D0277112

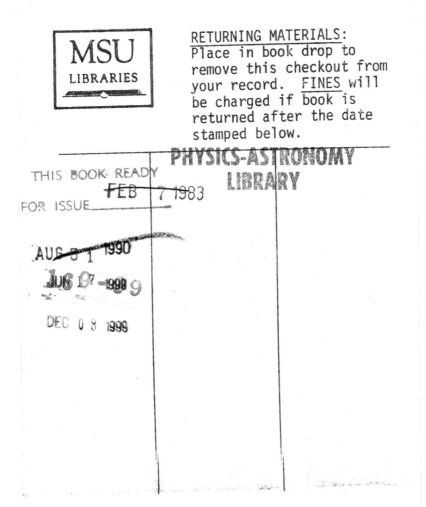

Molecular Interactions

Volume 2

Molecular Interactions

Volume 2

Edited by
H. Ratajczak
Institute of Chemistry, University of Wroclaw, Poland

and

W. J. Orville-Thomas
Department of Chemistry and Applied Chemistry, University of Salford, UK

Assistant Editor: **Mavis Redshaw**
Department of Chemistry and Applied Chemistry, University of Salford, UK

A Wiley-Interscience Publication

JOHN WILEY & SONS
Chichester · New York · Brisbane · Toronto

British Library Cataloguing in Publication Data:

Molecular interactions.
  Vol. 2
  1. Chemical bonds
  2. Molecules
  I. Ratajczak, H.
  II. Orville-Thomas, William James
  539'.6   QC461   79-40825

  ISBN 0 471 27681 2

Filmset in Northern Ireland at The Universities Press (Belfast) Limited
Printed and bound in Great Britain at The Pitman Press, Bath

List of contributors

BLINC, R.

Institut 'Josef Stefan' Ljubljana, University of Ljubljana, Ljubljana, 61001, Jamova 39, P.P. 199-IV, Yugoslavia.

BUREIKO, S. F.

Laboratory of Molecular Spectroscopy, Institute of Physics, Leningrad State University, 199164 Leningrad, B-164, U.S.S.R.

CHOJNACKI, H.

Institute of Organic and Physical Chemistry, Technical University, 50–370 Wroclaw, Wyb. Wyspianskiego 27, Poland.

DENISOV, G. S.

Laboratory of Molecular Spectroscopy, Institute of Physics, Leningrad State University, 199164 Leningrad, B-164, U.S.S.R.

GETTINS, W. J.

Department of Chemistry and Applied Chemistry, University of Salford, Salford M5 4WT, England.

GOLUBEV, N. S.

Laboratory of Molecular Spectroscopy, Institute of Physics, Leningrad State University, 199164 Leningrad, B-164, U.S.S.R.

GORMALLY, J.

Department of Chemistry and Applied Chemistry, University of Salford, Salford M5 4WT, England.

GUIBÉ, L.

Université Paris XI, Institut D'Électronique Fondamentale, Bâtiment 220, F-91405 Orsay, Cedex, France.

HILLIER, I. H.

Department of Chemistry, The University, Manchester, M13 9PL, England.

HUYSKENS, P.

Universiteit Te Leuven, Department Scheikunde, Laboratorium voor Fysicochemie en Stralingschemie, Celestijnenlaan 200 F, 3030 Heverlee, Belgium.

JUGIE, G.

Laboratoire de Chimie de Coordination du C.N.R.S. 205 Route de Narbonne, B.P. 4142, Toulouse 31030. France.

MATAGA, N.   *Department of Chemistry, Faculty of Engineering Science, Osaka University, Toyonaka, Osaka, Japan.*

PASTERNAK, M.   *Department of Physics and Astronomy, Tel-Aviv University, Ramat-Aviv, Israel.*

PIGON, K.   *Institute of Organic and Physical Chemistry, Technical University, 50–370 Wroclaw, Wyb, Wyspianskiego 27, Poland.*

SLIFKIN, M. A.   *Department of Biochemistry, University of Salford, Salford, M5 4WT, England.*

SMYTH, C. P.   *Department of Chemistry, Princeton University, Princeton, New Jersey, 08540, U.S.A.*

SONNINO, T.   *Israel Atomic Energy Commission, Soreq Nuclear Research Center, Yavne, Israel.*

TOKHADZE, K. G.   *Laboratory of Molecular Spectroscopy, Institute of Physics, Leningrad State University, 199164 Leningrad, B-164, U.S.S.R.*

VINOGRADOV, S. N.   *Department of Biochemistry, Wayne State University, Michigan 48201, U.S.A.*

WYN-JONES, E.   *Department of Chemistry and Applied Chemistry, University of Salford, Salford, M5 4WT, England.*

ZEEGERS-HUYSKENS, TH.   *Universiteit Te Leuven, Department Scheikunde, Laboratorium voor Fysicochemie en Stralingschemie, Celestijnenlaan 200 F, 3030 Heverlee, Belgium.*

Contents

List of contributors v

Preface to the Series xvii

Preface to Volume 2. xix

The International System of Units xxi

1 Proton transfer and ion transfer complexes **1**

Th. ZEEGERS-HUYSKENS AND P. HUYSKENS

1.1 Introduction 1
1.2 Experimental studies of proton or ion transfer complexes . 5
   1.2.1 X-Ray and neutron diffraction 5
   1.2.2 Vibrational spectroscopy 7
      1.2.2.1 Proton transfer complexes 7
          General considerations 7
          More specific systems 16
      1.2.2.2 Electron transfer complexes 21
   1.2.3 Electronic spectroscopy 25
      1.2.3.1 Proton transfer complexes 25
   1.2.4 Dipole moments 30
   1.2.5 Nuclear magnetic resonance spectroscopy 36
      1.2.5.1 ^{1}H spectra 37
      1.2.5.2 ^{13}C spectra 38
      1.2.5.3 General comment 40
   1.2.6 Some selected thermodynamic data for proton transfer
       complexes 40
1.3 Tautomerism between ion or proton transfer complexes and
   normal complexes 40
   1.3.1 Requirements for the existence of a tautomer-
      ism—Environmental effects 40
   1.3.2 Experimental methods for the study of the tautomer-
      ism 50

1.3.2.1 Spectroscopy   50
1.3.2.2 Isotope effect on the R distance and ν_{AH}/ν_{AD} frequency ratio   58
1.3.2.3 Dipole moments   60
1.3.2.4 Nuclear magnetic resonance spectroscopy   64
1.3.3 Experimental values of the transfer constant K_t and transfer enthalpy ΔH_t   65
1.3.4 Factors which influence the transfer constant   65
1.3.4.1 Influence of ΔpK_a   65
1.3.4.2 Influence of the proton affinity of the base   73
1.3.4.3 Influence of the dielectric constant of the medium .   74
1.3.4.4 Influence of other specific bonds formed by the complex with molecules of the environment . . .   74
1.4 Dissociation of proton and ion transfer complexes   77
1.4.1 Experimental determination of the dissociation constants of ion-pairs   78
1.4.2 Selected experimental values of the dissociation constant K_d and of the dissociation enthalpy ΔH_d of hydrogen-bonded and E.D.A.-bonded ion-pairs . . .   83
1.4.3 Factors governing the dissociation of hydrogen-bonded or E.D.A.-bonded ion-pairs into free ions   83
1.4.3.1 Dielectric constant   83
1.4.3.2 Specific interactions of the solvent   90
1.4.3.3 pK_a of the partners in water   91
1.4.3.4 Homoconjugation and heteroconjugation of the ions   92
1.4.3.5 Complexation of the ion-pairs   96
1.5 References   97

2 The kinetics of exchange and proton transfer processes in hydrogen-bonded systems in inert media **107**

G. S. DENISOV, S. F. BUREIKO,
N. S. GOLUBEV, AND K. G. TOKHADZE

2.1 Introduction   107
2.1.1 Formation and breaking of hydrogen bonds   108
2.1.2 Reversible proton transfer inside a complex, or 'molecular–ionic tautomerism'   109
2.1.3 Proton exchange between molecules or complexes . .   110
2.2 Lifetimes of hydrogen-bonded complexes   111
2.3 Reversible proton transfer inside hydrogen-bonded complexes   122

2.4   Proton exchange in hydrogen-bonded systems 130
2.5   References 139

3   Kinetic studies of micelle formation in surfactants **143**

J. GORMALLY, W. J. GETTINS,
AND E. WYN-JONES

3.1   Introduction 143
3.2   Properties of surfactants 143
3.3   Experimental techniques 148
3.4   Kinetic studies of micellization 151
   3.4.1   Introduction 151
   3.4.2   The Aniansson and Wall Model 152
      3.4.2.1   The rate equations 155
      3.4.2.2   The fast relaxation time 159
      3.4.2.3   The slow relaxation time 161
   3.4.3   The model of Sams, Wyn-Jones, and Rassing 163
3.5   The evaluation of experimental data 165
   3.5.1   The fast relaxation time 165
   3.5.2   The slow relaxation time 172
3.6   Other aggregation phenomena 173
3.7   References 174

4   Structural aspects of hydrogen bonding in amino acids, peptides, proteins, and model systems **179**

S. N. VINOGRADOV

4.1   Introduction 179
4.2   Hydrogen bonds in crystal structures of amino acids and peptides 181
   4.2.1   Types of hydrogen bonds 181
   4.2.2   Carbonyl–carboxyl hydrogen bonding 185
   4.2.3   Hydrogen-bonded ion-pair interactions 187
   4.2.4   Weak interactions: bifurcated hydrogen bonds and C—H $\cdots$ O contacts 189
4.3   Models of hydrogen bonds in proteins 191
   4.3.1   Monocation $(N—H—N)^+$ complexes 191
   4.3.2   Relationship between O—H $\cdots$ N and $N^+—H \cdots O^-$ bonds 195
   4.3.3   Hydrogen bonding in amides 204

4.3.4 Hydrogen bonds in peptides and peptide–water interactions 205
4.4 Hydrogen bonds in proteins 207
  4.4.1 Types of hydrogen bonding interactions 207
  4.4.2 Possible roles for ion-pair interactions 211
  4.4.3 N—H · · · S interactions 216
  4.4.4 Water–protein interactions 217
4.5 Conclusion 221
4.6 Acknowledgements 223
4.7 References 223

5 Hydrogen-bonded ferroelectrics and lattice dimensionality . . . 231

R. BLINC

5.1 Introduction 231
5.2 Lattice dimensionality and critical behaviour 231
5.3 The pseudo-spin Hamiltonian dynamics 236
  5.3.1 The Hamiltonian 236
  5.3.2 The pseudo-spin dynamics 240
    5.3.2.1 The molecular field approximation (MFA) 240
    5.3.2.2 The random phase approximation (RPA) 241
5.4 Pseudo-one-dimensional ferroelectricity 243
  5.4.1 $PbHPO_4$ and $PbHAsO_4$ 243
    5.4.1.1 Structure 243
    5.4.1.2 Theoretical Model for $PbDPO_4$ 244
  5.4.2 CsH_2PO_4 and CsD_2PO_4 247
5.5 Pseudo-two-dimensional ferroelectricity 248
  5.5.1 Introduction 248
  5.5.2 Squaric acid 250
    5.5.2.1 Structure 250
    5.5.2.2 Theoretical model 250
5.6 Three-dimensional uniaxial ferroelectricity in KH_2PO_4-type systems and proton–lattice interactions 254
  5.6.1 Structure 254
  5.6.2 Proton–lattice interactions in hydrogen-bonded systems 255
  5.6.3 The model Hamiltonian 256
    5.6.3.1 Proton–phonon coupling 256
  5.6.4 Direct $S_i^x S_j^x$ coupling 259
  5.6.5 $S_{-q}^x F_q^x Q_q$ coupling 262
  5.6.6 $S_j^x D_j Q_j^2$ 266
5.7 References 269

6   **The significance of charge transfer interactions in biology** . . . **271**

M. A. SLIFKIN

6.1   Introduction 271
6.2   Charge transfer complexes 271
6.3   Amino acids, proteins, and analogous molecules 278
6.4   Nucleic acid bases 285
6.5   Flavins 289
6.6   Phenothiazines and tranquillizers 293
6.7   Porphyrins 296
6.8   Charge transfer interaction and biological regulation . . . 299
6.9   Concluding remarks 301
6.10 References 301

7   **Dipole moment, dielectric loss, and molecular interactions** . . . **305**

C. P. SMYTH

7.1   Introduction 305
7.2   Dipole moment and dielectric permittivity 305
7.3   Solvent effect 309
7.4   Reaction field of a polarizable point dipole 312
7.5   Permanent dipoles, cohesion, boiling point, and solubility . 312
7.6   Dielectric loss and relaxation time 314
7.7   Effect of internal field on relaxation time 316
7.8   Rate theory of dielectric relaxation and viscosity 320
7.9   Relaxation time, inner friction, and viscosity 320
7.10 Individual dipole–dipole interactions 329
7.11 Hydrogen bonding, dipole moment, and relaxation time . . 331
7.12 Millimetre and submillimetre absorption 336
7.13 References 340

8   **Nuclear quadrupole resonance studies of molecular complexes** . **343**

L. GUIBÉ AND G. JUGIE

8.1   General introduction 343
8.2   Nuclear quadrupole resonance spectroscopy 344
8.2.1 Introduction 344
8.2.2 Basic principles 345
8.2.3 Experimental 348

8.2.4   Interpretation of quadrupole coupling constants in complexes 349
8.3   Electron transfer in molecular complexes as studied by NQRS 351
8.4   Complexes containing elements of group I 352
8.5   Complexes containing elements of group II 354
  8.5.1   Magnesium-containing complexes 354
  8.5.2   Cadmium- and zinc-containing complexes 355
  8.5.3   Mercury-containing complexes 355
8.6   Complexes containing elements of group III 359
  8.6.1   Boron-containing complexes 360
  8.6.2   Aluminium; gallium-, and indium-containing complexes 362
8.7   Complexes containing elements of group IV 365
  8.7.1   Carbon-containing complexes 365
  8.7.2   Tin-containing complexes 368
8.8   Complexes containing elements of group V 371
  8.8.1   Nitrogen-containing complexes 372
    8.8.1.1   Complexes of nitriles 372
    8.8.1.2   Complexes of amines 375
  8.8.2   Phosphorus-containing complexes 376
  8.8.3   Arsenic- and antimony-containing complexes 377
  8.8.4   Bismuth-containing complexes 380
  8.8.5   Complexes containing elements Vb (Nb and Ta) . . . 380
8.9   Complexes containing elements of group VI 381
8.10   Complexes containing elements of group VII 383
  8.10.1   Complexes of pure halogens 384
  8.10.2   Polyhalides 385
  8.10.3   Carbon chloride complexes (CCl_4, $CHCl_3$) 387
8.11   Complexes containing elements of Group 0 388
8.12   Chemical index 389
8.13   Formulae index 393
8.14   List of complexes 396
8.15   List of abbreviations 432
8.16   References 433

9   Mössbauer effect studies of molecular complexes **439**

M. PASTERNAK AND T. SONNINO

9.1   Introduction 439
9.2   The Mössbauer effect or nuclear gamma resonance 440
9.3   Hyperfine interactions and the Mössbauer parameters . . 442

9.3.1 The monopole interaction and the isomer shift 443
9.3.2 The electric quadrupole interaction 444
9.4 Chemical information from isomer shift and quadrupole splitting data 445
9.5 Review of selected experimental data 446
9.6 References 450

10 Electrical conductivity of solid molecular complexes **451**

K. PIGON AND H. CHOJNACKI

10.1 Introduction 451
10.2 Crystal structures of molecular complexes 451
    10.2.1 Phase diagrams of systems with CT interaction and stoichiometry of CT complexes 451
    10.2.2 Crystal structures of CT complexes 455
    10.2.3 Crystal structures of hydrogen-bonded crystals . . . 459
10.3 Survey of experimental results on electrical properties of molecular complexes 464
    10.3.1 Electrical conductivity and photoconductivity in CT complexes 464
    10.3.2 Current carrier mobility 470
    10.3.3 Ionic and electronic conductivity 473
10.4 Possible mechanism of the electrical conductivity 476
    10.4.1 Problem of interaction in the solid state 476
    10.4.2 Band structure considerations 479
    10.4.3 Tunnelling and hopping model 486
10.5 References 488

11 Photoelectron spectroscopic studies of molecular complexes . . **493**

I. H. HILLIER

11.1 Introduction 493
11.2 Theoretical background 493
    11.2.1 Interpretation of core ionization energies 495
11.3 ESCA studies of donor–acceptor complexes 496
    11.3.1 Pyridine–iodomonochloride 496
    11.3.2 Chloranil–donor adducts 497
    11.3.3 Determination of charge transfer in TTF–TCNQ . . 499
    11.3.4 Donor–acceptor complexes involving BF_3 501
11.4 Valence photoelectron spectra 506
    11.4.1 Complexes of borane with Lewis bases 506

11.5 Conclusions 507
11.6 References 508

12 Properties of molecular complexes in the electronic excited
    states 509

N. MATAGA

12.1 Introduction 509
12.2 Interaction of excited polar systems with polar solvent
      molecules 510
      12.2.1 Molecular emission process and solvent relaxation . 511
      12.2.2 Solvent dependence of electronic and geometrical
              structures of excited polar systems 513
      12.2.3 Kinetics of solvation processes 522
12.3 Mechanisms of intermolecular charge transfer processes in
      the excited electronic state 525
      12.3.1 Excitation of ground-state aggregates—change of
              structure due to intracomplex relaxation 526
      12.3.2 Possible different types of charge transfer processes in
              the excited state 528
12.4 Kinetics of exciplex formation and photochemical charge
      transfer processes 531
      12.4.1 Kinetic fluorescence studies of exciplex formation . 531
      12.4.2 Transient absorption studies by means of picosecond
              laser photolysis 533
      12.4.3 Relation between the fluorescence quenching rate
              constant and the free energy change of the charge
              transfer processes 536
12.5 Structures of exciplexes 539
      12.5.1 Electronic and geometrical structures of some EDA
              complexes in the fluorescent and phosphorescent
              state 540
      12.5.2 Structures of exciplexes as revealed from the studies
              upon intramolecular complex systems 543
      12.5.3 Solvent effects upon structures of exciplexes . . . 545
           12.5.3.1 Pyrene–DEA intermolecular heteroexcimer . . 546
           12.5.3.2 Intermolecular heteroexcimers, P_n 547
           12.5.3.3 Cyano-substituted layered cyclophanes . . . 547
12.6 Ionic photodissociation and related processes 548
      12.6.1 Effects of strength of CT interaction, spin multiplicity,
              and solvent polarity upon the dissociation yield . . 549

12.6.2   Local triplet formation from ion-pairs and relevant states 552

12.6.3   Ionic photodissociation in micellar solutions 554

12.7   Proton transfer and related processes 558

12.7.1   Charge transfer followed by proton transfer in exciplex systems 559

12.7.2   Hydrogen bonding interaction in the excited electronic state 564

12.8   References 566

Author index **571**

Subject index **591**

Subject index–Volume 1 . . . . **615**

Preface to the series

Currently the field of intermolecular interactions, or to use the current phrase, 'molecular interactions' represents one of the most important branches of molecular science. One reason for this is that the topic is important in chemistry, physics, and biology, and even more so at the interfaces of these subjects.

This series will cover both theoretical progress made in the field of molecular interactions as well as the development and applications of new experimental methods and techniques to study interactions between molecules. Much attention will be paid to the role and significance of molecular interactions in determining specific properties and structure of molecular systems as well as molecular phenomena related to interacting molecules.

The articles in this series will range from weak (van der Waal's type), through medium to strong molecular interactions in a variety of environments.

An attempt will be made to tailor the contents of each volume around one or two main themes. In this way it is hoped to avoid the production of disjointed accounts of very different topics.

Finally we do hope that this series will stimulate further progress in this fascinating field.

Institute of Chemistry
University of Wroclaw
Poland

and

Department of Chemistry
and Applied Chemistry
University of Salford
England

HENRYK RATAJCZAK
W. J. ORVILLE-THOMAS

Preface to volume 2

This volume is devoted to some specific properties and the significance of strong molecular interactions of the hydrogen bond and electron donor acceptor types in molecular biology and solid state physics, to the applications of selected physical methods to the study of electronic structure and properties of molecular complexes, and to electronic excited states of molecular systems.

Much attention has been paid to the phenomenon of proton transfer and its relation to ion-transfer in molecular complexes. These problems are extensively reviewed and discussed by Zeegers-Huyskens and Huyskens. The kinetics of proton exchange and proton transfer in hydrogen bonded systems in inert solvents are discussed by Denisov and his co-workers. In addition some problems of the kinetics of micelle formation in surfactants are described by Wyn-Jones and his colleagues.

Two articles presented in this volume emphasise the role and significance of molecular interactions in the biological sciences Vinogradov extensively reviews the structural aspects of hydrogen bonding in amino acids, peptides and proteins; Slifkin critically describes the role of electron donor-acceptor (charge-transfer) interactions in biological systems. In some way related to these is a chapter on the electric conductivity of solid molecular complexes of the hydrogen bond and charge-transfer types written by Pigon and Chojnacki.

The significance and role of hydrogen bonding in ferroelectric phase transitions of some hydrogen-bonded ferroelectrics is clearly demonstrated by Blinc.

The following four articles deal with applications of physical methods to the study of electronic structure and properties of molecular complexes: the dielectric method is described by Smyth; nuclear quadrupole resonance is extensively reviewed by Guibé and Jugie; Mössbauer effect by Pasternak and Sonnino and the photoelectron spectroscopic method is discussed by Hillier. Finally the properties of molecular complexes in the electronic excited states are reviewed and extensively discussed by Mataga.

It is a great pleasure for the Editors to thank all authors for their contributions and for their ready collaboration in preparing this book.

We are grateful to the copyright owners for their permission to reproduce various figures and tables.

Institute of Chemistry   HENRYK RATAJCZAK
University of Wroclaw   W. J. ORVILLE-THOMAS
Poland

and

Department of Chemistry and
Applied Chemistry
University of Salford
England
July 1979

The international system of units (SI)

Physical quantity	Name of unit	Symbol for unit
SI Base Units		
length	metre	m
mass	kilogram	kg
time	second	s
electric current	ampere	A
thermodynamic temperature	kelvin	K
amount of substance	mole	mol
SI Supplementary Units		
plane angle	radian	rad
solid angle	steradian	sr
SI Derived having Special Names and Symbols		
energy	joule	$J = m^2\,kg\,s^{-2}$
force	newton	$N = m\,kg\,s^{-2} = J\,m^{-1}$
pressure	pascal	$Pa = m^{-1}\,kg\,s^{-2} = N\,m^{-2} = J\,m^{-3}$
power	watt	$W = m^2\,kg\,s^{-3} = J\,s^{-1}$
electric charge	coulomb	$C = s\,A$
electric potential difference	volt	$V = m^2\,kg\,s^{-3}\,A^{-1} = J\,A^{-1}\,s^{-1}$
electric resistance	ohm	$\Omega = m^2\,kg\,s^{-3}\,A^{-2} = V\,A^{-1}$
electric conductance	siemens	$S = m^{-2}\,kg^{-1}\,s^3\,A^2 = \Omega^{-1}$
electric capacitance	farad	$F = m^{-2}\,kg^{-1}\,s^4\,A^2 = C\,V^{-1}$
magnetic flux	weber	$Wb = m^2\,kg\,s^{-2}\,A^{-1} = V\,s$
inductance	henry	$H = m^2\,kg\,s^{-2}\,A^{-2} = V\,s\,A^{-1}$
magnetic flux density	tesla	$T = kg\,s^{-2}\,A^{-1} = V\,s\,m^{-2}$
frequency	hertz	$Hz = s^{-1}$

SOME NON-SI UNITS

Physical quantity	Name of unit	Symbol and definition
Decimal Multiples of SI Units, Some having Special Names and Symbols		
length	ångström	$\text{Å} = 10^{-10}\,\text{m} = 0.1\,\text{nm}$
		$= 100\,\text{pm}$
length	micron	$\mu\text{m} = 10^{-6}\,\text{m}$
area	are	$\text{a} = 100\,\text{m}^2$
area	barn	$\text{b} = 10^{-28}\,\text{m}^2$
volume	litre	$1 = 10^{-3}\,\text{m}^3 = \text{dm}^3$
		$= 1000\,\text{cm}^3$
energy	erg	$\text{erg} = 10^{-7}\,\text{J}$
force	dyne	$\text{dyn} = 10^{-5}\,\text{N}$
force constant	dyne per centimetre	$\text{dyn cm}^{-1} = 10^{-3}\,\text{N m}^{-1}$
force constant	millidyne per ångström	$\text{mdyn Å}^{-1} = 10^2\,\text{N m}^{-1}$
force constant	attojoule per ångström squared	$\text{aJ Å}^{-2} = 10^2\,\text{N m}^{-1}$
pressure	bar	$\text{bar} = 10^5\,\text{Pa}$
concentration	—	$\text{M} = 10^3\,\text{mol m}^{-3}$
		$= \text{mol dm}^{-3}$
Units Defined Exactly in Terms of SI Units		
length	inch	$\text{in} = 0.0254\,\text{m}$
mass	pound	$\text{lb} = 0.453\,592\,37\,\text{kg}$
force	kilogram-force	$\text{kgf} = 9.806\,65\,\text{N}$
pressure	standard atmosphere	$\text{atm} = 101\,325\,\text{Pa}$
pressure	torr	$\text{Torr} = 1\,\text{mm Hg}$
		$= 133\,322\,\text{N m}^{-2}$
energy	kilowatt hour	$\text{kWh} = 3.6 \times 10^6\,\text{J}$
energy	thermochemical calorie	$\text{cal}_{\text{th}} = 4.184\,\text{J}$
thermodynamic temperature	degree Celsius*	$°\text{C} = \text{K}$

* Celsius or 'Centigrade' temperature θ_C is defined in terms of the thermodynamic temperature T by the relation $\theta_C/°\text{C} = T/\text{K} - 273.15$.

OTHER RELATIONS

1. The physical quantity, the wavenumber (units cm^{-1}), is related to frequency as follows:

$$cm^{-1} \approx (2.998 \times 10^{10})^{-1} s^{-1}$$

2. The physical quantity, the molar decadic absorption coefficient (symbol ε) has the SI units $m^2 \, mol^{-1}$. The relation between the usual non-SI and SI units is as follows:

$$M^{-1} cm^{-1} = 1 \, mol^{-1} cm^{-1} = 10^{-1} \, m^2 \, mol^{-1}$$

The SI Prefixes

Multiplication factor	Prefix	Symbol	Multiplication factor	Prefix	Symbol
10^{-1}	deci	d	10^{1}	deca	da
10^{-2}	centi	c	10^{2}	hecto	h
10^{-3}	milli	m	10^{3}	kilo	k
10^{-6}	micro	μ	10^{6}	mega	M
10^{-9}	nano	n	10^{9}	giga	G
10^{-12}	pico	p	10^{12}	tera	T
10^{-15}	femto	f	10^{15}	pera	P
10^{-18}	atto	a	10^{18}	exa	E

Molecular Interactions, Volume 2
Edited by H. Ratajczak and W. J. Orville-Thomas
© 1980 John Wiley & Sons

1 Proton transfer and ion transfer complexes

Th. Zeegers-Huyskens and P. Huyskens

1.1 Introduction

Theoretically, any interaction between molecules or ions modifies all the internuclear distances in the partners.

In most cases, however, these changes are very limited and remain below 1%. This is the reason why the positions of most stretching bands of molecules are not strongly affected by a change of the medium. Even in the case of fairly strong electrostatic interactions, as for example those between tetraalkylammonium ions and anions, the C—H stretching vibration bands are not much shifted from their position for the free ions.

However there are two types of interactions for which this rule no longer holds and where modifications brought about in some internuclear distances usually exceed 1%, in some cases even being very important. These interactions are the hydrogen bonds and the $n–\sigma$ electron donor–acceptor (E.D.A.) bonds. For instance in the complex between bromine and 1,4-dioxan in the solid state:

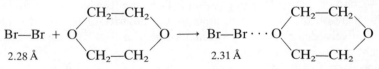

the distance between the two bromine nuclei is 2.31 Å (Hassel and Hvoslef, 1954) compared with 2.28 Å in the free bromine molecule. In hydrogen bonds in heavy ice, the O—D distance is lengthened from 0.970 Å in the gas phase to 0.982 Å (La Placa $et\ al.$, 1973). In hydrogen bonds between acetic acid molecules in the solid state the O—H interdistance reaches 1.011 Å (Jönsson, 1971) whereas in the monohydrate of trifluoromethane sulphonic acid, the hydrogen nucleus of CF_3SO_3H is removed as far as 1.50 Å from the corresponding oxygen nucleus of the acid molecule (Lundgren and Olovsson, 1976) by the hydrogen bond with the water molecule.

Let us consider the atom p of the acceptor molecule which comes into closest contact with the donor molecule. In the hydrogen bond p is a hydrogen atom, in the E.D.A. bond considered above, p is a bromine atom. In the acceptor molecule or in the acid, this atom is bound to another atom a.

A common characteristic of both the hydrogen bonds and charge transfer bonds is that the distance a—p *increases upon bond formation.*

$$A + D \rightarrow A \cdots D$$
$$-a-p+d \rightarrow -a-p \cdots d$$
$$r_{ap}^0 \qquad\qquad r_{ap}$$
$$r_{ap}^0 < r_{ap}$$

Usually this increase exceeds 1%.

It is also reflected by a decrease of the ν_{ap} stretching frequency and of the corresponding force constant k_{ap}. For instance in the complex between iodine and dimethylacetamide in benzene solution, the frequency of the I—I stretching band decreases from 207 cm^{-1} in the free molecule to 195 cm^{-1} (Dorval and Zeegers-Huyskens, 1973). In the stronger complex with triethylamine $\nu_{\text{I-I}}$ is 185 cm^{-1} (Yada et al., 1962). Similar observations were made for the complexes of bromine with pyridines of increasing basicity (D'Hondt and Zeegers-Huyskens, 1971).

In the case of deuterium bonds of the O—D $\cdots$ O type, the shift of the ν_{OD} frequency is correlated with the O—D distance determined by neutron diffraction techniques (La Placa et al., 1973). Plots of calculated O—H distances versus $\Delta\nu_{\text{OH}}$ in several O—H $\cdots$ O and O—H $\cdots$ N bonds seem to belong as a first approximation to the same curve (Zeegers-Huyskens, 1977), showing the correlation between $\Delta\nu_{\text{OH}}$ and Δr_{OH} for given a and p atoms to be rather insensitive to the nature of the electron donor.

A second property of hydrogen and E.D.A. bonds which is closely related to the increase of the a—p distance is the polarity of the bond, which is quantitatively described by the enhancement $\Delta\mu$ of the dipole moment brought about by its formation.

In a given family of complexes, *an increase of* r_{ap}, *and consequently a greater shift* $-\Delta\nu_{ap}$ *is always accompanied by an increase of* $\Delta\mu$. Here the correlation depends on the nature of the electron donor: according to our results the value of the ratio $\Delta\mu/\Delta r_{\text{OH}}$ is of the order of 10 Debye Å^{-1} in O—H $\cdots$ N bonds between phenols and pyridines or anilines (Huyskens and Hernandez, 1973). This ratio is even greater in O—H $\cdots$ O=C bonds (Pauwels and Huyskens, 1974). The ratio's $\Delta\mu/\Delta r_{ap}$ are also very important in E.D.A. bonds: in the complex between iodine and trimethylamine in the crystalline state, Δr_{II} is 0.17 Å (Strømme, 1959), whereas its dipole increment in dioxan solution reaches 5.2 Debye (Toyoda and Nagakura, 1966) and will probably be still greater in the solid state. This leads to a ratio $\Delta\mu/\Delta r_{II}$ which exceeds 30 Debye $Å^{-1}$.

A common characteristic of both E.D.A. and hydrogen bonds is therefore that the displacement of the nucleus of the p atom increases the negative charge on the remainder of the molecule.

It must be noted here that, if the only phenomenon is a displacement of *the positive ion* p^+, (which derives from the atom p by the loss of one electron), the ratio $\Delta\mu/\Delta r_{ap}$ equals 4.8 Debye $Å^{-1}$, which is less than half the experimental value.

The displacement of p *must therefore be considered rather as a moving of the positive ion* p^+, accompanied by some additional net displacement of electrons in the opposite direction which further improves the dipole increment.

If in a hydrogen bond or an E.D.A. bond the $a—p$ distance is greater than in the free acid or acceptor molecule, *an increase of this distance* r_{ap} *in a given family of complexes, is accompanied by a decrease of the distance* p—d between the nucleus p and the nearest nucleus d of the donor.

$$A+D \rightarrow A \cdots D$$
$$—a—p+d— \rightarrow a—p \cdots d—$$
$$r_{ap}^0 \qquad r_{ap} \quad r_{pd}$$

In the case of O—H $\cdots$ O hydrogen bonds, where numerous neutron diffraction data are available, Olovsson and Jönsson recently presented a correlation between r_{OH} and $r_{H \cdots O}$ (Olovsson and Jönsson, 1976). The experimental plots show some scatter due to the differing environment, the thermal motion, and of course, the experimental errors, but the deviations from the curve do not generally exceed a few per cent. The same authors give also a curve correlating the distances $r_{H \cdots O}$ and r_{NH} for N—H $\cdots$ O bonds. In charge transfer complexes, a similar regularity between the increase of the r_{ap} distance is observed. For instance in the charge transfer complex between bromine and methanol in the solid state:

$$Br—Br \cdots O—CH_3$$
$$\mid$$
$$H$$

the Br—Br distance is 2.29 Å and the Br $\cdots$ O distance is 2.78 Å, whereas one finds 2.31 Å and 2.71 Å in the complex with 1,4-dioxan (Hassel and Hvoslef, 1954).

The correlation between the two r_{ap} and r_{pd} distances is also reflected by the corresponding stretching frequencies and force constants. For the complexes between bromine and pyridines of increasing basicity, in benzene solution, the force constant k_{BrBr} decreases whereas that of the Br $\cdots$ N vibration increases (D'Hondt and Zeegers-Huyskens, 1971).

Let us now consider the ion p^+ which corresponds to the atom p when it has lost an electron.

A molecule D can only act as electron donor in an E.D.A. bond or as a base in a hydrogen bond when it can form *a chemical bond*, with this ion p^+ giving:

$$p^+ + D \rightarrow p^+ - D$$

For instance

$$Br^+ + NR_3 \rightarrow Br^+ - NR_3$$

$$H^+ + NR_3 \rightarrow H^+ - NR_3$$

Let $r_{p^+d}^0$ be the internuclear distance of $p-d$ in the free ion p^+-d which is formed in this way. (In the case of the first ion r_{NBr}^0 must be of the order of 2.15 Å, whereas in the second case $r_{NH^+}^0$ is of the order of 1.02 Å.)

It now appears that in both hydrogen bonds and charge transfer bonds, *the distance* p—d *is always intermediate between the sum of the Van der Waals radii of both atoms* p *and* d, *and the distance* $r_{p^+d}^0$ *in the free positive ion* p$^+$—d.

Furthermore, a decrease of the $p-d$ distance corresponds to an enhancement of the polarity of the bond.

One can thus consider the $n-\sigma$ charge transfer complexes and the hydrogen-bonded complexes as intermediate steps in the transfer of the positive ion p^+ from one entity to another.

Let A'^- be the negative ion formed from the acceptor or the acid when p^+ is completely removed. One can write for the overall *ionogenic* process:

$$A'-p+D \rightarrow A'-p \cdots D \rightarrow A'^- + p^+ - D$$

$$r_{ap}^0 \qquad\qquad \text{complex} \qquad\qquad r_{p^+d}^0$$

Hydrogen bonds and n—σ *E.D.A. bonds both involve a partial transfer of a positive ion from one given molecular entity to another.* For example:

$$Br-Br + NR_3 \rightarrow Br-Br \cdots NR_3 \rightarrow Br^- + Br^+ - NR_3$$

$$\phi\text{-}O-H + NR_3 \rightarrow \phi\text{-}O-H \cdots NR_3 \rightarrow \phi\text{-}O^- + H^+ - NR_3$$

In the first case the E.D.A. bond is an intermediate in the transfer of a positive Br^+ ion from the acceptor molecule Br_2 to the electron donor molecule. In the second case, the hydrogen bond is an intermediate in the transfer of a proton from the acid to the base.

One can consider as the limiting case

$$\frac{r_{ap}}{r_{ap}^0} = \frac{r_{pd}}{r_{p^+d}^0}.$$

Such complexes correspond to the class of *middle complexes* in the terminology of Mulliken (Mulliken and Person, 1969). However, in most complexes between neutral molecules one observes that

$$\frac{r_{ap}}{r_{ap}^0} < \frac{r_{pd}}{r_{p^+d}^0}.$$

Such complexes will be called normal complexes (*outer complexes* in the terminology of Mulliken). One also finds numerous complexes where

$$\frac{r_{ap}}{r_{ap}^0} > \frac{r_{pd}}{r_{p^+d}^0}.$$

In the case of hydrogen bonds we can call these complexes: *proton transfer complexes*. In the case of E.D.A. bonds they belong to the class of *inner complexes* of Mulliken. They can also be called '*ion transfer*' complexes.

Let us consider the quantity ζ defined by the relation

$$\zeta \equiv \frac{r_{ap}}{r_{pd}} \times \frac{r_{p^+d}^0}{r_{ap}^0}.$$

This function can be considered as a quantitative expression describing the magnitude of the ion or proton transfer in a given bond.

According to the proposed definition, an ion transfer or proton transfer complex between two entities A'—p and D is thus a complex where for the nuclei a, p, and d directly involved in the specific interaction the following inequality holds:

$$\zeta > 1$$

In the following sections, the experimental studies of proton or ion transfer complexes and the tautomerism between normal and proton transfer complexes will be successively reviewed and discussed.

1.2 Experimental Studies of Proton or Ion Transfer Complexes

1.2.1 *X-Ray and neutron diffraction*

For complexes in the crystalline state neutron and X-ray diffraction techniques constitute the most adequate tools in establishing the position of the nuclei and the mean interdistances between them.

Discussions about the accuracy of both methods will be found in the work of Hamilton and Ibers (1968) and that of Olovsson and Jönsson (1976). Reviews have been presented on the crystal structure of charge transfer complexes by Prout and Kamenar (1973) and on X-ray diffraction results for hydrogen-bonded complexes by Donohue (1968).

The main difference between the two methods lies in the fact that X-rays are primarily scattered by electrons whereas the nuclei themselves are responsible for the scattering of neutrons.

As a consequence, owing to the fact that the hydrogen atom does not possess inner electrons, X-ray crystallography is less useful for the determination of the position of the proton for which neutron diffraction is much more suitable.

The accuracy of both methods is influenced by the thermal motion of the nuclei. This thermal motion not only limits the precision in determining the *position* of the nuclei but has an additional unfavourable influence on the accuracy of the determination of the internuclear distances.

The reason is that *as a consequence of the thermal motion of the nuclei, the average distance between two nuclei does not correspond to the distance between their mean positions.*

The average bond length is always greater than the distance between the mean positions. In order to determine the mean bond length accurately, it would be necessary to know the correlations between the motions of the various nuclei. The lack of precision in this knowledge introduces an additional inaccuracy when determining the bond lengths from the mean positions which are given by diffraction methods.

This effect can reach several hundredths of an Ångström. As pointed out by Hamilton and Ibers (1968) it is therefore unrealistic to attribute an accuracy better than 0.02 Å to the mean internuclear distances determined by diffraction methods.

Table 1.1 reports the experimental data needed for the computation of the function

$$\zeta \equiv \frac{r_{ap}}{r_{pd}} \times \frac{r_{p^+d}^0}{r_{ap}^0}$$

which, according to the considerations of the introduction, describes the magnitude of the proton transfer or the ion transfer in the complexes.

For the hydrogen-bonded complexes the distances r_{ap} and r_{pd} obtained by neutron diffraction are not corrected for thermal motion. Although this sometimes leads to values which are smaller than r_{ap}^0 or r_{pd}^0, the effect of neglecting this correction is, to a given extent, cancelled out when computing the ratio r_{ap}/r_{pd}.

For the unperturbed distances r_{ap}^0 sufficient information exists in the literature to obtain values which are accurate within a few hundredths of an Ångström. Furthermore these distances do not change very much in a given family of compounds. In the case of hydrogen bonds there is also no major difficulty in finding the appropriate $r_{p^+d}^0$ value. For these complexes therefore one can deduce values of the function ζ which are accurate within a few per cent.

In the case of charge transfer complexes experimental data concerning the unperturbed $r_{p^+d}^0$ distance is lacking. For instance in the case of the I^+—N≡

bond, we take the value of 2.16 Å, whereas on the basis of the sum of atomic radii a value of 2.03 Å could be proposed, and conversely, considering the N—Br distance in nitrosyl bromide (2.14 Å) a higher value could be expected. The ζ value may differ from the proposed value by as much as 10% because of the lack of accuracy in $r^0_{p^+d}$ in the charge transfer complexes.

The data of Table 1.1 show that complexes in the solid state are characterized by values of the function ζ which differ significantly from each other and this is so for both hydrogen bonds and charge transfer complexes. One can thus conclude that, in the solid state, the complexes are characterized by different magnitudes of ion transfer, as described by their function ζ.

For numerous hydrogen bonds ζ is much greater than one, and complexes such as those of strong acids with water or ammonia clearly belong to the proton transfer class. From the data of Tables 1.2 and 1.3 it appears that the E.D.A. complexes of bromine and iodine with aliphatic amines could be classified among the 'ion transfer' complexes. However it may be more realistic to classify these complexes among the 'middle' class, because of the lack of accuracy in the $r^0_{p^+d}$ value and the fact that ζ is nearly one.

From the data of Table 1.1 it appears that *an important factor governing the transfer function ζ in the complexes in the solid phase is the strength of the acid*, as reflected by its pK_a in water: the OH $\cdots$ O bonds of oxalic acid with water still belong to the normal class whereas for the stronger acids the hydrates belong to the proton transfer class.

A similar remark also holds for the strength of the base: in the solid phase the bonds between oxalic acid and NH_3 are of the proton transfer type as a consequence of the much higher basicity of ammonia compared to water.

It must be emphasized that environmental effects may also play an important role. Differences in environmental effects are responsible for the fact that the distances in hydrogen bonds between H_3O^+ and H_2O differ from one case to another: in some cases the bond is nearly symmetrical and $r_{ap} \simeq r_{pd}$ (Attig and Williams, 1977), whereas in other cases marked differences appear between these two distances (Lundgren and Olovsson, 1976).

The coordination in the crystal lattice can enhance the magnitude of ion or proton transfer in the solid state. As a consequence of this, some complexes which already belong to the ion transfer class in the solid state, may still not fulfil the requirement in solution.

1.2.2 *Vibrational spectroscopy*

1.2.2.1 *Proton transfer complexes*

General considerations. The formation of a normal hydrogen-bonded complex brings about characteristic features in a vibrational spectrum and these

Table 1.2 Transfer of the bromide ion (X-ray diffraction data)

Electron acceptor	Electron donor	Compound	Bond	r_{ap} (Å)	r_{ap}^{0} (Å)	r_{pd} (Å)	$r_{p^+d}^{0}$ (Å)	ζ	Class
Oxalyl bromide	Oxalyl bromide	Oxalyl bromide	C—Br···O	1.84[a]	1.89[b]	3.27[a]	1.90[c]	0.57	Normal
Oxalyl bromide	Dioxan	Oxalyl bromide-1,4-dioxan	C—Br···O	1.96[d]	1.89[b]	3.21[d]	1.90[c]	0.61	Normal
C_2Br_4	Pyrazine	Tetrabromoethylene-pyrazine	C—Br···N	1.88[e]	1.89[b]	3.02[e]	2.14[f]	0.70	Normal
Br_2	Acetone	Br_2-acetone	Br—Br···O	2.28[g]	2.28[h]	2.82[g]	1.90[c]	0.67	Normal
Br_2	Methanol	Br_2-methanol	Br—Br···O	2.39[i]	2.28[h]	2.78[i]	1.90[c]	0.69	Normal
Br_2	Dioxan	Br_2-dioxan	Br—Br···O	2.31[j]	2.28[h]	2.71[j]	1.90[c]	0.71	Normal
Br_2	Hexamethylene-tetramine	$2Br_2$-hexamethylenetetramine	Br—Br···N	2.43[k]	2.28[h]	2.16[k]	2.14[f]	1.06	Middle

[a] Groth and Hassel (1963).
[b] Mean value in olefinic compounds, Weast (1967).
[c] Estimated from O—Cl = 1.70 Å in Cl_2O adding 0.20 Å when passing from Cl to Br.
[d] Damm et al. (1965).
[e] Dahl and Hassel (1966).
[f] Distance N—Br in nitrosyl bromide, ref. b.
[g] Hassel and Strømme (1959).
[h] Jesson and Muetterties (1969).
[i] Groth and Hassel (1963).
[j] Hassel and Hvoslef (1954).
[k] Eia and Hassel (1956).

Table 1.3 Transfer of the iodide ion (X-ray diffraction data)

Electron acceptor	Electron donor	Compound	Bond	r_{ap} (Å)	r_{ap}^o (Å)	r_{pd} (Å)	$r_{p^+d}^o$ (Å)	ζ	Class
CHI₃	1,4-Dioxan	CHI₃–1,4-dioxan	C—I···O	2.13[a]	2.13[b]	3.04[a]	2.00[c]	0.66	Normal
Iodocyano-acetylene	Iodocyano-acetylene	Iodocyanoacetylene (self complex)	C—I···N	1.79[d]	2.13[b]	2.93[d]	2.16[e]	0.62	Normal
ICl	1,4-Dioxan	2ICl–1,4-dioxan	Cl—I···O	2.3[f]	2.32[g]	2.57[f]	2.00[c]	0.78	Normal
I₂	1,4-Dithiane	2I₂–dithiane	I—I···S	2.79[h]	2.67[g]	2.87[h]	2.39[i]	0.88	Normal
I₂	4-Methylpyridine	I₂–4-picoline	I—I···N	2.83[i]	2.67[g]	2.31[i]	2.16[e]	0.99	Middle
ICl	Trimethylamine	ICl–trimethylamine	Cl—I···N	2.52[k]	2.32[g]	2.30[k]	2.16[e]	1.02	Middle
I₂	Trimethylamine	I₂–trimethylamine	I—I···N	2.83[l]	2.67[g]	2.27[l]	2.16[e]	1.01	Middle

[a] Bjorvatten (1969).

[b] Weast (1967).

[c] Estimated from O—Cl = 1.70 Å in Cl₂O and adding 0.30 Å when passing from Cl to I.

[d] Borgen et al. (1962).

[e] Distance in the Py₂—I⁺ ion determined by Hassel and Hope (1961).

[f] Hassel and Hvsoslef (1956).

[g] Jesson and Muetterties (1969).

[h] Chao and McCullough (1960).

[i] Estimated from the previous I···O distance, adding 0.39 Å when passing from O to S. The proposed value can be compared to P—I = 2.43 Å in PI₃ (ref. g).

[j] Hassel et al. (1961).

[k] Hassel and Hope (1960).

[l] Strømme (1959).

Other data can be found in Prout and Kamenar (1973).

are discussed in several books or reviews (Cannon, 1958; Hadzi, 1959; Pimentel and McClellan, 1960, 1971; Zeegers-Huyskens, 1968; Vinogradov and Linnell, 1971; Joesten and Schaad, 1974). The perturbations from the free molecules are mainly the frequency lowering and intensity enhancement of the ν_{AH} stretching vibration, a frequency increase of the AH in-plane deformation, a frequency increase and intensity enhancement of the γ_{AH} out-of-plane deformation and the appearance in the low-frequency region of new absorptions due to intermolecular modes.

When a proton transfer complex $A^- \cdots H^+B$ is formed, the spectrum will consist of a *superposition of the characteristics of the species* A^- *and* BH^+ with very little change (Hadzi, 1963). The main vibrational modes of this hydrogen-bonded system may be described as follows, assuming that the principal displacements involved in the oscillatory movements of atoms or groups are 'pure' vibrational modes:

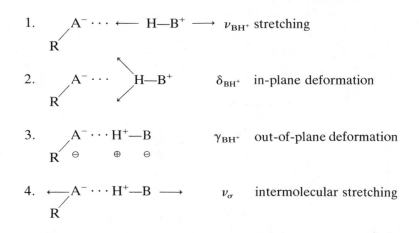

1. $A^- \cdots \longleftarrow H{-}B^+ \longrightarrow \quad \nu_{BH^+}$ stretching

2. $A^- \cdots \searrow H{-}B^+ \quad\quad \delta_{BH^+}$ in-plane deformation

3. $A^- \cdots H^+{-}B \quad\quad \gamma_{BH^+}$ out-of-plane deformation

4. $\longleftarrow A^- \cdots H^+{-}B \longrightarrow \quad \nu_\sigma$ intermolecular stretching

In these asymmetric systems where the proton remains closer to the base B, the four vibrational modes are infrared and Raman active. The main perturbations from the free ions are the following:

1. The absorption band due to the $B^+{-}H$ stretching vibrations is shifted to lower frequencies. This effect is related to the weakening of the force constant of the $B^+{-}H$ bond, arising principally from the charge transfer taking place from the non-bonded electrons of RA^- species to the antibonding orbital of the $B^+{-}H$ bond. The integrated intensity of fundamental ν_{B^+H} bands strongly increases. In contrast with normal hydrogen bonds, the complexation shift $\Delta\nu$ is more difficult to establish because the wavenumber for the free BH^+ ion is not known with great precision and must be evaluated by extrapolation methods. (The

wavenumber of the free ν_{NH^+} vibration of trialkylammonium salts may be located at about 3240 cm^{-1} (Moreno-Gonzales *et al.*, 1977).)

2. The in-plane deformation modes are shifted to higher frequencies, indicating that formation of hydrogen bonds constrains the deformation; the intensity of these vibrations does not change markedly.

3. The out-of-plane deformation modes are enhanced in frequency and in intensity.

4. New absorption bands appear in the far-infrared region below 200 cm^{-1}. This may be ascribed to intermolecular stretching or bending vibrations. These vibrations are generally studied by infrared spectrometry.

5. Some other vibrational modes not directly involved in the formation of the hydrogen-bonded ion (the internal modes of RA^- and BH^+) may be perturbed in frequency and intensity. These changes are usually greater than for normal hydrogen-bonded complexes and arise mainly from electronic redistribution effects. The shifts may be either to longer or shorter wavelengths.

Table 1.4 reports the observed wavenumber for the three vibrational modes involving the proton oscillation and for the intermolecular stretching mode of three pyridinium halides.

In this case, the strength of the hydrogen bond varies in the order $N^+H \cdots Cl^- > N^+H \cdots Br^- > NH^+ \cdots I^-$; the frequency ν_σ follows a similar trend. It must be pointed out, however, that the δ_{NH^+} vibrations show no regular variation with the strength of the bond. These vibrations are often coupled with skeletal modes in polyatomic molecules.

Some perturbations of the skeletal modes of the pyridine have been observed, mainly for the A_1 (1, 2, 8a, 19a) and B_1 (8b, 19b, 15) vibrations which are shifted to higher frequencies. These shifts can be explained by the electronic redistribution arising from protonation, mainly an increase of the σ character of the C—N bond (Filimonov and Bystrov, 1962) and of the overlap population of the ring (Adam *et al.*, 1968). The perturbation of the vibrational modes of the ring is more important in the pyridinium ion than in pyridine involved in a normal hydrogen bond (as for example with water or with aliphatic alcohols) and this fact can be explained by a greater charge

Table 1.4   Infrared data (cm^{-1}) relative to the pyridinium halides in the solid state (from Foglizzo and Novak, 1969a, b)

Vibrational mode	$C_6H_5NH^+Cl^-$	$C_6H_5NH^+Br^-$	$C_6H_5NH^+I^-$
ν_{NH^+}	2450	2800	3000
δ_{NH^+}	1249	1240	1245
γ_{NH^+}	945	912	871
$\nu_{\sigma(N^+\cdots X^-)}$	192	133	110

transfer to the proton: 0.69 e (Jordan, 1975) than to the water molecule; 0.03 e (Del Bene, 1975).

For normal hydrogen bonds, numerous correlations have been established between the spectroscopic parameters (frequency lowering or intensity enhancement of the ν_{AH} vibration) and the thermodynamic data (equilibrium constant, enthalpy of complex formation). For the ion-pair complexes, some difficulties arise in establishing such relations.

For a normal $AH \cdots B$ bond of weak to medium strength (15–25 kJ mol^{-1}), the ν_{AH} absorption appears as a broad ($\Delta\nu_{1/2} = 300$ cm^{-1}) definite band, reasonably free from structural effects and with a maximum which is easy to measure. This is the case for complexes between alcohols or phenol derivatives with ethers, aldehydes, amides, thioamides, etc. For such complexes, the band position and integrated intensity per complex are only weakly temperature dependent and the isotopic ratio ν_{AH}/ν_{AD} is close to $\sqrt{2}$ (Bratoz, 1975).

This is not generally the case for hydrogen-bonded ion-pairs because of the strength of the bands; the profile of the stretching band exhibits a more complicated pattern here and presents several submaxima, the spectral resolution of which is better at low temperature. This band is very broad with a half width of the order of 300–1000 cm^{-1} and the deuteration shift is usually lower than $\sqrt{2}$. For example, isotopic ratios ν_{NH^+}/ν_{ND^+} of 1.19 and 1.08 have been found for the ionic adducts of 3-methylpyridine with CCl_3COOH and $CHCl_2COOH$ respectively (Glazunov et al., 1976).

The presence of several submaxima in the ν_{AH} (or ν_{BH^+}) band has been extensively discussed (for a recent discussion, see Hadzi and Bratoz, 1976). The phenomena have been attributed to two mechanisms. The first is the anharmonic coupling between the high-frequency mode ν_{AH} (or $\nu_{(BH^+)}$) and the low external mode ν_{σ}. The second mechanism is the Fermi resonance between the stretching mode and some internal modes perturbed by the hydrogen bond which is at the present time proved beyond doubt (see for example Asselin et al., 1969). A typical absorption pattern is shown in Figure 1.1 where several submaxima are observed between 2750 and 2100 cm^{-1} for the ν_{NH^+} vibration of triethylammonium bromide in the solid state.

Several submaxima in the ν_{BH^+} vibration of hydrogen-bonded ion-pairs have been observed for $N^+H \cdots X^-$ bonds such as those in amine hydrohalides (Chenon and Sandorfy 1958; Cabana and Sandorfy, 1962) and hexamethylenetetramine hydrohalides (Marzocchi et al., 1965), or for $NH^+ \cdots O^-$ bonds such as those in aniline picrate (Briegleb and Delle, 1960) or triethylammonium dinitrophenolate (Zeegers-Huyskens, 1969). Recently, Huong and Schlaak (1974) observed a doublet in the Raman spectra of trimethylammonium halides, at 2600 and 2470 cm^{-1}; the disappearance of the second band in $(CD_3)_3NH^+Cl^-$ suggested that this band is

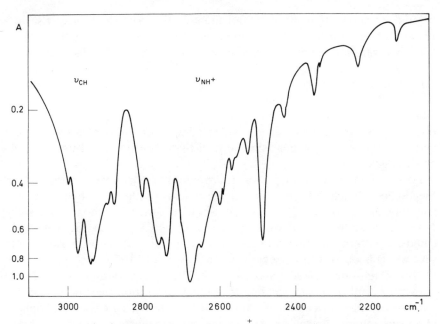

Figure 1.1   Infrared spectrum of solid $(C_2H_5)_3\overset{+}{N}HBr^-$ in the ν_{NH^+} region. KBr pellet; spectrophotometer P.E. 180. [Reproduced by permission of J. B. Rulinda from J. B. Rulinda, *Ph.D. Thesis*, Leuven, 1978]

due to the oscillation of the methyl groups and more specifically to the rocking vibration of these groups.

For strong hydrogen bonds a new type of absorption pattern can be observed. The characteristic features are the three bands called the *A*, *B*, *C* bands, usually observed between 2700 and 1600 cm^{-1} but sometimes at lower wavenumbers. These bands are connected with proton motion since they are affected by deuteration. As shown by Hadzi and Kobilarov (1966), the total integrated intensity of the *A*, *B*, *C* bands increases with increasing strength of the hydrogen bond and can be related to the wavenumber shift of the centre of gravity of the bands (Detoni *et al.*, 1970). The maxima of the bands are not temperature sensitive but show some narrowing at low temperature. The possible origin of this multiplicity has been discussed and Hadzi (1965) concluded that these three bands originate in an unusually strong Fermi resonance between the ν_{AH} and the second harmonic of the in-plane and out-of-plane deformation vibrations.

Another explanation was proposed by Claydon and Sheppard (1969) for a number of phosphinic and arsenic acids; these authors observed that the wavenumbers of the *minima* coincide better with the wavenumbers of the overtones. The three components of the bands are separated by broad transmission windows called 'Evans holes'. Evans (Evans and Wright, 1960) has indeed shown that when a fundamental vibration which gives rise to a *broad* IR band overlaps a *sharp* fundamental of the same symmetry, this can lead to a transmission window, the missing intensity being redistributed into the absorption region on either side of the window by resonance repulsion. For strongly hydrogen-bonded systems, where the stretching bands are intrinsically broad, these phenomena give rise to 'pseudo-maxima' in a vibrational spectrum. Several examples of transmission windows have been observed for strongly hydrogen-bonded complexes without proton transfer and for symmetrical or near-symmetrical hydrogen bonds. (This point will be discussed in section 3.3.)

For the complexes with proton transfer, three bands were observed for the ionic adducts of some dihalogeno-acids with trimethylarsine and trimethylamine oxide (Hadzi, 1970) where additional protonic bands are observed close to the antisymmetric band of the CO_2^- group. Some complicated IR patterns have been observed for the ionic adducts of pyridine with several acids of different strength (Odinokov *et al.*, 1976).

As pointed out by Hadzi and Bratoz (1976), it is very difficult 'to decide between Fermi resonance leading to several absorption maxima in a broad band or resonance cleaving windows in it by considering only the argument of wavenumbers'. Because of the strong anharmonicity involved, the overtones or combination bands will not be exactly twice the sum of the wavenumbers of their fundamentals and if the maxima and minima are separated by 50 or $100 \, \text{cm}^{-1}$, it is difficult to decide between the two proposed mechanisms. Moreover, it has been shown by these authors that these two descriptions are intimately related to each other.

The relationship between the structural and spectroscopic parameters is thus difficult to establish owing to the multiplicity of the observed bands and, moreover, the thermodynamic quantities relative to the formation of ionic complexes are rather limited.

Recently, Glazunov *et al.* (1976) and Odinokov *et al.* (1976) have correlated the enthalpy of complex formation and the frequency of the ν_{NH^+} stretching vibration for some ionic adducts of nitrogen bases and several acids. Numerical integration of the spectrogram data was performed to estimate the frequency of the centre of gravity of the ν_{NH^+} absorption.

More specific systems. The adducts of *carboxylic acids* with *aliphatic amines* generally appear to be of the ion-pair type in solvents of low dielectric constant. These complexes have been studied by several authors as, for example, the adduct of trichloroacetic acid with triethylamine (Jasinski and Kokot, 1967), pivalic, phenylacetic, formic acid with n-butylamine (Hibbert and Satchell, 1968), benzoic acid with cyclohexylamine or benzylamine (Bruckenstein and Saito, 1965; Bruckenstein and Unterecker, 1969), heptafluorobutyric acid and tri-n-octylamine (Gusakova *et al.*, 1971).

These $1:1$ adducts show no free $\nu_{C=O}$ absorption between 1780 and 1700 cm^{-1} and bands characteristic of the carboxylate ion appear at 1680–1560 cm^{-1} (ν_{as} CO$_2^-$) and at 1400–1300 cm^{-1} (ν_s CO$_2^-$). Some additional bands are observed: the $\nu_{NH^+\cdots O^-}$ vibration ranging from 2800 to 2200 cm^{-1} analogous to the salts in the solid state, the $\delta_{NH_2^+}$ or $\delta_{NH_3^+}$ bands at about 1620–1600 cm^{-1} and the $\delta_{CO_2^-}$ at vibration 670 cm^{-1}.

As pointed out by several authors (Barrow, 1956; Smith and Vitoria, 1968; Denisov et al., 1973b; Jasinski and Kokot, 1967), on decreasing the amine concentrations, a $2:1$ complex can be formed by addition of a second molecule of acid to the ion-pair; for this complex the absorption band of the carboxylate group can be lower than for the $1:1$ complex. This is the case for the adduct of trichloroacetic acid with tri-n-butylamine where the ν_{as} CO$_2^-$ absorption was observed at 1680 cm^{-1} for the $1:1$ adduct

$$CCl_3-C \overset{O}{\underset{O\cdots HN^+Bu_3}{\quad(-)}}$$

while this absorption band was shifted to 1640 cm^{-1} for the $2:1$ adduct (Jasinski and Kokot, 1967)

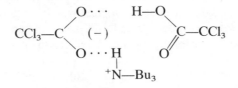

The formation of an ion-pair complex between pentachlorophenol and aniline (Zeegers-Huyskens, 1967) or secondary aliphatic amines (Denisov and Schreiber, 1972) was also indicated by IR spectroscopy by the appearance of the $\nu_{NH^+\cdots O^-}$ or by the ν_{C-O^-} vibration of the pentachlorophenolate ion.

A special case of complex formation was investigated by Matsunaga and Inoue (1972) for the 2,4-dinitrophenol adducts with various anilines. The complex is of the proton transfer type when the pK_a of aniline is higher than the pK_a of 2,4-dinitrophenol (4.09); this was shown by comparison with the spectra of the corresponding anilinium chloride. The adduct is of the charge transfer type when the pK_a is lower than 4.09.

The protonation of amides has been studied by several authors and vibrational spectroscopy clearly shows that the protonation site is the oxygen of the carbonyl function. The main arguments are the appearance of the ν_{OH^+}, δ_{OH^+}, and γ_{OH^+} bands, the low-frequency shift of the $\nu_{C=O}$ band, and the presence of a band characteristic of the amidium ion, lying between 1660 and 1725 cm^{-1}.

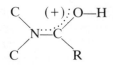

Some of these studies have been performed on the solid salts: N,N-dimethylacetamide adducts of HBr, HI, HCl (Cook, 1964) and N,N-dimethylformamide adducts with $HClO_4$ (Belin and Potier, 1976a). Table 1.5 reports the wavelengths of the ν_{OH^+}, δ_{OH^+}, and γ_{OH^+} vibrations. The ν_{OH^+} values suggest that the hydrogen bond strengths are in the following order: $Cl^- > Br^- > I^- > ClO_4^- > SbCl_6^-$. This sequence is confirmed by the isotopic ratio (ν_{OH^+}/ν_{OD^+}) which has the highest value for the $SbCl_6^-$ salt and the lowest for the Br^- salt. The correlation of the hydrogen bond strength with δ_{OH^+} and γ_{OH^+} frequencies appears to be less clear, probably due to coupling of these vibrations with other vibrational modes.

In concentrated aqueous solution, Combelas et al. (1970) studied the protonation of propionamide and polyacrylopiperidide by dichloroacetic acid; the formation of ionic species can be shown by the appearance of an absorption band characteristic of the acetate ion at $1607\ cm^{-1}$ and of the amidium cation at $1660\ cm^{-1}$; the latter band was observed at $1690\ cm^{-1}$ for the adduct of N,N-dimethylacetamide with HCl in aqueous solution (de Lozé et al., 1972). The IR spectra of mixtures of trifluoroacetic acid and some simple amides in carbon tetrachloride solution can also be interpreted in terms of proton transfer to the carbonyl group of the amide molecule (Tam and Klotz, 1973). The IR spectra of protonated acetone have been recently examined by Huong and Noel (1976); the major features observed in the spectrum are the appearance of the ν_{OH^+}, δ_{OH^+} and γ_{OH^+} vibrations. The displacement by only $130\ cm^{-1}$ of the $\nu_{C=O}$ vibration suggests that protonated acetone is an oxonium ion $(CH_3)_2C=O^+$—H and not a carbonium ion $(CH_3)_2C^+$—O—H.

The homoconjugated salts $(BHB)^+$ may be obtained in excess of base. The spectra of these adducts present a very broad absorption band, starting in the $1700\ cm^{-1}$ region, sometimes interrupted by transmission windows; this

Table 1.5   OH$^+$ vibrations of some acetamide salts (cm^{-1})

Adduct	ν_{OH^+}	ν_{OH^+}/ν_{OD^+}	δ_{OH^+}	γ_{OH^+}	References
N,N-Dimethylacetamide·HCl	2095		1352	928	Cook (1964a)
N,N-Dimethylacetamide·HBr	2460	1.25	1342	839	
N,N-Dimethylacetamide·HI	2720	1.29	1310	761	
N,N-Dimethylacetamide·HSbCl$_6$	3360	1.35	1300		
N,N-Dimethylformamide·HClO$_4$	2750	1.31	1372	790	Belin and Potier (1976a)

pattern is characteristic of very strong, quasi-symmetrical OHO bonds and will be discussed in section 3.3.

The $\nu_{C=O}$ vibration is shifted to low frequencies by protonation and an interesting feature of this frequency lowering is that it can be related to the charge transfer taking place from the amide (or ketone) molecule to the H^+ ion. The charge transfer has been calculated by *ab initio* SCF calculations for protonated acetone (Del Bene and Vaccaro, 1976) and protonated formamide (Hopkinson and Csizmadia, 1974). These values are listed in Table 1.6. Some values relative to normal hydrogen bonds are reported for comparison; for these $C=O \cdots HO$ bonds, the value of the charge transfer has been calculated by Johansson and Kollman (1972) for the formamide–water dimer or evaluated from the $\Delta\nu_{OH}$ values for some complexes of dimethylacetamide (DMA) with phenols (Zeegers-Huyskens, 1975).

Other spectroscopic studies have been performed in the protonation of trimethylamine oxide (Cook, 1963), ethers (Cook, 1964b); Lassègues *et al.*, 1971), trimethylamine, and trimethylarsine oxide (Hadzi, 1970).

The infrared and Raman spectra of pyridinium salts (Foglizzo and Novak, 1970), mono- and diprotonated purine salts (Lautie and Novak, 1971) and N,N-dimethylanilinium hydrochlorides (Perrier-Datin *et al.*, 1973) have been analysed and the effect of protonation on the ring vibrations discussed. For these salts, the isotopic ratio ν_{NH}/ν_{ND} ranges from 1.29 to 1.20, indicating medium to strong hydrogen bonds, and the ν_{NH} absorption often presents a complicated pattern with several submaxima.

Thus characteristic differences can be found between the spectra of normal molecular complexes and ionic complexes resulting from proton

Table 1.6   Comparison between $\Delta\nu_{C=O}$ (cm^{-1}) and charge transfer (e^-) for some normal hydrogen bonds and protonated species involving $C=O$ bonds

System	$\Delta\nu_{C=O}$ (cm^{-1})	References	Charge transfer (e^-)	References
DMA·H$_2$O	16	de Lozé *et al.* (1972)	0.036*	Johansson and Kollman (1972)
DMA·dimethylphenol	20	Dorval and Zeegers-Huyskens (1973)	0.069	Zeegers-Huskens (1975)
DMA·phenol	23	Dorval and Zeegers-Huyskens (1973)	0.073	Zeegers-Huyskens (1975)
DMA·3-nitrophenol	31	Dorval and Zeegers-Huyskens (1973)	0.093	Zeegers-Huyskens (1975)
Protonated acetone	130	Huong and Noel (1976)	0.476	Del Bene and Vaccaro (1976)
Protonated DMA	270	de Lozé *et al.* (1972)	0.621*	Hopkinson and Csizmadia (1974)

* Value relative to formamide.

transfer and typical examples have been given for the internal vibrations of the proton donor or the proton acceptor. Now, the low-frequency spectra show the intermolecular vibrational modes, the stretching (ν_σ) and generally below 100 cm^{-1}, the in-plane and out-of-plane bending vibrations. The limited amount of experimental data already obtained from these investigations does not enable any distinction to be made between the molecular complexes and the proton transfer species. As, for example, the ν_σ ($\text{OH} \cdots \text{N}$) vibration is observed at 168 cm^{-1} for the pyridine acetic acid complex (Denisov et al., 1973b) and at 134 cm^{-1} for the pyridine–phenol complex (Ginn and Wood, 1967). For the ionic adducts of trichloro- and trifluoroacetic acid with pyridine, the ν_σ ($\text{NH}^+ \cdots \text{O}^-$) are observed at 137 and 158 cm^{-1} respectively (Denisov et al., 1973b).

Table 1.7 lists the value of ν_σ for some ion-pair complexes and the corresponding force constant k_σ; the k_σ values are calculated by the diatomic model approximation (Foglizzo and Novak, 1969a) or a triatomic model (Marzocchi et al., 1965) using the entire masses of the cation and the anion in both cases. These results show that the low-frequency vibrations

Table 1.7 Intermolecular stretching vibration ν_σ (cm^{-1}) and force constant k_σ (N m^{-1}) of some proton transfer complexes

Complex	Bond	ν_σ (cm^{-1})	k_σ (N m^{-1})	State or solvent	References
Pyridine oxide·HCl	$\text{O—H}^+ \cdots \text{Cl}^-$	214		Solid	Hadzi (1962)
Pyridine oxide·HBr	$\text{O—H}^+ \cdots \text{Br}^-$	180			
Picoline oxide·HCl	$\text{O—H}^+ \cdots \text{Cl}^-$	241			
Hexamethylene-tetramine·HCl	$\text{N—H}^+ \cdots \text{Cl}^-$	187	75	Solid	Marzocchi et al. (1965)
Hexamethylene-tetramine·HBr	$\text{N—H}^+ \cdots \text{Br}^-$	121	50		
Hexamethylene-tetramine·HI	$\text{N—H}^+ \cdots \text{I}^-$	100	44		
Hexamethylene-tetramine·HClO$_4$	$\text{N—H}^+ \cdots \text{ClO}_4^-$	83	25		
Pyridine·HCl	$\text{N—H}^+ \cdots \text{Cl}^-$	192	53	Solid	Foglizzo and Novak (1969a)
Pyridine·HBr	$\text{N—H}^+ \cdots \text{Br}^-$	133	42		
Pyridine·HI	$\text{N—H}^+ \cdots \text{I}^-$	110	35	CHCl$_3$	Kludt et al. (1972)
Triethylamine·HCl	$\text{N—H}^+ \cdots \text{Cl}^-$	180		CHCl$_3$	
Triethylamine·HBr	$\text{N—H}^+ \cdots \text{Br}^-$	133			
Pyrimidine·HCl	$\text{N—H}^+ \cdots \text{Cl}^-$	198	57	Solid	Foglizzo and Novak (1971)
Pyrimidine·HI	$\text{N—H}^+ \cdots \text{I}^-$	112	37		
Trifluoroacetic acid·pyridine	$\text{N—H}^+ \cdots \text{O}^-$	158		CHCl$_3$	Denisov et al. (1973b)
Trichloroacetic acid·pyridine	$\text{N—H}^+ \cdots \text{O}^-$	137			

are sensitive to the nature of the cation and probably to its mass. The use of the entire masses of the ions leads to an overestimation of the k_σ values and a complete normal coordinate analysis is greatly needed to obtain the real values of the force constants. These calculations are, however, difficult because the usual method of fixing the interaction constants by the isotopic substitution method is inapplicable since deuteration in the bridge position causes a very small shift. For example an isotopic ratio of 1.01 was found for the ν_σ vibration of $C_6H_5NH^+Cl^-$ and $C_6D_5ND^+Cl^-$ (Foglizzo and Novak, 1969a).

1.2.2.2 *Electron transfer complexes*   If the proton transfer complexes have been intensively studied by vibrational spectroscopy, the data relative to $n-\sigma$ inner complexes in charge transfer interactions are rather limited. The formation of inner complexes leading to the formation of A^- and D^+ species can perturb the vibrational spectra and in this case, the spectra is *close* to a superposition of the spectra of the ionized components A^- and D^+.

The interaction of halogen with bases in solvents of medium to strong polarity can lead to the formation of $(B_2X)^+$ and X_3^- entities and the formation of these ion-pair complexes is indicated by the appearance of the typical vibration of the X_3^- ions and of the skeletal vibrations arising from the formation of the B—X bond. The presence of I_3^- ions can be inferred from bands arising from the asymmetric stretching vibration at 130–140 cm^{-1} (infrared) (Ginn and Wood, 1966) and from the symmetric stretching vibration at 114 cm^{-1} (Raman) (Stammreich *et al.*, 1961). The symmetric and asymmetric stretching modes of the Br_3^- ion have been found to occur near 160 cm^{-1} (Raman) and 190 cm^{-1} (infrared) respectively, in a number of tribromide compounds (Person *et al.*, 1961). These values depend on the nature of the associated cation and on the polarity of the solvent. Bands at 114 cm^{-1} and 160 cm^{-1} were observed in the Raman spectrum of moderately strong complexes formed between bases and iodine or bromine respectively; these were attributed to the formation of ion transfer complexes (Klaboe, 1967).

An interesting series of complexes formed between pyridine (Pyr) and halogens or mixed halogens, has been studied by spectroscopic and theoretical methods. X-Ray diffraction has shown that the solid substance recovered from solutions of iodine in pyridine contains the $(Pyr_2I)^+$ ion characterized by a linear, centrosymmetric N—I—N bond (Hassel and Hope, 1961).

In solvents of low polarity such as benzene, cyclohexane, chloroform, these complexes are un-ionized (form **1**) and their vibrational spectra show bands characteristic of the ν_{BrBr} and $\nu_{(N\cdots Br)}$ vibrations, in the low-frequency range. Normal complexes between pyridine and chlorine or bromine are also formed at low temperature (~ 77 K) as evidenced by the Raman spectrum (Kimel'fel'd *et al.*, 1975). In more polar solvents such as CH_2Cl_2,

CH_2ClCH_2Cl, the pyridine–halogen complexes ionize according to the following equilibrium:

$$2\,Pyr \cdots X\!\!-\!\!X \rightleftharpoons (Pyr_2X)^+ + X_3^-$$
$$\quad\;\; \mathbf{1} \qquad\qquad\qquad \mathbf{2}$$

When form (**2**) is formed in appreciable concentration, the vibrational spectra show typical bands of the $\nu_{(X_3^-)}$ and $\nu_{(N\!-\!X\!-\!N)}$ vibrations. For example, the iodine complex in CH_2Cl_2 solution is characterized by bands lying at $172\,cm^{-1}$ (IR) and $180\,cm^{-1}$ (R) attributed to the asymmetric and symmetric $\nu_{(N\!-\!I\!-\!N)}$ vibrations; the $\nu(I_3^-)$ band is observed at $137\,cm^{-1}$ (IR). Some bands are, however, difficult to observe individually owing to their overlapping; the frequency of the N—I—N stretching coincides with the I—I stretched of the un-ionized complex (Haque and Wood, 1967).

In a similar way, for the $(Pyr_2Br)^+$ ion, the $\nu(N\!-\!Br\!-\!N)$ vibrations are situated at $195\,cm^{-1}$ (R) and $170\,cm^{-1}$ (IR) respectively (Haque and Wood, 1968); here there is also some overlapping of the $195\,cm^{-1}$ band with the Br—Br stretch of the normal complex. Figure 1.2 shows the Raman spectrum of the pyridine–Br_2 complex in the 250—$150\,cm^{-1}$ region where the presence of the Br_3^- ion is shown by the band at $161\,cm^{-1}$; the relative intensity of this band is stronger in CH_3CN than in CH_2Cl_2, indicating greater concentration of form (**2**) in the first solvent.

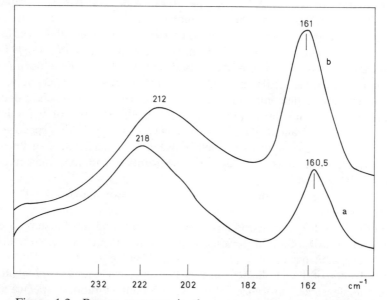

Figure 1.2   Raman spectrum in the $\nu_{Br\!-\!Br}$ region of pyridine and bromine solutions. Solvents: (a) CH_2Cl_2; (b) CH_3CN ($c_{pyridine} = 1\,mol\,l.^{-1}$; $c_{bromine} \approx 0.5\,mol\,l.^{-1}$); $T = 298\,K$.   Spectrophotometer Coderg T 800 equipped with a Kr^+ laser. [Reproduced by permission of G. Maes from G. Maes, *Ph.D. Thesis*, Leuven, 1978]

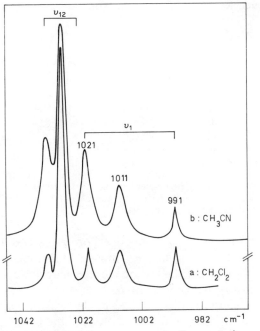

Figure 1.3 Raman spectrum in the ν_1 region of pyridine and bromine solutions. Solvents: (a) CH_2Cl_2; (b) CH_3CN ($c_{\text{pyridine}} = 1\ \text{mol}\ \text{l}.^{-1}$; $c_{\text{bromine}} \simeq 1\ \text{mol}\ \text{l}.^{-1}$); $T = 298$ K. (Reproduced by permission of G. Maes from G. Maes, *Ph.D. Thesis*, Leuven, 1978)

An interesting spectroscopic effect of such complexes is the perturbation of the internal modes of the pyridine molecule, mainly the ring modes. The Raman spectrum of the pyridine–Br_2 complex in the ν_1 region (ring breathing) is reproduced in Figure 1.3. The ν_1 vibration occurs at 991 cm^{-1} in free pyridine, at 1011 cm^{-1} in the normal complex and at 1021 cm^{-1} in the ionized complex; it must be pointed out that the relative intensity of the 1021 cm^{-1} band is greater in CH_3CN than in CH_2Cl_2 suggesting again greater contribution of form (2) in CH_3CN.

Table 1.8 reports the frequencies of some ring modes sensitive to the complex formation; the values relative to free pyridine, the normal pyridine–Br_2 complex, and the $(Pyr_2H)^+$ ion are also listed in this table. These results suggest that charge transfer occurring from the pyridine molecule(s) to Br_2 (or Br^+ ion) has proceeded even further in the three $(Pyr_2X)^+$ ions than in the normal pyridine $\cdots Br_2$ complex.

The extent of charge transfer is of the same order of magnitude for the $(Pyr_2Br)^+$ and $(Pyr_2I)^+$ ions for which the perturbations of the ring frequencies are very similar. Comparison of these two ions with $(Pyr_2H)^+$ seems

Table 1.8   Frequencies (cm^{-1}) of some ring modes sensitive to the complex formation

Vibrational modes	Free pyridine[a]	Pyr $\cdots$ Br$_2$[b]	(Pyr$_2$H)$^{+}$ [c]	(Pyr$_2$Br)$^{+}$ [d]	(Pyr$_2$I)$^{+}$ [d]
ν_{8a}	1583	1590	1639	1605	1605
ν_{19b}	1439	1447	1549	1455	1452
ν_1	992	1009	1007	1016	1013
ν_{6a}	605	626	628	644	640
ν_{16b}	405	418		450	438

[a] Cummings and Wood (1973), IR frequencies.
[b] Ginn et al. (1968), results relative to benzene solutions where no ionization occurs.
[c] Clements and Wood (1973a), IR frequencies, the Raman frequencies differ by 2–5 cm^{-1}.
[d] Haque and Wood (1968), results relative to CH$_2$Cl$_2$ solutions. The frequencies are the mean values of the IR and Raman frequencies.

more difficult: the ν_{8a} and ν_{19b} modes are more perturbed for (Pyr$_2$H)$^{+}$ and the reverse is true for the ν_1 and ν_{6a} modes. Direct comparision of these frequencies and particularly the low frequencies may however be affected by the change in mass of the central atom.

The frequency shifts of the pyridine modes can be explained, at least qualitatively, by a variation of the bond order of the N—C and C—C bonds of the ring. CNDO/2 calculations performed by Sabin (1972) have shown that the greatest changes in going from free pyridine to (Pyr$_2$I)$^{+}$ are for the 1,2 bond

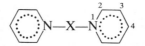

which decreases from 0.62 to 0.47 and the 2.3 bond which increases from 0.68 to 0.71. Similar calculations have been performed on the (Pyr$_2$X)$^{+}$ ions where X = H, F, Cl. The π-bond orders of the three bonds of the pyridine ring are listed in Table 1.9; this table also reports the values relative to free pyridine (Sabin, 1972). These results show clearly that the C$_{(2)}$—C$_{(3)}$ π-bond order increases and these changes should be reflected by an increase in

Table 1.9 π-Bond order in free pyridine and in (Pyr$_2$X)$^{+}$ complexes (modified after Sabin, 1972)

Molecule or ion	N$_{(1)}$—C$_{(2)}$	C$_{(2)}$—C$_{(3)}$	C$_{(3)}$—C$_{(4)}$
Free pyridine	0.6585	0.6795	0.6595
(Pyr$_2$X)$^{+}$, X = H	0.6308	0.6940	0.6484
X = F	0.6313	0.6929	0.6494
X = Cl	0.6334	0.6920	0.6501

frequency of the pyridine modes involving mainly displacements of the $C_{(2)}$ and $C_{(3)}$ atoms (as for example the ν_{19b}, ν_{6a}, and ν_{16b} modes).

Like the spectroscopic results, the calculations do not show definitive trends in the series of the H, F, and Cl complexes. The bond energies are high and the order of stability is Cl $(200 \text{ kJ mol}^{-1}) > H$ $(161 \text{ kJ mol}^{-1}) > F$ (65 kJ mol^{-1}). On the other hand, the values of the Mulliken atomic populations of the X atom are 0.7098, 6.9629, and 6.7678 for H, F, and Cl respectively, indicating that the extent of charge transfer from the two pyridine entities to the X atom is greatest for the F complex.

Sabin's calculations indicate also that the potential wells for the N—I—N asymmetric vibrations are single minimum potentials and this was observed experimentally by Haque and Wood (1968). The main spectroscopic arguments in favour of a linear, centrosymmetrical structure of the N—I—N chain are:

1. The infrared inactivity of the symmetric (N—I—N) vibration and the Raman activity of the asymmetric vibration;
2. The observation of only one band for each internal mole of the base. This implies that the two pyridine molecules are equivalent. The infrared bands coincide only accidentally with the Raman bands. As for example, the ν_1 ring vibration of the pyridine in the $(Pyr_2Br)^+$ ion was observed at 1011 cm^{-1} in the infrared and at 1021 cm^{-1} in the Raman spectrum. The values listed in Table 1.9 are the mean values of the two frequencies.

It must be pointed out however that the $(N—H—N)^+$ ion is characterized by an asymmetric potential where the proton remains closer to one nitrogen atom and some important spectroscopic features suggest a non-centrosymmetric structure in solution (Clements and Wood, 1973b; Dean and Wood, 1975a). These structural and spectroscopic properties will be discussed in more detail in section 3.3.

1.2.3 *Electronic spectroscopy*

1.2.3.1 *Proton transfer complexes*   The effects of hydrogen bond formation on the electronic spectra have been discussed extensively by Pimentel and McClellan (1960), Mataga and Kubota (1970), and by Davis (1968). When a normal hydrogen bond is formed, the absorption spectra resulting from electronic transitions are perturbed in frequency and intensity but the variations from the free molecules generally remain moderate. Absorption spectra are due to electronic transitions from the ground state to the Franck–Condon excited state and therefore the frequency shift should correspond to the difference between the hydrogen bond energies in the ground state (E^G) and in the excited state (E^E). A red shift is observed when $E^G < E^E$ and a blue shift arises if $E^E < E^G$. As, for example, the ^1L_b transition of a phenol derivative complexed with an aliphatic amine undergoes a bathochromic shift accompanied by a hyperchromic effect but the $n \rightarrow \sigma^{\pm}$ transition is shifted to shorter wavelengths (Baba and Nagakura, 1951; Nagakura and Baba 1952; Kubota, 1966). The same effect can sometimes be observed for the ^1B and ^1L_a transitions of the phenol derivative (Dupont *et al.*, 1971).

When a proton transfer complex is formed, the electronic spectrum resembles that of the RA^- or BH^+ ions. From the neutral molecules RAH and B, the shifts may either be to shorter or longer wavelengths and typical examples are the aniline and the *p*-nitrophenol molecules whose ^1L_b transitions are respectively shifted to the blue and to the red through ionization. (For a general discussion see Wheland, 1955.)

The most investigated systems are those formed between the phenol derivatives and aliphatic bases where no overlapping occurs between the $\pi \rightarrow \pi^*$ transitions of the acid and the $n \rightarrow \sigma^{\pm}$ transitions of the base, lying in the short-wavelength region (<230 nm).

The effect of solvent and added bases on the colour of bromophthalein magenta, which can be considered as a phenol derivative, has been studied extensively by Davis and coworkers (Davis and Schuhmann, 1947; Davis and Hetzer, 1952, 1954; Davis and Paabo, 1960). The yellow colour of the free base ($\lambda_{max} = 400$ nm) is explained by the quinoid structure of ring 1:

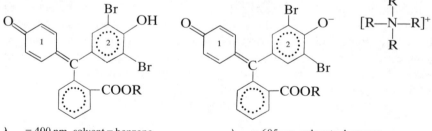

$\lambda_{max} = 400$ nm, solvent = benzene $\lambda_{max} = 605$ nm, solvent = benzene

The yellow tone of the free base can be converted at will to practically any colour by adding a proper base in a suitable amount. Solutions containing quaternary ammonium salts in benzene are found to give blue solutions ($\lambda_{max} = 605$ nm); the shift to a deep colour is explained by resonance structures in which the electronic patterns of ring 1 and 2 are quinoid and phenolate respectively and a second important structure is the one in which these electronic patterns are reversed. With tertiary amines, the limiting colour is magenta ($\lambda_{max} = 540$ nm) and in this case the proton remains hydrogen bonded to the cation:

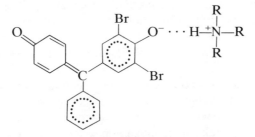

$\lambda_{max} = 540$ nm, solvent = benzene

In such a structure, a part of the charge of the O^- atom is retained by the ammonium cation and the delocalization of the charges in the phenolate ion is less pronounced.

A remarkable departure from indicator behaviour was observed in aqueous or ethanol solutions where all bases cause similar colour changes. In these hydroxylic solvents, the ion-pairs are strongly dissociated and further solvated by one or several hydroxyl groups; the blue colour is, in this case, independent of the nature of the base.

The formation of hydrogen-bonded ion-pairs was regarded as being an 'incipient ionization . . . an intermediate step in the complete transfer of a proton' (Davis, 1968). Similar studies were performed on bromocresol green, iodophenol blue, and tetrabromophenol blue (Davis *et al.*, 1948).

The electronic spectra of adducts of mono-, di-, or trinitrophenol derivatives with aliphatic or cyclic amines have been extensively studied and the existence of ion-pair complexes is generally shown by the appearance of a band lying at about the same wavelength as that of the corresponding phenoxide ion. For example, the ultraviolet spectrum of mixtures of 2,4-dinitrophenol and aliphatic or cyclic amines in aromatic solvents, indicates that the phenol derivative is ionized in a similar way as in water (Bayles and Chetwyn, 1958; Bayles and Taylor, 1961; Bayles and Evans, 1965). The spectrum of mixtures of 4-nitrophenol and 2-bromo-5-nitrophenol and triethylamine in toluene solution is not very different from that of the corresponding free phenol and it may be inferred that normal OH · · · N bonds are formed; but the 2,5- and 2,6-dinitrophenol complexes exhibit a shift very similar to that in alkaline water medium, indicating that these adducts form ion-pair complexes (Bell and Crooks, 1962). In a similar way, the spectrum of solutions of pentachlorophenol and triethylamine resembles the spectrum of the pentachlorophenoxide ion, and so, the formation of a proton transfer species can be deduced (Pilyugin, 1975).

Some other systems including mono- or dinitrophenol derivatives with aliphatic or aromatic nitrogen bases have been further studied by Pearson and Vogelsong (1958), Davis (1962), Jasinski and coworkers (1965a, b, c; 1967, 1968, 1969), Jannakoudakis and Moumtzis (1968), Steigman and Lorenz (1966), Baba et al. (1969), and more recently by Hudson and Scott (1972), Bron and Simmons (1976), Moulik et al. (1976), Kraft et al. (1975), and Romanowski and Sobczyk (1975). For all these systems, the band characteristic of the phenoxide ion is always observed near 400 nm but the wavelength and the extinction coefficient are dependent on the solvent. Table 1.10 reports the value of λ_{max} and ε_{max} of the $\pi \to \pi^{\pm}$ band for the p-nitrophenol (or p-nitrophenoxide ion) molecule in different environments.

The λ_{max} and ε_{max} take their maximum values for the ion-pair complex dissolved in basic solvents of high dielectric constants such as dimethylsulphoxide leading to the formation of 'solvent-separated ion-pairs', where the N^+H groups are solvated by the basic solvent but where the phenoxide ion remains essentially free (Baba et al., 1969). In dimethylformamide, or acetonitrile, the λ_{max} values are observed respectively at 430 nm (Scott et al., 1968) and 429 nm (Baba et al., 1969). For these free p-nitrophenoxide ions, the conjugation of the oxygen $2p$ π electrons to the aromatic ring has a maximum value. In hydroxylic solvents (water, butanol, or mixed hydroxylic solvents) the band is observed near 400–405 nm (Scott et al., 1968; Scott and Vinogradov, 1969); in these solvents, the phenoxide ion is solvated by one or more hydroxyl groups and a part of the charge of the oxygen is

Table 1.10   Characteristics of the electronic spectrum of the p-nitrophenol molecule in different environments

$O_2N-\langle \rangle-OH$	$O_2N-\langle \rangle-OH\cdots N$	$O_2N-\langle \rangle-O^-\cdots H^+N$
Free molecule	Normal hydrogen bond with triethylamine	Hydrogen-bonded ion-pair with piperidine
$\lambda_{max} = 286$ nm	$\lambda_{max} = 307$ nm	$\lambda_{max} = 390$ nm
$\varepsilon_{max} = 10{,}300$	$\varepsilon_{max} = 12{,}000$	$\varepsilon_{max} = 18{,}900$
Solvent = cyclohexane	Solvent = cyclohexane	Solvent = methanol
(Dupont et al., 1971)	(Dupont et al., 1971)	(Beier, 1975)

$O_2N-\langle \rangle-O^-\cdots n\,H_2O$	$O_2N-\langle \rangle-O^-\cdots$ Solvent
p-Nitrophenoxide ion in water	'Free' p-nitrophenoxide ion
$\lambda_{max} = 400$ nm	$\lambda_{max} = 436$ nm
$\varepsilon_{max} = 18{,}000$	$\varepsilon_{max} = 34{,}000$
Solvent = water	Solvent = dimethylsulphoxide
(Beier, 1975)	(Beier, 1975)

Table 1.11   Spectral characteristics of some amine 2,6-dinitrophenolates and amine picrates

	2,6-Dinitrophenolate[a]				Picrate[b]		
base	pK_a	λ_{max} (nm)	ε_{max}	base	pK_a	λ_{max} (nm)	ε_{max}
sym-Collidine	7.45	415	—	sym-Collidine	7.45	347	18,500
n-Butylamine	10.61	425	8700	Pyridine	5.13	346	17,400
Di-n-butylamine	11.31	425	7800	Aniline	4.57	348	15,000
Tri-n-butylamine	10.83	430	8600	Triethylamine	10.75	356	18,000
Triethylamine	10.75	430	9060	$(C_4H_9)_4N$	—	373^c	$19,200^c$

[a] Jasinski and Widernikova (1969); solvent = n-hexane, $T = 298$ K.
[b] Chantooni and Kolthoff (1968); solvent = acetonitrile.
[c] Since $(C_4H_9)_4N$ picrate is almost completely dissociated in 0.01 M solution, its spectrum is that of the simple picrate ion.

transferred to the $\sigma_{OH}^{\ddagger}$ orbitals, causing a lowering of the electron delocalization in the ring. This lowering is a little more pronounced when the phenoxide ion is bound to a H^+NR_3 cation and the band is observed near 390 nm.

In contrast with the behaviour of bromophthalein magenta which contains two aromatic rings, the spectra of mono-or polynitrophenoxide ions are not very sensitive to the nature of the associated cations. Table 1.11 reports the spectral characteristics of some amine picrates and amine 2,6-dinitrophenolates and the pK_a of the corresponding amine. These results show that the λ_{max} values are somewhat lower for the weaker bases such as the pyridine or aniline derivatives than for the stronger aliphatic bases but no definite relation exists between the wavelengths or the absorption coefficient and the pK_a values of the nitrogen bases. The lack of correlation may be due—at least partly—to the presence of multiple hydrogen bonds in the primary or secondary bases:

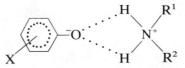

A special case of complexes formed by pyridine-1-oxide and carboxylic or sulphuric acids was investigated recently by Kreevoy and Chang (1976). The UV spectra are not combinations of the spectra of the protonated or unprotonated base, but show bands at intermediate positions, indicating that neither fully protonated nor unprotonated base are present. These complexes seem to have two partial OH bonds but the results do not indicate whether there is one, or two, minima between the two basic oxygens in the potential function for the hydrogen.

One of the great interests of electronic spectroscopy is the precise determination of the equilibrium constants and the related thermodynamic functions. Some recent values of these parameters are tabulated in section 2.6 of this chapter.

1.2.4   *Dipole moments*

The displacement Δr_{ap} of the positive ion p^+ brings about a change of the dipole moment, which, owing to the charge of the ion is given by the expression

$$\Delta \mu_{p^+} = 4.8(r_{ap} - r_{ap}^0) = 4.8\Delta r_{ap}$$
$$\text{(D)} \qquad\qquad\qquad\qquad\qquad \text{(\AA)}$$

However as mentioned in the introduction, this is not the only change in the dipole moment originated by the formation of the hydrogen bond or E.D.A. bond. The motion of electrons in the opposite direction makes some contribution to the change of dipole moment.

From the data of Table 1.1, it can be seen that, for typical proton transfer complexes, where ζ is markedly greater than one, Δr_{ap} exceeds 0.5 Å. Thus it appears that the formation of proton transfer complexes is accompanied by a strong enhancement of the dipole moment, $\Delta \mu$ by, in general, more than 5 D. A similar statement could be made for ion transfer E.D.A. complexes.

The appearance of proton transfer (or ion transfer) complexes can thus be inferred from the high value of the dipole increment brought about by complexation.

This overall dipole increment $\Delta \mu$ is, in effect, a *vector* difference defined by

$$\Delta \mu \equiv \mu_{AD} - \mu_A - \mu_D$$

where μ_{AD}, μ_A, and μ_D are the dipole moments of the complex and of the separate components respectively. In order to determine the value of $\Delta \mu$ it is therefore necessary to know, not only the values of the moments[†] but also the angles these vectors form with each other.

In a good approximation it can be assumed that $\Delta \mu$ has the direction of the hydrogen bond. Let us consider the plane formed by the dipole moment μ_D of the donor and the hydrogen bond (Figure 1.4). The moment μ_D forms with the hydrogen bond an angle θ_D. The moment μ_A forms with this plane

---

[†] These dipole moments are generally determined from the permittivities, the refractive indexes, and the densities of the solutions. A discussion of the methods was recently given by Sobczyk (Sobczyk *et al.*, 1976).

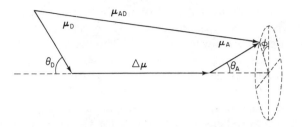

Figure 1.4 Dipole moments μ_{AD}, μ_A, and μ_D of the
complex and of the partners. Dipole increment $\Delta\mu$

an azimuthal angle ϕ and an angle θ_A with the direction of the hydrogen
bond. It is convenient to set ϕ equal to zero when the steric groups of both
partners lie in the *cis* position with respect to each other.

The vector equation can then be transformed in the expression

$$\Delta\mu = \sqrt{\mu_{AD}^2 - \mu_A^2 \sin^2 \theta_A - \mu_D^2 \sin^2 \theta_D - 2\mu_A\mu_D \sin \theta_A \sin \theta_D \langle \cos \phi \rangle} \cdot$$
$$- \mu_A \cos \theta_A - \mu_D \cos \theta_D$$

For many O—$H \cdots N$ hydrogen bonds the mean value $\langle \cos \phi \rangle$ is
zero, which means that all the rotamers around the axis of the bond are
equally represented. For O—$H \cdots O$ bonds the mean value $\langle \cos \phi \rangle$ is
generally negative because steric hindrance does not favour the rotamers in
the *cis* position (Huyskens *et al.*, 1980a). A similar effect was predicted by
Gerbier and Baraton (Baraton, 1979) for complexes between phenols and
acetonitrile.

The problem of finding a correct value for the angles θ_A and θ_D can be
more or less complicated. In $Et_3N \cdots I$—Cl for instance, both angles can be
taken equal to zero for reasons of symmetry. In the hydrogen-bonded
complexes of Et_3N with phenols θ_D remains equal to zero but θ_A depends
on the nature of the phenol. It can be found to a good approximation by
comparing the dipole moments of the substituted phenols and assuming the
additivity of the group moments. For symmetric nitrogen bases $\theta_D = 0$ but
this is no longer the case for symmetric oxygen bases as Et_2O or
$(CH_3)_2C$=O where θ_D is of the order of $55°$ and $60°$ respectively.‡

Errors as large as 0.3 D for $\Delta\mu$ are very common owing to uncertainties
about the values of the angles.

When, for a given family of hydrogen-bonded complexes in a given
solvent, $\Delta\mu$ is plotted against the difference

$$\Delta pK_a = pK_{a_{BH^+}} - pK_{a_{HA}}$$

‡ In a given family of complexes the most probable values of the angles are those which give
the best correlation between the computed values of $\Delta\mu$ and another characteristic of the bond
such as ΔH or ΔpK_a (Huyskens, 1974; Huyskens *et al.*, 1980a)

where BH^+ is the conjugated acid of the base, the pK_a's being taken in water, and when ΔpK_a extends over a large range, *sigmoidal* curves as shown in Figure 1.5 are always obtained. (The figure refers to complexes of substituted phenols with pyridines or imidazoles.) This behaviour was first observed by Ratajczak and Sobczyk (1969, 1970) for complexes between triethylamine and phenols, but it is quite general for $O—H \cdots N$ bonds (Debecker and Huyskens, 1971; Jadzyn and Malecki, 1972; Nouwen and Huyskens, 1973; Huyskens *et al.*, 1980b), for $O—H \cdots O$ bonds (Pauwels and Huyskens, 1974; Huyskens *et al.*, 1980a), and also for $Cl—H \cdots O$ bonds (Rospenk *et al.*, 1977). This clearly demonstrates that *above a given* ΔpK_a *value*, the dipole increment becomes very high and that in the organic solutions considered *most of the hydrogen bonds belong to the proton transfer type.*

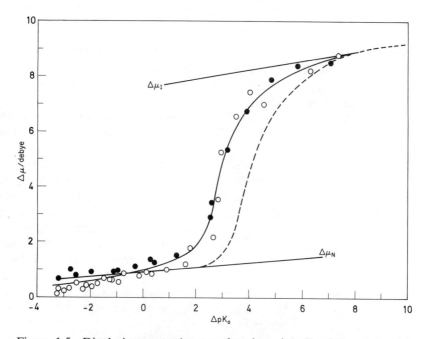

Figure 1.5   Dipole increment $\Delta\mu$ as a function of ΔpK_a of the partners in water. $\Delta\mu_N$ is the increment which should be found if all the complexes belonged to the normal type. $\Delta\mu_I$ is the increment which should be observed if all the complexes were ion-pairs. ●: phenols + pyridines in C_6H_6 (Nouwen and Huyskens, 1973); ○: phenols + imidazoles in C_6H_6 (Huyskens *et al.*, 1979); dashed line: phenols + triethylamine in C_6H_6 (Ratajczak and Sobczyk, 1970).

One can distinguish for a given family of complex three regions of ΔpK_a values:

1. A first one characterized by a low polarity of the complexes, $\Delta\mu$ remaining lower than 2 D: in this region $\Delta\mu$ does not vary much with increasing ΔpK_a. Obviously the majority of the complexes belong to the normal type with $\zeta \ll 1$.
2. A transient region where $\Delta\mu$ rapidly increases with increasing ΔpK_a.
3. A region of higher ΔpK_a where again $\Delta\mu$ does not vary much but where the polarity of the complexes is very high, $\Delta\mu$ being greater than 7 D. In this last region the majority of the bonds are therefore of the proton transfer type where $\zeta \gg 1$.

In the transient region either an equilibrium exists between the two forms of the complex which are present in significant amounts, or the complexes must belong to the 'middle' class. Therefore on the basis of these dipolar results it may be concluded that, in a given medium, ΔpK_a is an important factor governing the appearance of proton transfer complexes, and marked differences are observed between the families of complexes. For example the sigmoidal curve $\Delta\mu/\Delta pK_a$ is markedly shifted to the right for the complexes of triethylamine compared with those of pyridines or imidazoles with the same phenols (Figure 1.5). The curves also depend on the nature of the medium (Jadzyn and Malecki, 1972).

From this point of view it is very interesting to consider the influence on the dipole increment, $\Delta\mu$ of a change in the enthalpy of the bond $-\Delta H$. Such a function has the advantage of passing necessarily through the origin because if replacement of the molecule of base in the complex by solvent molecules has absolutely no influence on the relative positions of the nuclei and the electrons, such replacement would also be accompanied by an enthalpy change strictly equal to zero.

In some cases, as for instance for the complexes of *ortho*-substituted chloro- or nitrophenols an intramolecular hydrogen bond has to be broken before complexation takes place. It is of course necessary to take the enthalpy of the intramolecular bond into account in estimating from the complexation enthalpy. This leads to some inaccuracy in determining the enthalpy of the bond $-\Delta H$.

Figure 1.6 assembles data for different families of $O—H \cdots N$ bonded complexes (amines + phenols, pyridines + phenols) in benzene, toluene or, on occasion in CCl_4. It is clear that for all these hydrogen bonds the $\Delta\mu/\Delta H$ correlation can be considered, to a good approximation, as unique for all the families in these media (Huyskens *et al.*, 1980b), despite a lack of precision in determining both $\Delta\mu$ and ΔH.

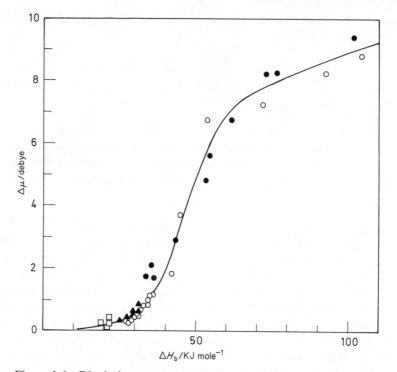

Figure 1.6   Dipole increment $\Delta\mu$ as a function of the enthalpy $-\Delta H_b$ of O—H $\cdots$ N hydrogen bonds. Solvents: benzene, toluene, or CCl_4. Proton donors: phenols. Bases: $\square$: anilines, $\Delta\mu$ from Debecker and Huyskens (1971), ΔH from Neerinck and Lamberts (1966); $\blacktriangle$: pyridines, $\Delta\mu$ from Nouwen and Huyskens (1973), ΔH from Neerinck and Lamberts (1966); $\bigcirc$: imidazoles, $\Delta\mu$ and ΔH from Huyskens *et al.* (1980b); $\bullet$ trialkylamines, $\Delta\mu$ from Ratajczak and Sobczyk (1969), ΔH from Van Audenhaege (1970). For 2-chloro- and 2-nitro-substituted phenols an amount of 12.5 kJ mol^{-1} and 29 kJ mol^{-1}, respectively was added to the experimental $-\Delta H$ values in order to take into account the breaking of the intramolecular bond

Here also three $-\Delta H$ regions can be distinguished:

1. from 0 to 30 kJ mol^{-1} a region of complexes of low polarity;
2. from 30 to approximately 70 kJ mol^{-1} a transient region where $\Delta\mu$ strongly increases with $-\Delta H$;
3. a region of higher polarity where an increase of $-\Delta H$ causes only a moderate enhancement of the polarity of the complexes.

From Figure 1.6 it can be seen that if $-\Delta H$ exceeds a value of about 50 kJ mol^{-1} in C_6H_6 or CCl_4, the majority of the bonds belong to the proton transfer type. Above 80 kJ mol^{-1} proton transfer occurs in the great majority of the bonds.

In the transient area, the points approximately obey the relation

$$-\Delta H(\text{kJ mol}^{-1}) = 30 + 5.0\ \Delta\mu(\text{D})$$

This relation resembles that proposed by Gur'yanova and coworkers (1976)

$$-\Delta H(\text{kJ mol}^{-1}) = 16 + 160\left(\frac{\Delta\mu}{er_{\text{DA}}}\right)$$

However it is worthwhile noting that such a relation no longer holds at values of $-\Delta H$ below 30 or above 80 kJ mol^{-1}.

For O—H $\cdots$ O bonds in *cyclohexane* a similar sigmoidal relationship between $\Delta\mu$ and $-\Delta H$ is found (Huyskens *et al.*, 1980a). However the inflexion point of the sigmoid lies at a larger value of $-\Delta H$ (about 80 kJ mole^{-1}) whereas the initial slope is more important.

Although this is partly due to the change of solvent, it seems that the difference is also related to the change of the nature of the bond and to the existence of a second lone pair of electrons on the oxygen atom of the base.

In the case of E.D.A. complexes the $\Delta\mu/\Delta H$ correlation differs greatly from the sigmoidal functions found for hydrogen bonds.

In Figure 1.7 we give the data relative to iodine–nitrogen base complexes

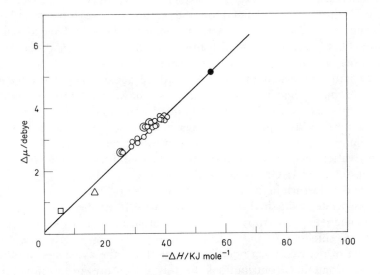

Figure 1.7 Dipole increment $\Delta\mu$ as a function of the enthalpy $-\Delta H$ of E.D.A. complexes. □: iodine–benzene (Keefer and Andrews, 1955; Fukui *et al.*, 1962); △: iodine–naphthalene (Kuroda *et al.*, 1967); ○: iodine–pyridines (Huyskens and Mahillon 1980); ●: iodine–triethylamine (Huyskens and Mahillon, 1980)

determined by Mahillon in C_6H_{12}, C_6H_6, and CCl_4. *All the points lie approximately on a straight line which passes through the origin.* The ratio $-\Delta\mu/\Delta H$ thus remains constant, with a mean value of 0.093 D/kJ mol^{-1} and from 0 to 55 kJ mol^{-1} the following relationship holds.

$$-\Delta H = 10.7\,\Delta\mu$$

$$(kJ\ mol^{-1})\ (D)$$

A similar relation had already been proposed in 1968 by Gol'dshtein and Gur'yanova for a wide variety of E.D.A. complexes, the enthalpies of formation of which vary from 0 to 80 kJ mol^{-1} (Gol'dshtein *et al.*, 1968). Their relationship is of the form:

$$-\Delta H = 148\left(\frac{\Delta\mu}{er_{DA}}\right)$$

The origin of the marked differences between the $\Delta\mu/\Delta H$ relationships of the hydrogen bonds and of the E.D.A. bonds will be discussed later.

1.2.5   *Nuclear magnetic resonance spectroscopy*

Protons such as NH, OH, SH, involved in hydrogen bonds, usually have chemical shifts which depend on the nature of the solvent and of the associated base. The chemical shifts are generally large and negative towards a lower field. At room temperature the lifetimes of hydrogen-bonded complexes in solution are usually too short ($<10^{-4}$ s) for the chemical shifts of free and complexed proton donor to be separately resolved and one single resonance line is observed; as first shown by Gutowski and Saika (1953), this line is a composite of bonded and non-bonded protons. As a consequence, NMR spectroscopy has been widely used as a tool in the determination of kinetic parameters such as the rate of proton exchange or the activation energy for proton transfer (see for example Connor and Loewenstein, 1961; Grunwald *et al.*, 1960).

Schneider, Bernstein, and Pople (1958) have considered that the chemical shifts associated with hydrogen bonding arise from the electrostatic effect and diamagnetic anisotropy of the donor atom and from the change in the long-range shielding of the acceptor group. It is now widely accepted that the effect of the hydrogen bonding is to deshield the associated proton, and that the dominant contributions to this effect are proportional to the reduction of the charge density around the bridging proton. For example, the effect of hydrogen bonding in the HCl–dimethyl ether system is to deshield the HCl proton by about 3 p.p.m. (Govil *et al.*, 1968) and in the bihalide ions $(XHX)^-$, the proton shieldings are about 14 p.p.m. less than those in the corresponding gas-phase hydrogen halide molecules (Martin and Fujiwara, 1972). This effect has also been discussed recently for hydrogen bonds involving O or N atoms (Lutskii *et al.*, 1976a).

Most high-resolution NMR studies have been conducted by monitoring ^{1}H shifts but recently, the use of nuclei such as ^{13}C, ^{15}N, or ^{19}F has been increasing rapidly.

1.2.5.1 *^{1}H spectra* Well-defined changes occur on protonation of carbonyl bases, amides, aliphatic amines, or nitrogen-containing heterocyclic systems and generally a pronounced downfield shift is observed.

The protonation of amides has been extensively studied and the results will not be discussed in detail here. From these studies, it has been concluded that amides are protonated predominantly on the oxygen atom in non-aqueous acidic solutions and that the barrier to rotation around the C—N bond increases by about 12 kJ mol^{-1} on going from free amide to the protonated species (see for example Fraenkel and Franconi, 1960; Berger *et al.*, 1959). At room temperature, no definite peak could be assigned to the OH proton because of the fast exchange of protons; this exchange catalysed by hydronium or hydroxide ions has been studied more recently by Cox (1970), Schleich *et al.* (1971), Molday and Kallen (1972), Liler (1972a,b), and Whidby and Morgan (1973). Gillespie and Birchall (1963) however, recorded the proton resonance spectra of acetamide, dimethylacetamide, and dimethylformamide in fluorosulphonic acid at low temperature ($\sim$190 K) and were able to observe the signal of the captured proton at low field (about -10 p.p.m. from tetramethylsilane (TMS)). For the *N*-methylformamide HBr and HCl complexes, Kuhn and McIntyre (1965) also observed a downfield shift of the formyl and NH protons. The proton spectrum of the conjugated acids of some ketones has been further studied at low temperatures (228–188 K) in fluorosulphuric acid and chemical shifts ranging from -12 to -15 p.p.m. (from TMS) have been observed for the proton involved in the —OH$^+$ bond (Birchall and Gillespie, 1965). The complexes of dimethylacetamide and tetramethylurea with HCl have been studied more recently by Bernander and Olofsson (1972), in the less polar solvent 1,2-dichloroethane; in this medium the acidic proton chemical shifts were observed at a considerably lower field than those reported for the protonated species, indicating the formation of hydrogen-bonded complexes rather than protonated species. For tertiary amides, the presence of two methyl peaks indicates the retention of the conjugated structure in the protonated bases but sometimes strong protonating solvents bring about the collapse of the *N*-methyl doublet, indicating that the two methyl groups have a similar time-averaged chemical environment. A small amount of *N*-protonation should be responsible for the unrestricted rotation. This collapse of the methyl signals was observed by Nawrot and Veis (1970) for some primary amides dissolved in anhydrous formic acid.

The protonation of nitrogen bases has also been extensively studied. Generally, the proton exchange of NH groups in trifluoroacetic is slow in the NMR time scale and characteristic changes in chemical shift occur when the NH group is transformed to a NH_2^+ group (see for example Jackman and Sternhell, 1969). So, the majority of saturated aliphatic amines ($pK_a \approx 10$) dissolved in trifluoroacetic acid show a downfield shift of the $NH(NH_2)$ group upon formation; moreover, the decreased shielding results in a downfield shift of the N-methyl resonance. At room temperature, the NH exchange rate is slow enough to permit observation of coupling of H(N) with the N-methyl protons (Ma and Warnhoff, 1965, Kohler, 1974). However, substituted anilinium and pyridinium ions with pK_a values below 4 show coalescence for the signals of the N^+H protons and disappearance of the coupling with the α protons (Giger and Simon, 1970). The chemical shift of the bonded proton of some tri-n-octylammonium salts dissolved in dry CCl_4 ranges from -6 to -10 p.p.m. (from the CH_3 group) and depends on the nature of the associated cation ($Cl^- > Br^- > I^-$) (Keder and Burger, 1965).

The protonation of pyridine in trifluoroacetic solution is shown by the position of the N—H signal to the low-field side of the carboxyl group; the splitting of this signal in triplet arises from spin coupling with the ^{14}N nucleus. A low-field displacement of the signals of the γ ring protons also occurs when the pyridinium is formed; the α proton resonance is hardly affected (Smith and Schneider, 1961). These effects have been further studied by Chakrabarty et al. (1973) for substituted pyridinium ions. As reported recently by Iogansen et al. (1976), the absorption signal of the proton of the NH^+ group of some pyridinium salts in CH_3CN solution, is observed at very weak field (-13.6 to -17.7 p.p.m. from TMS) and depends on the nature of the associated cation. The concentration dependence of the chemical shift indicates that pyridinium perchlorate is partly dissociated in nitromethane and the dissociation constant has been estimated as $80 \ dm^3 \ mol^{-1}$. The protonation of quinoline by trifluoroacetic acid in dichloromethane has been studied by Krakower and Reeves (1964). At 300 K, the peaks of the NH^+ and COOH protons are not resolved; this contrasts with the results obtained with pyridine in the same medium. At 255 K however, these two peaks are distinct but the NH^+ signal, observed at very low field (-18 p.p.m. from TMS), is not resolved into the triplet structure. A displacement of the ring protons was also observed.

1.2.5.2 ^{13}C spectra   It seems well established that the ^{13}C chemical shift is quite a sensitive experimental method for detecting the structure and electronic perturbations around a carbon atom. Pople (1964) predicted a low-field shift for molecules with low excitation energy and a high-field shift if the carbon has high electron density. It is not widely accepted that protonation of carbonyl or nitrogen bases brings about a variation of the charges of the different carbon atoms of the molecule and so, a variation of the chemical shift of the ^{13}C signal is to be expected. For example, it has been shown that protonation of acetone in sulphuric acid–water media brings about a downfield shift of the carbonyl ^{13}C while the methyl ^{13}C remains almost constant; these experimental results can be explained at least qualitatively by an electron displacement away from the carbonyl carbon while the charge distribution of the methyl carbon is hardly affected. The results for the coupling constants $^1J_{CH}$ and $^4J_{NH}$ appear to be less satisfactory (de Jeu, 1970).

Most of the studies have been performed on nitrogen bases such as nitrogen-containing heterocycles or aliphatic amines. Very interesting results have been obtained by Grant and Pugmire (1965) on the protonation shifts of pyridine, and by Mathias and Gil (1965) on the protonation shifts of pyrazine and pyrimidine. These studies show that most of the ring carbons are more shielded than in the neutral species. The NMR results can be compared with the total charges on the α, β, and γ-carbon atoms calculated by the extended Hückel theory (Adam et al., 1968); these charges and the ^{13}C chemical shifts for pyridine and the pyridinium ion are repeated in Table 1.12. The general effect of protonation of a nitrogen atom is to increase the total electron density at the adjacent carbon atom and this result is consistent with an increase of 8.5 p.p.m. of the shielding of the α-carbon atom. The results relative to the γ-carbon atom are less reliable. However, Adam and coworkers (1967) have shown that, for forty 6-membered ring heterocycles, the correlation between the ^{13}C chemical shifts and the total electron densities at the carbon atom is linear.

Further studies have been performed on the protonation of aliphatic amine and N-heterocyclic six-membered ring compounds and it was shown that the ^{13}C shifts exhibit marked structural and conformational dependences and are well interpreted in terms of charge densities of the carbon skeleton. The upfield shifts for most of the ^{13}C signals suggest that, upon protonation, the C—H bond could be polarized to produce the C^-—H^+

Table 1.12 ^{13}C chemical shifts and total charges of the carbon atoms in pyridine and pyridinium ions

	^{13}C chemical shift*			Total charges $(e)^\dagger$		
	$C_{(\alpha)}$	$C_{(\beta)}$	$C_{(\gamma)}$	$C_{(\alpha)}$	$C_{(\beta)}$	$C_{(\gamma)}$
Pyridine	−21.7	+4.6	−7.4	+0.369	−0.046	+0.110
Pyridinium ion	−13.2	+0.4	−19.2	+0.338	−0.038	+0.109

* In p.p.m. relative to benzene (Grant and Pugmire, 1965).
† Results of Adam et al. (1968).

structure (Morishima *et al.* 1973). The effects of important structural features on observed ^{13}C protonation shifts of aqueous amines have been discussed recently by Sarneski *et al.* (1975) but comparison of experimental ^{13}C shifts and calculated CNDO charge density changes failed to yield useful results in contrast with the work of Morishima *et al.* (1973). The use of protonation-induced changes in ^{13}C shifts has also been used to investigate the structure of partially protonated polybasic molecules such as imidazole (Reynolds *et al.*, 1973) or γ-pyridone (Sarneski *et al.*, 1976).

Although most of the studies have been performed on ^{13}C chemical shifts of the base, it has also been shown that ionization of an aliphatic acid results in a downfield shift of its ^{13}C resonance (Hagen and Roberts 1969).

Remark 1.2.5.3 *General comment*   The formation of proton transfer complexes also has a strong influence on other properties of the solutions. It causes, for example, an important decrease in the volume and thus affects the density. This volume decrease is not restricted to the area of the specific bond but extends to more remote regions of the molecules of the partners (Felix and Huyskens, 1975). The dissociation of the proton transfer complex into free ions brings about a further decrease in the volume.

Another characteristic of proton transfer complexes is their tendency to aggregate, forming crystalline lattices. As a consequence, the solubility of such complexes in apolar solvents is strongly reduced.

1.2.6   *Some selected thermodynamic data for proton transfer complexes*

In Table 1.13, some selected thermodynamic data for proton transfer complexes are reported; these data only refer to systems not already quoted in Joesten and Schaad's book (1974).

1.3   Tautomerism Between Ion or Proton Transfer Complexes and Normal Complexes

The possibility of simultaneous appearance of two kinds of complexes, one of the normal type (with $\zeta < 1$) and one of the proton transfer type (with $\zeta > 1$) was envisaged by Fuoss in 1939 (Elliott and Fuoss, 1939). An equilibrium between the two forms was postulated by Barrow and Bell (Barrow, 1956; Bell and Barrow, 1959a) in order to explain the spectroscopic results. Zimmerman (1964) used the term 'protomerism' to characterize this type of tautomerism.

1.3.1   *Requirements for the existence of a tautomerism—environmental effects*

Let us consider the three nuclei *a*, *p*, and *d* which are directly involved in the specific bond A $\cdots$ D. Let us assume that for given values of r_{ap} and r_{pd} the

Table 1.13　Selected thermodynamic data for the reaction $AH + B \rightleftharpoons A^- \cdots H^+B$

Acid	Base	Solvent	$T°$ (K)	K (dm³ mol⁻¹)	$-\Delta H$ (kJ mol⁻¹)	$-\Delta S$ (J mol⁻¹ deg⁻¹)	Method	References
2,4-Dinitrophenol	Piperidine	Dioxan	298	14,500			UV	Jasinski et al., 1965a
		Ethyl acetate	298	1760			UV	Jasinski et al., 1965a
		Isopropyl ether	298	8060			UV	Jasinski et al., 1965a
		C₆H₅Cl	298	6250			UV	Jasinski et al., 1965a
		CCl₄	298	640			UV	Jasinski, et al. 1965a
	N-Ethylpiperidine	C₆H₅Cl	298	4890			UV	Jasinski, et al. 1965a
	Di-n-butylamine	CCl₄	298	858			UV	Jasinski and Widernikova (1969)
	Tri-n-butylamine	CCl₄	298	417			UV	Jasinski and Widernikova (1969)
		n-Hexane	298	1003			UV	Jasinski and Widernikova (1969)
	Triethylamine	CCl₄	298	813			UV	Jasinski and Widernikova (1969)
		C₆H₅Cl	298	13,400	52.9*	98.4	UV	Bron and Simmons (1976)
		C₆H₅Cl	293	23,500	69 *	152.3	UV	Ivin et al. (1973)
	Tri-n-propylamine	C₆H₅Cl	298	3310	51.5*	105.4	UV	Bron and Simmons (1976)
		C₆H₅Cl	298	4900	57.3*	124.3	UV	Ivin et al. (1973)
	Tri-n-butylamine	C₆H₅Cl	298	4460	52.4*	106	UV	Ivin et al. (1973)
		C₆H₅Cl	298	3350	63.4*	144.6	UV	Caldin and Crooks (1967)
	Tri-n-pentylamine	C₆H₅Cl	293	5700	57.7*	124.7	UV	Ivin et al. (1973)
	Tri-n-octylamine	C₆H₅Cl	297	4800			UV	Caldin et al. (1973)
	Quinuclidine	C₆H₅Cl	298	119,000			UV	Caldin et al. (1973)
	2-Methylaniline	1,3-Dioxan	298	247			UV	Jasinski and Kokot (1967)
	3-Methoxyaniline	1,3-Dioxan	298	316			UV	Jasinski and Kokot (1967)
	Pyridine	C₆H₆	298	4210			UV	Jasinski et al. (1965c)
		C₆H₅Cl	298	20,000			UV	Jasinski et al. (1965c)
		CCl₄	298	5400			UV	Jasinski et al. (1965c)
	2,4,6-Trimethylpyridine	C₆H₆	298	111,000			UV	Jaskinski et al. 1968
	2,6-Dimethylpyridine	C₆H₆	298	48,800	51.9*	62	UV	Jasinski et al. (1968)
		C₆H₆	298				UV	Parbhoo et al. (1975)
	2-Methylpyridine	C₆H₆	298	20,830			UV	Jasinski et al. (1968)
	3-Methylpyridine	C₆H₆	298	11,350	46.5*	59.9	UV	Jasinski et al. (1968)
		C₆H₆	298					Parbhoo et al. (1975)
	4-Methylpyridine	C₆H₆	298	20,410			UV	Jasinski et al. 1968
	3-Chloropyridine	C₆H₆	298		40.4*	91.3		Parbhoo et al. 1975
	3-Chloropyridine	C₆H₅Cl	298		39.2*	80.4		Parbhoo et al. (1975)
	Tri-n-butylamine	C₆H₅CH₃	298		82.4*		Calorimetry	Van Audenhaege (1970)
	N,N,N',N'-TMPD⁺	C₆H₆	298	133,000			UV	Steigman and Cronkright (1970)

Table 1.13 (continued)

Acid	Base	Solvent	$T°$ (K)	K (dm^3 mol^{-1})	$-\Delta H$ (kJ mol^{-1})	$-\Delta S$ (J mol^{-1} deg^{-1})	Method	References
Acetic acid	n-Butylamine	Ether	298	69			IR	Hibbert and Satchell (1968)
Formic acid	n-Butylamine	Ether	298	3000			IR	Hibbert and Satchell (1968)
3-Chloropropionic acid	n-Butylamine	Ether	298	481			IR	Hibbert and Satchell (1968)
Trichloroacetic acid	Tri-n-butylamine	CCl$_4$	298	118			IR	Jasinski and Modro (1967)
Benzoic acid	n-Butylamine	Ether	298	371			IR	Hibbert and Satchell (1968)
2,5-Dinitrobenzoic acid	N,N,N',N'-TMPD†	C$_6$H$_6$	298	13,000			UV	Steigman and Cronkright (1970)
3,5-Dinitrobenzoic	N,N,N',N'-TMPD†	C$_6$H$_6$	298	4300			UV	Steigman and Cronkright (1970)
2-Nitrobenzoic acid	N,N,N',N'-TMPD†	C$_6$H$_6$	298	1470			UV	Steigman and Cronkright (1970)
4-Nitrobenzoic	N,N,N',N'-TMPD†	C$_6$H$_6$	298	280			UV	Steigman and Cronkright (1970)
2,5-Dinitrophenol	Di-n-butylamine	CCl$_4$	298	270			UV	Jasinski and Widernikova (1969)
		n-Hexane	298	312			UV	Jasinski and Widernikova (1969)
	Tri-n-butylamine	CCl$_4$	298	50.8			UV	Jasinski and Widernikova (1969)
		n-Hexane	298	17.5			UV	Jasinski and Widernikova (1969)
	Triethylamine	CCl$_4$	298	91			UV	Jasinski and Widernikova (1969)
		n-Hexane	298	28.5			UV	Jasinski and Widernikova (1969)
		C$_6$H$_5$Cl	298	1270	45.8	95	UV	Bron and Simmons (1976)
	Tri-n-propylamine	C$_6$H$_5$Cl	298		43.8	100.2	UV	Bron and Simmons (1976)
2,6-Dinitrophenol	n-Butylamine	CCl$_4$	298	6250			UV	Jasinski and Widernikova (1969)
	Di-n-butylamine	CCl$_4$	298	3590			UV	Jasinski and Widernikova (1969)
		n-Hexane	298	6250			UV	Jasinski and Widernikova (1969)
	Triethylamine	CCl$_4$	298	20,000			UV	Jasinski and Widernikova (1969)
		n-Hexane	298	12,500			UV	Jasinski and Widernikova (1969)
		C$_6$H$_5$Cl	298	158,000	60.4*	103.1	UV	Bron and Simmons (1976)
	Tri-n-propylamine	C$_6$H$_5$Cl	298	30,900	61.1*	119.1	UV	Bron and Simmons (1976)
	Tri-n-butylamine	CCl$_4$	298	9340			UV	Jasinski and Widernikova (1969)
		n-Hexane	298	5400			UV	Jasinski and Widernikova (1969)
		C$_6$H$_5$Cl	298	5750	65.6*	129.1	UV	Bron and Simmons (1976)
2,3-Dinitrophenol	n-Butylamine	CCl$_4$	298	500			UV	Jasinski and Widernikova (1969)
	Di-n-butylamine	CCl$_4$	298	5000			UV	Jasinski and Widernikova (1969)
	Tri-n-butylamine	CCl$_4$	298	666			UV	Jasinski and Widernikova (1969)
	Triethylamine	CCl$_4$	298	2000			UV	Jasinski and Widernikova (1969)

Acid	Base	Solvent	T				Method	Reference
3,5-Dinitrophenol	Triethylamine	C_6H_{12}	298	15,600			UV	Dupont et al. (1971)
2,6-Dichloro-4-nitrophenol	n-Butylamine	CCl_4		2520			UV	Romanowski and Sobczyk (1975)
	Diethylamine	$C_2H_4Cl_2$		30,200			UV	Romanowski and Sobczyk (1975)
		CCl_4		53,600			UV	Romanowski and Sobczyk (1975)
	Triethylamine	$C_2H_4Cl_2$		380,000			UV	Romanowski and Sobczyk (1975)
		CCl_4		16,900			UV	Romanowski and Sobczyk (1975)
		$C_2H_4Cl_2$		316,500			UV	Romanowski and Sobczyk (1975)
		C_6H_{12}	298	28,000			UV	Dupont et al. (1971)
Pentachlorophenol	Di-n-propylamine	$CHCl_3$	293	7000	35.10*	46	IR	Denisov and Schreiber (1972)
	Di-n-butylamine	$CHCl_3$	293	11,000	36.8*	47.6	IR	Denisov and Schreiber (1972)
	Triethylamine	CCl_4	293	28,000	48.1*	79.4	IR	Denisov and Schreiber (1972)
	Triethylamine	CCl_4	299	1445			IR	Clotman et al. (1970)
Picric acid	Aniline	Tetrahydrofuran	298	1370				Jasinski et al. (1965b)
		1,3-Dioxan	298	416				Jasinski and Kokot (1967)
		Isopropyl ether	298	3120				Jasinski and Kokot (1967)
	4-Methylaniline	Tetrahydrofuran	298	3150				Jasinski et al. (1965b)
		1,3-Dioxan	298	1260				Jasinski and Kokot (1967)
		Isopropyl ether	298	5710				Jasinski et al. (1965b)
	4-Methoxyaniline	Tetrahydrofuran	298	3570				Jasinski et al. (1965b)
		1,3-Dioxan	298	4450				Jasinski and Kokot (1967)
		Isopropyl ether	298	35,800				Jasinski and Kokot (1967)
	4-Chloroaniline	Tetrahydrofuran	298	138				Jasinski and Kokot (1967)
		1,3-Dioxan	298	52.5				Jasinski and Kokot (1967)
		Isopropyl ether	298	286				Jasinski (1965b)
	4-Bromoaniline	Tetrahydrofuran	298	85.4				Jasinski (1965b)
		1,3-Dioxan	298	222				Jasinski (1965b)
	3-Methylaniline	1,3-Dioxan	298	457				Jasinski and Kokot (1967)
	2-Methoxyaniline	1,3-Dioxan	298	282				Jasinski and Kokot (1967)
Bromophenol blue	Pyridine	C_6H_5Cl	298	2120	43.2	81.3	UV	Crooks and Robinson (1971)
	2-Methylpyridine	C_6H_5Cl	298	30,600	58	108.3	UV	Crooks and Robinson (1971)
	Trimethylamine	C_6H_5Cl	298	3400			UV	Crooks and Robinson (1975)
	Triethylamine	C_6H_5Cl	298	33,000			UV	Crooks and Robinson (1975)
	Tri-n-propylamine	C_6H_5Cl	298	4900			UV	Crooks and Robinson (1975)
	Tri-n-butylamine	C_6H_5Cl	298	7200			UV	Crooks and Robinson (1975)
	Tri-n-octylamine	C_6H_5Cl	298	6300			UV	Crooks and Robinson (1975)

* This value refers to the enthalpy of the reaction between the acid and the base. In order to get the true enthalpy of the hydrogen bond, it is necessary to add to this value the enthalpy needed for breaking the intramolecular hydrogen bond of the acid;
† N,N,N',N'-Tetramethyl-p-phenylenediamine.

other particles of the system have acquired their equilibrium in the spatial distribution. Each couple of values r_{ap} and r_{pd} will then correspond to a given potential energy $\mathscr{V}(r_{ap}, r_{pd})$ with respect to some reference state. This potential energy surface can be represented in a three-dimensional map (Figure 1.8).

A first requirement for the existence of tautomerism of the type

$$-a-p \cdots d- \rightleftarrows -a^- \cdots p^+-d-$$

will be that the potential surface shows two minima. A second requirement is that the depth of the two minima will be sufficient to lead to the existence of two stationary wavefunctions $\Psi(r_{ap}, r_{pd})$, the maxima of which corresponds to the minima in the potential surface.

The two stationary states will then show different spectral characteristics and if both are experimentally observed it can be concluded that the two conditions above are fulfilled. However, the lack of simultaneous appearance of the bands of both forms in the spectrum cannot be considered as proof of the existence of a single minimum. The second minimum may be located at such a high level that its population becomes too small to be detected.

Theoretical calculations can be used to estimate the shape of the potential surface. Generally the energies are computed at constant values of

$$R \equiv r_{ap} + r_{pd}$$

(This corresponds to sections at 45° in Figure 1.8.)

When dealing with the main factors which can influence the shape of the potential surface, semi-empirical potential functions can provide good qualitative information.

A common result of all these methods however is that at *sufficiently high values of R two minima appear in the curve of potential energy whereas below a given interdistance R a single minimum is obtained.*

The most used semi-empirical potential function is that of Lippincott and Schroeder (1955). This model was extended by Reid (1959) and more recently by Spencer, Schreiber, and their coworkers who studied hydrogen bonds involving sulphur atoms, intramolecular hydrogen bonds, very strong hydrogen bonds, etc. (Snijder *et al.*, 1973; Robinson *et al.*, 1972; Spencer *et al.*, 1974).

According to Lippincott's equation, at a given interdistance R where two minima occur, the depth of the two minima will depend on D_0, the depth of the potential well in the *free* acid molecule, r_{ap}^0 the corresponding interdistance, and k_{ap}^0 the force constant and on similar terms D_0^*, r_{pd}^0, and k_{pd}^0 characterizing the proton acceptor.

We may note here that, in hydrogen bonds D_0^*, depends on the proton affinity of the base (and on the electron affinity of the radical derived from

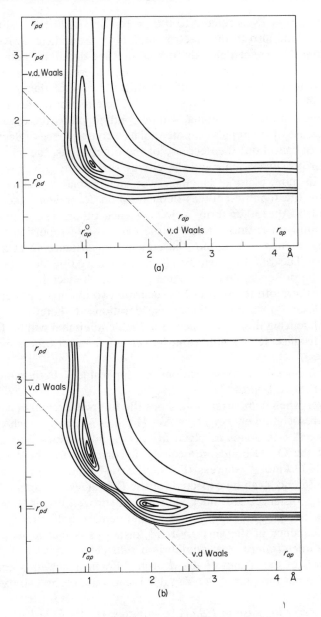

Figure 1.8 Upper figure: isopotential lines with weak R-depending repulsion. Single-well potential surface. Dotted line: constant R distance at 2.4 Å). Lower figure: isopotential lines with stronger R-depending repulsion. Double-well potential surface. Dotted line: constant R distance at 2.8 Å

the acid whereas D_0 is related to the proton affinity of the conjugated base of the acid (and also to the electron affinity of the radical derived from the acid). From this one can already foresee that the difference between the two levels will depend in a given family on the difference between the proton affinity of the base and that of the conjugated base of the acid.

Let us now change the interdistance R. The effect of decreasing R, on the energy levels in the two minima, will be conditioned by the repulsion forces. If the increase of the repulsion potential as a consequence of decreasing R is sufficient to cancel out the decrease of the other terms, the minima will rise from their previous values and the energy surface $\mathscr{V}(r_{ap}, r_{pd})$ will show two minima. In more pictorial language we can say that, in this case, the dike formed by the repulsion forces depending on R prevents the two valleys originated by the other terms rejoining each other. The appearance of a double minimum is thus related to the existence of repulsion forces which are strong enough to maintain the equilibrium distance R at a sufficiently high value. This repulsion can be strongly enhanced by steric factors, as for instance the presence of cumbersome groups in the vicinity of the atoms 'a' or 'd'. Furthermore the barrier between the two minima will depend on the depths D_0 and D_0^*. From these considerations it therefore appears that factors influencing the appearance of a double minimum will be D_0, D_0^*, and the interdistance R at equilibrium (governed amongst others by the repulsion forces).

As a principle these considerations are applicable to both hydrogen bonds and n–σ E.D.A. bonds.

However when comparing both cases the following remarks can be made; first, the dissociation energy D_0 of A'—H bonds is usually much greater than that of the A'—X bonds involved in E.D.A. interactions. For instance the energy of the O—H bonds in alcohols (uncorrected for the zero vibration) reaches $465 \, kJ \, mol^{-1}$ whereas that of the I—I bond is only $151 \, kJ \, mol^{-1}$. Similar differences can be expected for the D_0^* values. A second difference is that the r_{ap}^0 distances are much shorter in hydrogen bonds (0.96 Å against 2.67 Å in the chosen examples). A similar remark holds for r_{pd}^0.

A consequence of the very short r_{ap}^0 distances is that in *many hydrogen bonds the interdistance* R at equilibrium *falls below the sum of the Van der Waals radii* of the a and of the d atom. (For instance in water R equals 2.76 Å whereas the sum of the Van der Waals radii of the two oxygen atoms is 2.80 Å.)

This is *never the case in E.D.A. complexes.* (In $R_3N \cdots$ I—I the distance between the nucleus N and the *second* iodine nucleus is 5.11 Å whereas the sum of their Van der Waals radii is only 3.65 Å.) These differences between the two types of specific bonds must have an influence on the repulsion term. One can thus expect that the repulsion 'dike' is more efficient in the hydrogen bonds. As a consequence the occurrence of a double minimum and of tautomerism between the two forms seems *a priori* more probable for hydrogen bonds than for E.D.A. bonds.

Theoretical calculations generally concern 'free' complexes *in vacuo*. Clementi (1967a; Clementi and Gayles 1967b) carried out LCAO-MO calculations on NH_4Cl (neglecting electron correlations). This calculation shows that, in the gas phase, the potential shows a minimum for a distance r_{N-H} of 1.24 Å (compared with 1.01 Å in the free ion) and $r_{H-Cl} = 1.62$ Å (compared with 1.32 Å in the free molecule). Clementi computes an excess electronic charge of 1.11 on the nitrogen whereas the chlorine excess electronic charge is 0.64. It can be therefore concluded that 'the NH_4Cl molecule at its equilibrium distance is only partially an ion-pair'.

According to the definition proposed above, the value of ζ describing the magnitude of proton transfer in the gas phase complex is 1.02. Such a complex belongs therefore to the 'middle' class. This conclusion is also supported by the infrared study of this complex isolated in a nitrogen matrix at 15 K. This study was made by Ault and Pimentel (1973a) who concluded that the complex is bound by hydrogen bond in which the bridge proton is preferentially attached neither to the chlorine nor to the nitrogen atom. According to Haney and Franklin (1969) the proton affinity of ammonia in the gas phase is 875 kJ mol^{-1}. Ault and Pimentel predict proton transfer between HCl and a nitrogen base when the proton affinity of the last exceeds 940 kJ mol^{-1}.

$$H-N \cdots\cdots H \cdots\cdots Cl$$

The situation is completely different in crystalline NH_4Cl where the four N—H distances are equal and which clearly belongs to the proton transfer class.

The environment thus has a strong influence on the potential energy surface.

Similar remarks hold for the system F—H $\cdots$ NH$_3$. Theoretical calculations on the isolated systems were performed by several authors who compute R_{NF} distances between 2.68 and 2.77 Å. The calculated bonding energy strongly depends on the basis set used: (STO-3G) yields 36 kJ mol^{-1} (Kollman *et al.*, 1974), (4-31G) 75 kJ mol^{-1}, and (6-31 G^x) 50 kJ mol^{-1} (Dill *et al.*, 1975). The *ab initio* calculations of Kollman and Allen (1971) give a single minimum at a r_{FH} distance of the order of 1 Å. This distance is thus only stretched by 0.06 Å from the isolated H—F bond length. *Thus in the gas base H$_3$N—HF belongs to the normal type of hydrogen bonded complexes with $\zeta < 1$.*

$$H-N \cdots\cdots H-F$$

This was also found by the CNDO/2 studies performed by Schuster *et al.* (1974).

However CNDO/2 calculations by the last authors on a system formed by $NH_3 \cdots HF$ *and six water molecules*:

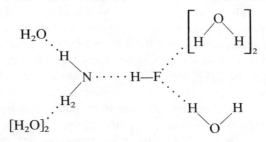

show an *evolution* towards the appearance of a double minimum (Figure 1.9).

In the crystalline state, one finds NH_4^+ ions tetrahedrally surrounded by F^- ions. Here again, the environment strongly influences the structure of the complex.

Another example is that of the complex $H_2O \cdots HCl$ studied by Ault and Pimentel (1973b). The infrared spectrum of the 1 : 1 complex of HCl and H_2O isolated in a nitrogen matrix at 15 K shows that this complex does not involve a hydronium ion but corresponds to a normal complex $H_2O \cdots HCl$.

On the contrary, when an equimolar mixture of HCl and H_2O is condensed to the crystalline form, this crystal is made up of H_3O^+ and Cl^- ions.

Many other complexes which are ionic in the crystalline state appear to belong to the normal form in the isolated state. This is so for the complexes

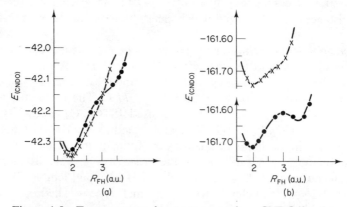

Figure 1.9 Energy curves for proton transfer—CNDO/2 calculations.   (a) $NH_3.HF$ without   hydratation   shell.   (b) $H_3N.HF].H_2O$. $\times$—$\times$—$\times$—$\times$: $R_{NF} = 2.7$ Å; ·—·—·—·: $R_{NF} = 3$ Å. [Reproduced by permission of the authors from P. Schuster *et al.*, 1974]

of aliphatic amines with HF which have been studied recently by Romanowska and coworkers (1977) using a CNDO/2 method. For the complex between pyridine and HCl (STO-3G) calculations give in the isolated state a single-well potential with a $N \cdots Cl$ distance of 2.8 Å, a N—H distance of 1.4 Å, and a H—Cl distance of the order of 1.4 Å. This complex in the gas phase resembles even more the separated molecules than the separated ions (Jordan, 1975).

It must be noted here that, as early as 1965, Sokolov pointed out that 'In the gas phase proton transfer according to the scheme $AH \cdots B \rightarrow A^- \cdots H^+B$ does not occur in practice, because the process is highly endothermic' (Pshenichnov and Sokolov, 1965).

From all these data it is clear that environmental effects can modify the potential energy surface of the proton to a marked extent. This conclusion can also been extended to $n-\sigma$ bonds.

Several environmental effects can be envisaged *a priori*:

1. Formation of *additional specific bonds* by the molecules of the partners of the complex with the neighbouring molecules (as in the case studied by Schuster, cited above). Such interactions will increase the proton donor character of the acid (thus diminishing the value of D_0 in the Lippincott–Schroeder equation) and strengthen the electron donor power of the base (increasing D_0^*).
2. *Polarization of the solvent* molecules around the complex. This polarization will be more important for the polar form. An increase of the dielectric constant of the medium will therefore modify the potential energy surface in favour of the ion-pair.
3. A similar influence in the crystalline state by the *Madelung effect* of the neighbours, stabilizing the polar form.

It must be noted here that the last two effects will increase the equilibrium value of the interdistance R with respect to the situation of minimal energy in the gas phase.

As pointed out by Kollman and Allen (1972), this equilibrium interdistance in the gas phase is generally so low that only one position of proton well exists. Moreover, with the exception of complexes between very strong acids with sufficiently strong bases, this minimum lies near the acid molecule.

In the crystalline state and in liquid media, environmental effects (1) increase the equilibrium value of R so that double-minimum potential surfaces become much more current, (2) stabilize the ionic form.

Such environmental effects are much weaker in a nitrogen or an argon matrix, where the situation resembles more that of the gas phase.

To conclude we can say that *although proton transfer and ion transfer complexes exist very seldom in the gas phase they are expected to be much more important in solution and in the crystalline state.* Furthermore, *tautomerism between two forms is expected to be much more current in solution than in the gas phase.*

1.3.2   *Experimental methods for the study of the tautomerism*

1.3.2.1   *Spectroscopy*   The most convincing proof of the existence of a double minimum is perhaps the simultaneous appearance in the electronic spectrum of bands characteristic of the ionic form near bands which characterize the normal form.

The difficulty however lies in the assignment of a given absorption band to one or the other form. One of the most rigorous ways of obtaining this evidence consists in changing a factor which can modify the transfer constant K_t

$$A\!-\!H \cdots B \overset{K_t}{\rightleftarrows} A^- \cdots H\!-\!B^+$$

As shown above, such a factor may be the dielectric constant of the medium.

One can therefore expect that increasing the permittivity of the medium would enhance the relative intensity of the bands characteristic of the polar form whereas it would decrease the surface of the bands corresponding to the normal form.

This was indeed observed by Bell and Barrow (1959a) for the complex of p-nitrophenol with triethylamine. In C_6H_{12}, the hydrogen-bonded phenol gives rise to an absorption centred near 308 nm while in the more polar $CHCl_3$ a second band characteristic of the p-nitrophenolate ion appears near 400 nm. The simultaneous presence of the two peaks at 308 and 400 nm can be ascribed to the two tautomeric forms of the hydrogen bonds. Figure 1.10 gives the electronic spectrum of the same complex studied quantitatively by Baba *et al.* (1969) who found a value for K_t of 0.68 at 293 K in 1,2-dichloroethane. The enthalpy change, determined from the spectral data indicates that the ion-pair is more stable by 13 kJ mol^{-1} in the ground state.

$$NO_2C_6H_4\!-\!O\!-\!H \cdots NR_3 \overset{K}{\rightleftarrows} NO_2C_6H_4\!-\!O^- \cdots H\!-^+\!N\!-\!R_3$$

For the adducts of the same phenol with aliphatic amines, Scott *et al.* (1968) showed that, in binary solvents like dimethylformamide–benzene mixtures, the tautomeric equilibrium could be shifted, changing the proportion of the polar component. The role of solvent polarity (and also of the solvation of the base) was investigated for the 3,4-dinitrophenol–amine system by Hudson *et al.* (1972). Kraft *et al.* (1975) studied the interaction of

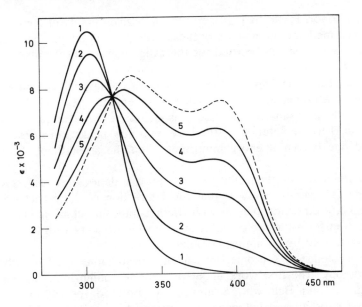

Figure 1.10 Electronic spectrum of 4-nitrophenol in 1,2-dichloroethane at 293 K with various amounts of triethylamine. $c_{amine} = (1)$ 0; (2) 2.4×10^{-4}; (3) 9.5×10^{-4}; (4) 2.1×10^{-3}; (5) 5.9×10^{-3} mol l.$^{-1}$ Broken curve: spectrum of the equilibrium mixture of complexes. [Reproduced by permission of Pergamon Press from Baba *et al.*, 1969]

2,6-dihalo-substituted phenols with aliphatic amines by electronic spectroscopy. They showed that in solvents of low dielectric constant (2–7) the equilibrium is displaced in favour of the non-ionic form. In alcoholic solution however, the presence of the ionic form may be inferred from the presence of a band lying at 385–410 nm. The role of the solvent was also studied by Romanowski and Sobczyk (1975) for the proton transfer in complexes between 2,6-dichloro-4-nitrophenol and aliphatic or aromatic nitrogen bases.

It is worthwhile noting that the increase of the dielectric constant not only favours the most polar form but also exerts a favourable influence on the dissociation of the ion-pairs into free ions. It can also favour the formation of 'solvent-separated' ion-pairs. It can then happen that in solvents of higher dielectric constant, such as nitrobenzene or methanol, the entities present in the solution are mainly free ions or solvent-separated ion-pairs. In this case the observed spectrum is characteristic of these new entities and the corresponding band can no longer be used to determine the transfer constant K_t.

Another factor which can influence the relative proportions of the two forms is the acidic strength of the proton donor or the basicity of the electron donor. It can be expected that, in a given family of complexes and in a given medium, an increase of these strengths will favour the polar form. This characteristic can be used for the assignment of the bands.

This was done, for instance, by Nasielski and Van der Donckt (1963), who examined the electronic spectra of mixtures of bases like quinoline, isoquinoline, acridine, and 1-, 4-, and 9-azaphenanthrenes with halogenated acetic acids in hexane. In the case of isoquinoline, for example, on addition of the acid to the solution of the base, two new peaks were observed, the second one becoming more important when the strength of the acid increases.

The influence of the acidity of the proton donor was also studied by Pilyugin (1975) who showed that the interaction of 2,4,6-trichlorophenol and 2,3,4,6-tetrachlorophenol with triethylamine in octane leads to partial proton transfer while the complex of pentachlorophenol with the same base is characterized by an almost complete proton transfer.

The existence of tautomerism can also be demonstrated by vibrational spectroscopy. The first work about protomeric equilibrium was performed by Barrow and Bell who examined the infrared spectra of complexes between pyridine and halogenated acetic acids in solvents of low polarity (Barrow, 1956; Bell and Barrow, 1959b). They showed the weakest acid (CH_3COOH, $pK_a = 4.75$) forms a hydrogen-bonded complex and the strongest acid (CF_3COOH, $pK_a = 0$) an ion-pair salt. The IR spectrum of the adducts of acids of intermediate strength indicates a tautomeric equilibrium between the two forms. Barrow's studies have been extended by Johnson and Rumon (1965) to the solid adducts between pyridine and benzoic acids. The appearance of two widely separated ν_{NH} and OH bonds was taken as evidence for a double-minimum potential for the proton stretching motion whereas a single broad NH or ν_{OH} bond below 1700 cm^{-1} was regarded as an indication of a single-minimum potential function.

A similar study was made by Zeegers-Huyskens, (1969) of the complexes of methanol and substituted phenols with triethylamine in $CHCl_3$ ($D = 4.8$). For 4-nitrophenol ($pK_a = 7.15$) two absorption bands are observed: one at about 3050 cm^{-1} which can be ascribed to the O—H stretching vibration in the normal complexes and a broad band near 2500 cm^{-1} characteristic of the ν_{NH^+} vibration of hydrogen-bonded ion-pairs. For methanol ($pK_a \approx 18$), one band only is observed at 3220 cm^{-1}, assigned to normal OH $\cdots$ N bonds but for the more acidic 2,4-dinitrophenol ($pK_a = 4.1$), the spectrum is characterized by a broad absorption band with several submaxima at 2670, 2560, and 2420 cm^{-1} and is analogous with the spectrum of the ammonium picrates (Figure 1.11). This suggests that with this phenol, hydrogen-bonded ion-pairs are formed almost-exclusively. Similar observations have been reported by Lutskii and coworkers for complexes of substituted phenols with amines (1976b) and with monoazoaromatic bases (1976c).

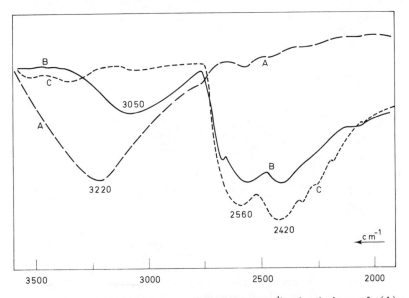

Figure 1.11   Infrared spectrum (3500–$2500\ cm^{-1}$) of solutions of: (A) Methanol–triethylamine;   (B)   4-nitrophenol–triethylamine;   (C)   2,4-dinitrophenol–triethylamine. Solvent = chloroform; $T^{\circ} = 298$ K. [from Th. Zeegers-Huyskens, *Thesis*, Louvain, 1969]

Pyridine complexes with carboxylic acids have been studied by other spectroscopists. When a normal hydrogen bond is formed, the bands of pyridine at 1585, 991, and 605 cm^{-1} shift to higher frequencies. This is observed for the adduct with acetic acid or with isobutyric acid in $CHCl_3$. With CCl_3COOH, bands characteristic of the pyridinium ion appear at 1635 and 1490 cm^{-1}. The presence of the CO_2^- bands at 1675 cm^{-1} ($\nu_{as}\,CO_2^-$) and at 1320 cm^{-1} ($\nu_s\,CO_2^-$) also brings confirmation of the formation of ion-pairs (Gusakova *et al.* 1970). With acids of intermediate strength like $CHCl_2COOH$, the presence of bands both at 1490 cm^{-1} and 1675 cm^{-1}, provides evidence for tautomeric equilibrium. Both $\nu_{C=O}$ and $\nu_{CO_2^-}$ vibration bands have also been observed by Gusakova and coworkers (1972a) in the system isobutyric acid–piperidine in dioxan solution. These authors have calculated the transfer constant K_t and the corresponding enthalpy ΔH_t from the intensity of the bands (Gusakova *et al.*, 1972b).

Extensive quantitative work using infrared spectroscopy was performed by Zundel and his coworkers; they studied the tautomeric equilibrium in carboxylic acid–nitrogen bases systems using the $\nu_{C=O}$ band characterizing the normal form and the ($\nu_s\,CO_2^-$) band appearing from the ion-pair. The percentage of ion-pairs is obtained by comparing the integral absorbance of the $\nu_{C=O}$ band in a given system with that of the same band of a system in which no proton transfer is detectable (formic acid–2–methylpyrazine, for

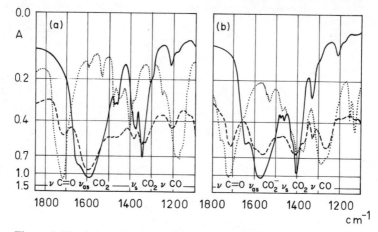

Figure 1.12   Infrared spectra: (a) —— formic acid + n-propylamine; ---- formic acid + imidazole; ··· formic acid + 2-methylpyrazine; (b) —— acetic acid + n-propylamine; ---- acetic acid + imidazole; ··· acetic acid + pyrazole. [Reproduced by permission of G. Zundel from Lindemann and Zundel, 1977]

instance) as demonstrated by the absence of ν_{CO_2} bands. It is assumed that the extinction coefficient does not change appreciably from one system to another but this is confirmed by the fact that the transfer constants obtained from $(\nu_s\,CO_2^-)$ are in good agreement with those obtained from $\nu_{C=O}$. In Figure 1.12 we give examples of the spectra determined by Zundel and his coworkers illustrating the principle of the method (Lindemann and Zundel, 1977). These authors also observed that the continuum absorption between 3500 and 1500 cm^{-1} becomes very important for those complexes for which K_t lies in the vicinity of unity. This was ascribed to the *high polarizability* of such bonds.

Proton transfer can also occur in *intra*molecular hydrogen bonds. Such bonds were investigated by Brzezinski and Zundel (1976) in 1-piperidine carboxylic acids:

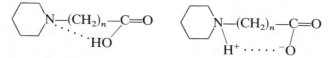

The authors came to the conclusion that a double-minimum energy surface exists in the system studied. The strong infrared continuum also indicates that intramolecular bonds are easily polarizable.

The previous examples all refer to the transfer equilibrium in hydrogen bonds of the type

$$O\text{---}H \cdots N \rightleftharpoons O^- \cdots H\text{---}^+N$$

For O—H $\cdots$ O bonds an example of tautomerism has been detected by Kirszenbaum, Corset and Josien (1970) in the adduct of trichloroacetic acid with tetrahydrofuran (THF)

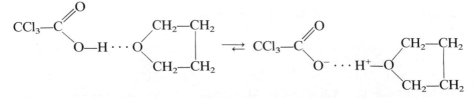

In a solvent like 1,2-dichloroethane the acid is characterized by a $\nu_{C=O}$ vibration at 1804 cm^{-1}. When THF is added, a new band corresponding to the normal hydrogen bond is found at 1781 cm^{-1}; in nitromethane solution, two bands are observed, at 1780 and 1762 cm^{-1}, the first corresponds to the normal complex and the second to the proton transfer complex.

Spectroscopic methods are thus very adequate tools to detect tautomerism and to determine the transfer constant K_t.

The examples discussed above show clearly the existence of a double minimum in the potential curve of the proton, and the tautomerism in A—H $\cdots$ B $\rightleftarrows$ A$^-$ $\cdots$ H$^+$—B where A and B represent two different entities, seems well established in solution.

Vibrational spectroscopy has also been widely used to deduce the nature of the proton potential in formally symmetric systems, such as the (BHB)$^+$ or (AHA)$^-$ systems. The (NH $\cdots$ N)$^+$ bonds have been extensively studied by Wood and coworkers, as, for example, the bonds in (pyridine)$_2$H$^+$ (Clements and Wood, 1973a), (4-methylpyridine)$_2$H$^+$ (Dean and Wood, 1975a), (imidazole)$_2$H$^+$ (Bonsor et al., 1976), and (benzimidazole)$_2$H$^+$ (Borah and Wood, 1976). All these systems, studied in solutions, are characterized by two broad bands lying at $\sim$2500 and $\sim$2100 cm^{-1}, associated with the N—H stretching motion; in the corresponding deuterium-bridged cations only one band was observed at $\sim$1900 cm^{-1}. The origin of this doublet has been interpreted in terms of Fermi resonance involving the δ_{NH} overtone and this interpretation has been confirmed recently for the spectrum of amine semiperchlorate: Grech et al. (1976) have indeed shown that the position of the minimum dividing the ν_s band into two components remains constant with decrease in temperature.

In the far-infrared region, Dean and Wood (1975a) observed a broad band, lying between 100 and 150 cm^{-1}, the maximum of which depended on the nature of the associated cation. This absorption could arise from the tunnelling transition between the two levels of the split ground state in a double-minimum potential, but the small deuteration shift rules out this possibility. The far-infrared absorption has been assigned to an intermolecular N $\cdots$ N stretching motion. For these formally symmetric systems, the proton is not located at the middle of the bridge and this fact is suggested by the following spectroscopic arguments:

1. the infrared activity of the intermolecular ν_σ (N $\cdots$ N) mode;
2. the Raman activity of the δ_{NH} in-plane bending mode;
3. the observation of the vibrational modes derived from both the proton donor (as for example the pyridinium ion) and the proton acceptor (as for example the pyridine).

The fact that the hydrogen-bonded proton is not located at the middle of the N $\cdots$ N bridge has also been confirmed by X-ray studies of the benzimidazolium semitetrafluoroborate in which the N $\cdots$ N bond is found to be 2.787 Å and the proton is located nearest a nitrogen atom (Quick *et al.*, 1974). The asymmetry is also reflected in corresponding differences in the bond lengths within the two imidazole rings. The ν_s doublet remains in the infrared spectrum of the solid state and it is therefore very probable that this doublet is not indicative of proton tunnelling. The N $\cdots$ N bond length of 2.853 Å in pyrazine crystal (Glowiak *et al.*, 1975a) and of 2.84 Å in triethylenediamine perchlorate (Glowiak *et al.*, 1975b) and the N—H distance of about 1 Å also indicates the presence of an asymmetric double-minimum potential for the proton compounds.

Some other studies have been performed on asymmetric $(N—H \cdots N)^+$ bonds, containing two different bases such as 2,6-trimethylpyridine, 3,5-dimethylpyridine, 2-fluoropyridine and pyridine (Clements *et al.*, 1973) or 4-methylpyridine, N-methylimidazole and pyridine (Dean and Wood, 1975b). In the hydrogen bridged cations of these systems, the ν_s NH band is a doublet, the ratio of the intensity of the upper/lower component varying regularly with the ΔpK_a values of the bases. Only a single ν_s ND band was observed for the deuterium bridged cations. The absence of any observable splitting of the internal modes of the base indicates that the bridging proton potential is asymmetric. It thus seems that nitrogen atoms are unable to form N—H $\cdots$ N hydrogen bonds with a single minimum and this fact was pointed out by Pawlak and Sobczyk (1973) for some other (NHN)$^+$ cations. A double-minimum potential energy curve with a low barrier for the $(N—H \cdots N)^+$ bond in the pyridinium–pyridine complex was also predicted from LCAO–MO calculations (Sabin, 1968, 1972).

In some cases however, the assignment of the vibrational bands may become difficult, especially when the R distance is small (<2.6 Å) and in this last decade both diffraction and infrared spectroscopic studies have been performed on these very short hydrogen bonds existing for example in many salts of carboxylic acids containing the (COOHCOO)$^-$ groups. These (OHO)$^-$ bonds can be represented by a symmetric *single* minimum, or a symmetric or asymmetric *double* minimum. Many of these salts have been studied by diffraction methods and Speakman (1972) distinguishes A and B types of structure. The B structure corresponds to asymmetric hydrogen bonds where the acid residues are neither symmetry related nor equivalent and the distances between the C and O atoms are characteristic of ionized and non-ionized carbonyl groups.

In A-type salts, the centre of the hydrogen bond coincides with a crystallographic symmetry element, both carboxylic residues being equivalent; the A structure corresponds to symmetric hydrogen bonds. Diffraction methods cannot differentiate between two minima separated by less than 0.1 Å and a single centred well (Hamilton and Ibers, 1968) but vibrational spectroscopy may provide some information about the symmetry of the hydrogen bond. For A-type salts, the stretching vibrations of the $(OHO)^-$ groups are observed at very low wavenumbers, as low as $700\ cm^{-1}$ for the asymmetric stretching vibrations active in the infrared; the symmetric stretching vibration active in the Raman spectra is observed between 100 and $320\ cm^{-1}$ (Novak, 1972). The (OHO) bending frequencies have been assigned to bands at $\sim 1500\ cm^{-1}$ (δ_{OH}) and $1285\ cm^{-1}$ (γ_{OH}) and as pointed out by Novak (1974), this is one of the rare instances where the bending frequencies are higher than those of stretching. For these A-type salts, the two carboxylic residues are equivalent and one should observe two C—O stretching bands, the first one corresponding to the shorter bond (at $\sim 1750\ cm^{-1}$) and the second one corresponding to the longer bond (at $\sim 1400\ cm^{-1}$).

For the B-type salts, the ν_{OH} stretching band is usually observed at wavenumbers higher than $1000\ cm^{-1}$. The carboxylic residues are not equivalent and one should observe two stretching bands corresponding to the un-ionized residue (at about 1700 and $1200\ cm^{-1}$) and two stretching bands (at about 1600 and $1400\ cm^{-1}$) corresponding to the asymmetric and symmetric vibrations of the ionized residues. Table 1.14 compares the spectroscopic parameters for two types of salts. Potassium hydrogen bis(trifluoroacetate) is characterized by a very short hydrogen bond; the $R(O\cdots O)$ distance is equal to 2.437 Å and a twofold axis passes through the centre of

Table 1.14 Frequencies of some infrared and Raman bands due to the $(COOHCOO)^-$ group (in cm^{-1})

Compound	$\nu_{as}(OH\cdots O)$ (IR)	$\nu_s(OH\cdots O)$ (R)	δ_{OH} (IR)	γ_{OH} (IR)	$\nu_{C=O}$ and ν_{C-O} (IR)
$KH(CF_3COO)_2$	800	130	1480	1285	1790 1420
					(Hadzi *et al.*, 1973)
$NaHC_2O_4\cdot H_2O$	1800	—	1455	1042	1740, 1600, 1420, 1230
					(de Villepin and Novak, 1971a)

the $O \cdots O$ bond (Macdonald *et al.*, 1972). Sodium hydrogen oxalate monohydrate is of B type, the $HC_2O_4^-$ anions being associated by short $OH \cdots O$ hydrogen bonds (2.571 Å) (Tellgren and Olovsson, 1971). The crystallographic and spectroscopic data are, in this case in good agreement about the equivalence, or non-equivalence of the carboxylic residues but the assignments of the OH stretching or bending vibrations are sometimes uncertain or unknown. The infrared and Raman spectra of homoconjugated salts of acetone (Huong and Noel, 1976) and *N,N*-dimethylformamide (Belin and Potier, 1976b) also suggest that the homoconjugated salts contain a very strong and quasi-symmetrical $\diagup C{=}O{-}H{-}O{=}C \diagdown$ hydrogen bond. The spectra generally present a complicated pattern; the bands are usually very broad and may be interrupted by resonance windows.

Other criteria have been used tentatively to decide between single- or double-minimum potential; these criteria are the isotope effect on the *R* distance and the values of the ν_{AH}/ν_{AD} frequency ratio.

1.3.2.2 *Isotope effect on the* R *distance and* ν_{AH}/ν_{AD} *frequency ratio* The change in hydrogen bond length on replacing hydrogen by deuterium, first investigated by Robertson and Ubbelohde (1939) has been further discussed by Gallagher (1959) on the basis of two terms contributing to the change in the bond length: a contraction term arising from the increased dipole moment and an expansion term depending on the hydrogen bond energy. Using Stepanov's approximation, Singh and Wood (1969) calculated the effect of isotopic substitution on the $R(A \cdots B)$ distance and found that these distances *increase* on deuteration if the potential energy function is characterized by a symmetric or asymmetric *double* minimum. According to Savelev and Sokolov (1970), the isotope expansion is related to the bond length and should vary between 2×10^{-3} and 6×10^{-2} Å for a double-minimum potential. Other examples of this expansion have been reviewed by Hamilton and Ibers (1968) and more recently by Olovsson and Jönsson (1976). For example, an extension of 0.022 Å was found in $NaHC_2O_4 \cdot H_2O$ (Tellgren and Olovsson, 1971) where the $O{-}H \cdots O$ bond is asymmetric.

Now, according to Singh and Wood (1969), a *small contraction* is to be expected when the proton potential curve shows a *single* symmetric *minimum* (although Rundle (1964) predicted a *zero isotope* effect in this case). A small contraction has been found indeed in the $(FHF)^-$ bond of KHF_2 (Ibers, 1964) and this bond has long been recognized as symmetric and very short (2.26 Å). A zero isotope effect has been observed recently for the symmetric hydrogen bond of $KH(CF_3COO)_2$ where the two $O \cdots H \cdots O$ and $O \cdots D \cdots O$ distances are both strictly equal (2.437 Å) (Macdonald *et al.*, 1972).

The isotope effect on the distance can thus be considered as an indication of the shape of the potential function, but, as pointed out by Hamilton and Ibers (1968) and Olovsson and Jönsson (1976), these indications must be regarded with caution because it was generally assumed 'that any changes brought about by isotopic substitution were confined in the hydrogen bond'. The isotope effect is more likely to be distributed in the whole unit cell; some difficulties can also arise when several hydrogen bonds are present (usually one short and one or two long hydrogen bonds).

Another indication of the shape of the potential function is the value of the isotopic ratio ν_{AH}/ν_{AD}; it was pointed out in section 2.2.1 that these ratios are close to 1.355 for weak complexes but may become lower for stronger complexes. Most of the studies have been performed on the ν_{OH} stretching vibration and it has been shown that in solution (Odinokov and Iogansen, 1972) or in the crystalline state (Novak, 1974) the isotopic ratio ν_{OH}/ν_{OD} systematically decreases with increased strength of the hydrogen bond. Novak (1974) has shown that when the ν_{OH}/ν_{OD} frequency ratio is plotted against the $R(O \cdots O)$ distance, the resulting curve shows that this ratio decreases slowly for weak to medium strong hydrogen bonds ($R > 2.6$ Å). The potential barrier separating the two wells remains relatively high and the proton remains preferentially in one well. When the $R(O \cdots O)$ distance is between 2.60 and 2.48 Å, the isotopic ratio decreases rapidly and values of about 1 are observed; this range should correspond to a double minimum with a relatively low barrier. Laane (1971) has calculated the isotopic ratio for barriers of various shapes including quartic terms in the potential function. He found that the ratio approaches 1.4 for a high barrier and falls as the barrier decreases; these calculations are qualitatively in agreement with observations. For example, a ratio ν_{OH}/ν_{OD} of 1 has been found for $NaHC_2O_4 \cdot H_2O$; the existence of a double minimum is also compatible with an extension of 0.022 Å on replacing hydrogen by deuterium (Tellgren and Olovsson, 1971). This extension implies a relatively higher ν_{OD} and this effect combined with a higher value of the mass should eventually explain the similar values of ν_{OH} and ν_{OD} (de Villepin and Novak, 1971b). As pointed out by Somorjai and Hornig (1962) abnormally low ν_{OH}/ν_{OD} ratios can also be observed for barriers whose top is in the vicinity of the first excited OH vibrational level.

Lastly, the ν_{OH}/ν_{OD} ratio increases suddenly for distances less than 2.46 Å and this range should correspond to a single central minimum (Novak, 1974). The stochastic model of the hydrogen bond proposed by Romanowski and Sobczyk (1977) also predicts an increase of the ν_{OH}/ν_{OD} values for short distances. For these short and symmetrical hydrogen bonds, values between 1.2 and 1.45 have been reported by Hadzi and Orel (1973). Table 1.15 reports some of these values and the $O \cdots O$ and $O-H$ distances measured by neutron diffraction. Now, ν_{OH}/ν_{OD} ratios near unity have also been found for very short hydrogen bonds as in $NH_4^+ CH_2ClCOO^-$

Table 1.15 $R(O \cdots O)$ and r_{OH} distances, $\nu_{as}OHO$ and ν_{OHO}/ν_{ODO} for symmetrical (OHO) bonds

Compound	$R(O \cdots O)$ (Å)	r_{OH} (Å)	$r_{H \cdots O}$ (Å)	References	$\nu_{as}OHO$ (cm^{-1}) -83 K	$\dfrac{\nu_{OHO}}{\nu_{ODO}}$	References
KH bistri-fluoro-acetate	2.437	1.218	1.218	Macdonald *et al.* (1972)	800	1.35	Hadzi and Orel (1973)
KH malonate	2.468	1.234	1.234	Currie and Speakman (1970)	600	1.4	Hadzi and Orel (1973)
KH succinate	2.444	1.227	1.227	McAdam *et al.* (1971)	850	1.4	Hadzi and Orel (1973)

where the $O \cdots O$ distance is 2.43 Å (Ichikawa, 1972). Hadzi and Orel (1973) further distinguish the pseudo-A acid salts where no symmetry is imposed on the hydrogen bond and the carboxylic residues show only small differences in dimensions. These salts should be characterized by an isotopic ratio ν_{OH}/ν_{OD} near unity.

It thus seems that the values of the ν_{OH}/ν_{OD} ratio alone do not provide sufficient indications about the potential curve of the proton. These data should be compared with the variation of the R distance on substituting hydrogen by deuterium; for symmetrical hydrogen bonds, the only indication of a zero isotope effect on the distance has been found for $KH(CCl_3COO)_2$. The decrease of the ν_{OH}/ν_{OD} ratio with increasing bond strength (or decreasing distance) seems well established for usual bonds but for very short hydrogen bonds, some difficulties remain in the interpretation of the experimental data.

Now, for ($NH^+ \cdots X^-$) or ($NH^+ \cdots N^+$) hydrogen bonds, the ν_{NH^+}/ν_{ND^+} ratio regularly decreases with the N—H stretching frequency (Novak, 1974) or with the enthalpy of complex formation (Glazunov *et al.*, 1976). This indicates that the proton remains fixed to the nitrogen in full agreement with the previously discussed results on the hydrogen bonds involving nitrogen atoms.

1.3.2.3   *Dipole moments*

In a previous section it was shown that the dipole increment $\Delta\mu$ brought about by the formation of a specific bond increases when the enthalpy of the bond, $-\Delta H$, becomes stronger (Figures 1.6 and 1.7).

When tautomerism exists between the normal form characterized by the dipole moment μ_N and the ion-pair with the dipole moment μ_I, the experimental dipole moment μ_{AD} is intermediate between the two values.

One can write

$$\mu_{AD}^2 = \frac{C_N}{C_{AD}}\mu_N^2 + \frac{C_I}{C_{AD}}\mu_I^2 = \frac{1}{1+K_t}\mu_N^2 + \frac{K_t}{1+K_t}\mu_I^2 \qquad (1.1)$$

where C_N is the concentration of the normal form, C_I that of the ion-pair, and C_{AD} the sum of both. This equation holds when the exchange time between the two tautomers is longer than the dielectric relaxation time. When a broad potential barrier separates the two minima, the displacement of the nucleus p is much more important in the ion-pair and $\mu_I \gg \mu_N$.

Let us now consider the complex for which the transfer constant K_t is unity. In such a complex the concentrations of both forms are equal and μ_{AD}^2 is the arithmetic mean of μ_I^2 and μ_N^2. This complex is characterized by the enthalpy $-\Delta H_e$ of the bond.

Complexes in the given series for which $-\Delta H$ is markedly smaller than $-\Delta H_e$ have a transfer constant which is much less than one. For these complexes μ_{AD} will be close to μ_N.

In contrast, complexes with an enthalpy $-\Delta H$ sufficiently greater than $-\Delta H_e$ will show high transfer constants and their dipole moment μ_{AD} will be close to μ_I.

In the vicinity of $K_t = 1$ there will thus be a sharp increase in $\Delta\mu$ when $(-\Delta H)$ increases supposing that $\Delta\mu_I$ remains much greater than $\Delta\mu_N$ and when a broad potential barrier separates the two minima.

The populations of the two levels depend on their vertical separation and on the entropy change. The passage of the supremacy of the normal type to the ion-pair type when a broad potential barrier exists between the minima is illustrated by Figure 1.13.

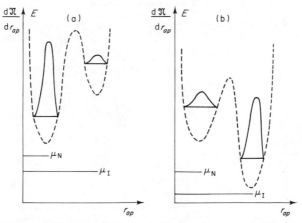

Figure 1.13 Populations of the levels separated by a broad potential barrier: (a) μ_{AD} resembles μ_N; (b) μ_{AD} resembles μ_I

In such a case a sudden change in $\Delta\mu$ is thus expected in the vicinity of $K_t = 1$ and a *sigmoidal* curve is expected for the $\Delta\mu/\Delta H$ relation (Figure 1.14):

If now the potential barrier between the two minima is suppressed, this will cause the disappearance of the highest level and a flattening of the bottom of the energy surface. Such flattening will displace the maximum of the wavefunction towards the centre. Passing from the first case illustrated by Figure 1.15 to the second one will thus be accompanied by a smaller change in the dipole moment than in the cases illustrated by Figure 1.13. In the sigmoidal curve, the change around the inflexion point will be less abrupt. At the limit, when a potential barrier does not appear in the whole range of $-\Delta H$, the inflexion point will disappear (Figure 1.16).

The spectacular difference between the $\Delta\mu/\Delta H$ relations for hydrogen bonds and for E.D.A. bonds has been mentioned in a previous section. In solution O—H $\cdots$ N and O—H $\cdots$ O bonds give sigmoidal $\Delta\mu/\Delta H$ relations whereas more or less linear relations are obtained for E.D.A. complexes.

From this behaviour it can be inferred that for O—H $\cdots$ N and O—H $\cdots$ O bonds *in solution* potential barriers which are large enough to assure the appearance of tautomerism separate the two wells in the whole range of $(-\Delta H)$ values. This is confirmed by the spectrometric data given above.

In contrast, the $\Delta\mu/\Delta H$ relationships for E.D.A. bonds belong rather to the type described by Figure 1.16. This suggests that over a broad ΔH range such a barrier does not exist for these bonds in solution.

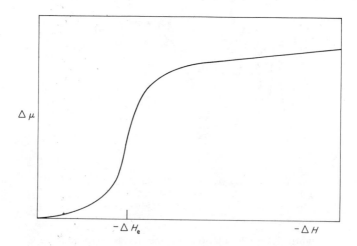

Figure 1.14   Expected $\Delta\mu/\Delta H$ relation when the minima remain separated by broad potential barriers

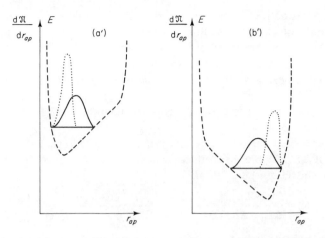

Figure 1.15 Situation when the potential barriers of Figure 1.13 are suppressed (... curves of Figure 1.13); (a') is more polar than (a); (b') is less polar than (b)

The reasons for this difference in behaviour between E.D.A. and hydrogen bonds have already been discussed.

When $\Delta\mu/\Delta H$ or $\Delta\mu/\Delta pK_a$ relationships are available over a sufficiently large range to cover regions where either the normal bonds or the ion-pairs are dominant, it is possible to extract values of K_t from these data.

As a matter of fact the values of $\Delta\mu_N$ and $\Delta\mu_I$ for a given complex can be determined by extrapolation of those parts of the $\Delta\mu/\Delta pK_a$ curve where practically only normal complexes or only ion-pairs appear (Figure 1.5). To

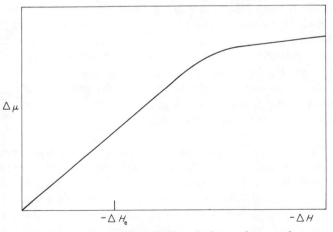

Figure 1.16 Expected $\Delta\mu/\Delta H$ relation when only one minimum exists in the whole ΔH range

a first approximation these values may even be taken as constant for the whole family of complexes. For O—H $\cdots$ N bonds involving phenols and amines, $\Delta\mu_N$ is of the order of 0.4 D whereas $\Delta\mu_I$ is of the order of 9.4 D (Ratajczak and Sobczyk, 1969). For O—H $\cdots$ N bonds involving carboxylic acids and amines, a value of 7.0 D was proposed for $\Delta\mu_I$ (Sobczyk and Pawelka, 1974). By extrapolation in the transient region for pyridine–phenol complexes we found for $\Delta\mu_N$ an order of magnitude of 1.0 D and for $\Delta\mu_I$ approximately 8 D.

In the transient region where $\Delta\mu$ is sufficiently different from either $\Delta\mu_N$ or $\Delta\mu_I$, it is possible to deduce the value of K_t from the relationship (Nouwen and Huyskens, 1973):

$$K_t = \frac{\Delta\mu - \Delta\mu_N}{\Delta\mu_I - \Delta\mu} \left\{ \frac{\Delta\mu + \Delta\mu_N + 2(\mu_A \cos\theta_A + \mu_D \cos\theta_D)}{\Delta\mu + \Delta\mu_I + 2(\mu_A \cos\theta_A + \mu_D \cos\theta_D)} \right\} \qquad (1.2)$$

When the correction factor equals unity, one obtains the equation used by Sobczyk. (See for instance Sobczyk and Pawelka, 1974.) As a matter of fact, for the O—H $\cdots$ N bonds studied in the literature, this factor lies between 0.30 and 0.70. The validity of this expression is based on the fact that the lifetime of the tautomers is longer than the dielectric relaxation time. This may not be the case for the complexes for which the value of K_t lies in the vicinity of unity. An indication of this may be found in the work of Malecki, determining non-linear dielectric effects in mixtures of 2,4,6-trichlorophenol and triethylamine in various solvents (Malecki, 1976). In this case the correction factor approaches unity.

Owing to the lack of accuracy in determining the dipole increments, the experimental values of $\Delta\mu$ and $\Delta\mu_N$ or $\Delta\mu$ and $\Delta\mu_I$ must differ by at least 0.2 D to make the method relevant. As a consequence, its application is limited to the range of K_t values going from 0.01 to 30. Furthermore the accuracy of dipolemetric determinations of K_t remains very poor when K_t is smaller than 0.1 or greater than 10. It is worthwhile noting that, contrary to spectroscopic methods, dipolemetry does not provide proof of the existence of tautomerism. The equation above can only be used if such a tautomerism exists. In the cases with a single minimum in the potential energy surface, the equation above (without the correcting factor) would only give some information on the position of this minimum.

1.3.2.4 *Nuclear magnetic resonance spectroscopy* NMR spectroscopy can also provide some indirect information about tautomerism in a given family of complexes. It has been shown by Odinokov *et al.* (1975) that for complexes of hydroxy acids with triethylamine, the enthalpy of complex formation regularly increases with the acid strength but the proton shift takes a maximum of 15 p.p.m. and further decreases with the acid strength.

Similar observations have been reported by Le-van and Ratajczak (1977) for complexes of substituted phenols with triethylamine in benzene solution. When the shift of the proton is plotted against the pK_a of the phenol, a transition point is observed for a pK_a of the order of six. The existence of these transition points has been attributed to the increase in importance of ion-pair complexes relative to normal hydrogen bonds.

Blinc *et al.* (1967) have also shown that useful information on potential curves of hydrogen bonds might be obtained from nuclear quadrupole resonance studies; some acid salts of chloroacetic acids have been studied by this technique (Ratajczak *et al.*, 1977).

1.3.3 *Experimental values of the transfer constant K_t and transfer enthalpy ΔH_t.*

In Table 1.16 we give a list of selected values of K_t and ΔH_t computed from the literature data. Most of the data refer to $O—H \cdots N$ hydrogen bonds which were the most studied from this point of view. A few data concern $O—H \cdots O$ bonds.

It is worthwhile noting here that when the two levels of *free energy* are the same for both forms, and thus when K_t is unity, there is already a marked difference in the *energy* levels. It appears indeed that for the systems where K_t is approximately equal to unity, the energy of the proton transfer form in $O—H \cdots N$ bonds lies about 13–15 kJ mol^{-1} *lower* than that of the normal form.

Thus for the $O—H \cdots N$ complexes for which $K_t = 1$, the double well of the energy curve is not symmetric: the well near the nitrogen nucleus is markedly deeper than that near the oxygen nucleus.

The reason for this behaviour lies in the fact that the proton transfer form is not favoured from the point of view of entropy. The polar form polarizes the solvent molecules around the complex much more. These solvent molecules are thus more preferentially orientated when the transfer takes place and this increase of order in the solution corresponds, of course, to a decrease of the entropy. At $K_t = 1$, this decrease of the entropy $-\Delta S_t^{\ominus}$ just cancels out the decrease of the enthalpy $-\Delta H_t$ in the standard free energy of transfer.

1.3.4 *Factors which influence the transfer constant*

1.3.4.1 *Influence of ΔpK_a* From the data of Table 1.16, it is clear that an important factor governing the transfer constant in a complex is the difference between the pK_a of the conjugated acid of the base and the pK_a of the acid in water

$$\Delta pK_a = pK_{a_{BH^+}} - pK_{a_{AH}}.$$

Table 1.16 Selected values of tautomerization constants and enthalpies at 298 K

Acid	$pK_{a_{AH}}$	Base	$pK_{a_{BH^+}}$	ΔpK_a water	Medium [dielectric constant]	K_t (l mol^{-1})	log K_t	$-\Delta H_t$ (kJ mol^{-1})	Method
Phenol derivatives		*Aliphatic amines*							
3,4,5-(CH$_3$)$_3$-	10.80	Triethylamine	10.65	-0.25	Toluene [2.39]	0.01^a	-2.00		Dipol.
3,4,5-(CH$_3$)$_3$-	10.80	Triethylamine	10.65	-0.25	Trichloroethylene [3.56]	0.04^a	-1.40		Dipol.
4-CH$_3$-	10.26	Triethylamine	10.65	0.39	Trichloroethylene	0.06^a	-1.22		Dipol.
Phenol	9.25	Triethylamine	10.65	0.70	C$_6$H$_6$ [2.27]	0.01^b	-2.00		Dipol.
Phenol	9.95	Triethylamine	10.65	0.70	Toluene	0.02^a	-1.70		Dipol.
Phenol	9.95	Triethylamine	10.65	0.70	Trichloroethylene	0.04^a	-1.40		Dipol.
4-Cl-	9.42	Propylamine	10.53	1.11	CD$_3$CN [36.0]	0.01^c	-2.00		I.R.
4-Cl-	9.42	Triethylamine	10.65	1.23	C$_6$H$_6$	0.01^b	-2.00		Dipol.
4-Br-	9.36	Triethylamine	10.65	1.29	C$_6$H$_6$	0.01^b	-2.00		Dipol.
4-I-	9.31	Triethylamine	10.65	1.34	C$_6$H$_6$	0.01^b	-2.00		Dipol.
3-Cl-	9.02	Propylamine	10.53	1.51	CD$_3$CN	0.03^c	-1.52		I.R.
2-Cl-	8.66	Propylamine	10.53	1.87	CD$_3$CN	0.06^c	-1.22		I.R.
3,5-Cl$_2$-	8.09	Propylamine	10.53	2.44	CD$_3$CN	0.18^c	-0.74		I.R.
2,4-Cl$_2$-	7.92	Propylamine	10.53	2.61	CD$_3$CN	0.30^c	-0.52		I.R.
2,3-Cl$_2$	7.61	Propylamine	10.53	2.92	CD$_3$CN	5.23^c	+0.72		I.R.
4-CN-	7.52	Triethylamine	10.65	3.13	C$_6$H$_6$	0.09^b	-1.05		Dipol.
4-NO$_2$-	7.15	Triethylamine	10.65	3.50	C$_6$H$_6$	0.12^b	-0.92		Dipol.
4-NO$_2$-	7.15	Triethylamine	10.65	3.50	Cyclohexane [2.02]	0.07^a	-1.15		Dipol.
4-NO$_2$-	7.15	Triethylamine	10.65	3.50	Toluene	0.18^a	-0.74		Dipol.
4-NO$_2$-	7.15	Triethylamine	10.65	3.50	Trichloroethylene	0.55^a	-0.26		Dipol.
4-NO$_2$-	7.15	Triethylamine	10.65.	3.50	1,2-Dichloroethane [10.4]	0.68^d	-0.17	12.9^d	U.V.
4-NO$_2$-	7.15	Piperidine	11.10	3.95	CH$_3$CN	10.3^e	+1.01		U.V.
3,5-(NO$_2$)$_2$-	6.70	Triethylamine	10.65	3.95	C$_6$H$_6$	0.27^b	-0.57		Dipol.
2,4,5-Cl$_3$-	6.17	Propylamine	10.53	4.36	CD$_3$CN	99^c	+2.00		I.R.
2,4,6-Cl$_3$	6.00	Triethylamine	10.65	4.65	C$_6$H$_6$	0.46^b	-0.34		Dipol.
2,4,6-Cl$_3$-	6.00	Triethylamine	10.65	4.65	Cyclohexane	0.15^a	-0.82		Dipol.
2,4,6-Cl$_3$-	6.00	Triethylamine	10.65	4.65	Toluene	0.57^a	-0.24		Dipol.
2,4,6-Cl$_3$-	6.00	Triethylamine	10.65	4.65	Trichloroethylene	1.38^a	+0.14		Dipol.
2,4,5-Cl$_3$	6.17	Dibutylamine	11.25	5.08	Tetrachloroethylene	2.1^r	+0.32		I.R., U.V.
2,5-(NO$_2$)$_2$-	5.22	Triethylamine	10.65	5.43	C$_6$H$_6$	1.53^b	+0.18		Dipol.
Cl$_5$-	5.26	Triethylamine	10.65	5.39	Cyclohexane	0.39^a	-0.41		Dipol.

Donor		Acceptor			Solvent				Method
Cl$_5$-	5.26	Triethylamine	10.65	5.39	Toluene	1.53^a	+0.18		Dipol.
Cl$_5$-	5.26	Triethylamine	10.65	5.39	Trichloroethylene	3.05^a	+0.48		Dipol.
Br$_5$-	5.26	Tributylamine	10.89	5.63	C$_6$H$_6$	1.75^f	+0.24		Dipol.
2,6-Cl$_2$-4-NO$_2$-	3.68	Triethylamine	10.65	6.97	C$_6$H$_6$	8.96^f	+0.95		Dipol.
2,6-Br$_2$-4-NO$_2$-	3.39	Triethylamine	10.65	7.26	Toluene	4.52^a	+0.66		Dipol.
2,6-Br$_2$-4-NO$_2$-	3.39	Triethylamine	10.65	7.26	Trichloroethylene	28.6^a	+1.46		Dipol.
Phenol derivatives		*Pyridines*							
2,6-Cl$_2$-4-NO$_2$-	3.68	Pyridine	5.23	1.55	CCl$_4$ [2.23]	0.14^g	-0.85		Dipol.
2,6-Cl$_2$-4-NO$_2$-	3.68	Pyridine	5.23	1.55	C$_6$H$_6$	0.11^h	-0.96		Dipol.
Cl$_5$-	5.26	2,4,6-(CH$_3$)$_3$-	7.43	2.17	C$_6$H$_6$	0.15^h	-0.82		Dipol.
2,6-Cl$_2$-4-NO$_2$-	3.68	4-CH$_3$-	6.02	2.34	CCl$_4$	0.15^g	-0.82		Dipol.
2,6-Cl$_2$-4-NO$_2$-	3.68	3,5-(CH$_3$)$_2$-	6.29	2.61	C$_6$H$_6$	0.25^h	-0.60		Dipol.
2,6-Cl$_2$-4-NO$_2$-	3.68	2,6-(CH$_3$)$_2$-	6.81	3.13	C$_6$H$_6$	0.85^h	-0.07		Dipol.
2,6-Cl$_2$-4-NO$_2$-	3.68	2,4,6-(CH$_3$)$_3$-	7.43	3.75	CCl$_4$	0.52^g	-0.28		Dipol.
2,6-Cl$_2$-4-NO$_2$-	3.68	2,4,6-(CH$_3$)$_3$-	7.43	3.83	C$_6$H$_6$	2.29^h	+0.36		Dipol.
2,4,6-(NO$_2$)$_3$-	0.44	Pyridine	5.23	4.79	C$_6$H$_6$	9.00^h	+0.95		Dipol.
Phenol derivatives		*Imidazole derivatives*							
Cl$_5$-	5.26	N-CH$_3$-	7.04	1.78	C$_6$H$_6$	0.11^i	-0.96		Dipol.
3,4-(NO$_2$)$_2$-	5.42	1,2-(CH$_3$)$_2$-	8.07	2.65	C$_6$H$_6$	0.16^i	-0.80		Dipol.
Cl$_5$-	5.26	1,2-(CH$_3$)$_2$-	8.07	2.81	C$_6$H$_6$	0.39^i	-0.41		Dipol.
2,4-(NO$_2$)$_2$-	4.07	N-CH$_3$-	7.04	2.97	C$_6$H$_6$	1.02^i	+0.01		Dipol.
2,6-Cl$_2$-4-NO$_2$-	3.68	N-CH$_3$-	7.04	3.36	C$_6$H$_6$	2.41^i	+0.38		Dipol.
2,4-(NO$_2$)$_2$-	4.07	1,2-(CH$_3$)$_2$-	8.07	4.00	C$_6$H$_6$	3.39^i	+0.53		Dipol.
2,6-Cl$_2$-4-NO$_2$-	3.68	1,2-(CH$_3$)$_2$-	8.07	4.39	C$_6$H$_6$	3.15^i	+0.50		Dipol.
Bromophenol blue		*Pyridines*							
Bromophenol blue	4.10	3-Cl-	2.84	-1.26	C$_6$H$_5$Cl [5.62]	1.2^j	+0.08	14^j	U.V.
Bromophenol blue	4.10	Pyridine	5.23	1.13	C$_6$H$_5$Cl	300^j	+2.48	23^j	U.V.
Bromophenol blue	4.10	2-CH$_3$-	5.97	1.87	C$_6$H$_5$Cl	4200^j	+3.62	37^j	U.V.
Bromophenol blue	4.10	2,6-(CH$_3$)$_2$-	6.75	2.65	C$_6$H$_5$Cl	18,000^j	+4.26	29^j	U.V.
Bromophenol blue	4.10	2,4,6-(CH$_3$)$_3$-	7.43	3.33	C$_6$H$_5$Cl	92,000^j	+4.96	37^j	U.V.
Carboxylic acids		*Amines*							
Isobutyric	4.80	Piperidine	11.12	6.32	Dioxan [2.21]	100^k	+2.00	30.6^k	I.R.
Chloroacetic	2.86	Trioctylamine	11.19	8.33	C$_2$Cl$_4$ [2.30]			15.8^l	I.R.

Table 1.16 (continued)

Acid	$pK_{a_{AH}}$	Base	$pK_{a_{BH^+}}$	ΔpK_a water	Medium [dielectric constant]	K_t (l mol^{-1})	log K_t	$-\Delta H_t$ (kJ mol^{-1})	Method
Carboxylic and benzoic acids									
		Pyridines							
Acetic	4.76	2-Methanolpyridine	5.01	0.25	1:1 mixture	0.01^m	-2.00		I.R.
Palmitic	4.90	Pyridine	5.23	0.33	CCl$_4$	0.57^n	-0.24		I.R.
Myristic	4.90	Pyridine	5.23	0.33	CCl$_4$	0.56^n	-0.25		I.R.
Benzoic	4.20	Pyridine	5.23	1.03	C$_6$H$_6$	0.06^o	-1.22		Dipol.
4-Chlorobenzoic	3.99	Pyridine	5.23	1.24	C$_6$H$_6$	0.12^o	-0.92		Dipol.
4-Bromobenzoic	3.97	Pyridine	5.23	1.26	C$_6$H$_6$	0.12^o	-0.92		Dipol.
Formic	3.75	2-Methanolpyridine	5.01	1.26	1:1 mixture	0.11^m	-0.96		I.R.
3,5-Dichlorobenzoic	3.70	Pyridine	5.23	1.53	C$_6$H$_6$	0.14^o	-0.85		Dipol.
Iodoacetic	3.15	Pyridine	5.23	2.08	C$_6$H$_6$	0.06^o	-1.22		Dipol.
Bromoacetic	2.90	Pyridine	5.23	2.33	C$_6$H$_6$	0.11^o	-0.96		Dipol.
Chloroacetic	2.86	Pyridine	5.23	2.37	C$_6$H$_6$	0.10^o	-1.00		Dipol.
3,5-Dinitrobenzoic	2.82	Pyridine	5.23	2.41	C$_6$H$_6$	0.37^o	-0.43		Dipol.
Acetic	4.76	2,4,6-(CH$_3$)$_3$-	7.43	2.67	1:1 mixture	0.10^m	-1.00		I.R.
Chloroacetic	2.86	4-CH$_3$-	6.02	3.16	C$_6$H$_6$	0.11^o	-0.96		Dipol.
Dichloroacetic	1.30	Pyridine	5.23	3.93	C$_6$H$_6$	0.35^o	-0.46		Dipol.
Dichloroacetic	1.30	Pyridine	5.23	3.93	CHCl$_3$	1.2^p	+0.08	15.1	I.R.
Trichloroacetic	0.63	Pyridine	5.23	4.60	C$_6$H$_6$	1.33^o	+0.12		Dipol.
Trifluoroacetic	0.23	Pyridine	5.23	5.00	C$_6$H$_6$	1.94^o	+0.29		Dipol.
Trichloroacetic	0.63	4-(CH$_3$)-	6.02	5.39	C$_6$H$_6$	2.57^o	+0.41		Dipol.
Trichloroacetic	0.63	2,4,6-(CH$_3$)$_3$-	7.43	6.80	C$_6$H$_6$	11.50^o	+1.06		Dipol.
Carboxylic acids									
		Imidazole							
Formic	3.75	Pyrazole	2.47	-1.28	1:1 mixture	0.09^m	-1.05		I.R.
Acetic	4.76	Imidazole	6.96	2.20	1:1 mixture	0.87^m	-0.06		I.R.
Formic	3.75	Imidazole	6.96	3.21	1:1 mixture	1.78^m	+0.25		I.R.
Formic	3.75	N-CH$_3$-	7.04	3.29	1:1 mixture	0.20^m	-0.70		I.R.
Acetic	4.76	2-C$_2$H$_5$-	8.06	3.30	1:1 mixture	2.21^m	+0.34		I.R.
Chloroacetic	2.86	N-CH$_3$-	7.04	4.18	1:1 mixture	1.22^m	+0.09		I.R.
Formic	3.75	2-C$_2$H$_5$-	8.06	4.31	1:1 mixture	8.52^m	+0.93		I.R.
Dichloroacetic	1.30	Imidazole	6.96	5.66	1:1 mixture	99^m	+2.00		I.R.
Dichloroacetic	1.30	N-CH$_3$-	6.96	5.66	1:1 mixture	99^m	+2.00		I.R.

Phenols		Ethers			I.R.
4-NO$_2$-	7.15	Dibutyl	0.02[a]	-1.7	Dipol.

Phenols		Phosphine oxides			I.R.
4-NO$_2$-	7.15	Trioctyl	0.17[a]	-0.77	Dipol.
3,4-(NO$_2$)$_2$-	5.42	Trioctyl	0.26[a]	-0.59	Dipol.
Cl$_5$-	5.26	Trioctyl	0.08[a]	-0.15	Dipol.
2,4-(NO$_2$)$_2$-	4.07	Trioctyl	0.58[a]	-0.24	Dipol.
2,6-Cl$_2$-4-NO$_2$-	3.68	Trioctyl	0.19[a]	-0.72	Dipol.
2,4,6-(NO$_2$)$_3$-	0.44	Trioctyl	2.88[a]	+0.46	Dipol.

Phenols		Amides			I.R.
4-NO$_2$-	7.15	N,N-Diethylacetamide	0.01[a]	-2	Dipol.
3,4-(NO$_2$)$_2$-	5.42	N,N-Diethylacetamide	0.01[a]	-2	Dipol.
Cl$_5$-	5.26	N,N-Dimethylacetamide	0.02[a]	-1.7	Dipol.
2,4-(NO$_2$)$_2$-	4.07	N,N-Dimethylacetamide	0.15[a]	-0.8	Dipol.
2,6-Cl$_2$-4-NO$_2$-	3.68	N,N-Dimethylacetamide	0.09[a]	-1.1	Dipol.
2,4,6-(NO$_2$)$_3$-	0.44	N,N-Dimethylacetamide	1.17[a]	0.1	Dipol.

[a] Computed from the data of Jadzyn and Malecki (1972) using equation (1.2).
[b] Computed from the data of Ratajczak and Sobczyk (1969) using equation (1.2).
[c] Zundel and Nagyrevi (1978).
[d] Baba et al. (1969).
[e] Beier (1975).
[f] Janusek et al. (1975).
[g] Calculated from the data of Hawranek et al. (1972) using the Sobczyk formula.
[h] Nouwen and Huyskens (1973).
[i] Franz (1978).
[j] Crooks and Robinson (1970, 1975). The transfer of the proton implies here also a change of the position of the hydrogen bond.
[k] Gusakova et al. (1972b), $T = 293$ K.
[l] Denisov et al. (1973a).
[m] Lindemann and Zundel (1977).
[n] Roy (1966). These values are probably too high.
[o] Calculated from the data of Sobczyk and Pawelka (1974) using equation (1.2).
[p] Gusakova et al. (1970), $T = 293$ K.
[q] Cleuren (1979).
[r] Denisov and Schreiber (1976).

When only the terms of a given family of complexes in a given solvent are compared, it appears from the above data that $\log K_t$ is, to a good approximation, linearly related to the difference in the pK_a

$$\log K_t = \xi \, \Delta pK_a - \delta \tag{1.3}$$

ξ and δ being constants for the family and medium considered. Figure 1.17, for example, gives a plot of $\log K_t$ against ΔpK_a for some complexes of phenols with pyridine and triethylamine. Equation (1.3) was used by Ratajczak and Sobczyk (1969) to improve an equation which we deduced in 1964 (Huyskens and Zeegers-Huyskens, 1964) and where as a first approximation the factor ξ was put equal to unity. Already in 1961 Gordon had proposed a linear relation to relate the logarithm of the transfer constant of the complexes of a series of acids with a given base to the pK_a of the acids (Gordon, 1961).

As a matter of fact, the experimental values of ξ lie between 0.2 and 1, depending on the systems and solvents considered. This implies that in a rather narrow ΔpK_a range, the percentage of the proton transfer complexes passes from 1% to 99%. The more or less linear dependence of $\log K_t$ on ΔpK_a in water for a given family of complexes and in a given medium can be expected on the basis of the following considerations:

The transfer of the proton in the complex which takes place in the medium S and which is described by the equation

$$(A\!-\!H \cdots B)_s \overset{K_t}{\rightleftharpoons} (A^- \cdots H\!-\!{}^+B)_s$$

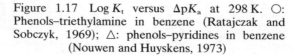

Figure 1.17 Log K_t versus ΔpK_a at 298 K. $\bigcirc$: Phenols–triethylamine in benzene (Ratajczak and Sobczyk, 1969); $\triangle$: phenols–pyridines in benzene (Nouwen and Huyskens, 1973)

can also be realized passing through the following steps:

1. Bringing the normal complex from the solvent S to water:

$$(A—H \cdots B)_s \overset{P_{Nws}}{\rightleftharpoons} (A—H \cdots B)_w$$

P_{Nws} is the distribution coefficient of the complex between water and the solvent.

2. Rupture of the hydrogen bond of the normal complex in water:

$$(A—H \cdots B)_w \overset{1/K_{abw}}{\rightleftharpoons} A_w—H_w + B_w$$

The equilibrium constant of this reaction is the reverse of the complexation constant of the acid and of the base in water.

3. Formation of the free ions in water, starting from the dissolved free molecules:

$$A_w—H_w + B_w \overset{K_{aAH}/K_{aBH^+}}{\rightleftharpoons} A_w^- + H_w—B_w^+$$

The logarithm of the equilibrium constant of this reaction is equal to ΔpK_a.

4. Formation of a hydrogen bond between the ions in water:

$$A_w^- + H_w—B_w^+ \overset{1/K_{dw}}{\rightleftharpoons} (A^- \cdots H—B^+)_w$$

The reverse of the equilibrium constant of this reaction is the dissociation constant of the ion-pair in water K_{dw}.

5. Transfer of the ion-pair from water to the solvent S:

$$(A^- \cdots H—B^+)_w \overset{1/P_{Iws}}{\rightleftharpoons} (A^- \cdots H—B^+)_s$$

Let P_{Iws} be the distribution coefficient of the ion-pair between water and the medium S.

The transfer constant K_t in the medium S is thus related to the various equilibrium constants considered above by the expression:

$$\log K_t = \{\log P_{Nws} - \log P_{Iws}\} - \log K_{abw} - \log K_{dw} + \Delta pK_a$$

The term within brackets can be expected to remain roughly constant in a given series of complexes. A substitution in the partners will indeed alter the distribution coefficients between water and the solvent of both forms of the complex approximately in the same manner.

On the other hand, it was established in a wide variety of cases (Dierckx *et al.*, 1965; Zeegers-Huyskens, 1969) that, in solvents like CCl_4, linear relations hold between $\log K_{ab}$ and the pK_a of the acid in a given family of complexes with a given base. Similar relations are found with the pK_a of the base in a family of complexes with a given acid. The slopes of the lines in CCl_4 lie between 0.2 and 0.7. In water one can expect that such relations also hold but that the slopes will be markedly smaller as a consequence of the effect this solvent exerts against the complexation: the K_{ab} values are indeed much lower in water because of the formation of hydrogen bonds of the free acid and of the free base with water. However, these hydrogen bonds are stronger when the acid or the base are stronger and the reduction of K_{ab} with respect to the value in CCl_4 is more important. As a consequence $K_{ab\,w}$ in water will increase more slowly when changing the pK_a's. On the basis of these considerations and neglecting secondary effects a relation of the type:

$$\log K_{ab\,w} = \rho_{ab\,w} \Delta pK_a + \text{constant}$$

with $\rho_{ab\,w}$ values markedly less than one, can be expected.

A rough linear relation can also be found between $\log K_d$ of the ion-pairs between alkylammonium and substituted phenolates ions determined by Kolthoff et al. (1966) in *acetonitrile* and the ΔpK_a of the partners in water. The slope here is of the order of 0.2. The effect of water will be similar to that in the previous case and will reduce the slope. One can thus expect a relation of the form

$$\log K_{dw} = \rho_{dw} \Delta pK_a + \text{constant}$$

with a ρ_{dw} value much smaller than one.

On the basis of the previous approximations the expected relation between $\log K_t$ and ΔpK_a takes the form:

$$\log K_t = (1 - \rho_{ab\,w} - \rho_{dw}) \Delta pK_a + \text{constant}$$

which corresponds well to the observed data.

It must be noted that the value of δ in the equation

$$\log K_t = \xi \Delta pK_a - \delta$$

which corresponds to the value of ΔpK_a for which K_t is unity depends both on the nature of the family of complexes considered and on the medium. For instance, in benzene solution, it can be seen from Figure 1.17 that δ is of the order of 5.2 for the complexes between triethylamine and the phenols whereas the value falls to 3.4 for the complexes of these phenols with pyridines. In pure $1:1$ carboxylic acid–nitrogen base mixtures, Zundel and his coworkers found 50% proton transfer at $\Delta pK_a = 2.3$ (Lindemann and Zundel, 1977). Substituted benzoic acid–pyridine systems, studied by Johnson and Rumon (1965) give 50% proton transfer at $\Delta pK_a = 3.7$. For the carboxylic acid–pyridine systems studied by Sobczyk and Pawelka (1974), δ is of the order of 4.7. According to the results of Jadzyn and Malecki (1972) the values of δ for the systems phenols–triethylamine are respectively 6.5 in cyclohexane, 5.2 in toluene, and 4.1 in trichloroethylene. In the case of intramolecular hydrogen bonds in Mannich bases, Sucharda-Sobczyk and Sobczyk (1978) showed that the ΔpK_a value at the point where K_t is unity is somewhat shifted to higher values in comparison with the intermolecular interactions. According to the most recent results (Cleuren, 1979; Sobczyk, private communication) it appears that the coefficient ξ is lower for O—H $\cdots$ O hydrogen bonds (where it has values between 0.2 and 0.3) than for O—H $\cdots$ N hydrogen bonds (where it generally lies between 0.4 and 1).

1.3.4.2 *Influence of the proton affinity of the base on the transfer constant* A parameter which is still more adequate than ΔpK_a when dealing with proton transfer and related problems, is the proton affinity of the base. Of course this can only be used for complexes involving the same acid. For instance, it was shown by one of us (Zeegers-Huyskens, 1976) that when the enthalpies of complex formation of O—H $\cdots$ N bonds involving the phenol molecule are plotted against the proton affinity of nitrogen bases as different as pyridines and aliphatic amines (Aue *et al.*, 1976a, b), all the points lie on the same curve (with the exception of a few deviations caused by steric hindrance). This is not the case when ΔH_{ab} is plotted against the pK_a. In this case several curves are found depending on the family to which the case belongs.

A similar observation can be made for the transfer constant K_t of the complexes of 2,6-dichloro-4-nitrophenol for which the data are given in Table 1.16.

When $\log K_t$ is plotted versus the proton affinity of the base, the point relative to triethylamine lies on the same curve as those relative to the pyridines (Figure 1.18). Thus when considering the proton affinity instead of

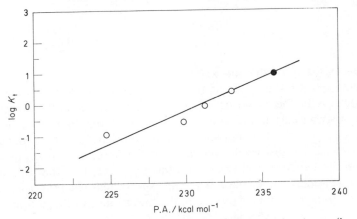

Figure 1.18 Log K_t versus proton affinity of the base (in kcal mol^{-1}) for complexes of 2,6-dichloro-4-nitro-phenol. O: pyridines; ●: triethylamine. (Proton affinities from Aue *et al.*, 1972, 1975, 1976a and b)

the pK_a the differentiation into several families of complexes which appears in Figure 1.17 vanishes within the limits of the experimental errors.

1.3.4.3 Influence of the dielectric constant of the medium on the transfer constant K_t

As already mentioned above, the polarization of the molecules of the medium by the complex is an essential factor governing not only the transfer constant but even the appearance of tautomerism. As seen from the values of Table 1.16, increasing the dielectric constant of the solutions has generally a positive influence on the value of K_t, but other effects may also play a role.

Attempts were made by Baba *et al.* (1969) and by Jadzyn and Malecki (1972) to describe, in a more quantitative way, the effect of the dielectric constant on K_t. These treatments are based on the reaction field which, according to the theory of Onsager (1936), the polarized solvent molecules exert on the electric dipole of the complex. This reaction field is, of course, more important for ion-pairs with a high dipole moment μ_I than for the normal complexes characterized by dipole moments μ_N.

The equation of Jadzyn and Malecki can be written as

$$\log K_t = \log K_t^0 + \left(\frac{4\pi N^2}{6.9\,RT\bar{V}}\right)(\mu_I^2 - \mu_N^2)\left(\frac{D-1}{2D+1}\right)$$

K_t^0 is the transfer constant in a medium of unit dielectric constant and $\bar{V}$ is the molar volume of the complex. The other symbols have their usual meaning.

Baba *et al.* take the polarization of the dipoles into account. Their equation can be formulated as

$$\log K_t = \log K_t^0 + \left(\frac{4\mu N^2}{6.9\,RT\bar{V}}\right)(\mu_I^2 - \mu_N^2)\left(\frac{n^2+2}{3}\right)\left(\frac{D-1}{2D+n^2}\right)$$

where n is the refractive index.

In both cases the complex is treated as a *point dipole* at the centre of a spherical cavity the value of which is that of the complex. Treating the charge distribution in the complex as a point dipole is a rather crude approximation, especially in the case of the ion-pair. Moreover, the shape of the complexes is distorted from a perfect sphere. Hence the agreement between the experimental values and those predicted by the two equations above is generally rather poor. The most spectacular changes of K_t with the dielectric constant can be expected in the lower range of the permittivities.

1.3.4.4 Influence of other specific bonds formed by the complex with molecules of the environment

From a quantitative point of view (Sobczyk *et*

al., 1976) the effect of varying the dielectric constant of the medium may often be overshadowed by that of another factor, i.e. the formation of specific bonds between the complex and molecules of the solvent or other dissolved substances.

Complexes of secondary or primary amines still bear external NH groups which can act as proton donor and which can bind a second amine molecule:

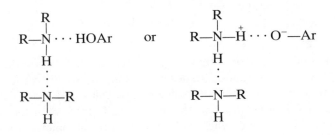

Using a calorimetric procedure, we showed that the proton-transferred complexes are much more able to bind a second amine molecule than the normal complexes (Huyskens, 1972). The reasons for such behaviour have been discussed recently (Huyskens, 1977). If the transfer of the proton favours binding of a second molecule, it can be expected that, reciprocally, the formation of such an additional specific bond will improve the proton transfer (Huyskens and Hernandez, 1973). This prediction has been confirmed by Lindemann and Zundel (1977), who pointed out that the pK_a where 50% proton transfer occurs is lowered by 1.7 units in carboxylic acid–nitrogen base systems where the nitrogen base possess additional NH groups.

Similar effects can be expected when the complex forms specific bonds with surrounding water molecules. The theoretical work of Schuster on this topic has already been mentioned above (Schuster *et al.*, 1974). Extensive experimental work on the effect of the presence of water on the transfer constant has been performed by Zundel *et al.* who studied the systems acetic acid + pyridine, acetic acid + imidazole, and other systems (Lindemann and Zundel, 1977, 1978). In the case of 1:1 mixtures acetic acid–pyridine, for instance, the percentage proton transfer is very small. When two water molecules are added per acid–nitrogen base pair, this percentage remains still negligible. It begins to increase when the ratio water molecules/complexes reaches 3. With 30 water molecules per associate the percentage proton transfer lies not far from 100%. It must be noted that, in this case, the effect also involves the change in K_t brought about by the increase of the dielectric constant.

The influence of water can be ascribed to the stabilization of the ion-pair form by additional hydrogen bonds involving the water molecule. As for example, for the pyridinium–acetate ion-pair:

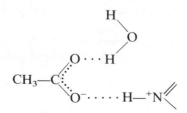

From all these data it is clear that the influence of a solvent on the transfer constant of a complex cannot be described only by its dielectric constant. Specific interactions between the solvent molecules and the complex can also play a determining role and this is especially true when the dielectric constant exceeds 10 and when variation of this parameter causes only limited changes of K_t.

Several parameters have been proposed to describe the specific donor–acceptor properties of a solvent. A widely used parameter is the so-called Z value, introduced by Kosower (1965). This value is based on the charge transfer transition of 1-ethyl-4-carbomethoxypyridinium iodide in a given solvent. Another parameter is the E_T value introduced by Reichardt (Dimroth et al., 1963; Reichardt, 1965) and based on the light absorption of pyridinium phenol betaine. There is a correlation between the two parameters. With respect to the influence of specific solvent effects on the transfer constant it should be noted here that Sobczyk found good correlation between the dipole moment of the 2,4-dinitrophenol–triethylamine complex in various solvents and the E_T value of Reichardt characterizing the solvent (Sobczyk et al., 1976).

When dealing with the donor–acceptor properties of a solvent it must be remembered that it is necessary to make a distinction between the donor (or nucleophilic) properties and the acceptor (or electrophilic) power. This was emphasized by Gutmann (1976) who introduced two parameters to describe both properties: the donor properties of a solvent are related to its 'donicity' or donor number (Gutmann, 1969). The acceptor power of a solvent is characterized, according to Gutmann, by its 'acceptor number' which is based on the ^{31}P NMR shift of triethylphosphine in the given solvent (Mayer et al., 1975). Beier pointed out that the percentage of proton transfer in the complex p-nitrophenol–morpholine in various solvents such as acetonitrile, ethylene carbonate, dimethylformamide, and dimethylsulphoxide is much better correlated with Gutmann's donicity for these solvents than with their dielectric constant, indicating the predominance of specific solvent effects over polarization effects in this range of dielectric constants (Beier, 1975).

1.4 Dissociation of Proton and Ion Transfer Complexes

The complexation of an acceptor or an acid molecule A by the electron donor D

$$A + D \rightleftharpoons A \cdots D$$

has repercussions on the *strengths* of the various chemical *bonds* in the partners. Some of these bonds are reinforced but others become weaker. This is especially the case for the bond a—p of the acceptor molecule where p is the atom which comes into the closest contact with the donor.

This bond is more extended when the interaction becomes stronger

$$-a-p + d- \rightleftharpoons -a-p \cdots d-$$

This increase of r_{ap} becomes very important in the proton or ion transfer complexes

$$-a-p + d- \rightleftharpoons -a^- \cdots p-^+d-$$

where

$$\frac{r_{ap}}{r^0_{ap}} > \frac{r_{pd}}{r^0_{pd^+}}$$

r^0_{ap} and $r^0_{pd^+}$ being the *bond lengths* in the free acceptor molecule and the free p—^+d ion respectively.

A consequence of this important lengthening of r_{ap} is that in proton or ion transfer complexes the energy needed for the rupture of the bond a—p is smaller than that required to break the bond p—d. In such complexes the a—p bond is thus the weakest bond between the nuclei and can be considered as a hydrogen bond or an E.D.A. bond between the corresponding ions A^- and p—D^+.

The rupture of this labile a—p bond brings about the formation of free ions:

$$A^- \cdots p^+ -D \overset{K_d}{\rightleftharpoons} A^- + p^+ -D$$

as for example:

$$(NO_2)_3C_6H_3O^- \cdots H^+ -NR_3 \rightleftharpoons (NO_2)_3C_6H_3O^- + H^+ -NR_3$$

or

$$I^- \cdots I-^+NR_3 \rightleftharpoons I^- + I-^+NR_3$$

In this chapter we deal with the experimental determinations of the dissociation constants K_d of proton transfer and ion transfer complexes and we examine the factors which govern the stability of the hydrogen bond or E.D.A. bond in these complexes.

1.4.1 *Experimental determination of the dissociation constants of ion pairs*

The most widely used method for the determination of K_d is based on the measurement of *the electric conductivity* κ of the solutions.

The possibility of migration towards its appropriate electrode of an ion bound with the counterion in a proton or ion transfer complex is drastically reduced by the labile bond between the two ions. As a consequence ion-pairs contribute practically nothing to the passage of electric current in solutions.

The conductance of solutions is thus essentially due to the migration of '*free*' ions. The electric conductivity κ of the solutions is related to the concentrations C_i of the various ions by the relation

$$\kappa = \frac{\sum \lambda_i |Z_i| C_i}{1000}$$

where Z_i is the charge number of the ion and λ_i its equivalent ionic conductance. (The factor 1000 refers to the concentration units generally used: $mol\,dm^{-3}$.)

In the case where all the ions derive from the same univalent–univalent ion-pair this relation can be written:

$$[i^+] = [i^-] = \frac{1000\kappa}{(\lambda_+ + \lambda_-)}$$

$[i^+]$ and $[i^-]$ being the concentrations of the free cations and anions and λ_+, λ_- their ionic conductances.

The concentration of the 'free' ions can thus be determined from the experimental value of the conductivity κ provided the ionic conductances are known. It should be noted that conductivity measurements cover a wide range of κ values so that conductimetry can be used for the determination of very small ion concentrations. The limiting factor is the presence of ionic impurities in the solution. The residual conductivity κ_0 of the pure organic solvents is, in most cases, mainly caused by these ionic impurities.

The main problem however, when ion concentrations must be determined from conductivity data, generally lies in the exact evaluation of the *ionic conductances* λ_i. Three factors may influence the values of λ_i: 1. the viscosity of the medium, 2. the presence of other ions, and 3. complexation of the ions by ligand molecules or the solvent itself.

1. The *viscosity* of the medium. When the viscosity η increases, the mobility is reduced. As a first approximation the relation between λ_i and η is given by '*Walden's rule*' (1906):

$$\lambda_i = \frac{w_i}{\eta}$$

where the *Walden product* w_i is a constant for a given ion. As a principle w_i should be independent of the nature of the medium and of the temperature. However, as can be seen from Table 1.17 strong deviations can occur in water, especially for ions of a small size. For the other solvents the constancy is much better.

2. A second factor which can reduce the mobility of the ions is the effect of the presence of other ions around it. These other ions form around the given ion what is called an 'ionic atmosphere'. The ionic atmosphere exerts two main effects on the migration of an ion in an external field: the *relaxation effect* and the *electrophoretic effect*.

The relaxation effect arises from the fact that the ionic atmosphere requires a finite time to recover the spherical distribution around the given ion when it is moved from one point to another. This phenomenon produces an additional force ΔX acting on the moving ion and working in the opposite direction to the external field X.

Table 1.17   Walden products w_i^0 (in $ohm^{-1}\ cm^2\ mol^{-1}\ cP$) of various ions in various solvents (at zero ionic strength) at 298 K

	Solvents			
Ion	acetonitrile	nitromethane	nitrobenzene	water
$(i\text{-}Am)_4B^-$	19.9^a	19.5^a	19.9^a	—
$(i\text{-}Am)_4N^+$	19.9^a	19.5^a	19.9^a	—
Ph_4B^-	20.0^a	19.6^a	20.1^a	—
K^+	31.0^b	—	32.6^a	65.4^c
Cl^-	32.7^d	38.8^a	42.2^a	68.0^c
Br^-	35.2^d	39.0^a	40.8^a	69.5^c
I^-	34.6^d	—	39.5^a	68.4^c
$Picr^-$	26.8^a	—	30.3^a	26.9^b
HSO_4^-	28.9^d	—	—	—
NO_3^-	36.5^d	—	—	63.6^c
ClO_4^-	35.9^a	—	32.9^a	59.9^c
Me_4N^+	32.8^a	33.8^a	31.6^a	40.0^c
Et_4N^+	29.5^a	29.5^a	29.9^a	29.1^c
Pr_4N^+	24.5^a	24.3^a	24.6^d	20.8^c
Bu_4N^+	21.5^a	21.1^a	21.5^a	17.3^c
Et_3NH^+	30.2^e	—	32.0^f	17.3^c
$Et_2NH_2^+$	32.6^g	—	35.6^f	—

[a] Coetzee and Cunningham (1965a).
[b] Harned and Owen (1958).
[c] Robinson and Stokes (1959).
[d] Huyskens and Lambeau (1978b).
[e] Haulait and Huyskens (1975a).
[f] Walden and Birr (1933).
[g] Haulait and Huyskens (1975b).

The electrophoretic effect is the consequence of the moving of the ionic atmosphere, with its associated molecules of solvent, in the opposite direction to that in which the ion is moving.

The net result of both effects on the ionic conductivity can be represented by the following equation:

$$\lambda_i = \lambda_i^0 \left(1 - \frac{\Delta X}{X}\right) - \Delta \lambda_{\text{electrophoretic}}$$

λ_i^0 is the equivalent conductance of the ion in the given medium at zero ionic strength, when all the other ions are kept at infinite distances. It can be expected that both effects disappear when this is the case and that they depend on the concentration $[i]$ of the free ions present in the solution.

Theoretical approaches have been made to establish relations between $\Delta X/X$ and $\Delta \lambda_{\text{electrophoretic}}$ and the potential due to the ionic atmosphere. This has led to various equations relating λ_i and the concentration of the free ions $[i]$. One of the most developed of these *equations* is that of *Fuoss* and his coworkers, which can be written in the form:

$$\lambda_i = \lambda_i^0 - s[i]^{1/2} + c[i] \log [i] + a[i] + b[i]^{3/2}$$

(see for instance James and Fuoss, 1975) wherein s, c, a, and b are constants. A discussion of the coefficients of the higher terms was made by Barthel *et al.* (1973). See also Treiner and Justice (1971). When the concentrations of the free ions remain low enough, as is often the case in organic solutions, the higher terms may be neglected and the equation above reduces to the *limiting equation of Onsager* (Onsager, 1927):

$$\lambda_i = \lambda_i^0 - \{\alpha \lambda_i^0 + \beta\}[i]^{1/2}$$

with

$$\alpha = \frac{0.8204 \times 10^6}{(DT)^{3/2}} \quad \text{and} \quad \beta = \frac{41.25}{\eta (DT)^{1/2}}$$

λ_i^0 being expressed in ohm^{-1} cm^2 mol^{-1} and η in centipoise. D is the dielectric constant and T is the temperature. From the above equations, it follows that the concentration $[i]$ can be found by solving the equation

$$[i] = \frac{(\kappa - \kappa_0) \times 1000}{(\lambda_+^0 + \lambda_-^0) - [\alpha(\lambda_+^0 + \lambda_-^0) + 2\beta][i]^{1/2}}.$$

An easy way to solve this equation was proposed by Fuoss and Accascina (1959) introducing the variable z defined by

$$z = \frac{\sqrt{1000}[\alpha(\lambda_+^0 + \lambda_-^0) + 2\beta](\kappa - \kappa_0)^{1/2}}{(\lambda_+^0 + \lambda_-^0)^{3/2}}$$

and calculating the function

$$\mathscr{F}(z) = \tfrac{4}{3}\cos^2\left[(\tfrac{1}{3})\cos^{-1}\left(-3^{3/2}\times\frac{z}{2}\right)\right].$$

One finds:

$$[i] = \frac{1000(\kappa - \kappa_0)}{(\lambda_+^0 + \lambda_-^0)} \times \frac{1}{\mathscr{F}(z)}.$$

3. A third effect which can influence the mobility of the ions is their *complexation* by ligand molecules or by the solvent itself. When an external field is applied, the entity which moves towards the electrode is not always the bare ion. When the ion forms sufficiently strong bonds with solvent molecules or with ligands present in the solution, the molecules may accompany the ion in its migration. Of course the presence of these 'passengers' reduces the mobility of the moving particle. As a consequence the values of λ_i and of the Walden product of a solvated or complexed ion are smaller.

In Table 1.18 we give values of the Walden products of some singly or doubly complexed anions in nitrobenzene or acetonitrile. It is, of course,

Table 1.18 Walden products w_i^0 (in $ohm^{-1}\,cm^2\,mol^{-1}\,cP$) of complexed anions at zero ionic strength in acetonitrile and in nitrobenzene at 298 K

Complexed ion	w_i^0 (in acetonitrile)	w_i^0 (in nitrobenzene)
Cl^-–benzoic acid[a]	23.7	—
Cl^-–p-methylbenzoic acid[a]	22.5	—
Cl^-–m-chlorobenzoic acid[a]	22.7	—
Cl^-–m-nitrobenzoic acid[a]	22.0	—
Cl^-–3,5-dinitrobenzoic acid[a]	19.1	—
Cl^-–citric acid[b]	17.2	—
Cl^-–(citric acid)$_2$[b]	9.7	—
Br^-–benzoic acid[a]	23.5	28.2
Br^-–p-methylbenzoic acid[a]	22.0	—
Br^-–m-chlorobenzoic acid[a]	21.9	25.9
Br^-–m-nitrobenzoic acid[a]	20.7	24.3
Br^-–3,5-dinitrobenzoic acid[a]	18.0	19.9
Br^-–citric acid[a]	17.5	—
Br^-–(benzoic acid)$_2$[a]	—	13
Br^-–(m-methylbenzoic acid)$_2$[a]	—	8
Br^-–(m-chlorobenzoic acid)$_2$[a]	—	11
Br^-–(m-nitrobenzoic acid)[a]	—	15
HSO_4^-–citric acid[b]	15.4	—
HSO_4^-–(citric acid)$_2$[b]	9.6	—

[a] Pirson and Huyskens (1974).
[b] Huyskens and Lambeau (1978b).

necessary, when using the previous equation, to utilize the correct values of λ^0_+ and λ^0_- which correspond to the actual state of solvation of the ions and to the proper viscosity of the medium.

The dissociation constant K_d of the ion-pair can then be computed from the equation

$$K_d = \left(\frac{[i]^2 y^2}{F - [i]} \right)$$

where F is the formal concentration of the ionophore (ion-pairs + cations) and y the activity coefficient of the ions in the medium of ionic strength $[i]$. These activity coefficients can be computed from the Debye–Hückel equation but it must be borne in mind that no reliable equations exist for this purpose when the concentration of the ions becomes too great.

A common procedure is that given by Fuoss (Fuoss and Accascina, 1959), based on the following equation deduced from the previous ones:

$$\left(\frac{\mathcal{F}(z)F}{1000(\kappa - \kappa_0)} \right) = \left(\frac{1}{(\lambda^0_+ + \lambda^0_-)} \right) + \left(\frac{1}{K_d(\lambda^0_+ + \lambda^0_-)^2} \right) \left\{ \frac{1000(\kappa - \kappa_0)y^2}{\mathcal{F}(z)} \right\}$$

wherein κ_0 is the conductivity in absence of the ionophore.

One plots the experimental quantity corresponding to the left hand side versus the experimental quantity within the brackets { } for several values of the formal concentration F. One must obtain a straight line, the intercept of which gives $(\lambda^0_+ + \lambda^0_-)^{-1}$. The dissociation constant can be deduced from the slope.

It must be noted that when the ion concentration exceeds 10^{-3} mol l^{-1}, it is necessary to take the higher terms of the Fuoss equation into account.

When using the Fuoss equation it is necessary that the ions should remain in the same state of solvation in the full concentration range of the ionophore. This is required to maintain the values λ^0_+, w^0_+ and λ^0_-, w^0_- constant. As a matter of fact all the ions must not appear in the same state of solvation, but the proportion of the ions in the various solvation or complexation state must remain constant. The previous equation had thus to be used for such solutions where the concentrations of free ligands, if they are present, do not change.

Acetonitrile ($D = 36.0$ at 298 K) and nitrobenzene ($D = 34.8$ at 298 K) are very convenient solvents for the determination of the dissociation constants of dissolved hydrogen-bonded ion pairs. In this range of dielectric constants the undissociated hydrogen-bonded ion-pairs remain important at the formal concentrations of ionophore which are adequate for the measurements. In solvents of higher dielectric constants, such as water, the dissociation of the ionophores is often almost complete. Furthermore the *donicities* D.N. and the *acceptor numbers* A.N. of acetonitrile (D.N. = 14.1; A.N. = 18.9) and of nitrobenzene (D.N. = 4.4; A.N. = 14.8) which according to

Gutmann describe their abilities to form specific bonds with the solutes (Gutmann, 1976) are moderate. The donicity D.N. is defined as the negative ΔH value in kcal mol^{-1} for the interaction of the electron donor, with SbCl$_5$ in a diluted solution of dichloroethane. The 'acceptor number' A.N. is derived from the ^{31}P NMR shift produced in triethylphosphine oxide by the solvent. The A.N. is a hundred times the ratio between the shift relative to the peak in 1,2-dichloroethane and that of the (Et)$_3$PO$\rightarrow$SbCl$_5$ adduct, dissolved in dichloroethane. In solvents for which either the donicity or the acceptor number are large, as for instance diethyl ether (D.N. = 19.2), tetrahydrofuran (D.N. = 20), dimethylsulphoxide (D.N. = 29.8), or ethyl alcohol (A.N. = 37.1) the specific interactions between the ions and the surrounding solvent molecules become more important.

On the other hand, in solvents of low dielectric constant, low donicity, and low acceptor number, like benzene ($D = 2.4$; D.N. = 0.1; A.N. = 8.2) difficulties can arise from the lack of solubility of the ion-pairs. Great care must be taken to reduce the presence of ionic impurities which increase the residual conductivity κ_0 of the solvent. Furthermore technical difficulties in the measurement of the conductivities are encountered because of the polarization of the electrodes in the d.c. bridge method which is generally used in this case. A last obstacle in the range of low dielectric constants is the formation of aggregates between the free ions and the ion-pairs ('*triple ions*').

1.4.2 *Selected experimental values of the dissociation constant* K$_d$ *and of the dissociation enthalpy* ΔH$_d$ *of hydrogen-bonded and* E.D.A.-*bonded ion-pairs*

1.4.3 *Factors governing the dissociation of hydrogen-bonded or* E.D.A.-*bonded ion-pairs into free ions*

1.4.3.1 *Dielectric constant* From Table 1.19 it can be seen that the dielectric constant is one of the most important factors governing the dissociation of the ion-pairs.

The dielectric constant D is, as a matter of fact, a *macroscopic* quantity which is related to the dipole moment μ of the molecules of the medium and to their concentration C. To a good approximation the quantitative relation for pure solvents can be written as:

$$\frac{(D - n^2)(2D + n^2)}{D(n^2 + 2)^2} = \left(\frac{4\pi N}{9kT}\right) g\mu_s^2 C_s$$

where μ_s is the *dipole moment* of the solvent, C_s the concentration of the solvent molecules, g the correlation parameter of *Kirkwood*, and n the

Table 1.19 Dissociation constants K_d and dissociation enthalpies ΔH_d of hydrogen-bonded and E.D.A.-bonded ion-pairs at 298 K

Solvent: acetonitrile ($D = 36.00$)

Cation	pK_a (in water)	Anion	pK_a (acid in water)	K_d^0 (mol l^{-1})	log K_d^0	ΔH_d (kJ mol^{-1})
Triethylammonium	10.65	3-Nitrophenolate	8.35	8.3×10^{-6}	-5.08^a	
Triethylammonium	10.65	4-Nitrophenolate	7.15	1.5×10^{-4}	-3.8^a	
Triethylammonium	10.65	3,5-Dinitrophenolate	6.70	6.3×10^{-5}	-4.2^b	
Anilinium	4.63	Picrate	0.38	6.3×10^{-5}	-4.2^a	
Triethylammonium	10.65	2,4-Dinitrophenolate	4.07	8.5×10^{-4}	-3.1^c	
Triethylammonium	10.65	2,6-Dinitrophenolate	3.71	5.0×10^{-4}	-3.3^a	
Triethylammonium	10.65	3,5-Dinitrobenzoate	2.82	2.0×10^{-4}	-3.7^a	
Tributylammonium	10.89	Salicylate	2.75	1.2×10^{-5}	-4.9^d	
Dibutylammonium	11.25	Salicylate	2.75	4.3×10^{-5}	-4.4^c	
Butylammonium	10.59	2,6-Dihydroxybenzoate	1.22	5.5×10^{-5}	-4.3^c	
Tributylammonium	10.89	2,6-Dihydroxybenzoate	1.22	1.5×10^{-3}	-2.8^c	
Dibutylammonium	11.25	2,6-Dihydroxybenzoate	1.22	4.7×10^{-4}	-3.3^c	
Butylammonium	10.59	Picrate	0.38	5.3×10^{-4}	-3.3^c	
Tributylammonium	10.89	Picrate	0.38	3.4×10^{-3}	-2.5^c	
Dibutylammonium	11.25	Picrate	0.38	4.6×10^{-3}	-2.7^c	
Triethylammonium	10.65	Cl$^-$	-7.0^a	3.4×10^{-3}	-2.5^c	
Triethylammonium	10.65	HSO$_4^-$	-3.0^a	3.0×10^{-5}	-4.5^e	
n-Butylammonium	10.59	HSO$_4^-$	-3.0^a	7.9×10^{-4}	-3.1^f	$+2.3^f$
Triethylammonium	10.65	NO$_3^-$	-1.4^a	5.0×10^{-4}	-3.3^d	
Triethylammonium	10.65	Br$^-$	-9.0^a	7.8×10^{-4}	-3.1^f	$+3.8^f$
Triethylammonium	10.65	I$^-$	$\sim 10^9$	2.7×10^{-4}	-3.6^f	
Tributylammonium	10.89	ClO$_4^-$	$\sim 10^9$	3.1×10^{-4}	-2.5^f	
Dibutylammonium	11.25	ClO$_4^-$	$\sim 10^9$	9.5×10^{-2}	-1.0^c	
Butylammonium	10.59	ClO$_4^-$	$\sim 10^9$	1.4×10^{-2}	-1.9^c	
Butylammonium	11.25	ClO$_4^-$	$\sim 10^9$	1.5×10^{-2}	-1.9^c	
(Tetrabutylammonium)	—	ClO$_4^-$		5.6×10^{-2}	-1.3^c	

Solvent: nitrobenzene ($D = 34.82$)

Triethylammonium	10.65	4NO$_2$ phenolate	7.15	1.3×10^{-6}	-5.8^b +8.2^l
Anilinium	4.63	Picrate	0.38	2×10^{-5}	-4.7^g
N,N'-Dimethylanilinium	5.15	Picrate	0.38	4.1×10^{-5}	-4.4^g
Pyridinium	5.21	Picrate	0.38	5.9×10^{-5}	-4.2^h
3-Methylpyridinium	5.52	Picrate	0.38	7.0×10^{-5}	-4.15^h
3,4-Dimethylpyridine	6.46	Picrate	0.38	1.0×10^{-4}	-4.00^h
Triethanolammonium	7.77	Picrate	0.38	4.9×10^{-4}	-3.3^i
Diethanolammonium	8.88	Picrate	0.38	1.9×10^{-4}	-3.7^i
N,N',N''-Triphenylguanidinium	9.1	Picrate	0.38	2.0×10^{-4}	-3.7^i
Dialkylammonium	9.29	Picrate	0.38	3.7×10^{-4}	-3.4^i
Alkylammonium	9.49	Picrate	0.38	2.3×10^{-4}	-3.6^i
Ethanolammonium	9.50	Picrate	0.38	2.6×10^{-4}	-3.6^i
N,N'-Diphenylguanidinium	10.12	Picrate	0.38	3.14×10^{-5}	-4.5^i
n-Butylammonium	10.59	Picrate	0.38	1.5×10^{-4}	$-3.8^{g,k}$
n-Pentylammonium	10.63	Picrate	0.38	1.5×10^{-4}	-3.8^g
Triethylammonium	10.65	Picrate	0.38	2.0×10^{-4}	-3.7^l
Tributylammonium	10.89	Picrate	0.38	1.9×10^{-4}	$-3.7^{a,g}$
Piperidinium	11.12	Picrate	0.38	1.5×10^{-4}	-3.8^i
Dibutylammonium	11.25	Picrate	0.38	1.6×10^{-4}	$-3.8^{i,g}$
Guanidinium	13.6	Picrate	0.38	2.0×10^{-4}	-3.7^l
Tetramethylguanidinium	13.6	Picrate	0.38	5.44×10^{-3}	-4.3^l
Triethylammonium	10.65	Cl$^-$	-7.00	1.1×10^{-6}	-5.9^e
Triethylammonium	10.65	Br$^-$	-9.00	6.1×10^{-6}	$-5.2^{e,m}$
Tributylammonium	10.89	I$^-$	-10.00	9.5×10^{-5}	-4.0^g
Benzylammonium	9.34	ClO$_4^-$	-10.00	3.6×10^{-4}	-3.4^i
Butylammonium	10.59	ClO$_4^-$	-10.00	2.5×10^{-3}	-2.6^g
C$_5$H$_5$N—I$^+$		I$^-$		1.5×10^{-12}	-11.8^n
4-CH$_3$C$_5$H$_4$N—I$^+$		I$^-$		7.0×10^{-12}	-11.1^n
3,4-(CH$_3$)$_2$C$_5$H$_3$N—I$^+$		I$^-$		2.0×10^{-11}	-10.6^n

Table 1.19 (continued)

Cation	pK_a (in water)	Anion	pK_a (acid in water)	K_d^0 (mol l^{-1})	log K_d^0	ΔH_d (kJ mol^{-1})
Solvent: 1,2-dichloroethane (D = 10.36)						
Butylammonium	10.59	Picrate	0.38	2.1×10^{-8}	-7.7^o	
Tributylammonium	10.89	Picrate	0.38	2.03×10^{-8}	-7.7^p	
Tributylammonium	10.59	Br$^-$	-9.00	8.07×10^{-10}	-9.1^p	
Tributylammonium	10.59	I$^-$	-10.00	1.57×10^{-8}	-7.8^p	
Solvent: 1,2-dichlorobenzene (D = 9.93)						
Pyridinium	5.21	Picrate	0.38	3.5×10^{-11}	-10.5^p	
Triethylammonium	10.65	Picrate	0.38	2.1×10^{-10}	-9.7^p	
Tributylammonium	10.89	Picrate	0.38	2.9×10^{-10}	-9.5^p	
Tributylammonium	10.89	Br$^-$	-9.00	2.2×10^{-12}	-11.5^p	
Solvent: chlorobenzene (D = 5.62)						
Tributylammonium	10.89	Picrate	0.38	4.8×10^{-13}	-12.3^p	

[a] Kolthoff et al. (1966).
[b] Neven (1978).
[c] Coetzee and Cunningham (1965b).
[d] Kolthoff and Chantooni (1963b).
[e] Pirson and Huyskens (1974).
[f] Huyskens and Lambeau (1978).
[g] Witschonke and Kraus (1947).
[h] Haulait-Pirson (1980).
[i] Jasinski and Pawlak [1970].
[j] Collaer (1980).
[k] Macau et al. (1971).
[l] Haulait and Huyskens (1975a).
[m] Delcoigne and Haulait (1976).
[n] Estimated from the results of Poskin and Huyskens (1976), assuming a value of 5×10^6 l mol^{-1} for the homoconjugation constant of I$^-$ and 10^5 mol^{-1} for the homoconjugation constant of C_5H_5N—I$^+$ and of the other pyridine–iodinium ions. On account of these approximations, the cited value gives only an order of magnitude.
[o] Nead et al. (1939).
[p] Ralph and Gilkerson (1964).
[q] Values of pK_a given by Bell (1959).

internal refractive index, the other symbols having their usual meaning. This equation is derived from the theory of Onsager (1936). The refractive indices of liquids do not differ very much from each other.

Thus when the liquid medium possesses a high dielectric constant this means, from a molecular point of view, that the molecules of the medium have a high dipole moment or that the number of the dipoles per unit volume is important.

From a macroscopic point of view the force acting between two point charges in a medium of dielectric constant D is reduced by a factor $1/D$ compared to the value *in vacuo*.

From a molecular point of view this force is not reduced but additional interactions occur between the point charges and the dipoles of the molecules of the medium. The net effect is an apparent reduction of the interaction energy between the point charges.

The free ions derived from the ion-pair polarize the solvent molecules around themselves by *ion–dipole interaction*. For spherical ions this phenomenon takes place in the same way in all directions.

Around the contact ion-pair the polarization of the solvent molecules brought about by one ion is partially annihilated by the polarization in a different direction caused by the other ion. Furthermore in the contact zone solvent molecules are no longer present.

The ion–dipole interactions are thus less important for the contact ion-pair than for the sum of the two separated ions (Figure 1.19).

It can be concluded that dipole–ion interactions favour the dissociation of the ion-pairs. The dipole–ion interactions between the ions and the dipolar

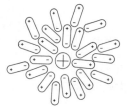

Figure 1.19 Polarization of the dipoles of the solvent molecules by the separate ions and by a contact ion-pair

solvent molecules increase when the dipole moment of these molecules becomes larger and when their number per unit volume increases. According to the previous equation such an increase also enhances the dielectric constant.

Thus when the dielectric constant is stronger, this means that the concentration of the dipoles or the magnitude of the dipole moment of the solvent molecules become greater. As a consequence the ion–dipole interactions are more important.

This is the reason of the correlation between the macroscopic dielectric constant and the dissociation constant of the ion-pairs. When the ions in the ion-pair are not bound by additional hydrogen bonds or E.D.A. bonds, so that their binding results only from the electrostatic Coulomb interactions, the ion–dipole interactions between the separate ions and the solvent molecules are often so large that the dissociation of the ion-pairs in these solvents becomes *exothermic* and the corresponding entropy change negative. This entropy effect results from the orientation of the solvent molecules around the ions, increasing the order in the solution.

This was observed, for instance, by Denison and Ramsey (1955) for the dissociation of phenyltrimethylammonium perchlorates in 1,2-di-chloroethane and in ethylidene chloride (ΔH_d being of the order of $-14 \, \text{kJ mol}^{-1}$). In this case the exothermic attraction of the solvent molecules by the ions surpasses thus the electrostatic attraction between the ions in the ion-pairs.

In the case of *hydrogen-bonded* ion-pairs, the energy of the labile bond is still strong enough to maintain the *dissociation* endothermic, but as can be seen from the data above, the values of ΔH_d although positive are rather small.

The first thermodynamic approach to take into account the effect of the dielectric constant on the dissociation of ion-pairs was made by Denison and Ramsey in 1955. It is based on the difference in energy of two point charges in a continuous medium having a dielectric constant D, when they are at a distance a and when they are infinitely far from each other. Denison and Ramsey's equation can be written

$$\log K_d^0 = \log K_d^{0x} - \left(\frac{e^2}{2.3aDkT} \right)$$

K_d^{0x} is the dissociation constant at infinite dielectric constant where the attraction between the two point charges is completely annihilated by the polarization effects. According to Denison and Ramsey the logarithm of the dissociation constant should be a linear function of the reciprocal of the dielectric constant of the medium.

As a matter of fact, the experimental curves of $\log K_d$ as a function of $1/D$ in mixed solvents of decreasing dielectric constant show a slight upwards

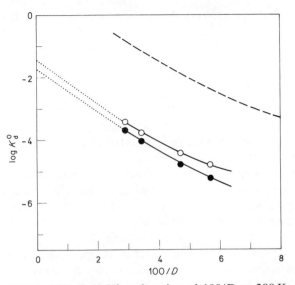

Figure 1.20 Log K_d^o as function of $100/D$ at 298 K.
○: Triethylammonium picrate; ●: methylimidazolium
picrate in nitrobenzene–benzene mixtures [Repro-
duced by permission of Plenum Publishing Corp. from
Haulait-Pirson and Van Even, 1977]. Dashed curve:
tetrabutylammonium salts in nitrobenzene–CCl₄ mixtures after the data of Hirsch and Fuoss (1960)

curvature (Figure 1.20). This was shown for non-hydrogen-bonded ion-pairs (tetraalkylammonium salts) in nitrobenzene–CCl₄ mixtures by Hirsch and Fuoss (1960) and for hydrogen-bonded ion-pairs by Haulait (Haulait-Pirson and Van Even, 1977) and others.

The slight curvature was ascribed by Hirsch and Fuoss to an accumulation of the more polar component of the mixed solvent near the ions, increasing the local dielectric constant in the less polar media. The slopes of the curves lie between 30 and 70. From these values one deduces values for the parameter a lying between 8 and 3.5 Å on the basis of Denison and Ramsey's equation. This is an acceptable order of magnitude for the interdistance between the charges in contact pairs. Generally the second term of the equation has a value between 1 and 2. Thus if the electrostatic attraction between the counterions was annihilated, the logarithms of the dissociation constants in acetonitrile and in nitrobenzene would be 1–2 units higher.

For non-hydrogen-bonded ion-pairs, where

$$\log K_d^{00} = \log K_d^{00x} - \left(\frac{e^2}{2.3aDkT}\right).$$

Fuoss and Accascina (1959) deduced an equation relating the first term to the parameter a:

$$K_d^{00x} = \left(\frac{3000}{4\pi Na^3}\right).$$

With a value of 6 Å for a, this yields a value of 1.8 mol l^{-1} for K_d^{00x}, leading to a value of -0.9 mol l^{-1} for K_d^{00} in acetonitrile and in nitrobenzene. This figure is in good agreement with the experimental value of -1.3 for tetraethylammonium perchlorate in acetonitrile, found by Coetzee and Cunningham (1965b).

For hydrogen-bonded ion-pairs, as can be seen from Table 1.19 K_d^0 *generally lies several orders of magnitude lower* than the above figure. Log K_d^{0x} often differs by several units from log K_d^{00x} and this is due to the stabilizing effect of the hydrogen bond.

This stabilizing effect is due to a great extent to the covalent character of the hydrogen bond also involving some electronic charge transfer from the anion to the cation. It can not therefore be described in an adequate quantitative manner on the basis of simple models considering, for instance, ion–dipole or dipole–dipole interactions.

This remark holds also for *E.D.A.-bonded ion-pairs*. Although the figures given for $C_5H_5I^+ \cdots I^-$ are only orders of magnitude, it is quite clear that the *dissociation constant* here is many orders of magnitude less than that which would be expected on the basis of the sole electrostatic attraction between the ions. This is due to the very important electronic charge transfer which takes place from the iodide ion to the other parts of the complex.

For non-hydrogen-bonded or non-E.D.A.-bonded ion-pairs K_d^{00x} is practically independent of the temperature. Using the Van't Hoff relation, one finds for the dissociation enthalpy of such ion-pairs

$$\Delta H_d^{00} = \frac{e^2 N}{aD}\left[1 + \frac{d(\ln D)}{d(\ln T)}\right].$$

For many liquids, the value of $d(\ln D)/d(\ln T)$ is negative and less than one (see for instance Szwarc, 1972). This explains why in these media the dissociation of non-hydrogen-bonded ion-pairs is endothermic. In the case of the hydrogen-bonded or E.D.A.-bonded ion-pairs an additional term, related to the specific bond, has to be taken into account.

1.4.3.2 *Specific interactions of the solvent* The simple treatment of Denison and Ramsey predicts that the dissociation constant should remain identical in all the solvents with the same dielectric constant. As early as 1956 it was pointed out by Gilkerson (1956) that this is not the case. From Table 1.19 it can be seen for instance that log K_d^0 differs generally by more than one for the same ion-pair in passing from nitrobenzene to acetonitrile. The slight difference in dielectric constant should only account for a difference of 0.05. The reason lies in the fact that acetonitrile can act as a weak electron donor. This was shown by Joesten and Drago (1962), White and Thompson (1966), Gramstad and Sandström (1969), and Gerbier and Baraton (Baraton, 1971; Baraton et al., 1971, 1973) amongst others. The hydrogen bond between phenol and CH_3CN is markedly weaker than with the pyridines, for instance. In CCl_4 the enthalpy of the bond is only $13 \, kJ \, mol^{-1}$ according to Joesten and Drago, $\Delta\nu_{OH}$ reaches only $160 \, cm^{-1}$. The equilibrium constant according to Gerbier and Baraton and other authors is of the order of $5 \, l \, mol^{-1}$. Acetonitrile can thus form weak hydrogen bonds with the trialkylammonium ion.

$$CH_3CN\cdots\cdots H\overset{+}{-}\underset{\underset{R}{|}}{\overset{\overset{R}{|}}{N}}-R$$

According to the determinations of Huyskens and Lambeau (1978a), this interaction lowers the standard free energy of the ion dissolved in acetonitrile by approximately $9.4 \, kJ \, mol^{-1}$ with respect to nitrobenzene.

The specific interactions of the solvent with the ions were taken into account in Gilkerson's treatment (Gilkerson, 1956). Gilkerson introduces a term E_s/RT into the expression for $\log K_d$, which is related to the difference between the specific interaction energies of ions and ion-pairs with the nearest solvent molecules. In the case of non-hydrogen-bonded ion-pairs, Gilkerson's expression can be written:

$$K_d^{00} = \left(\frac{2\pi\mu kT}{h^2}\right)^{3/2} g\nu\delta \exp\left(\frac{E_s}{RT}\right)\exp\left(\frac{-e^2}{aDkT}\right)$$

where μ is the reduced mass of the pair; g, ν, terms related to the partition functions and to the available free volume per particle; δ is a constant slightly larger than unity.

1.4.3.3 pK$_a$ of the partners in water When one restricts the comparison to a given series of complexes in a given solvent, there is, as a rule, a relationship between the dissociation constant and the difference in pK$_a$ of the cation and of the conjugated acid of the anion in water.

In a given series of ion pairs, the dissociation constant K$_d$ generally increases when either the base from which the cation derives or the acid from which the anion derives becomes stronger. This rule holds for complexes where the ion-pairs are predominant.

For instance, from the data of Table 1.19 it appears that to a first approximation a linear relation holds between log K_d and pK_a of the acid for the complexes between amines and substituted phenols in acetonitrile (Figure 1.21) (The slope is approximately 0.2.) However, the correlations can be altered by other factors, for example, the presence of steric groups which does not favour the formation of a hydrogen bond between the ions, thus increasing K_d. Furthermore, marked differences appear when passing from one family to another. This is the case for instance when comparing the ion-pairs involving the halide ions with the other groups. Their dissociation constants in the organic solvents are much lower than would be expected on the basis of the pK_a values in water. This effect is related to the strong stabilization of the halide ions by hydrogen bonds with the surrounding water molecules in water.

A similar remark holds when comparing the inorganic anions HSO_4^-/NO_3^- with the picrate ion.

Furthermore differences also exist between the various groups of organic ions.

1.4.3.4 *Homoconjugation and heteroconjugation of the ions*
Free ions derived from hydrogen-bonded or E.D.A.-bonded ion-pairs possess specific sites which make them able to form specific bonds with electron donor or electron acceptor neutral molecules present in the medium. A consequence of the formation of such specific bonds with the ligands is an increase of the

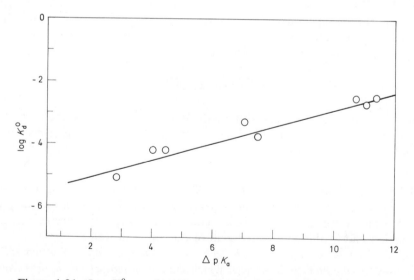

Figure 1.21 Log K_d^0 as a function of ΔpK_a of the partners for ion pairs between alkylammonium and substituted phenolate ions in acetonitrile. Values of Table 1.19

dissociation constant K_d with respect to the value K_d^0 in absence of the ligand.

When only one kind of ion is complexed by the ligand, the following equation may be written:

$$\frac{K_d}{K_d^0} = R = (1 + k_1 L + k_1 k_2 L^2 + \cdots)$$

where $k_1 k_2 \cdots$ are the addition constants of the first and second ligand molecules on the ion and L is the concentration of the free ligand.

The ligand may be the parent acid or base molecule and in this case, Kolthoff and Chantooni (1963b) introduced the denomination of 'homoconjugation' to describe this kind of complexation:

$$R_3N^+—H \cdots NR_3 \qquad PiO—H \cdots {}^-O—Pi$$

When the ligand is not the parent molecule, Kolthoff and Chantooni use the term 'heteroconjugation'.

As early as 1947 Witschonke and Kraus ascribed the dependence of the dissociation constant of pyridinium picrate in nitrobenzene on the concentration of pyridine, to the complexation of the cation by the last molecule.

Several methods were used to determine the homoconjugation and heteroconjugation constants between ions and ligands. They are based on the influence of the ligand concentration on the electric conductivity of the solutions, on the electrode potentials, on the solubility of the ionophores, etc. Important and extensive work has been carried out in this field by Kolthoff, Chantooni, and their coworkers (Kolthoff et al., 1952, 1961, 1966, 1968; Kolthoff and Chantooni, 1962, 1965, 1966, 1969, 1970, 1971; Chantooni and Kolthoff, 1968, 1973; etc.); Coetzee and his coworkers (Coetzee et al. (1964); Coetzee and Cunningham, 1965a, b; Coetzee and Padmanabhan, 1965; Coetzee, 1967; Gilkerson and his coworkers (Ralph and Gilkerson, 1964; Gilkerson and Ralph, 1965; Gilkerson and Ezell, 1965; Ezell and Gilkerson, 1968; Flora and Gilkerson, 1973; Aitken and Gilkerson, 1973; Junker and Gilkerson, 1975; Gilkerson, 1976; etc.); Pawlak (1973); Haulait-Pirson (Delcoigne and Haulait, 1976; Haulait-Pirson, 1976); and others.

It lies beyond the scope of this chapter to enter into the details of these works. Table 1.20 gives values of the homoconjugation constants of some selected anions and cations in two solvents in order to detail the factors which seem to play a role in this phenomenon. Some of these factors have already been discussed by Davis (1968).

First, there is a marked solvent effect which is due to specific interactions of the solvent with either the ion or the neutral molecule rather than to a change in dielectric constant. As can be seen from Table 1.20, the homoconjugation constants are generally at least one order of magnitude higher in

Table 1.20   Homoconjugation constants k_1 in $1\,\text{mol}^{-1}$ of anions and cations in acetonitrile (A.N.) and in nitrobenzene (N.B.) at 298 K

Anion	pK_a (acid)	$\log k_1$ (A.N.)	$\log k_1$ (N.B.)	Cation	pK_b (base)	$\log k_1$ (A.N.)	$\log k_1$ (N.B.)
Phenolate	9.95	5.8[a]	—	Pyridinium	8.77	0.6[h]	
2-Nitrophenolate	7.21	2.2[b]	—	—	—	—	—
3-Nitrophenolate	7.15	4.0[c]	—	—	—	—	—
4-Nitrophenolate	7.15	3.5[b]	—	—	—	—	—
4-Nitro-3-chloro-phenolate	~6.25	3.6[c]	—	—	—	—	—
3,5-Dinitro-phenolate	6.70	4.2[d]	—	Morpholinium	5.60	1.0[h]	—
2,4-Dinitro-phenolate	4.07	2.1[b]	—	—	—	—	—
Picrate	0.38	0.3[b]	—	—	—	—	—
Benzoate	4.20	3.6[e]	—	—	—	—	—
Hydroxybenzoate	4.61	3.1[e]	—	Benzylammonium	4.65	1.2[h]	—
3-Bromobenzoate	5.00	3.8	—	Ethylammonium	3.33	1.4[h]	—
3-Nitrobenzoate	3.49	4.0	—	Diethylammonium	3.00	0.4[h]	—
4-Nitrobenzoate	3.42	3.8	—	Triethylammonium	3.35	[i]	—
2-Nitrobenzoate	2.21	4.0	—	Butylammonium	3.41	1.4[h]	2.45[j]
3-Nitro-4-chloro-benzoate	2.00	4.0	—	Dibutylammonium	2.75	[i]	1.22[j]
3,5-Dinitrobenzoate	2.82	4.0	6.0[g]	Tributylammonium	3.11	[i]	-0.4[k]
Chloroacetate	1.30		6.4[g]	Piperidinium	3.75	1.4[h]	
Methanesulphonate		3.8	—	Triphenyl-guanidinium	4.9	—	0.8[l]
2,5-Dichloro-benzene-sulphonate		2.7[b]	—	Diphenyl-guanidinium	3.9	—	1.8[l]
Cl^-	-7	2.2[f]	—	Tetramethyl-guanidinium	0.4	—	1.5[l]
NO_3^-	-1.4	2.3[f]	—	—	—	—	—
HSO_4^-	-3.0	3.0[f]	—	—	—	—	—
Br^-	-9	2.4[f]	—	—	—	—	—
ClO_4^-	-10	0[f]	—	—	—	—	—

[a] Coetzee and Padmanabhan (1965).
[b] Kolthoff and Chantooni (1965).
[c] Kolthoff and Chantooni (1969).
[d] Kolthoff and Chantooni (1963b).
[e] Kolthoff and Chantooni (1966).
[f] Kolthoff et al. (1961).
[g] Jasinski and Pawlak (1970).
[h] Coetzee et al. (1964), Coetzee and Padmanabhan (1965), and Coetzee (1967).
[i] Not observable (ref. h).
[j] Haulait and Huyskens (1975b).
[k] Macau et al. (1971).
[l] Collaer (1980).

nitrobenzene, than in acetonitrile, and this for the homoconjugation of both anions or cations.

The effect of changing the *dielectric constant on homo- and heteroconjugation constants* was studied by Haulait-Pirson and Van Even (1977). The constants increase with decreasing dielectric constant of the mixed solvent, but this effect is less pronounced than that on the dissociation constant. This effect was interpreted as related to the change of the size of the complexed ions.

As a rule, homoconjungation constants for anions with hydrogen bonds of the type $O—H \cdots {}^-O$ are stronger than those for cations with hydrogen-bonds of the type $N^+—H \cdots N$.

Furthermore, a general trend seems to exist for the constants to decrease when the *acidity* of the acid increases in $O—H \cdots {}^-O$ bonds or when the *basicity* of the base decreases in $N^+—H \cdots N$ bonds. This may be related to the depth of the potential well in the $N^+—H$ bond which increases with increasing basicity, whereas the depth in the $O—H$ bond increases when the acidity decreases.

However, an important factor which can completely overshadow the effect of changing the pK_a, is the *steric hindrance*: there is for instance a difference of two orders of magnitude between the homoconjugation constants of the o-nitrophenolate and m-nitrophenolate ions. Similar remarks hold for trialkylammonium ions compared with the monoalkylammonium ions.

Homoconjugation is perhaps still more important in the case of *ions derived from E.D.A. bonds*. For instance the homoconjugation constant of I^-:

$$I^- + I_2 \rightleftarrows I_3^-$$

shows a value as high as $6.3 \times 10^8 \, l \, mol^{-1}$ in dichlorobenzene (Uruska and Szpakowska, 1972). The values reported in acetonitrile lie between 4 and $15 \times 10^6 \, l \, mol^{-1}$ (Nelson and Iwamoto, 1964; Barraque *et al.*, 1969; Baucke *et al.*, 1971). In nitromethane Nelson and Iwamoto found a value of $5 \times 10^6 \, l \, mol^{-1}$. Only in amphiprotic solvents, such as ethanol or water, does the homoconjugation constant drop to values as low as $1.4 \times 10^4 \, l \, mol^{-1}$ (Barraque *et al.*, 1969) and $7.8 \times 10^2 \, l \, mol^{-1}$ (Davies and Gwynne, 1952). Higher homoconjugation

$$I_3^- + I_2 \rightleftarrows I_5^-$$

is still very important in solvents such as nitrobenzene or o-dichlorobenzene (Huyskens and Poskin, 1975). The homoconjugation of cations such as

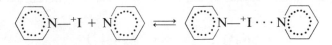

is also very important in these solvents (Poskin and Huyskens, 1976). It can also be noted here that in the crystalline phase, the complex between iodine and pyridine comprises I_3^- ions and $C_5H_5N—I^+ \cdots NC_5H_5$ cations.

A consequence of the high homoconjugation constants of both anions and cations derived from E.D.A. bonds such as those considered before, is that the ions which are observed in solution are mainly homoconjugate ions.

Another type of interaction is that between an ion and the parent ion-pair. Such entities have been called 'triple ions' by Kraus and Fuoss (Fuoss and Kraus, 1933; Kraus, 1939). When the ion-pair does not bear additional electron donor or electron acceptor sites, such *triple ions* are held together only through electrostatic interactions. In this case they are as a rule observed only when the dielectric constant falls below 20. However, when the ion-pair has additional sites so that it can form supplementary hydrogen bonds with the ion, the formation of triple ions may also be detected in solvents of much higher dielectric constant such as nitrobenzene (Collaer, 1980).

1.4.3.5 *Complexation of the ion-pairs*   A homoconjugated cation such as

$$Bu_3N—{}^+H \cdots NBu_3$$

had no additional sites which could allow it to bind molecules or other ions. However, an electrostatic attraction remains between this homoconjugated ion and the anion, so that the formation of ion-pairs of the type

$$(Bu_3N—{}^+H \cdots NBu_3)X^-$$

can be envisaged. As a matter of fact the appearance of such ion-pairs has been observed by Gilkerson and Ralph (Gilkerson and Ralph, 1965).

However, the stability of such complexed ion-pairs becomes much more important, when the original ion-pair bears additional specific sites as for instance in

$$\underset{\overset{|}{H}}{Bu_2N^+}—H \cdots Pi^- + Bu_2—\underset{\overset{|}{H}}{N} \xrightarrow{K_1} Bu_2—\underset{\overset{|}{\underset{\vdots}{H}}}{N^+} \cdots H \cdots Pi^-$$

$$H—N—Bu_2$$

where the hydrogen bond reinforces the electrostatic interactions. In such a case, the presence of the ligand L (here the free amine $Bu_2—N—H$) changes the overall dissociation constant of the ionophore in a manner given by the following equation (Macau *et al.*, 1971) which is a generalization of that used by Gilkerson

$$\frac{K_d}{K_d^0} \equiv R = \frac{(1 + k_1^+ L + k_1^+ k_2^+ L^2)}{1 + K_1 L}.$$

The curve of R versus the concentration of free ligand L shows often a downward curvature. *This indicates not only that the ion-pair is complexed but also that the corresponding constant K_1 is greater than the addition constant k_2^+ of a second ligand molecule on the once-complexed ion.*

This is the case for instance for the ion-pairs $BuN^+H_3Pi^-$ complexed by pyridines and amines, on the additional NH sites, and $Et_3NH^+Br^-$ complexed on the additional electron donor sites of the bromide ion by proton donors such as benzoic acids or by electron acceptors such as Br_2 or I_2 (Huyskens *et al.*, 1973). The reason why K_1 may be greater than k_2^+ lies in the fact that the perturbation, in the ability of the second site to form a specific bond, brought about by the first bond, is here of a more covalent than electrostatic nature. This has been discussed in a recent paper (Huyskens, 1977).

In conclusion we can say that additional specific interactions of the ion-pair with the surrounding molecules generally favours the proton or ion transfer in the complex, whereas, such interactions do not favour the dissociation of the ion-pairs in free ions.

1.5 References

Adam, W., Grimison, A., and Rodriguez, G. (1967). *Tetrahedron*, **23**, 2513.

Adam, W., Grimison, A., Hoffmann, R., and de Ortiz, C. Z. (1968). *J. Amer. Chem. Soc.*, **90**, 1509.

Aitken, H. W. and Gilkerson, W. R. (1973). *J. Amer. Chem. Soc.*, **95**, 8551.

Almlöf, J., Lindgren, J., and Tegenfeldt, J. (1972). *J. Mol. Struct.*, **14**, 427.

Asselin, M., Bélanger, G., and Sandorfy, C. (1969). *J. Mol. Spectrosc.*, **30**, 96.

Attig, R. and Williams, J. M. (1977). *J. Chem. Phys.*, **66**, 1389.

Aue, D. H., Webb, H. M., and Bowers, M. T. (1972). *J. Amer. Chem. Soc.*, **94**, 4726.

Aue, D. H., Webb, H. M., and Bowers, M. T. (1975). *J. Amer. Chem. Soc.*, **97**, 4137.

Aue, D. H., Webb, H. M., and Bowers, M. T. (1976a). *J. Amer. Chem. Soc.*, **98**, 318.

Aue, D. H., Webb, H. M., Bowers, M. T., Liotta, C. L., Alexander, C. J., and Hopkins, M. P. (1976b). *J. Amer. Chem. Soc.*, **98**, 854.

Ault, B. S. and Pimentel, G. C. (1973a). *J. Phys. Chem.*, **77**, 1649.

Ault, B. S. and Pimentel, G. C. (1973b). *J. Phys. Chem.*, **77**, 57.

Baba, H. and Nagakura, S. (1951). *J. Chem. Soc. Japan*, **72**, 3.

Baba, H., Matsuyama, A., and Kokobun, H. (1969). *Spectrochim. Acta*, **25A**, 1709.

Baraton, M. I. (1971). *J. Mol. Struct.*, **10**, 231.

Baraton, M. I. (1979). Thesis, Limoges

Baraton, M. I., Gerbier, J., and Besnaïou, S. (1973). *C. R. Acad. Sci. Paris*, **276**, 797.

Baraton, M. I., Gerbier, J., and Lorenzelli, V. (1971). *C. R. Acad. Sci. Paris*, **272**, 750.

Barraque, C., Vedel, J., and Tremillon, R. (1969). *Anal. Chim. Acta*, **46**, 263.

Barrow, G. M. (1956). *J. Amer. Chem. Soc.*, **78**, 5802.

Barthel, J., Justice, J. C., and Wachter, R. (1973). *Phys. Chem. (NF)*, **84**, 100.

Baucke, F. G. K., Bertram, R., and Cruse, K. (1971). *J. Electroanal. Chem.*, **32**, 247.

Bayles, J. W. and Chetwyn, A. (1958). *J. Chem. Soc.*, 2328.

Bayles, J. W. and Evans, B. (1965). *J. Chem. Soc.*, 6984.
Bayles, J. W. and Taylor, A. F. (1961). *J. Chem. Soc.*, 417.
Beier, G. (1975). *Ph.D. Thesis*, Universität Wien.
Belin, C. and Potier, J. (1976a). *J. Chim. Phys.*, **73**, 117.
Belin, C. and Potier, J. (1976b). *J. Chim. Phys.*, **73**, 122.
Bell, C. L. and Barrow, G. M. (1959a). *J. Chem. Phys.*, **31**, 1158.
Bell, C. L. and Barrow, G. M. (1959b). *J. Chem. Phys.*, **31**, 300.
Bell, R. P. (1959). *The Proton in Chemistry*, Cornell University Press, New York.
Bell, R. P. and Crooks, J. E. (1962). *J. Chem. Soc.*, 3513.
Berger, A., Loewenstein, A., and Meiboom, S. (1959). *J. Amer. Chem. Soc.*, **81**, 62.
Bernander, L. and Olovsson, G. (1972). *Tetrahedron*, **28**, 3521.
Birchall, T. and Gillespie, R. J. (1965). *Canad. J. Chem.*, **43**, 1045.
Bjorvatten, T. (1969). *Acta Chem. Scand.*, **23**, 1109.
Blinc, R., Mali, M., and Trontelj, Z. (1967). *Phys. Lett.*, **25A**, 289.
Bonsor, D., Borah, B., Dean, R. L., and Wood, J. L. (1976). *Canad. J. Chem.*, **54**, 2458.
Borah, B. and Wood, J. L. (1976). *Canad. J. Chem.*, **54**, 2470.
Borgen, B., Hassel, O., and Rømming, C. (1962). *Acta Chem. Scand.*, **16**, 2469.
Bratoz, S. (1975). *J. Chem. Phys.*, **63**, 3499.
Briegleb, G. and Delle, H. (1960). *Z. Phys. Chem.*, **24**, 359.
Bron, J. and Simmons, E. L. (1976). *J. Phys. Chem.*, **80**, 17.
Bruckenstein, S. and Saito, A. (1965). *J. Amer. Chem. Soc.*, **87**, 698.
Bruckenstein, S. and Unterecker, D. F. (1969). *J. Amer. Chem. Soc.*, **91**, 5741.
Brzezinski, B. and Zundel, G. (1976). *Chem. Phys. Lett.*, **44**, 521.
Cabana, A. and Sandorfy, C. (1962). *Spectrochim. Acta*, **18**, 843.
Caldin, E. F. and Crooks, J. E. (1967). *J. Chem. Soc. B*, 959.
Caldin, E. F., Crooks, J. E., and Donnell, D. O. (1973). *J. Chem. Soc. Faraday I*, **69**, 993.
Cannon, C. G. (1958). *Spectrochim. Acta*, **10**, 341.
Chakrabarty, M. R., Handloser, C. S., and Mosher, M. W. (1973). *J. Chem. Soc. Perkin II*, 938.
Chantooni, M. K. and Kolthoff, I. M. (1968). *J. Amer. Chem. Soc.* **90**, 3005.
Chantooni, M. K. and Kolthoff, I. M. (1973). *J. Phys. Chem.*, **77**, 1.
Chao, G. Y. and McCullough, J. O. (1960). *Acta Cryst.*, **13**, 727.
Chenon, B. and Sandorfy, C. (1958). *Canad. J. Chem.*, **36**, 1181.
Choi, C. S., Mapes, J. E., and Prince, E. (1972). *Acta Cryst.*, **B28**, 1357.
Claydon, M. F. and Sheppard, N. (1969). *Chem. Commun.*, 1431.
Clementi, E. (1967a). *J. Chem. Phys.*, **47**, 2323.
Clementi, E. and Gayles, J. N. (1976b). *J. Chem. Phys.*, **47**, 3837.
Clements, R., Dean, R. L., and Wood, J. L. (1973). *J. Mol. Struct.*, **17**, 291.
Clements, R. and Wood, J. L. (1973a). *J. Mol. Struct.*, **17**, 265.
Clements, R. and Wood, J. L. (1973b). *J. Mol. Struct.*, **17**, 283.
Cleuren, W. (1979). *Thesis*, Leuven.
Clotman, D., Van Lerberghe, D., and Zeegers-Huyskens, Th. (1970). *Spectrochim. Acta*, **26A**, 1621.
Coetzee, J. F. (1967). *Prog. Phys. Org. Chem.*, **4**, 45.
Coetzee, J. F. and Cunningham, G. P. (1965a). *J. Amer. Chem. Soc.*, **87**, 2580.
Coetzee, J. F. and Cunningham, G. P. (1965b). *J. Amer. Chem. Soc.*, **87**, 2534.
Coetzee, J. F. and Padmanabhan, G. R. (1965). *J. Phys. Chem.*, **69**, 3193.
Coetzee, J. F., Padmanabhan, G. R., and Cunningham, G. (1964). *Talanta*, **11**, 93.
Collaer, H. (1980). Thesis, Leuven.

Combelas, P., Cruege, F., Lascombe, J., Quivoron, C., Rey-Lafon, M., and Sebille, B. (1970). *Spectrochim. Acta*, **26A,** 1323.
Conor, T. M. and Loewenstein, A. (1961). *J. Amer. Chem. Soc.*, **83,** 560.
Cook, D. (1963). *Canad. J. Chem.*, **41,** 1127.
Cook, D. (1964a). *Canad. J. Chem.*, **42,** 2721.
Cook, D. (1964b). *Canad. J. Chem.*, **42,** 2292.
Coppens, P. and Sabine, T. M. (1969). *Acta Cryst.*, **B25,** 2442.
Cox, B. G. (1970). *J. Chem. Soc. B*, 1780.
Crooks, J. E. and Robinson, B. H. (1970). *Chem. Commun.*, 979.
Crooks, J. E. and Robinson, B. H. (1971). *Trans. Faraday Soc.*, **67,** 1707.
Crooks, J. E. and Robinson, B. H. (1975). *Faraday Symp. Chem. Soc.*, **10,** 29.
Cummings, D. L. and Wood, J. L. (1973). *J. Mol. Struct.*, **17,** 257.
Currie, M. and Speakman, J. C. (1970). *J. Chem. Soc. A*, 1923.
Dahl, T. and Hassel, O. (1966). *Acta Chem. Scand.*, **20,** 2009.
Damm, E., Hassel, O., and Rømming, C. (1965). *Acta Chem. Scand.*, **19,** 1159.
Davies, M. and Gwynne, E. (1952). *J. Amer. Chem. Soc.*, **74,** 2748.
Davis, M. M. (1962). *J. Amer. Chem. Soc.*, **84,** 3623.
Davis, M. M. (1968). *Acid-Base Behaviour in Aprotic Organic Solvents*, Natl. Bur. Std. Monograph 105.
Davis, M. M. and Hetzer, H. B. (1952). *J. Res. Natl. Bur. Std.* **48,** 381.
Davis, M. M. and Hetzer, H. B. (1954). *J. Amer. Chem. Soc.*, **76,** 4247.
Davis, M. M. and Paabo, M. (1960). *J. Amer. Chem. Soc.*, **82,** 5081.
Davis, M. M. and Schuhmann, P. J. (1947). *J. Res. Natl. Bur. Std.*, **39,** 221.
Davis, M. M., Schuhmann, P. J., and Lovelace, M. I. (1948). *J. Res. Natl. Bur. Std.*, **41,** 27.
Dean, R. L. and Wood, J. L. (1975a). *J. Mol. Struct.*, **26,** 197.
Dean, R. L. and Wood, J. L. (1975b). *J. Mol. Struct.*, **26,** 215.
Debecker, G. and Huyskens, P. (1971). *J. Chim. Phys.*, **68,** 287.
de Jeu, W. H. (1970). *J. Phys. Chem.*, **74,** 822.
Del Bene, J. E. (1975). *J. Amer. Chem. Soc.*, **97,** 5330.
Del Bene, J. E. and Vaccaro, A. (1976). *J. Amer. Chem. Soc.*, **98,** 7526.
Delcoigne, V. and Haulait, M. C. (1976). *J. Solut. Chem.*, **5,** 47.
de Lozé, C., Combellas, P., Bacelon, P., and Garrigou-Lagrange, C. (1972). *J. Chim. Phys.*, **69,** 397–403.
Denison, J. and Ramsey, J. (1955). *J. Amer. Chem. Soc.*, **77,** 2615.
Denisov, G. S., Gusakova, G. V., and Smolyansky, A. L. (1973a). *J. Mol. Struct.*, **15,** 377.
Denisov, G. S. and Schreiber, V. M. (1972). *Spectrosc. Lett.*, **5,** 377.
Denisov, G. S. and Schreiber, V. M. (1976). *Vestnik LGU*, **4,** 61.
Denisov, G. S., Starosta, Ya., and Shraiber, V. M. (1973b). *Opt. Spektrosk.*, **35,** 447.
Detoni, S., Hadzi, D., Smerkolj, R., Hawranek, J., and Sobczyk, L. (1970). *J. Chem. Soc. A*, 2851.
de Villepin, J. and Novak, A. (1971a). *Spectrochim. Acta*, **27A,** 1259.
de Villepin, J. and Novak, A. (1971b). *Spectrosc. Lett.*, **4,** 1.
D'Hondt, J. and Zeegers-Huyskens, Th. (1971). *J. Mol. Struct.*, **10,** 135.
Dierckx, A. M., Huyskens, P., and Zeegers-Huyskens, Th. (1965). *J. Chim. Phys.*, **62,** 336.
Dill, J., Allen, L. C., Topp, W. C., and Pople, J. A. (1975). *J. Amer. Chem. Soc.*, **97,** 7220.
Dimroth, K., Reichardt, C., Siepmann, T., and Bohlmann, F. (1963). *Ann. Chem.*, 661.

Donohue, J. (1968). Selected topics in hydrogen bonding in: *Structural Chemistry and Molecular Biology* (Eds. Rich, A. and Davidson, N.), Freeman, San Francisco.

Dorval, C. and Zeegers-Huyskens, Th. (1973). *Spectrochim. Acta*, **29A**, 1805.

Dupont, J. P., D'Hondt, J., and Zeegers-Huyskens, Th. (1971). *Bull. Soc. Chim. Belg.*, **80**, 369.

Eia, G. and Hassel, O. (1956). *Acta Chem. Scand.*, **10**, 139.

Elliott, M. A. and Fuoss, R. M. (1939). *J. Amer. Chem. Soc.*, **61**, 294.

Evans, J. C. and Wright, N. (1960). *Spectrochim. Acta*, **16**, 352.

Ezell, J. B. and Gilkerson, W. R. (1968). *J. Phys. Chem.*, **72**, 144.

Felix, N. and Huyskens, P. (1975). *J. Amer. Chem. Soc.*, **79**, 2316.

Filimonov, V. N. and Bystrov, D. S. (1962). *Optics and Spectroscopy*, **12**, 31.

Flora, H. B. and Gilkerson, W. R. (1973). *J. Phys. Chem.*, **77**, 1421.

Foglizzo, R. and Novak, A. (1969a). *J. Chem. Phys.*, **50**, 5366.

Foglizzo, R. and Novak, A. (1969b). *J. Chim. Phys.*, **66**, 1539.

Foglizzo, R. and Novak, A. (1970). *Spectrochim. Acta*, **26A**, 2281.

Foglizzo, R. and Novak, A. (1971). *J. Mol. Struct.*, **7**, 205.

Fraenkel, G. and Franconi, C. (1960). *J. Amer. Chem. Soc.*, **82**, 4478.

Franz, M. (1978). *Thesis*, Leuven.

Fukui, K., Imamura, A., Yonezawa, T., and Nagato, C. (1962). *Bull. Chem. Soc. Japan*, **35**, 33.

Fuoss, R. M. and Accascina, F. (1959). *Electrolyte Conductance*, Interscience Publishers, New York.

Fuoss, R. M. and Kraus, C. A. (1933). *J. Amer. Chem. Soc.*, **55**, 3614.

Gallagher, K. J. (1959). In *Hydrogen Bonding* (Eds. Hadzi, D. and Thompson, H. W.), Pergamon Press.

Giger, W. and Simon, W. (1970). *Helv. Chem. Acta*, **53**, 1609.

Gilkerson, W. R. (1956). *J. Chem. Phys.*, **25**, 1199.

Gilkerson, W. R. (1976). *J. Phys. Chem.*, **80**, 2488.

Gilkerson, W. R. and Ezell, J. B. (1965). *J. Amer. Chem. Soc.*, **87**, 175.

Gilkerson, W. R. and Ralph, III. E. (1965). *J. Amer. Chem. Soc.*, **87**, 175.

Gillespie, R. and Birchall, T. (1963). *Canad. J. Chem.*, **41**, 148.

Ginn, S. G. W., Haque, I., and Wood, J. L. (1968). *Spectrochim. Acta*, **24A**, 1531.

Ginn, S. G. W. and Wood, J. L. (1966). *Trans. Faraday Soc.*, **62**, 777.

Ginn, S. G. W. and Wood, J. L. (1967). *Spectrochim. Acta*, **23A**, 611.

Glazunov, V. P., Mashkovsky, A. A., and Odinokov, S. E. (1976). *Spectrosc. Lett.*, **9**, 391.

Glowiak, T., Sobczyk, L., and Grech, E. (1975a). *Chem. Phys. Lett.*, **34**, 292.

Glowiak, T., Sobczyk, L., and Grech, E. (1975b). *Chem. Phys. Lett.*, **36**, 106.

Gol'dshtein, I. P., Kharlamova, E. N., and Gur'yanova, E. N. (1968). *Zh. Obshch. Khim.*, **38**, 1984.

Gordon, J. E. (1961). *J. Org. Chem.*, **26**, 738.

Govil, G., Clague, A. D. H., and Bernstein, H. J. (1968). *J. Chem. Phys.*, **49**, 2821.

Gramstad, T. and Sandström, J. (1969). *Spectrochim. Acta*, **25**, 31.

Grant, D. M. and Pugmire, R. J. (1965). Cited by Mathias, A. and Gil, V. M. S.

Grech, E., Malarski, Z., and Sobczyk, L. (1976). *Spectrosc. Lett.*, **9**, 749.

Groth, P. and Hassel, O. (1963). *Mol. Phys.*, **6**, 543.

Grunwald, E., Karabatsos, P. J., Kromhout, R. A., and Purlee, E. L. (1960). *J. Chem. Phys.*, **33**, 556.

Gur'yanova, E. N., Gol'dshtein, I. P., and Perepelkova, T. I. (1976). *Uspekhi*, **45**, 1568.

Gusakova, G. V., Denisov, G. S., and Smolyanskii, A. L. (1971). *Zh. Prikl. Spektr.*, **14**, 992.

Gusakova, G. V., Denisov, G. S., and Smolyanskii, A. L. (1972a). *Zh. Prikl. Spektr.*, **16**, 320.

Gusakova, G. V., Denisov, G. S., and Smolyanskii, A. L. (1972b). *Zh. Prikl. Spektr.*, **16**, 503.

Gusakova, G. V., Denisov, G. S., Smolyanskii, A. L., and Schraiber, V. M. (1970). *Dokl. Akad. Nauk SSSR*, **193**, 1065.

Gutmann, V. (1969). *Coordination Chemistry in Non-aqueous Solutions*, Springer-Verlag, Wien–New York.

Gutmann, V. (1976). *Coordination Chem. Rev.*, **18**, 225.

Gutowski, H. and Saika, A. (1953). *J. Chem. Phys.*, **21**, 1688.

Hadzi, D. (1959). *Hydrogen Bonding*, Pergamon Press.

Hadzi, D. (1962). *J. Chem. Soc. A*, 5128.

Hadzi, D. (1963). *Boll. Sc. Fac. Chim. Ind. Bologna*, **XXI**, 23.

Hadzi, D. (1965). *Pure Appl. Chem.*, **11**, 435.

Hadzi, D. (1970). *J. Chem. Soc. A*, 418.

Hadzi, D. and Bratoz, S. (1976). In *The theory of Hydrogen Bond* (Eds. Schuster, P., Zundel, G., and Sandorfy, C.) Part II, North-Holland.

Hadzi, D. and Kobilarov, N. (1966). *J. Chem. Soc. A*, 439.

Hadzi, D., Orel, B., and Novak, A. (1973). *Spectrochim. Acta*, **29A**, 1745.

Hagen, R. and Roberts, Y. D. (1969). *J. Amer. Chem. Soc.*, **91**, 4504.

Hamilton, W. C. and Ibers, J. A. (1968). *Hydrogen Bonding in Solids*, W. A. Benjamin Inc., New York.

Haney, M. A. and Franklin, J. L. (1969). *J. Chem. Phys.*, **50**, 2028.

Haque, I. and Wood, J. L. (1967). *Spectrochim. Acta*, **23A**, 2523.

Haque, I. and Wood, J. L. (1968). *J. Mol. Struct.*, **2**, 217.

Harned, H. S. and Owen, B. (1958). *The Physical Chemistry of Electrolyte Solutions*, Reinhold Publ. Corp., New York.

Hassel, O. and Hope, H. (1960). *Acta Chem. Scand.*, **14**, 341.

Hassel, O. and Hope, H. (1961). *Acta Chem. Scand.*, **15**, 407.

Hassel, O. and Hvoslef, J. (1954). *Acta Chem. Scand.* **8**, 873.

Hassel, O. and Hvoslef, J. (1956). *Acta Chem. Scand.*, **10**, 138.

Hassel, O., Rømming, C., and Tufte, T. (1961). *Acta Chem. Scand.*, **15**, 967.

Hassel, O. and Strømme, K. O. (1959). *Acta Chem. Scand.*, **13**, 275.

Haulait, M. C. and Huyskens, P. (1975a). *J. Solut. Chem.*, **4**, 853.

Haulait, M. C. and Huyskens, P. (1975b). *J. Phys. Chem.*, **79**, 1812.

Haulait-Pirson, M. C. (1976). *Bull. Soc. Chim. Belg.*, **85**, 639.

Haulait-Pirson, M. C. (1980). To be published.

Haulait-Pirson, M. C. and Van Even, V. (1977). *J. Solut. Chem.*, **6**, 757.

Hawranek, J. P., Oszust, J., and Sobczyk, L. (1972). *J. Phys. Chem.*, **76**, 1989.

Hibbert, F. and Satchell, D. N. P. (1968). *J. Chem. Soc. B*, 1967.

Hirsch, E. and Fuoss, R. M. (1960). *J. Amer. Chem. Soc.*, **82**, 1018.

Hopkinson, A. C. and Csizmadia, I. G. (1974). *Canad. J. Chem.*, **52**, 546.

Hudson, R. A., Scott, R. M., and Vinogradov, S. N. (1972). *J. Phys. Chem.*, **76**, 1989.

Huong, P. V. and Noel, G. (1976). *Spectrochim. Acta*, **32A**, 831.

Huong, P. V. and Schlaak, M. (1974). *Chem. Phys. Lett.*, **27**, 111.

Huyskens, P. (1972). *Ind. Chim. Belg.*, **37**, 15.

Huyskens, P. (1974). *Bull. Soc. Chim. Belg.*, **83**, 239.

Huyskens, P. (1977). *J. Amer. Chem. Soc.*, **99**, 2578.

Huyskens, P., Cleuren, W., and Vanbrabant-Govaerts, H. and Vuylsteke, A. (1980a). To be published.

Huyskens, P., Cleuren, W., and Franz, M., and Vuylsteke, A. (1980b). To be published.

Huyskens, P. and Hernandez, G. (1973). *Ind. Chim. Belg.*, **38**, 1237.
Huyskens, P. and Lambeau, Y. (1978a). *J. Phys. Chem.*, **82**, 1886.
Huyskens, P. and Lambeau, Y. (1978b). *J. Phys. Chem.*, **82**, 1892.
Huyskens, P. and Mahillon, P. (1980). To be published.
Huyskens, P., Pirson, D., Haulait, M. C., and Poskin, G. (1973). *Proceedings of the Third International Conference on Chemical Thermodynamics*, Baden, Vienna.
Huyskens, P. and Poskin, G. (1975). *Bull. Soc. Chim. Belg.*, **84**, 947.
Huyskens, P. and Zeegers-Huyskens, Th. (1964). *J. Chim. Phys.*, **61**, 81.
Ibers, J. A. (1964). *J. Chem. Phys.*, **40**, 402.
Ichikawa, M. (1972). *Acta Cryst.*, **B28**, 755.
Iogansen, A. V., Kiselev, S. A., Rassadin, B. V., and Samoilenko, A. A. (1976). *Zh. Strukt. Khim.*, **17**, 629.
Ivin, K. J., McGarvey, J. J., Simmons, E. L., and Small, R. (1973). *J. Chem. Soc. Faraday I*, **69**, 1016.
Jackman, L. M. and Sternhell, S. (1969). *Applications of Nuclear Magnetic Resonance Spectroscopy in Organic Chemistry*, 2nd edn., Pergamon Press.
Jadzyn, J. and Malecki, J. (1972). *Acta Phys. Pol.*, **A41**, 599.
James, C. J. and Fuoss, R. M. (1975). *J. Solut. Chem.*, **4**, 91.
Jannakoudakis, D. and Moumtzis, J. (1968). *Z. Naturforsch*, **23B**, 1303.
Janusek, E., Pajdowska, M., and Sobczyk, L. (1975). *Chem. Phys.*, **9**, 205.
Jasinski, T. and Kokot, Z. (1967). *Roczn. Chem.*, **41**, 139.
Jasinski, T., Misiak, T., and Skarzynska, T. (1965a). *Roczn. Chem.*, **39**, 79.
Jasinski, T., Misiak, T., and Skarzynska, T. (1965b). *Roczn. Chem.*, **39**, 1485.
Jasinski, T., Misiak, T., and Skarzynska, T. (1950). *Roczn. Chem.*, **39**, 1549.
Jasinski, T., Misiak, T., and Skarzynska-Klentak, T. (1968). *Roczn. Chem.*, **42**, 875.
Jasinski, T. and Modro, A. (1967). *Roczn. Chem.*, **41**, 2115.
Jasinski, T. and Pawlak, Z. (1970). *Roczn. Chem.*, **44**, 1577.
Jasinski, T. and Widernikova, T. (1969). *Roczn. Chem.*, **43**, 1253.
Jesson, J. P. and Muetterties, E. L. (1969). *Chemist's Guide. Basic Chemical and Physical Data*, Marcel Dekker Inc., New York.
Joesten, M. D. and Drago, R. S. (1962). *J. Amer. Chem. Soc.*, **84**, 3817.
Joesten, M. D. and Schaad, L. J. (1974). *Hydrogen Bonding*, M. Dekker Inc., New York.
Johansson, A. and Kollman, P. A. (1972). *J. Amer. Chem. Soc.*, **94**, 6196.
Johnson, S. L. and Rumon, K. A. (1965). *J. Phys. Chem.*, **69**, 74.
Jönsson, P. G. (1971). *Acta Cryst.*, **B27**, 893.
Jordan, F. (1975). *J. Amer. Chem. Soc.*, **97**, 3330.
Junker, M. L. and Gilkerson, W. R. (1975). *J. Amer. Chem. Soc.*, **97**, 493.
Keder, W. E. and Burger, L. L. (1965). *J. Phys. Chem.*, **69**, 3075.
Keefer, R. M. and Andrews, L. J. (1955). *J. Amer. Chem. Soc.*, **75**, 4538.
Kimel'fel'd, Ja M., Mostovoy, A. B., and Mostovaja, L. M. (1975). *Chem. Phys. Lett.*, **33**, 114.
Kirszenbaum, M., Corset, J., and Josien, M. L. (1970). *C. R. Acad. Sci. Paris*, **271**, 630.
Klaboe, P. (1967). *J. Amer. Chem. Soc.*, **89**, 3667.
Kludt, J. R., Kwong, G. Y. W., and McDonald, R. L. (1972). *J. Phys. Chem.*, **76**, 339.
Kohler, F. (1974). *Proceedings of the Symposium on Specific Interactions, Louvain.*
Kollman, P. and Allen, L. C. (1971). *J. Amer. Chem. Soc.*, **93**, 4991.
Kollman, P. and Allen, L. C. (1972). *Chem. Rev.*, **72**, 283.
Kollman, P. A. and Bender, C. F. (1973). *Chem. Phys. Lett.*, **21**, 271.
Kollman, P., Johansson, A., and Rothenberg, S. (1974). *Chem. Phys. Lett.*, **24**, 199.

Kolthoff, I. M., Bruckenstein, S., and Chantooni, M. K. (1961). *J. Amer. Chem. Soc.*, **83**, 3927.

Kolthoff, I. M. and Chantooni, M. K. (1962). *J. Phys. Chem.*, **66**, 1675.

Kolthoff, I. M. and Chantooni, M. K. (1963a). *J. Amer. Chem. Soc.*, **85**, 426.

Kolthoff, M. I. and Chantooni, Jr. M. K. (1963b). *J. Amer. Chem. Soc.*, **85**, 2195.

Kolthoff, M. I. and Chantooni, Jr. M. K. (1965). *J. Amer. Soc.*, **87**, 4428.

Kolthoff, M. I. and Chantooni, M. K. (1966). *J. Phys. Chem.*, **70**, 856.

Kolthoff, M. I. and Chantooni, M. K. (1969). *J. Amer. Soc.*, **91**, 4621.

Kolthoff, M. I. and Chantooni, M. K. (1970). *J. Amer. Chem. Soc.*, **92**, 7025.

Kolthoff, M. I. and Chantooni, M. K. (1971). *J. Amer. Chem. Soc.*, **93**, 3943.

Kolthoff, M. I., Chantooni, M. K., and Bhowmik, S. (1966). *J. Amer. Chem. Soc.*, **88**, 5430.

Kolthoff, M. I., Chantooni, M. K., and Bhowmik, S. (1968). *J. Amer. Chem. Soc.*, **90**, 23.

Kolthoff, I. M., Stocesoca, D., and Lee, T. S. (1952). *J. Amer. Chem. Soc.*, **75**, 1834.

Kosower, E. M. (1965). *Progr. Phys. Org. Chem.*, **3**, 81.

Kraft, J., Walker, S., and Magee, M. D. (1975). *J. Phys. Chem.*, **79**, 881.

Krakower, E. and Reeves, L. W. (1964). *Spectrochim. Acta*, **20**, 71.

Kraus, C. A. (1939). *Science*, **90**, 281.

Kreevoy, M. M. and Chang, K. C. (1976). *J. Phys. Chem.*, **80**, 259.

Kubota, T. (1966). *J. Amer. Chem. Soc.*, **88**, 211.

Kuhn, S. T. and McIntyre, J. S. (1965). *Canad. J. Chem.*, **43**, 995.

Kuroda, H., Amano, T., Ikemoto, I., and Akamatu, H. (1967). *J. Amer. Chem. Soc.*, **89**, 6056.

Laane, J. (1971). *J. Chem. Phys.*, **55**, 2514.

La Placa, S. J., Hamilton, W. C., Kamb, B., and Prakash, A. (1973). *J. Chem. Phys.*, **58**, 567.

Lassègues, J. C., Cornut, J. C., and Huong, P. V. (1971). *Spectrochim. Acta*, **27A**, 73.

Lautie, A. and Novak, A. (1971). *J. Chim. Phys.*, **68**, 1492.

Leuchs, M. and Zundel, G. (1978). *J. Phys. Chem.*, **82**, 1632.

Le-van, L. and Ratajczak, H. (1977). Personal communication.

Liler, M. (1972a). *J. Chem. Soc. Perkin II*, 816.

Liler, M. (1972b). *J. Chem. Soc. Perkin II*, 720.

Lindemann, R. and Zundel, G. (1977). *J. Chem. Soc. Faraday II*, 788.

Lindemann, R. and Zundel, G. (1978). *Biopolymers*, **17**, 1285.

Lippincott, E. R. and Schroeder, R. (1955). *J. Chem. Phys.*, **23**, 1099.

Lundgren, J. O. and Olovsson, I. (1976). *The Hydrogen Bond* (Eds. Schuster, P., Zundel, G., and Sandorfy, C., North-Holland, Amsterdam–New York.)

Lundgren, J. O. and Williams, J. M. (1973). *J. Chem. Phys.*, **58**, 788.

Lutskii, A. E., Gordienko, V. G., and Beilis, Y. I. (1976a). *Z. Obshch. Khim.*, **46**, 2335.

Lutskii, A. E., Klepanda, T. I., Sheina, G. G., and Batrakova, L. P. (1976b). *Zh. Obshch. Khim.* **46**, 1356.

Lutskii, A. E., Klepanda, T. I., Sheina, G. G., and Batrakova, L. P. (1976c). *Zh. Obshch. Khim.*, **46**, 1361.

Ma, J. C. N. and Warnhoff, E. W. (1965). *Canad. J. Chem.*, **43**, 1849.

McAdam, A. Currie, M., and Speakman, J. C. (1971). *J. Chem. Soc. A*, 1994.

Macau, J., Lamberts, L., and Huyskens, P. (1971). *Bull. Soc. Chim. France*, 2387.

Macdonald, A. L., Speakman, J. C., and Hadzi, D. (1972). *J. Chem. Soc. Perkin II*, 825.

Maes, G. (1978). Thesis, Leuven.

Mahillon, Ph. and Huyskens, P. (1978). To be published.
Malecki, J. (1976). *J. Chem. Soc. Faraday II*, **72**, 1214.
Martin, J. S. and Fujiwara, F. (1972). *J. Chem. Phys.*, **56**, 4098.
Marzocchi, M. P., Fryar, C. W., and Bambagiotti, M. (1965). *Spectrochim. Acta*, **21**, 155.
Mataga, N. and Kubota, T. (1970). *Molecular Interactions and Electronic Spectra*, M. Dekker, Inc., New York.
Mathias, A. and Gil, V. M. S. (1965). *Tetrahedron Lett.*, **35**, 3163.
Matsunaga, Y. and Inoue, N. (1972). *Bull. Chem. Soc. Japan*, 45.
Mayer, V., Gutmann, V., and Geiger, W. (1975). *Mh. Chem.*, **106**, 1235.
Molday, R. S. and Kallen, R. G. (1972). *J. Amer. Chem. Soc.*, **94**, 6739.
Moreno-Gonzales, E., Marraud, M., and Neel, J. (1977). *J. Chim. Phys.*, **74**, 563.
Morishima, I., Yoshikawa, K., Okada, K., Yonezawa, T., and Goto, K. (1973). *J. Amer. Chem. Soc.*, **95**, 165.
Moulik, S. P., Ray, S., and Das, A. R. (1976). *J. Phys. Chem.*, **80**, 157.
Mulliken, R. S. and Person, W. B. (1969). *Molecular Complexes*, Wiley–Interscience, New York–London.
Nagakura, S. and Baba, H. (1952). *J. Amer. Chem. Soc.*, **74**, 5693.
Nasielski, J. and Van der Donckt, E. (1963). *Spectrochim. Acta*, **19**, 1989.
Nawrot, C. F. and Veis, A. (1970). *J. Amer. Chem. Soc.*, **92**, 3903.
Nead, D. J., Fuoss, R. M., and Kraus, C. A. (1939). *J. Amer. Chem. Soc.*, **61**, 3257.
Neerinck, D. and Lamberts, L. (1966). *Bull. Soc. Chim. Belg.*, **75**, 473.
Nelson, I. V. and Iwamoto, R. (1964). *Electroanal. Chem.*, **7**, 218.
Neven, L. (1978). Thesis, Leuven.
Nouwen, R. and Huyskens, P. (1973). *J. Mol. Struct.*, **16**, 459.
Novak, A. (1972). *J. Chim. Phys.*, **69**, 1615.
Novak, A. (1974). *Structure and Bonding*, **18**, 177.
Odinokov, S. E. and Iogansen, A. V. (1972). *Spectrochim. Acta*, **28A**, 2343.
Odinokov, S. E., Mashkovsky, A. A., Dzizenko, A. K., and Glazunov, V. P. (1975). *Spectrosc. Lett.*, **8**, 157.
Odinokov, S. E., Mashkovsky, A. A., Glazunov, V. P., Iogansen, A. V., and Rassadin, B. V. (1976). *Spectrochim. Acta*, **32A**, 1355.
Olovsson, I. and Jönsson, P. (1976). In *The Hydrogen Bond*, Vol. II, *Structure and Spectroscopy* (Eds. Schuster, P., Zundel, G., and Sandorfy, C.), North-Holland, 395.
Onsager, L. (1927). *Physik. Z.*, **28**, 277.
Onsager, L. (1936). *J. Amer. Chem. Soc.*, **58**, 1486.
Parbhoo, D. M., Deenadayalan, K. C., and Bayles, J. W. (1975). *J. Chem. Soc. Perkin II*, 1057.
Pauwels, H. and Huyskens, P. (1974). *Bull. Soc. Chim. Belg.*, **83**, 427.
Pawlak, Z. (1973). *Roczn. Chem.*, **47**, 347.
Pawlak, I. and Sobczyk, L. (1973). *Adv. Mol. Relax. Proc.*, **5**, 99.
Pearson, R. G. and Vogelsong, D. C. (1958). *J. Amer. Chem. Soc.*, **80**, 1038.
Perrier-Datin, A., Julien-Laferrière, S., and Lebas, J. M. (1973). *J. Chim. Phys.*, **70**, 1684.
Person, W. B., Anderson, G. R., Fordemwalt, J. N., Stammreich, H., and Forneris, R. (1961). *J. Chem. Phys.*, **35**, 908.
Pilyugin, V. S. (1975). *Zh. Obshch. Khim.*, **45**, 136.
Pimentel, G. C. and McClellan, A. L. (1960). *The Hydrogen Bond*, W. H. Freeman.
Pimentel, G. C. and McClellan, A. L. (1971). *Ann. Rev. Phys. Chem.*, **22**, 347.
Pirson, D. and Huyskens, P. (1974). *J. Solut. Chem.*, **3**, 503.
Pople, J. A. (1964). *Mol. Phys.*, **7**, 301.

Poskin, G. and Huyskens, P. (1976). *Bull. Soc. Chim. France*, 337.
Prout, C. K. and Kamenar, B. (1973). *Molecular Complexes* (Ed. Foster, R.) Elek. Science, London, p. 151.
Pshenichnov, E. A. and Sokolov, N. D. (1965). *Kinetika i Kataliz*, **6**, 724.
Quick, A., Williams, D. J., Borah, B., and Wood, J. L. (1974). *J. Chem. Soc. Chem. Commun.*, 891.
Ralph III, E. and Gilkerson, W. R. (1964). *J. Amer. Chem. Soc.*, **86**, 4783.
Ratajczak, H., Orville-Thomas, W. J., and Choloniewska, I. (1977). *Chem. Phys. Lett.*, **45**, 208.
Ratajczak, H. and Sobczyk, L. (1969). *J. Chem. Phys.*, **50**, 556.
Ratajczak, H. and Sobczyk, L. (1970). *Bull. Acad. Pol. Sci. Ser. Sci. Chim.*, **18**, 93.
Reed, J. W. and Harris, P. M. (1961). *J. Chem. Phys.*, **35**, 1730.
Reichardt, C. (1965). *Angew. Chem.*, **77**, 30.
Reid, C. (1959). *J. Chem. Phys.*, **30**, 182.
Reynolds, W. F., Peat, I. R., Freedman, M. H., and Lyerla, J. R. (1973). *J. Amer. Chem. Soc.*, **95**, 328.
Robertson, J. M. and Ubbelohde, A. R. (1939). *Proc. Roy. Soc. (London) A*, **170**, 222.
Robinson, E. A., Schreiber, H. D., and Spencer, J. N. (1972). *Spectrochim. Acta*, **28A**, 397.
Robinson, R. A. and Stokes, R. H. (1959). *Electrolyte Solutions*, Butterworths, London.
Romanowska, K., Ratajczak, H., and Orville-Thomas, W. J. (1977). *Adv. Mol. Relax. Proc.*
Romanowski, H. and Sobczyk, L. (1975). *J. Phys. Chem.*, **79**, 2535.
Romanowski, H. and Sobczyk, L. (1977). *Chem. Phys.*, **19**, 361.
Rospenk, M., Koll, A., and Sobczyk, L. (1977). *Adv. Mol. Relaxation and Interaction Proc.*, **11**, 129.
Roy, R. S. (1966). *Spectrochim. Acta*, **22**, 1877.
Rundle, R. E. (1964). *Physique*, **25**, 487.
Sabin, J. R. (1968). *J. Quantum Chem.*, **2**, 23.
Sabin, J. R. (1972). *J. Mol. Struct.*, **11**, 33.
Sarneski, J. E., Surprenant, H. L., and Reilley, C. N. (1975). *Anal. Chem.*, **47**, 2116.
Sarneski, J. E., Suprenant, H. L., and Reilley, C. N. (1976). *Spectrosc. Lett.*, **9**, 885.
Savelev, V. A. and Sokolov, N. D. (1970). *Teor. Eksp. Khim.*, **6**, 174.
Schleich, T., Rollefson, B., and von Hippel, P. H. (1971). *J. Amer. Chem. Soc.*, **93**, 7070.
Schlemper, E. O. and Hamilton, W. C. (1966). *J. Chem. Phys.*, **44**, 4498.
Schneider, W. G., Bernstein, H. J., and Pople, J. A. (1958). *J. Amer. Chem. Soc.*, **80**, 3497.
Schuster, P., Jakubetz, W., Beier, G., Meyer, W., and Rode, B. M. (1974). *Jerusalem Symposia on Quantum Chemistry and Biochemistry*, **VI**, 257.
Scott, R., De Palma, D., and Vinogradov, S. (1968). *J. Phys. Chem.*, **72**, 3192.
Scott, R. and Vinogradov, S. (1969). *J. Phys. Chem.*, **73**, 1890.
Singh, T. R. and Wood, J. L. (1969). *J. Chem. Phys.*, **50**, 3572.
Smith, I. C. and Schneider, W. G. (1961). *Canad. J. Chem.*, **39**, 1158.
Smith, J. W. and Vitoria, M. C. (1968). *J. Chem. Soc. A*, 2468.
Snijder, W. R., Schreiber, H. D., and Spencer, J. N. (1973). *Spectrochim. Acta*, **29A**, 1225.
Sobczyk, L., Engelhardt, H., and Bunze, K. (1976). *The Hydrogen Bond* (Eds. Schuster, P., Zundel, G., and Sandorfy, C.), North-Holland, Amsterdam–New York, p. 937.

Sobczyk, L. and Pawelka, Z. (1974). *J. Chem. Soc. Faraday I.*, **70**, 832.

Somorjai, R. L. and Hornig, D. F. (1962). *J. Chem. Phys.*, **36**, 1980.

Speakman, J. C. (1972). *Structure and Bonding*, **12**, 141.

Spencer, J. N., Casey, G. J., Buckfelder, J., and Schreiber, H. D. (1974). *J. Phys. Chem.*, **78**, 1415.

Stammreich, H., Forneris, R., and Tavares, Y. (1961). *Spectrochim. Acta*, **17**, 1173.

Steigman, J. and Cronkright, W. (1970). *J. Amer. Chem. Soc.*, **92**, 6729.

Steigman, J. and Lorenz, P. M. (1966). *J. Amer. Chem. Soc.*, **88**, 2083.

Strømme, K. O. (1959). *Acta Chem. Scand.*, **13**, 268.

Sucharda-Sobczyk, A. and Sobczyk, L. (1978). *Bull. Acad. Pol. Sci.*, **26**, 549.

Szwarc, M. (1972). *Ions and Ion Pairs in Organic Reactions*, Wiley–Interscience.

Tam, J. W. O. and Klotz, I. M. (1973). *Spectrochim. Acta*, **29A**, 633.

Taylor, J. C. and Sabine, T. M. (1972). *Acta Cryst.*, **B28**, 2340.

Tellgren, R. and Olovsson, I. (1971). *J. Chem. Phys.*, **54**, 127.

Tenzer, L., Frazer, B. C., and Pepinsky, R. (1958). *Acta Cryst.*, **11**, 505.

Toyoda, K. and Nagakura, S. (1966). *J. Amer. Chem. Soc.*, **88**, 3905.

Treiner, C. and Justice, J. C. (1971). *J. Chim. Phys.*, **68**, 56.

Uruska, I. and Szpakowska, M. (1972). *Bull. Acad. Pol. Sci.*, **20**, 449.

Van Audenhaege, A. (1970). *Thesis*, Leuven.

Vinogradov, S. N. and Linnell, R. H. (1971). *Hydrogen Bonding*, Van Nostrand Reinhold.

Walden, P. (1906). *Z. Phys. Chem.*, **55**, 207.

Walden, P. and Birr, E. J. (1933). *Z. Phys. Chem.*, **A163**, 263.

Weast, R. C. (1967). *Handbook of Chemistry and Physics*, The Chemical Rubber Co., Cleveland.

Wheland, G. W. (1955). *Resonance in Organic Chemistry*, John Wiley.

Whidby, J. F. and Morgan, W. R. (1973). *J. Phys. Chem.*, **77**, 2999.

White, S. C. and Thompson, H. W. (1966). *Proc. Roy. Soc.* (*London*) A, **291**, 460.

Witschonke, C. R. and Kraus, C. A. (1947). *J. Amer. Chem. Soc.*, **69**, 2472.

Yada, M., Tanaka, J., and Nagakura, S. (1962). *J. Mol. Spectrosc.*, **9**, 461.

Zeegers-Huyskens, Th. (1967). *Spectrochim. Acta*, **23A**, 855.

Zeegers-Huyskens, Th. (1968). *Ind. Chim. Belg.*, **33**, 525.

Zeegers-Huyskens, Th. (1969). *Thesis*, Louvain.

Zeegers-Huyskens, Th. (1975). *J. Mol. Struct.*, **26**, 329.

Zeegers-Huyskens, Th. (1976). *Ann. Soc. Sci. Brux.*, **90**, 263.

Zeegers-Huyskens, Th. (1977). *Bull. Soc. Chim. Belg.*, **86**, 921.

Zimmerman, H. (1964). *Angew. Chem. Intern. Ed. Engl.*, **3**, 157.

Zundel, G. and Nagyrevi, A. (1978). *J. Phys. Chem.*, **82**, 685.

Molecular Interactions, Volume 2
Edited by H. Ratajczak and W. J. Orville-Thomas
© 1980 John Wiley & Sons Ltd

2 The kinetics of exchange and proton transfer processes in hydrogen-bonded systems in inert media

G. S. Denisov, S. F. Bureiko, N. S. Golubev, and K. G. Tokhadze

2.1 Introduction

The traditional approach to the investigation of hydrogen bonding is the study of static characteristics of complexes. The general tendencies in the development of this approach are connected with the study of optical and NMR spectra which makes it possible to determine geometry, electrooptical and thermodynamic characteristics of complexes, and to obtain information on parameters of the potential surface of proton donor and proton acceptor interaction (Schuster *et al.*, 1976). Among the conventional problems, that of correlation between electronic structure of partner molecules, on the one hand, and thermodynamic and spectral properties of hydrogen-bonded systems, on the other, is of significance.

In treating some static characteristics of complexes the question of dynamic properties of hydrogen-bonding already arises, the possible influence of fast exchange processes upon the shape of bands in optical and NMR spectra being of interest. The study of the dynamics of systems with hydrogen bonding is interesting in itself, since hydrogen bonding plays a decisive role in the kinetics of a number of processes, in particular, the processes of proton transfer and proton exchange. By investigation of the kinetics of these processes we may gain information on the values of the barriers separating the minima on the potential energy surface, and give answers to some fundamental questions concerning the nature and spectral properties of hydrogen bonding. Because of high speed of the reactions at room temperature, most of the problems referring to the kinetics of hydrogen bonded systems depend, to a large extent, on the development of special techniques for the study of fast reactions (Hammes, 1974).

Although the interest in hydrogen bond dynamics has increased considerably in recent years (a number of papers on measurements of the lifetime of complexes in a condensed phase testify to that: Rassing, 1975; Tabuchi, 1976; Hopmann, 1976; Denisov and Golubev, 1978) yet it is still not clear, how to approach these problems. Besides, most results have been obtained by studying concentrated solutions and pure liquids, in cases when the interaction with the surrounding medium, which influence the mechanism of the process qualitatively, not only cannot be excluded but becomes decisive.

This fact makes investigation of the dynamics of hydrogen-bonded systems expedient under conditions of minimum interaction with the surroundings, i.e. in the gas phase, or at a low concentration in inert solvents whose energy of interaction with the molecules under investigation is apparently lower than the energy of interaction between the partner molecules. The kinetic study of processes in solvents, whose molecules do not stimulate electrolytic dissociation, allows one to neglect consideration of acid–base catalysis and is of greatest interest for finding the mechanism of the initial interaction between the molecules concerned. However, it is not improbable that even a neutral solvent may play a vital role in the kinetics of processes involving hydrogen bonding. Therefore, it would be desirable to carry out experiments under conditions where it is possible to separate the influence of the surrounding medium from that of the partner molecules, on the dynamics of hydrogen bonding. Interesting conclusions may be drawn by comparing the results of experiments in solution, where the energy exchange between the molecules of a solvent and complexes is almost uninterrupted, with the data obtained in the gaseous phase.

The kinetic characteristics of hydrogen bonding, a well as the rates of conventional chemical reactions, depend largely on the kind of potential surface of interaction between a proton donor and a proton acceptor. As a generalization, one can consider three types of processes occurring in hydrogen-bonded systems.

2.1.1 *Formation and breaking of hydrogen bonds*

The process of complex formation in the gaseous phase requires no activation energy and, apparently, proceeds with the participation of a third particle:

$$AH + B + M \underset{k_{diss}}{\overset{k_{ass}}{\rightleftharpoons}} AH \cdots B + M$$

The particle M removes the excess energy, consisting of the energy of the hydrogen bond ΔH_0 and part of the kinetic energy of the colliding partners AH and B; the probability of transfer of this energy onto the vibrational degrees of freedom of the $AH \cdots B$ complex is very small, as in the processes

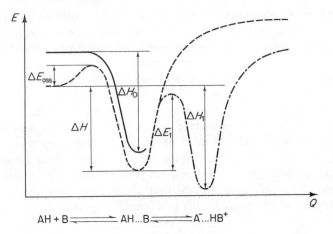

Figure 2.1 Reaction path profiles for the formation of a complex containing a hydrogen bond: in the gaseous phase (——); in solution (– – – –). Reaction path profile for proton transfer (– · – · – · –)

of recombination of free radicals and atoms. Interaction with a solvent affects the thermodynamic characteristics of a complex ($\Delta H_0 \neq \Delta H$, Figure 2.1), as well as the lifetime of an individual hydrogen bond, the lifetime in solution being decreased apparently due to a more effective energy transfer in the condensed phase. In this case complex formation does require some activation energy ΔE_{ass}, which is obtained by the interaction of donor and acceptor molecules with those of the solvent. Directly connected with these processes is the process of molecular exchange between hydrogen-bonded complexes:

$$AH \cdots B + A'H \cdots B' \rightleftarrows AH \cdots B' + A'H \cdots B,$$

AH and A'H, B and B' being either the same, or different molecules.

2.1.2 Reversible proton transfer inside a complex, or 'molecular–ionic tautomerism'

Reversible transfer of a proton inside a hydrogen-bonded complex, $AH \cdots B \rightleftarrows A^- \cdots HB^+$, occurs in systems having two minima on the potential surface of proton donor–proton acceptor interaction, separated by a barrier ΔE_1 and corresponding to two kinds of a complex, ionic and molecular, which are in equilibrium with free molecules. Such states exist for only a limited number of systems. In the case of one-minimum potential surface, it is possible to observe the process of setting up of an equilibrium either between free molecules and an ionic pair, $AH + B \rightleftarrows A^- \cdots HB^+$

(here, there is no molecular complex), or between free molecules and a molecular complex, $AH + B \rightleftarrows AH \cdots B$. In some cases a complex has an intermediate composition. Investigation of the structure of complexes and the types of equilibria in systems with different values of proton donor and proton acceptor ability under static conditions makes it possible to obtain an idea of the shape of the potential surface and to determine the depth of the minima, ΔH and ΔH_1. The value of the potential barrier, ΔE_1, determining the rate of transfer from one state to the other, may be obtained from kinetic measurements only. It must be emphasized that up to the present day no one has succeeded in recording the equilibria between a molecular complex and an ionic pair in the gaseous phase. Therefore, in this respect, the question of the influence of a medium on the parameters of the surface and the kinetics of the process acquires special importance.

2.1.3 *Proton exchange between molecules or complexes*

The decisive role of hydrogen bonding in proton exchange kinetics $AH + BH^* \rightleftarrows AH^* + BH$, between molecules capable of acting both as proton donor and as proton acceptor, has been indicated in a number of papers (Brodsky, 1964, Denisov and Tokhadze, 1973; Bureiko *et al.*, 1976, 1977b). A study of the kinetics of this process provides some information on proton transfer in systems having one minimum on the potential surface, belonging to a molecular complex, $AH \cdots B$. As a rule, proton exchange is a complicated process involving several stages, in particular, the process of complex formation. In the gaseous phase and in dilute solution in inert solvents, proton exchange between individual molecules occurs in hydrogen-bonded complexes. In (see section 2.4) the simplest case, the first step of exchange is cooperative proton transfer in a cyclic hydrogen-bonded complex. The existence of a minimum on the potential surface of interaction between AH and BH, referring to a cyclic complex with two hydrogen bonds, $AH \cdots B$ and $BH \cdots A$ (A and B may be the same), indicates the existence of another minimum, which results from the symmetry (Figure 2.2). These two minima refer to two physically indistinguishable states of a complex, and the transfer of a proton from one state to the other requires some activation energy, $\Delta E^{\neq}$ (Denisov and Tokhadze, 1973).

The present chapter discusses the results of experimental investigations of the kinetics of these processes in inert solvents and in the gaseous phase. Results have been obtained mainly by means of NMR and kinetic IR spectroscopy. As to the theoretical aspects of the problem, they have not been adequately developed so far. The few papers referring to these problems are concerned with the simplest model processes. Here, special attention is devoted to the influence of the electronic structure of the interacting molecules on the rate of the process; and comparison is made of static and dynamic characteristics of hydrogen bonding.

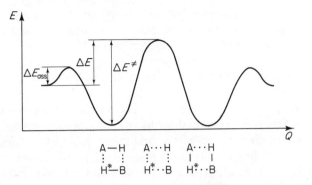

Figure 2.2   Reaction path profile for proton exchange
(the case of cooperative proton transfer)

2.2 Lifetimes of Hydrogen-Bonded Complexes

Experimental study of the kinetic characteristics of hydrogen-bonded complexes is of great importance for understanding some fundamental questions connected with the nature of hydrogen bonding and its spectral manifestations. The main kinetic characteristic of a complex is its lifetime. The lifetime of an individual complex can be determined by various processes of molecular exchange between complexes. Such processes may be, for example, the exchange of proton acceptor molecules:

$$AH \cdots B' + B'' \rightleftharpoons AH \cdots B'' + B'; \qquad (2.1)$$

or proton donor molecules:

$$AH' \cdots B + AH'' \rightleftharpoons AH'' \cdots B + AH'; \qquad (2.2)$$

B' and B'', AH' and AH'' being either different, or identical molecules. Some exchange processes occur with the help of self-associates, e.g.:

$$(AH)_2 + 2B \rightleftharpoons 2AH \cdots B, \qquad (2.3)$$

or

$$2AH \rightleftharpoons (AH)_2 \overset{AH}{\rightleftharpoons} (AH)_3 \rightleftharpoons \cdots \rightleftharpoons (AH)_n \qquad (2.4)$$

The lifetimes of complexes, determined by these equilibria, have a complicated dependence on the concentrations of species. Of particular importance is the lifetime, determined by the monomolecular dissociation process:

$$AH \cdots B \rightleftharpoons AH + B, \qquad (2.5)$$

which is a characteristic of the kinetic stability of the $AH \cdots B$ complex. This value could be found by studying more complicated processes such as (2.1) or (2.2), provided the simple process (2.5) was one of the stages.

The lifetimes of hydrogen-bonded complexes have been measured by various techniques—ultrasonic relaxation and dynamic NMR being most effective. Investigation of hydrogen-bonded systems by ultrasonic methods shows that in most cases, at room temperature, the rate of hydrogen bond formation is close to the rate of molecular collisions in liquids and is limited by diffusion. As a rule, the bimolecular rate constants, k_{ass}, are within the range $10^8–10^{11}\,l\,mol^{-1}\,s^{-1}$. The fact that they are less than the diffusion limit may be accounted for by the entropy factor, since this process requires the collision of definitely orientated molecules. Such rates were observed for the dimerization of benzoic acid (Maier, 1960), caprolactam (Bergman *et al.*, 1963), pyridone (Hammes and Lillford, 1970), and other compounds. The activation energy in all these systems does not exceed $\Delta E_{ass} \sim 1–3\,kcal\,mol^{-1}$, and coincides with the activation energy of diffusion. To a first approximation, the rate of dissociation of complexes is determined by the energy of hydrogen bonding, and generally decreases as the energy increases, with the lifetime of the complex getting longer. For example k_{diss} of the $NH \cdots O$ bond in 2-pyridone is less than that of the $NH \cdots S$ bond in 2-thiopyridone $(2.2 \times 10^7$ and $4.4 \times 10^7\,s^{-1})$ due to the greater strength of the former complex. The difference in the activation energies for dissociation and association coincides, as a rule, with the energy of the complex measured under static conditons. For intramolecular hydrogen bonds, like $OH \cdots O{=}C$ in salicylates and in salicylic aldehyde, the process

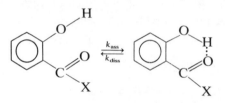

is monomolecular, and for $X = H$, OCH_3 $k_{ass} = 2.2 \times 10^7$, 2.6×10^7, and $k_{diss} = 3 \times 10^5$, $9.5 \times 10^5\,s^{-1}$, respectively (Yasunaga *et al.*, 1969).

In some cases, ultrasonic methods can be used for a more detailed investigation of the mechanism of the monomer–dimer relaxation. The results obtained for acetic acid solution (Carsaro and Atkinson, 1971) are in accordance with the assumption of successive breaking of two hydrogen bonds, i.e. monomolecular transformation of a cyclic dimer into an open dimer with one hydrogen bond, followed by dissociation of the open dimer:

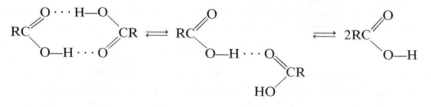

Carsaro and Atkinson (1971) suggested that the rate of cyclic complex formation, $k_{ass} = 3.4 \times 10^6\,s^{-1}$, is determined by the internal rotation around the hydrogen bond, and depends upon viscosity. Analysis of the ultrasonic absorption in the range 300–1500 MHz has shown that the lifetime of the open dimer is $\tau = 2 \times 10^{-10}\,s$ (Bader and Plass, 1971). The step-by-step mechanism of dissociation has been found also for propionic acid in cyclohexane solution at concentrations of 0.3–1 mol l^{-1}, the rate constants being $k_{diss} \sim 10^4\,s^{-1}$ and $k_{ass} \sim 10^7\,s^{-1}$ (Tatsumoto *et al.*, 1972).

A study of the kinetics of hydrogen bonding in various solvents has led to some interesting conclusions. In a more active solvent, the activation energy of dissociation of the complex, ΔE_{diss}, decreases, the value of ΔE_{ass} remaining unchanged. The decrease of ΔE_{diss} results in an increase in the dissociation rate constant. Thus for caprolactam at 22 °C in CCl$_4$ $k_{diss} = 4.6 \times 10^7\,s^{-1}$, and in, C$_6H_6$, $k_{diss} = 26 \times 10^7\,s^{-1}$ (Bergman *et al.*, 1963; Maeyer *et al.*, 1968). Consequently, a decrease of the equilibrium constant takes place, as is well known from static measurements. This statement may be illustrated by the results for benzoic acid in various solvents: C$_6$H$_{12}$, CCl$_4$, C$_6$H$_5$Cl, C$_6$H$_5$CH$_3$ (Borucki, 1967). For these solvents k_{diss} increases from 0.22×10^6 to $3.3 \times 10^6\,s^{-1}$, and ΔH_{diss} decreases from 13.9 to 9.9 kcal mol^{-1}. In this series the solvent activity rises according to the rise of its proton acceptor ability. The energy of a hydrogen bond measured under static conditions is equal to the difference in energies of the initial and final states: $\Delta H = (\Delta H_0 + E_{AH\cdots B}^S) - (E_{AH}^S + E_B^S)$, where $\Delta H_0 + E_{AH\cdots B}^S$ is the total energy of the complex in a solvent, $(E_{AH}^S + E_B^S)$ is the energy of the initial state, corresponding to the free molecules of donor and acceptor in solution, ΔH_0 is the energy of the hydrogen bond in the gaseous phase, $E_{AH\cdots B}^S$, E_{AH}^S, and E_B^S are the interaction energies of the complex, free donor, and acceptor molecules with a solvent (Figure 2.3). Since the active centres of the complex are blocked ($E_{AH\cdots B}^S$ in different solvents are very close), the total

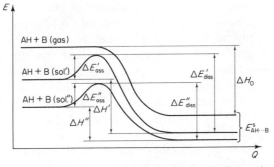

Figure 2.3 Reaction path profile for the formation of a complex, AH . . . B, in the gaseous phase, and in various solvents

energy of a complex remains almost unchanged, on its transfer from gaseous to liquid phase, or in different solvents. The interaction energy of free molecules grows with the growth of the solvent's activity, which results in $\Delta H'' < \Delta H' < \Delta H_0$. The value, ΔE_{ass}, is almost unaffected by the medium, which means that interaction with the medium has existed in the transitional state.

It is necessary to emphasize that, in ultrasonic experiments, there is some degree of uncertainty in establishing correspondence between the absorption observed and the process under investigation. The registered dependence of the absorption coefficient on the frequency permits determination of the relaxation time only in those cases where one process is going on. However, most results are available for concentrated solutions and pure liquids, where more complicated processes are possible.

In the liquid phase, due to the high rate of energy transfer, the mechanism of the processes discussed above cannot be studied in detail. In particular, a study of the role of triple collisions and of intramolecular energy redistribution during the formation of hydrogen-bonds is possible only in the gaseous phase. For this reason measurement of the lifetime of complexes in the gaseous phase is of special interest. An effort has been made to study monomer–dimer relaxation in a gas by means of the shockwave technique using IR monitoring (Tokhadze, 1975; Gerasimov et al., 1981). This technique studies relaxation processes taking $\sim 10^{-6}$ s, at concentrations $\sim 10^{-5}$ mol l.$^{-1}$, by measuring the changes in intensity of IR absorption bands. Strongly hydrogen-bonded complexes have been used: cyclic dimers of trifluoroacetic and acetic acids ($\Delta H \sim 12$–14 kcal mol^{-1}) and mixed complexes $CF_3COOH \cdots O(C_2H_5)_2$ ($\Delta H = 8.4$ kcal mol^{-1}) and $CF_3CF_2CH_2OH \cdots N(CH_3)_3$ ($\Delta H = 8.8$ kcal mol^{-1}). Unfortunately, in all the systems within the temperature range 40–100 °C, the decay of the complexes comes to an end at the front of the shock wave, and an estimate of the upper limit of the lifetimes, $\tau < (1$–$2) \times 10^{-6}$ s, only, is possible. However, even such estimates suggest a possible mechanism for the process in the gas phase. Thus in the case of cyclic dimers, the rate of dissociation observed can be explained only on the supposition that the two hydrogen bonds do not break simultaneously. This is in accordance with the ultrasonic absorption data, where two relaxation processes are observed, corresponding to the conversion of the cyclic dimer to an open dimer, and of the open dimer to the monomer. The high rate of dissociation of a single hydrogen bond ($\Delta H \sim 9$ kcal mol^{-1}) may be explained in terms of the model, taking into account the consequent exitation of the energy levels of the complex, corresponding to low-frequency vibrations of the hydrogen bond, $\Delta \nu \sim 20$–100 cm^{-1}, the probabilities of the vibrational translational energy transfer for the low-frequency vibrations of polyatomic molecules being used in the model.

Dynamic nuclear magnetic resonance (DNMR) is one of the most convenient and reliable experimental methods for studying the kinetics of fast exchange processes (see, for example, Jackman and Cotton, 1975). This method can be used to investigate reactions within a large time range (1–10^{-8} s, if pulse techniques are used), and is of considerably higher sensitivity compared to the ultrasonic method. Indeed, at present, the Fourier-transform ^{1}H NMR spectra suitable for carrying out the lineshape analysis can be obtained easily at concentrations of about 10^{-4} mol l^{-1}. One of the benefits of DNMR is that the composition of an equilibrium mixture and the structure of the molecular forms involved can be ascertained simultaneously. This makes it considerably easier to attribute the rate measured to a certain process. Proton chemical shifts are very susceptible to the formation of even weak hydrogen bonds. Strong hydrogen bonding is apt to shift the signal of the proton involved by up to 20 p.p.m., which exceeds the interval of proton chemical shifts (for molecules with no hydrogen bonding). Therefore, it would be natural to assume that the application of DNMR to a study of the kinetics of hydrogen bonding could be rather fruitful. However, attempts to determine the lifetimes of hydrogen-bonded complexes by DNMR in a direct way have not been successful until recently. Among the earlier works it is worth mentioning a number of papers by Grunwald and coworkers (see Grunwald and Ralf, 1971), where information on the kinetics of dissociation of a complex was obtained indirectly, through a study of proton exchange kinetics. Having assumed a certain mechanism, the authors evaluated the lifetimes of the complexes of ammonia and mono- and trimethylamine with water at ~ 300 K as 10^{-12}, 10^{-11}, and 10^{-10} s, respectively. This order of values is in accordance with the increase of bond strength in this series, although such a large effect is unexpected and can hardly be accounted for by the difference in hydrogen bond energy only. Since the lifetimes mentioned above are of the same order as the characteristic diffusion times in liquids, diffusion must play an important part in the kinetics of breaking and re-forming of these bonds. Fratiello et $al.$ (1973) found that at temperatures below $-100\,°$C the rate of exchange processes in systems with strong hydrogen bonding may be so retarded that separate signals belonging to different complexes are observed. Thus velocities of molecular exchange processes, under conditions where they are not limited by diffusion (i.e. the lifetime of the complexes being 10^{-6}–10^{-2} s), may be estimated.

The lifetimes of complexes determined by different kinds of equilibria were obtained recently by a study of separated signals. As mentioned above, the simplest equilibrium (2.5) is between a binary complex, AH $\cdots$ B, and molecules with no bonding. Unfortunately, study of such a simple process is possible only rarely, for most proton donor molecules are capable of self-association. At low temperature this leads to the formation of various complexes involving more than two molecules. Moreover, between molecules containing hydroxyl groups intensive proton exchange would occur which, as well as the process under investigation, could result in the averaging of signals. (In section 2.4 these phenomena will be shown to accompany each other, because both self-association and proton exchange are affected by the ability of an AH molecule to form hydrogen bonds as a proton donor and as a proton acceptor, simultaneously.) Therefore, it would be worth choosing a proton donor which had very little acceptor ability. Equilibrium (2.5) involving a complex of CHF_3 with triethylamine in liquid argon was described by Golubev *et al.* (1977b). In the ${}^1H$ NMR spectra, at 90 K, broadening of the resonance CHF_3 signal was seen, which was at a maximum when the binary complex and free CHF_3 molecule concentrations were comparable. This broadening was not observed in the absence of the acceptor, or if it were great excess. In the latter case equilibrium (2.5) was completely shifted to the left. Thus it was concluded that the linewidth was determined by the process of hydrogen bond breaking and re-forming. Estimation of the lifetime of the complex at 90 K gave the value $\tau = 7 \times 10^{-4}$ s, which is much greater than the characteristic diffusion time in liquid argon.

The complex of $(CF_3)_3COH$ with hexamethylphosphoroustriamide (HMPT) is another example (Golubev and Denisov, 1981). Since the acceptor ability of this alcohol is greatly lowered, at concentrations less than 10^{-2} mol l^{-1} no self-associates or complicated complexes have been revealed even at 110 K. Figure 2.4 shows the ${}^1H$ NMR spectra of a solution containing HMPT and $(CF_3)_3COH$ (the latter in excess) in CDF_2Cl. At 110 K, two signals are seen,

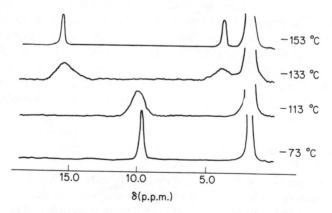

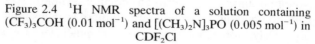

Figure 2.4 ${}^1H$ NMR spectra of a solution containing $(CF_3)_3COH$ (0.01 mol^{-1}) and $[(CH_3)_2N]_3PO$ (0.005 mol^{-1}) in CDF_2Cl

the high-field signal belonging to the free hydroxyl groups, and the low-field one attributed to a binary complex with HMPT. As relative concentrations are varied, so are their relative intensities, but not their chemical shifts. As the temperature rises, the signals broaden, overlap, and collapse. The averaging of the free molecule and complex signals could be caused either by the breaking and re-forming process (2.5), or by the bimolecular exchange (2.2), where AH' and AH'' are identical molecules, or by proton exchange between these forms. However, the lineshapes have been shown to be unaffected by dilution of a solution with a given $[AH]_0/[BH]_0$ concentration ratio. The authors consider this fact to be an argument in favour of the monomolecular process (2.5). Indeed, in this case,

$$\tau_{AH\cdots B} = k_{diss}^{-1}$$

$$\tau_{AH} = k_{ass}^{-1}[B]^{-1} = Kk_{ass}^{-1}\frac{[AH]}{[AH\cdots B]} \approx Kk_{ass}^{-1}\left(\frac{[AH]_0}{[B]_0} - 1\right) \qquad (2.6)$$

because equilibrium (2.5) is strongly shifted to the left, and, with an excess of AH, the concentration [B] is small compared to [AH] and [AH $\cdots$ B]. It is clear that information could be obtained only about the dissociation velocity, since finding k_{ass} would require a very accurate value of the equilibrium constant, K, to be known. The lifetime of the complex proved to depend strongly on temperature. Thus at 130 K, $\tau_{AH\cdots B} = 3.4 \times 10^{-2}$ s; at 150 K, $\tau_{AH\cdots B} = 4 \times 10^{-4}$ s; at 170 K, $\tau_{AH\cdots B} = 1.8 \times 10^{-5}$ s. The activation energy was found to be $\Delta E_{diss} = 8.7$ kcal mol^{-1}, which is close to the enthalpy of this complex (9.6 kcal mol^{-1} in CCl$_4$, Kuopio et al., 1976). It is worth emphasizing that if the data obtained were extrapolated to room temperature, the lifetime of the complex would be estimated as 10^{-9} s, which means that it is likely to be determined by diffusion.

The equilibrium between ortho-substituted phenols with strong intramolecular hydrogen bonding and their complexes with proton acceptors has been studied at 110–200 K in CHF$_2$Cl (Denisov et al., 1977). Intramolecular hydrogen bonding in a proton donor molecule hinders self-association and proton exchange, which makes it possible to observe the simple process (2.5). Figure 2.5 shows the spectra of a solution of o-nitrophenol and 2,4,6-trimethylpyridine. As the temperature falls, the hydroxyl signal is shifted to the lower field (which corresponds to the equilibrium shift towards the intermolecular hydrogen bonding) and, then, is split into two. With no acceptor, only the high-field signal is observed, while with excess of acceptor only the low-field signal is left. With an excess of the phenol the lifetimes of the complex measured are independent of concentrations, and the lifetimes of the intramolecular hydrogen bond are inversely proportional to the acceptor concentration. A two-stage mechanism was

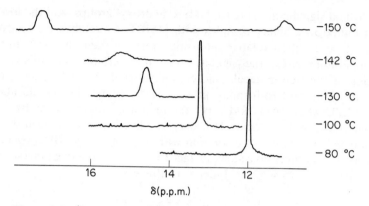

Figure 2.5 ^{1}H NMR spectra of the o-nitrophenol (0.06 mol l^{-1})
+ 2,4,6-trimethylpyridine (0.04 mol l^{-1}) system in CHF$_2$Cl

suggested, involving the total breaking of one hydrogen bond followed by
the formation of the other:

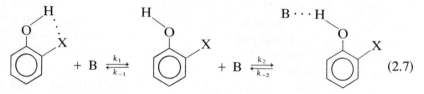

$$\tag{2.7}$$

The lifetimes of the forms with intra- and intermolecular bonding are

$$\tau_{\text{intra}} = k_1^{-1}\left(1 + \frac{k_{-1}}{k_2[\text{B}]}\right), \qquad \tau_{\text{inter}} = k_{-2}^{-1}\left(1 + \frac{k_2[\text{B}]}{k_{-1}}\right) \tag{2.8}$$

the observed reaction orders being in accordance with (2.8), if $k_2[\text{B}] \ll k_{-1}$.

If salicylic aldehyde were taken as proton donor, the kinetics of the
process could be studied using both OH and CHO group signals. The CHO
signal is very sensitive to the breaking of the intramolecular hydrogen bond,
due to the simultaneous rotation of this group through 180°:

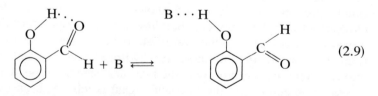

$$\tag{2.9}$$

The lifetimes measured by these two signals proved practically the same,
which confirms that proton exchange has no influence upon the spectra. The
activation parameters are given in Table 2.1. The difference between the
activation energy values of the forward and backward reactions is close to

Table 2.1   Enthalpy change ΔH and activation parameters ΔE_{diss}, ΔS_{diss} of the forward and backward reaction (2.9).*

No.	Phenol + acceptor	$-\Delta H$ (kcal mol^{-1})	ΔE_{diss}^{inter} (kcal mol^{-1})	ΔE_{diss}^{intra} (kcal mol^{-1})	ΔS_{diss}^{inter} (e.u.)	ΔS_{diss}^{intra} (e.u.)
1.	2-Nitrophenol + collidine	6.0±0.3	13.2±0.6	6.8±0.6	+20	+6
2.	2-Nitrophenol + hexamethapole	5.1±0.3	11.9±0.6	6.7±0.6	+19	+6
3.	2-Formylphenol + collidine	5.0±0.3	12.8±0.6	8.0±0.5	+17	+2.5
4.	2-Formylphenol + hexamethapole	3.2±0.3	11.4±0.7	8.1±0.6	+16	+2.3
5.	2-Acetylphenol + hexamethapole	−0.3±0.1	10.6±0.6	10.8±0.6	+19	+5

* Nos. 1–4 in CHF$_2$Cl, No. 5 in a mixture of C$_2$H$_5$Cl + CH$_2$Cl$_2$

the enthalpy of the process determined independently. In addition, the activation energy of breaking an intramolecular hydrogen bond is unchanged, when various acceptors are used, which confirms mechanism (2.7).

The hindered internal rotation in molecules of 2,6-disubstituted phenols, involving the breaking of intramolecular hydrogen bonds, was examined by Koelle and Forsen (1974), and Golubev and Denisov (1975):

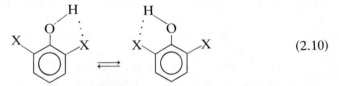

$$(2.10)$$

The substituents X, become non-equivalent when process (2.10) is slow (Figure 2.6), which makes it possible to use their signals for determining the rate. The activation parameters of the internal rotation, given in Table 2.2,

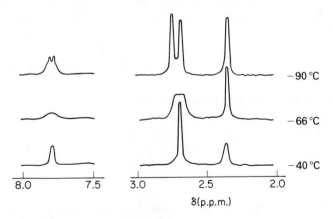

Figure 2.6 ^{1}H NMR spectra of 2,6-diacetyl-4-methylphenol in CHF$_2$Cl at various temperatures

Table 2.2   Kinetic characteristics of the hindred internal rotation in 2,6-disubstituted phenols

No.	Phenol	ΔE_{diss} (kcal mol^{-1})	ΔS_{diss} (e.u.)
1.	2,6-Dicarbomethoxyphenol	9.0 ± 0.4	-1.4 ± 0.5
2.	2,6-Diacetyl-4-methylphenol	10.5 ± 0.4	-2.0 ± 0.5
3.	2,6-Diformylphenol	7.1 ± 0.2	-21 ± 1

turn out to be close to the respective values for the OH $\cdots$ X bonds, breaking in the presence of an acceptor. Thus the conclusion that an acceptor molecule does not take part in the first stage of process (2.7) is supported.

Molecular exchange of the type (2.1) has been thoroughly investigated by means of DNMR under conditions of excess of one of the proton acceptors. This excess can often depress the formation of more complicated complexes. In the paper by Fratiello *et al.*, (1973), spectra of mixtures of $(CF_3)_2CHOH$ or p-FC$_6$H$_4$OH molecules with triethylamine in diethyl ether were obtained. The complexes of the donor with amine and ether were in equilibrium. The activation energy of the process leading to collapse of the signals was very high (~ 15 kcal mol^{-1}), which proves that proton exchange does not occur. The kinetics of molecular exchange (reaction 2.1) between complexes of carboxylic acids with different proton acceptors have been investigated, one of the acceptors being varied and the other being a molecule of diethyl ether (Golubev and Denisov, 1976). Splitting of the signals was observed in the temperature range 120–60 K. It was found that the lifetime of the complex, RCOOH $\cdots$ O(C$_2$H$_5$)$_2$, is independent of the nature of the second acceptor, and is determined by its concentration only. Therefore, it was supposed that this exchange is monomolecular, dissociation of the complex being the velocity-limiting stage:

$$AH \cdots B' + B'' \underset{k_{-1}}{\overset{k_1}{\rightleftarrows}} AH + B' + B'' \underset{k_{-2}}{\overset{k_2}{\rightleftarrows}} AH \cdots B'' + B' \qquad (2.11)$$

The rate of the forward reaction is

$$V = k_1[AH \cdots B'] \frac{k_2[B'']}{k_2[B''] + k_{-1}[B']} = k_1[AH \cdots B']\left(1 + \frac{k_1[B']}{Kk_{-2}[B'']}\right)^{-1}$$

$$= k_1[AH \cdots B']\left(1 + \frac{k_1}{k_{-2}} \frac{[AH \cdots B']}{[AH \cdots B'']}\right)^{-1}, \qquad (2.12)$$

where K is the equilibrium constant of the exchange (2.1). A similar equation may be written for the rate of the inverse process. The average lifetime of a molecular form X being equal to $[X]V^{-1}$, the monomolecular dissociation velocity constants k_1 and k_{-2} can be obtained from the lifetimes observed, $\tau_{AH\cdots B'}$, $\tau_{AH\cdots B''}$, by means of the equations:

$$\tau_{AH\cdots B'} = k_1^{-1}\left(1 + \frac{k_1[AH\cdots B']}{k_{-2}[AH\cdots B'']}\right),$$

$$\tau_{AH\cdots B''} = k_{-2}^{-1}\left(1 + \frac{k_{-2}[AH\cdots B'']}{k_1[AH\cdots B']}\right). \tag{2.13}$$

The values k_1 and k_{-2} can be determined from the plots $\tau_{AH\cdots B'} = f([AH\cdots B']/[AH\cdots B''])$. The lifetimes of complexes determined by hydrogen bond dissociation and formation processes, are $t_{AH\cdots B'} = k_1^{-1}$; $t_{AH\cdots B''} = k_{-2}^{-1}$. In Table 2.3 t values of HCOOH and CF_3COOH complexes with various acceptors, at 170 K, are given together with activation parameters of the exchange. Denisov *et al.* (1976) emphasized that the activation entropy values are positive, resulting in very high values of the pre-exponential factors in the Arrhenius equation (10^{15}–$10^{18}\,s^{-1}$). This means that the hydrogen bond in an activated complex is much weaker compared to the initial state. In the case of some bimolecular mechanisms, e.g.

$$AH\cdots B' + B'' \rightleftarrows AH : \genfrac{.}{.}{0pt}{}{\cdot B'}{\cdot B''} \rightleftarrows AH\cdots B'' + B'$$

a decrease of entropy in the transitional state would be expected rather than an increase.

Golubev (1977) examined the possible contribution of proton exchange to the observed averaging of signals of trifluoroacetic acid complexes when an excess of one of the two acceptors was present. The CHO signal in the $(CH_3)_2NCHO$ molecule appeared to be rather sensitive to the formation of a hydrogen bond with strong proton acceptors. The kinetics of the molecular

Table 2.3   Kinetic characteristics of hydrogen-bonded complexes in CHF_2Cl

No.	Acceptor	$t^{170\,K} \times 10^5$ (s)	ΔE_{diss} (kcal mol^{-1})	ΔS_{diss} (e.u.)
	1. Complexes of HCOOH			
1.	$(C_2H_5)_2O$	0.17 ± 0.03	7.6 ± 0.3	12.0 ± 0.6
2.	$(CH_3)_2SO$	0.9 ± 0.2	8.9 ± 0.6	16 ± 1
3.	$(CH_3)_2NCHO$	3.0 ± 0.5	9.6 ± 0.6	18 ± 1
4.	$[(CH_3)_2N]_3PO$	3.5 ± 0.7	10.2 ± 0.8	21 ± 1.5
	2. Complexes of CF_3COOH			
1.	$(CH_3)_2O$	0.6 ± 0.1	8.7 ± 0.6	16 ± 1
2.	$(C_2H_5)_2O$	1.7 ± 0.2	9.8 ± 0.3	20 ± 1
3.	$(CH_2)_4O$	5.6 ± 0.7	10.2 ± 0.8	20 ± 1
4.	$(CH_3)_2SO$	12 ± 2	10.8 ± 0.8	22 ± 1.5
5.	$(CH_3)_2NCHO$	16 ± 4	11.2 ± 1	24 ± 1.5
6.	$[(CH_3)_2N]_3PO$	43 ± 8	11.9 ± 1	26 ± 1.5

exchange (2.1) between the complexes of CF_3COOH with $(CH_3)_2NCHO$ and $(CH_3)_2O$ was studied, the lifetimes being determined using either OH or CHO signal shape. As the proton exchange cannot affect the CHO signal shape, the good correspondence of the times obtained proved the proton exchange rate in this case to be negligibly small. The step-like mechanism of the molecular exchange was confirmed.

Summing up, one can say, that the monomolecular process, $AH \cdots B \rightleftarrows AH + B$, while requiring comparatively high activation energy, is the most profitable way of breaking either inter- or intramolecular hydrogen bonds.

2.3 Reversible Proton Transfer Inside Hydrogen-Bonded Complexes

Proton transfer is the simplest and, at the same time, one of the most important chemical reactions. The study of this process, as well as of more complicated reactions, involving proton transfer as one of the stages, has attracted the attention of a great many investigators (Caldin and Gold, 1975). However, a similar process in inert solvents has not been well studied so far, probably, in view of its minor practical importance. Since reactions with the participation, or formation, of ions are, to a much greater extent, affected by the solvent in comparison with pure molecular processes, the acid–base interaction in inert media may differ considerably from the 'common' reaction in solvents strongly solvating the ions. It is known that, in this case, interaction between comparatively weak partners results in the formation of a molecular complex with a hydrogen bond, $AH \cdots B$, while interaction between strong partners results in the formation of a close ionic pair, $A^- \cdots HB^+$, with interionic hydrogen bonding. Dissociation of an ionic pair into free ions in solvents, like hydrocarbons and their halogen derivatives, is extremely disadvantageous. Thus the reaction is always a process of complex formation, with a hydrogen bond between molecules contributing much to the energy balance of the reaction. Hydrogen bonding affects the kinetics of proton transfer even more considerably.

The structure of the binary complex in the case of intermediate interaction energy and the gradual alteration of this structure with increasing complex formation energy are problems of special importance. These problems have been dealt with in a number of experimental works, carried out mainly by IR, UV, and NMR spectroscopy. It was in 1956 that Barrow, having analysed the IR spectra of complexes of pyridine with carboxylic acids in $CHCl_3$ in the region of skeleton vibrations, stated that the interaction with CH_3COOH proceeds no further than the formation of molecular complexes, but, with CCl_3COOH and CF_3COOH, results in proton transfer and the formation of an ionic pair. In the intermediate cases, $CH_2ClCOOH$ and $CHCl_2COOH$, the complex formed is tautomeric: $AH \cdots B \rightleftarrows A^- \cdots HB^+$. This conclusion was, at first, doubted (Davis, 1968), but later confirmed by Gusakova *et al.* (1970). In a number of papers (e.g. Vinogradov and Linnel, 1971) the assumption of a tautomeric proton transfer inside a complex was made in order to explain the structure of the stretching vibration band, ν_{XH}. However, the question of the shape of this band seems to be rather obscure in the spectroscopy of hydrogen bonding (Hadzi and Bratos, 1976). It may be determined by many factors having no direct relation to proton migration (Fermi resonance, interaction with low-frequency vibrations of hydrogen bond, etc.). At present, there is not a single case known when the ν_{AH} band structure observed could be referred without any doubt to proton tunnelling in a double-minimum potential well.

In later works it has been shown that reliable identification of a tautomeric structure of a complex is a difficult task requiring a study of each complex over a wide range of the spectrum. Equilibrium between a molecular complex and an ionic pair of $OH \cdots N \rightleftharpoons O^- \cdots HN^+$ type has been found from the IR spectra in the region of skeletal vibrations of partner molecules (Gusakova *et al.*, 1972, Lindemann and Zundel, 1977), in the region of low-frequency intermolecular vibrations of the hydrogen bond (Denisov *et al.*, 1973), from UV spectra (Baba *et al.*, 1969; Hudson *et al.*, 1972), and from NMR spectra at low temperatures (Golubev *et al.*, 1977a). In the last work, by the way, it has been stated that the frequency of the reversible proton transfer in the $OH \cdots N$ complexes is above $10^5 \, s^{-1}$, even at a temperature of $-170\,°C$. Proton transfer in complexes with $CH \cdots N$ bonding, studied by IR spectroscopy (Golubev and Denisov, 1977) and NMR spectroscopy (Golubev *et al.*, 1978), appeared to be much slower.

Thus the fact of proton migration in hydrogen-bonded complexes in inert media is not in doubt. It has been safely established for a comparatively large number of compounds, in particular, for complexes of acids of rather different types with aliphatic and aromatic amines. However, not even a single case of molecular–ionic tautomerism has been reported for a complex with an oxygen atom as a proton acceptor. In analysing some data obtained by a number of authors (Hadzi, 1965; Matrosov and Kabachnik, 1977), one can suppose that, in the case of complexes with strong $OH \cdots O$ hydrogen bonding, their structure is altered gradually with the rise of the interaction energy, from the molecular, $AH \cdots B$, to the ionic, $A^- \cdots HB^+$. In the intermediate case, the spectral characteristics of such complexes resemble those of 'symmetrical' hydrogen bonding which have been observed in complex ions, like FHF^-, and bimaleate ion, whose potential surface of interaction has only one minimum corresponding to the central position of the proton. Thus for such complexes (e.g. complexes of CF_3COOH with some oxides) a very low ν_{OH} frequency and a very high chemical shift of the OH proton are typical. A number of frequencies in these complexes are intermediate between those characteristic of the intrinsic molecular and ionic complexes. One may assume the structure of complexes with extremely strong manifestation of hydrogen bonding to be intermediate between molecular and ionic, $A^{\delta-} \cdots H \cdots B^{\delta+}$, with a 'quasi-symmetrical' hydrogen bond.

So, from the experimental data available, one may conclude that, as the interaction energy grows, the potential energy surface of proton donor–acceptor interaction may be altered at least in two ways. Figure 2.7 shows a schematic cross-section of this surface along the reaction coordinate (note, that this coordinate does not coincide with that corresponding to the position of the proton between the two nearest nuclei, since, on proton transfer, a change occurs in the bond lengths and the angles of A and B fragments). To the left of the hachured area there is a region, corresponding to a molecular complex; to the right is the region of ionic pairs. In case I, at low energy, only one minimum is seen. When interaction is intensified, another minimum appears, corresponding to the ionic pair, which deepens more rapidly than the first one (Denisov and Schreiber, 1974). At very high energy of complex formation, the potential barrier, dividing these two wells, disappears, and the complex formed will be of ionic structure. In case II, along with the deepening of the only well, a simultaneous shift to the region of an ionic complex occurs. In the intermediate case the hydrogen bond is quasi-symmetrical. The primary criterion of the type of proton potential function is to be found in equilibria observed in the system. However, if the presence of the molecular–ionic tautomerism proves the existence of at least two minima, the absence of an experimentally observed tautomerism is not sufficient proof of the absence of the second well. For example no equilibria have ever been reported for which an ionic pair would be more profitable

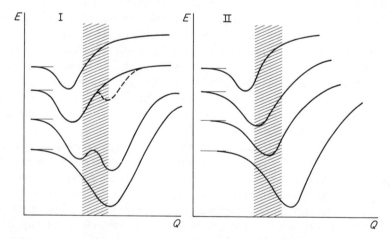

Figure 2.7   Two types of cross-sections of potential surfaces of interaction between a proton donor and a proton acceptor along the reaction coordinate Q

than a molecular complex. This is only natural, since the ionic complex with a large dipole moment, 10–15 D, strongly affects the arrangement of the solvent molecules around itself. This accounts for the fact that proton transfer is accompanied by a strong decrease in entropy (of 20–30 e.u.). Therefore, the equilibrium constant of the endothermal proton transfer would be comparatively small, hampering detection of an ionic pair. However, the existence of a second minimum, lying above the first, may strongly influence the kinetics of proton exchange (Bureiko *et al.*, 1979). To identify the potential function type correctly in cases where the tautomerism (i.e. two groups of bands related to molecular and ionic complexes, respectively) is not clearly seen in the spectra, it is necessary to turn to the finer spectroscopic features of the complexes, either with a migrating proton, or with a quasi-symmetrical hydrogen bond. These involve, for example, isotopic effects in IR and NMR spectra (Gunnarson *et al.*, 1976; Shapet'ko *et al.*, 1976), as well as a considerable decrease in the spin–spin coupling constants of the proton, $A^{\delta-} \cdots H \cdots B^{\delta+}$, with nuclei of the fragments, A and B, in comparison with the structures, AH and HB$^+$ (Fujiwara and Martin, 1974).

Where a double-well potential function exists, the question naturally arises of the frequency of proton migration inside a complex. As early as 1948, Sokolov made an assumption that this process is an elementary act of acid–base interaction. Formation of an intermediate hydrogen-bonded complex was considered to be the first stage of proton transfer, though such complex cannot always be observed immediately, in active solvents. However, repeated attempts to determine the rates of exothermal proton transfer inside any OH $\cdots$ N complex in studying the kinetics of interaction between two molecules AH and B, have failed. (It is worth mentioning that the rate of an analogous endothermal process is not so interesting in terms of kinetics, since its potential barrier cannot be lower than the energy of the process, and, consequently, its rate may be hindered by purely thermodynamic causes. Most reactions, known as 'reactions, controlled by proton transfer', are slow enough, since this transfer is unprofitable in terms of thermodynamics.)

Caldin *et al.* (1973) studied the kinetics of the interaction between a number of phenols and amines in chlorobenzene by means of microwave jump and detected the ionic pair by UV absorption. The velocities varied from 0.01 to 0.1 of the diffusion limit; the rate constants did not correlate with the equilibrium constants, and were determined by steric factors. This has led the authors to the conclusion that formation of an ionic pair is limited by the stage of hydrogen-bonded complex formation (which has been found by its UV absorption) rather than by proton transfer. The rate of this process is less than the diffusion limit due to an entropy factor, i.e. due to the need for collision of two definitely oriented polyatomic molecules. A similar conclusion was drawn by Ivin *et al.* (1971), in studying the influence of solvent viscosity on the kinetics of ion-pair formation from 2,4-dinitrophenol and amines.

Recently, the kinetics of proton transfer from free radicals, like **1**, to various amines, in a low-polar solvent, has been intensively investigated by means of ESR (Masalimov *et al.*, 1976, 1977a,b):

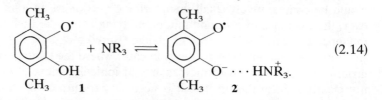

(2.14)

In molecule, **1,** superfine coupling with the OH proton can be observed, and this disappears in ionic pair, **2**. When both forms are present in solution, some broadening of the signal is observed, which can be used to determine the rate of the process. Unfortunately, the monomeric free radical, **1,** cannot be distinguished from the molecular hydrogen-bonded complex by ESR, and the authors have not discussed the possibility of its intermediate formation. Thus it is hard to say, to which stage the kinetic characteristics are related (specific rates being about $10^8 \, \text{l mol}^{-1} \text{s}^{-1}$). However, the activation parameters of the forward and backward reactions being rather close to the similar values for forming and breaking of complexes with strong hydrogen-bonding (section 2), one may suppose that it is this process which is rate limiting. This conclusion is supported by a weak dependence of the forward reaction velocity upon the basicity of the amines used.

Crooks and Robinson (1970, 1971) determined the rates of interaction between the bromphenol blue molecule and amines:

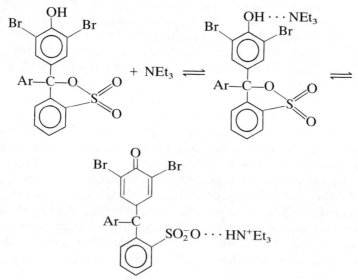

This process was found to proceed via two stages. The formation of a hydrogen-bonded complex (seen in the UV spectrum) was accomplished at the usual rate, while the proton transfer inside the complex was extremely retarded (down to $10^2 \, s^{-1}$!). The authors tried to explain this strange fact by the unfavourable orientation of the molecules in the reaction complex, and by the weak solvation of the transitional state. However, in this case, absorption in the visible spectral region cannot appear immediately on proton transfer, but appears only after the formation of a quinoide structure, which requires the sulton ring to be broken. Thus it is obvious, that the rates observed have no relation to proton transfer, but are determined by the C—O bond breaking.

We believe that so far there have been no direct measurements made of proton migration frequency inside any OH $\cdots$ N complex. In Grunwald's (1966) review an analysis of a good deal of indirect evidence concerning the rates of various processes involving extremely fast proton transfer has been made. The conclusion was drawn that, in some cases, the frequency of proton migration inside a complex can reach $10^{13} \, s^{-1}$. Indeed Kreevoy (1965), for example, observed broadening of some bands in Raman spectra of solutions, containing CF_3COOH and the CF_3COO^- ion, compared with the same bands in solutions of pure CF_3COOH or CF_3COONa. This could be accounted for by proton migration with a frequency of about $10^{13} \, s^{-1}$, in an extremely strong complex, $CF_3COOH \cdots \overline{O}COCF_3$. However, in the IR spectra of complexes of carboxylic acids with amines, referred to above, these bands are not broadened. Thus it can be stated that, in these cases, the frequency must be below $10^{12} \, s^{-1}$ (but above $10^9 \, s^{-1}$, since the bimolecular proton transfer is controlled by diffusion). It is obvious, that such an immense frequency value could only be reached either due to a very small potential barrier or a high probability of proton tunnelling. It may be supposed, that in complexes with a strong OH $\cdots$ N hydrogen bond both possibilities are combined. Indeed, the formation of a strong hydrogen bond must constrict strongly interacting molecules, which results in a lowering and narrowing of the barrier. It must be mentioned, that so far there has been no strict proof of proton tunnelling in OH $\cdots$ O or OH $\cdots$ N complexes.

In tautomeric complexes with weaker hydrogen bonding the proton migration frequency is much lower. Ulashkevich *et al.* (1977) and Golubev *et al.* (1980) have used ^1H NMR to study the kinetics of such a process in complexes with NH $\cdots$ N, SH $\cdots$ N, and CH $\cdots$ N hydrogen bonds. Figure 2.8 shows the spectrum of a solution containing CH_3NHNO_2 ($pK_a = 5$), and triethylamine in CHF_2Cl. (The occurrence of reversible proton transfer in the system is proved by IR spectra.) At low temperatures, splitting of the signal of the α-methylene group of the amine is observed, as well as that of the movable proton. The relative intensity and the lineshape of NH signals do not depend on concentration, therefore, the averaging of the signal involved is sure to be determined by the monomolecular process. The

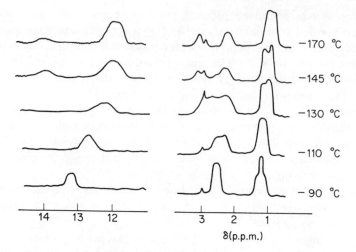

Figure 2.8 ^{1}H NMR spectra of a solution containing CH$_3$NHNO$_2$ (0.03 mol l^{-1}) and (CH$_3$CH$_2$)$_3$N (0.1 mol l^{-1}) in CHF$_2$Cl at various temperatures

low-field signal belongs to the NH group of the molecular complex (with excess amine, at low temperatures, there are practically no free CH$_3$NHNO$_2$ molecules). The high-field signal belongs to the NH$^+$ group of the ionic pair, its intensity increasing as the temperature decreases. The monomolecular rate constant for proton transfer ranges from 500 s^{-1}, at 120 K, to 30,000 s^{-1}, at 170 K. The spectrum of the thiophenol–triethylamine complex (Figure 2.9)

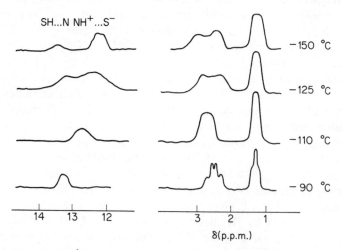

Figure 2.9 ^{1}H NMR spectra of a solution containing C$_6$H$_5$SH and (CH$_3$CH$_2$)$_3$N in CHF$_2$Cl at various temperatures

has a similar appearance, though the splitting of the $SH \cdots N \rightleftarrows S^- \cdots HN^+$ signal takes place at a higher temperature. The rate of proton transfer inside complexes with $CH \cdots N$ hydrogen bonding may be varied within a wide range, but, generally, it is lower than in the similar complexes of NH and SH acids. Table 2.4 gives data concerning complexes of different types of acids with triethylamine. These acids have very similar acidities ($pK_a = 5$–6), but differ most strikingly in their proton–donor abilities, the value of which may be judged by the energy of their hydrogen bonding with dimethylsulphoxide. Since the rates of proton transfer in the tautomeric complexes of these acids are within the interval of characteristic NMR frequencies at various temperatures, they cannot be directly compared with each other. So, the temperature is given, at which the monomolecular rate constant is equal to $10^3 \, s^{-1}$. It can be seen that, as the energy of hydrogen bonding increases, the rate increases greatly, while the activation energy falls. It is noteworthy, that CH acids do not seem to be a particular case of 'pseudo-acids', and low rates of deprotonization appear to be directly related to their weakness as proton donors in hydrogen bonding. The considerable influence of the strength of hydrogen bonding on the rate of proton migration seems very natural. At a fixed enthalpy of formation, ΔH_2, for the ion pair, and with an increase in hydrogen bonding energy, ΔH_1, the energy barrier falls since the transitional state, $[A^{\delta-} \cdots H \cdots B^{\delta+}]$, is stabilized (Table 2.4). Also, the minima on the potential energy surface are brought together on account of 'constriction' of A and B fragments by the proton (Figure 2.10). As a result, the probability of proton migration must increase considerably. One can suppose that in the limiting case of very strong hydrogen bonding, the potential barrier may disappear altogether, with the resulting potential function having only one minimum. At intermediate values of ΔH_2, the corresponding complexes could be described by the structure, $A^{\delta-} \cdots H \cdots B^{\delta+}$, with a 'quasi-symmetrical' hydrogen bond.

Table 2.4 The enthalpy values of hydrogen-bonded complexes of acids with $(CD_3)_2SO$ and kinetic characteristics of proton transfer inside the complexes formed with $(CH_3CH_2)_3N$

No.	Acid	pK_a	$-\Delta H_{DMSO}$ (kcal mol^{-1})	$t(°C)$	ΔE_1 (kcal mol^{-1})
1.	$(CF_3)_3COH$	5.5	11.8	<-180	
2.	$(CH_3)_3CCOOH$	5.1	9.6	<-180	
3.	CH_3NHNO_2	5.1	7.0	-145	2.2
4.	C_6H_5SH	6.2	5.2	-115	3.4
5.	$(CF_3)_2CHNO$	5.3	4.4	-55	4.7
6.	$CH_3CH(NO_2)_2$	5.1	2.2	$+42$	7.9
7.	$(CF_3)_2CHCOOCH_3$	6.2	<1	$+70$	10

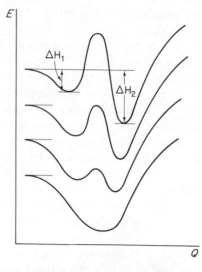

Figure 2.10 The possible evolution of a reaction path profile for proton transfer with an increase of the hydrogen bonding energy, ΔH_1

2.4 Proton Exchange in Hydrogen-Bonded Systems

In systems containing molecules with AH and BH groups capable of hydrogen bond formation as proton donors and proton acceptors, proton exchange takes place in hydrogen-bonded complexes. Therefore, some useful information about the processes of forming and breaking of these complexes, and the proton transfer processes in them, can be expected from a study of the proton exchange kinetics. It should be noted that here, as in previous sections, special attention is paid to proton exchange between molecules in inert media, where it is possible to exclude, or at least to minimize, all other reaction paths except the proton exchange in the hydrogen-bonded complexes.

The information available on the kinetics of proton exchange processes embraces practically all classes of molecules capable of forming hydrogen bonds. The following methods are often used for the investigation of proton exchange: chemical methods and optical spectroscopy for the study of labelled compounds, NMR spectroscopy; some work has been carried out using ESR (Prokofiev *et al.*, 1974). For an investigation of proton exchange under the conditions mentioned above, kinetic IR spectroscopy (Denisov *et al.*, 1968) is very convenient. Combined with the stopped-flow method (Bureiko and Denisov, 1974) it permits investigation of proton exchange in solutions (see Figure 2.11) with a half-exchange period as little as a few milliseconds.

In spite of the broad variety of physical and chemical properties of the molecules studied, and the wide range of characteristic times of proton exchange (from some milliseconds to several hours), one may conclude from the available experimental data that the ability of a molecule to form a hydrogen bond determines the kinetic characteristics of proton exchange with its participation. Tables 2.5 and 2.6 list some values of rate constants, k, and activation energies, ΔE.

In comparing different classes of compounds, it will be obvious that the exchange of the proton of the thiohydrylic group of thiols with all the partners studied is accomplished much more slowly than exchange of the proton of the hydroxylic group of similar alcohols. The ability of the SH group to form hydrogen bonds as a proton donor and a proton acceptor is considerably lower than that of the hydroxylic group. Since the acidity of thiols is greater than that of alcohols, one may conclude that, in this case, the proton exchange rate is determined by the ability of the molecule to form hydrogen bonds rather than by its acidic properties.

Supposing (see section 2.1) that proton exchange takes place in inter-mediate cyclic complexes, of the type, $R{-}A{\diagdown}_{\cdot\cdot H}^{H\cdot\cdot}{\diagup}B{-}R'$, then the kinetics of the process must depend on the strengths of both $AH \cdots B$ and $BH \cdots A$ bonds, i.e. on the proton donor and proton acceptor abilities of

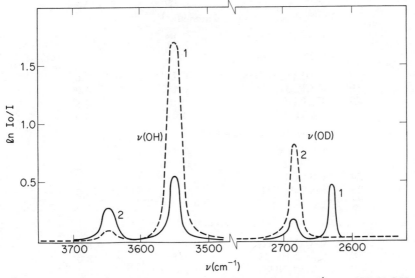

Figure 2.11 IR spectra of the CH_3OD $(0.05 \text{ mol l}^{-1}) + o\text{-ClC}_6H_4OH$ $(0.02 \text{ mol l}^{-1})$ system in CCl_4 before (– – –) and after (——) proton exchange. 1. Bands of the phenol; 2. bands of the alcohol

Table 2.5   Kinetic parameters of proton exchange processes in
CCl$_4$ solutions obtained by kinetic IR spectroscopy*

No.	System	k^{**}(l mol^{-1} s^{-1})	ΔE(kcal mol^{-1})
	I. Carboxylic acid–ethanol		
1.	CF$_3$COOH	4.8×10^4	4.2
2.	CCl$_3$COOH	3.6×10^4	4.8
3.	CHCl$_2$COOH	2.4×10^4	3.4
4.	CH$_2$ClCOOH	6.8×10^3	4.3
5.	CH$_3$COOH	3.5×10^3	4.5
6.	HCOOH	1.3×10^3	5.0
7.	(CH$_3$)$_3$CCOOH	1.0×10^3	5.7
	II. Phenol–methanol		
8.	4-ClC$_6$H$_4$OH	320	2.0
9.	C$_6$H$_5$OH	240	2.1
10.	2-ClC$_6$H$_4$OH	65	3.3
11.	2,4,6-Cl$_3$C$_6$H$_2$OH	37	3.1
	III. Alcohol–water		
12.	CD$_3$OH	270	1.7
13.	C$_2$H$_5$OH	240	1.9
	IV. Secondary amine–ethanol		
14.	C$_6$H$_5$NHCH$_3$	240	1.8
15.	(C$_2$H$_5$)$_2$NH	150	1.5
16.	(iso-C$_4$H$_9$)$_2$NH	80	3.1
	V. Carboxylic acid–isobutanethiol		
17.	iso-C$_3$H$_7$COOH	2.1	13.0
	VI. Phenol–butanethiol		
18.	4-NO$_2$C$_6$H$_4$OH	0.21	3.0
19.	4-ClC$_6$H$_4$OH	0.13	4.2
20.	C$_6$H$_5$OH	0.10	4.5
21.	4-CH$_3$C$_6$H$_4$OH	0.023	5.0
	VII. Alcohol–butanethiol		
22.	CF$_3$CH$_2$OH	3.0×10^{-2}	—
23.	CH$_3$OH	7.0×10^{-2}	3.0
24.	C$_2$H$_5$OH	2.2×10^{-2}	5.0
25.	iso-C$_3$H$_7$OH	0.9×10^{-2}	5.5
26.	t-C$_4$H$_9$OH	0.4×10^{-2}	5.0
	VIII. Amine–t-butanethiol		
27.	C$_6$H$_5$NHCH$_3$	0.1	8.0
28.	C$_6$H$_5$NHC$_2$H$_5$	0.08	8.0
29.	(C$_6$H$_5$)$_2$NH	0.1×10^{-2}	6.0
30.	(C$_2$H$_5$)$_2$NH	1.1×10^{-2}	6.0
31.	(C$_3$H$_7$)$_2$NH	0.8×10^{-2}	8.0
32.	(iso-C$_4$H$_9$)$_2$NH	0.4×10^{-2}	8.0

Table 2.5 (*Continued*)

No.	System	$k^{**}(1\,\text{mol}^{-1}\,\text{s}^{-1})$	$\Delta E(\text{kcal}\,\text{mol}^{-1})$
	IX. Heterocyclic amine–methanol		
33.	Pyrrole	1.2×10^{-4}	8.0
34.	Indole	4.5×10^{-4}	11.0
35.	Carbazole	0.3	14.0

* R.m.s. deviations in Tables 2.5 and 2.6 are 5–20%.
** At 20 °C.

each molecule. Experimental data have shown (Table 2.5) that the maximum values of the rate constants of the proton exchange with alcohols, or thiols, were found for carboxylic acids (Denisov and Smolyansky, 1968, Bureiko and Lange, 1978). The reaction becomes slower for phenol derivatives (Bureiko *et al.*, 1972), still slower when water (Bureiko *et al.*, 1977a) or alcohols are used, and is further retarded for secondary amines (Bureiko *et al.*, 1971; Bureiko and Denisov, 1973a). In C_6H_6, the proton exchange rate of $(C_3H_7)_2PH$ with methanol ($k = 0.011\,\text{mol}^{-1}\,\text{s}^{-1}$) is greater than the k value for the $(C_3H_7)_2PH + (C_2H_5)_2NH$ system (Ryltsev *et al.*, 1980). As a rule, the sequence of decreasing proton donor ability in a series of compounds is the same. A similar dependence has been obtained for the gas phase by Denisov and Tokhadze (1972) and Bureiko and Denisov (1973b).

The same regularity is observed in studying the influence of proton donor ability on the rate of proton exchange in a series of compounds of one class. As a rule, an increase in the proton donor ability of the AH group in such a series results in an increase of k. This means that the influence of the growth of proton donor ability on the exchange kinetics is greater than the simultaneous decrease of proton acceptor ability. In a series of RAH molecules,

Table 2.6   Solvent effect on the proton exchange rate involving the participation of methanol (obtained by ^{1}H NMR spectroscopy)

No.	Solvent	$\varepsilon_{\text{solv}}$	$k^*(1\,\text{mol}^{-1}\,\text{s}^{-1})$	
			$(CH_3)_3COH$	CH_3COOH
1.	C_6H_{12}	2.02	900	
2.	CCl_4	2.28	850	4.8×10^3
3.	C_6H_6	2.28	480	
4.	$CHCl_3$	4.70	420	
5.	C_6H_5Cl	5.62	80	1.1×10^4
6.	CH_2Cl_2	8.90	150	6.5×10^4
7.	$C_6H_4Cl_2$	9.93	120	9.0×10^4

* At 27 °C.

the proton donor and proton acceptor abilities are changed in different directions by varitaion of the substituent, R. Therefore, in cyclic complexes formed by the BH molecule with a number of partners, RAH, an increase in the strength of one hydrogen bond will be accompanied by a decrease in the strength of the other. At the same time, the total energy of the complex, determining the depth of the potential well (Figure 2.2), does not change in the series as obviously as the energy of each of the two bonds.

The dominating influence of the proton donor ability on the rate of proton exchange is not yet fully understood. There are deviations from this experimental dependence, for example, the rate constant for proton exchange in the trifluoroethanol–butanethiol system (Table 2.5) is less than that for methanol-butanethiol, although the proton donor ability of CF_3CH_2OH is considerably greater than that of CH_3OH since it contains electronegative substituents. A decrease in k is also observed, on the transfer from N-alkylanilines to $(C_6H_5)_2NH$, a still stronger proton donor (Table 2.5). These facts may be considered as an indication of the influence of decrease in proton acceptor ability of the A atom in the AH group, and of the cyclic structure of proton exchange intermediates. It should be noted that, contrary to alcohols and aliphatic amines, an increase of the proton donor ability of diphenylamine and N-alkylanilines is brought about not by the inductive effect, but by conjugation of an electron lone pair of the N atom with the ring π electrons. This conjugation, and that in amide molecules also, is responsible for the decrease of the proton acceptor ability of the N atom, which, in the case of heterocyclic amines (systems IX, Table 2.5), results in a decrease of the exchange rate and even in a change of mechanism for the process (Belozerskaya et al., 1970).

In the proton exchange of phenol derivatives with methanol (9–11 in Table 2.5), the reaction was retarded as proton donor ability of the molecule increased. This series of molecules is of interest also for studying the influence of intramolecular hydrogen bonding (in 2-chlorophenol and 2,4,6-trichlorophenol) on the rate of the process. The results of Forsen and Hoffman (1963) and Bureiko et al. (1975) testify to the fact that formation of an intramolecular hydrogen bond by the AH group proton is accompanied by a considerable decrease in the rate of proton exchange.

The experimental data available confirm the conclusion that the rate of the molecular proton exchange processes of the type considered is determined by the same peculiarities of electronic structure which control the hydrogen bonding ability of the functional groups of the molecules (Bureiko et al., 1976, 1977b).

The value of another kinetic parameter of proton exchange process, i.e. the activation energy, ΔE, is not as sensitive as the k values to a change in molecular structure. ΔE for various classes of molecules ranges from 1.5 to 14 kcal mol^{-1}, but in most cases, for molecules of one class, the activation energy varies only slightly. Most surprising is the fact that ΔE values are so small compared to the energy of breaking bonds. However, nothing can be said so far about the influence of the electronic structure of the molecules on the $\Delta E^{\neq}$ value, i.e. the barrier height measured from the bottom of the potential well (Figure 2.2). This value might be found experimentally through direct measurement of the cyclic complex concentration, and this may be looked upon as one of the most interesting problems to be studied in the future.

The importance of hydrogen bonding in proton exchange kinetics is supported also by a study of the influence, on the kinetics of the process, of solvents capable of forming hydrogen bonded complexes with the molecules concerned. The addition of proton acceptor compounds (acetone, DMSO, THF, dioxane) to inert solvents results in a decrease in the proton exchange rate (Tewari and Li, 1970; Denisov and Semenova, 1972). This has been discussed by Denisov and Tokhadze (1975) in terms of the dynamic characteristics of hydrogen bonding (see section 2.1). The influence of hydrogen bonding is seen clearly by comparing the results of a proton exchange study in alcohol–water systems in dilute CCl_4 solutions (Bureiko et al., (1977a), and in binary mixtures (e.g. Paterson and Spedding, 1963; Tewari and Li, 1970). In the latter case, the k values are smaller by two orders of magnitude, while ΔE is markedly higher than in the case of proton exchange, at component concentrations of 10^{-2}–10^{-3} mol l^{-1}. This can be attributed to the effect of a network of hydrogen bonds in the binary mixtures hampering the formation of cyclic complexes between alcohol and water molecules.

In investigating the role of hydrogen bonding in proton exchange processes, consideration of the reaction mechanism and its limiting stage are extremely important. The data above are consistent with the supposition that the first stage involves proton transfer in the cyclic intermediate. For such a complex to form, the A atom and the B atoms must have lone pairs of electrons. The NMR investigation of amine–dinitroethane systems has shown that proton transfer in them is rapid, while the rate of proton exchange is very low. This fact is naturally related to the electronic structure of a carbon acid molecule. We have obtained another remarkable result in studying proton exchange involving the (3-aminopropyl)dibutylborane molecule, by means of kinetic IR spectroscopy. This molecule possesses a proton donor ability comparable with that of aliphatic alcohols (Iogansen et

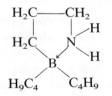

al., 1971), but, on account of the lone pair of the N atom involved in coordination to the boron atom, it loses its proton acceptor function completely. While the k values for alcohol–alcohol and alcohol–amine systems in CCl_4 are of the order of 100–$400 \ l \ mol^{-1} \ s^{-1}$, the rate constants for proton exchange between aminoborane and methanol, or secondary amines, in the same solvent are 3–4 orders of magnitude lower. Such a result points to a cyclic rather than linear structure for the intermediate.

Experimental measurements of the order of reaction with respect to each component a supposition as to the number of molecules involved in the first stage of the process to be made. In all cases of proton exchange in solution, the reaction order is close to unity. Hence, the process can be seen as bimolecular, i.e. the first step of proton exchange takes place in the cyclic complex formed by two hydrogen bonded molecules. That the reaction proceeds in the cyclic bimolecular intermediates has been accepted by Huyskens and Zeegers-Huyskens (1961) and Limbach (1977), also.

What can be said as to the mechanism of such a process? An assumption of synchronous transfer of two protons in the cyclic complex was made by Brodsky (1949). The mechanism of such cooperative process may be represented by the scheme:

$$AH + BH^* \rightleftharpoons A\begin{matrix} H \\ \cdots \\ H^* \end{matrix}B \rightleftharpoons A\begin{matrix} H \\ \cdots \\ H^* \end{matrix}B \rightleftharpoons AH^* + BH \quad (2.15)$$

In solutions, other reaction mechanisms via binary complexes are possible (Grunwald and Meiboom, 1963). An alternative mechanism is a sequential transfer of two protons in a linear complex, where the intermediate has the form of a cyclic ionic pair with two equal hydrogen bonds:

$$AH + BH^* \rightleftharpoons AH \cdots BH^* \rightleftharpoons$$

$$A^-\begin{matrix} H \\ \cdots \\ H^* \end{matrix}B^+ \rightleftharpoons AH^* \cdots BH \rightleftharpoons AH^* + BH \quad (2.16)$$

Such a process would not involve the stage of breaking of the ionic pair (i.e. the electrolytic dissociation) and, therefore, could go on in an inert medium (section 2.3). Investigation of the possible proton exchange mechanisms is a promising field.

A study of the influence of the dielectric permeability of a solvent on the proton exchange rate may be a help in choosing between mechanisms (2.15) and (2.16). The results reported by Bureiko *et al.* (1979) indicate that, as ε_{solv} increases, the rate of proton exchange in alcohol–alcohol system decreases, while, for alcohol–acetic acid system, the proton exchange rate rises considerably (Table 2.6). The authors explain this dependence by the cooperative mechanism (2.15) in the former case, and by the ion-pair mechanism (2.16) in the latter. On formation of the transitional state of mechanism (2.16), a great increase of the dipole moment, as compared to the initial state, must follow. Therefore, when ε is increased, the rate of proton exchange must be also increased. However, on formation of the transitional state of the cooperative mechanism, which resembles in structure the symmetrical cyclic complex, the dipole moment decreases, and this would result in the opposite effect. This very dependence has been observed experimentally.

Speaking of intermediate cyclic structures, it should be clearly understood that the cyclic model of the binary complex, formed by two non-linear hydrogen bonds, is less expedient than the linear structures, in terms of energy and entropy. Although, so far there is no direct experimental evidence to support the existence of four-membered cyclic hydrogen-bonded dimers, still the results of quantum mechanical calculations (Kollman and Allen, 1972) show that, in some systems, such a complex does possess energy, considerably exceeding the thermal energy, and is stable under some variations of geometrical parameters.

Very few experimental investigations of proton exchange processes have been carried out in the gas phase. The available data show that the reaction mechanism may be quite different for different systems, though the dependence of the rate of the process on the hydrogen-bonding abilities of the molecules seems to be the same. In all cases in solution, and for amines, alcohols, hydrogen halides, and phosphines (Ryltsev et al., 1980) in the gas phase, proton exchange processes are bimolecular. In the homogeneous reaction of thiol a trimolecular reaction was observed with one thiol molecule and two molecules of the second component participating in the first step (Bureiko et al., 1977b). The results obtained were interpreted by Denisov and Tokhadze (1972) who postulated a two-step reaction: (a) formation of a bimolecular hydrogen-bonded complex, and (b) proton transfer either in this complex activated by a third molecule, or in the trimolecular cyclic complex. In the condensed phase, the energy, ΔE, required for transition over the potential barrier, is obtained by the system from interaction with the solvent. Although such a two-step mechanism must occur in a solvent as well, it does not influence the kinetic characteristics because of almost continuous energy exchange of the complexes with the medium.

The problem of the limiting stage of proton exchange processes is still not clear. As seen from (2.15) and (2.16), the rate of the reaction may be determined by the time taken to form or break the intermediate, or by the time for proton transfer in this complex. So one can imagine that all the cases may be realized for different molecular systems under various conditions.

We know of only one system where breaking of the intermediate could be the determining stage of proton exchange: it is the proton exchange between carboxylic acid dimers forming strong eight-membered rings. In studying proton exchange of HCOOH, Denisov and Golubev (1977) concluded that the rate of the process was limited by the breaking of hydrogen bonds.

Large activation energies, ΔE, for proton exchange reactions involving thiol are not in accordance with the supposition that in this case the formation of hydrogen-bonded intermediates is the limiting stage, the activation energy of this process being very low (Tokhadze, 1975). The opinion was expressed (Denisov and Smolyansky, 1968) that the first step of cooperative proton transfer in the intermediate complex determines the kinetics of proton exchange in such systems. For proton exchange between alcohols, amines, water, which are probably also characterized by the cooperative mechanism, the rate constants, k, are considerably greater, while the activation energies, ΔE, are close to ΔE of the diffusion. By studying the kinetic isotope effect in a methanol–water system in CCl_4, Bureiko et al. (1977a) deduced that the proton exchange rate is determined by the stage of cyclic complex formation. This process may be very slow due to the extremely small value of the equilibrium constant. In this case, although the rate of the reaction is not determined by diffusion, the activation energy can, in fact, be determined by it. The problem of the limiting stage in different mechanisms of proton exchange processes is open to further investigation.

Summing up, one can say that we are still far from a full understanding of the nature of the proton exchange processes in hydrogen-bonded systems. There is little doubt that the ability of exchanging molecules to form hydrogen bonds influences the kinetic characteristics of the proton exchange. The evidence available testifies to a molecular mechanism via formation of cyclic intermediates (mostly, bimolecular ones) in an inert medium. The cooperative mechanism of proton transfer is the simplest model of the reaction. Its realization in a pure form is most probable in systems with symmetrical intermediates. If the hydrogen-bond forming abilities of component molecules differ greatly, then the step-like mechanism via formation of the hydrogen-bonded ion pair may be correct. In nature, a whole variety of intermediate reaction paths may exist for different molecular systems.

A better insight into the mechanism of proton exchange molecular processes may be obtained by a further spectroscopic study of the cyclic complexes, determination of their lifetimes, investigation of the dynamics of successive steps of the process by various physical and chemical methods and techniques, and by theoretical calculations of the potential surfaces of the interaction.

2.5 References

Baba, H., Matsuyama, A., and Kokubun, J. (1969). *Spectrochim. Acta*, **25A**, 1703.

Bader, F. and Plass, K. G. (1971). *Ber. Bunsenges. Phys. Chem.*, **75**, 353.

Barrow, G. M. (1956), *J. Amer. Chem. Soc.* **78**, 5802.

Belozerskaya, L. P., Denisov, G. S., and Tupitsin, I. F. (1970). *Teor. Eksp. Khim.*, **6**, 408.

Bergman, K., Eigen, M., and Maeyer, L. (1963). *Ber. Bunsenges. Phys. Chem.*, **67**, 819.

Borucki, L. (1967). *Ber. Bunsenges. Phys. Chem.*, **71**, 504.

Brodsky, A. I. (1949). *Izv. Akad. Nauk SSSR, Ser. Khim.*, 3.

Brodsky, A. I. (1964). *Vodorodnaja Sviaz* (Eds. Sokolov, N. D. and Chulanovsky, V. M.) Nauka, Moscow, p. 115.

Bureiko, S. F. and Denisov, G. S. (1973a). *Organic Reactivity*, **10**, 959.

Bureiko, S. F. and Denisov, G. S. (1973b). *Kinetika i Kataliz*, **14**, 1384.

Bureiko, S. F. and Denisov, G. S. (1974). *React. Kinet. Catal. Lett.*, **1**, 283.

Bureiko, S. F., Denisov, G. S., Golubev, N. S., and Lange, I. J. (1977a). *React. Kinet. Catal. Lett.*, **7**, 139.

Bureiko, S. F., Denisov, G. S., Golubev, N. S., and Lange, I. J. (1979). *React. Kinet. Catal. Lett.*, **11**, 35.

Bureiko, S. F., Denisov, G. S., and Martsinkovsky, R. (1975). *React. Kinet. Catal. Lett.*, **2**, 343.

Bureiko, S. F., Denisov, G. S., and Tokhadze, K. G. (1971). *Kinetika i Kataliz*, **12**, 62.

Bureiko, S. F., Denisov, G. S., and Tokhadze, K. G. (1976). *Studia Biophysica*, **57**, 205.

Bureiko, S. F., Denisov, G. S., and Tokhadze, K. G. (1977b). *Molekuliarnaja Spektroskopija*, No. 4, Leningrad Univ. Publ., p. 74.

Bureiko, S. F., Denisov, G. S., and Tupitsyn, I. F. (1972). *Organic Reactivity*, **9**, 773.

Bureiko, S. F. and Lange, I. J. (1978). *Vestnik Leningr. Univ.*, No. 4, p. 52.

Caldin, E. F., Crooks, J. E., and O'Donnell, D. (1973). *J.C.S. Faraday Tr. II*, **69**, 993, 1000.

Caldin, E. and Gold, V. (Eds.) (1975). *Proton Transfer Reactions*, Chapman and Hall, London.

Carsaro, R. D. and Atkinson, G. (1971). *J. Chem. Phys.*, **54**, 4090.

Crooks, J. E. and Robinson, B. U. (1970). *Trans. Faraday Soc.*, **66**, 1496.

Crooks, J. E. and Robinson, B. U. (1971). *Trans. Faraday Soc.*, **67**, 1707.

Davis, M. M. (1968). *Natl. Bur. Std. U.S. Monogr.*, 105.

Denisov, G. S. and Golubev, N. S. (1977). *Abstr. XIII Europ. Congr. Mol. Spectr.*, Wroclaw, Poland, p. 52.

Denisov, G. S. and Golubev, N. S. (1981). *Adv. Mol. Relax. Interact. Proc.*

Denisov, G. S., Golubev, N. S., and Koltsov, A. I. (1977). *Adv. Molec. Relax. Interact. Proc.*, **11**, 283.

Denisov, G. S., Golubev, N. S., and Schreiber, V. M. (1976). *Studia Biophysica*, **57**, 25.

Denisov, G. S., Kazakova, E. M., and Ryltsev, E. V. (1968). *Zh. Prikl. Spektr.*, **8**, 690.

Denisov, G. S. and Schreiber, V. M. (1974). *Dokl. Akad. Nauk SSSR*, **215**, 627.

Denisov, G. S. and Semenova, A. E. (1972). *Teor. Eksp. Khim.*, **8**, 822.

Denisov, G. S. and Smolyansky, A. L. (1968). *Kinetika i Kataliz*, **9**, 902.

Denisov, G. S., Starosta, J., and Schreiber, V. M. (1973). *Opt. Spektrosk.*, **35**, 447.

Denisov, G. S. and Tokhadze, K. G. (1972). *Dokl. Akad. Nauk SSSR*, **207**, 1387.

Denisov, G. S. and Tokhadze, K. G. (1973). *Molekuliarnaja Spektroskopija*, No. 2, Leningrad Univ. Publ., p. 65.

Denisov, G. S. and Tokhadze, K. G. (1975). *React. Kinet. Catal. Lett.*, **2**, 457.

Forsen, S. and Hoffman, R. A. (1963). *J. Chem. Phys.*, **39**, 2892.

Fratiello, A., Schuster, R. E., Vidulich, G. A., Bragin, J., and Lin, D. (1973). *J. Amer. Chem. Soc.*, **95**, 631.

Fujiwara, F. J. and Martin, J. S. (1974). *J. Amer. Chem. Soc.*, **96**, 7625.

Gerasimov, I. V., Denisov, G. S., and Tokhadze, K. G. (1981). *React. Kinet. Catal. Lett.*

Golubev, N. S. (1977). *React. Kinet. Catal. Lett.*, **7**, 225.

Golubev, N. S. and Denisov, G. S. (1975). *Dokl. Akad. Nauk SSSR*, **220**, 1352.

Golubev, N. S. and Denisov, G. S. (1976). *React. Kinet. Catal. Lett.*, **4**, 87.

Golubev, N. S. and Denisov, G. S. (1977). *Molekuliarnaja Spektroskopija*, No. 4, Leningrad Univ. Publ., p. 62.

Golubev, N. S. and Denisov, G. S. (1981). *React. Kinet. Catal. Lett.*

Golubev, N. S., Denisov, G. S., and Koltsov, A. I. (1977a). *Dokl. Akad. Nauk SSSR*, **232**, 841.

Golubev, N. S., Denisov, G. S., and Schreiber, V. M. (1980). *Vodorodnaja Sviaz* Ed. Sokolov, N. D. Nauka, Moscow.

Golubev, N. S., Kolomijtsova, T. D., Melikova, S. M., and Shchepkin, D. N. (1977b). *Teoreticheskaya Spektroskopija*, AN SSSR Publ., Moscow, p. 78.

Golubev, N. S., Safarov, N. A., and Tanasijchuk, A. S. (1978). *Kinetika i Kataliz*, **19**, 302.

Grunwald, E. (1966). *Advances in Physical Organic Chemistry* (Ed. Gold, V.) Vol. 4, Academic Press, NY.

Grunwald, E. and Meiboom, S. (1963). *J. Amer. Chem. Soc.*, **85**, 2047.

Grunwald, E. and Ralf, E. K. (1971). *Acc. Chem. Res.*, **4**, 107.

Gunnarson, G., Wennerstrom, H., Egan, W., and Forsen, S. (1976). *Chem. Phys. Lett.*, **38**, 96.

Gusakova, G. V., Denisov, G. S., Smolyansky, A. L., and Schreiber, V. M. (1970). *Dokl. Akad. Nauk. SSSR*, **193**, 1056.

Gusakova, G. V., Denisov, G. S. and Smolyansky, A. L. (1972). *Zh. Prikl. Spektr.*, **17**, 666.

Hadzi, D. (1965). *Pure Appl. Chem.*, **11**, 435.

Hadzi, D. and Bratos, S. (1976). In *The Hydrogen Bond* (Ed. P. Schuster, G. Zundel and C. Sandorly) Vol. 2. North Holland. p. 566.

Hammes, G. G. (Ed.) (1974). *Investigation of Rates and Mechanisms of Reactions*, 3rd edn., John Wiley, NY.

Hammes, G. G. and Lillford, P. J. (1970). *J. Amer. Chem. Soc.*, **92**, 7578.

Hopmann, R. F. W. (1976). *Proc. III Intern. Symp. on Specific Interactions*, Wroclaw-Karpacz, Poland, p. 209.

Hudson, R. A., Scott, R. M., and Vinogradov, S. N. (1972). *J. Phys. Chem.*, **76**, 1989.

Huyskens, P. and Zeegers-Huyskens, T. (1961). *Bull. Soc. Chim. Belg.*, **70**, 511.

Iogansen, A. V., Rassadin, B. V., Botchkariova, M. N., Dorokhov, V. A., and Mikhailov, B. M. (1971). *Zh. Prikl. Spektr.*, **15**, 1047.

Ivin, K. J., McGravey, J. J., Simmons, L. L., and Small, R. (1971). *Trans. Faraday Soc.*, **67**, 104.

Jackman, L. M. and Cotton, F. A. (Eds.) (1975). *Dynamic Nuclear Magnetic Resonance Spectroscopy*, Academic Press, NY.

Koelle, U. and Forsen, S. (1974). *Acta Chem. Scand.*, **A28**, 531.

Kollman, P. A. and Allen, L. C. (1972). *Chem. Revs.*, **72**, 283.
Kreevoy, M. (1965). *Discuss. Faraday Soc.*, No. 39, 166.
Kuopio, R., Kivinen, A. and Murto, J. (1976). *Acta Chem. Scand.*, **A30**, 1.
Limbach, H. H. (1977). *Ber. Bunsenges. Phys. Chem.*, **81**, 1112.
Lindemann, R. and Zundel, G. (1977). *J.C.S. Faraday Tr. II*, **73**, 788.
Maeyer, L., Eigen, M., and Shares, J. (1968). *J. Amer. Chem. Soc.*, **90**, 3157.
Maier, W. (1960). *Z. Elektrochem.*, **64**, 145.
Masalimov, A. S., Prokofiev, A. I., Bubnov, N. N., Solodovnikov, S. P., and Kabachnik, M. I. (1976). *Izv. Akad. Nauk SSSR, Ser. Khim.*, 190.
Masalimov, A. S., Prokofiev, A. I., Bubnov, N. N. Solodovnikov, S. P., and Kabachnik, M. I. (1977a). *Izv. Akad. Nauk SSSR, Ser. Khim.*, 761.
Masalimov, A. S., Prokofiev, A. I., Bubnov, N. N., Solodovnikov, S. P., and Kabachnik, M. I. (1977b). *Izv. Akad. Nauk SSSR, Ser. Khim.*, 767.
Matrosov, E. I. and Kabachnik, M. I. (1977). *Dokl. Akad. Nauk SSSR*, **232**, 89.
Paterson, W. G. and Spedding, H. (1963). *Canad. J. Chem.*, **41**, 2477.
Prokofiev, A. I., Bubnov, N. N., Solodovnikov, S. P., Belostotskaya, I. S., Ershov, V. V., and Kabachnik, M. I. (1974). *Izv. Akad. Nauk SSSR, Ser. Khim.*, 2467.
Rassing, J. (1975). *Chem. and Biol. Appl. Relaxat. Spectrometry*, Dodrecht-Boston, p. I.
Ryltsev, E. V., Bureiko, S. F., and Shurubura, A. K. (1980), *Vodorodnaja Sviaz* (Ed. Sokolov, N. D.) Nauka, Moscow.
Schuster, P., Zundel, G., and Sandorfy, C. (Eds.) (1976). *The Hydrogen Bond*, Vols. II–III, North-Holland.
Shapet'ko, N. N., Bogachev, Y. S., Radushnova, I. L., and Shigorin, D. N. (1976). *Dokl. Akad. Nauk SSSR*, **231**, 401.
Sokolov, N. D. (1948). *Dokl. Akad. Nauk SSSR*, **60**, 825.
Tabuchi, T. (1976). *Mem. Inst. Sci. and Ind. Osaka Univ.*, **33**, 17.
Tatsumoto, N., Sano, T., and Yasunaga, T. (1972). *Bull. Chem. Soc. Japan*, **45**, 3096.
Tewari, K. C. and Li, N. C. (1970). *Canad. J. Chem.*, **48**, 1616.
Tokhadze, K. G. (1975). *Molekuliarnaja Spektroskopija*, No. 3, Leningrad Univ. Publ., p. 81.
Ulashkevich, Y. V., Golubev, N. S., Basanov, A. G., Denisov, G. S., and Zelinski, I. V. (1977). *Dokl. Akad. Nauk SSSR*, **234**, 872.
Yasunaga, T., Tatsumoto, N., Inoue, H., and Miura, M. (1969). *J. Phys. Chem.*, **73**, 477.
Vinogradov, S. N. and Linnel, R. H. (1971). *Hydrogen Bonding*. Van Nostrand, p. 163.

Molecular Interactions, Volume 2
Edited by H. Ratajczak and W. J. Orville-Thomas
© 1980 John Wiley & Sons Ltd.

3 Kinetic studies of micelle formation in surfactants

J. Gormally, W. J. Gettins, and E. Wyn-Jones

3.1 Introduction

As a result of the pioneering efforts of Eigen and De Maeyer (Eigen and De Maeyer, 1963; Claesson, 1967) the kinetics of extremely fast reactions in solution can now be successfully studied by the use of chemical relaxation methods as well as continuous- and stopped-flow techniques (Caldin, 1964; Britton Chance *et al.*, 1964; Czerlinski, 1966; Kustin, 1969; Hague, 1971; Hammes, 1974; Bradley, 1975; Wyn-Jones, 1975; Bernasconi, 1976; Strehlow, and Knoche, 1977; Gettins and Wyn-Jones, 1979). By using different techniques it is now possible to 'tune in' to the time scale of the individual steps of complex reactions and through studies over the extensive time domain, $1-10^{-11}$ s, detailed reaction mechanisms can be evaluated. This in turn can lead to the acquisition of much new information on systems which may be of industrial, biological, or biochemical significance.

In the past decade, one of the success stories in this field of fast reactions in solution has been in gaining an understanding of the kinetics of micelle formation in surfactants. In these studies the experimental techniques, stopped-flow, pressure-jump, temperature-jump, shock-wave, and ultrasonics, have been extensively used to measure the relaxation processes which occur in micellar solutions (Wyn-Jones, 1975; Mittal, 1977; Hoffmann, 1978; and references therein). The purpose of this chapter is to review the progress which has been made in this field up to 1979.

3.2 Properties of Surfactants

The fundamental characteristic of surface-active molecules or ions is their amphiphilicity or amphipathicity, that is, the presence in the molecule or ion of both polar and non-polar moieties (Shinoda, 1967). In aqueous solutions, the polar head groups are hydrophilic and the non-polar moieties are usually relatively long hydrocarbon tails and are therefore hydrophobic.

Depending on the chemical structure, a surfactant can be anionic, cationic, ampholytic, or non-ionic. Examples of these types of surfactants are:

(a) *Anionic.* The anion is the surface-active species and the cation, known as the counterion has no surface-active properties.
Example: sodium nonyl sulphate,

$$CH_3(CH_2)_8OSO_3^-Na^+$$

(b) *Cationic.* The cation is the surface-active species.
Example: cetyltrimethylammonium bromide

$$CH_3(CH_2)_{15}\overset{+}{N}(CH_3)_3Br^-$$

(c) *Ampholytic.* These surfactants may be anionic, cationic, or non-ionic, depending on the pH of the solution.
Example: *N*-dodecyl-*N,N*-dimethyl betaine

$$CH_3(CH_2)_{11}\overset{+}{N}(CH_3)_2CH_2COO^-$$

(d) *Non-ionic.* The hydrophilic group may be either a hydroxyl or a poly(oxyethylene) group.
Example: tetraethylene glycol-*p*-iso octyl phenyl ether

$$(CH_3)C\cdot CH_2\cdot C(CH_3)_2\cdot C_6H_4O(CH_2CH_2O)_4H$$

In aqueous solution these molecules have a tendency to collect at interfaces and their surface activity is illustrated in Figure 3.1. These molecules have the ability to adsorb to air–water and oil–water interfaces or to the surfaces of hydrophobic solids such as polymers and proteins. The hydrophilic head groups remain in the aqueous phase whereas the hydrophobic moieties can be partially or completely removed from the aqueous phase to be in contact with a more hydrophobic environment. The dual properties of these surfactants also allow them to form organized structures such as soap films, bilayers, and micelles.

In general when water is added to solid surfactants, three types of behaviour can occur:

(a) Some of the surfactant dissolves to form an aqueous micellar solution.
(b) The surfactant is practically insoluble, and remains as a solid crystal plus an aqueous solution of surfactant monomers.
(c) A lyotropic liquid crystal is formed, which may dissolve in more water to form an aqueous micellar solution.

This chapter is concerned with the kinetic studies of surfactant solutions

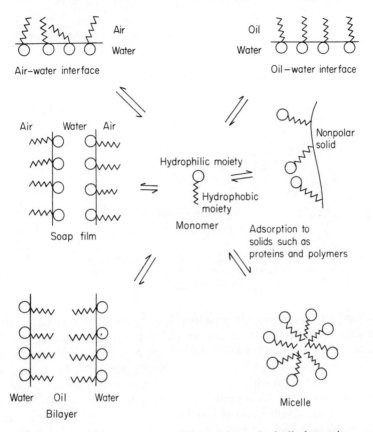

Figure 3.1   An illustration of surface activity and micelle formation

which form micelles. At very low surfactant concentrations, 10^{-4}–10^{-2} mol dm^{-3}, the behaviour of ionic surface-active molecules in aqueous solution resembles that of strong electrolytes whilst the properties of non-ionic surfactants are similar to those of simple organic molecules. On increasing the concentration of surfactant however, a pronounced deviation from 'ideal' behaviour is observed. Figure 3.2 is a typical diagram of the observed variation of equivalent conductivity which occurs as a function of surfactant concentration. The well-defined, but not abrupt, change in this physical parameter is attributable to the association of the surfactant monomers to form high molecular weight aggregates, known as micelles. The concentration at which micelles form is known as the critical micelle concentration (c.m.c.) and corresponds to the concentration at which the change in slope of, for example, the physical parameter shown in Figure 3.2, occurs. It can be seen that this change usually occurs over a narrow

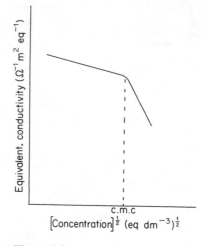

Figure 3.2 A plot of equivalent conductivity as a function of concentration

surfactant concentration range and the magnitude of this range depends somewhat on the physical parameter which is being measured.

Generally surfactants form micelles in a narrow temperature region ($\approx 10\ K$) above their Krafft points. The Krafft point may be defined as that temperature below which micelles are insoluble, that is where the monomer solubility is too small for micelle formation. For the surfactants described in this chapter the Krafft point is usually below room temperature.

When micelles are formed in aqueous solution the c.m.c. is one of the most readily obtainable parameters (Mukerjee and Mysels, 1971; Mukerjee, 1974) and useful results such as the concentrations of monomers and micelles can be evaluated from it. It is now generally accepted that below the c.m.c. only monomers are present in solution whereas above the c.m.c. both monomers and micelles are present in solution and the monomeric surfactant concentration is reasonably constant (Mukerjee, 1974). Micelles are not themselves very surface active and the significance of the c.m.c. is that it is the equilibrium concentration at which surface activity ends and colloid chemistry begins.

Micelles usually have very high aggregation numbers (50–100) and in micellar solutions it is generally thought that the abundant species are monomers and micelles: there is very little experimental evidence concerning significant concentrations of the intermediate species and these are usually neglected. Micelles are in fact polydisperse, having a distribution of aggregation numbers about a mean value. Many experimental techniques have been employed to determine c.m.c.'s (Mukerjee and Mysels, 1971) and aggregation numbers (Shinoda, 1967). In addition, many theories have been developed to account for this behaviour (Shinoda, 1967; Mittal, 1977). C.m.c.'s and aggregation numbers can be affected by the addition of salts and also neutral molecules.

A schematic two-dimensional representation of an ionic micelle in aqueous solution is shown in Figure 3.3. A variety of evidence (Balmabra *et al.*, 1962; Courcheene, 1964; Shinoda, 1967; Mittal, 1977), in support of the idealized spherical shape in the vicinity of the c.m.c. has been reported for cationic, anionic, and non-ionic micelles. The hydrophobic part of the aggregate forms the core of the micelle while the polar head groups are located at the micelle–water interface in contact with, and hydrated by a number of water molecules. There is much evidence that the micelle interior resembles a liquid hydrocarbon (Clifford, 1965; Mukerjee, 1965) and this means that they can solubilize water-insoluble molecules.

In ionic micelles the charged head groups, along with some of the counterions are located in a very compact region, known as the Stern layer. This layer is in turn surrounded by the Gouy–Chapman electrical double layer which comprises water and counterions. The rest of the solution consists of water, free surfactant ions, and free counterions.

When the surfactant concentration markedly exceeds the c.m.c., the shape of the spherical micelle is thought to undergo gradual changes until a point is reached where larger, rod-shaped micelles are probably formed (Ekwall, 1964; Ekwall and Holmberg, 1965). The formation of these larger micelles is thought to be responsible for the 'second c.m.c.' that has been reported in certain systems.

Micellar solutions have a diverse range of applications. For example, they can be used as model colloid systems (Mukerjee, 1967) and in many cases they can catalyse chemical reactions (Fendler, 1976). In addition they have many industrial, technological (Weisstuck and Lange, 1971; Franklin and

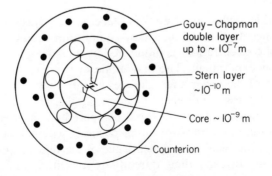

Figure 3.3   A two-dimensional schematic representation of a spherical micelle

Honda, 1977), and pharmaceutical (Mittal, 1977) uses as well as topical applications in connection with energy storage (Geffen, 1975; Alkaitis and Gratzel, 1976; Bansal and Shah, 1977).

3.3 Experimental Techniques

The experimental techniques that have been used in the study of micellar kinetics are stopped-flow, temperature- and pressure-jump, shock-wave and various ultrasonic methods.

In the stopped-flow technique the micellar solution is subjected to an abrupt concentration change by rapid mixing (Caldin, 1964; Wyn-Jones, 1975), generally in 1–0.1 ms. The ensuing changes within the diluted solution are then monitored by a fast-detection method usually spectrophotometry or conductivity and the resulting concentration–time curves are subsequently analysed using normal kinetic methods.

Temperature-jump, pressure-jump, and ultrasonics (Eigen and De Maeyer, 1963; Caldin, 1964; Czerlinski, 1966; Claesson, 1967; Kustin, 1969; Hague, 1971; Hammes, 1974; Bradley, 1975; Wyn-Jones, 1975; Bernasconi, 1976; Strehlow and Knoche, 1977; Gettins and Wyn-Jones, 1979) form a group of techniques known as relaxation methods which were developed by Eigen and de Maeyer for kinetic studies of fast reactions in solution. The general principle of the 'jump' methods is that we start with a reaction at equilibrium. This system is then rapidly perturbed by means of a temperature or pressure change and the equilibrium constant is changed to its new position in accordance with the van't Hoff isochore:

$$\left[\frac{d(\ln K)}{dT}\right]_P = \frac{\Delta H^0}{RT^2} \quad \text{(Temperature jump)}$$

$$\left[\frac{d(\ln K)}{dP}\right]_T = -\frac{\Delta V^0}{RT} \quad \text{(Pressure jump)}$$

The reaction which subsequently takes place is the adjustment of reactant and product concentrations to their new equilibrium values. This process of readjustment of concentrations is monitored by a fast-response method as in the stopped-flow experiment. The magnitude of the temperature and pressure 'jumps' are of the order a few degrees and up to 300 atm, respectively. The principle of these methods demands that the rate of change of the temperature or pressure must be rapid compared to the reaction rate. In practice temperature and pressure jumps occuring within 1 μs are achieved. Figure 3.4 shows the ideal time dependence of the temperature or pressure change and of a corresponding quantity representing the extent of the reaction, such as the concentration of a product. In practice the displacement is small so that a first-order exponential is obtained. This is important

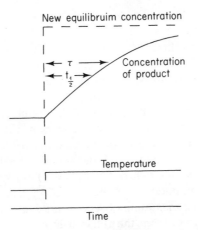

Figure 3.4 The principle of the temperature-jump method

in connection with the evaluation of rate constants from the mechanism (Czerlinski, 1966; Bernasconi, 1976).

We can derive rate constants from reaction-time curves such as that in Figure 3.4 as follows. The rate–concentration relationship depends on the reaction type. Consider the equilibrium

$$A + B \underset{k_b}{\overset{k_f}{\rightleftharpoons}} C$$

which applies to a great many reactions, such as ligand substitution, dissociation of a weak acid, or recombination of atoms or radicals.

Simple mathematics shows that when a small perturbation occurs the system moves to the new equilibrium position according to a first-order rate law, with a rate constant, k, given by

$$k = k_f(\bar{a} + \bar{b}) + k_b$$

where $\bar{a}$ and $\bar{b}$ are the concentrations of A and B at the new equilibrium, while k_f and k_b are the forward and backward rate constants in the balanced reaction shown. The first-order rate constant k is often expressed as τ^{-1}, where τ is the relaxation time which may be defined as the time required for the deviation Δc of the concentration c from its new equilibrium values to be reduced by a factor e (≈ 2.718); its relation to the half life of the reaction, $t_{1/2}$, is

$$t_{1/2}/\tau = k t_{1/2} = \ln 2 \approx 0.7$$

In addition, information concerning the equilibrium/thermodynamic parameters can be obtained from the relaxation amplitudes (Wyn-Jones, 1975; Bernasconi, 1976).

In the ultrasonic method the perturbation of the equilibrium is achieved by means of the periodic variation of pressure and temperature accompanying the passage of a sound wave through the system. The response of the equilibrium to such periodic variations is shown in Figure 3.5. When conditions are such that the frequency of the sound wave is the same order of magnitude as the reciprocal of the relaxation time, relaxational effects give rise to characteristic changes in the quantity α/f^2, where α is the sound absorption coefficient at frequency f, the sound velocity and the absorption, per wavelength, $\alpha\lambda$ as functions of frequency (f) as shown in Figure 3.6. The relaxation time is obtained from the inflection point of the α/f^2 against frequency curve or the maximum observed in the absorption per wavelength against frequency curve, from the relationship:

$$f_c = \frac{1}{2\pi\tau}$$

where f_c is the relaxation frequency. The amplitude of the relaxation process is related to the equilibrium/thermodynamic parameters.

The ranges of τ values which can be investigated by these techniques are: stopped-flow $\tau > 1$ ms, temperature-jump $5\,\mu s < \tau < 0.2$ s, pressure-jump 0.1 ms $< \tau < 100$ s, shock-wave $1\,\mu s < \tau < 2$ ms. Ultrasonic techniques are commonly available which enable absorption and velocity measurements to be made in the frequency range 0.5–250 MHz representing a relaxation time range of 10^{-6}–10^{-9} s.

In general these techniques can detect more than one relaxation time. However, a confident quantitative analysis of the experimental relaxation

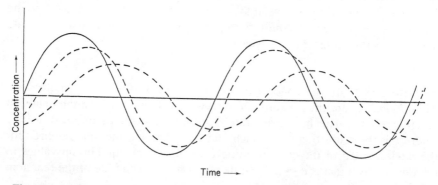

Figure 3.5 Periodic disturbance of a chemical equilibrium by a sound wave where —— represents the change in the equilibrium concentration and ·-~-·-~ represents the actual concentrations caused by a pressure perturbation for a reaction with $1/\tau \approx 2\pi f_c$

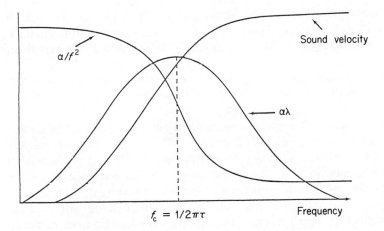

Figure 3.6 Plots of α/f^2, $\alpha\lambda$, and sound velocity against frequency for a single relaxation process

data is only possible if the relaxation times are separated by a factor of 4 or more and the relaxations have comparable amplitudes. The analysis of complex mechanisms in terms of relaxation spectra is in principle possible (Czerlinski, 1966; Bernasconi, 1976) and many have been considered in detail. In relaxation methods rates are considered close to equilibrium and consequently the relationship between rate constants and formal concentrations are not quite the same as for initial rates. The principles invoked in these calculations have been considered in detail (Czerlinski, 1966; Wyn-Jones, 1975; Bernasconi, 1976).

3.4 Kinetic Studies of Micellization

3.4.1 Introduction

The initial attempts to investigate the kinetics associated with micelle formation were carried out in the mid-sixties using stopped-flow (Jaycock and Ottewill, 1964) and pressure-jump (Mijnlieff and Ditmarsch, 1965) methods. These investigations were followed by temperature-jump (Kresheck et al., 1966; Bennion et al., 1969; Bennion and Eyring, 1970) and ultrasonic (Yasunaga et al., 1967, 1969) experiments. As a result of these experiments it was evident that the relaxation spectra of micellar solutions were characterized by time constants ranging from seconds to nanoseconds. In the early nineteen seventies, further ultrasonic measurements (Graber and Zana, 1970; Graber et al., 1970; Sams et al., 1972, 1973) showed that the relaxation time which occurs in time range 10^{-6}–10^{-9} s is associated with monomer–micelle exchange and a simple kinetic model (Sams et al., 1972) to describe this process was formulated in 1972. During this time it was inferred (Muller, 1972; Takeda and Yasunaga, 1972; Hoffmann and Janjic, 1973; Rassing *et al.*, 1974; Nakagawa, 1974) and also shown experimentally (Folger *et al.*, 1974) that the relaxation spectra of micellar solutions are actually characterized by two distinct relaxation times, the so called slow time, τ_2, in the second to millisecond range and a time, τ_1, which is up to three orders of magnitude smaller.

Of the various theoretical treatments (Kresheck *et al.*, 1966; Graber and Zana, 1970; Sams *et al.*, 1972; Muller, 1972; Colen, 1974; Aniansson and Wall, 1974, 1975; Nakagawa, 1974; Wyn-Jones, 1975; Takeda *et al.*, 1977) involving models for micelle formation that have been presented to date to account for the kinetic behaviour of surfactant solutions, that described by Aniansson and Wall is of particular interest. In common with other kinetic treatments this model assumes that micelles are formed from monomers by a series of step-wise bimolecular equilibria according to the aggregation scheme:

$$A_1 + A_1 \underset{k_2^-}{\overset{k_2^+}{\rightleftharpoons}} A_2$$

$$A_1 + A_2 \underset{k_3^-}{\overset{k_3^+}{\rightleftharpoons}} A_3$$

$$\cdots\cdots\cdots\cdots$$

$$A_1 + A_{s-1} \underset{k_s^-}{\overset{k_s^+}{\rightleftharpoons}} A_s$$

$$A_1 + A_s \underset{k_{s+1}^-}{\overset{k_{s+1}^+}{\rightleftharpoons}} A_{s+1}$$

(3.1)

In this scheme only equilibria which involve an interaction between aggregates and monomers are considered to make a contribution. This means that the bimolecular equilibria involving only aggregates are neglected. The theoretical treatment by Aniansson and Wall represents a novel way of looking at the problem and predicts the two relaxation times which are observed experimentally (Aniansson *et al.*, 1976). This model also allows the calculation of the width of the micellar distribution curve from kinetic data. We present below an account of the model but suggest that details be sought by reference to the original papers.

3.4.2 *The Aniansson and Wall model*

The model presupposes that in a surfactant solution at equilibrium and above the c.m.c., the micelles and smaller aggregates are distributed according to the aggregation number, s, in the manner depicted in Figure 3.7.

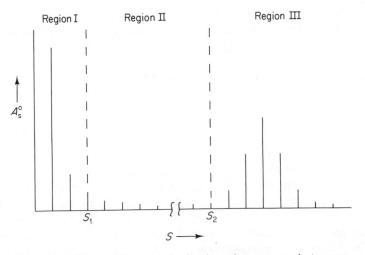

Figure 3.7   The equilibrium distribution of aggregates in 'aggregation space'

The concentration of aggregates with aggregation number s is A_s, A_s^0 being its equilibrium value. One should note that the symbol A_s is used to denote a particular species and also the concentration of that species. The diagram shows the equilibrium distribution of aggregates in 'aggregation space'; this being the one-dimensional space represented by the horizontal axis.

The aggregation space is divided into the three regions, I, II, and III. Region I contains monomer and the lower oligomers with aggregation numbers $1 \leqslant s \leqslant s_1$. For all practical purposes the value of s_1 can be taken as 1 since the concentration of such species as dimers and trimers is assumed to be very low. Region II contains aggregates of intermediate size. The distinguishing feature of these species is that their concentration is also assumed to be very much less than the concentration of either the monomer or the micelles proper (Mukerjee, 1974, 1975; Mukerjee and Cardinal, 1976). The existence of these aggregates cannot be verified directly by experiment. Region III contains micelles proper with aggregation numbers $s \geqslant s_2 + 1$. The aggregates are considered to be of detectable size in the sense that they give rise to the usual physical effects associated with the presence of micelles, such as the increased intensity of scattered light. The location of the boundary s_2 is clearly related to the way in which A_s^0 depends upon s in the region III, that is, it depends upon the form of the micelle distribution curve. In the model it is the form of this dependence that is assumed rather than the value of s_2.

Surfactant systems of this sort generally exhibit relaxation behaviour with two relaxation times, τ_1 and τ_2 with $\tau_2 > \tau_1$. Typically, τ_2 may be two or three orders of magnitude larger than τ_1. The faster process characterized by τ_1 is identified with a monomer–micelle exchange represented by the equations:

$$\left.\begin{array}{l} A_1 + A_{s-1} \underset{k_s^-}{\overset{k_s^+}{\rightleftharpoons}} A_s \\[2ex] A_1 + A_s \underset{k_{s+1}^-}{\overset{k_{s+1}^+}{\rightleftharpoons}} A_{s+1} \\[2ex] \text{etc. for } s > s_2 \end{array}\right\} \begin{array}{l} \text{fast} \\ \text{process} \end{array} \qquad (3.2)$$

where the values of $s-1$, s, $s+1$, etc. are values corresponding to micelles in region III. Similar steps corresponding to values of s in region II do not in themselves give rise to observable relaxational effects because of the assumed low concentrations in this region. After disturbing the equilibrium, the monomer–micelle exchange will give rise to a redistribution of micelles within region III of aggregation space but it will not change the total number of micelles. This redistribution will be accompanied by a corresponding change in the monomer concentration in region I (Aniansson, 1978) as indicated schematically in Figure 3.8.

It should be noted that this fast process alone does not return the system to its final equilibrium state. This is accomplished by the inclusion of the slow process which is identified with the step-wise build-up of micelles from monomers and the dispersion of micelles into monomers according to:

$$\left.\begin{array}{l} A_1 + A_1 \underset{k_2^-}{\overset{k_2^+}{\rightleftharpoons}} A_2 \\[2ex] A_1 + A_2 \underset{k_3^-}{\overset{k_3^+}{\rightleftharpoons}} A_3 \\[2ex] \cdots\cdots\cdots\cdots \\[2ex] A_1 + A_{s-1} \underset{k_s^-}{\overset{k_s^+}{\rightleftharpoons}} A_s \end{array}\right\} \begin{array}{l} \text{slow} \\ \text{process} \end{array} \qquad (3.3)$$

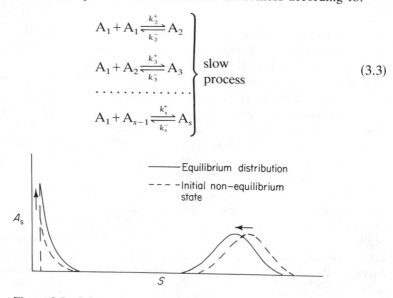

Figure 3.8   Schematic representation of the fast relaxation
process alone

It is clear that a shift in these equilibria to the left or the right, will result in an increase or decrease in the number of micelles in region III. It will be shown that it is possible to consider this process as a flow through aggregation space between regions I and III through region II. In general, both the fast and the slow processes contribute to the return of the system to its final equilibrium state and both involve changes in the monomer concentration. It may be noted that no consideration is given to the rearrangement of counterions which must take place during association. This is assumed to be a much faster process than either of the two processes referred to and would not appear in the experiments normally used to investigate these systems.

From the interpretation presented it is clear that an account of the fast process must be sought by considering the return to equilibrium in regions I and III alone, whereas the slow process can be described by considering the subsequent flow through region II.

3.4.2.1 *The rate equations* For any of the species appearing in the reaction steps depicted in (3.1) the relative displacement from equilibrium is denoted by the dimensionless quantity:

$$\xi_s(t) = \frac{A_s(t) - A_s^0}{A_s^0}. \tag{3.4}$$

In general, the displacements $\xi_s(t)$ will be much less than unity. For the single step

$$A_1 + A_{s-1} \underset{k_s^-}{\overset{k_s^+}{\rightleftharpoons}} A_s$$

the rate equation for $s > 1$ will be,

$$\frac{dA_s}{dt} = -k_s^- A_s + k_s^+ A_1 A_{s-1} \tag{3.5}$$

and at equilibrium,

$$k_s^- A_s^0 = k_s^+ A_1^0 A_{s-1}^0. \tag{3.6}$$

Elimination of k_s^+ from (3.5) using (3.6) allows one to write (3.5) in the form

$$\frac{dA_s}{dt} = A_s^0 \frac{d\xi_s}{dt} = +J_s \tag{3.7}$$

where

$$J_s = -k_s^- A_s^0 [\xi_s - \xi_{s-1} - \xi_1 (1 + \xi_{s-1})]. \tag{3.8}$$

It can be seen from (3.7) that J_s is the rate of increase of the concentration A_s due to the reaction step considered. J_s can, therefore, be regarded as a flow in aggregation space.

However, the rate of change of A_s is also governed by the step

$$A_1 + A_s \underset{k_{s+1}^-}{\overset{k_{s+1}^+}{\rightleftharpoons}} A_{s+1}.$$

In a similar way it is found that for this step alone,

$$\frac{dA_s}{dt} = A_s^0 \frac{d\xi_s}{dt} = -J_{s+1}. \tag{3.9}$$

Combining the effects of both steps (equations 3.7 and 3.9) gives the overall rate for A_s,

$$\frac{dA_s}{dt} = A_s^0 \frac{d\xi_s}{dt} = -(J_{s+1} - J_s) \tag{3.10}$$

for $s > 1$.

The case when $s = 1$ requires special consideration because the dimerization reaction is the first in the sequence and also because A_1 is involved in every step. The rate for A_1 can be found by using the material balance condition

$$\sum_s s\xi_s A_s^0 = 0 \tag{3.11}$$

from which

$$A_1^0 \dot{\xi}_1 = -\sum_{s>1} s\dot{\xi}_s A_s^0 \tag{3.12}$$

The equations (3.8) and (3.10) are very similar to the transport equations which describe flow processes such as heat conduction or diffusion. The correspondence becomes closer if we imagine quantities such as A_s, ξ_s, k_s^-, and J_s to be continuous functions of s; $A(s)$, $\xi(s)$, $k^-(s)$, and $J(s)$.

We can then write,

$$\xi_s - \xi_{s-1} \simeq \frac{\delta\xi(s)}{\delta s} \Delta s = \frac{\delta\xi(s)}{\delta s} \quad \text{when} \quad \Delta s = 1 \tag{3.13}$$

with a similar expression for $J_{s+1} - J_s$.

Equation (3.8) then becomes,

$$J(s) = -k^-(s)A^0(s)\left[\frac{\delta\xi(s)}{\delta s} - \xi_1(1 + \xi(s))\right]. \tag{3.14}$$

Similarly, equation (3.10) becomes,

$$A^0(s)\frac{\delta\xi(s)}{\delta t} = -\frac{\delta J(s)}{\delta s}. \tag{3.15}$$

Equations (3.14) and (3.15) can be compared with the equations which describe heat conduction along a bar of cross-sectional area A.

$$Q(x) = -K\mathcal{A}\frac{\delta\theta}{\delta x} \qquad (3.16)$$

and

$$\mathcal{A}\frac{\delta\theta}{\delta t} = -\frac{\delta Q(x)}{\delta x}. \qquad (3.17)$$

In (3.16) and (3.17), x denotes distance along the bar and corresponds to s; $Q(x)$ is the rate of flow of heat at position x and corresponds to $J(s)$; $\theta(x)$ denotes temperature and corresponds to $\xi(s)$; K is the thermal diffusivity and it corresponds to $k^-(s)$; $\mathcal{A}$ is the product $A\rho c$ where A is the cross-sectional area of the bar, ρ is its density, and c its specific heat. This product corresponds to $A^0(s)$. The analogy is exact except for the term $\xi_1(1+\xi(s))$ in (3.14), there being nothing which corresponds to this in the case of heat conduction. This difference leads to the kinetic equations having some interesting features. In a flow process as described by (3.16) and (3.17), equilibrium is attained when the flow becomes zero, assuming that the system is closed. Inspection of equations (3.8) or (3.14) shows that in the case considered, this can be achieved in two ways. From (3.8), J_s will be zero when

$$\xi_s = \xi_{s-1} = \xi_1 = 0. \qquad (3.18)$$

This corresponds to the final equilibrium state when both the fast and the slow processes have reached completion. However, J_s will also be zero when,

$$\xi_s - \xi_{s-1} = \xi_1(1+\xi_{s-1}) \qquad (3.19)$$
$$\simeq \xi_1.$$

The last approximation results from the assumed smallness of the ξ_s which allows the neglect of ξ_{s-1} on the right-hand side of (3.19). The state corresponding to (3.19) is referred to as a pseudo-equilibrium state. It is not a state of true equilibrium as such a state is characterized by zero values of the ξ_s. Because the values of $k_s^- A_s^0$ are expected to be much larger in regions I and III than in region II, there will be a tendency towards the rapid establishment of pseudo-equilibria in these regions due to the fast monomer–micelle exchange process. The complete return to equilibrium in the sense denoted by (3.18) is then accomplished by the slow flow process through region II. The equations which characterize the slow process are then,

$$\xi_s - \xi_{s-1} - \xi_1 \simeq 0 \quad \text{in regions I and III} \qquad (3.20)$$

and

$$\xi_s - \xi_{s-1} - \xi_1 = -\frac{J_s}{k_s^- A_s^0} \quad \text{in region II.} \tag{3.21}$$

The left-hand side of (3.20) is written as being approximately equal to zero since the value of J_s in regions I and III cannot be identically zero until the system has reached the final equilibrium state.

The overall process has been likened to the attainment of thermal equilibrium in a narrow metal bar with two large ends as indicated in Figure 3.9. The ends, L and N, are analogous to regions I and III and the narrow connecting bar, M, is analogous to region II.

Whilst this analogy has some instructive features, it should not be pressed too far as the process of heat conduction is not really described by an equation such as (3.14). However, bearing in mind that $A\rho c$ is equivalent to A_s^0, A being the cross-sectional area, ρ the density, and c the specific heat of the bar, it is apparent that the cross-sectional areas of the left-hand end L and the right-hand end N bear analogy with the concentrations A_1^0 and A_s^0 in regions I and III of aggregation space. The thinness of the connecting portion M corresponds to the assumed low concentration of oligomers in region II of aggregation space.

To pursue the analogy, we note that a relaxation experiment involves disturbing the equilibrium such that the total amount of material expressed as monomer remains constant,

$$\sum_s s\xi_s A_s^0 = 0.$$

In terms of the thermal flow analogy, this means that when thermal equilibrium is disturbed, the amount of heat $q(x)$ which flows into a thin section of thickness dx must be such as to satisfy,

$$\int q(x)\, dx = 0. \tag{3.22}$$

In other words, if heat is gained in one part of the bar, there must be a corresponding loss elsewhere.

The start of a relaxation experiment then corresponds to the situation in which the temperature in end L is raised say, and the temperature distribution in end N is altered in a non-uniform way but such that (3.22) is

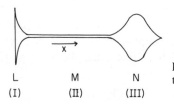

L   M   N
(I)   (II)   (III)

Figure 3.9   The attainment of thermal equilibrium in a narrow metal bar with two large ends

satisfied. Immediately following this disturbance there will be a rapid movement towards the establishment of a uniform temperature in end N accompanied by a corresponding change in temperature in end L so as to satisfy (3.22). This is the fast process which leaves the bar in a pseudo-equilibrium state in which end N has a uniform distribution of temperature but this temperature may be different from that in L. Note that because of (3.19), the pseudo-equilibrium state referred to here is not exactly analogous to that which occurs in micellar systems. The two ends are then brought to the same temperature by a relatively slow transfer of heat through the centre section M. This transfer is slow because of the thinness of the centre section and it corresponds to the slow process observed in many surfactant systems.

By means of this analogy we can understand some of the features of the kinetic behaviour of surfactants in a qualitative way. For example, if the overall concentration of surfactant is increased we can view this as something which increases the cross-sectional area of end N, that is, it increases the concentration of micelles but keeps the monomer concentration constant at the c.m.c. In this case the attainment of pseudo-equilibrium would take place more rapidly due to the increase in cross-sectional area in N. It is, in fact, the case that an increase in concentration shortens the fast relaxation time, τ_1. The effect upon the slow time would be quite different because now heat would have to flow through the centre section for a longer period to establish equilibrium because of the increased thermal capacity of the end N. Again it is observed that increasing concentration tends to lengthen the slow relaxation time, τ_2.

We now indicate briefly how expressions for the fast and slow relaxation times were derived by Aniansson and Wall.

3.4.2.2 *The fast relaxation time*   The fast process has been treated by writing the s-dependent quantities appearing in (3.10) as continuous functions of s, assuming $k^-(s)$ to be independent of s. Equations (3.8) and (3.10) then lead to the second-order differential equation

$$A_s^0 \frac{\delta \xi(s, t)}{\delta t} = k^- \frac{\delta}{\delta s} \left[A^0(s) \left\{ \frac{\delta \xi(s, t)}{\delta s} - \xi_1(t)(1 + \xi(s, t)) \right\} \right]. \qquad (3.23)$$

It should be noted that, as with (3.10), this equation can only describe the behaviour of ξ_s for $s > 1$.

To solve (3.23) some form for the equilibrium concentrations $A^0(s)$ must be assumed. Aniansson and Wall assumed a Gaussian form,

$$A^0(s) = A_0 \cdot \exp(-z^2) \qquad (3.24)$$

where

$$z = \frac{s-n}{\sigma\sqrt{2}} \tag{3.25}$$

n being the mean aggregation number and σ^2 the variance of the distribution.

Using this form for $A^0(s)$ it is possible to find solutions for (3.23) in terms of Hermite polynomials, $H_m(z)$,

$$\xi(z, t) = \sum_{m=0}^{\infty} c_m(t)H_m(z). \tag{3.26}$$

Inserting (3.26) into (3.23) and using the condition for material balance (3.11) leads to the following expression for $\xi_1(t)$

$$\xi_1(t) = -ac_0(t) - \left(\frac{a\sigma\sqrt{2}}{n}\right)c_1(t) \tag{3.27}$$

where

$$c_0(t) = c_0, \text{ independent of } t \tag{3.28}$$

$$c_1(t) = (c_1(0) + B) \cdot \exp(-t/\tau_1) - B \tag{3.29}$$

where B is a constant containing a, c_0, n, and σ, and

$$a = \frac{(A_{\text{tot}} - A_1^0)}{A_1^0} \tag{3.30}$$

A_{tot} being the overall surfactant concentration expressed as monomer. The relaxation time τ_1 appearing in (3.29) is given by

$$\frac{1}{\tau_1} = \frac{k^-}{\sigma^2} + \frac{k^-}{n}(a(1+c_0)). \tag{3.31}$$

It is not difficult to show that c_0 is equal to the average value of $\xi(s, t)$ in the micellar region and so is small compared to unity. To a good approximation then, (3.31) can be written in the form:

$$\frac{1}{\tau_1} = \frac{k^-}{\sigma^2} + \frac{k^-}{n}\left(\frac{(A_{\text{tot}} - A_1^0)}{A_1^0}\right). \tag{3.32}$$

It may not be obvious that the relaxation time found by this method is in fact the fast relaxation time. It should be noted, however, that the process described will approach equilibrium as the expression in brackets { } in (3.23) approaches zero. This is precisely the pseudo-equilibrium condition (3.19). The final state attained by this process is then a state of pseudo-equilibrium and this identifies it as the fast process. This is also borne out by considering the behaviour of (3.27) and (3.29) as $t \to \infty$ which shows that $\xi_1(\infty) \neq 0$, whereas for the final equilibrium state, $\xi_1(\infty) = 0$. It may also be noted that the time independence of c_0 as expressed by (3.28) implies that the average value of $\xi(s, t)$ in the micellar region is independent of time. That is,

$$\frac{\sum_{s>s_2} \xi_s A_s^0}{\sum_{s>s_2} A_s^0} = c_0, \text{ independent of } t. \tag{3.33}$$

Bearing in mind the definition of ξ_s, (3.4), this means that the number of micelles remains constant during the fast process, as expected.

In using (3.32), A_1^0 is taken as being equal to the c.m.c. (Shinoda, 1967). A plot of $1/\tau_1$ against $(A_{tot} - A_1^0)/A_1^0$ then gives k^-/σ^2 as the intercept on the $1/\tau_1$ axis and k^-/n as the gradient. An independent determination of n by light scattering, for example, allows the calculation of k^- and σ. Further, from the expression for the equilibrium constant,

$$\frac{k^+}{k^-} = \frac{A_s^0}{A_1^0 A_{s-1}^0} \tag{3.34}$$

we can obtain k^+ using the approximate relationship,

$$\frac{k^+}{k^-} = \frac{1}{A_1^0} = \frac{1}{\text{c.m.c.}} \tag{3.35}$$

3.4.2.3 The slow relaxation time

To derive an expression for the slow relaxation time, one starts with equations (3.20) and (3.21) which characterize the slow process. In order to proceed, it is necessary to assume that J_s is a slowly varying function of s in region II so that J_s can be replaced by J.

An expression for $\xi_1(t)$ is then sought in the form

$$\xi_1(t) = \xi_1(0) \exp(-t/\tau_2) \tag{3.36}$$

where τ_2 is the slow relaxation time. This can be done by noting that the slow process corresponds to a flow into the micellar region which changes the number of micelles present. The instantaneous value of $\xi_1(t)$ must then be related to the excess number of micelles in region III, i.e. it must be related to the difference in the number of micelles present at t and the number present at equilibrium. This latter quantity is simply given by,

$$\text{excess number of micelles} = \sum_{s>s_2} \xi_s A_s^0. \tag{3.37}$$

Since the change in this quantity results from the flow J into region III we can write,

$$\frac{d}{dt} \sum_{s>s_2} \xi_s A_s^0 = J \tag{3.38}$$

$\sum_{s>s_2} \xi_s A_s^0$ and J are then expressed as linear functions of $\xi_1(t)$ to give a first-order differential equation to which (3.36) is the solution. This is achieved by summing (3.20) and (3.21) over values of s in the regions to which these equations apply giving

$$\xi_s = s\xi_1 \qquad \text{for} \quad s \leqslant s_1 \tag{3.39}$$

$$\xi_s = s\xi_1 - RJ \quad \text{for} \quad s_1 < s \leqslant s_2 \tag{3.40}$$

and

$$\xi_s = s\xi_1 - RJ \quad \text{for} \quad s > s_2 \tag{3.41}$$

where

$$R = \sum_{s=s_1+1}^{s_2} \frac{1}{k_s^- A_s^0}. \tag{3.42}$$

Following some straightforward algebra which entails using the material balance equation (3.11), one then finds,

$$\frac{1}{\tau_2} = \frac{1}{Rc_3} \left[\frac{\overline{n_1^2}c_1 + \overline{n_3^2}c_3}{\overline{n_1^2}c_1 + \sigma^2 c_3} \right] \tag{3.43}$$

where n_1 and n_3 represent aggregation numbers in region I and III respectively and c_1 and c_3 are total concentrations of aggregates in these regions.

$$c_1 = \sum_{s=1}^{s_1} A_s^0 \tag{3.44}$$

and

$$c_3 = \sum_{s>s_2} A_s^0. \tag{3.45}$$

The bars denote average values defined by

$$\overline{n^m} = \frac{\sum_s s^m A_s^0}{\sum_s A_s^0} \tag{3.46}$$

where the summations are taken over s values in the region concerned.

Also,

$$\sigma^2 = \overline{n_3^2} - \bar{n}_3^2. \tag{3.47}$$

Equation (3.43) can be simplified for surfactant solutions sufficiently above the c.m.c. in the following way,

$$\text{(a) } \overline{n_1^2} = 1$$
$$\text{(b) } c_1 = A_1^0 = \text{c.m.c.} \tag{3.48}$$

These assume that the concentrations of dimers, trimers, etc. in region I are negligible.

$$\text{(c) } \overline{n_3^2 c_3} \simeq \bar{n}_3^2 c_3 \gg A_1^0. \tag{3.49}$$

This assumes that, σ^2, is small in comparison to the mean aggregation number $\bar{n}_3$. Writing $\bar{n}_3$ as n, (3.43) then becomes,

$$\frac{1}{\tau_2} = \frac{n^2}{R}\left[A_1^0 + \frac{\sigma^2}{n}(A_{\text{tot}} - A_1^0)\right]^{-1}. \tag{3.50}$$

From (3.50), τ_2 appears to increase with increasing A_{tot}. However, experimental results indicate that variations in R with concentration must also be considered (Aniansson 1976). This corresponds to changing the cross-sectional area at the region M in Figure 3.9. These interpretations assume that σ and n do not vary with concentration.

The rate constants which enter into the slow process are contained in R as indicated in (3.42). The value of τ_2 has been found to be very sensitive to ionic strength and also to the presence of small amounts of impurity in the surfactant solution. This sensitivity has been interpreted as resulting from changes in the value of R (Aniansson et al., 1976).

The model described above is most suitable as a description of surfactant solutions in which τ_1 and τ_2 differ by several orders of magnitude. The treatment is expected to be less satisfactory for systems in which τ_1 and τ_2 are of similar magnitudes. In such cases, the coupling between the fast and the slow processes would have to be taken into account and they could no longer be treated separately as is implied by equations (3.20) and (3.21) for example.

3.4.3 The model of Sams, Wyn-Jones, and Rassing

At this point attention should be drawn to a much simpler and earlier model (Sams et al., 1972) for the fast process.

As in the model of Aniansson and Wall, the fast process is identified with a monomer–micelle exchange interaction as indicated by (3.2), but no attempt was made to restrict the values of the aggregation number s to large values. This corresponds to assuming a large value for σ^2, the variance of the micelle distribution function. The rate for A_1 was written as

$$\frac{d}{dt}A_1 = -k_f A_1 \sum_{s>1} sA_s + k_b \sum_{s>1} sA_s. \tag{3.51}$$

It is here assumed that the rate for A_1 is proportional to the number of monomers in aggregated form. The relaxation time derived from (3.51) is

$$\frac{1}{\tau_1} = k_b\left(\frac{(A_{tot} - A_1^0)}{A_1^0}\right). \tag{3.52}$$

It should be noted, however, that the constants k_f and k_b which appear in (3.51) are not true rate constants because of the factor s which appears before the concentration A_s in the summations. In terms of true rate constants k^+ and k^-, (3.51) should be written as

$$\frac{dA_1}{dt} = -k^+ A_1 \sum_{s>1} A_s + k^- \sum_{s>1} A_s. \tag{3.53}$$

Noting that the mean aggregation number, n, is defined by:

$$n = \frac{\displaystyle\sum_{s>1} sA_s^0}{\displaystyle\sum_{s>1} A_s^0} \tag{3.54}$$

we see that

$$k_f = \frac{k^+}{n} \quad \text{and} \quad k_b = \frac{k^-}{n}. \tag{3.55}$$

Equation (3.52) then becomes

$$\frac{1}{\tau_1} = \frac{k^-}{n}\left(\frac{(A_{tot} - A_1^0)}{A_1^0}\right). \tag{3.56}$$

This is to be compared with (3.32). The omission of the term k^-/σ^2 is to be expected due to the wide distribution curve assumed in this simpler model.

If one distinguishes between proportionality constants and true rate constants then it is evident that this model of Sams and coworkers yields very similar rate constants to the Aniansson and Wall model. The kinetic principles of this simple model have also been successfully applied to investigations of the fast relaxation process in mixed micelles (Hall *et al.*, 1977), the exchange process between neutral molecules and micelles (Gettins *et al.*, 1978), and the exchange process between water molecules in bulk solution and those bound to surfactant aggregates (Tiddy *et al.*, 1979).

Finally, Aniansson has further developed his model to consider mixed micelles (Aniansson, 1979).

3.5 The evaluation of Experimental Data

3.5.1 *The fast relaxation time*

The fast relaxation time, τ_1, which is associated with the monomer–micelle exchange process, has been investigated mainly using sound absorption and pressure-jump techniques. The magnitude of this time depends on the length of the hydrocarbon tail of the surfactant: the shorter the chain length, the faster is the relaxation time. Figure 3.10 shows the variation of the

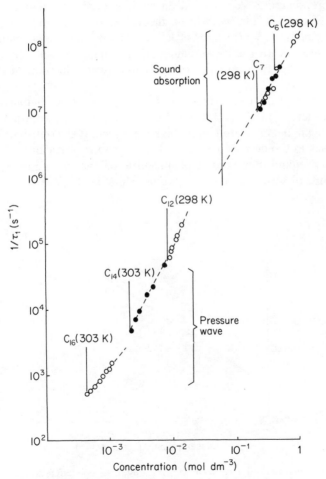

Figure 3.10 A plot of $1/\tau_1$ against concentration for a series of sodium alkyl sulphates (Aniansson *et al.*, 1976)

reciprocal of this fast relaxation time with chain length for a homologous series of sodium alkyl sulphates.

An examination of the concentration dependence of the fast relaxation time shows that, in general, as the surfactant concentration, A_{tot}, increases, $1/\tau_1$ increases linearly as is shown in Figures 3.11 and 3.12. Using the relationship $k^+/k^- = 1/A_1^0$, equation (3.32) may be expressed in the form:

$$\frac{1}{\tau_1} = \frac{k^+}{n} \cdot A_{tot} + \frac{k^-}{n} \left[\frac{n}{\sigma^2} - 1 \right]. \tag{3.57}$$

In many micellar systems the mean aggregation number, n, has been determined from independent measurements, usually light scattering. Making the approximation $A_1^0 = $ c.m.c., an analysis of the concentration dependence of $1/\tau_1$ is possible by plotting the values of $1/\tau_1$ against A_{tot} and by the use of equation (3.57) values of k^+, k^-, and σ^2 may be estimated. These quantities along with the mean aggregation number, n, and the c.m.c. are listed in Table 3.1 for a number of surfactants which have been studied by various authors.

The accuracy of the derived values of these parameters obviously depends on several factors. In general, it has been found that the accuracy to which the c.m.c. can be measured may make a substantial contribution to the uncertainties in the derived values of k^+, k^-, and in particular σ^2. It is well known from equilibrium studies of micellization that the c.m.c. is in fact a narrow range of surfactant concentrations (Shinoda, 1967, Mittal, 1977) and

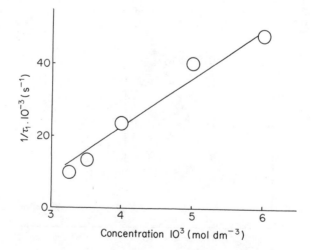

Figure 3.11   A plot of $1/\tau_1$ against concentration for dodecylpyridinium perfluorbutyrate at 298 K (Hoffmann *et al.*, 1979)

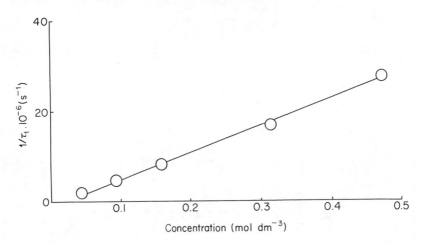

Figure 3.12   A plot of $1/\tau_1$ against concentration for caesium perfluoro-ctanoate at 298 K (Gettins *et al.*, 1979)

in general the different methods which are used to estimate this quantity do not always give identical results. For long chain length surfactants this range of concentrations is small enough to allow precise results to be obtained but on the other hand the c.m.c. for short chain length surfactants is not so well defined. Of all of the parameters which may be obtained from kinetic data it is the tolerance in the values of σ^2 which is most sensitive to the c.m.c. value used.

The order of magnitude of the rate constants for monomer–micelle exchange, listed in Table 3.1 indicates that the forward rate constant, k^+ has a value within the range $\sim(1-55)\times10^8$ mol^{-1} dm^3 s^{-1} and is almost diffusion controlled (Aniansson *et al.*, 1976). This has recently been supported by further experiments measuring the fast relaxation times of surfactants as solutions in both water and deuterium oxide (Gettins *et al.*, 1979a). An example of the $1/\tau_1$ against surfactant concentration plots for decyl-trimethylammonium bromide (DTAB) is shown in Figure 3.13. The most significant feature of these data is that the slopes of the relaxation plots are greater when the solvent is water. The forward rate constants for DTAB obtained from the plots shown in Figure 3.13 are $k^+_{H_2O} = 5.5\times10^9$ mol^{-1} dm^3 s^{-1}, $k^+_{D_2O} = 5.1\times10^9$ mol^{-1} dm^3 s^{-1} and the ratio of these, $k^+_{H_2O}/k^+_{D_2O} = 1.1$ compares extremely well with the ratio of the viscosities of the solvents $\eta_{D_2O}/\eta_{H_2O} = 1.2$ confirming that the process is diffusion controlled. The dissociation rate constant, k^-, is found to decrease with increasing chain length for a homologous series of surfactants. This observation has been correlated with the hydrophobic energy of the CH_2 groups in the hydrocarbon chains (Aniansson *et al.*, 1976).

Table 3.1

Surfactant	C.m.c. (mol dm⁻³)	n	k^+ (mol⁻¹ dm³ s⁻¹)	k^- (s⁻¹)	σ^2	Temperature (K)	References
Sodium hexyl sulphate	0.42	17	3.2×10^9	1.32×10^9	36	298	Aniansson et al. (1976)
Sodium heptyl sulphate	0.22	22	3.3×10^9	7.3×10^8	100	298	Aniansson et al. (1976)
Sodium octyl sulphate	0.13	27	7.7×10^8	1.0×10^8		298	Aniansson et al. (1976)
Sodium octyl sulphate (solvent, water)	0.11	25	2.7×10^9	2.9×10^8	312	298	Gettins et al. (1979a)
Sodium octyl sulphate (solvent, deuterium oxide)	0.11	19	1.5×10^9	1.6×10^8	136	298	Gettins et al. (1979a)
Sodium nonyl sulphate	6.10^{-2}	33	2.3×10^9	1.4×10^8		298	Aniansson et al. (1976)
Sodium decyl sulphate	3.3×10^{-2}	41	2.7×10^9	9×10^7		313	Aniansson et al. (1976)
Sodium undecyl sulphate	1.6×10^{-2}	52	2.6×10^9	4×10^7		298	Aniansson et al. (1976)
Sodium dodecyl sulphate	8.2×10^{-3}	64	1.2×10^9	1.0×10^7	169	298	Aniansson et al. (1976)
Cobalt dodecyl sulphate	2.46×10^{-3}	108	1.05×10^9	2.6×10^6	121	298	Baumuller et al. (1978)
Cobalt dodecyl sulphate	2.46×10^{-3}	103	1.5×10^9	3.7×10^6	144	303	Baumuller et al. (1978)
Cobalt dodecyl sulphate	2.46×10^{-3}	98	2.0×10^9	5×10^6	144	308	Baumuller et al. (1978)
Nickel dodecyl sulphate	2.48×10^{-3}	112	0.6×10^9	1.6×10^6	169	283	Baumuller et al. (1978)
Nickel dodecyl sulphate	2.48×10^{-3}	107	0.9×10^9	2.2×10^6	185	288	Baumuller et al. (1978)
Nickel dodecyl sulphate	2.48×10^{-3}	102	1.2×10^9	2.9×10^6	188	293	Baumuller et al. (1978)
Nickel dodecyl sulphate	2.48×10^{-3}	97	1.6×10^9	3.9×10^6	169	298	Baumuller et al. (1978)
Nickel dodecyl sulphate	2.48×10^{-3}	92	2.3×10^9	5.7×10^6	185	303	Baumuller et al. (1978)
Sodium tetradecyl sulphate	2.05×10^{-3}	80	4.7×10^8	9.6×10^5	272	298	Aniansson et al. (1976)
Sodium tetradecyl sulphate	2.1×10^{-3}	80	4.6×10^8	9.6×10^5	784	298	Inoue et al. (1978)
Sodium tetradecyl sulphate	2.2×10^{-3}	80	5.4×10^8	1.2×10^6	289	303	Inoue et al. (1978)
Sodium tetradecyl sulphate	2.2×10^{-3}	80	6.7×10^8	1.5×10^6	256	308	Inoue et al. (1978)
Sodium hexadecyl sulphate	4.5×10^{-4}	100	1.3×10^8	6×10^4	121	303	Aniansson et al. (1976)
Octylpyridinium chloride	0.21	27	2.5×10^9	5.3×10^8	48	298	Hoffmann et al. (1976)
Dodecylpyridinium iodide	4.96×10^{-3}	76	9.8×10^8	4.8×10^6	369	278	Hoffmann et al. (1976)
Dodecylpyridinium iodide	4.85×10^{-3}	73	1.4×10^9	6.7×10^6	419	283	Hoffmann et al. (1976)
Dodecylpyridinium iodide	5.01×10^{-3}	70	1.9×10^9	9.4×10^6	348	288	Hoffmann et al. (1976)
Dodecylpyridinium iodide	5.13×10^{-3}	67	2.3×10^9	1.2×10^7	293	293	Hoffmann et al. (1976)

Dodecylpyridinium iodide	5.25×10^{-3}	64	3.0×10^{9}	1.6×10^{7}	225	298	Hoffmann et al. (1976)
Dodecylpyridinium perfluorpropionate	5.69×10^{-3}	65	1.4×10^{9}	7.8×10^{6}	100	278	Hoffmann et al. (1979)
Dodecylpyridinium perfluorpropionate	4.88×10^{-3}	62	1.9×10^{9}	9.3×10^{6}	92	288	Hoffmann et al. (1979)
Dodecylpyridinium perfluorbutyrate	3.38×10^{-3}	189	1.1×10^{9}	3.8×10^{6}	1225	278	Hoffmann et al. (1979)
Dodecylpyridinium perfluorbutyrate	2.93×10^{-3}	193	1.9×10^{9}	5.6×10^{6}	1332	288	Hoffmann et al. (1979)
Dodecylpyridinium perfluorbutyrate	2.86×10^{-3}	189	2.8×10^{9}	7.9×10^{6}	1369	289	Hoffmann et al. (1979)
Dodecylpyridinium methylbenzoate	3.80×10^{-3}	71	1.3×10^{9}	5.0×10^{6}	144	288	Hoffmann et al. (1979)
Tetradecylpyridinium chloride	4.48×10^{-3}	96	1.5×10^{9}	6.7×10^{6}	239	278	Hoffmann et al. (1976)
Tetradecylpyridinium chloride	4.2×10^{-3}	92	2.0×10^{9}	8.6×10^{6}	232	283	Hoffmann et al. (1976)
Tetradecylpyridinium chloride	4.1×10^{-3}	88	2.4×10^{9}	9.7×10^{6}	180	288	Hoffmann et al. (1976)
Tetradecylpyridinium bromide	2.63×10^{-3}	96	6.7×10^{8}	1.8×10^{6}	150	278	Hoffmann et al. (1976)
Tetradecylpyridinium bromide	2.57×10^{-3}	92	1.0×10^{9}	2.6×10^{6}	137	283	Hoffmann et al. (1976)
Tetradecylpyridinium bromide	2.49×10^{-3}	88	1.3×10^{9}	3.2×10^{6}	123	288	Hoffmann et al. (1976)
Tetradecylpyridinium bromide	2.45×10^{-3}	84	1.8×10^{9}	4.3×10^{6}	116	293	Hoffmann et al. (1976)
Tetradecylpyridinium bromide	2.55×10^{-3}	80	2.5×10^{9}	6.3×10^{6}	107	298	Hoffmann et al. (1976)
Tetradecylpyridinium bromide	2.63×10^{-3}	76	3.9×10^{9}	1.0×10^{7}	106	303	Hoffmann et al. (1976)
Tetradecylpyridinium iodide	1.5×10^{-3}	72	2.0×10^{9}	3.1×10^{6}	564	308	Hoffmann et al. (1976)
Tetradecylpyridinium iodide	1.5×10^{-3}	68	3.3×10^{9}	5.1×10^{6}	699	313	Hoffmann et al. (1976)
Tetradecylpyridinium perfluoracetate	1.93×10^{-3}	41	3.0×10^{8}	5.7×10^{5}	31	288	Hoffmann et al. (1979)
Tetradecylpyridinium perfluoracetate	1.90×10^{-3}	44	4.7×10^{8}	9.0×10^{5}	41	298	Hoffmann et al. (1979)
Tetradecylpyridinium perfluoracetate	2.00×10^{-3}	35	5.3×10^{8}	1.1×10^{6}	28	308	Hoffmann et al. (1979)
Tetradecylpyridinium perfluorpropionate	1.30×10^{-3}	137	2.1×10^{9}	2.7×10^{6}	361	278	Hoffmann et al. (1979)
Tetradecylpyridinium perfluorpropionate	1.10×10^{-3}	153	3.2×10^{9}	3.5×10^{6}	441	288	Hoffmann et al. (1979)
Tetradecylpyridinium perfluorpropionate	1.15×10^{-3}	103	2.5×10^{9}	2.8×10^{6}	225	298	Hoffmann et al. (1979)
Tetradecylpyridinium perfluorpropionate	1.20×10^{-3}	98	3.0×10^{9}	3.7×10^{6}	225	308	Hoffmann et al. (1979)
Tetradecylpyridinium perfluorbutyrate	8.40×10^{-4}	27	8.0×10^{7}	6.8×10^{4}	27	278	Hoffmann et al. (1979)
Tetradecylpyridinium perfluorbutyrate	7.60×10^{-4}	20	1.2×10^{8}	9.0×10^{4}	11	288	Hoffmann et al. (1979)
Tetradecylpyridinium perfluorbutyrate	6.50×10^{-4}	18	2.8×10^{8}	1.8×10^{5}	10	298	Hoffmann et al. (1979)
Tetradecylpyridinium butanesulphonate	2.2×10^{-3}	56	1.1×10^{9}	2.5×10^{6}	81	288	Hoffmann et al. (1977)
Tetradecylpyridinium butanesulphonate	2.2×10^{-3}	47	1.8×10^{9}	4.0×10^{6}	64	298	Hoffmann et al. (1977)
Tetradecylpyridinium benzenesulphonate	2.3×10^{-3}	53	8.2×10^{8}	1.9×10^{6}	64	288	Hoffmann et al. (1977)
Tetradecylpyridinium benzenesulphonate	2.3×10^{-3}	42	1.1×10^{9}	2.4×10^{6}	42	298	Hoffmann et al. (1977)
Tetradecylpyridinium methanesulphonate	4.7×10^{-3}	50	1.4×10^{9}	6.5×10^{6}	108	283	Hoffmann (1978)
Tetradecylpyridinium ethanesulphonate	4.1×10^{-3}	48	9.6×10^{8}	3.9×10^{6}	86	283	Hoffmann (1978)

Table 3.1 (Continued)

Surfactant	C.m.c. (mol dm⁻³)	n	k^+ (mol⁻¹ dm³ s⁻¹)	k^- (s⁻¹)	σ^2	Temperature (K)	References
Tetradecylpyridinium propanesulphonate	3.1×10^{-3}	53	8.7×10^{8}	2.7×10^{6}	85	283	Hoffmann (1978)
Tetradecylpyridinium butanesulphonate	2.4×10^{-3}	56	7.2×10^{8}	1.7×10^{6}	76	283	Hoffmann (1978)
Tetradecylpyridinium pentanesulphonate	1.55×10^{-3}	49	4.7×10^{8}	7.3×10^{5}	52	283	Hoffmann (1978)
Tetradecylpyridinium hexanesulphonate	9.5×10^{-4}	46	4.4×10^{8}	4.1×10^{5}	41	283	Hoffmann (1978)
Tetradecylpyridinium heptanesulphonate	5.5×10^{-4}	21	9.5×10^{7}	5.3×10^{4}	18	283	Hoffmann (1978)
Hexadecylpyridinium chloride	9×10^{-4}	120	2.4×10^{8}	2.3×10^{5}	74	278	Hoffmann et al. (1976)
Hexadecylpyridinium chloride	9×10^{-4}	115	3.8×10^{8}	3.6×10^{5}	113	283	Hoffmann et al. (1976)
Hexadecylpyridinium chloride	9×10^{-4}	110	4.6×10^{8}	4.4×10^{5}	113	288	Hoffmann et al. (1976)
Hexadecylpyridinium chloride	9×10^{-4}	105	5.2×10^{8}	5.0×10^{5}	109	293	Hoffmann et al. (1976)
Hexadecylpyridinium chloride	9×10^{-4}	100	6.3×10^{8}	6.0×10^{5}	118	298	Hoffmann et al. (1976)
Hexadecylpyridinium chloride	9×10^{-4}	95	7.7×10^{8}	7.3×10^{5}	120	303	Hoffmann et al. (1976)
Hexadecylpyridinium chloride	9×10^{-4}	90	9.8×10^{8}	9.3×10^{5}	116	308	Hoffmann et al. (1976)
Hexadecylpyridinium bromide	6.0×10^{-4}	115	2.1×10^{8}	1.2×10^{5}	60	283	Hoffmann et al. (1976)
Hexadecylpyridinium bromide	6.3×10^{-4}	110	3.4×10^{8}	2.1×10^{5}	70	288	Hoffmann et al. (1976)
Hexadecylpyridinium bromide	6.5×10^{-4}	105	4.5×10^{8}	2.7×10^{5}	77	293	Hoffmann et al. (1976)
Hexadecylpyridinium bromide	6.8×10^{-4}	100	5.5×10^{8}	3.3×10^{5}	92	298	Hoffmann et al. (1976)
Hexadecylpyridinium bromide	7.1×10^{-4}	95	6.7×10^{8}	4.7×10^{5}	98	303	Hoffmann et al. (1976)
Hexadecylpyridinium bromide	7.6×10^{-4}	90	7.6×10^{8}	5.8×10^{5}	83	308	Hoffmann et al. (1976)
Hexadecylpyridinium bromide	7.8×10^{-4}	85	9.4×10^{8}	7.4×10^{5}	83	313	Hoffmann et al. (1976)
Hexadecylpyridinium bromide	8.0×10^{-4}	80	1.1×10^{9}	8.4×10^{5}	88	318	Hoffmann et al. (1976)
Dimethylammonium perfluoroctanesulphonate	2.08×10^{-3}	753	2.3×10^{9}	4.8×10^{6}	12,100	278	Hoffmann (1978)
Dimethylammonium perfluoroctanesulphonate	2.17×10^{-3}	183	9.2×10^{8}	2.0×10^{6}	784	288	Hoffmann (1978)
Dimethylammonium perfluoroctanesulphonate	2.32×10^{-3}	136	1.3×10^{9}	2.9×10^{6}	484	298	Hoffmann (1978)
Dimethylammonium perfluoroctanesulphonate	2.42×10^{-3}	126	2.0×10^{9}	4.9×10^{6}	484	308	Hoffmann (1978)
Decyltrimethylammonium bromide (solvent, water)	6.6×10^{-2}	36	5.5×10^{9}	3.6×10^{8}	900	298	Gettins et al. (1979a)
Decyltrimethylammonium bromide (solvent, deuterium oxide)	6.6×10^{-2}	42	5.1×10^{9}	3.4×10^{8}	840	298	Gettins et al. (1979a)

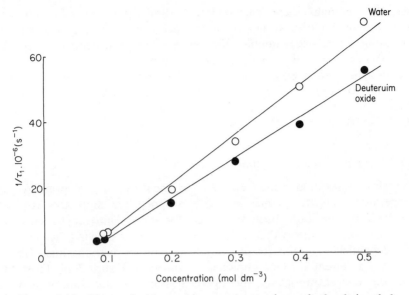

Figure 3.13 Plots of $1/\tau_1$ against concentration of decyltrimethyl-ammonium bromide as solutions in water and deuterium oxide at 298 K (Gettins *et al.*, 1979)

In both ultrasonic and pressure-jump experiments, deviations from the linearity of the $1/\tau_1$ against A_{tot} plots predicted by Aniansson and Wall, have been observed (Rassing *et al.*, 1974; Hoffmann *et al.*, 1976). $1/\tau_1$ exhibits curvature at higher surfactant concentrations, which has been explained in terms of possible changes in n and σ^2. An interesting observation has also been reported in a narrow surfactant concentration region just above the c.m.c. in ultrasonic relaxation studies of short chain length surfactants (Sams *et al.*, 1975; Reinsborough, 1975). On increasing the surfactant concentration it is found that $1/\tau_1$ decreases with increasing surfactant concentration before the normal kinetic behaviour according to equation (3.57) is observed. It has been shown that this apparently unusual type of behaviour is probably associated with pre-micellar aggregates (Gettins *et al.*, 1979a). At the initial surfactant concentration in this narrow range it is conceivable that pre-micellar aggregates (submicelles) are present and that they continually 'grow' with increasing surfactant concentration until 'micelles proper' are formed. During the growth of these aggregates it is possible for their molar concentration to actually decrease with increasing surfactant concentration which accounts for the decrease in the observed values of $1/\tau_1$.

The magnitude of the fast relaxation time has been found to be fairly insensitive to impurities and moderate ionic strength (Wyn-Jones, 1975).

Activation enthalpies and entropies and volume differences for monomers associating with micelles can also be obtained from measurements of the fast relaxation time. For example, the temperature dependence of the rate constants can be used to evaluate the activation energies and enthalpies for this process (Aniansson *et al.*, 1976; Hoffmann *et al.*, 1976) and the concentration dependence of the relaxation amplitude yields the volume difference between a monomer in free solution and in a micelle (Rassing *et al.*, 1974; Teubner, 1979).

3.5.2 *The slow relaxation time*

The slow relaxation time, τ_2, can be measured using pressure- and temperature-jump techniques as well as stopped-flow methods. At present this quantity has only been measured for long chain length surfactants (hydrocarbon tail longer than C_{10}). A relaxation plot showing the concentration dependence of the reciprocal of the slow relaxation time, $1/\tau_2$, for dodecylpyridinium perfluoropropionate is shown in Figure 3.14. From these data the concentrations of the aggregates in the distribution minimum (region II, Figure 3.9) can be calculated from the term R in equation (3.50), (Aniansson *et al.*, 1976; Hoffmann *et al.*, 1976):

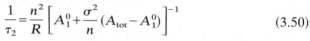

$$\frac{1}{\tau_2} = \frac{n^2}{R}\left[A_1^0 + \frac{\sigma^2}{n}(A_{tot} - A_1^0)\right]^{-1} \qquad (3.50)$$

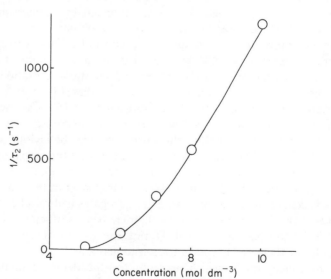

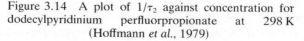

Figure 3.14 A plot of $1/\tau_2$ against concentration for dodecylpyridinium perfluoropropionate at 298 K (Hoffmann *et al.*, 1979)

where

$$R = \sum_{s=s_1+1}^{s_2} \left(\frac{1}{k_s^- A_s^0}\right). \tag{3.42}$$

In some surfactants the aggregation number has not been determined from equilibrium measurements but for certain systems both the fast and slow relaxation times have been measured and a value of R can be calculated by assuming a Gaussian curve for the micellar distribution (Hoffmann, 1978). In these circumstances n and σ^2 can be evaluated from the relaxation data. These data are also listed in Table 3.1. An alternative approach for the evaluation of the width of the micellar distribution curve has also been described by the analysis of the amplitude of the slow process (Teubner et al., 1978).

Kinetic and thermodynamic energy parameters associated with the formation of aggregates at the distribution minimum can also be obtained from studies of the slow relaxation time (Aniansson et al., 1976; Hoffmann et al., 1976). From studies of the temperature dependence of the relaxation time activation enthalpies and entropies can be derived and the relaxation amplitude yields the appropriate volume changes (Teubner et al., 1978).

It has also been observed that the slow relaxation time is very sensitive to additives (Wyn-Jones, 1975).

3.6 Other Aggregation Phenomena

This chapter has so far been exclusively concerned with aggregation phenomena in aqueous micellar solutions of surfactants which have definitive aggregation distributions as depicted in Figure 3.7. Other modes of aggregation have been postulated to account for the association which occurs in solutions of solutes such as alcohols, amides, purine, and pyrimidine derivatives, proteins, and dyes. In some cases, hydrogen bonding is thought to be the cause of the association, whereas in the case of the nucleoside bases, hydrophobic interactions are thought to be of prime importance. The nucleoside bases and dyes are essentially planar aromatic compounds and aggregation is described in terms of the stacking of molecules, one on top of the other. The term, 'base-stacking' is often used to describe this process. These systems do not exhibit a critical concentration at which large aggregates are formed but in some instances aggregates of comparable sizes to those existing in typical surfactant micellar solutions are found. It must be noted that it cannot be assumed that small aggregates are present in solution only in negligible quantities as is done in connection with micelle-forming surfactants. The relaxation spectra of all of these systems are characterized by either a well-defined single relaxation time or by a distribution of relaxation times which are very similar. Kinetic models involving sequences of step-wise bimolecular equilibria have been considered for these aggregation phenomena (Hammes and Hubbard, 1966; Rassing, 1972; Porschke and Eggers, 1972; Turner et al., 1972; Garland and Patel, 1975; Garland and Christian, 1975; Robinson et al., 1975; Thusius, et al., 1975; Thusius, 1975; Heyn et al., 1977).

It is interesting to note that drugs also exhibit colloidal behaviour (Mukerjee, 1974; Mukerjee and Cardinal, 1976; Mittal, 1977; Attwood, 1979). The structure of many of these molecules is such that they contain hydrophobic moieties as well as aromatic bases. This means that two distinct mechanisms for self-association are possible: association of the hydrophobic tails of the molecules and also, stacking of the planar aromatic bases. Equilibrium studies on these systems have shown that some molecules definitely form micelles similar to those of the surfactants listed in Table 3.1. Other systems have been found to contain monomers and aggregates with sometimes the aggregates having a small aggregation number. There are reports of relaxation studies on these types of systems and in general a 'fast' relaxation time is observed (Gettins et al. 1976; Gettins et al., 1979b). The relaxation spectra are characterized either by a well-defined single relaxation time or a distribution of relaxation times which are very similar. We are not aware of any reports of the existence of a 'slow' relaxation time in these solutions. There can be no doubt that the aggregates are formed in these systems by a step-wise process as illustrated in reaction (3.1), with perhaps additional random bimolecular equilibria between aggregates. Up to the present time, very little effort has been concentrated on the interpretation of the available relaxation data. Clearly, a knowledge of the form of the aggregation distribution function is required and in order to understand the kinetic behaviour the relaxation data must be complemented by equilibrium measurements.

Finally, there is also a paucity of kinetic data on the aggregation phenomena which occur in systems which contain reversed micelles, that is, a 'normal' surfactant dissolved in a non-aqueous medium.

3.7 References

Alkaitis, S. A. and Gratzel, M. (1976). *J. Amer. Chem. Soc.*, **98,** 3549.
Aniansson, E. A. G. (1978). *Ber. Bunsenges Phys. Chem.*, **82,** 981.
Aniansson, E. A. G. (1979). In *Techniques and Applications of Fast Reactions in Solution* (Eds. Gettins, W. J. and Wyn-Jones, E.), Reidel, Dordrecht.
Aniansson, E. A. G. and Wall, S. N. (1974). *J. Phys. Chem.*, **78,** 1024.
Aniansson, E. A. G. and Wall, S. N. (1975). *J. Phys. Chem.*, **79,** 857.
Aniansson, E. A. G., Wall, S. N., Almgren, M., Hoffmann, H., Kielmann, I., Ulbricht, W., Zana, R., Lang, J., and Tondre, C. (1976). *J. Phys. Chem.*, **80,** 905.
Attwood, D. (1979). In *Techniques and Applications of Fast Reactions in Solution* (Eds. Gettins, W. J. and Wyn-Jones, E.), Reidel, Dordrecht.

Balmbra, R., Clunie, J., Corkill, J., and Goodman, J. (1962). *Trans. Faraday Soc.*, **58,** 1661.

Bansal, V. M. and Shah, D. O. (1977). In *Micellization, Solubilization and Microemulsions* (Ed. Mittal, K. L.), Vol. 1, Plenum Press, p. 87.

Baumuller, W., Hoffmann, H., Ulbricht, W., Tondre, C., and Zana, R. (1978). *J. Colloid Interface Sci.*, **64,** 418.

Bennion, B. C. and Eyring, E. M. (1970). *J. Colloid Interface Sci.*, **32,** 286.

Bennion, B. C., Tong, L. K., Holmes, L. P., and Eyring, E. M. (1969). *J. Phys. Chem.*, **73,** 3288.

Bernasconi, C. F. (1976). *Relaxation Kinetics*, Academic Press, New York.

Bradley, J. N. (1975). *Fast Reactions*, Clarendon Press, Oxford.

Britton Chance, Eisenhardt, R. H., Gibson, R. H., and Lonberg-Holm, K. K. (Eds.). (1964). *Rapid Mixing and Sampling Techniques in Biochemistry*, Academic Press, New York.

Caldin, E. F. (1964). *Fast Reactions in Solution*, Blackwell, Oxford.

Claesson, S. (Ed.). (1967). *Fast Reactions and Primary Processes in Chemical Kinetics*, Interscience.

Clifford, J. (1965). *Trans. Faraday Soc.*, **61,** 1267.

Colen, A. (1974). *J. Phys. Chem.*, **78,** 1676.

Courcheene, W. L. (1964). *J. Phys. Chem.*, **68,** 1870.

Czerlinski, G. (1966). *Chemical Relaxation*, Arnold.

Eigen, M. and De Maeyer, L. (1963). *Technique of Organic Chemistry*, (Ed. Weissberger, A.), Vol. VIII, Part II, John Wiley, New York.

Ekwall, P. (1964). *Chemistry, Physics and the Application of Surface-Active Substances*, **2,** 651.

Ekwall, P. and Holmberg, P. (1965). *Acta Chem. Scand.*, **19,** 455.

Fendler, J. H. (1976). *Acc. Chem. Res.*, **9,** 153.

Folger, R., Hoffmann, H., and Ulbricht, W. (1974). *Ber. Bunsenges. Phys. Chem.*, **78,** 986.

Franklin, T. C. and Honda, T. (1977). In *Micellization, Solubilization and Microemulsions*, (Ed. Mittal, K. L.), Vol. 2, Plenum Press, 617.

Garland, F. and Christian, S. D. (1975). *J. Phys. Chem.*, **79,** 1247.

Garland, F. and Patel, R. C. (1975). *J. Phys. Chem.*, **78,** 848.

Geffen, T. M. (1975). *World Oil*, March.

Gettins, J., Gould, C., Hall, D., Jobling, P. L., Rassing, J. E., and Wyn-Jones, E. (1979a), *J. Chem. Soc. Faraday II*, in press.

Gettins, W. J., Greenwood, R. C., Natarajan, R., and Wyn-Jones, E. (1979b). Unpublished results.

Gettins, J., Greenwood, R., Rassing, J., and Wyn-Jones, E. (1976). *Chem. Commun.*, 1030.

Gettins, J., Hall, D., Jobling, P. L., Rassing, J. E., and Wyn-Jones, E. (1978). *J. Chem. Soc. Faraday II*, **74,** 1957.

Gettins, J., Jobling, P. L., Walsh, M. F., and Wyn-Jones, E. (1979c). *J. Chem. Soc. Faraday II*, in press.

Gettins, W. J. and Wyn-Jones, E. (Eds.) (1979). *Techniques and Applications of Fast Reactions in Solution*, Reidel, Dordrecht.

Graber, E., Lang, J., and Zana, R. (1970). *Kolloid Z. Z. Polymere*, **238,** 470.

Graber, E. and Zana, R. (1970). *Kolloid Z. Z. Polymere*, **238,** 479.

Hague, D. N. (1971). *Fast Reactions*, John Wiley.

Hall, D., Jobling, P. L., Wyn-Jones, E., and Rassing, J. E. (1977). *J. Chem. Soc. Faraday II*, **73,** 1582.

Hammes, G. G. (Ed.) (1974). *Investigation of Rates and Mechanisms of Reactions, Techniques of Chemistry*, Vol. VI, part II, Interscience, New York.

Hammes, G. G. and Hubbard, C. D. (1966). *J. Phys. Chem.*, **70**, 1615.

Heyn, M. P., Nicola, C. U., and Schwarz, G. (1977). *J. Phys. Chem.*, **81**, 1611.

Hoffmann, H. (1978). *Ber. Bunsenges Phys. Chem.*, **82**, 988.

Hoffmann, H. and Janjic, T. (1973). *Z. Phys. Chem. NF*, **86**, 322.

Hoffmann, H., Nagel, R., Platz, G., and Ulbricht, W. (1976). *Colloid Polym. Sci.*, **254**, 812.

Hoffmann, H., Nusslein, H., and Ulbricht, W. (1977). In *Micellization, Solubilization and Microemulsions* (Ed. Mittal, K. L.), Vol. 1., Plenum Press, p. 263.

Hoffmann, H., Tagesson, B., and Ulbricht, W. (1979). *Ber. Bunsenges Phys. Chem.*, **83**, 148.

Inoue, T., Shibuya, Y., and Shimozawa, R. (1978). *J. Colloid Interface Sci.*, **65**, 370.

Jaycock, M. J. and Ottewill, R. H. (1964). *Proc. Int. Cong. Surface-Active Substances*, 4th, Section B, paper 8.

Kresheck, G. C., Hamori, E., Davenport, G., and Scheraga, H. A. (1966). *J. Amer. Chem. Soc.*, **88**, 246.

Kustin, K. (Ed.) (1969). *Fast Reactions*, (*Methods in Enzymology*, Vol. XVI), Academic Press, New York.

Mijnlieff, P. F. and Ditmarsch, R. (1965). *Nature*, **208**, 889.

Mittal, K. L. (Ed.) (1977). *Micellization Solubilization and Microemulsions* (Ed. Mittal, K. L.), Vol. 1, Plenum Press.

Mukerjee, P. (1965). *J. Phys. Chem.*, **69**, 2821.

Mukerjee, P. (1967). *Adv. Colloid Interface Sci.*, **1**, 241.

Mukerjee, P. (1974). *J. Pharm. Sci.*, **63**, 972.

Mukerjee, P. (1975). In *Physical Chemistry: Enriching Topics from Colloid and Surface Chemistry* (Eds. van Olphen, H. and Mysels, K. J.), IUPAC, Theorex, California.

Mukerjee, P. and Cardinal, J. R. (1976). *J. Pharm. Sci.*, **65**, 882.

Mukerjee, P. and Mysels, K. J. (1971). *Critical Micelle Concentrations of Aqueous Surfactant Systems*, (NSRDS-NBS-36, Washington D.C.).

Muller, N. (1972). *J. Phys. Chem.*, **76**, 3017.

Nakagawa, T. (1974). *Colloid Polym. Sci.*, **252**, 56.

Porschke, D. and Eggers, F. (1972). *Eur. J. Biochem.*, **26**, 490.

Rassing, J. (1972). *Adv. Mol. Relax. Proc.*, **4**, 55.

Rassing, J., Sams, P. J., and Wyn-Jones, E. (1974). *J. Chem. Soc. Faraday II*, **70**, 1247.

Reinsborough, V. C. (1975). In *Chemical and Biological Applications of Relaxation Spectrometry* (Ed. Wyn-Jones, E.), Reidel, Dordrecht, p. 159.

Robinson, B. H., Seelig-Loffler, A., and Schwarz, G. (1975). *J. Chem. Soc. Faraday I*, **71**, 815.

Sams, P. J., Rassing, J. E., and Wyn-Jones, E. (1975). In *Chemical and Biological Applications of Relaxation Spectrometry* (Ed. Wyn-Jones, E.), Reidel, Dordrecht, p. 163.

Sams, P. J., Wyn-Jones, E., and Rassing, J. (1972). *Chem. Phys. Lett.*, **13**, 233.

Sams, P. J., Wyn-Jones, E., and Rassing, J. (1973). *J. Chem. Soc. Faraday II*, **69**, 180.

Shinoda, K. (Ed.) (1967). *Solvent Properties of Surfactant Solutions*, M. Dekker, New York.

Strehlow, H. and Knoche, W. (1977). *Fundamentals of Chemical Relaxation*, Verlag Chemie, Weinheim.

Takeda, K. and Yasunaga, T., (1972). *J. Colloid Interface Sci.*, **40**, 127.

Takeda, K., Yasunaga, T., and Uehara, H. (1977). In *Micellization, Solubilization and Microemulsions* (Ed. Mittal, K. L.), Vol. 1, Plenum Press, p. 305.

Teubner, M. (1979). *J. Phys. Chem.*, in press.

Teubner, M., Diekmann, S., and Kahlweit, M. (1978). *Ber. Bunsenges Phys. Chem.*, **82,** 1278.

Thusius, D. (1975). *J. Mol. Biol.*, **94,** 367.

Thusius, D., Dessen, P., and Jallon, J.-M. (1975). *J. Mol. Biol.*, **92,** 413.

Tiddy, G. T., Walsh, M. F., and Wyn-Jones, E. (1979). Unpublished results.

Turner, D. H., Flynn, G. W., Lundberg, S. K., Faller, L. D., and Sutin, N. (1972). *Nature*, **239,** 215.

Turner, D. H., Yuan, R., Flynn, G. W., and Sutin, N. (1974). *Biophys. Chem.*, **2,** 385.

Weisstuck, A. and Lange, K. R. (1971). *Materials Protection,* **10,** 29.

Wyn-Jones, E. (Ed.) (1975). *Chemical and Biological Applications of Relaxation Spectrometry*, Reidel, Dordrecht.

Yasunaga, T., Fujii, S., and Miura, M. (1969). *J. Colloid Interface Sci.*, **30,** 399.

Yasunaga, T., Oguri, H., and Miura, M. (1967). *J. Colloid Interface Sci.*, **23,** 352.

Molecular Interactions, Volume 2
Edited by H. Ratajczak and W. J. Orville-Thomas
© 1980 John Wiley & Sons Ltd.

4 Structural aspects of hydrogen bonding in amino acids, peptides, proteins, and model systems

S. N. Vinogradov

4.1 Introduction

A hydrogen bond is an interaction between a group X—H and an atom or group Y in which the energy of complex formation is larger than dipolar and London dispersion energies (Kollman and Allen, 1972). In biological materials X is usually N or O, and Y can be N, O, or an anion. Proof of the existence of a hydrogen bond $H \cdots Y$ depends on the demonstration that, (1) the distance $H \cdots Y$ is shorter than the sum of the vad der Waals radii of H and Y and (2), the potential energy of the X—Y group is influenced by the acceptor atom. It is important to realize that *both* criteria must be satisfied in order to prove the existence of a specific hydrogen bond.

An X—Y $\cdots$ Y interaction is considered to be a hydrogen bond if the distance $H \cdots Y < W_H + W_Y$ and $110° < \theta < 180°$, where W_H and W_Y are the van der Waals radii of the H and acceptor atoms, respectively, and θ is the angle X—Y $\cdots$ Y. Hamilton and Ibers (1968) have proposed that the existence of a hydrogen bond is certain when the distance $H \cdots Y < W_H + W_Y - 0.2$ Å. The accepted van der Waals radii of interest are listed in Table 4.1. The neutron diffraction data on a variety of crystalline hydrogen-containing substances accumulated over the last decade have shown that H—H distances of 2.2 Å are common. Consequently, it has been proposed that the van der Waals radius for hydrogen should be revised to 1.0 Å

Table 4.1   van der Waals Radii, Å (Bondi, 1964)

H	C	N	O	S	F	Cl	Br
1.20	1.70	1.55	1.52	1.80	1.47	1.75	1.85

(Baur, 1972; Koetzle *et al.*, 1972). If the new value is accepted, the distance H $\cdots$ Y in cases where Y is O or N, must be less than 2.5 Å in order to indicate unequivocally the presence of a hydrogen bond.

Single-crystal neutron diffraction studies of small molecules have shown that hydrogen atoms can be located to a precision† of better than 0.02 Å. In contrast, X-ray diffraction studies of crystals of biological macromolecules have rarely reached resolutions below 2.0 Å. The average error in a set of atomic coordinates obtained from a 2.0–2.8 Å resolution electron density map is about 0.2–0.5 Å (Matthews, 1977). Thus structural evidence concerning the presence of specific hydrogen bonds in proteins can be somewhat tentative.

Generally, spectroscopic methods are used to demonstrate whether the second criterion for the existence of a hydrogen bond is satisfied in a given case. The shift of the X—H stretching vibration frequency is widely used in studies of small molecules in solution, to demonstrate the effect of an acceptor atom or group on a group X—H and thus prove the existence of the hydrogen bond X—H $\cdots$ Y. Individual O—H and N—H stretching vibration frequencies are difficult to observe in infrared spectra of aqueous solutions of polypeptides and proteins due to the twin obstacles of resolving broad absorption bands and of overcoming the absorption background due to the strongly hydrogen-bonded solvent. The development of Raman spectroscopy promises to provide the means of surmounting these obstacles. Characteristic chemical shifts of either the H or the X resonances in NMR spectra of small molecules possessing a hydrogen-bonded X—H group are also used to demonstrate the effect of an acceptor atom or group Y. Although significant advances have been made in using NMR spectroscopy to determine the presence of hydrogen bonds in peptides in solution, the NMR spectra of proteins at the available resolutions, are of limited usefulness insofar as identification of specific hydrogen bonds is concerned.

The overwhelming majority of hydrogen bonds inferred to exist in amino acids, peptides, and proteins are based on the occurrence of appropriately short contacts between likely donor and acceptor groups in crystal structures determined by X-ray diffraction. Spectroscopic evidence satisfying the second of the two criteria for the existence of a hydrogen bond is generally either unavailable or controversial. Consequently, this review will include a limited discussion of hydrogen bonding in small-molecule systems which can be viewed as possible models for interactions occurring in proteins.

---

† In many cases where a crystal structure has been determined accurately by both X-ray and neutron diffraction the difference between the hydrogen atom positions found by the two methods is about 0.25 Å (Baur, 1972). The X—H distances observed in X-ray determinations tend to be 0.1–0.2 Å shorter than in neutron diffraction determinations (Olovsson and Jönsson, 1976).

4.2   Hydrogen Bonds in Crystal Structures of Amino Acids and Peptides

4.2.1   Types of hydrogen bonds

A survey of hydrogen bonds in 28 structures of amino acids, peptides, and related molecules obtained by neutron diffraction and 68 structures obtained by X-ray diffraction, was carried out as part of the present review For details see Vinogradov (1979). The results of early X-ray diffraction studies of amino acids and peptides have been reviewed by Marsh and Donohue (1967), Gurskaya (1968), and by Ramakrishnan and Prasad (1971). The X-ray diffraction studies selected from the period 1969–78 were those in which the hydrogen atom positions were determined. Since Koetzle and Lehmann (1976) have shown that the N—H $\cdots$ O angle distribution exhibits a maximum at 165° analogous to the O—H $\cdots$ O angle distribution (Olovsson and Jönsson, 1976), only the data for linear hydrogen bonds ($\theta > 150°$) were employed to calculate the average lengths of the various types of hydrogen bonds. The results are shown in Table 4.2. Some 55 different types of hydrogen bonding occur among the 439 hydrogen bonds counted. In 325 of these N—H groups are the donor, while in 114 O—H groups are the donor; oxygen is the acceptor in 342 hydrogen bonds and halogen ions in 97 hydrogen bonds.

The protonated amine groups (imidazolium, ammonium, and guanidinium) interact with negatively charged groups such as carboxylate, oxygen anions, and halide ions and with neutral acceptor groups, such as water oxygens and the carbonyl oxygen of unionized carboxyl groups, but not with the carbonyl oxygen of amide groups. The neutral N—H of the amide group interacts predominantly with the carbonyl oxygen of acyl groups, of amide groups, and of unionized carboxyl groups; only a few interactions with carboxylate groups and halogen ions were observed. The paucity of data pertaining to the uncharged imidazole and indole N—H groups precludes any generalizations concerning their interactions. The hydroxyl group, whether of a side-chain group, an unionized carboxyl group, or a water molecule, tends to interact with carboxylate groups, oxygen, and halogen anions, carbonyl oxygens and the oxygen atom of water molecules. In particular, the carboxyl hydroxyl group appears to have a preference for Cl⁻ ions, while the water hydroxyl tends to interact with carboxylate groups and the carbonyl oxygen of amide groups.

In an analysis of N—H $\cdots$ O bonds in 84 structures of amino acids and peptides available prior to 1970, Ramakrishnan and Prasad (1971) have shown that the distributions of hydrogen bond lengths exhibit a maximum between 2.8 and 2.9 Å for the charged N—H group and between 2.9 and 3.0 Å for the uncharged N—H group. These findings were in accord with the statement by Marsh and Donohue (1967), that typical values for N⁺—H $\cdots$ O distances are 2.80–2.85 Å, while peptide N—H $\cdots$ O distances are

    Molecular Interactions Volume 2

Table 4.2 Average hydrogen bond lengths* in crystal structures of amino acids, peptides, and related molecules (Vinogradov, 1979)

Acceptor group	Donor group								
	(Im)⁺N—H	(NH₃)⁺N—H	(Gnd)⁺N—H	CO—N—H	(Im)N—H	(In)N—H	CO—OH	(W)O—H	O—H
COO⁻	2.72 (6) ±0.07	2.82 (88) ±0.08	2.95 (17) ±0.13	2.96 (6) ±0.10	2.71 (2)	2.79 (2)	2.50 (4) ±0.06	2.81 (26) ±0.08	2.74 (5) ±0.07
(H₂PO₄)⁻O	2.69 (3) ±0.02	2.82 (12) ±0.03	2.92 (20) ±0.12				2.55 (2)	2.78 (5) ±0.05	
(NO₃)⁻O		2.84 (3) ±0.04							
(SO₄)⁼O	2.73 (2)	2.85 (13) ±0.07					2.65 (1)	2.78 (6) ±0.07	
(—SO₃)⁻O		2.83 (5) ±0.06					2.59 (3) ±0.03		
Cl⁻	3.11 (2)	3.22 (50) ±0.07	3.31 (2)	3.26 (1)		3.28 (1)	2.99 (14) ±0.03	3.16 (11) ±0.05	3.11 (2)
Br⁻		3.37 (4) ±0.05						3.40 (3) ±0.10	3.29 (1)
OH \ C=O (—C=O)		2.89 (12) ±0.09		2.99 (4) ±0.11			2.62 (4) ±0.08		
C=O	2.71 (1)			2.97 (32) ±0.11	2.93 (1)		2.62 (2)	2.81 (5) ±0.12	2.90 (1)
(W)O	2.73 (1)	2.86 (12) ±0.04	2.93 (11) ±0.05	2.90 (2)	2.87 (3) ±0.01	2.87 (1)		2.79 (9) ±0.08	
H—O		2.87 (2)					2.59 (5) ±0.05	2.84 (3) ±0.07	
(Im)N	2.86 (1)	2.75 (1)		3.00 (1)	2.86 (1)			2.82 (2)	
Total	16	202	50	46	7	4	35	70	9

* And standard deviations for hydrogen bonds with X—H—Y angles >150°. The number of values averaged is given in parentheses.

about 2.9 Å. The average hydrogen bond lengths collected in Table 4.2 suggest that several additional generalizations can be made, subject in many cases to the uncertainty due to a small number of observations (Vinogradov, 1979). A comparison of the values in the first three columns of Table 4.2 indicates that there exists the following order of hydrogen bond lengths in cases where protonated amine groups are donors: imidazolium < ammonium < guanidinium. This order is observed with all the acceptor groups for which data is available. Although the number of hydrogen bond lengths pertaining to the imidazolium and guanidinium groups is appreciably smaller than for ammonium groups, and although the distributions of bond lengths for the three types overlap, the order is the same whether the acceptor group is a carboxylate, an oxygen anion, a Cl^- ion, or the oxygen of a water molecule. Furthermore, the order of imidazolium, ammonium, and guanidinium hydrogen bond lengths remains unaltered whether the neutron diffraction and X-ray diffraction results are considered separately or not, whether all oxygen acceptors are considered together or whether N—H $\cdots$ O bonds with $\theta < 150°$ are included. Figure 4.1 presents a histogram of the three types of hydrogen bonds with carboxylate, phosphate, sulphate, and nitrate anions. There exists considerable overlap between the three distributions; nevertheless, the order of the position of their maxima is unambiguous. The observed order is in agreement with the order of the free energies and enthalpies of protonation of imidazole (Paiva et al., 1976), primary amines (Jones, III and, Arnett, 1974), and guanidine (Fabbrizzi et al., 1977) in aqueous solution (9.5, ~15, and 19 kcal mol^{-1} and 8.8, ~14, and 18 kcal mol^{-1}, respectively).

The amide N—H group appears to form longer hydrogen bonds than the imidazole N—H for all the acceptors listed in Table 4.2, subject to the uncertainty due to the small number of available observations. The indole N—H appears to be similar to the imidazole N—H.

The carboxyl O—H forms much shorter hydrogen bonds with all the acceptors listed in Table 4.2 than the side-chain and water O—H groups. The shortest hydrogen bonds are formed with carboxylate groups. The (W)O—H $\cdots$ O bond lengths in Table 4.2 fall within the range of hydrogen bond lengths observed in crystalline hydrates (Ferraris and Franchini-Angela, 1972; Falk and Knop, 1973). The average (W)O—H $\cdots$ O(W) distance calculated using the values for 189 linear hydrogen bonds in inorganic hydrates taken from Brown (1976) is 2.79 Å, in excellent agreement with the value in Table 4.2.

The interactions involving Cl^- and Br^- are relatively long hydrogen bonds. If their lengths are corrected for the difference between the van der Waals radii of Cl and Br (1.75 and 1.85 Å, respectively) and those for O and N (1.52 and 1.55 Å, respectively), they remain appreciably longer and thus probably weaker, than the hydrogen bonds with the other acceptor groups

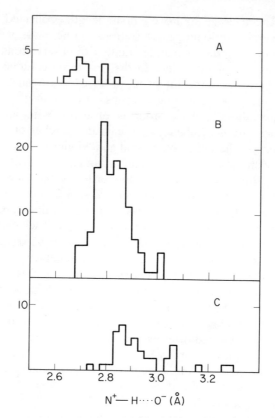

Figure 4.1 Histograms of $N^+ - H \cdots O^-$ bond lengths for (A) imidazolium, (B) ammonium, and (C) guanidinium hydrogen bonds with oxygen anions (Vinogradov, 1979)

listed in Table 4.2. For example, in the case of (W)O—H as donor, the corrected lengths of the hydrogen bonds with Cl^- and Br^- are 2.92 and 3.07 Å, respectively. In contrast, the average (W)O—H $\cdots$ F^- bond length in metal fluorides is 2.68 Å (Simonov and Bukvetsky, 1978); since the van der Waals radii of O and F are about equal, this hydrogen bond is probably much stronger than the (W)O—H $\cdots$ Cl^- and (W)O—H $\cdots$ Br^- bonds. This conclusion is in agreement with the experimental vapour-phase hydration energies of F^- and Cl^- of -23 and -13 kcal mol^{-1}, respectively (Arshadi *et al.*, 1970) and the hydration energies calculated by an *ab initio* molecular orbital method: -22 and -12 kcal mol^{-1}, respectively (Kollman and Kuntz, 1976).

Oxygen is the dominant acceptor among the hydrogen bonds listed in Table 4.2. Examination of the hydrogen bond lengths for the 9 different types of oxygen acceptors, of which 5 are negatively charged and 4 are neutral, with the donors $(NH_3)^+N$—H and (W)O—H, shows that no effect can be discerned between the two groups. This conclusion is in agreement with the findings of a previous survey (Ramakrishnan and Prasad, 1971). Investigations of normal hydrogen bonds within the framework of the molecular orbital theory, have shown over the last few years, that the equilibrium distance $X \cdots Y$ depends sensitively on the proton donor group X—H but is not strongly dependent on the proton acceptor group Y (Allen, 1975; Umeyama and Morokuma, 1977).

The foregoing survey of recent crystallographic results suggests that there are differences between the lengths of hydrogen bonds formed by the various donor and acceptor groups in amino acids and peptides and by implication, in the energetics of their formation. The differences between the various types of hydrogen bonds observed in the crystal structures of the selected group of amino acids and peptides, are also likely to be encountered in proteins.

4.2.2 *Carboxyl–carboxyl hydrogen bonding*

The strongest O—H $\cdots$ O bonds known occur when the O $\cdots$ O distance is below 2.50 Å (Speakman, 1972; Novak, 1974). The feature common to the shortest O—H $\cdots$ O bonds is that the proton interacts with two chemically very similar acceptor groups forming monoanion $(O \cdots H \cdots O)^-$ or monocation $(O \cdots H \cdots O)^+$ complexes where the charge is shared by both moieties of the complex. In the former case one of the two groups is negatively charged, e.g. carboxylate groups in acid salts such as sodium hydrogen diacetate (Barrow *et al.*, 1975), pyridine-2,3-dicarboxylic acid (Kvick *et al.*, 1974), etc. Monocation complexes are the result of protonation of a neutral molecule such as H_2O, e.g. in toluenesulphonic acid monohydrate (Lundgren and Williams, 1973), trinitrosulphonic acid tetrahydrate (Lundgren and Tellgren, 1974), and other hydrates of strong acids. See Lundgren and Olovsson (1976) for a recent review. The monocation $(X \cdots H \cdots Y)^+$ type of complex can of course be also formed by amines and will be discussed later.

In a number of amino acid crystal structures, relatively short and by inference, moderately strong O—H $\cdots$ O bonds are found, in which the proton donor is a carboxyl group and the acceptor is a carboxylate (Table 4.3). The O—H $\cdots$ O distances vary from 2.43 Å in di-L-leucine·HCl (Golič and Hamilton, 1972) to 2.58 Å in L-aspartic acid (Derissen *et al.*, 1968). The very short hydrogen bond in di-L-leucine·HCl is probably symmetric because the OH absorption band is broad, intense and lies between 400 cm^{-1} and 1500 cm^{-1} (Hadzi and Marciszewski, 1967).) In contrast, the carboxyl–carboxyl hydrogen bond distance in formic acid and acetic acid dimers is 2.625 Å (Nahringbauer, 1978).

Table 4.3 Carboxyl hydrogen bonds

Compound	Structure	Study[a]	H···Y (Å)	X···Y (Å)	<X—H—Y (deg.)	References
Di-L-leucine·HCl[b]	(COOHOOC)⁻	X	1.37	2.429	175	Golič and Hamilton (1972)
L-Glutamic acid[c]	—C—O—H···⁻OOC (O=)	N	1.475	2.519	172	Lehmann et al. (1972b)
L-Aspartic acid[c]	—C—O—H···⁻OOC (O=)	X		2.577	175	Derissen et al. (1968)
N-Chloroacetyl-DL-alanine	COOH···O(acyl)	X		2.68	166	Cole (1970)
N-Acetylnorvaline	COOH···O(acyl)	X		2.55	132	Lovas et al. (1974).
N-Formyl-L-methionine	COOH···O(acyl)	X		2.555		Chen and Parthasarathy (1977)
N-Acetyl-L-glutamine	C—O—H···O(amide) (O=)	X	1.64	2.511	174	Narasimhamurthy et al. (1976)
N-Acetyl-L-phenylalanyl-L-tyrosine	COOH···O=C(amide)	X	1.85	2.633	171	Stenkamp and Jensen (1973)
Hippuric acid	—C—O—H···O=C(amide) (O=)	N	1.692	2.679	168	Currie and Macdonald (1974)
Pyridine-2,3-dicarboxylic acid[d]	(COOHOOC)⁻	N	1.238	2.398	174	Kvick et al. (1974)

[a] N—neutron diffraction; X—X-ray diffraction.
[b] Intermolecular carboxyl–carboxylate monoanion hydrogen bond.
[c] The proton donor group is the side-chain carboxyl.
[d] Intramolecular carboxyl–carboxylate monoanion hydrogen bond.

It is appropriate to mention here a recent review of the structures of small carboxylic acids in the vapour and solid states (Derissen, 1977). The average deformation of the structure of a carboxylic acid group on crystallization, consists of an expansion of the O—H bond (0.04 Å), the C=O bond (0.01 Å) and the C—C—O bond angle (1.5°), and a shrinkage of the C—O bond (0.041 Å), the $C_{(\alpha)}$—C bond (0.012 Å), and the $C_{(\alpha)}$—C=O bond angle (2.0°). The energy necessary for these deformations to occur was estimated to be 1.8 kcal mol^{-1}. This is not negligible relative to the total hydrogen bond energy in carboxylic acid dimers.

An interesting fact noted by Chen and Parthasarathy (1978) and also seen in Table 4.3, is the presence of intermediate (2.55–2.68 Å) hydrogen bonds, between the carboxyl OH and the acyl oxygen in most N-acylamino acids and N-acyl peptides. These carboxyl–acyl oxygen hydrogen bonds appear to be intermediate between the carboxyl–carboxylate hydrogen bonds on one hand and the carboxyl–peptide carbonyl oxygen hydrogen bonds as in hippuric acid, on the other.

Although strong carboxyl–carboxylate hydrogen bonds occur in the crystalline state, evidence for their occurrence in solution is tenuous. Kinetic evidence for the existence of a (COOHOOC)$^-$ complex in concentrated aqueous formate buffers, possessing an association constant $K = 0.25$ M^{-1}, has been obtained recently by Hand and Jencks (1975). As the authors point out, association constants of this magnitude require careful definition and examination with respect to the particular experimental property and technique used for their determination, since equilibrium constants for the formation of random encounter complexes between two molecules are in the range of 0.017–0.3 M^{-1}.

Carboxyl–carboxylate interactions are found in proteins. An example is the complex formed between the side-chain groups of Asp(32) and Asp(215) in the active site of penicillopepsin, an acid protease of 323 residues, whose structure was recently determined by X-ray diffraction (Hsu *et al.*, 1977). A carboxylate anion and a protonated carboxyl group have been implicated in the catalytic mechanism of acid proteases (Fruton, 1976). Furthermore, in penicillopepsin, the hydroxyl group of Ser(35) is hydrogen bonded to the other carboxyl oxygen of Asp(32). These three hydrogen-bonded residues, Ser(35), Asp(32), and Asp(215), are also conserved in the known sequences of acid proteases and are reminiscent of the charge relay system of the serine proteases. The hydrogen bonding could easily explain the low pK_a of Asp(32) ($\sim$2) and the slightly higher pK_a of Asp(215) ($\sim$4.5) (James *et al.*, 1977).

4.2.3 Hydrogen-bonded ion-pair interactions

Historically, interactions between a protonated amine group (ammonium, imidazolium, guanidinium) and an ionic group, whether a carboxylate or a halide ion, have been called 'salt bridges' or 'salt linkages' by protein chemists. This terminology, by emphasizing the cation–anion aspect of the interactions, has served to obfuscate the basic similarity between the hydrogen-bonded ion-pairs and the hydrogen-bonded complexes formed by neutral donor and acceptor groups.

Our knowledge concerning the properties of hydrogen bonding interactions involving the carboxylate anion on one hand, and the ammonium, imidazolium, or guanidinium cation on the other, is still very limited. It is clear that ammonium–carboxylate hydrogen bonds are long when compared to the monoanion $(O—H \cdots O)^-$ bonds. A recent careful study of valence electron distribution in perdeuteroglycylglycine by a combination of X-ray and neutron diffraction data (Griffin and Coppens, 1975), has provided the following observations concerning the hydrogen bonds formed by the ND_3^+ group with three different carboxylate groups: (1) there is no buildup of density near the midpoint of the line between the D atom and the oxygen acceptor, (2) the N—D vectors point toward the lone-pair density rather than toward the centre of the acceptor oxygen, and (3), the lone-pair density accumulates close to the line connecting the D and O atoms. Furthermore, net charges on the COO^- and ND_3^+ groups are about -0.4 to -0.5 and $+0.4$ electrons, respectively. Such a picture is in accordance with an electrostatic model for the $NH_3^+ \cdots {}^-OOC$ interaction. In contrast, the buildup of electron density observed in a typical, short, symmetrical $(O \cdots H \cdots O)^-$ bond of 2.408 Å in sodium hydrogen diacetate is incompatible with an ionic model (Stevens *et al.*, 1977).

While ammonium or imidazolium groups form single hydrogen bonds with carboxylate groups, guanidinium groups can form a cyclic eight-membered complex involving two hydrogen bonds. Carboxyl groups also form cyclic complexes with the ureido group $(—C(—N—H)=O)$ in biotin (DeTitta *et al.*, 1976), dethiobiotin (Chen *et al.*, 1976) and in $N_{(6)}$-(N-glycylcarbonyl)-adenosine (Parthasarathy *et al.*, 1977). Figure 4.2 shows the structure of the cyclic guanidinium–carboxylate complex in the salt arginine glutamate (Bhat and Vijayan, 1977). A similar cyclic complex also occurs in the complex of glycylglycine with guanidine hydrochloride (Cotton *et al.*, 1974b).

It is interesting to note that carboxylate groups of N-benzoylglycine and N-formylglycine can also form cyclic complexes with nucleic acid bases such as cytosine, protonated at $N_{(3)}$ (Tamura *et al.*, 1972; Ohki *et al.*, 1975). Recent high-resolution PMR studies of the association in chloroform of butyric acid, as a model for glutamic or aspartic acid side chain, with nucleic acid bases, have led to the following preferential order of associations: 2-dimethylamino-9-methylguanine $(K = 660 \text{ M}^{-1}) > 1$-cyclohexylcytosine $(K = 270 \text{ M}^{-1}) > 9$-ethyladenine $(K = 160 \text{ M}^{-1}) >$ cyclohexyluracil $(K = 80 \text{ M}^{-1})$ (Lancelot, 1977). It is clear that there exists in such associations a potential mechanism of recognition of specific nucleic acid bases by side-chain groups of protein residues.

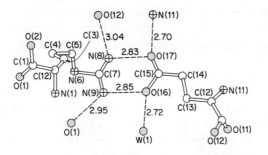

Figure 4.2 Disposition of the arginine and the glutamate ions involved in the specific ion-pair interaction between the guanidinium group and the γ-carboxylate group in arginine -L-glutamate (Bhat and Vijayan, 1977). The broken lines indicate hydrogen bonds

The association of carboxylate anion with guanidinium and n-butyl-ammonium cations in aqueous solution was determined recently by Haake and his collaborators using potentiomeric titration and kinetic measurements (Springs and Haake, 1977; Knier and Haake, 1977). The association equilibrium constants of guanidinium with carboxylates varied from $0.37\,\text{M}^{-1}$ for acetate, to $0.43\,\text{M}^{-1}$ for chloroacetate, and $0.55\,\text{M}^{-1}$ for succinate, indicating that the basicity of the oxyanion had little or no effect. The association of n-butylammonium with acetate was found to have an equilibrium constant of $0.31\,\text{M}^{-1}$.

A Raman spectroscopic study of L-alanine crystals showed that the spectral width of the NH_3^+ torsional vibration band at $484\,\text{cm}^{-1}$ when measured as a function of temperature, exhibited an unusual transition at 220–230 K (Wang and Storms, 1971). The mechanism for this transition was suggested to be the breaking of $(NH_3)^+N$—H $\cdots$ ^-OOC bonds; an activation energy of $3.5\pm0.6\,\text{kcal mol}^{-1}$ was determined from an Arrhenius-type plot. Similar transitions were observed in the temperature dependence of the bandwidths of the first overtones of the NH_3^+ torsional vibration at 972 and $975\,\text{cm}^{-1}$ in the infrared absorption spectra of L- and DL-alanine, respectively (Kaneko et al., 1975). The activation energy for the hydrogen bond breaking process was found to be $3.6\pm0.6\,\text{kcal mol}^{-1}$.

4.2.4 Weak interactions: bifurcated hydrogen bonds and C—H $\cdots$ O contacts

In a number of amino acids and peptides the NH_3^+ group is surrounded by more than three acceptors. For example, in glycylglycine monohydrochloride monohydrate (Parthasarathy, 1969; Koetzle et al., 1972), the

N-terminal NH_3^+ group is surrounded by three oxygens and three Cl^- ions (Figure 4.3). The contacts $H_{(1)} \cdots Cl'$ and $H_{(2)} \cdots O_{(2'')}$ are only very slightly shorter than van der Waals distances, and should not be called hydrogen bonds, although weak electrostatic interactions may contribute significantly to the stability of the crystal structure. $H_{(3)}$ forms a hydrogen bond $N_{(1)}$—$H_{(3)} \cdots O_{(4)}$ to the water molecule and is also involved in a short intramolecular contact with the oxygen $O_{(1)}$ of the peptide unit. The $H_{(3)} \cdots O_{(1)}$ distance is 2.360 Å. Although such a short intramolecular contact has occasionally been called a hydrogen bond, usually as the weaker member of a bifurcated pair, the utility of this appellation can be questioned. The distance is less than 2.4 Å, the criterion suggested by Hamilton and Ibers (1968) for an X—H $\cdots$ O hydrogen bond but the N—H $\cdots$ O angle of 97° is at the extreme of the range normally found in hydrogen bonds. Distances between α-amino hydrogen atoms and the adjacent carbonyl oxygen as short as this are not uncommon in amino acids. Distances of 2.327, 2.291, and 2.467 Å occur in L-histidine (Lehmann et al., 1972a), L-asparagine monohydrate (Verbist et al., 1972), and L-arginine dihydrate (Lehmann et al., 1973), respectively. A survey of over 100 crystal structures

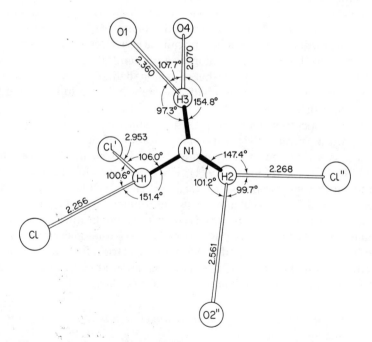

Figure 4.3 Environment of the terminal nitrogen atom in the neutron diffraction structure of glycylglycine monohydrochloride monohydrate (Koetzle et al., 1972)

has shown that bifurcated interactions occur for the following hydrogen donors besides the NH_3^+ group in amino acids: the charged cationic hetero nitrogen in nucleic acid bases and an HN_2 or NH bond when they are highly polarized due to neighbouring positive charges (Parthasarathy, 1977). When these bifurcated hydrogen bonds are formed, the donor, the hydrogen, and the two acceptors are usually all in almost the same plane.

In a recent review Leiserowitz (1976) has discussed the possible role of C—H $\cdots$ O=C interactions in influencing the intralayer and interlayer packing of carboxylic acids and amides in the solid state. There is a general tendency to regard a C—H $\cdots$ O contact as not being significant if the C $\cdots$ O distance is greater than about 3.3 Å. Rather conclusive crystallographic and spectroscopic evidence for C—H $\cdots$ Y interactions exists where Y is a strong proton acceptor and the C—H is a constituent of halogen-substituted alkane (e.g. X_3CH, X_2CH_2, XCH_3, $CH_3(X_2)CH$, etc.), alkene ($RCH=CH_2$, $RCH=(X)CH$, etc.), and alkyne ($N\equiv CH$, $RCH\equiv CH$, etc.) derivatives (Green, 1974). These is little evidence for hydrogen bonding between an (sp^3)C—H group attached to other (sp^3)C atoms, and an acceptor oxygen or nitrogen atom. However, a weak interaction can occur between an acceptor and a C—H group attached to a nitrogen atom. Such C—H $\cdots$ O contacts of 3.2–3.3 Å are observed in crystals of amino acids and peptides. Examples are the two C—H $\cdots$ O interactions of 3.325 and 3.265 Å in glycyl-DL-phenylalanine (Marsh et al., 1976), where both carbon atoms are each bonded to a nitrogen atom.

Studies of hydrogen bonding in pyridine carboxylic acids have revealed that although N^+—H $\cdots$ $^-$O bonds are found in the pyridine dicarboxylic acids, quinolinic acid (Takusagawa et al., 1973a), and cinchomeronic acid (Takusagawa et al., 1973b), O—H $\cdots$ N bonds occur in pyridine monocarboxylic acids, nicotinic (Takusagawa and Shimada, 1973), pyrazinic (Takusagawa et al., 1974), and isonicotinic acids (Takusagawa and Shimada, 1976). The molecular packing in the monocarboxylic acids is such that the carboxyl oxygen atom is surrounded by two hydrogen atoms at distances close enough to indicate appreciable C—H $\cdots$ O interactions. In dicarboxylic acids there is never more than one hydrogen atom in the vicinity of the carboxyl oxygen. Thus two weak C—H $\cdots$ O interactions could play a role in inhibiting the formation of N^+—H $\cdots$ $^-$O bonds.

4.3 Models of Hydrogen Bonds in Proteins

4.3.1 *Monocation* $(N—H—N)^+$ *complexes*

Monocation $(N—H—N)^+$ complexes are formed by amines in the presence of a strong acid. Wood and his collaborators have examined an extensive series of hydrogen-bonded cations of the type $(B_1HB_2)^+$ where B_1 and B_2

can be the same or dissimilar bases. Trimethylammonium–trimethylamine (Masri and Wood, 1972a), pyridinium–pyridine (Clements and Wood, 1973a), imidazolium–imidazole (Bonsor *et al.*, 1976), triethylammonium–pyridine (Masri and Wood, 1972b), benzimidazolium–benzimidazole (Borah and Wood, 1976), and other complexes (Clements and Wood, 1973b; Clements *et al.*, 1973). The infrared absorption spectra are characteristic of the cation and are independent of the anion (which can be ClO_4^-, BF_4^-, etc.) provided it is not strongly hydrogen bonding, such as Cl^-. Furthermore, the cation spectra are independent of the solvent, which can be excess amine or for example acetonitrile, and the N—H stretching vibration band shows a pronounced doublet structure in the 2000–2700 cm^{-1} region (Clements *et al.*, 1971). In the case of symmetrical $(N—H—N)^+$ complexes, the percentage of complex formed in acetonitrile solution increases with the pK_a of the amine, and there is present in the infrared absorption spectra, an absorption continuum in the region below 2000 cm^{-1}; the continuum is also found in the spectra of asymmetrical $(N—H—N)^+$ complexes; however, the continuum decreases with increase in ΔpK_a, the difference in the pK_a's of the two amine moieties (Brzezinski and Zundel, 1976).

The investigations of Zundel and his collaborators over the last few years have demonstrated that monocation complexes $(X \cdots H \cdots Y)^+$ and monoanion complexes $(X \cdots H \cdots Y)^-$ $(X, Y = O, N)$ possess similar spectroscopic properties (Zundel, 1976). In particular, the absorption continuum present in the infrared spectra of both types of complexes, is characteristic of a highly polarizable hydrogen bond. The potential energy surface describing the location of the proton in both types of complexes appears to be either a single very broad and flat potential well, or a double-minimum potential well with a low barrier with tunnelling of the proton. The high polarizability of the proton has three consequences. Firstly, the electrical fields of anions and the dipole fields of solvent molecules in the neighbourhood of the hydrogen bonds polarize these bonds. Secondly, the hydrogen bonds exert a mutual influence on each other. Thirdly, the X—H stretching vibrations in the hydrogen bonds couple with low-frequency intermolecular vibrations. These three interactions shift the energy levels of the protons, and because the strength of the interactions passes through a variety of values in a fluid, owing to the various distances and orientations between the groupings, the resultant continuity of energy-level shifts and energy-level differences manifests itself as an absorption continuum in the infrared region (Janoschek *et al.*, 1972).

The N $\cdots$ N bond lengths observed in some of the monocation complexes are listed in Table 4.4. The $(Im—H—Im)^+$ bond length is appreciably shorter than the other N $\cdots$ N distances, in keeping with the other hydrogen bonds formed by imidazolium N—H groups (Table 4.2). The hydrogen bond length in 9-ethylguanine hemihydrochloride (Mandel and Marsh, 1975)

Table 4.4 $N \cdots N$ distances in some $(N—H—N)^+$ complexes

Base	Complex	$N \cdots N$ (Å)	References
Imidazole	$(Im—H—Im)^+ClO_4^-$	2.73	Quick and Williams (1976a)
Benzimidazole	$(Bzim—H—Bzim)^+BF_4^-$	2.787	Quick and Williams (1976b)
Pyrazine	$(Pyr—H—Pyr)^+ClO_4^-$	2.853	Glowiak et al. (1975a)
Triethylenediamine	$(Tea—H—Tea)^+ClO_4^-$	2.84	Glowiak et al. (1975b)
9-Ethylguanine hemihydrochloride	$(Gu—H—Gu)^+Cl^-$	2.637	Mandel and Marsh (1975)

included for comparison, is nearly 0.4 Å shorter than the van der Waals $N \cdots N$ contact distance. A shortening of 0.4 Å in on $O—H \cdots O$ system below the $O \cdots O$ van der Waals distance of 2.8 Å, would probably result in a symmetric or almost symmetric hydrogen bond. The hydrogen bonds observed in acid metal cyanides $H_3M(CN)_6$ $(M = Fe, Co)$ with $N \cdots N$ distances of 2.665 Å (Fe) and 2.582 Å (Co) (Haser et al., 1977) are also among the shortest known $N—H—N$ bonds.

The increase in conductivity of solutions in aprotic solvents of ammonium salts in the presence of added amines and other ligands has been used by several investigators to study the formation of monocation complexes and of cation–neutral acceptor complexes. Table 4.5 lists some representative results and compares them to results obtained by other methods. It is seen that the equilibrium constants for the formation of the $(N—H—N)^+$ complexes are similar to those for strong hydrogen bonds and that they are sensitive to the polarity of the solvent. Furthermore, there are present in these solutions other equilibria involving the cation–amine complex and the ion-pair (Haulait and Huyskens, 1975). Gilkerson and his collaborators (Flora and Gilkerson, 1970; Aitken and Gilkerson, 1973; Junker and Gilkerson, 1975) have measured the thermodynamic parameters of association of the n-Bu_3NH^+ cation with a number of ligands. Some of their results shown in Table 4.6, demonstrate that the NH_3^+ group forms moderately strong hydrogen bonds with neutral oxygen acceptors.

Wood (1974) has drawn attention recently to the pH dependence of a monocation complex $(BHB)^+$. When only a small proportion of the total base is complexed, the amount of complex formed is proportional to $pK_a - pH$ and is thus sharply pH dependent. As the proportion complexed increases, the curve broadens and becomes flat topped, with more than half the base complexed over the range of pH values $pK_a \pm \log KC$, approximately (K is the equilibrium constant and C is the total concentration). In the case of complexes formed by two bases B_1 and B_2, the formation of the complex $(B_1HB_2)^+$ will be optimum over a pH range between pK_1 and pK_2.

The foregoing results suggest that monocation complexes can occur in

Table 4.5 Equilibrium constants for the formation of monocation hydrogen-bonded complexes[a] obtained from conductance measurements

Cation	Anion	Ligand	Solvent	Permittivity	$K_1(M^{-1})$	$K_2(M^{-1})$	$K(M^{-1})$	References
Pyr^+	Perchlorate	Pyridine	Acetonitrile	3.60	2.2^b			Brzezinski and Szafran (1977)
Im^+	BF_4^-	Imidazole	Acetonitrile	3.60	8^c			Bonsor et al. (1976)
$BuNH_3^+$	Picrate	$BuNH_2$	Nitrobenzene	34.82	279 ± 10	10 ± 1	50 ± 2	Haulait and Huyskens (1975)
$Bu_2NH_2^+$	Picrate	Bu_2NH			16.5 ± 0.5	0.5	4 ± 0.2	
$Et_2NH_2^+$	Perchlorate	3,4-Lutidine			60 ± 2	5.1 ± 0.6	22.7 ± 2	
$Et_2NH_2^+$	Picrate	3,4-Lutidine			61 ± 2	5 ± 0.5	16.5 ± 2	
$BuNH_3^+$	Chloride	$BuNH_2$	Nitromethane	35.9	382 ± 18	1.1		Pawlak (1973)
$BuNH_3^+$	Chloride	$BuNH_2$	Nitrobenzene	34.82	3350 ± 100	1.4		
$Bu_2NH_2^+$	Chloride	Bu_2NH	Nitromethane		150 ± 10			
Bu_3NH^+	Chloride	Bu_3N	Nitrobenzene		352 ± 10			
$n\text{-}Bu_3NH^+$	Picrate	Pyridine	Chlorobenzene	5.62	1380			Junker and Gilkerson (1975)
			o-Dichlorobenzene	10.07	1130			
$n\text{-}BuNH_3^+$	Perchlorate	$n\text{-}BuNH_2$	Acetonitrile	36.0	26^d	2.5		Coetzee and Padmanabhan (1965)

[a] $R_1NH_3^+ + R_2NH_2 \overset{K_1}{\rightleftharpoons} R_1NH_3^+NH_2R_2$; $R_1NH_3^+NH_2R_2 + R_2NH_2 \overset{K_2}{\rightleftharpoons} R_1NH_3^+(NH_2R_2)_2$; $R_1NH_3^+Y^- + R_2NH_2 \overset{K}{\rightleftharpoons} R_1NH_3^+Y^-NH_2R_2$.
[b] PMR measurements.
[c] Infrared absorption spectroscopic measurements.
[d] Potentiometric measurements.

Table 4.6 Thermodynamics of association of tri-n-butylammonium cation with ligands in chlorobenzene (Junker and Gilkerson, 1975)

Ligand	$K_1 (M^{-1})$ 25°	$-\Delta H$ (kcal mol^{-1})	$-\Delta S$ (cal deg^{-1} mol^{-1})
Tetrahydrofuran	149	5.50±0.17	8.56±0.57
Acetonitrile	290	4.51±0.28	3.95±0.95
Acetone	270	4.52±0.07	4.03±0.24
Pyridine	1380	7.43±0.19	10.6±0.62
Pyridine-N-oxide	76,000	7.10±0.18	1.55±0.59

proteins as easily as do the monoanion carboxyl–carboxylate complexes. The NMR studies of Rüterjans et al. (1969, 1971) have provided evidence for the existence of a $(N—H \cdots N)^+$ interaction between two histidine side-chain groups in the active site of pancreatic ribonuclease. Close approach of amino side-chain groups has been observed in protein crystals. In papain, two lysine side-chain groups interact with the same water molecule (Berendsen, 1975). The amino groups of Lys(99) of the α subunits of deoxyhaemoglobin lie in the internal cavity and are only 5 Å apart (Ladner et al., 1977). In triclinic lysozyme, the guanidinium groups of Arg(14) and Arg(61) are within 2.7–3.8 Å of one another (Moult et al., 1976).

4.3.2 Relationship between O—H $\cdots$ N and N$^+$—H $\cdots$ $^-$O bonds

The dissociation constants of quaternary ammonium salts obtained from conductance measurements in aprotic non-polar solvents, are much greater than the dissociation constants of other ammonium salts (Kraus, 1956). This difference is attributed to the formation of a hydrogen bond between the cation and anion moieties of the ion-pair. Numerous investigations over the last 30 years have shown that amines and proton donor groups capable of ionization, such as phenols and carboxylic acids, form ion-pairs in solvents of low dielectric constant (Davis, 1968; Kohler and Huyskens, 1976). Under appropriate conditions, there can exist an equilibrium between a hydrogen-bonded complex and a hydrogen-bonded ion-pair:

$$R_1—O—H \cdots N—R_2 \overset{K_{pt}}{\rightleftharpoons} R_1—O^- \cdots H^+NR_2 \qquad (4.1)$$

The position of this equilibrium depends on two factors. One is ΔpK_a. Huyskens and Zeegers-Huskens (1964) proposed the following relation between K_{pt}, the equilibrium constant for the proton transfer process and ΔpK_a, for a given family of similar complexes:

$$\log K_{pt} = \Delta pK_a + B \qquad (4.2)$$

where B is a constant. Clerbaux *et al.* (1967) and Ratajczak and Sobczyk (1969) showed that this relation was obeyed in phenol–triethylamine complexes. Formation of the more polar ion-pair is accompanied by a large change in the dipole moment of the hydrogen–bonded complex. Figure 4.4 shows the variation in the dipole moment increment and enthalpy of complex formation between phenols and trialkylamines with the pK_a of phenol. The enthalpies of formation of hydrogen-bonded ion-pairs between phenols with $pK_a < 5$ and amines vary between 10 and 25 kcal mol^{-1} and increase with decrease in pK_a (Neerinck *et al.*, 1968; Elegant *et al.*, 1977). The break in the enthalpy curve appears to coincide with the pK_a region 5–6 of the transition between hydrogen-bonded complex and hydrogen-bonded ion pair. A detailed study of phenol–trialkylamine complexes from the viewpoint of Mulliken's charge transfer theory by Ratajczak (1972), showed that the expected transition should occur at a pK_a of about 6.

The region of ΔpK_a where 50% proton transfer occurs varies with the system and other factors (Table 4.7). An interesting property of complexes which fall within that region has been brought to light recently in studies

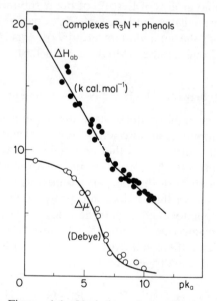

Figure 4.4 Variation of the dipole moment increment $\Delta\mu$ (Ratajczak and Sobczyk, 1969, 1970) and enthalpy of complex formation between phenols and trialkylamines (Neerinck *et al.*, 1968) as function of phenol pK_a. Taken from Kohler and Huyskens (1976)

Table 4.7 ΔpK_a for 50% proton transfer from infrared spectroscopic studies

Proton donor	Proton acceptor	State	ΔpK_a	References
Chlorophenols	n-Propylamine	Acetonitrile soln.	3.25	Zundel and Nagyrevi (1978)
Carboxylic acids	Polyhistidine	Film	2.8	Lindemann and Zundel (1978)
Trichloroacetic acid	Amines	Solid	~0[a]	Pietrazak et al. (1977)
Polyglutamic acid	Amines	Film	2.0	Lindemann and Zundel (1977a)
Carboxylic acids	Amines[b]	1:1 molar ratio (pure liquids)	2.3 (0.9)[c]	Lindemann and Zundel (1977b)
			4.0[d]	Odinokov et al. (1976)
Carboxylic acids	Pyridine	Pyridine	4.5	Sobczyk and Pawelka (1974)
Carboxylic acids	Pyridine	Benzene soln.	4.5[e]	Sobczyk and Pawelka (1973)
Carboxylic acids	Trialklamines	Benzene soln.	7.5[e]	Lindemann and Zundel (1972)
Carboxylic acids	Amines	Pure liquid	~4.0	Ratajczak and Sobczyk (1969, 1970)
	Triethylamine		~5.0[e]	
Phenols	Polylysine	Film	1.9	Kristof et al. (1978)
Phenols	Polyarginine	Film	2.3	Kristof et al. (1978)
Benzoic acids	Pyridine	Solid state	3.75	Johnson and Rumon (1965)

[a] Nuclear quadrupole resonance measurements.
[b] Amines with more than one NH group.
[c] In the presence of four water molecules per acid–base pair.
[d] Amines with only one NH group.
[e] Permittivity measurements.

carried out by Zundel and his collaborators. These ion-pair complexes exhibit infrared absorption continua below $3000 \, cm^{-1}$, which are similar to those observed in monocation and monoanion systems. Furthermore, the continuum decreases on either side of the transition in pK_a (Figure 4.5), i.e. as the hydrogen-bonded complex tends to be entirely neutral or completely ion-pair, the hydrogen bond becomes less polarizable. This phenomenon was observed in phenol–n-propylamine (Zundel and Nagyrevi, 1978), poly-histidine–carboxylic acids (Lindemann and Zundel, 1978), polyglutamic acid–amine (Lindemann and Zundel, 1977a), carboxylic acid–amine (Lindemann and Zundel, 1977b), and phosphate–amine (Matthies and Zundel, 1977) systems. It is possible that the large polarizability of such types of hydrogen bonds is responsible for the function and regulation of the charge relay system in serine protease enzymes such as chymotrypsin (Zundel, 1978).

The other factor affecting the position of equilibrium (4.1) is the environment of the hydrogen-bonded complex. The effect of the environment can

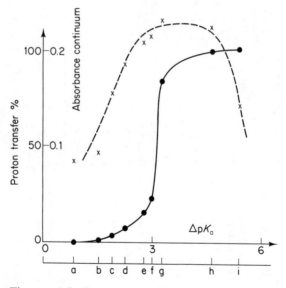

Figure 4.5 Proton transfer in phenol +n-propylamine systems, 1:1 molar ratio in trideuteroacetonitrile. Solid curve: per cent proton transfer determined from the absorbance of the NH_3^+ antisymmetric bending frequency at $1631 \, cm^{-1}$. Dashed curve: absorbance of the continuum at $1750 \, cm^{-1}$; (a) phenol, (b) 4-chloro-, (c) 3-chloro-, (d) 2-chloro-, (e) 3,5-dichloro-, (f) 2,4-dichloro-, (g) 2,3-dichloro-, (h) 2,3,5-trichloro-, and (i) pentach-lorophenol (Zundel and Nagyrevi, 1978)

be due to: (i) its polarity, i.e. the permittivity (dielectric constant) of the surrounding medium, (ii) non-specific interactions, and (iii) specific interactions of the complexes with either the solvent molecules or ligand molecules. Being a polar interaction hydrogen bonding is maximum in the most apolar solvents. The largest known enthalpy of formation of an O—H $\cdots$ N bond between neutral proton donor and acceptor is the -10.5 kcal mol^{-1} recorded for p-nitrophenol–triethylamine in cyclohexane (Hudson et $al.$, 1970). Increase in solvent polarity can be expected to favour the formation of an ion-pair, all other factors being equal. However, beyond a certain optimum polarity, further increase in solvent polarity leads to dissociation of the ion-pair.

$$R_1—O^- \cdots H^+—NR_2 \xrightleftharpoons{K_d} R_1—O^- + N^+—NR_2 \qquad (4.3)$$

This effect occurs irrespective of whether an ion-pair is hydrogen bonded or not. Extensive conductance measurements by Fuoss and his collaborators in mixed solvents have shown that $\log K_d$ for uni-univalent electrolytes is inversely proportional to the permittivity (Fuoss, 1968). Representative values of the equilibrium constants for the formation of ion-pairs K_{pt}, and the dissociation of hydrogen-bonded ion-pairs K_d, taken from recent literature are given in Table 4.8.

The enhancement of proton transfer by a small increase in solvent polarity was first demonstrated for phenol–amine systems by Vinogradov and coworkers (Scott et $al.$, 1968; Scott and Vinogradov, 1969; Hudson et $al.$, 1970). Jadzyn and Malecki (1972) have employed Onsager's theory to obtain an expression for the potential energy of the proton in the tautomeric equilibrium (4.1) as a function of the permittivity of the surrounding medium. According to them the change in potential energy minima distribution with change in permittivity reflecting the transition from an O—H $\cdots$ N complex to an O$^-$ $\cdots$ H—$^+$N complex can be depicted schematically as in Figure 4.6.

The effect of solvent polarity is a function of ΔpK_a; the larger the ΔpK_a the smaller the range of solvent polarity over which the enhancement of proton transfer can be demonstrated (Wojtowicz and Malecki, 1977). In the extreme case of 3,4-dinitrophenol–triethylamine ($\Delta pK_a \sim 5.1$) one observes an example of a non-specific solvent effect. In cyclohexane the O—H $\cdots$ N complex is uncharged; addition of benzene leads to progressive increase in the proportion of the ion-pair which is the only species in pure benzene (Hudson et $al.$, 1972; Wojtowicz and Malecki, 1976).

Specific interaction of a polar molecule with the hydrogen-bonded ion-pair can also lead to enhancement of proton transfer. Pawelka and Sobczyk (1975) have demonstrated that specific interaction with solvent affects the dipole moments and thermodynamic parameters of formation of nitrophenolate–amine complexes. Dramatic increases in per cent proton transfer were

Table 4.8   Representative equilibrium constants for the formation, K_{pt}, and dissociation, K_d, of hydrogen-bonded ion-pairs from conductance and dielectric measurements

Ion-pair	Solvent	Permittivity	K_{pt} (M^{-1})	$K_d \times 10^6$ (M)	Dipole moment (D)	References
L-phe Me ester + N-acetyl-D-phe	1-Octanol	9.88	22±2	2.68±0.24	9.9	Highsmith (1975)
L-phe Me ester + N-acetyl-L-phe			19.5±2	2.93±0.25	10.6	
n-BuNH$_3^+$-picrate	Nitrobenzene	34.8		152		Haulait and Huyskens (1975)
(n-Bu)$_2$NH$_2^+$-picrate				156	11.5[b]	
Et$_2$NH$_2^+$-picrate				165		
Et$_2$NH$_2^+$-perchlorate				2170		
n-BuNH$_3^+$-picrate	Nitrobenzene	34.8		155		Macau et al. (1971)
	Methanol	32.6		555,000		
	n-Butanol	17.7		943		
n-Bu$_3$NH$^+$-picrate	Nitrobenzene	34.8		258	11.4[c]	
	Methanol	32.6		90,900		
	n-Butanol	17.7		552		

Compound	Solvent		K		Reference
Bu$_4$N$^+$-picrate	Nitrobenzene	34.8	143,000	15.0[c]	
	Methanol	32.6	100,000		
	n-Butanol	17.7	1320		
n-BuNH$^+$-2,6-dinitrophenolate	Dioxan	2.21	1.35×10^7		Vinogradov et al. (1970)
n-BuNH$^+$-2,4-dinitrophenolate			1.55×10^5		
n-BuNH$_3$-acetate	Diethyl ether	4.24	69 ± 8		Hibbert and Satchell (1968)[a]
n-BuNH$_3$-benzoate			371 ± 50		
n-BuNH$_3$-formate			3000 ± 500		
EtNH$_3^+$-dichloroacetate	Acetone	20.7	8.1		Jasinski and Pawlak (1967)
n-Bu$_3$NH$^+$-dichloroacetate			5.7		

[a] Infrared spectroscopic study.
[b] In benzene (Maryott, 1948).
[c] In benzene (Davies and Williams, 1960).

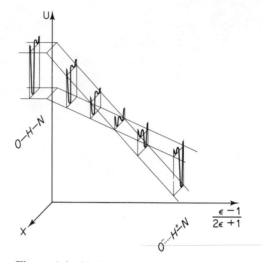

Figure 4.6 Variation in the proton potential energy curve of a hydrogen bond related to the permittivity (Jadzyn and
Malecki, 1972)

observed in chlorophenol–n-propylamine systems in acetonitrile with addition of water (Zundel and Nagyrevi, 1978). In carboxylic acid–amine systems, the presence of four water molecules per acid–base pair reduces the
ΔpK_a for 50% proton transfer from 2.3 to 0.9 in the case of primary or
secondary amines and from 4 to 2 in the case of tertiary amines (Figure 4.7)
(Lindemann and Zundel, 1977b). Likewise, increases in proton transfer
were observed in polyglutamic acid–amine and polyhistidine–carboxylic acid
systems with increase in hydration (Lindemann and Zundel, 1977a, 1978).
In polyamino acid systems particularly in polyhistidine–carboxylic acid systems, there is also an interdependence of proton transfer and polymer
conformation (Lindemann and Zundel, 1978).

Huyskens (1977) has reviewed recently the thermodynamic and spectroscopic data obtained by his group and other investigators, dealing with the
effect of a first hydrogen bond on the formation of a second hydrogen bond
by the same molecule or ion. In a hydrogen bond h_1 the proton donor or
acid group XH and the base Y can each possess additional proton acceptor
(n) and proton donor (h) sites:

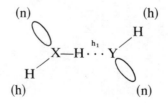

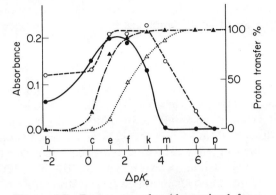

Figure 4.7 Proton transfer (determined from $\nu(C{=}O)$ and $\nu_s(COO^-)$) and absorbance of the continuum at 1700–2500 cm^{-1} in carboxylic acid–amine systems in the case of amines possessing additional hydrogen bond donor groups. Per cent proton transfer: .... pure systems, -.-.-. systems with four water molecules per acid–amine pair. Absorbance of the continuum: - - - - pure systems, ——— systems with four water molecules per acid–amine pair (Lindemann and Zundel, 1977b)

The ratio of equilibrium constants for the formation of a hydrogen bond with a ligand, K_2 in the presence of bond h_1 and K_2^0 in the absence of bond h_1, was examined as a function of the formation equilibrium constant K_h of bond h_1 and of the X—H stretching vibration frequency shift $\Delta\nu$ caused by its formation. Although in a given family a correlation often exists between $\log K_2/K_2^0$ and $\log K_h$, the correlations between $\log K_2/K_2^0$ and $\Delta\nu$ are more general. They were found in the following cases: (1) the interaction of a second molecule with one of the lone electron pairs of a first phenol molecule hydrogen bonded to triethylamine or tetramethylurea; (2) the interaction of a second phenol molecule with a halogen ion involved in an O—H $\cdots$ X$^-$ band with a first phenol molecule, and (3), the interaction of amines with the second N—H site of a dialkyl or monoalkylammonium ion, the first N—H group of which is involved in a N$^+$—H $\cdots$ X or N$^+$—H $\cdots$ X$^-$ bond. Thus the leading factor governing the ratios K_2/K_2^0 for the additional sites is not the hydrogen bond h_1 equilibrium constant K_h, but rather the perturbation of the X—H distance brought about by the hydrogen bond h_1 and reflected in the frequency shift of the X—H stretching vibration. The effect appears to be independent of the charge of the partners in the bond h_1 (Huyskens, 1977).

It is likely that specific interactions of water molecules, anions, cations, etc., with proton acceptor and proton donor sites on O—H $\cdots$ N and

$O^- \cdots H-^+N$ complexes formed side-chain groups are important in determining the properties of such complexes in proteins.

4.3.3   *Hydrogen bonding in amides*

In the crystal structures of simple aliphatic amides both amide hydrogen atoms are usually involved in $N-H \cdots O=C$ bonds. Often, amide molecules form cyclic dimers linked by pairs of hydrogen bonds which involve the hydrogen atom *cis* to the carbonyl oxygen atom. The atom which is *trans* to the oxygen atom forms hydrogen bonds which further link dimers to give ribbons (Hsu and Craven, 1974a). The $N \cdots O$ distances for the two kinds of hydrogen bonds in amide crystal structures are very similar. The average of 20 *cis* $N \cdots O$ distances is 2.97 Å and of 21 *trans* distances is 2.95 Å.

On the basis of X-ray diffraction studies of several small amides and a series of theoretical studies, Ottersen (1976) has found that formation of one intermolecular $N-H \cdots O=C$ bond results in a shortening of the $C-N$ bond by ~0.025 Å and a lengthening of the $C=O$ bond by ~0.014 Å. Similar consequences of hydrogen bonding appear to occur also in the hydrogen-bonded complexes of barbiturates: a systematic lengthening of the exocyclic $C=O$ bond (~0.01 Å) when it participates in hydrogen bonding (Hsu and Craven, 1974b). These consequences of hydrogen bonding can be explained by the stabilization of tautomers **2** relative to **1** in the scheme below. Formation of a hydrogen bond would decrease the negative charge at

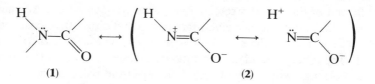

the carbonyl and cause an electron density shift from the nitrogen lone-pair orbital. The electron density shift from the $N-H$ to the carbonyl results in a decreased magnetic shielding for the amide proton leading to a shift of its resonance to lower fields in PMR spectra of peptides (Llinas and Klein, 1975).

Urea is exceptional among amides in that it possesses four $N-H$ groups per molecule in addition to the one oxygen. The $N \cdots O$ distances for hydrogen bonds in which urea is both donor and acceptor have an average value of 2.98 Å and range between 2.81 and 3.13 Å. The hydrogen bonds in which the carbonyl of urea is the acceptor, are all short ranging from 2.70 to 2.80 Å. The available data for the $N-H \cdots O=C$ bonding of urea suggest that urea is more effective as an acceptor than as a donor (Gartland and Craven, 1974). Measurements of the heat capacities of ureas, alkyl ureas and water in N,N-dimethylformamide have shown that the relative hydrogen bonding strengths of protons to the peptide carbonyl oxygen are in the order $NH_2 >$ water $> NH$ (Bonner et al., 1977).

Amides self-associate through $N—H \cdots O{=}C$ bonds in non-polar solvents such as CCl_4. They have been studied extensively as models of amide hydrogen bonds in polypeptides and proteins. The enthalpy of interamide hydrogen bonding in N-methylacetamide in CCl_4 is 4–5 kcal mol^{-1} (Klotz and Franzen, 1962; Znaskar and Rao, 1967; Kreschek and Klotz, 1969; Lowenstein et al., 1970; Graham and Chang, 1971; Lindheimer et al., 1974). Calorimetric studies of enthalpy changes for the transfer of short-chain N-alkylamides between water and CCl_4 have shown that it is an almost constant 7.4 ± 0.5 kcal mol^{-1} (Ojelund et al., 1976). This result suggests that amide–water interactions are stronger than amide–amide interactions. An infrared and Raman spectroscopic study of amide–water–salt solutions showed that the hydrogen bond donor strength of the water proton is greater than that of the amide $N—H$ group and that the acceptor strength varied in the order $Cl^- >$ amide carbonyl oxygen $>$ water oxygen (Bonner and Jordan, 1976). Recent vapour pressure measurements of aqueous solutions of acetamide, N-methylacetamide and N,N-dimethylacetamide (Wolfenden, 1978), showed that N-methylation of acetamide has little effect on its hydrophilic character, suggesting that the hydrophilic character of the peptide bond is mainly associated with the carbonyl oxygen rather than the $N—H$ group. These findings are in agreement with *ab initio* molecular orbital calculations of amide–water interactions (Johansson et al., 1974; Alagona et al., 1973; Pullman et al., 1974; Hinton and Harpool, 1977).

4.3.4   *Hydrogen bonds in peptides and peptide–water interactions*

Various types of intramolecular $N—H \cdots O{=}C$ bonds are found in the crystal structures of small peptides. They are designated $m \rightarrow n$, where m is the sequence number of the amino acid residue containing the $N—H$ group and n is the sequence number of the residue containing the $C{=}O$ group. Figure 4.8 shows the seven types of intramolecular $N—H \cdots O{=}C$ bonds observed in crystalline peptides. The $N \cdots O$ distances cover a range of 2.80–3.20 Å in cyclic polypeptides (Karle et al., 1970; Karle, 1974, 1975, 1977). In the cyclic dodecadepsipeptide valinomycin there are four $4 \rightarrow 1$ type and two $5 \rightarrow 1$ type hydrogen bonds (Karle, 1975; Smith et al., 1975). The latter are weaker than the former: the average $N \cdots O$ distance is 3.05 Å as compared to 2.95 Å and the $N—H \cdots O$ angles are 120–125°. On the other hand, the single intramolecular $N—H \cdots O{=}C$ bond in

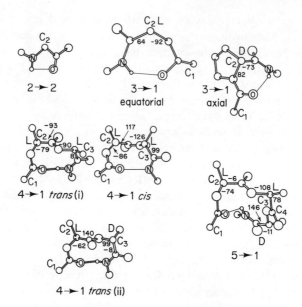

Figure 4.8   Types of intramolecular hydrogen bonds found in crystal structures of peptides (Benedetti, 1977)

(Phe^{4}Val^6)antamanide dodecahydrate, though of the $5 \rightarrow 1$ type, has an N $\cdots$ O distance of 2.91 Å (Karle and Duesler, 1977). Thus it appears that a wide variation in hydrogen bond lengths is possible with all the types of intramolecular N—H $\cdots$ O=C bonds.

In amino acids and small peptides the water molecules whenever present in the crystal structure, tend to have approximately tetrahedral environments, i.e. to have their two proton donor and two proton acceptor sites occupied, e.g. in L-alanyl-L-alanyl-L-alanine monohydrate (Fawcett *et al.*, 1975). In the larger cyclic peptides such as the cyclic decapeptide (Phe^{4}Val^6)antamanide dodecahydrate, it is possible to distinguish between three distinct roles for the water molecules (Karle and Duesler, 1977). Three water molecules appear to be an integral part of the peptide, in that they reside in the interior of the molecule and stabilize the conformation of the 30-membered ring by forming hydrogen bonds with N—H groups. Water molecules hydrogen bonded to the carbonyl oxygens which are directed to the outer surface of the molecule, represent 'bound' water. The remaining water molecules hydrogen bond to each other. The structural roles of water in (Phe^{4}Val^6)antamanide may well serve as a model for the unusual properties and arrangements of water molecules in proteins.

4.4   Hydrogen Bonds in Proteins

4.4.1   *Types of hydrogen bonding interactions*

The hydrogen bonds in proteins are usually classified into three groups:

1. main-chain–main-chain interactions, involving N—$H \cdots O$=O bonds between the peptide groups;
2. main-chain–side-chain interactions, between the peptide N—H and C=O groups on one hand and the side-chain groups on the other; and
3. side-chain–side-chain interactions.

The uncertainty in the position of an individual atom in a protein crystal structure varies depending on whether it is part of the polypeptide backbone or of a side-chain group and in the latter case, also on the location of the side-chain group, i.e. whether it is in the protein interior, at the surface of, or exterior to the protein. In high-resolution crystal structure determinations the $H \cdots Y$ distances in hydrogen bonds range from 1.5 to 2.5 Å. If the maximum error is 0.2 Å in main-chain nitrogen or calculated hydrogen atom position, and is 0.25 Å in peptide oxygen atom position then the error in $N \cdots O$ or $H \cdots O$ distance should not exceed 0.35 Å $((0.2^2 + 0.25^2)^{1/2} = 0.32)$ (Moews and Kretsinger, 1975). For main-chain–main-chain hydrogen bonds the generally accepted limits are an $H \cdots O$ distance of about 2.5 Å and an N—$H \cdots O$ angle of greater than 120–135°. Although the uncertainty in the position of the polypeptide backbone atoms is usually less than that in the positions of the side-chain atoms and consequently, the delineation of main-chain–main-chain hydrogen bonds is usually the most reliable, there are cases where regions of the polypeptide chain can be mobile in the crystal structure of a protein. Such segmental flexibility occurs in the 'activation domain' of trypsinogen, where the four segments, the nitrogen terminus to Gly(19), Gly(142) to Pro(152), Gly(184) to Gly(193), and Gly(216) to Asn(223), either are flexibly wagging in the crystal structure or adapt several different conformations statistically (Huber and Bode, 1978).

There can be an appreciable uncertainty in the position of an atom of a side-chain group exterior to the molecular surface of a protein. This is seen in Table 4.9 which compares the values of B and σ position for the side-chain nitrogen atom of Lys(3) and Lys(46) in the structure of the small non-haem iron protein rubredoxin (54 amino acids) determined by Jensen and his group at 1.5 Å resolution (Watenpaugh *et al.*, 1973). B is the isotropic thermal parameter proportional to the mean-square amplitude of vibration (Stout and Jensen, 1968). For protein structures this parameter includes contributions from statistical disorder in the atomic position as well as simple thermal vibration. σ is the standard deviation of the positional

Table 4.9 B values and σ positions for atoms in Lys(3) and Lys(46) of rubredoxin (Watenpaugh *et al.*, 1973)

Lys(3)			Lys(46)		
Atom	B (Å^2)	σ position (Å)	Atom	B (Å^2)	σ position (Å)
C_β	14	0.12	C_β	9	0.09
C_γ	15	0.14	C_γ	8	0.07
C_δ	18	0.15	C_δ	8	0.08
C_ε	31	0.18	C_ε	7	0.08
N_ζ	58	0.30	N_ζ	5	0.06

parameters. While the side-chain group of Lys(3) is fully extended, project-ing from the molecular surface into the solution, the N_ζ of Lys(46) is firmly bound by hydrogen bonds to $O_{(30)}$ and $O_{(33)}$. The uncertainty in position of N_ζ of Lys(3) is thus 5–10 times that in the position of N_ζ of Lys(46).

The main-chain–main-chain N—H $\cdots$ O=C bonds are responsible for the maintenance of secondary structures in proteins—helices, pleated sheets, and turns. It is therefore not surprising to find that the number of main-chain–main-chain interactions is approximately proportional to the number of amino acids residues (Figure 4.9). The number of hydrogen bonds belonging to the other two categories is usually much smaller and varies widely from protein to protein. Likewise, the number of water molecules found inside the protein varies widely: none in parvalbumin (Kretsinger and Nockolds, 1973) myoglobin or haemoglobin (Perutz, 1978), one in *Chromatium* high-potential iron protein (Carter *et al.*, 1974), four in pan-creatic trypsin inhibitor (Deisenhofer and Steigmann, 1975), ten in carboxy-peptidase A (Quiocho and Lipscomb, 1971), 25 in elastase (Sawyer *et al.*, 1978) and trypsin (Bode and Schwager, 1975), and 18 in trypsinogen (Fehlhammer *et al.*, 1977).

Elastase is one of the larger proteins whose crystal structure has been refined recently. Its hydrogen bonding has been described in great detail by Sawyer *et al.* (1978) and compared to that observed in chymotrypsin (Birktoft and Blow, 1972). The different types of hydrogen bond were counted and the results are shown in Table 4.10. Several points of interest emerge, particularly in contrast to the types of hydrogen bond observed in amino acids and peptides (Table 4.2):

1. The numerical preponderance of peptide–peptide hydrogen bonds over main-chain–side-chain and side-chain–side-chain hydrogen bonds;
2. the substantial number of hydrogen bonds formed by side-chain O—H groups with peptide C=O, hydrogen bonding which is not observed in the amino acid and peptide structures (Table 4.2);

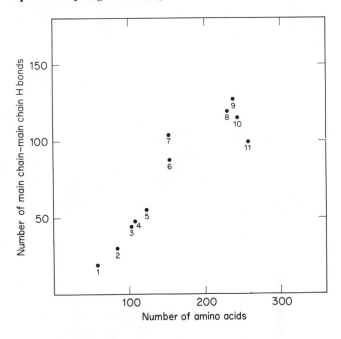

Figure 4.9 Plot of the number of main-chain–main-chain N—H $\cdots$ O$=$C bonds versus number of amino acids for: 1. pancreatic trypsin inhibitor (Deisenhofer and Steigmann, 1975). 2. *Chromatium* high-potential iron protein (Carter *et al.*, 1974), 3. cytochrome *c* (Takano *et al.*, 1977), 4. parvalbumin (Kretsinger and Nockolds, 1973), 5. ribonuclease S (Wyckoff *et al.*, 1970), 6. myoglobin (Schoenborn, 1971, 7. metmyoglobin (Takano, 1977), 8. elastase (Sawyer *et al.*, 1978), 9. concanavalin (Reeke *et al.*, 1975), 10. tosyl-α-chymotrypsin (Birktoft and Blow, 1972), and 11. carbonic anhydrase C (Liljas *et al.*, 1972)

3. the tendency of the internal water molecules to hydrogen bond with peptide carbonyls and side-chain hydroxyl groups, and
4. the tendency by N—H groups of the imidazolium, ammonium, and guanidinium side-chains to hydrogen bond to peptide carbonyl groups and external water molecules.

Elastase and α-chymotrypsin exhibit only 39% identity and 51% homology in their amino acid sequence (Shotton and Hartley, 1970). Homologous residues are those which are identical or chemically similar to their counterparts in corresponding positions along the polypeptide chain. The homologous residues are mostly located in the internal regions of each molecule. This fact probably accounts for the similarity between the hydrogen bonding

Table 4.10 Hydrogen bonds in elastase[a] (Sawyer et al., 1978)

	(Im)⁺N—H	(NH₃)⁺N—H	(Gnd)⁺N—H	Peptide N—H	Amide[b] N—H	(In)N—H	O—H	Int(W)O—H	Ext(W)O—H
COO⁻	1	1	2	8		1	2	2	6
Peptide C=O	1	5	4	119	5	3	13	21	77[a]
Amide C=O[b]			1	6	1	1	2		22
H—O				2	1				
(Im)N				1					
Int(W)O	2		1	14	1		8[e]		
Ext(W)O	4	4[c]	15[c]	16[d]	24[d]	1	24[d]	17	
Sulphate			4						

[a] Crystals grown at pH 5.0.
[b] Of glutamine and asparagine side chains.
[c] Counting 2 hydrogen bonds for extended side-chain groups of lysine and arginine.
[d] Counting one hydrogen bond for surface N—H, C=O, and O—H groups.
[e] Includes O—H···O(W) and (W)O—H···O—H interactions.

patterns made by the side-chain groups in the two proteins. Of the 60 side-chain hydrogen bonds found in elastase, 26 are also found to be present in α-chymotrypsin (Sawyer *et al.*, 1978).

4.4.2 *Possible roles for ion-pair interactions*

The many different roles (see also the recent discussion of electrostatic effects in proteins by Perutz (1978)) that hydrogen-bonded ion-pairs are likely to perform in proteins can be roughly divided into three groups:

1. conformational roles, contribution to the maintenance of a specific conformation,
2. recognition roles, provisions of specificity in the association or binding of proteins, and
3. roles necessary to the function of proteins.

The contribution of 'salt bridges' to protein conformational stability has been generally neglected for a long time, due to the following two arguments: (i) 'salt bridges' should require a low local dielectric constant environment and hence can only exist in the interior of proteins, and (ii), acid–base titrations of protein molecules should result in the neutralization of the two moieties of a 'salt bridge' and liberation of the water molecules forming its solvation shell, leading to a volume change which is not observed. These objections do not appear to be too serious (Arvidsson, 1972). Hydrogen-bonded ion-pair bonds have been observed in external regions of several protein crystal structures. For example, in subtilisin (275 residues), charged groups form ion-pairs at several locations on the surface of the molecule: Lys(136)-Asp(140), Lys(141)-Glu(112) and Lys(237)-carboxyl terminal (Drenth *et al.*, 1971). The water molecules which interact with polar side-chain groups involved in hydrogen-bonded ion-pairs in the interior of proteins can, subsequent to neutralization of the two moieties of the ion-pair, participate in hydrogen bonding with the now uncharged moieties of the original ion-pair as well as with other neighbouring polar groups.

A possible conformational role for hydrogen-bonded ion-pairs in proteins is the enhancement of heat stability. Perutz and Raidt (1975) have compared the amino acid sequences of ferredoxins from mesophile and thermophile bacteria and used an atomic model to compare the effect of the known amino acid replacements, assuming the structures of all the ferredoxins to be identical. Their results indicate clearly that possible formation of external ion-pairs, linking residues near the amino terminus to others near the carboxyl terminus could be responsible for the greater heat stabilities of the thermophile ferredoxins. They also pointed out that large (20–fold) differences in the rates of thermal denaturation correspond to differences in free energy of activation of less than 2.5 kcal mol^{-1}, a value in accord with the total extra free energy of stabilization provided by several additional (2–4) hydrogen-bonded ion-pairs. The structure of thermolysin, a heat-stable protease of 316 residues, incorporates an unusually large number of ionic interactions: 15 ion pairs formed by aspartate, glutamate, lysine, and arginine residues and 6 interactions between the four Ca^{2+} ions and the side-chain groups of aspartate and glutamate residues (Colman et al., 1972). Unique intersubunit Arg-Glu and Arg-Asp ion pairs in B.stearothermophilus glyceraldehyde phosphate dehydrogenase may also be responsible for the heat stability of this enzyme (Harris and Walker, 1977; Perutz, 1978).

Stabilization of elastase and protection against digestion by carboxypeptidase A, in the presence of which elastase has to function in vivo, has been suggested for the surface ion-pair Arg(170)-Asn(245) C-terminal carboxylate which could hold in place the C-terminal helix (Sawyer et al., 1978). The analogous ion pairs Lys(107)-C-terminal carboxylate group are observed in α-chymotrypsin (Birktoft and Blow, 1972) and in β-trypsin (Bode and Schwager, 1975).

There are several examples of the role of ion-pairs in the binding of proteins or fragments thereof, to one another. One is the Arg(10)-Glu(2) ion pair in ribonuclease that is formed in a hydrophobic environment when S peptide complexes with S protein (Finn et al., 1972). Another example is the interaction(s) of Glu(35) with His(15) and/or Arg(14) in the concentration- and pH-dependent self-association of lysozyme studied by Rupley and coworkers (Banerjee et al., 1975, Shindo et al., 1977). A third example is found in the associations of trypsin with its inhibitors, the pancreatic trypsin inhibitor (58 amino acids) and soybean trypsin inhibitor (181 amino acids), highly specific interactions resulting in the formation of stable, enzymatically inactive complexes with a free energy of dissociation of 15 kcal mol^{-1} or more. Only a small proportion of both molecules is in contact in the complex: 14 amino acid residues in the case of the pancreatic trypsin inhibitor and 24 residues of trypsin (out or 224) (Huber and Bode, 1978). Although the contribution of polar interactions to the free energy of dissociation is probably small compared to that of hydrophobic interactions (Janin and Chothia, 1976), an essential part in the specificity of trypsin–inhibitor interactions is a complex network of hydrogen bonds involving Lys(15) of the pancreatic inhibitor and Arg(63) of soybean inhibitor, on one hand, and Ser(190), Asp(189) of trypsin and two water molecules on the other (Janin et al., 1974; Bode et al., 1976, 1978). A final example, would be the involvement of an ion-pair in a receptor binding site Tsernoglou et al. (1978) have proposed that formation of an Arg-Asp ion-pair upon bonding of a snake venom curarimimetic neurotoxin to the acetylcholine receptor protein results in specific occupation of the membrane acetylcholine binding site.

Ion-pairs may also play a role in the interaction of charged denaturants with proteins. A recent crystallographic study of the interaction of sodium dodecyl sulphate with lysozyme showed that the molecules of the detergent bound at three different locations each representing a different type of interaction (Yonath *et al.*, 1977). All of them possessed one feature in common which is probably characteristic of sodium dodecyl sulphate binding: the hydrocarbon chain makes hydrophobic contacts while the sulphate groups form ion-pairs with protonated amine groups.

There are numerous examples of the involvement of ion pairs in the function of proteins. Chemical and crystallographic data have shown that arginyl residues occur at active sites of many enzymes where they interact with negatively charged phosphate or carboxylate moieties of substrates or cofactors. The work of Cotton *et al.* (1973, 1974) and of Riordan *et al.* (1977) has shown that a major biological function of arginine might be to interact with phosphorylated metabolites, since the guanidinium group is ideally suited for interaction with phosphate-containing substances by virtue of its planar structure and ability to form multiple hydrogen bonds with the phosphate moiety. In guanidinium phosphates, almost all of the guanidinium N—H groups participate in hydrogen bonding: 18 of 19 in trisguanidinium hydrogen pyrophosphate (Adams and Ramdas, 1976) and 13 of 15 in bisguanidinium hydrogen phosphate monohydrate (Adams and Small, 1976). The methylguanidinium phosphates were investigated by Cotton and his group as models of the complex observed in the binding of deoxy-thymidine-3″,5″-diphosphate to staphylococcal nuclease whose structure (Arnone *et al.*, 1971) is shown in Figure 4.10. Two cyclic complexes are formed by each guanidyl group of the two arginine residues (35 and 37) to a phosphate group. In methylguanidinium dihydrogen orthophosphate (Cotton *et al.*, 1973) one such structure is observed, while in bis(methyl-guanidinium) monohydrogen orthophosphate (Cotton *et al.*, 1974a) two cyclic complexes occur in a very similar arrangement, though not precisely the same, as that in the enzyme–inhibitor complex. The hydrogen bonding versatility of the arginine side-chain group is further demonstrated in the nuclease, by the formation of additional hydrogen bonds by Arg(35) to one or two peptide carbonyls and by Arg(87) to the carboxylate of Asp(83) (Hazen, 1978).

The crystal structures of the serine proteases contain only two buried ion pairs. Both play crucial roles in the function of these enzymes. One is part of the hydrogen-bonded complex involving imidazole in the 'charge relay system' of Asp(102), His(57), and Ser(195) of α-chymotrypsin (Blow *et al.*, 1969). See Blow (1976) and Kraut (1977) for recent reviews. An almost identical geometrical arrangement of the Asp(32), His(64), and Ser(221) side-chain groups was found in subtilisin whose main chain is, however, folded in a very different way from that of α-chymotrypsin (Alden *et al.*, 1970). Subsequently, the same geometrical arrangement of Asp, His, and

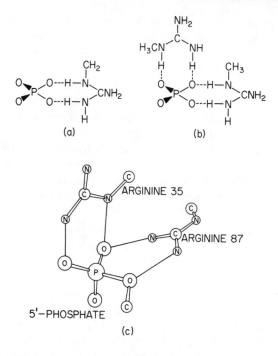

(a)   (b)

(c)

Figure 4.10 Cyclic hydrogen-bonded complexes in (A) methylguanidinium dihydrogen
orthophosphate (Cotton *et al.*, 1973), (B)
bis(methylguanidinium) monohydrogen orthophosphate (Cotton *et al.*, 1974b), and (C) in
the ternary complex of the staphylococcal nuclease, deoxy-thymidine-3′,5′-diphosphate and
calcium ion (Arnone *et al.*, 1971)

Ser was also found in other chymotryspin-homologous enzymes including
elastase (Shotton and Watson, 1970), trypsin (Bode and Schwager, 1975),
and *Streptomyces griseus* protease B (Codding *et al.*, 1974). In all cases, the
side-chain groups of the three residues were thought to form a hydrogen-
bonded system. Recent refinement of the crystal structure of subtilisin
(Matthews *et al.*, 1977) has revealed that the hydroxyl oxygen of the reactive
serine (Ser(221)) is in fact about 3.7 Å from NE2 of His(64), although the
His(64)-(Asp(32) ion-pair bond remains a strong hydrogen bond of about
2.7 Å. A similar situation was observed also in elastase where the N · · · O
distance between His(57) and Ser(195) is 3.4 Å (Sawyer *et al.*, 1978).† It can
be seen from Figure 4.1 that a distance of 3.4–3.7 Å is significantly outside

† In trypsin this hydrogen bond is 3.3 Å and bent at pH 8, while at pH 5 it is 2.9 Å and more
linear (Huber and Bode, 1978).

the range of $(Im)^+N—H \cdots O^-$ bonds. The alteration in the view of the structure of the three residues at the active site of serine proteases, illustrates the difficulty associated with the unequivocal identification of specific hydrogen bonds in protein crystals.

The other internal ion pair in serine proteases is formed subsequent to a conformational alteration attendant upon activation of the zymogen proenzyme by proteolytic removal of a small peptide N-terminal to residue (16), between the carboxylate group of Asp(194) and the newly created ammonium group of residue (16). The careful studies of Fersht and Requena (1971) and Fersht (1972) have shown that in α-chymotrypsin, this ion-pair‡ contributes a stabilization energy of $2.9 \, \text{kcal mol}^{-1}$ to the active conformation of the enzyme. In trypsin, Huber and Bode (1978) has described the Asp(194)-(Ile(16) ion-pair as the clamp of a complex hydrogen bonding system consisting of more than twenty hydrogen bonds cross-linking the segments of the 'activation domain', which are either absent or are replaced by hydrogen bonds to water molecules in trypsinogen. The flexibility of the four segments of the 'activation domain' in trypsinogen stands in contrast to their rigidity in trypsin due to extensive hydrogen bonding.

In 1970 Perutz proposed that the cooperative effects in the oxygenation of haemoglobin arise from an equilibrium between the liganded or R structure whose oxygen affinity is comparable to the average of the affinities of the subunits, and the deoxy or T structure whose oxygen affinity is lowered by the presence of constraining ion-pair interactions, shown below (Perutz, 1976).

$$\text{Lys H10(127)}\alpha \, NH_3^+ \cdots {}^-OOC \text{ Asp A4(6)}\alpha$$

$$\text{C-terminal Arg HC3(141)}\alpha \overset{\displaystyle COO^- \cdots {}^+H_3N \text{ Val A1(1)}\alpha}{\underset{\displaystyle Gnd^+ \cdots}{\diagdown}} \cdots Cl^-$$

$${}^-OOC \text{ Asp H9(126)}\alpha$$

$$\text{C-terminal His HC3(146)}\beta \overset{\displaystyle COO^- \cdots {}^+H_3N \text{ Lys C6(40)}\alpha}{\underset{\displaystyle Im^+ \cdots {}^-OOC \text{ Asp FG1(94)}\beta}{\diagdown}}$$

$$\text{Arg FG4(92)}\alpha \, Gnd^+ \cdots {}^-OOC \text{ Glu CD2(43)}\beta$$

Although the extent to which these hydrogen-bonded ion-pairs are involved in the mechanism of oxygenation of human haemoglobin remain to

‡ As expected, the Ile(16) ammonium group possesses a high pK_a of 10 (Fersht, 1972).

be completely determined, Kilmartin *et al.* (1978) have recently provided substantial evidence for the role played by the Bohr group 'salt bridges', namely the His(146)β—Im$^+$ $\cdots$ $^-$OOC Asp(94)β ion-pair and the complex formed by the inorganic anion positioned between Val(1)α and Arg(141α side-chain groups.

In conclusion, ion-pair interactions that are conserved in the evolution of a protein family must obviously be important in maintaining the structural or functional integrity of proteins, or both. There are a number of such ion-pair interactions whose precise roles have yet to be determined. An example are the two ion pairs Lys(14A)-Asp or Glu(4GH) and Lys(10H-Glu(4A) found in all myoglobins and mammalian haemoglobins (Takano, 1977).

4.4.3   N—H $\cdots$ S *Interactions*

Hydrogen bonds involving sulphur are long and by implication, weak in amino acids. In the crystal structure of cysteine the S—H group of one molecule forms weak hydrogen bonds of 3.84 Å and 3.39 Å with the S—H and carboxylate groups of another molecule (Kerr *et al.*, 1975). Two bent N—H $\cdots$ S bonds of 3.00 Å and 3.08 Å are found in 3,3,3',3'-tetramethyl-D-cystine dihydrochloride (Rosenfield, Jr. and Parthasarathy, 1975). In a survey of N—H $\cdots$ S interactions in crystals of small compounds Donohue (1969) observed the N $\cdots$ S distance to occur in the range of 3.25–3.66 Å, with H $\cdots$ S distances varying from 2.3 Å to 2.8 Å, the sum of the van der Waals radii. Recent work has shown that the van der Waals radius of sulphur should be reduced from the commonly accepted values of 1.8–1.9 Å (Van Wart *et al.*, 1975; Boyd, 1978). In addition, a divalent sulphur (sulphur bonded to two ligands, neither one being hydrogen) is not spheroidal on its non-bonded side. Studies by Parthasarathy and his collaborators of non-bonded atomic contacts with sulphur in crystal structures of sulphur-containing amino acids and other molecules, have shown that the electron density is up to 0.2 Å longer perpendicular to the Y—S—Z plane than in the plane (Rosenfield, Jr. and Parthasarathy, 1974, 1975; Rosenfield, Jr. *et al.*, 1977). Furthermore, electrophilic atoms prefer approaches from a direction perpendicular to the Y—S—Z plane, while nucleophilic atoms prefer approaches in this plane along extensions of the S—Y or S—Z bonds (Rosenfield, Jr. *et al.*, 1977).

Several kinds of N—H $\cdots$ S bonds are found in proteins. In α-chymotrypsin the sulphur atom of Met(180) is 3.7 and 3.8 Å from the peptide N—H groups of Lys(177) and Trp(215), respectively (Birktoft and Blow, 1972). Five N—H $\cdots$ S bonds occur between peptide N—H groups and the Fe$_4$S$_4$ cluster of *Chromatium* high-potential iron protein (Carter *et al.*, 1974). A large number of N—H $\cdots$ S bonds involving peptide N—H groups and both inorganic and cysteine sulphur atoms were found by Jensen and his collaborators in the iron proteins ferredoxin and rubredoxin (Adman *et al.*, 1975). Fifteen to eighteen such bonds occur in ferredoxin and 6 in rubredoxin with lengths in the range of 3.1–3.9 Å. Four types of N—H $\cdots$ S interactions were observed: three involving cysteinyl sulphur atoms and one involving inorganic sulphur atoms. Two types use the cysteinyl sulphur atom of residue n and the peptide N—H group of residue $n+2$. The third type of N—H $\cdots$ S bond involves a peptide N—H group distant in the sequence. In the fourth type, the peptide N—H group is directed towards an inorganic sulphur atom.

4.4.4 *Water–protein interactions*

The hydration of biological macromolecules has been studied extensively.[†] The water molecules associated with a protein can be divided into two types: internal and external. The internal water molecules are a small number of tightly bound water molecules in the protein interior, that may be considered an integral part of the protein. Less tightly bound and consequently more mobile, external water molecules are distributed over the protein molecular surface. These include molecules associated with the charged and uncharged hydrophilic side-chain groups at the protein surface as well as more distant water molecules which form a transitional layer merging into the bulk water. Some of the external water molecules make more than one hydrogen bond to the same protein molecule while others are hydrogen bonded to only one group and presumably serve to connect the protein with the surrounding solvent. Some water molecules bridge across to neighbouring molecules in the crystal and others are found in active-site clefts and along channels running into the protein.

Watenpaugh *et al.* (1978) have recently obtained a model of rubredoxin based on X-ray diffraction data of 1.2 Å resolution, which includes 127 water sites with B values ranging from 15–60 Å^2 and occupancies from unity down to 0.3. They find the minimum distances from water to protein oxygen and nitrogen atoms to be distributed about two distinct maxima, the major one occurring at 2.5–3.0 Å and the minor one at 4.0–4.5 Å. Thus two thirds of the water sites occur in the first shell of water extending to 3.5 Å and the rest occur in the second shell between 3.5 and 5.5 Å. The B values and occupancies increase and decrease, respectively, with increase in the distance from the protein surface. Only one water molecule is so tightly bound that it can be considered a structural part of the protein. The detailed picture obtained by Jensen *et al.* is consistent with the concept of the external water molecules comprising two layers, one of molecules hydrogen

[†] See reviews by Kuntz and Kauzmann (1974), Cooke and Kuntz (1974), Berendsen (1975), Hopfinger (1977), and Finney (1977).

bonded to the molecular surface of rubredoxin, and the other consisting of more distant water molecules blending into the bulk water. Of the water molecules in the proximal layer those hydrogen bonded to the main chain C=O and N—H groups occur at better defined sites than those hydrogen bonded to the hydrophilic side chains, presumably because of the greater mobility of the latter compared to the main-chain atoms.

In Table 4.11 are collected the frequencies of hydrogen bonding interactions between water molecules and main-chain amide and carbonyl groups in several recent protein crystal structures. It is evident that the number of (W)O—H $\cdots$ O=C bonds is always greater than that of N—H $\cdots$ O(W) bonds, irrespective of whether only the internal water molecules or all the water molecules located in the protein crystal structure, are considered. This pattern suggests that the energy of interaction of water with a peptide carbonyl group is higher than with a peptide amide group, in agreement with the results of studies with amides mentioned earlier. Of greatest structural interest are the water molecules found in the protein interior. Detailed descriptions of interior water hydrogen bonding have been given in tosyl-α-chymotrypsin (Birktoft and Blow, 1972), β-trypsin (Bode and Schwager, 1975), trypsin-pancreatic trypsin inhibitor (trypsin–PTI) complex (Bode et al., 1976), and elastase (Sawyer et al., 1978). Figure 4.11 provides a diagrammatic representation of the hydrogen bonding engaged in by the internal 25 water molecules in elastase. The internal water molecules occur singly as well as in clusters of two to seven; when single, they form at least two hydrogen bonds and in some caes, as many as five hydrogen bonds, e.g. W-25 in elastase (Figure 4.11 (H)). In rubredoxin 11 of the 127 water oxygen atoms are involved in five hydrogen bonds. This is understandable in view of the dynamic nature of water structure, with most water sites being

Table 4.11   Frequency of interaction of the main-chain N—H and C=O groups with water molecules

Protein	Number of water molecules	N—H	C=O	References
Myoglobin	106[a]	8	30	Schoenborn (1971)
Tosyl-α-chymotrypsin	14[b]	13	18	Birktoft and Blow (1972)
Trypsin	25[b]	13	28	Bode and Schwager (1975)
Trypsin–PTI	34[b]	17	45	Bode et al. (1976)
Metmyoglobin	80[a]	7	10	Takano (1977)
Elastase	{ 25[b]	14	21	Sawyer et al. (1978)
	111[c]	36	39	
Rubredoxin	127[a]	11	26	Watenpaugh et al. (1978)

[a] Total number located in the structure.
[b] Internal water molecules.
[c] External water molecules.

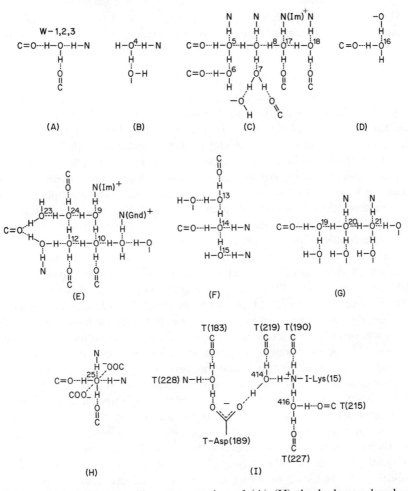

Figure 4.11 Diagrammatic representation of (A)–(H) the hydrogen bonds formed by the 25 internal water molecules of elastase (Sawyer *et al.*, 1978), and (I) of the hydrogen bonding in the specificity pocket of the trypsin–PTI complex (Bode *et al.*, 1976). Water molecule 25 in elastase (H) may be a Ca^{2+} ion

only partially, i.e. intermittently, occupied (Watenpaugh *et al.*, 1978). The internal water molecules overwhelmingly form hydrogen bonds with neutral hydrophilic groups, main-chain and side-chain amide groups, and side-chain hydroxyl groups in the foregoing proteins. Only a small proportion of hydrogen bonds are formed to protonated amine and carboxylate side-chain groups. Furthermore, there do not appear to be any water molecules directly associated with an ion-pair in the protein interior.

The presence of clusters of two and more internal water molecules must provide several advantages insofar as maintenance of a protein conformation is concerned: (1) they help accommodate polar hydrophilic side-chain groups in the relatively hydrophobic protein interior, (2) the donor and acceptor groups do not have to be sterically paired, (3) there can be odd numbers of donor and acceptor groups, (4) the whole hydrogen-bonded network is flexible, and (5) the additional hydrogen bends formed enhance the stability of the whole structure. For example, Figure 4.11 (C) shows a cluster of 6 water molecules forming 16 hydrogen bonds with each other and 10 protein main-chain and side-chain groups; in the absence of the water molecules and assuming correct pairing of the groups only 5 hydrogen bonds could be formed. In elastase (Sawyer et al., 1978) 48 hydrophilic groups form at least 65 hydrogen bonds with the 25 internal water molecules. In the trypsin–PTI complex (Bode et al., 1976) 84 groups form at least 94 hydrogen bonds with the 34 water molecules, 23 in the interior of the trypsin, 3 inside the inhibitor and 8 between the two moieties of the complex. In β-trypsin (Bode and Schwager, 1975) some 53 groups form 64 hydrogen bonds with 25 water molecules and in tosyl-α-chymotrypsin (Birktoft and Blow, 1972) 41 groups form about 43 hydrogen bonds with 13 water molecules. The results of counting the various hydrogen bonding configurations of the internal water molecules in these proteins are shown in Table 4.12. The commonest configurations, 54 out of 61 involving 68 out of 97 water molecules, involve one or two water molecules; the average number of hydrogen bonds per water molecule varies from 2.6 to 2.9.

To the list of possible structural roles played by internal water molecules mentioned above, two more should be added. In some regions of the protein, internal water clusters hydrogen bond to external water molecules and thereby provide communication between the protein interior and the surrounding aqueous environment. Furthermore, water molecules can apparently play the role of hydrogen bonding 'fill-ins'. An example is the interesting hydrogen bonding pattern found in the specificity pocket of the trypsin–pancreatic trypsin inhibitor complex (Figure 4.11 (I)). The Lys(15) of the inhibitor interacts with the carboxylate group of Asp(189) of trypsin through a water molecule (water 414) (Bode et al., 1976). In the case of the soybean inhibitor, the Arg(63) side-chain group forms an ion pair directly with the Asp(189) carboxylate (Janin et al., 1974). Likewise, water 414 is absent in benzamidine-inhibited trypsin, the guanidine group interacting directly with the Asp(189) carboxylate group (Huber and Bode, 1978). Thus water 414 helps to accommodate the different hydrogen bonding requirement of the ammonium side-chain group of Lys(15) in the trypsin–PTI complex.

In conclusion, internal water molecules may well participate in the function of proteins. An example is the water molecule hydrogen bonded

Table 4.12   Frequencies of distribution of internal water molecules[a]

No. of protein groups	No. of water molecules	No. of hydrogen bonds[b]	Frequency	Average number hydrogen bonds per water molecule
2	1	2	12	
3	1	3	20	2.9
4	1	4	8	
2	2	3	1	
3	2	4	2	
4	2	5	5	2.6
5	2	6	4	
6	2	7	2	
5	3	7	1	
5	3	8	1	2.8
7	3	9	1	
8	3	10	1	
5	4	11	1	
10	6	16	1	
9	7	14	1	

[a] In tosyl-α-chymotrypsin (Birktoft and Blow, 1972), β-trypsin (Bode and Schwager, 1975), trypsin–PTI complex (Bode et al., 1976), and elastase (Sawyer et al., 1978).
[b] In counting the number of hydrogen bonds in a cluster any hydrogen bonds made by internal water molecules to external water molecules were not counted.

between Glu(43) and the 5′-phosphate in the staphylococcal nuclease–pdTp complex, which is believed to act as the nucleophile in the mechanism of action of the enzyme (Hazen, 1978).

4.5   Conclusion

The attempt has been made to review briefly different types of hydrogen bonding interactions in order to emphasize their possible occurrence and importance in proteins. It is worthwhile summarizing the salient points that have emerged.

The survey of hydrogen bonding in crystal structures of amino acids, peptides, and related molecules suggests that imidazolium, ammonium, and guanidinium groups form with anionic and neutral acceptors hydrogen bonds whose lengths increase in the order given. Very little is known about the energetics of such interactions. Further work on hydrogen-bonded ion-pairs is very necessary.

Monoanion and monocation hydrogen bonds occur extensively in small molecules. Although there appears to be good evidence for the existence and importance of carboxyl–carboxylate interactions in some proteins, the existence of monocation complexes remains questionable.

In amides and small peptides N—H $\cdots$ O=C bond lengths can vary from 2.8 to 3.2 Å. Although both groups interact readily with water molecules, the C=O group appears to be favoured.

In proteins the dominant hydrogen bonding interactions are those involving main-chain and side-chain amide groups and the side-chain hydroxyl groups. Hydrogen-bonded ion-pairs are relatively rare, particularly in the protein interior. Examples of the different types of possible roles played by ion-pairs include:

1. stabilization of a specific conformation, e.g. increasing resistance to thermal denaturation or to proteolytic degradation,
2. provision of specificity in protein–protein association or recognition, e.g. in protease–inhibitor association, and
3. involvement in protein function, e.g. the role of buried ion-pairs in serine proteases.

Ion-pairs conserved in the evolution of homologous proteins are likely to be important: in many cases their role remains unknown.

The water associated with protein crystal structures can be broadly divided into:

1. external water molecules which hydrogen bond specifically to the molecular surface and which also form part of the transitional layer between the protein and the bulk solvent, and
2. internal water molecules which can be considered part of the protein structure.

Both types of water molecules interact extensively with main-chain and side-chain amide groups, the interactions with carbonyl groups being numerically preponderant. Examination of the structure of internal water molecules in chymotrypsin, trypsin, trypsin–pancreatic trypsin inhibitor complex, and elastase reveals that:

1. the dominant majority form hydrogen bonds to amide N—H and C=O groups and to side-chain hydroxyl groups,
2. over two thirds occur as single molecules or as clusters of two molecules,
3. clusters of up to seven water molecules can occur, and
4. the average number of hydrogen bonds per water molecule varies between 2.6 and 2.9.

The presence of internal water molecules in a protein may present several structural as well as functional advantages.

The paucity of knowledge concerning hydrogen bonding in proteins is due to the complexity of the structures involved no less than to the favour enjoyed by hydrophobic interactions as determinants of protein structure and protein–protein association. It is perhaps time for a rehabilitation of the hydrogen bond and particularly the hydrogen-bonded ion-pair, as important factors in protein structure and function.

4.6 Acknowledgements

Criticisms and suggestions made by Drs. E. Hazen, Jr., R. Parthasarathy, and G. Zundel are gratefully acknowledged.

4.7 References

Adams, J. M. and Ramdas, V. (1976). *Acta Cryst.*, **B32**, 3224.

Adams, J. M. and Small, R. W. H. (1976). *Acta Cryst.*, **B32**, 832.

Adman, E., Watenpaugh, D. D., and Jensen, L. H. (1975). *Proc. Natl. Acad. Sci. U.S.*, **72**, 4854.

Aitken, H. W. and Gilkerson, W. R. (1973). *J. Amer. Chem. Soc.*, **95**, 8551.

Alagona, G., Pullman, A., Scrocco, E., and Tomasi, J. (1973). *Int. J. Peptide Prot. Res.*, **5**, 25.

Alden, R. A., Wright, C. S. and Kraut, J. (1970). *Phil. Trans. Roy. Soc. (London)*, **B257**, 119.

Allen, L. C. (1975). *J. Amer. Chem. Soc.*, **97**, 6921.

Arnone, A., Bier, C. J., Cotton, F. A., Day, V. W., Hazen, E. E., Jr., Richardson, D. C., Richardson, J. S., and Yonath, A. (1971). *J. Biol. Chem.*, **246**, 2302.

Arshadi, M., Yamdagni, R., and Kebarle, P. (1970). *J. Phys. Chem.*, **74**, 1475.

Arvidsson, E. D. (1972). *Biopolymers*, **11**, 2197.

Banerjee, S. K. Pogolotti, A., Jr., and Rupley, J. A. (1975). *J. Biol. Chem.*, **250**, 8260.

Barrow, M. J., Currie, M., Muir, K. W., Speakman, J. C., and White, D. N. J. (1975). *J. Chem. Soc. Perkin II*, **71**, 15.

Baur, W. H. (1972). *Acta Cryst.*, **B28**, 1456.

Benedetti, E. (1977). In *Peptides: Proceedings of the Fifth American Peptide Symposium* (Eds. Goodman, M. and Meienhofer, J.), John Wiley New York, p. 257.

Berendsen, H. J. C. (1975). In *Water: A Comprehensive Treatise* (Ed. Franks, F.), Vol. 5, Plenum Press, New York, p. 293.

Bhat, T. N. and Vijayan, M. (1977). *Acta Cryst.*, **B33**, 1754.

Birktoft, J. J. and Blow, D. M. (1972). *J. Mol. Biol.*, **68**, 187.

Blow, D. M. (1976). *Acc. Chem. Res.*, **9**, 145.

Blow, D. M., Birktoft, J. J., and Hartley, B. S. (1969). *Nature*, **221**, 337.

Bode, W. and Schwager, P. (1975). *J. Mol. Biol.*, **98**, 693.

Bode, W., Schwager, P., and Huber, R. (1976). In *Proteolysis and Physiological Regulation*, Academic Press, New York, p. 43.

Bode, W., Schwager, P., and Huber, R. (1978). *J. Mol. Biol.*, **118**, 99.

Bondi, A. (1964). *J. Phys. Chem.*, **68**, 441.

Bonner, O. D., Bednarek, J. M., and Arisman, R. K. (1977). *J. Amer. Chem. Soc.*, **99**, 2898.

Bonner, O. D. and Jordan, C. F. (1976). *Physiol. Chem. Phys.*, **8**, 293.

Bonsor, D. H., Borah, B., Dean, R. L., and Wood, J. L. (1976). *Canad. J. Chem.*, **54**, 2458.

Borah, B. and Wood, J. L. (1976). *Canad. J. Chem.*, **54**, 2470.

Boyd, D. B. (1978). *J. Phys. Chem.*, **82**, 1407.

Brown, J. D. (1976). *Acta Cryst.*, **B32**, 24.

Brzezinski, B. and Szafran, M. (1977). *Roczn. Chem.*, **51**, 589.

Brzezinski, B. and Zundel, G. (1976). *J. Chem. Soc. Faraday II*, **72**, 2127.

Carter, C. W., Jr., Kraut, J., Freer, S. T., Alden, R. A. (1974). *J. Biol. Chem.*, **249**, 6339.

Chen, C. S. and Parthasarathy, R. (1977). *Acta Cryst.*, **B33**, 3332.

Chen, C. S. and Parthasarathy, R. (1978). *Int. J. Peptide Prot. Res.*, **11**, 9.

Chen, C. S., Parthasarathy, R., and DeTitta, G. T. (1976). *J. Amer. Chem. Soc.*, **98**, 4983.

Clements, R., Dean, R. L., Singh, T. R., and Wood, J. L. (1971). *Chem. Commun.*, 1125, 1127.

Clements, R., Dean, R. L., and Wood, J. L. (1973). *J. Mol. Struct.*, **17**, 291.

Clements, R. and Wood, J. L. (1973a). *J. Mol. Struct.*, **17**, 265.

Clements, R. and Wood, J. L. (1973b). *J. Mol. Struct.*, **17**, 283.

Clerbaux, T., Duterme, P., Zeegers-Huyskens, T., and Huyskens, P. (1967). *J. Chim. Phys.*, **64**, 1325.

Codding, P. W., Delbaere, L. T. J., Hayakawa, K., Hutcheon, W., James, M. N. G., and Jursek, L. (1974). *Canad. J. Biochem.*, **52**, 208.

Coetzee, J. F. and Padmanabhan, G. R. (1965). *J. Amer. Chem. Soc.*, **87**, 5005.

Cole, F. C. (1970). *Acta Cryst.*, **B26**, 622.

Colman, P. M., Jansonius, J. N., and Matthews, B. W. (1972). *J. Mol. Biol.*, **70**, 701.

Cooke, R. and Kuntz, I. D. (1974). *Ann. Rev. Biophys. Bioeng.* **3**, 95.

Cotton, F. A., Day, V. W., Hazen, E. E., Jr., and Larsen, S. (1973). *J. Amer. Chem. Soc.*, **95**, 4834.

Cotton, F. A., Day, V. W., Haxen, E. E., Jr., Larsen, S., and Wong, S. T. K. (1974a). *J. Amer. Chem. Soc.*, **96**, 4471.

Cotton, F. A., LaCour, T., Hazen, E. E., Jr., and Legg, M. (1974b). *Biochim. Biophys. Acta*, **359**, 7.

Currie, M. and Macdonald, A. L. (1974). *J. Chem. Soc. Perkin II*, 784.

Davies, M. and Williams, G. (1960). *Trans. Faraday Sco.*, **56**, 1619.

Davis, M. M. (1968). *Natl. Bur. St. U.S. Monogr.*, No. 105.

Deisenhofer, J. and Steigman, W. (1975). *Acta Cryst.*, **B31**, 238.

Derissen, J. L. (1977). *J. Mol. Struct.*, **38**, 177

Derissen, J. L., Enderman, H. J., and Peerdeman, A. F. (1968). *Acta Cryst.*, **B24**, 1349.

DeTitta, G. T., Edmonds, J. W., Stallings, W., and Donohue, J. (1976). *J. Amer. Chem. Soc.*, **98**, 1920.

Donohue, J. (1969). *J. Mol. Biol.*, **45**, 231.

Drenth, J., Hol, W. G. J., Jansonius, J. N., and Koekoek, R. (1971). *Cold Spring Harbor Symp.*, **36**, 107.

Elegant, L., Fidanza, J., Gal, J. F., and Azzaro, M. (1977). *J. Chim. Phys.*, **74**, 1015.

Fabbrizzi, L., Micheloni, M., Paoletti, P., and Schwarzenbach, G. (1977). *J. Amer. Chem. Soc.*, **99**, 5574.

Fair, C. K. and Schlemper, E. O. (1978). *Acta Cryst.*, **B34**, 436.

Falk, M. and Knop, D. (1973). In *Water: A Comprehensive Treatise* (Ed. Frank, F.), Vol. 2, Plenum Press, New York, p. 55.

Fawcett, J. K., Camerman, N., and Camerman, A. (1975). *Acta Cryst.*, **B31**, 658.

Fehlhammer, H., Bode, W., and Huber, R. (1977). *J. Mol. Biol.*, **111**, 415.

Ferraris, G. and Franchini-Angela, M. (1972). *Acta Cryst.*, **B28,** 3572.
Fersht, A. R. (1972). *J. Mol. Biol.*, **64,** 497.
Fersht, A. R. and Requena, Y. (1971). *J. Mol. Biol.*, **60,** 279.
Finn, F. M., Dadok, J., and Bothnerby, A. A. (1972). *Biochemistry*, **11,** 455.
Finney, J. L. (1977). *Phil. Trans. Roy. Soc. (London)*, **B278,** 3.
Flora, H. B. and Gilkerson, W. R. (1970). *J. Amer. Chem. Soc.*, **92,** 3273.
Fruton, J. S. (1976). *Adv. Enzymol.*, **44,** 1.
Fuoss, R. (1968). *Rev. Pure Appl. Chem.*, **18,** 125.
Gartland, G. L. and Craven, B. M. (1974). *Acta Cryst.*, **B30,** 980,
Glowiak, T., Sobczyk, L., and Grech, E. (1975a). *Chem. Phys. Lett.*, **34,** 292.
Glowiak, T., Sobczyk, L., and Grech, E. (1975b). *Chem. Phys. Lett.*, **36,** 106.
Golič, L. and Hamilton, W. C. (1972). *Acta Cryst.*, **B28,** 1265.
Graham, L. L. and Chang, C. Y. (1971). *J. Phys. Chem.*, **75,** 776.
Green, R. D. (1974). *Hydrogen Bonding by C—H Groups*, John Wiley, New York.
Griffin, J. R. and Coppens, P. (1975). *J. Amer. Chem. Soc.* **97,** 3496.
Gurskaya, G. V. (1968). *The Molecular Structure of Amino Acids*, Consultants Bureau, New York.
Hadzi, D and Marciszewski, H. (1967). *Chem. Commun.*, 2.
Hamilton, W. C. and J. A. Ibers, J. A. (1968). *Hydrogen Bonding in Solids*, W. A. Benjamin Inc., New York.
Hand, E. S. and Jencks, W. P. (1975). *J. Amer. Chem. Soc.*, **97,** 6221.
Harris, W. I. and Walker, J. E. (1977). In *Pyridine Nocleotide-Dependent Dehydrogenases* (Ed. Sund, H.), de Gruyter, Berlin.
Haser, R., Bonnet, B., and Roziere, J. (1977). *J. Mol. Struct.*, **40,** 177.
Haulait, M. C. and Huyskens, P. L. (1975). *J. Phys. Chem.*, **79,** 1812.
Hazen, E. E., Jr. (1978). Private communication.
Hibbert, F. and Satchell, D. N. P. (1968). *J. Chem. Soc.*, 573.
Highsmith, S. (1975). *J. Phys. Chem.*, **79,** 1456.
Hinton, J. F. and Harpool, R. D. (1977). *J. Amer. Chem. Soc.*, **99,** 349.
Hopfinger, A. J. (1977). *Intermolecular Interactions and Biomolecular Organization*, Wiley–Interscience, New York, p. 114.
Hsu, I. N. and Craven, B. M. (1974a). *Acta Cryst.*, **B30,** 974.
Hsu, I. N. and Craven, B. M. (1974b). *Acta Cryst.*, **B30,** 1299.
Hsu, I. N., Delbaere, L. T. J., James, M. N. G., and Hoffman, T. (1977). *Nature*, **226,** 140.
Huber, R. and Bode, W. (1978). *Acc. Chem. Res.*, **11,** 114.
Hudson, R. A., Scott, R. M., and Vinogradov, S. (1970). *Spectrochim. Acta*, **26A,** 337.
Hudson, R. A., Scott, R. M., and Vinogradov, S. (1972). *J. Phys. Chem.*, **76,** 1989.
Huyskens, P. (1977). *J. Amer. Chem. Soc.*, **99,** 2578.
Huyskens, P. and Zeegers-Huyskens, T. (1964). *J. Chim. Phys.*, **61,** 81.
Jadzyn, J. and Malecki, J. (1972). *Acta Phys. Pol.*, **A41,** 599.
James, M. N. G., Hsu, I. N. and Delbaere, L. T. J. (1977). *Nature*, **267,** 808.
Janin, J. and Chothia, C. (1976). *J. Mol. Biol.*, **100,** 197.
Janin, J., Sweet, R. M., and Blow, D. M. (1974). In *Proteinase Inhibitors* (Eds. Fritz, H., Tschesche, H., Greene, L. J., and Truscheit, E.), Springer-Verlag, Berlin, p. 513.
Janoschek, R., Weidemann, E. G., Pfeiffer, H., and Zundel, G. (1972). *J. Amer. Chem. Soc.*, **94,** 2387.
Jasinski, T. and Pawlak, Z. (1967). *Roczn. Chem.*, **41,** 1943.
Johansson, A., Kollman, P., Rothenberg, S., and McKelvey, J. (1974). *J. Amer. Chem. Soc.*, **96,** 3794.

Johnson, S. L. and Rumon, K. A. (1965). *J. Phys. Chem.*, **69**, 74.

Jones, F. M., III and Arnett, E. M. (1974). *Progr. Phys. Org. Chem.*, **11**. 263.

Junker, M. L. and Gilkerson, W. R. (1975). *J. Amer. Chem. Soc.*, **97**, 493.

Kaneko, N., Takahashi, H., and Higasi, K. (1975). *Bull. Chem. Soc. Japan*, **48**, 1961.

Karle, I. L. (1974). *J. Amer. Chem. Soc.*, **96**, 4000.

Karle, (1975). *J. Amer. Chem. Soc.*, **97**, 4379.

Karle, I. L. (1977). In *Peptides: Proceedings Fifth American Peptide Symposium* (Eds. Goodman, M. and Meinhofer, J.), John Wiley, New York, p. 274.

Karle, I. L. and Duesler, E. (1977). *Proc. Natl. Acad. Sci. U.S.*, **74**, 2602.

Karle, I. L., Gibson, J. W. and Karle, J. (1970). *J. Amer. Chem. Soc.*, **92**, 3755.

Kerr, K. A., Ashmore, J. P., and Koetzle, T. F. (1975). *Acta Cryst.*, **B31**, 2022.

Kilmartin, J. V., Imai, K., Jone, R. T., Faruqui, A. R., Fogg, J., and Baldwin, J. M. (1978). *Biochim. Biophys. Acta*, **534**, 15.

Klotz, I. M. and Franzen, J. S. (1962). *J. Amer. Chem. Soc.*, **84**, 3641.

Knier, B. L. and Haake, P. (1977). *Tetrahedron Lett.*, 3219.

Koetzle, T. F., Hamilton, W. C., and Parthasarathy, R. (1972). *Acta Cryst.*, **B28**, 2083.

Koetzle, T. F. and Lehmann, M. S. (1976). In *The Hydrogen Bond* (Eds. Schuster, P., Zundel, G., and Sandorfy, C.), Vol. II, North-Holland, Amsterdam, p. 457.

Kohler, F. and Huyskens, P. (1976). *Adv. Mol. Relax. Proc.*, **8**, 125.

Kollman, P. A. and Allen, L. C. (1972). *Chem. Rev.*, **72**, 283.

Kollman, P. A. and Kuntz, I. (1976). *J. Amer. Chem. Soc.*, **98**, 6820.

Kraus, C. A. (1956). *J. Phys. Chem.*, **60**, 129.

Kraut, J. (1977). *Ann. Rev. Biochem.*, **46**, 331.

Kresheck, G. C. and Klotz, I. M. (1969). *Biochemistry*, **8**, 8.

Kretsinger, R. H. and Nockolds, C. E. (1973). *J. Biol. Chem.*, **248**, 3313.

Kristof, W., Vogt, B. and Zundel, G. (1978). Private communication.

Kuntz, I. D. and Kauzmann, W. (1974). *Adv. Protein Chem.*, **28**, 239.

Kvick, A., Koetzle, T. F., Thomas, R., and Takusagawa, F. (1974). *J. Chem. Phys.*, **60**, 3866.

Ladner, R. C., Heidner, E. J., and Perutz, M. F. (1977). *J. Mol. Biol.*, **114**, 385.

Lancelot, G. (1977). *J. Amer. Chem. Soc.*, **99**, 7037.

Lehmann, M. S., Koetzle, T. F., and Hamilton, W. C. (1972a). *Int. J. Peptide Prot. Res.*, **4**, 229.

Lehmann, M. S., Koetzle, T. F., and Hamilton, W. C. (1972b). *J. Cryst. Mol. Struct.*, **2**, 225.

Lehmann, M. S., Verbist, J. J. Hamilton, W. C., and Koetzle, T. F. (1973). *J. Chem. Soc. Perkin II*, 1333.

Leiserowitz, L. (1976). *Acta Cryst.*, **B32**, 775.

Liljas, A., Kannan, K. K. Bergsten, P. C., Waara, I., Fridborg, K., Strandberg, B., Carlbom, V., Jarup, L., Lovgren, S., and Petif, M. (1972). *Nature New Biol.*, **235**, 131.

Lindemann, R. and Zundel, G. (1972). *J. Chem. Soc. Faraday. II*, **68**, 979.

Lindemann, R. and Zundel, G. (1977a). *Biopolymers*, **16**, 2407.

Lindemann, R. and Zundel, G. (1977b). *J. Chem. Soc. Faraday II*, **73**, 788.

Lindemann, R. and Zundel, G. (1978). *Biopolymers*, **17**, 1285.

Lindheimer, M., Etienne, G., and Brun, B. (1974). *J. Chim. Phys.*, **71**, 35.

Llinas, M. and Klein, M. P. (1975). *J. Amer. Chem. Soc.*, **97**, 4731.

Lovas, G., Kalman, A., and Argay, G. (1974). *Acta Cryst.*, **B30**, 2882.

Lowenstein, H., Lassen, H., and Hvidt, A. (1970). *Acta Chem. Scand.*, **24**, 13.

Lundgren, J. O. and Olovsson, I. (1976). In *The Hydrogen Bond* (Eds. Schuster, P., Zundel, G., and Sandorfy, C.), Vol. II, North-Holland, Amsterdam, p. 471.

Lundgren, J. O. and Tellgren, R. (1974). *Acta Cryst.*, **B30,** 1937.

Lundgren, J. O. and Williams, J. M. (1973). *J. Chem. Phys.*, **58,** 788.

Macau, J., Lamberts, L., and Huyskens, P. L. (1971). *Bull. Soc. Chim.*, 2387.

Mandel, G. S. and Marsh, R. E. (1975). *Acta Cryst.*, **B31,** 2862.

Marsh, R. E. and Donohue, J. (1967). *Adv. Protein Chem.*, **22,** 235.

Marsh, R. E., Ramakumar, S. and Venkatasan, K. (1976). *Acta Cryst.*, **B22,** 66.

Maryott, A. A. (1948). *J. Res. Natl. Bur. St.*, **41,** 7.

Masri, F. N. and Wood, J. L. (1972a). *J. Mol. Struct.*, **14,** 217.

Masri, F. N. and Wood, J. L. (1972b). *J. Mol. Struct.*, **11,** 201.

Matthews, B. D. (1977). In *Proteins* (Eds. Neurath, H. and Hill, R. L.), Vol. III, Academic Press, New York, p. 403.

Matthews, D., Alden, R. A., Birktoft, J. J., Freer, S. T., and Kraut, J. (1977). *J. Biol. Chem.*, **252,** 8875.

Matthies, M. and Zundel, G. (1977). *Biochem. Biophys. Res. Commun.*, **74,** 831.

Moews, P. C. and Kretsinger, R. H. (1975). *J. Mol. Biol.*, **91,** 201.

Moult, J., Yonath, A. Traub, W., Smilansky, A., Podjarny, A., Rabinovich, D., and Saya, A. (1976). *J. Mol. Biol.*, **100,** 179.

Nahringbauer, I. (1978). *Acta Cryst.*, **B34,** 315.

Narasimhamurthy, M. R., Venkatesan, K., and Winkler, F. (1976). *J. Chem. Soc. Perkin II*, 768.

Neerinck, D., Van Audenhaege, A., Lamberts, L., and Huyskens, P. (1968). *Nature,* **218,** 461.

Novak, A. (1974). *Structure and Bonding,* **18,** 177.

Odinokov, S. E., Mashkovsky, A. A., Glazunov, V. P., Iogansen, A. V., and Rassadin, B. V. (1976). *Spectrochim. Acta,* **32A,** 1355.

Ohki, M., Takenaka, A., Shimanouchi, H., and Sasada, Y. (1975). *Bull. Chem. Soc. Japan,* **48,** 848.

Ojelund, G., Skold, R., and Wadsö, I. (1976). *J. Chem. Thermod.*, **8,** 45.

Olovsson, I. and Jönsson, P. G. (1976). In *The Hydrogen Bond* (Eds. Schuster, P., Zundel, G., and Sandorfy, C.), Vol. II, North-Holland, Amsterdam, p. 395.

Ottersen, T. O. (1976). *Adv. Mol. Relax. Proc.*, **9,** 105.

Paiva, A. C. M., Juliano, L., Boschov, P. (1976). *J. Amer. Chem. Soc.*, **98,** 7645.

Parthasarathy, R. (1969). *Acta Cryst.*, **B25,** 509.

Parthasarathy, R. (1977). *Abstr. Amer. Cryst. Assoc. Mt.*, Series 2, Vol. 5, p. 38.

Parthasarathy, R., Ohrt, J. M., and Chheda, G. B. (1977). *Biochemistry,* **16,** 4999.

Pawelka, Z. and Sobczyk, L. (1975). *Roczn. Chem.*, **49,** 1383.

Pawlak, Z. (1973). *Roczn. Chem.*, **47,** 347.

Perutz, M. F. (1976). *Brit. Med. Bull.*, **32,** 195.

Perutz, M. F. (1978). *Science*, **201,** 1187.

Perutz, M. F. and Raidt, H. (1975). *Nature*, 255, 256.

Pietrazak, J. Nogaj, B., Dega-Szafran, Z., and Szafran, M. (1977). *Acta Phys. Pol.,* **A52,** 779.

Pullman, A., Alagona, G., and Tomasi, J. (1974). *Theor. Chim. Acta,* **33,** 87.

Quick, A. and Williams, D. J. (1976a). *Canad. J. Chem.*, **54,** 2465.

Quick, A. and Williams, D. J. (1976b). *Canad. J. Chem.*, **54,** 2482.

Quiocho, F. A. and Lipscomb, W. N. (1971). *Adv. Protein Chem.*, **25,** 1.

Ramakrishnan, C. and Prasad, N. (1971). *Int. J. Peptide Prot. Res.*, **3,** 209.

Ratajczak, H. (1972). *J. Phys. Chem.*, **76,** 3000, 3991.

Ratajczak, H. and Sobczyk, L. (1969). *J. Chem. Phys.*, **50,** 556.

Ratajczak, H. and Sobczyk, L. (1970). *Bull. Acad. Pol.*, **18,** 93.

Reeke, G. N., Jr., Becker, J. W., and Edelman, G. M. (1975). *J. Biol. Chem.*, **250,** 1525.

Riordan, J. F., MeElvany, K. D., and Bordua, C. L., Jr. (1977). *Science*, **195**, 884.

Rosenfield, R. E., Jr. and Parthasarathy, R. (1974). *J. Amer. Chem. Soc.*, **96**, 1925.

Rosenfield, R. E., Jr. and Parthasarathy, R. (1975). *Acta Cryst.*, **B31**, 462.

Rosenfield, R. E., Jr., Parthasarathy, R., and Dunitz, J. D. (1977). *J. Amer. Chem. Soc.*, **99**, 4860.

Rüterjans, H. and Pongs, O. (1971). *Eur. J. Biochem.*, **18**, 313.

Rüterjans, and Witzel, H. (1969). *Eur. J. Biochem.*, **9**, 118.

Sawyer, L., Shotton, D. M., Campbell, J. W., Wendell, P. L. Muirhead, H., Watson, H. E., Diamond, R., and Ladner, R. C. (1978). *J. Mol. Biol.*, **118**, 137.

Schoenborn, B. P. (1971). *Cold Spring Harbor Symp.*, **36**, 569.

Scott, R., DePalma, D., and Vinogradov, S. N. (1968). *J. Phys. Chem.*, **72**, 3192.

Scott, R. and Vinogradov, S. N. (1969). *J. Phys. Chem.*, **73**, 1890.

Shindo, H., Cohen, J. S., and Rupley, J. A.(1977). *Biochemistry*, **16**, 3879.

Shotton, D. M. and Hartley, B. S. (1970). *Nature*, **225**, 802.

Shotton, D. M. and Watson, H. C. (1970). *Nature*, **225**, 811.

Simonov, V. I. and Bukvetsky, B. V. (1978). *Acta Cryst.*, **B34**, 355.

Smith, G. D., Duax, W. L., Langs, D. A., DeTitta, G. T., Edmonds, J. W., Rohrer, D. C., and Weeks, C. M. (1975). *J. Amer. Chem. Soc.*, **97**, 7242.

Sobczyk, L. and Pawelka, Z. (1973). *Roczn. Chem.*, **47**, 1523.

Sobczyk, L. and Pawelka, Z. (1974). *J. Chem. Soc. Faraday I*, **70**, 832.

Speakman, J. C. (1972). *Structure and Bonding*, **12**, 114.

Springs, B. and Haake, P. (1977). *Bioorg. Chem.*, **6**, 181.

Stenkamp, R. E. and Jensen, L. H. (1973). *Acta Cryst.*, **B29**, 2872.

Stevens, E. D., Lehmann, M. S., and Coppens, P. (1977). *J. Amer. Chem. Soc.*, **99**, 2839.

Stout, G. H. and Jensen, L. H. (1968). *X-Ray Structure Determination*, Macmillan Publ. Co., New York.

Takano, T. (1977). *J. Mol. Biol.*, **110**, 537.

Takano, T., Trus, B. L., Mandel, N., Mandel, G., Kallai, O. B., Swanson, R., and Dickerson, R. E. (1977). *J. Biol. Chem.*, **252**, 776.

Takusagawa, F., Higuchi, T., Shimada, A., Tamura, C., and Sasada, Y. (1974). *Bull. Chem. Soc. Japan*, **47**, 1409.

Takusagawa, F., Hirotsu, K., and Shimada, A. (1973a). *Bull. Chem. Soc. Japan*, **46**, 2372.

Takusagawa, F., Hirotsu, K., and Shimada, A. (1973b). *Bull. Chem. Soc. Japan*, **46**, 2669.

Takusagawa, F. and Shimada, A. (1973). *Chem. Lett.*, 1089.

Takusagawa, F. and Shimada, A. (1976). *Acta Cryst.*, **B32**, 1925.

Tamura, C., Hata, T., Sato, S., and Sakurai, N. (1972). *Bull Chem. Soc. Japan*, **45**, 3254.

Tsernoglou, D., Petsko, G. A., and Hudson, R. A. (1978). *Mol. Pharmacol.*, **14**, 710.

Umeyama, H. and Morokuma, K. (1977). *J. Amer. Chem. Soc.*, **99**, 1316.

Van Wart, H. E., Shipman, L. L. and Scheraga, H. A. (1975). *J. Phys. Chem.*, **79**, 1436.

Verbist, J. J., Lehmann, M. S., Koetzlee, T. F., and Hamilton, W. C. (1972). *Acta Cryst.*, **B28**, 3006.

Vinogradov, S. N. (1979). *Int. J. Peptide Prot. Res.*, **14**, 281.

Vinogradov, S. N., Hudson, R. A., and Scott, R. M. (1970). *Biochim. Biophys. Acta*, **214**, 6.

Wang, C. H. and Storms, R. D. (1971). *J. Chem. Phys.*, **55**, 3291, 5110.

Watenpaugh, K. D., Margulis, T. N., Sieker, L. C., and Jensen, L. H. (1978). *J. Mol. Biol.*, **122**, 175.

Watenpaugh, K. D., Sieker, L. C., Herriott, J. R., and Jensen, L. H. (1973). *Acta Cryst.*, **B29**, 943.
Wojtowicz, A. and Malecki, J. (1976). *Roczn. Chem.*, **50**, 2121.
Wojtowicz, A. and Malecki, J. (1977). *Bull. Acad. Pol. Sci.*, **25**, 385.
Wolfenden, R. (1978). *Biochemistry*, **17**, 201.
Wood, J. L. (1974). *Biochem. J.*, **143**, 775.
Wyckoff, H. W., Tsernoglou, D., Hanson, A. W., Knox, J. R., Lee, B., and Richards, F. M. (1970). *J. Biol. Chem.*, **245**, 305.
Yonath, A., Podjarny, A., Honig, B., Sielecki, A., and Traub, W. (1977). *Biochemistry*, **16**, 1418.
Znaskar, K. T. and Rao, C. N. R. (1967). *Biochim. Biophys. Acta*, **136**, 561.
Zundel, G. (1976). In *The Hydrogen Bond, Recent Developments in Theory and Experiments* (Eds. Schuster, P., Zundel, G., and Sandorfy, C.), Vo.. II, North-Holland, p. 687.
Zundel, G. (1978). *J. Mol. Struct.*, **45**, 55.
Zundel, G. and Nagyrevi, A. (1978). *J. Phys. Chem.*, **82**, 685.

Molecular Interactions, Volume 2
Edited by H. Ratajczak and W. J. Orville-Thomas
© 1980 John Wiley & Sons Ltd.

5 Hydrogen-bonded ferroelectrics and lattice dimensionality

R. Blinc

5.1 Introduction

The interactions between hydrogen-bonded structural units and the resulting collective behaviour of protons or deuterons in systems of hydrogen-bonds represent one of the most fascinating problems in the physics of condensed matter (Blinc, 1960; De Gennes, 1963; Brout *et al.*, 1966; Kobayashi, 1968; Blinc and Žekš, 1974). They often lead (Blinc and Žekš, 1974) to order–disorder type ferroelectric, antiferroelectric, ferroelastic, and other structural phase transitions and play an important role in the conformation changes in certain biologically important macromolecules and perhaps also in DNA replication.

The importance of the role of lattice dimensionality in structural and other phase transitions has been realized a long time ago. Only recently however hydrogen-bonded systems with pronounced one- and two-dimensional character were found and their properties studied. It is the purpose of this chapter to review the collective behaviour of hydrogen bonds in the simplest possible examples of pseudo-one-dimensional (1D), pseudo-two-dimensional (2D), and three-dimensional (3D) hydrogen-bonded ferroelectrics and antiferroelectrics.

5.2 Lattice Dimensionality and Critical Behaviour

Ferroelectric (FE) and antiferroelectric (AFE) phase transitions represent a special class of structural phase changes where the transition from the high- to the low-symmetry phase is connected with the appearance of an order parameter η which is the spontaneous polarization in FE and the sublattice polarization in AFE systems (Blinc and Žekš, 1974) (Figure 5.1). The order parameter measures the extent to which the atomic configuration in the less

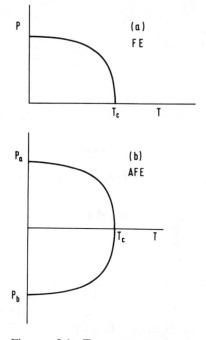

Figure 5.1 Temperature dependence of (a) the spontaneous polarization in ferroelectrics and (b) the
sublattice polarizations in antiferroelectrics

symmetrical phase departs from the configuration of the more symmetrical
phase. The occurrence of a spontaneous polarization is only a necessary and
not a sufficient condition for ferroelectricity. The spontaneous polarization
should also be reversible under the influence of an external electric field, i.e.
the change in the crystal structure at the phase transition should be relatively small. Phase transitions in ferroelectric and antiferroelectric systems
are thus connected only with the rearrangements of a few atoms in the unit
cell whereas the positions of all others remain unchanged. The phase
changes in hydrogen-bonded ferroelectrics are usually triggered (Slater,
1941; Silsbee *et al.*, 1964) by a rearrangement of the hydrogens in the
double-well-type hydrogen bond potentials.

In continuous—so called second-order type—phase transitions the order
parameter η goes continuously to zero as the transition temperature T_c is
approached from below. The temperature dependence of the order parameter can be expressed as

$$\eta \propto (T_c - T)^\beta \qquad\qquad (5.1)$$

where β is a critical exponent, which depends (Stanley, 1971; Fisher, 1974) on the dimensionality (d) of the interactions, the symmetry of the problem, and the number (n) of the components of the order parameter. It does not depend on the range of the interactions as long as it is finite. The critical behaviour of the order parameter susceptibility $\chi = (\partial\eta/\partial E)_T$—which measures the response of the system to the field E which is conjugate to the order parameter η—and the specific heat C_η are determined by the critical exponents γ and α:

$$\chi \propto (T - T_c)^{-\gamma} \tag{5.2a}$$

$$C_\eta \propto (T - T_c)^{-\alpha}. \tag{5.2b}$$

The hydrogen bond can be in many systems described as a two-position dipole (Slater, 1941; Silsbee *et al.*, 1964). The collective behaviour of such a system of hydrogen bonds can be approximated by the spin $\frac{1}{2}$ Ising model in a transverse tunnelling field (Blinc, 1960; De Gennes, 1963; Brout *et al.*, 1966; Blinc and Žekš, 1974). Since the transverse field does not change the critical properties of the system let us consider the simpler Hamiltonian

$$\mathcal{H} = -\sum_{i,j} J(r_i - r_j)S(r_i)S(r_j) \tag{5.3}$$

where $S(r_i) = [S_1(r_i), S_2(r_i) \cdots S_n(r_i)]$ is a n-component spin variable at the position $r_i = (r_1, r_2 \cdots r_d)$ in d-dimensional space. For $n = 1$ and $d = 2$, expression (5.3) reduces to the famous 2-dimensional Ising model which was exactly solved by Onsager (Onsager, 1944).

The critical exponents β, γ, and α are collected in Table 5.1 for isotropic spin $\frac{1}{2}$ models and a variety of n and d values (Stanley, 1971; Fisher, 1974).

As can be seen from Table 5.1, a one-dimensional ($d = 1$) Hamiltonian—describing for instance, infinite chains of structural units connected by hydrogen bonds with no contacts between the chains—does not exhibit any singularity in the thermodynamic functions at $T > 0$. If however weak interchain interactions are taken into account in addition to intrachain coupling, a spontaneous polarization and a divergence in the dielectric constant may occur.

Table 5.1 Critical exponents for some isotropic spin $\frac{1}{2}$ systems and lattice dimensionality

	Landau theory ($d = 1, 2, 3$)	$d = 3, n = 3$ (Heisenberg)	$d = 3$, $n = 2$(XY)	$d = 3, n = 1$ (Ising)	$d = 2, n = 1$ (Ising)	$d = 1$, $n = 1, 2, 3$
α	0 (discont.)	0.2	0 (log)	0.125	0 (log)	No
β	$\frac{1}{2}$	0.385	0.33	0.312	$\frac{1}{8}$	phase
γ	1	1.43	1.33	1.25	1.75	transition

In two-dimensional systems ($d = 2$, $n = 1$), where the spins interact only if they belong to the same layer with no interactions between the layers the critical exponent β is much smaller ($\frac{1}{8}$) than in the three-dimensional ($d = 3$, $n = 1$) Ising model ($\beta = 0.31$).

In the Landau theory—or, what is the same in the molecular field approximation (MFA)—the critical exponents have their classical values $\beta = \frac{1}{2}$, $\gamma = 1$, $\alpha = 0$ in any dimension. This approximation is equivalent to $d = \infty$ as fluctuations in the order parameter are neglected.

In the homogeneous case the Landau expansion of the non-equilibrium free energy density $g(\eta, T)$ in powers of the order parameter η can be written as:

$$g(\eta, T) = g_0(T) + \tfrac{1}{2}a(T)\eta^2 + \tfrac{1}{4}b\eta^4 + \cdots \qquad (5.4a)$$

where one assumes that

$$a = a'(T - T_c) \qquad (5.4b)$$

whereas b is a positive constant if the transition is to be of second order. The equilibrium value of the order parameter η_0 at a temperature T is obtained by a minimization of g:

$$\left(\frac{\partial g}{\partial \eta}\right)_{\eta_0} = 0, \qquad \left(\frac{\partial^2 g}{\partial \eta^2}\right)_{\eta_0} > 0. \qquad (5.5a)$$

From (5.5a), (5.4a), and (5.4b) one finds that

$$\eta_0 = 0, \qquad T > T_c \qquad (5.5b)$$

whereas

$$\eta_0 = [a'(T_c - T)/b]^{1/2}, \qquad T < T_c \qquad (5.5c)$$

i.e. $\beta = \frac{1}{2}$.

Similarly one finds from

$$\chi_T^{-1} = \left(\frac{\partial^2 g}{\partial \eta^2}\right)_{\eta_0} = a'(T - T_c), \qquad T > T_c \qquad (5.5d)$$

that $\gamma = 1$.

The Landau expansion breaks down in the critical region around the transition temperature T_c, where the fluctuations in the order parameter cannot be neglected and non-classical critical exponents—as given in Table 5.1—are expected to be found.

According to Ginzburg (1960), the critical region can be defined as that region where the fluctuations in the order parameter within a coherence length ξ are equal to or greater than the value of the order parameter itself:

$$\overline{(\Delta\eta^2)} \geq \eta_0^2. \qquad (5.6)$$

The above condition strictly applies only for $T < T_c$. If, however the assumption is made that the critical region is symmetrical around T_c then the above condition also determines the size of the critical region in the paraelectric phase.

When fluctuations are taken into account, the system becomes inhomogeneous and the Landau expansion of the free energy density is given by:

$$g[\eta(r), T] = g_0(T) + \tfrac{1}{2}a(T)\eta^2(r) + \tfrac{1}{4}b\eta^4(r) + \tfrac{1}{2}C(\nabla\eta(r))^2 + \cdots \qquad (5.7)$$

with $C > 0$.

The critical region in a three-dimensional system is then defined by

$$\varepsilon = |(T - T_c)/T_c| \leq \varepsilon_c \qquad (5.8a)$$

where

$$\varepsilon_c = \frac{1}{32\pi^2}\left(\frac{k_B}{\Delta c}\right)^2 \frac{1}{\xi_{T=0}^6}, \qquad d = 3. \qquad (5.8b)$$

Here Δc is the jump in the specific heat at T_c as derived by the MFA, k_B is the Boltzmann constant and $\xi_{T=0} = (c/2a'T_c)^{1/2}$ is the zero temperature coherence length, i.e. the range of the interactions (Ginzburg, 1960).

The critical exponents are thus expected to have their classical values if $|T - T_c| > \varepsilon_c T_c$ and depend on the dimensionality of the system (Table 5.1) only if $|T - T_c| < \varepsilon_c T_c$, i.e. inside the critical region.

In a two-dimensional system the Ginzburg criterion gives the width of the critical region as:

$$\varepsilon_c = \frac{1}{16\pi}\frac{k_B}{\Delta c}\left(\frac{1}{l\xi_{T=0}^2}\right), \qquad d = 2 \qquad (5.9a)$$

where l is the spacing of the planes. We see that

$$\varepsilon_c(2d) \approx \sqrt{\varepsilon_c(3d)} \qquad (5.9b)$$

i.e. the critical region for a two-dimensional system is much wider than for a three-dimensional system with comparable ranges of the coupling forces.

Most uniaxial ($n = 1$) 3D ferroelectric systems show a Landau-type classical critical behaviour (Figure 5.2). This is not surprising in view of the importance of long-range dipolar forces in these systems. Larkin and Khmelnitskii (1969) have indeed found that apart from logarithmic corrections—which are hard to observe—Landau exponents govern the critical behaviour of $d = 3$, $n = 1$ dipolar Ising systems.

Non-classical critical exponents have been, however, observed in ferroelectric liquid crystals where $d = 3$, $n = 2$, i.e. in systems where the order parameter is not uniaxial. Dipolar interactions as well become irrelevant

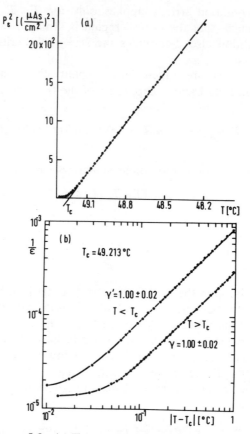

Figure 5.2   (a) Temperature dependence of the square of the spontaneous polarization in TGS. (b) Log–log plot of the inverse dielectric constant versus $T - T_c$ in TGS

near T_c in antiferroelectric systems so that non-classical exponents should be observable.

Non-classical behaviour should be, however, most easily studied in two-dimensional systems where the critical region is wide, or in pseudo-one-dimensional systems where—in the absence of three-dimensional interchain coupling—fluctuations suppress the phase transition down to $T = 0$.

5.3   The Pseudo-spin Hamiltonian Dynamics

5.3.1   *The Hamiltonian*

In many strongly anharmonic solids the single-particle potential for the motion of one or more atoms or atomic groups has more than one potential minimum (Blinc and Žekš, 1974). Such situations are often found in hydrogen-bonded systems where the proton can move between two equilibrium sites in the hydrogen bond potential. Quantum mechanical tunnelling of the proton through the potential barrier between the two wells produces a splitting of the ground state into a doublet. In addition to the ground state there are of course higher vibrational states. As the frequency separation between the ground and higher vibrational states is much larger than the splitting of the ground-state doublet, we can often neglect excited vibrational states and limit ourselves to the treatment of a two-level system (Figure 5.3).

For every two-level system, all particle operators can be expressed in terms of spin $\frac{1}{2}$ operators. The three spin $\frac{1}{2}$ Pauli matrices

$$S^x = \begin{vmatrix} 0 & \frac{1}{2} \\ \frac{1}{2} & 0 \end{vmatrix}, \qquad S^y = \begin{vmatrix} 0 & -i/2 \\ i/2 & 0 \end{vmatrix}, \qquad S^z = \begin{vmatrix} \frac{1}{2} & 0 \\ 0 & -\frac{1}{2} \end{vmatrix} \qquad (5.10)$$

and the unit matrix in fact form a complete set in the space of Hermitian 2×2 matrices. Such a pseudo-spin representation of the system is extremely useful both because the large anharmonicity of the problem has been transformed out and because of the analogy with magnetic systems.

To obtain a better feeling for the pseudo-spin formalism let us now derive the expression for the model Hamiltonian for the special case of

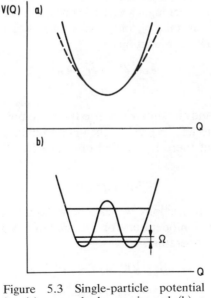

Figure 5.3 Single-particle potential for (a) a nearly harmonic and (b) a strongly anharmonic solid

hydrogen-bonded ferroelectrics where each hydrogen nucleus moves between two equilibrium sites in the hydrogen bonds (Figure 5.3).

The Hamiltonian $\mathscr{H}$ consists of a single-particle part $\mathscr{H}_1(i)$ and an interaction part $\mathscr{H}_2(i, j)$:

$$\mathscr{H} = \sum_i \mathscr{H}_1(i) + \sum_{i<j} \mathscr{H}_2(i, j). \tag{5.11}$$

In the representation in which the single-particle Hamiltonian $\mathscr{H}_1(i)$ is diagonal we obtain

$$\mathscr{H}_1(i) = \sum_\alpha E_\alpha a_\alpha^{i*} a_\alpha^i \tag{5.12}$$

and

$$\mathscr{H}_2(i, j) = \sum_{\alpha,\beta,\gamma,\delta} v_{\alpha\beta\gamma\delta}^{ij} a_\alpha^{i*} a_\beta^i a_\gamma^{j*} a_\delta^j, \tag{5.13}$$

where the operators a_α^{i*} and a_α^i are Fermi or Bose creation and annihilation operators for a proton or a deuteron at site i in the single-particle quantum state α.

Since there is always only one hydrogen nucleus in each hydrogen bond, the results are independent of the statistics. The matrix element v^{ij} represents a general ineraction between the ith and jth hydrogen bonds which depends not only on the distance between the positions of i and j but also on their quantum states. The ground state of a single hydrogen in a double well is a doublet, and, if we neglect higher excited states, the single-particle quantum numbers α, β, γ, δ can take on only two values, which we label $+$ and $-$.

The corresponding eigenfunctions

$$\psi_+ = 2^{-1/2}(\varphi_L + \varphi_R), \tag{5.14}$$

$$\psi_- = 2^{-1/2}(\varphi_L - \varphi_R) \tag{5.15}$$

are the symmetric and the antisymmetric linear combinations of wavefunctions localized in the left (φ_L) and right (φ_R) equilibrium sites.

The condition that there is one and only one hydrogen in each bond is now expressed as

$$a_+^{i*} a_+^i + a_-^{i*} a_-^i = 1. \tag{5.16}$$

Using the above restrictions we can now express the products of the creation and annihilation operators for a hydrogen in a given hydrogen bond with fictitious spin $\frac{1}{2}$ operators:

$$S_i^x = \tfrac{1}{2}(a_+^{i*} a_+^i - a_-^{i*} a_-^i), \tag{5.17}$$

$$S_i^y = \tfrac{1}{2}(a_+^{i*} a_-^i - a_-^{i*} a_+^i), \tag{5.18}$$

$$S_i^z = \tfrac{1}{2}(a_+^{i*} a_-^i + a_-^{i*} a_+^i), \tag{5.19}$$

which indeed obey the well-known commutation relations for spin operators:

$$[S_i^x, S_j^y] = i\,\delta_{ij}S_i^z,$$ (5.20a)

$$[S_i^y, S_j^z] = i\,\delta_{ij}S_i^x,$$ (5.20b)

$$[S_i^z, S_j^x] = i\,\delta_{ij}S_i^y.$$ (5.20c)

The physical meaning of the pseudo-spin operators becomes more transparent if one goes from the representation where the single-particle Hamiltonian is diagonal to the representation of localized states.

The operators a_α^* and a_α which create and annihilate a particle in the quantum state $\alpha = +, -$ are the symmetric ($\alpha = +$) and antisymmetric ($\alpha = -$) linear combinations of the corresponding operators which create or annihilate a particle in the left or the right equilibrium site in the hydrogen bond:

$$a_+^i = 2^{-1/2}(a_L^i + a_R^i), \qquad a_+^{i*} = 2^{-1/2}(a_L^{i*} + a_R^{i*}),$$ (5.21a)

$$a_-^i = 2^{-1/2}(a_L^i - a_R^i), \qquad a_-^{i*} = 2^{-1/2}(a_L^{i*} - a_R^{i*}).$$ (5.21b)

Using expressions (5.17)–(5.19) and (5.21), we obtain

$$S_i^x = \tfrac{1}{2}(a_L^{i*}a_R^i + a_R^{i*}a_L^i),$$ (5.22a)

$$S_i^y = \tfrac{1}{2}(a_L^{i*}a_R^i - a_R^{i*}a_L^i),$$ (5.22b)

and

$$S_i^z = \tfrac{1}{2}(a_L^{i*}a_L^i - a_R^{i*}a_R^i).$$ (5.22c)

The z component of the pseudo-spin thus represents the coordinate dipole moment operator, the y component the local current operator, and the x component the tunnelling operator. This is equivalent to saying that the expectation value of S^z measures the difference between the occupations of the 'left' and 'right' equilibrium sites, whereas the expectation value of S^x measures the difference between the occupations of the symmetric and antisymmetric energy states.

Introducing spin operators (5.22a) to (5.22c) into expressions (5.11) to (5.13) we obtain

$$\mathscr{H} = -\Omega\sum_i S_i^x - \tfrac{1}{2}\sum_{i,j} J_{ij}S_i^z S_j^z - \tfrac{1}{2}\sum_{i,j} B_{ij}S_i^x S_j^x,$$ (5.23)

where

$$\Omega = \left[E_- - E_+ + \sum_j (v_{----}^{ij} - v_{++++}^{ij})\right] \approx (E_- - E_+),$$ (5.24a)

is the tunnelling integral,

$$J_{ij} = -4v_{+-+-}^{ij},$$ (5.24b)

and

$$B_{ij} = 2v^{ij}_{++--} - v^{ij}_{++++} - v^{ij}_{----}. \tag{5.24c}$$

It should be noted that terms such as $S_i^x S_j^z$ vanish for symmetry reasons since in KDP-type structures the Hamiltonian has to be invariant with respect to the exchange $S_i^z \rightarrow -S_i^z$. It can be shown that the $B_{ij} S_i^x S_j^x$ term is small compared with the ΩS_i^x term and can usually be neglected. In this case expression (5.23) is identical with the Ising Hamiltonian in a transverse tunnelling field.

5.3.2 *The pseudo-spin dynamics*

It is well known that a crystal lattice is stable as long as all normal mode frequencies are positive and non-zero. Some 18 years ago Cochran and Anderson suggested (Blinc and Žekš, 1974) that phase transitions in certain ferroelectrics might result from an instability of one of the normal vibrational modes of the lattice. In this theory—which is by now well established—the frequency of the relevant soft phonon decreases on approaching the critical temperature T_c and the restoring force for the mode displacements tends to zero until the phonon has condensed out at the stability limit. The static atomic displacements on going from the paraelectric to the ferroelectric phase thus represent the frozen-in mode displacements of the unstable phonon. The order parameter of such a transition is the static component of the eigenvector of the unstable phonon. As the ferroelectric state is characterized by a macroscopic spontaneous polarization, the soft phonon must be both polar and of long wavelength; i.e. the instability of the infrared-active phonon must occur at the Brillouin zone centre. The antiferroelectric state, on the other hand, involves the instability of a phonon at the Brillouin zone boundary, as antiferroelectricity is characterized by the appearance of two opposite sublattice polarizations and an increase in the size of the unit cell.

Whereas the soft-mode theory was first applied only to 'displacive' phase transitions (such as $BaTiO_3$) it was later suggested that essentially the same ideas can be applied to the case of 'order–disorder' systems (such as KH_2PO_4), too. In such a case the soft collective excitations are not phonons, but rather unstable pseudo-spin waves (sometimes called configuration tunnelling excitations). Pseudo-spin modes occur in order–disorder systems in addition to all phonon modes predicted by the harmonic theory of crystal lattices.

5.3.2.1 *The molecular field approximation* (MFA) In the molecular field approximation (MFA)—or its extension to time-dependent problems, the random phase approximation (RPA)—the density matrix of the many-body

system is written as a product of single-particle density matrices and Hamiltonian (5.23) with $B_{ij} = 0$ is replaced by an effective Hamiltonian:

$$\mathcal{H}_i^{\text{MFA}} = -\boldsymbol{H}_i \boldsymbol{S}_i = \frac{\partial \langle \mathcal{H} \rangle}{\partial \langle \boldsymbol{S}_i \rangle} \boldsymbol{S}_i, \tag{5.25}$$

where $\boldsymbol{S}_i$ is a vector the components of which are the three spin $\frac{1}{2}$ Pauli matrices. The term $\boldsymbol{H}_i$ stands for the molecular field vector

$$\boldsymbol{H}_i = (H_{ix}, H_{iy}, H_{iz}) = \left(\Omega, 0, \sum_j J_{ij} \langle S_j^z \rangle \right) \tag{5.26}$$

in our pseudo-spin space which interacts with the pseudo-spin variables. The brackets designate the thermal expectation values:

$$\langle \boldsymbol{S}_i \rangle = \frac{\text{tr } \boldsymbol{S}_i \exp(-\beta \mathcal{H}_i^{\text{MFA}})}{\text{tr} \exp(-\beta \mathcal{H}_i^{\text{MFA}})} = \frac{1}{2} \frac{\boldsymbol{H}_i}{|\boldsymbol{H}_i|} \tanh \tfrac{1}{2}\beta \, |\boldsymbol{H}_i|. \tag{5.27}$$

Equation (5.27) represents the self-consistency equation for the order parameter $\langle S^z \rangle$ and $\beta_c = 1/kT_c$.

In the paraelectric phase above T_c, $\langle S^z \rangle = 0$, the two dipolar equilibrium sites are occupied with equal probability, and the molecular field points along the x direction. Below T_c, $\langle S^z \rangle \neq 0$ and $H_z \neq 0$. The dipoles order into one of the two possible sites and the molecular field points along a general direction in the x–z plane.

5.3.2.2 *The random phase approximation* (RPA)

The Heisenberg equations of motion for the expectation values of the pseudo-spin operators are

$$\frac{d\langle \boldsymbol{S}_i \rangle_t}{dt} = -\frac{i}{\hbar} \langle [\boldsymbol{S}_i, \mathcal{H}] \rangle_t = \frac{1}{\hbar} \langle \boldsymbol{S}_i \rangle_t \times \boldsymbol{H}_i(t). \tag{5.28}$$

These equations are equivalent to the classical free precession of the pseudo-spin 'magnetization' around the instantaneous value of the molecular field $\boldsymbol{H}_i(t)$ (Figure 5.4).

In the paraelectric phase, where $\langle S^z \rangle = 0$, we find in the long-wavelength limit $(\boldsymbol{q} \to 0)$ three eigenfrequencies of system (5.28):

$$\omega_1(\boldsymbol{q}) = 0, \tag{5.29a}$$

$$\omega_{2,3}^2(\boldsymbol{q}) = \Omega[\Omega - \tfrac{1}{2}J(\boldsymbol{q}) \tanh(\tfrac{1}{2}\beta\Omega)]. \tag{5.29b}$$

The solution $\omega_1(\boldsymbol{q}) = 0$ corresponds to the longitudinal mode, i.e. to a motion in the direction of the molecular field.

The two degenerate solutions $\omega_{2,3}(\boldsymbol{q})$ represent transverse excitations and describe the free precession of the pseudo-spins around the molecular field. As $T \to T_c$ the soft pseudo-spin mode condenses,

$$\omega_{2,3} \to 0, \qquad T \to T_c, \tag{5.29c}$$

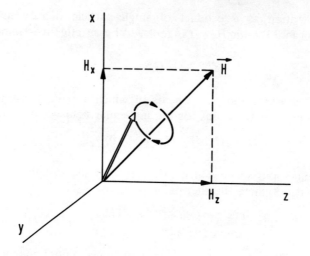

Figure 5.4   Free precession of the pseudo-spin 'magnetization' around the molecular field

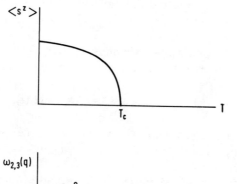

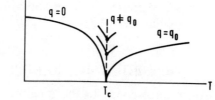

Figure 5.5   Order parameter $\langle S^z \rangle$ and soft
pseudo-spin precession frequency $\omega_{2,3}$ as a
function of temperature. Above T_c $\omega_{2,3}$ becomes critical for $q_0 = 0$ in the ferroelectric
case and for $q_0 \neq 0$ in the antiferroelectric
case

and the structure of the low-temperature phase is determined by the frozen-in mode displacements of the soft mode. The transition occurs when

$$\Omega - \tfrac{1}{2}J(\boldsymbol{q}) \tanh (\tfrac{1}{2}\beta_c \Omega) = 0, \qquad T = T_c. \tag{5.29d}$$

In the ferroelectric phase below T_c

$$\omega_1 = 0 \tag{5.30a}$$

and

$$\omega_{2,3}(q = 0) = J_0 \langle S^z \rangle, \tag{5.30b}$$

where $J_0 = \sum_j J_{ij}$.

The Larmor frequencies of the precessing pseudo-spin 'magnetization' are thus strongly temperature dependent both above and below T_c (Figure 5.5).

5.4 Pseudo-one-dimensional Ferroelectricity

5.4.1 $PbHPO_4$ and $PbHAsO_4$

5.4.1.1 *Structure* $PbHPO_4$ and $PbDPO_4$ undergo (Negran *et al.*, 1974) second-order ferroelectric phase transitions at $T_{c,H} = 37\,°C$ and $T_{c,D} = 179\,°C$, respectively. The rather large isotope shift of the paraelectric to ferroelectric transition temperature T_c on deuteration demonstrates that the $O—H \cdots O$ hydrogen bonds play an important role in the transition mechanism. This is further supported by the fact that the direction of the spontaneous polarization is nearly parallel to the direction of the $O—H \cdots O$ bonds.

$PbHPO_4$ is monoclinic at room termperature with $a = 4.688\,Å$, $b = 6.649\,Å$, $c = 5.781\,Å$, $\beta = 97.11°$. The ferroelectric space group is Pc, whereas the paraelectric space group is $P2/c$. The phase transition thus involves a change in the crystal point symmetry from $2/m$ to m and the condensation of a Bu soft mode at the centre of the Brillouin zone.

The PO_4 groups are linked together (Figure 5.6) by $O—H \cdots O$ bonds. The $O—H \cdots O$ distances are 2.52 Å. The hydrogen positions are 0.22 Å from the pseudo-centre of symmetry in the $O—H \cdots O$ bonds so that the $O—H$ bond length is 1.04 Å. The hydrogens are disordered above T_c and ordered below T_c.

$PbHAsO_4$ is presumably isomorphous with $PbHPO_4$. It also is monoclinic with $a = 4.85\,Å$, $b = 6.76\,Å$, $c = 5.83\,Å$ and $\beta = 95.5°$. The ferroelectric transition takes place (Blinc *et al.*, 1976a) at $T_c \approx 30\,°C$ and the isotope effect on deuteration is similar to the one in $PbHPO_4$.

The important feature of these crystal structures is the existence of infinite chains of PO_4 groups connected by $O—H \cdots O$ bonds with no direct

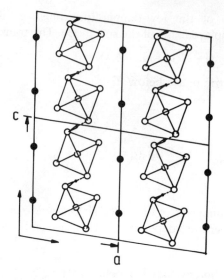

Figure 5.6   Projection of the structure
of PbHPO$_4$ on the a–c plane

contacts between the chains. The hydrogen bond network is thus one
dimensional. Interchain interactions, which are responsible for a finite T_c,
are weak as compared to intrachain interactions.

5.4.1.2   *Theoretical model for* PbDPO$_4$ If—in view of the heavier mass of
the deuteron as compared to the proton—tunnelling effects can be neglected
in PbDPO$_4$, one can write the pseudo-spin Hamiltonian in the presence of
an external field E for each chain of O—H $\cdots$ O bonded PO$_4$ groups in
PbDPO$_4$ as (de Carvalho and Salinas, 1978):

$$\mathscr{H}_0 = -J \sum_{i,j} \sigma_i \sigma_j - \mu E \sum_i \sigma_i. \tag{5.31}$$

Here $\sigma_i = \pm 1$, μ is the O—H $\cdots$ O bond dipole moment, the index i
labels a particular O—H $\cdots$ O bond along the chain and (ij) describes a pair
of neighbouring bonds. The two possible values of the pseudo-spin variable,
spin 'up' ($\sigma_i = +1$) and spin 'down' ($\sigma_i = -1$) represent the position of the
hydrogen in one of the two possible sites of the double-minimum O—
H $\cdots$ O potential. If no ions other than HPO$_4^{2-}$ are allowed, all 'spins' along
the chain are pointing in the same direction, either 'up' or 'down'. The
presence of PO$_4^{3-}$ or H$_2$PO$_4^-$ ions introduces disorder into the above model.
The coupling constant $J > 0$ measures the energy of formation of such a
defect.

One-dimensional intrachain coupling of the above type—which involves only short-range interactions—will never result in a transition to an ordered state with a non-zero spontaneous polarization $P \propto \left\langle \mu \sum_i \sigma_i \right\rangle$ at a finite temperature. For a ferroelectric transition to take place one has to add to expression (5.31) interchain interactions which are presumably weak and long range, and can be treated by the MFA.

The corresponding Helmholtz free energy is in the MFA given (de Carvalho and Salinas, 1978) by

$$F(T, P) = G_0(T, E_0) + PE_0 - \gamma P^2 \tag{5.32}$$

where $G_0(T, E)$ is the thermodynamic potential per 'pseudo-spin' associated with the chain Hamiltonian (5.31), which can be evaluated exactly

$$G_0(T, E) = -kT \ln \{e^{\beta J} \cosh \beta \mu E_0 + [e^{2\beta J} \sinh^2 (\beta \mu E_0) + e^{-2\beta J}]^{1/2}\} \tag{5.33}$$

and $\beta = 1/kT$. The field E_0 is eliminated from (5.32) with the help of the equation $P = -\partial G_0(T, E_0)/\partial E_0$. γ is a parameter characterizing the strength of the MFA-type long-range interchain coupling.

The free energy of the above model (de Carvalho and Salinas, 1978)—expression (5.32)—predicts the occurrence of a paraelectric to ferroelectric transition at a critical temperature T_c which is given by the solution of the equation

$$e^{-2\beta_c J} = 2\beta_c J\eta \tag{5.34}$$

where $\beta_c = (1/kT_c)$. Here $\eta = \gamma\mu^2/J$ measures the ratio between the strengths of the short-range (intrachain) and long-range (interchain) interactions. The reduced spontaneous polarization $p = P/\mu$ is given by

$$\frac{p}{e^{2\beta J}(1-p^2)^{1/2}} = \sinh (2\beta J\eta p), \qquad T < T_c \tag{5.35}$$

and the electric susceptibility $\chi = (\partial P/\partial E)_{E=0}$ by

$$\chi = \beta\mu^2/(e^{-2\beta J} - 2\beta J\eta), \qquad T > T_c \tag{5.36a}$$

and

$$\chi = \beta\mu^2 \Big/ \left\{\frac{1}{(1-p^2)[e^{4\beta J}(1-p^2)+p^2]} - 2\beta J\eta\right\}, \qquad T < T_c. \tag{5.36b}$$

The experimental results for T_c and the temperature dependences of χ and P in PbDPO$_4$ can be quantitatively described (de Carvalho and Salinas, 1978) by the above model with $J = 518.42$ K, $\eta = 0.0436$, and $\mu = 0.6$ D (Figures 5.7a and 5.7b). In view of the relatively small value of η, the

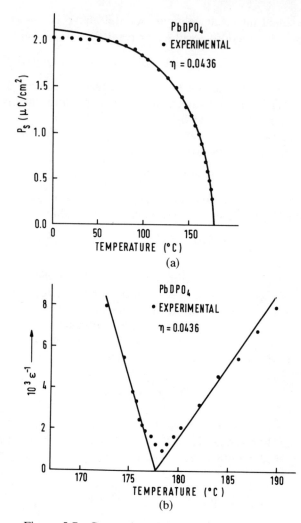

Figure 5.7 Comparison between the experimental and the theoretical temperature dependences of (a) the spontaneous polarization and (b) the inverse dielectric constant in PbDPO$_4$. [Reproduced, with permission, from de Carvalho and Salinas, 1978]

interchain coupling is indeed small as compared to the intrachain one. The MFA treatment of the interchain interactions should thus be a fairly good approximation.

Whereas the above model works well for PbDPO$_4$, this is not the case (de Carvalho and Salinas, 1978) for PbHPO$_4$. The reason for this might be the neglect of the tunnelling term in expression (5.31).

The isotope effects on deuteration in $PbHPO_4$ can be qualitatively understood from a MFA model (Blinc $et\ al.$, 1976a). In the usual pseudo-spin notation (section 3) one can rewrite the Hamiltonian (5.31) with the addition of a tunnelling term as

$$\mathcal{H} = -\Omega \sum_i (S_{i1}^x + S_{i2}^x) - \tfrac{1}{2} \sum_{i,j} [K_{ij}(S_{i1}^z S_{j1}^z + S_{i2}^z S_{j2}^z) + L_{ij} S_{i1}^z S_{j2}^z] \qquad [5.37]$$

where $\mathbf{S} = (S^x, S^y, S^z)$ are the Pauli spin matrices for a spin $\tfrac{1}{2}$, Ω is the proton tunnelling integral, K_{ij} describes proton–proton coupling within a chain, and L_{ij} coupling between protons on two neighbouring chains, designated here as 1 and 2 ($K_{ii} = 0$, $L_{ii} \neq 0$).

The two molecular fields, acting on the pseudo-spins 1 and 2 are

$$\mathbf{H}_1 = (\Omega, 0, K\langle S_1^z\rangle + \tfrac{1}{2} L\langle S_2^z\rangle) \qquad (5.38a)$$

$$\mathbf{H}_2 = (\Omega, 0, K\langle S_2^z\rangle + \tfrac{1}{2} L\langle S_1^z\rangle) \qquad (5.38b)$$

where

$$K = \sum_j K_{ij}, \qquad L = \sum_j L_{ij}. \qquad (5.38c)$$

The above Hamiltonian can be now in the MFA expressed as

$$\mathcal{H}_{\mathrm{MFA}} = -S_1 H_1 - S_2 H_2 \qquad (5.39)$$

and the spontaneous polarization $P/N\mu = \langle S_1^z\rangle + \langle S_2^z\rangle$ is obtained by solving the two self-consistency equations:

$$\langle S_1^z\rangle = \frac{1}{2} \frac{H_{z1}}{|H_1|} \tanh\left(\tfrac{1}{2}\beta\, |H_1|\right) \qquad (5.40a)$$

$$\langle S_2^z\rangle = \frac{1}{2} \frac{H_{z2}}{|H_2|} \tanh\left(\tfrac{1}{2}\beta\, |H_2|\right). \qquad (5.40b)$$

The transition temperature T_c is now obtained from the equation

$$\frac{2\Omega}{K + \tfrac{1}{2}L} = \tanh\left(\tfrac{1}{2}\beta_c\Omega\right) \qquad (5.41)$$

which predicts a significant isotope effect in T_c as $\Omega_D \ll \Omega_H$.

5.4.2 CsH_2PO_4 and CsD_2PO_4

A recent neutron scattering study (Semmingsen $et\ al.$, 1977) of the ferroelectric transition in CsD_2PO_4 has yielded a quasi-elastic diffuse distribution of intensity characteristic for a one-dimensional system with chain-like ordering parallel to the ferroelectric b axis of this monoclinic crystal. This is in sharp contrast to tetragonal KD_2PO_4 where the scattering exhibits characteristics of a three-dimensional dipolar system.

CsH_2PO_4 and CsD_2PO_4 crystallize in the monoclinic space group $P2_1/m$ with two formula units per unit cell. The lattice parameters for CsD_2PO_4 just above T_c are: $a = 7.91$ Å, $b = 6.36$ Å, $c = 4.89$ Å, and $\beta = 107.7°$. The transition temperatures (Levstik *et al.*, 1975) are $T_{c,H} = 153$ K and $T_{c,D} = 267$ K. The ferroelectric space group is $P2_1$ with the polar axis being the unique b axis. There are two chemically different O—H $\cdots$ O bonds with $R_{O \cdots O} = 2.43$ Å and $R_{O \cdots O} = 2.56$ Å. The shorter of these bonds is situated across the centre of inversion and joins the PO_4 groups into infinite zig-zag chains running along the b axis. The hydrogen in the longer O—H $\cdots$ O bond is ordered at all temperatures, whereas the hydrogen in the shorter O—H $\cdots$ O bond is disordered between the two possible equilibrium sites above T_c and ordered below T_c. As T_c is approached from above, the 1D correlation between the hydrogen positions within a chain develops. Three degrees K above T_c the 1D correlation length is about 250 Å and even 26 K above T_c it is still about 100 Å (Semmingsen *et al.*, 1977).

Dielectric measurements in CsH_2PO_4 showed (Levstik *et al.*, 1975) that close to T_c the dielectric properties are governed by classical critical exponents which presumably reflect the onset of 3D ordering. Dielectric measurements in a wider temperature range, on the other hand, seem to show non-classical values (Baranov, priv. comm.) of critical exponents:

$$\beta = 0.33 \tag{5.42a}$$

$$\gamma = 1.15 \pm 0.02 \tag{5.42b}$$

possibly reflecting the influence of 1D critical fluctuations.

5.5 Pseudo-two-dimensional Ferroelectricity

5.5.1 *Introduction*

Protonic order–disorder transitions have recently been discovered in several systems with a layered hydrogen-bonded network: $SnCl_2 \cdot 2H_2O$ (Kiriyama, 1970), $Cu(COOH)_2 \cdot 4H_2O$ (Kiriyama, 1962), $NaH_3(SeO_3)_2$ (Makita and Miki, 1970), and squaric acid, $C_4O_4H_2$ (Semingsen and Feder, 1974). Rigorous statistical calculations have been performed for some of these systems in two dimensions (Lieb and Wu, 1972).

The various vertex models used in exact statistical calculations of 2D ferroelectric models are illustrated in Table 5.2. In Figure 5.8 a two-dimensional square lattice with the various hydrogen arrangements around the lattice vertices is shown. A given hydrogen is assumed to be 'close' to a given vertex if the arrow points towards it whereas it is 'far' from the vertex if the arrow points away from it. The 'ice' rules—or, the Slater–Takagi rules (Slater, 1941; Silsbee *et al.*, 1964), which are the same in KH_2PO_4—request that there are only two hydrogens close to each vertex. There are only six

Table 5.2 Some 2D ferroelectric and antiferroelectric vertex models

Model	1.	2.	3.	4.	5.	6.	7.	8.
2D KDP	0	0	E	E	E	E	∞	∞
'F'	E	E	E	E	0	0	∞	∞
'IF'	$-E$	$-E$	$-E$	$-E$	0	0	∞	∞
'Baxter'	E_1	E_1	E_2	E_2	E_3	E_3	E_4	E_4

vertices (designated 1–6 in Table 5.2) which satisfy the 'ice' rules. The various models presented in Table 5.2 differ in the assignation of the energies to the allowed vertices. The 2D KDP model predicts a ferroelectric transition of the first order. The antiferroelectric 'F' model, used in the description of antiferroelectricity in $NH_4H_2PO_4$, predicts an infinite-order transition with no divergence in the specific heat at T_c. The 'IF' model, on the other hand, is nothing but the 'F' model with negative energies for the configurations 1–4 and exhibits no phase transition at $T > 0$ in the absence of an external electric field. The 'ice' rules are contained as a special case in the 8-vertex model, which has been solved exactly by Baxter (Baxter, 1972). Here the critical exponents depend on the values of the vertex energies.

For a review of the above exact results one should consult the text of Lieb and Wu (Lieb and Wu, 1972) and the article by Baxter (Baxter, 1972).

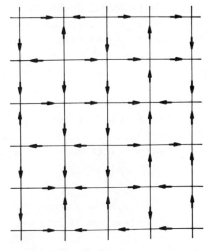

Figure 5.8 Hydrogen arrangements
on a square lattice

In the following section we shall illustrate 2D ferroelectricity using the example of squaric acid.

5.5.2 Squaric acid

5.5.2.1 *Structure* Squaric acid $(C_4O_4H_2)$ undergoes (Semmingsen and Feder, 1974; Feder, 1976) an antiferroelectric second-order phase transition at $T_{c,H} = 97\,°C$. The crystal structure at room temperature was determined by X-rays and neutron diffractions. The room temperature space group is monoclinic $(P2_1/m)$, but pseudo-body-centred tetragonal $(a = 6.31\,\text{Å},\ b = 5.27\,\text{Å},\ c = 6.14\,\text{Å},\ \beta = 90°,\ z = 2)$. The structure consists of planar layers of C_4O_4 subunits each of which is linked by $O—H \cdots O$ bonds to four neighbouring groups in the same layer (Figure 5.9). In the low-temperature phase there are two hydrogens located close to each C_4O_4 anion and the molecules in one layer are related to molecules in the layers above and below by screw axis. Each layer is thus ferroelectrically ordered whereas neighbouring layers are polarized in opposite directions. The coupling between the layers (separation $b/2 = 2.64\,\text{Å}$) is weak, whereas the interactions within the layers are strong, so that 2D behaviour is expected. Measurements of the temperature-dependent splitting of the ^{13}C NMR lines, $\Delta\sigma = ((T_c - T)/T_c)^\beta \Delta\sigma_0$, birefringence measurements $\Delta n \propto ((T_c - T)/T_c)^{2\beta}$, and neutron scattering data indeed show (Feder, 1976) that the temperature dependence of the order parameter is governed by a non-classical critical exponent $\beta = 0.14$, reflecting the two-dimensional character of this transition. It should be noted that in the region of 2D behaviour squaric acid is ferroelectric and only after cross-over to 3D behaviour near T_c do antiferroelectric correlations take over.

The high-temperature space group is tetragonal $(I4/m)$ with one molecule per unit cell $(z = 1)$. The transition is thus connected with the condensation of a soft mode at the Z point of the Brillouin zone of the body-centred tetragonal lattice. The ferroelectric order within the layers vanishes above T_c due to a dynamic disordering of the hydrogens between the two equilibrium sites in the $O—H \cdots O$ bonds. The large isotope effect in the transition temperature on deuteration—$T_{c,D}/T_{c,H} = 1.39$—also demonstrates the role of the hydrogens in the order–disorder transition mechanism (Suwelack *et al.*, 1977).

5.5.2.2 *Theoretical model* Though the Landau theory is a poor approximation for squaric acid it is nevertheless useful for a qualitative description of the essential features of the phase transition (Feder, 1976).

Let P_a and P_b be the in-plane spontaneous polarizations in two successive—'odd' and 'even'—layers of squaric acid. The non-equilibrium

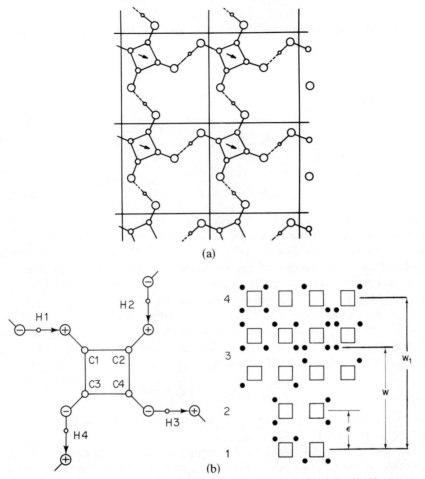

Figure 5.9 (a) One layer of $C_4O_4H_2$ molecules projected on the $\langle 010 \rangle$ plane. (b) Schematic drawing of a squaric acid molecule with the four nearest $O-H \cdots O$ bonds. The energies of the 16 possible hydrogen configurations around a C_4O_4 group are also shown. [Reproduced by permission of Akademie-Verlag from Zinenko, 1976]

free energy density $g(T, \boldsymbol{P}_a, \boldsymbol{P}_b)$ can be now written (Feder, 1976) in polar coordinates $\boldsymbol{P}_i = P_i(\cos\phi_i, \sin\phi_i)$, $i = a, b$ as:

$$g = \tfrac{1}{2}a(P_a^2 + P_b^2) + \tfrac{1}{4}(b_0 + b_1\cos 4\phi_a + b_2\sin 4\phi_a)P_a^4 +$$

$$+ \tfrac{1}{4}(b_0 + b_1\cos 4\phi_b + b_2\sin 4\phi_b)P_b^4 + cP_aP_b\cos(\phi_a - \phi_b) \quad (5.43)$$

where $a = a'(T - T_0)$ and a', b_0, b_1, b_2, and c are assumed to be temperature-independent positive constants.

The minimization of the free energy shows that the second-order transition between the disordered phase and the antiferroelectrically ordered phase $(P_a = -P_b \neq 0)$ occurs at

$$T_c = T_0 + c/a' \tag{5.44}$$

For $T > T_c$, $P_a = P_b = 0$, whereas one finds that for $T < T_c$

$$P_a = P_b = \sqrt{\frac{a'(T_c - T)}{b}} \tag{5.45a}$$

$$\phi_b = \phi_a + \pi \tag{5.45b}$$

$$\tanh 4\phi_a = b_2/b_1 \tag{5.45c}$$

The in-plane polarization in two successive layers is indeed rotated for 180°, so that the system is antiferroelectric. Expression (5.45a) predicts $\beta = \frac{1}{2}$ in sharp contrast to the experimental value $\beta \approx 0.14$. This discrepancy is due to the fact that the 2D properties are lost in the MFA-like Landau treatment.

A microscopic model of the phase transition in squaric acid has been recently developed by Zinenko (Zinenko, 1976). In analogy to the 3D cluster approximation of Blinc and Svetina (Blinc and Svetina, 1966; Vaks and Zinenko, 1973) and Vaks (1973) for KH_2PO_4 he developed a 2D cluster approximation treatment for the eigenvalues of the Ising model in a transverse tunnelling field Hamiltonian

$$\mathcal{H} = -\Omega \sum_i S_i^x - \frac{1}{2} \sum_{i,j} J_{ij} S_i^z S_j^z \tag{5.46}$$

describing the motion of the protons in the double well $O\!-\!H \cdots O$ potentials around the C_4O_4 groups.

The first cluster is formed by the group of four protons close to a C_4O_4 molecule in the layer at $y = \frac{1}{4}$ whereas the other cluster is the group of four protons close to a C_4O_4 molecule in the layer at $y = \frac{3}{4}$. The interactions within the clusters are evaluated exactly whereas the interactions between different layers and the interactions with the protons outside a given cluster are taken into account in a molecular field approximation.

The energies of the 16 possible arrangements of the four protons around a C_4O_4 group (Zinenko, 1976) are shown schematically in Figure 5.9.

These configuration energies are related to the constants of interaction between neighbouring 'pseudo-spins' (Figure 5.9) in the following way:

$$J_{13} = J_{24} = \frac{1}{2}(w - \varepsilon/2) \tag{5.47a}$$

$$J_{12} = J_{23} = J_{34} = J_{41} = \varepsilon/8 \tag{5.47b}$$

$$w_1 = 2w - \varepsilon. \tag{5.47c}$$

If interactions between the 'pseudo-spins' of different layers can be neglected ($J = 0$) as well as long-range interactions ($\gamma = 0$) the transition temperature T_c of $C_4O_4D_2$—where $\Omega_D = 0$—is determined by the following relation:

$$e^{-\beta_c \varepsilon} + 2e^{-\beta_c w_1} + 2e^{-\beta_c w} = 1, \qquad \beta_c = 1/(kT_c). \qquad (5.48)$$

In view of the antiferroelectric ordering of different layers the constant J has to be non-zero and negative, $J < 0$. The data for deuterated squaric acid have been fitted by: $\varepsilon = 32$ K, $w = 1070$ K, $\gamma = 38$ K, $J = -16$ K, $\Omega_D = 0$. With these parameters $T_c = 590$ K and the transition entropy, ΔS, equals 1.45 cal/K mol.

The qualitative features of the phase transition do not change if $\Omega \neq 0$, i.e. if the proton tunnelling is included. Zinenko found that with $\Omega_H = 105$ K and a 30% decrease in the coupling constants T_c becomes 390 K and $\Delta S \approx 136$ cal/K mol. The above cluster approximation yielded for the critical exponent β of the order parameter of the transition $\eta = (P_a - P_b)/2 \propto ((T_c - T)/T_c)^\beta$ a value $\beta \approx 0.3$ which is smaller than the MFA value $\beta = \frac{1}{2}$ but still higher than the experimental value. In addition one finds that $\beta_H(\Omega \neq 0) \leq \beta_D(\Omega = 0)$ in apparent agreement with the experimental data.

The temperature dependence of the order parameter η obtained with the above parameters for $C_4O_4D_2$ is shown in Figure 5.10. The increase in the order parameter η with decreasing temperature takes place in a much larger interval of T/T_c values than in 3D ferroelectrics. This reflects the fact that in 2D systems the critical region is much broader than in 3D ones.

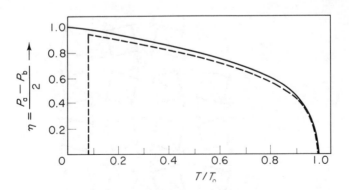

Figure 5.10   Temperature dependence of the order parameter $\eta = (P_a - P_b)/2$ in $C_4O_4D_2$. [Reproduced by permission of Akademie-Verlag from Zinenko, 1976]

5.6 Three-dimensional Uniaxial Ferroelectricity in KH_2PO_4-type Systems and Proton–Lattice Interactions

5.6.1 *Structure*

The structure of KH_2PO_4 is illustrated in Figure 5.11. Each PO_4 group is linked to four neighbouring PO_4 groups by O—H $\cdots$ O hydrogen bonds in such a way that there is a hydrogen bond between one 'upper' oxygen of a PO_4 group and one 'lower' oxygen of the neighbouring group. In such a way a 3D diamond lattice-type hydrogen-bonded network is formed (Slater, 1941; Silsbee *et al.*, 1964). All four hydrogen bonds around a given PO_4 group are nearly perpendicular to the ferroelectric *c* axis. In each O— H $\cdots$ O bond there are two equilibrium sites in which a hydrogen can be located. There are thus $2^4 = 16$ possible arrangements of the four hydrogens surrounding a given PO_4 group. If one involves the Slater 'ice' rules (Slater, 1941; Silsbee *et al.*, 1964) that two and only two hydrogens can be closely attached to any given PO_4 groups the 16 possible configurations are reduced to 6. According to the Slater model (Slater, 1941; Silsbee *et al.*, 1964), two out of these six H_2PO_4 configurations—with dipole moments along the $+c$ and $-c$ directions—represent states with an energy zero (Slater, 1941; Silsbee *et al.*, 1964; Blinc and Žekš, 1974) (Table 5.3). The other four H_2PO_4 configurations with dipole moments along the $\pm a$ and $\pm b$ crystal directions represent states with an energy $\varepsilon > 0$. Whereas the Slater KH_2PO_4 model has been exactly solved for a 2D square lattice by Lieb (Lieb and Wu, 1972), this is not the case for the 3D diamond lattice. The equilibrium properties of KH_2PO_4 and KD_2PO_4 however have been well described by the cluster expansion model of Blinc and Svetina (Blinc and Svetina, 1966; Vaks and Zinenko, 1973) and Vaks (Vaks, 1973) where

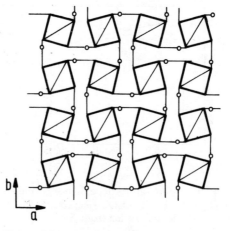

Figure 5.11   Structure of KH_2PO_4

Table 5.3 Energies of the 16 hydrogen configurations around a PO_4 group in KD_2PO_4 according to the Slater–Takagi model.

1	2	3	4	5	6	7	8	9	10	11	12	13	14	15	16
0	0	ε	ε	ε	ε	w	w	w	w	w	w	w	w	w_1	w_1

tunnelling effects as well as all 16 hydrogen configurations around a PO_4 group are taken into account. In this treatment the coupling between the four protons around a PO_4 group is taken into account exactly, whereas the coupling with all other protons is replaced by a molecular field.

As KH_2PO_4 type ferroelectrics have been recently reviewed by Blinc and Žekš (1974), we shall in the following limit ourselves to a discussion of proton–lattice interactions in these systems. Here we only wish to point out that the critical exponents in KH_2PO_4-type crystals are classical ($\gamma = 1$, $\beta = \frac{1}{2}$).

5.6.2 *Proton–lattice interactions in hydrogen-bonded systems*

The pseudo-spin $\frac{1}{2}$ Ising model in a transverse tunnelling field (IMTF) has been extensively used to describe structural—and, in particular, ferroelectric—phase transitions in hydrogen-bonded systems, where the transition is triggered by the ordering of protons into one out of the two equilibrium sites in the O—H $\cdots$ O bonds (Blinc, 1960; De Gennes, 1963; Brout *et al.*, 1966; Blinc and Žekš, 1974). The fact that the spontaneous polarization often appears in a direction which is perpendicular to the hydrogen bonds, has been accounted for by Kobayashi (1968) by introducing the coupling of the pseudo-spin proton tunnelling modes with polar optic phonons. Though the IMTF Hamiltonian is just a first approximation to the true physical situation in systems like KH_2PO_4, where four-body Slater-type interactions (Slater, 1941; Silsbee *et al.*, 1964; Blinc and Svetina, 1966; Vaks and Zinenko, 1973)—which are in KH_2PO_4 equivalent to introducing a pseudo-spin $\frac{15}{2}$ Hamiltonian—have to be included for a complete quantitative fit (Slater, 1941; Silsbee *et al.*, 1964; Blinc and Svetina, 1966; Vaks and Zinenko, 1973) of the thermal and dielectric properties, it has nevertheless been remarkably successful in providing a consistent, simple, and qualitatively correct description of both dynamic and static aspects of structural phase transitions in widely different classes of hydrogen-bonded solids such as KH_2PO_4 (Blinc and Žekš, 1974) and $NaH_3(SeO_3)_2$ (Makita and Miki, 1970), pseudo-two-dimensional squaric acid (Semmingsen and Feder, 1974) and pseudo-one-dimensional $PbHPO_4$ (Semmingsen *et al.*, 1977; de Carvalho and Salinas, 1978). One of the most remarkable features of the IMTF

Hamiltonian is the prediction of the simultaneous occurrence of:

(i)   a soft proton tunnelling mode (Kaminov and Damen, 1968; Lowndes *et al.*, 1974) the frequency of which decreases on deuteration and vanishes at T_c;
(ii) an increase in the Curie temperature T_c on deuteration (Blinc and Žekš, 1974);
(iii) a decrease in T_c with increasing hydrostatic pressure (Samara, 1967, 1971, 1973)—or, what is the same, increased tunnelling probability (Blinc and Žekš, 1968)—leading to a vanishing of ferroelectricity at high pressures (Samara, 1967, 1971, 1973; Blinc and Žekš, 1968). In deuterated isomorphs, where the tunnelling term is smaller, ferroelectricity should vanish at much higher hydrostatic pressure (Samara, 1967, 1971, 1973; Blinc and Žekš, 1968) than in undeuterated systems.

In all hydrogen-bonded systems studied so far these three phenomena, which result from the presence of the 'tunnelling' term, have been indeed found to coexist, though the proton tunnelling mode is—except at high pressures (Peercy, 1973, 1975a, 1975b)—usually overdamped.

Recent far-infrared (Kroupa and Petzelt, to be publ.), submillimetre dielectric (Volkov, 1977) and Raman (Lavrenčič and Petzelt, 1977) scattering experiments in $PbHPO_4$ and squaric acid (Nakashima and Balkanski, 1976), however, seem to show some new facts which cannot be, even qualitatively, explained within the simple IMTF framework or the modification introduced by (Kobayashi, 1968). Whereas it follows from the IMTF Hamiltonian that a significant isotope shift in T_c on deuteration should be found only in those systems where the frequency of the proton tunnelling mode is comparable—though smaller—than kT_c, this rule seems to be broken in $PbHPO_4$ (Kroupa and Petzelt, to be publ.; Volkov, 1977; Lavrenčič and Petzelt, 1977). Here T_c shifts from $T_{c,H} = 37\,°C$ to $T_{c,D} = 187\,°C$ on deuteration, whereas the frequency of the 'soft' proton mode ($\sim 1\,cm^{-1}$) is by more than two orders of magnitude smaller (Nakashima and Balkanski, 1976; Lavrenčič and Petzelt, 1977) than kT_c. At the same time another 'hard' temperature-dependent proton tunnelling mode has been observed below—but not above—T_c (Lavrenčič and Petzelt, 1977). The situation is similar in squaric acid, where $T_{c,H} = 97\,°C$, $T_{c,D} = 243\,°C$ and the soft mode frequency is rather low (Nakashima and Balkanski, 1976).

5.6.3   The model Hamiltonian

5.6.3.1   *Proton–phonon coupling*   The total Hamiltonian of the coupled proton–phonon system we are investigating is the sum of protonic term, a lattice term and a proton–lattice interaction term (Kobayashi, 1968; Blinc and Žekš, 1974)

$$H = H_P + H_L + H_I. \tag{5.49}$$

The Hamiltonian H_P of the 'bare' O—H $\cdots$ O proton system in a rigid lattice can be written in the usual pseudo-spin $\frac{1}{2}$ operator formalism (Blinc, 1960; De Gennes, 1963; Brout et al., 1966; Blinc and Žekš, 1974) as

$$H_P = -\Omega \sum_i S_i^x - \frac{1}{2} \sum_{i,j} J_{ij} S_i^z S_j^z - \frac{1}{2} \sum_{i,j} B_{ij} S_i^x S_j^x \tag{5.50}$$

where Ω is the proton tunnelling frequency, J_{ij} is the pseudo-spin–pseudo-spin Ising interaction constant, and B_{ij} measures the effect of the tunnelling motion of one proton on the tunnelling motion of another. It should be noted that the first two terms in expression (5.50) represent just the usual IMTF Hamiltonian (Blinc, 1960; De Gennes, 1963; Brout et al., 1966; Blinc and Žekš, 1974).

Whereas the tunnelling term Ω is proportional to the overlap integral $s < 1$ between the protonic wave functions at the two equilibrium sites in the O—H $\cdots$ O bond, B is proportional to the square of the overlap, s^2.

H_L represents the usual Hamiltonian of lattice vibrations

$$H_L = \frac{1}{2} \sum_{q,p} [P_{q,p} P_{-q,p} + \omega_{q,p}^2 Q_{q,p} Q_{-q,p}] \tag{5.51}$$

where $Q_{q,p}$, $P_{q,p}$, and $\omega_{q,p}$ are the normal coordinate, momentum, and frequency of the pth lattice mode with wave vector q.

The pseudo-spin–phonon interaction consists of three different contributions

$$H_I = H_I^{(a)} + H_I^{(b)} + H_I^{(c)}. \tag{5.52}$$

Here

$$H_I^{(a)} = -\sum_i S_i^z F_{i,p}^z Q_{p,i} \tag{5.53a}$$

describes the well-known interaction (Kobayashi, 1968) of the pseudo-spins with polar optical phonons, which makes the two potential wells for proton motion unequivalent.

$H_I^{(b)}$, on the other hand, describes the modulation of the distance 2ξ between the two equilibrium sites in the O—H $\cdots$ O bonds—and the resulting modulation of the tunnelling term Ω—by non-polar optic phonons:

$$H_I^{(b)} = -\sum_i S_i^x F_i^x Q_i. \tag{5.53b}$$

This term provides an indirect coupling between the tunnelling motion of one proton and the tunnelling motion of another, which may be larger than the direct $S_i^x S_j^x$ interaction.

Expanding the tunnelling integral Ω in powers of the lattice coordinates Q

$$\Omega(Q) = \Omega(0) + \left(\frac{\partial \Omega}{\partial Q}\right)_0 Q + \cdots \qquad (5.54)$$

we see that

$$F^x = \left(\frac{\partial \Omega}{\partial Q}\right)_0. \qquad (5.55)$$

If one, for sake of simplicity, represents the O—H $\cdots$ O bond potential by a sum of two harmonic potentials, displaced by 2ξ, one finds Ω as:

$$\Omega = E_0 \frac{4q}{\sqrt{\pi}} \exp(-q^2) \qquad (5.56a)$$

Here

$$q^2 = 2mE_0\xi^2/\hbar^2 \qquad (5.56b)$$

where m stands for the proton mass and E_0 is the protonic vibrational zero point energy in a given harmonic potential well in the absence of tunnelling. If we further assume that the distance between the two wells depends linearly on the phonon coordinates

$$\xi = \xi_0 + cQ \qquad (5.56c)$$

one finds that F^x is proportional to the tunnelling integral Ω:

$$F^x = \left(\frac{\partial \Omega}{\partial \xi}\right)\left(\frac{\partial \xi}{\partial Q}\right) = -(2q^2 - 1)\frac{\Omega c}{\xi} \qquad (5.57)$$

F^x is thus proportional to the overlap s and not to s^2 as the 'direct' S^x. S^x coupling constant $\cdot F^x$ is positive if the lattice vibration results in a compression of the O $\cdots$ O bond distance so that $C < 0$. It should be noted that $F^x Q$ may be in some cases comparable to $\Omega(0)$.

The third term in expression (5.52), $H_I^{(c)}$ represents the coupling of the tunnelling motion with polar optic phonons. For crystals, which are not polar above T_c, symmetry requires this coupling to be an even function of Q:

$$H_I^{(c)} = -\sum_i S_i^x D_{i,p} Q_{i,p}^2 \qquad (5.58a)$$

Polar optic phonons thus make the two potential minima in the O—H $\cdots$ O bond non-equivalent—equation (5.53a)—and in addition produce a deformation of the intervening potential barrier resulting in expression (5.58a).

The deformation in the shape of the potential well produces a fluctuation in E_0 such as

$$E_0 = E_0(0) + dQ_p^2 \qquad (5.58b)$$

so that one finds that $(\partial\Omega/\partial Q_p) = 0$, $(\partial^2\Omega/\partial Q_p^2) = 2D$ where

$$D = -(2q^2 - 3)\frac{\Omega d}{2E_0}. \qquad (5.58c)$$

5.6.4 Direct $S_i^x S_j^x$ coupling

Let us first investigate the IMTF Hamiltonian with the addition of the 'direct' $S_i^x S_j^x$ interaction term. The Hamiltonian given by expression (5.50), can be in the molecular field approximation (MFA) replaced by (Blinc and Žekš, 1974)

$$H_{p,i}^{MFA} = -\mathbf{H}_i \mathbf{S}_i \qquad (5.59)$$

The molecular field $\mathbf{H}_i$ represents a vector in the pseudo-spin space $\mathbf{H}_i = (H_x, 0, H_z)$ with components

$$H_x = \Omega + \sum_j B_{ij}\langle S_j^x \rangle = \Omega + B_0 \langle S^x \rangle \qquad (5.60a)$$

$$H_z = \sum_j J_{ij}\langle S_j^z \rangle = J_0 \langle S^z \rangle \qquad (5.60b)$$

In contrast to the usual IMTF not only H_z but also H_x is temperature dependent.

The thermal expectation values of the pseudo-spin operators are now determined from the self-consistency equations (Blinc and Žekš, 1974):

$$\langle S^x \rangle = \frac{1}{2}\frac{H_x}{H}\tanh\left(\frac{\beta H}{2}\right) \qquad (5.61a)$$

$$\langle S^z \rangle = \frac{1}{2}\frac{H_z}{H}\tanh\left(\frac{\beta H}{2}\right) \qquad (5.61b)$$

where $H = (H_x^2 + H_z^2)^{1/2}$ and $\beta = 1/kT$. Non-zero solutions for the spontaneous polarization, which is proportional to $\langle S^z \rangle$, are stable below the transition temperature T_c. This is given by

$$\frac{2(\Omega + B_0\langle S^x \rangle_c)}{J_0} = \tanh\left[\tfrac{1}{2}\beta_c(\Omega + B_0\langle S^x \rangle_c)\right] \qquad (5.62a)$$

where $\beta_c = 1/kT_c$ and $\langle S_x \rangle_c = \langle S_x \rangle_{T_c}$.

Eliminating $\langle S^x \rangle_c$ with the help of (5.61a) we can rewrite equation (5.62a) as

$$\frac{2\Omega}{J_0 - B_0} = \tanh\left(\frac{\beta_c\Omega}{2}\frac{J_0}{J_0 - B_0}\right). \qquad (5.62b)$$

The dependence of the transition temperature T_c on B_0 is shown in Figure 5.12 for two representative cases: (i) $2\Omega_H < J_0$ (undeuterated system) and (ii) $2\Omega_D \leqslant J_0$ (deuterated system). The transition temperature of the undeuterated system $T_{c,H}$ decreases continuously with increasing B_0 and ferroelectricity vanishes ($T_c \to 0$) for

$$B_0 \geqslant J_0 - 2\Omega \qquad (5.63)$$

In the same range of B_0 values $T_{c,D}$ does not depend on B_0 so that the isotope shift in the Curie temperatures $(T_{c,D} - T_{c,H})/T_{c,D}$ continuously increases with increasing B_0 and reaches the value of 1 for $B_0 = J_0 - 2\Omega_H$. It should be noted that Figure 5.12 closely resembles the experimentally observed dependence of T_c on hydrostatic pressure (Samara, 1967, 1971, 1973) in KH_2PO_4 and KD_2PO_4. In KH_2PO_4 $T_{c,H} \to 0$ at ~ 16 kbar, whereas no vanishing of $T_{c,D}$ was observed in KD_2PO_4 even at much higher pressures (Samara, 1967, 1971, 1973). We shall see later that a similar effect is produced by the indirect $S_i^x S_j^x$ coupling via phonons ($F^x \neq 0$).

Let us now study the dynamic properties of the above system.

Introducing collective variables as Fourier transform of the pseudo-spin operators (De Gennes, 1963; Brout *et al.*, 1966; Blinc and Žekš, 1974), we can rewrite expression (5.50) as:

$$H_P = -\Omega S_0^x - \frac{1}{2}\sum_q J_q S_q^z S_{-q}^z - \frac{1}{2}\sum_q B_q S_q^x S_{-q}^x \qquad (5.64)$$

and obtain the linearized random-phase approximation (RPA) Heisenberg

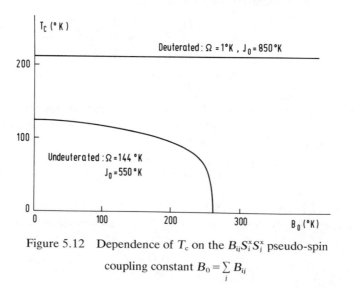

Figure 5.12   Dependence of T_c on the $B_{ij} S_i^x S_j^x$ pseudo-spin

coupling constant $B_0 = \sum_j B_{ij}$

equations of motion for the spin deviation operators $\delta S_q e^{i\omega t} = S_q - \langle S \rangle$ as:

$$i\omega \delta S_q^x = J_0 \langle S^z \rangle \delta S_q^y \tag{5.65a}$$

$$i\omega \delta S_q^y = [-J_0 \langle S^z \rangle + B_q \langle S^z \rangle] \delta S_q^x + [\Omega' - J_q \langle S^x \rangle] \delta S_q^z \tag{5.65b}$$

$$i\omega \delta S_q^z = -\Omega' \delta S_q^y \tag{5.65c}$$

where Ω' stands for the renormalized temperature-dependent tunnelling frequency:

$$\Omega' = \Omega + B_0 \langle S^x \rangle \tag{5.65d}$$

and $J_q = \sum_j J_{ij} e^{iq(r_i - r_j)}$ as well as $B_q = \sum_j B_{ij} e^{iq(r_i - r_j)}$.

The homogeneous system of linear equations (5.65a)–(5.65c) has a non-trivial solution only if the corresponding secular determinant is identically equal to zero. The resulting equation for the eigenfrequencies is:

$$i\omega[-\omega^2 + \Omega'(\Omega' - J_q \langle S^x \rangle) + J_0 \langle S^z \rangle^2 (J_0 - B_q)] = 0. \tag{5.66}$$

One solution—which corresponds to longitudinal excitations—is:

$$\omega_1 = 0 \tag{5.67a}$$

The frequency of transverse excitations which represent the free precession of the pseudo-spins around the molecular field, is, on the other hand, obtained as:

$$\omega_{2,3}^2 = \Omega'(\Omega' - J_q \langle S^x \rangle) + J_0 \langle S^z \rangle^2 (J_0 - B_q) \tag{5.67b}$$

and approaches zero as $q \to 0$, $T \to T_c$, where T_c is given by expression (5.62b). It should be noted that the $B_q S_q^x S_{-q}^x$ term renormalizes the soft mode frequency (5.67b). The renormalization effects are significant only close to T_c where they—as discussed above—increase the isotope shifts in T_c.

For $T > T_c$ we get

$$\omega_{2,3}^2 = \Omega'(\Omega' - J_q \langle S^x \rangle), \tag{5.67c}$$

which is the same as in the IMTF model except for Ω' instead of Ω. At high enough temperatures

$$\Omega' = \Omega \left(1 + \frac{B_0}{4kT}\right) \to \Omega \tag{5.68}$$

so that the soft mode frequency is the same as in the absence of the $BS_i^x S_j^x$ term though T_c is different.

5.6.5 $S_{-q}^x F_q^x Q_q$ *Coupling*

Let us now discuss the coupling of the pseudo-spin tunnelling motion with non-polar optical phonons, which results in an indirect $S_i^x S_j^x$ coupling. The coupling of S^z with polar optical phonons is as well taken into account (Kobayashi, 1968).

The Hamiltonian of our problem is now

$$H = -\Omega \sum_i S_i^x - \frac{1}{2} \sum_{i,j} J_{ij} S_i^z S_j^z - \sum_q S_{-q}^x F_q^x Q_q^{(1)} - \sum_q s_{-q}^z F_q^z Q_q^{(2)}$$
$$+ \frac{1}{2} \sum_q \sum_{p=1}^{2} (P_q^{(p)} P_{-q}^{(p)} + \omega_{q,p}^2 Q_q^{(p)} Q_{-q}^{(p)}) \tag{5.69}$$

where $Q_q^{(1)}$ is the normal coordinate of a non-polar optic phonon, interacting with S^x and $Q_q^{(2)}$ the normal coordinate of a polar optic phonon interacting with S^z.

The thermal expectation values of the pseudo-spin operators are—within the MFA—again determined by the self-consistency equations (5.61a) and (5.61b). The components H_x and H_z of the molecular field are, however, now given by:

$$H_x = \Omega + F_0^x \langle Q^{(1)} \rangle \tag{5.70a}$$

$$H_z = J_0 \langle S^z \rangle + F_0^z \langle Q^{(2)} \rangle \tag{5.70b}$$

where

$$\langle Q^{(1)} \rangle = \frac{F_0^x}{\omega_{1,0}^2} \langle S^x \rangle \tag{5.70c}$$

and

$$\langle Q^{(2)} \rangle = \frac{F_0^z}{\omega_{2,0}^2} \langle S^z \rangle \tag{5.70d}$$

Non-zero solutions for the spontaneous polarization, which is proportional to $\langle S^z \rangle$, are stable below the transition temperature T_c. This is given by

$$\frac{2\Omega}{J_0 + b - a} = \tanh \left(\frac{\beta_c \Omega}{1} \frac{J_0 + b}{J_0 + b - a} \right) \tag{5.71}$$

where $\beta_c = 1/kT_c$ and $a = |F^x|^2/\omega_{1,0}^2$. Equation (5.71) is analogous to (5.62b). Expression (5.71) shows that at T_c proton–phonon interactions renormalize the tunnelling integral: $\Omega \to \Omega(J_0 + b)/(J_0 + b - a)$. For $a = b = 0$, we obtain the equation for T_c of the IMTF Hamiltonian. Since a is proportional to the square of the tunnelling integral, the renormalization of Ω is mass and pressure dependent. A non-vanishing solution for T_c exists only if

$$2\Omega < J_0 + b - a \tag{5.72}$$

Introducing, as before, Fourier-transformed collective variables, we find—using expression (5.69)—for the linearized pseudo-spin and phonon deviation operators

$$i\omega\,\delta S_q^x = [J_0\langle S^z\rangle + F_0^z\langle Q^{(2)}\rangle]\,\delta S_q^y \tag{5.73a}$$

$$i\omega\,\delta S_q^y = -[J_0\langle S^z\rangle + F_0^z\langle Q^{(2)}\rangle]\,\delta S_q^x + [\Omega - J_q\langle S^x\rangle + F_0^x\langle Q^{(1)}\rangle]\,\delta S_q^z$$
$$- F_q^z\langle S^x\rangle\,\delta Q^{(2)} + F_q^x\langle S^z\rangle\,\delta Q^{(1)} \tag{5.73b}$$

$$i\omega\,\delta S_q^z = -[\Omega + F_0^x\langle Q^{(1)}\rangle]\,\delta S_q^y \tag{5.73c}$$

$$i\omega\,\delta Q_q^{(1)} = \delta P_q^{(1)} \tag{5.73d}$$

$$i\omega\,\delta Q_q^{(2)} = \delta P_q^{(2)} \tag{5.73e}$$

$$i\omega\,\delta P_q^{(1)} = -\omega_{1,q}^2\,\delta Q_q^{(1)} + F_{-q}^x\,\delta S_q^x \tag{5.73f}$$

$$i\omega\,\delta P_q^{(2)} = -\omega_{2,q}^2\,\delta Q_q^{(2)} + F_{-q}^z\,\delta S_q^z \tag{5.73g}$$

The above system of linear equations has non-trivial solutions if the corresponding secular determinant vanishes. The resulting equation for the eigenfrequencies of the coupled proton–phonon modes can be written in the form:

$$(\omega^2 - \omega_{1,q}^2)(\omega^2 - \omega_{2,q}^2)(\omega^2 - \omega_{B'}^2) - \omega_A^4(\omega^2 - \omega_{1,q}^2) - \omega_C^4(\omega^2 - \omega_{2,q}^2) = 0 \tag{5.73h}$$

where

$$\omega_{B'}^2 = \omega_B^2 + g^2 \tag{5.74a}$$

$$\omega_B^2 = \Omega''(\Omega'' - J_q\langle S^x\rangle) \tag{5.74b}$$

$$\Omega'' = \Omega + F_0^x\langle Q^{(1)}\rangle \tag{5.74c}$$

$$g = J_0\langle S^z\rangle + F_0^z\langle Q^{(2)}\rangle \tag{5.74d}$$

$$\omega_A^4 = \Omega''\,|F_q^z|^2\,\langle S^x\rangle \tag{5.74e}$$

$$\omega_C^4 = g\,|F_q^x|^2\,\langle S^z\rangle \tag{5.74f}$$

The frequency of the longitudinal excitations—i.e. fluctuations parallel to the molecular field—is zero both above and below T_c.

For $T > T_c$, $\langle S^z\rangle = \langle Q^{(2)}\rangle = \omega_C^4 = 0$, and we find

$$\omega^2 = \omega_{1,q}^2 \tag{5.75a}$$

and

$$\omega^4 - \omega^2(\omega_B^2 + \omega_{2,q}^2) + \omega_B^2\omega_{2,q}^2 - \omega_A^4 = 0 \qquad (5.75b)$$

The dispersion relation (5.75b) is equal to the Kobayashi result if Ω is replaced by the temperature-dependent renormalized frequency Ω''. The frequencies of the two coupled transverse modes are given by (Kobayashi, 1968):

$$\omega_{\pm}^2 = \tfrac{1}{2}[(\omega_{2,q}^2 + \omega_B^2) \pm \sqrt{(\omega_{2,q}^2 - \omega_B^2)^2 + 4\omega_A^4}]; \qquad T > T_c \qquad (5.76)$$

where ω_B is the soft mode frequency of the 'renormalized' protonic system given by equations (5.74b) and (5.74c).

The ω_- mode describes the in-phase motion of the pseudo-spin system and the lattice, whereas in the 'hard' ω_+ mode the two systems move with opposite phase. It should be noted that the coupling with the polar phonons occurs via the F^z term, whereas the non-polar phonons are above T_c decoupled from the proton modes.

The stability limit of the paraelectric phase is determined by the temperature T_c—expression (5.71)—where ω_- vanishes. The transitions are second order if $J_0/2\Gamma \geqslant 1$.

If $F_q^z = 0$, we find for $T > T_c$ that $\omega_+ = \omega_{2,q}$ and $\omega_- = \omega_B$. Below T_c we obtain in this case

$$\omega^2 = \omega_{2,q}^2 \qquad (5.77a)$$

and

$$\omega_{\pm}^2 = \tfrac{1}{2}[(\omega_{1,q}^2 + \omega_{B'}^2) \pm \sqrt{(\omega_{1,q}^2 - \omega_{B'}^2)^2 + 4\omega_C^4}], \qquad T < T_c \qquad (5.77b)$$

where $\omega_{B'}$ and ω_C are given by expressions (5.74a) and (5.74f). The frequency of the ω_- mode approaches zero at T_c for $q = 0$, whereas the ω_+ mode varies only weakly with temperature.

The above results show that the $S_{-q}^x F_q^x Q_q^{(1)}$ coupling renormalizes the frequency of the 'proton-like' mode ω_- both above and below T_c, but makes the 'non-polar optic phonon-like' mode frequency, ω_+ different from $\omega_{1,q}$ only below T_c where $\langle S^z \rangle$, and hence ω_C, are different from zero. This is quite different from the $S_{-q}^z F_q^z Q_q^{(2)}$ coupling (Blinc et al., 1976b) where the 'polar optic phonon-like' frequency ω_+ is different from $\omega_{2,q}$ for $T > T_c$ as well as $T < T_c$. The difference is the result of the fact that $\langle S^z \rangle \neq 0$ only below T_c, whereas $\langle S^x \rangle \neq 0$ both above and below T_c.

In looking for $S_{-q}^x F_q^x Q_q^{(1)}$ effects on the lattice vibrations we should thus study the temperature dependence of the lattice modes (Lavrenčič and Petzelt, 1977) below T_c and not, as in case of $S_{-q}^z F_q^z Q_q^{(2)}$ coupling, above T_c.

The temperature dependence of the 'soft' ω_- and 'hard' ω_+ modes is presented in Figures 5.13a and 5.13b for $F^z = 0$, whereas Figure 5.13c

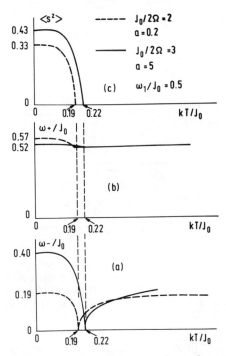

Figure 5.13 Temperature dependence of ω_-, ω_+, and $\langle S^z \rangle$ in case of $S_i^x F^x Q_i$ pseudo-spin phonon coupling. Here a is
$$a = J_0 |F^x|^2 / (4\omega_1^2 \Omega^2)$$

shows the temperature dependence of the spontaneous polarization, $\langle S^z \rangle$.

The dependence of the paraelectric soft mode frequency ω_- on the strength of the F^x coupling is shown in Figures 5.14a and 5.14b both for a deuterated $(\Omega_D = 1 \text{ K}, T_{c,D} = 213 \text{ K})$ and an undeuterated system $(\Omega_H = 144 \text{ K}, T_{c,H} = 122 \text{ K})$.

The strength of the F^x coupling is measured in terms of the parameter $a = J_0 |F^x|^2 / (4\omega_1^2 \Omega^2)$ which should be—in view of the proportionality of F^x to Ω—mass independent. Whereas the F^x coupling scarcely affects the soft mode frequency and T_c in the deuterated system, it affects ω_- of the undeuterated system close to $T_{c,H}$ and $T_{c,H}$ itself even for rather small values of a. For $a = 1$, $(\omega_-)_H$ is $\neq 0$ down to $T = 0$, whereas (ω_-) and $T_{c,D}$ are not changed at all. The high-temperature value of $(\omega_-)_H$ is as well only insignificantly higher than for $a = 0$.

The dependence of $T_{c,H}$ and $T_{c,D}$ on F^x is presented in Figure 5.15 whereas Figure 5.16 shows the dependence of the normalized isotope shift $(T_{c,D} - T_{c,H})/T_{c,D}$ on $\alpha = |F_x|^2 / (4\omega_1^2 \Omega^2)$ for $\Omega_D = 1 \text{ K}$, $J_H = J_D = 850 \text{ K}$ and different values of Ω_H.

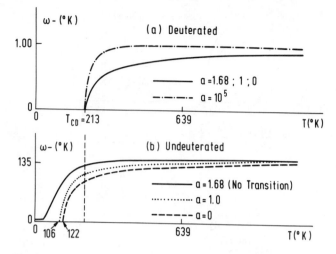

Figure 5.14 Dependence of ω_- on temperature and strength of the $S_i^x F^x Q_i$ pseudo-spin phonon coupling. $a = J_0 |F^x|^2/(4\omega_1^2\Omega^2)$ for a deuterated ($\Omega_D = 1$ K, $J_0 = 850$ K) and an undeuterated system ($\Omega_H = 144$ K, $J_0 = 544$ K)

The F_x coupling thus provides a mechanism which allows for large isotope shifts in T_c even in systems with relatively low Ω and low proton tunnelling mode frequencies. At the same time it could also account for the observed rather different dependences of $T_{c,H}$ and $T_{c,D}$ on pressure.

The effects on T_c are qualitatively similar to those provided by the 'direct' $S_i^x S_j^x$ coupling but are probably more significant in view of the larger magnitude of the $S_{-q}^x F_q^x Q_q^{(1)}$ term as compared to the $B S_q^x S_{-q}^x$ term.

5.6.6 $S_j^x D_j Q_j^2$ coupling

For sake of completeness let us now also look into the coupling of the tunnelling motion with polar optical phonons. In crystals with a non-polar high-temperature phase, this coupling is of the form $S_j^x D_j Q_j^2$. The linear coupling term vanishes due to symmetry as $S_j^x \to S_j^x$, $Q_j \to -Q_j$, $S_j^z \to -S_j^z$ on inversion.

The Hamiltonian of our problem can be thus written as:

$$H = -\Omega \sum_i S_i^x - \tfrac{1}{2} \sum_{i,j} J_{ij} S_i^x S_j^x - \sum_i S_i^z F_i^z Q_i - \sum_i S_i^x D_i Q_i^2 + \tfrac{1}{2} \sum_i (P_i^2 + \omega_{ph}^2 Q_i^2)$$

(5.78)

Within the molecular field approximation we now find that

$$H_i^{MFA} = -\boldsymbol{H}_i \boldsymbol{S}_i - F_i^z \langle S^z \rangle Q_i + \tfrac{1}{2}[P_i^2 + (\omega_{ph}^2 - 2D\langle S^x \rangle)Q_i^2] \qquad (5.79a)$$

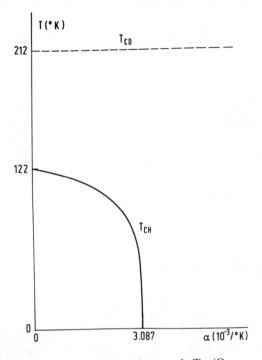

Figure 5.15 Dependence of $T_{c,H}(\Omega_H = 144\ K$, $J_0 = 544\ K)$ and $T_{c,D}(\Omega_D = 1\ K$, $J_0 = 850\ K)$ on the normalized strength of the $S_i^x F^x Q_i$ pseudo-spin phonon coupling $\alpha = |F_x^2|/(4\omega_1^2\Omega^2)$

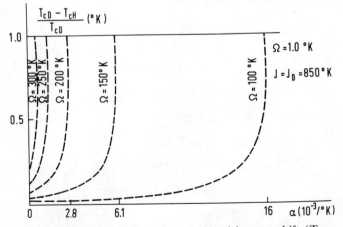

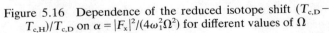

Figure 5.16 Dependence of the reduced isotope shift $(T_{c,D} - T_{c,H})/T_{c,D}$ on $\alpha = |F_x|^2/(4\omega_1^2\Omega^2)$ for different values of Ω

where

$$H_x = \Omega + D\langle Q^2 \rangle \approx \Omega + D\langle Q \rangle^2 \tag{5.79b}$$

and

$$H_z = J_0 \langle S^z \rangle + F_0^z \langle Q \rangle. \tag{5.79c}$$

In expression (5.79b) the fluctuations in the polar phonon coordinates are neglected. The expectation value of the polar phonon coordinate is now obtained as:

$$\langle Q \rangle = \frac{F_0^z \langle S^z \rangle}{\omega_{\mathrm{ph}}^2 - 2D\langle S^x \rangle}. \tag{5.79d}$$

The expression for T_c is given by:

$$\frac{2\Omega}{\tilde{J}_{0,c}} = \tanh \left(\tfrac{1}{2} \beta_c \Omega \right) \tag{5.80}$$

where $\tilde{J}_{0,c}$ stands for the renormalized pseudo-spin interaction constant

$$\tilde{J}_0 = J_0 + \frac{(F_0^z)^2}{\omega_{\mathrm{ph}}^2 - 2D\langle S^x \rangle} \tag{5.81}$$

at $T = T_c$. The proton–phonon interactions thus renormalize the Ising pseudo-spin interaction constant $J_0 \to \tilde{J}_0$. Since both D and $\langle S^x \rangle$ are proportional to Ω, the renormalization is strongly mass and temperature dependent. This might be at least in part responsible for the fact that $(J)_{\mathrm{H}} \neq (J)_{\mathrm{D}}$ in hydrogen-bonded ferroelectrics. The mass-dependent renormalization disappears in the Kobayashi model where $D = 0$.

The resulting equation for the eigenfrequencies of the coupled 'transverse' proton–polar optic phonon modes is:

$$\omega_{\pm}^2 = \tfrac{1}{2}[(\tilde{\omega}_{\mathrm{ph}}^2 + \tilde{\omega}_B^2) \pm \sqrt{(\tilde{\omega}_{\mathrm{ph}}^2 - \tilde{\omega}_B^2)^2 + 4(\omega_A^4 + \omega_C^4 - \omega_D^4)}] \tag{5.82a}$$

where

$$\tilde{\omega}_{\mathrm{ph}}^2 = \omega_{\mathrm{ph}}^2 - 2D\langle S^x \rangle \tag{5.82b}$$

$$\tilde{\omega}_B^2 = \tilde{\Omega}(\tilde{\Omega} - J_0\langle S^x \rangle) + (\tilde{J}_0\langle S^z \rangle)^2 \tag{5.82c}$$

$$\tilde{\Omega} = \Omega + D\langle Q \rangle^2 \tag{5.82d}$$

$$\omega_A^4 = \tilde{\Omega} \, |F_0^z|^2 \, \langle S^x \rangle \tag{5.82e}$$

$$\omega_C^4 = 4D^2 \hat{J}_0 \langle S^z \rangle^2 \langle Q \rangle^2 \tag{5.82f}$$

$$\omega_D^4 = 2DF_0^z(\tilde{\Omega} + \tilde{J}_0\langle S^x \rangle)\langle S^z \rangle\langle Q \rangle. \tag{5.82g}$$

Expression (5.82a) is valid both for $T > T_c$ and $T < T_c$. Above T_c our result is equivalent to the one obtained by Kobayashi except for the mass- and temperature-dependent renormalization of the phonon frequency

$\omega_{ph}^2 \rightarrow \omega_{ph}^2 - 2D\langle S^x \rangle$. There is, however, a considerable difference below T_c. The coupling term $\omega_A^4 + \omega_C^4 - \omega_D^4$ is in the Kobayashi model given by $\Omega |F_0^z|^2 \langle S^x \rangle$, whereas it is in the present case obtained as:

$$\omega_A^4 + \omega_C^4 - \omega_D^4 = \langle S^x \rangle \tilde{\Omega} \left[|F_0^z|^2 + \frac{4D^2}{\tilde{\Omega}\langle S^x \rangle} \tilde{J}_0 \langle S^z \rangle^2 \langle Q \rangle^2 \right.$$

$$\left. - \frac{2F_0^z D}{\tilde{\Omega}} \left(\tilde{J}_0 + \frac{\tilde{\Omega}}{\langle S^x \rangle} \right) \langle S^z \rangle \langle Q \rangle \right]. \quad (5.83)$$

The renormalization of the coupling term is critically temperature dependent and vanishes at $T \geq T_c$. In addition it is also dependent on the mass of the tunnelling particle and on hydrostatic pressure thus explaining the strong dependence of this term on pressure in KH_2PO_4-type crystals below T_c (Blinc et al., 1976b).

5.7 References

Baranov, A. I. (1972). Private communication.

Baxter, R. J. (1972). Ann. Phys., **70**, 193.

Blinc, R. (1960). J. Phys. Chem. Solids, **13**, 204.

Blinc, R., Arend, H., and Kandušer, A. (1976a). Phys. Stat. Sol. (b), **74**, 425.

Blinc, R., Pirc, R., and Žekš, B. (1976b). Phys. Rev., **B13**, 2943.

Blinc, R. and Svetina, S. (1966). Phys. Rev., **147**, 423.

Blinc, R. and Žekš, B. (1968). Helv. Phys. Acta, **41**, 700.

Blinc, R. and Žekš, B. (1974). Soft Modes in Ferroelectrics and Antiferroelectrics, North-Holland, Amsterdam, and references therein.

Brout, R. K., Müller, K. A., and Thomas, H. (1966), Solid State Commun., **4**, 507.

de Carvalho, A. V. and Salinas, S. R. (1978). J. Phys. Soc. Japan, **44**, 238.

De Gennes, P. G. (1963). Solid State Commun., **1**, 132.

Feder, J. (1976). Ferroelectrics, **12**, 71.

Fisher, M. E. (1974). Rev. Mod. Phys., **46**, 597.

Ginzburg, V. L. (1960). Fiz. Tverd. Tela, **2**, 2031.

Kaminov, I. P. and Damen, T. C. (1968). Phys. Rev. Lett., **20**, 1105.

Kiriyama, H. (1962). Bull. Chem. Soc. Japan, **35**, 1199.

Kiriyama, H. (1970). Proc. Int. Meeting Ferroelectricity, Kyoto; J. Phys. Soc. Japan, **28**, Suppl., p. 114.

Kobayashi, K. K. (1968). J. Phys. Soc. Japan, **24**, 497.

Kroupa, J. and Petzelt, J. To be published.

Larkin, A. I. and Khmelnitskii, D. E. (1969). Zh. Eksp. Teor. Fiz., **56**, 2087.

Lavrenčič, B. B. and Petzelt, J. (1977). J. Chem. Phys., **67**, 3890.

Levstik, A., Blinc, R., Kadaba, P., Čižikov, S., Levstik, I., and Filipič, C. (1975). Solid State Commun., **16**, 1339.

Lieb, E. M. and Wu, F. Y. (1972). In Phase Transitions and Critical Phenomena (Eds. Domb, C. and Green, M. S.), Vol. 1, Academic Press, New York, p. 332.

Lowndes, R. P., Tornberg, N. E., and Leung, R. C. (1974). Phys. Rev., **B10**, 911.

Makita, Y. and Miki, H. (1970). J. Phys. Soc. Japan, **28**, 1221.

Nakashima, S. and Balkanski, M. (1976). Solid State Commun., **19**, 1225.

Negran, T. J., Glass, A. M., Brickenkampf, C. S., Rosenstein, R. D., Osterheld, R. K., and Sussot, R. (1974). Ferroelectrics, **6**, 179.

Onsager, L. (1944). *Phys. Rev.*, **65,** 117.
Peercy, P. S. (1973). *Phys. Rev. Lett.*, **31,** 379.
Peercy, P. S. (1975a). *Solid State Commun.*, **16,** 439.
Peercy, P. S. (1975b). *Phys. Rev.*, **B12,** 2725.
Samara, G. A. (1967). *Phys. Lett.*, **25A,** 664.
Samara, G. A. (1971). *Phys. Rev. Lett.*, **27,** 103.
Samara, G. A. (1973). *Ferroelectrics,* **5,** 25.
Semmingsen, D., Ellenson, W. D., Frazer, B. C., and Shirane, G. (1977). *Phys. Rev. Lett.*, **38,** 1299.
Semmingsen, D. and Feder, J. (1974). *Solid State Commun.* **15,** 1369.
Silsbee, H. B., Uehling, E. A., and Schmidt, V. H. (1964). *Phys. Rev.*, **133,** A 165.
Slater, J. C. (1941). *J. Chem. Phys.*, **9,** 16.
Stanley, H. E. (1971). *Introduction to Phase Transitions and Critical Phenomena,* Clarendon Press, Oxford.
Suwelack, D., Becker, J., and Mehring, M. (1977). *Solid State Commun.*, **22,** 597.
Vaks, V. I. (1973). *Introduction to the Microscopic Theory of Ferroelectricity* (In Russian), Izd. Nauka, Moscow.
Vaks, V. I. and Zinenko, V. I. (1973). *Zh. Eksp. Teor. Fiz.*, **64,** 650.
Volkov, A. (1977). Paper presented at the IVth International Meeting on Ferroelectricity, Leningrad, Sept.
Zinenko, V. I. (1976). *Phys. Stat. Sol. (b)*, **78,** 721.

Molecular Interactions, Volume 2
Edited by H. Ratajczak and W. J. Orville-Thomas
© 1980 John Wiley & Sons Ltd.

6 *The significance of charge transfer interactions in biology*

M. A. Slifkin

6.1 Introduction

Discussions of charge transfer interactions in biology have usually been acrimonious and a wide spectrum of views is hotly propagated. The proponents of the arguments range from those who see charge transfer interactions as being of vital significance, to those who do not accept that charge transfer interactions exist and believe them to be the figment of a rather overactive imagination. The middle ground is occupied by those who believe in charge transfer interactions but not that they are particularly significant in biology. One of the difficulties in discussing the topic is that the definitions vary from author to author. However even allowing for this, the arguments are not ones of semantics but of acceptance of the basic concepts. Any discussion must therefore begin with an attempt at defining the terms to be used and a review albeit brief on the arguments for and against the very existence of charge transfer complexes and interactions. The ground has been well covered in the literature so that this is necessarily a rather brief survey (Mulliken and Person, 1969; Foster, 1969; Slifkin, 1971a, 1973).

6.2 Charge Transfer Complexes

The term charge transfer complex from which is derived charge transfer interaction was first used by Mulliken (1950) to describe a set of phenomena which hitherto had defied a satisfactory explanation. It had been known from the beginning of the 20th century that certain combinations of, usually, organic compounds formed brightly coloured addition products or complexes in solution. It was possible in many cases to crystallize out of these solutions, brightly coloured crystals in which the components were represented in simple stoichiometry. The solutions or dry material displayed basically the chemical behaviour of the individual components, although clearly the physical behaviour was quite different.

Many explanations were given of the phenomenon but the first really satisfactory explanation was that of Mulliken (1950).

In his model two components were an electron acceptor and an electron donor or Lewis acid and Lewis base. In the absence of electromagnetic energy they would coexist in a state in which there was some bonding due to the partial transfer of charge from the donor to the acceptor. On the absorption of electromagnetic radiation of suitable energy they could be excited into a state in which a much greater proportion of charge transfer took place. Mulliken used a valence bond description so that the two states of the complex were described thus:

the ground state $\psi_N = a\psi_0(D, A) + b\psi_1(D^+, A^-)$ with $a \gg b$
and the excited state $\psi_E = b^*\psi_1(D^+, A^-) - a^*\psi_0(D, A)$ with $b^* \gg a^*$.

The function ψ_0 describes the two molecules in close proximity where dispersion forces are operating and is called the non-bonding function, and the function ψ_1 describes a situation where charge has been wholly transferred between the donor and the acceptor. The transition between the two states gives rise to the characteristic charge transfer absorption band. The name charge transfer complex is given to this associated pair. This immediately raises problems. Firstly, as pointed out by Dewar and Thompson (1966) the valence bond approach is a mathematical description and does not necessarily bear any resemblance to the physical picture. Furthermore, they suggest that any combination of structure could be added to give a satisfactory description. They suggest that if sufficient locally excited structures were included in the valence bond model, i.e. structures of the form $c\psi(D^*, A) + d\psi(D, A^*)$ etc., then a satisfactory description could also be obtained. They further point to the obvious difficulty that the charge transfer is a phenomenon of the excited state and not the ground state, so that the name is itself misleading. In their view the so-called 'charge transfer complexes' are merely part of a much wider range of weak complexes bound by dispersion and van der Waals forces and there is no justification in attempting to pick out a special category. Furthermore, they claim that there is no experimental evidence for any detectable charge transfer in the ground state of the complexes so designated by Mulliken (1950).

Mulliken and Person (1969) have replied that firstly, the charge transfer description ranges from weak complexes of the kind discussed up to very strong salt-like complexes and that the very weak ones represent a limiting case of a recognizable category of interactions. Although the amount of charge transfer in the ground state of some of the weak complexes is very small, nevertheless it still exists. Whilst it might be possible to represent charge transfer complexes in a valence bond approach which included locally excited states, a large number of these would have to be included in the model and the simplicity of the approach would be lost. The role of dispersion forces in solutions of weak complexes is unlikely to be important as any interactions between donor and acceptor would be more or less cancelled out by interactions with the solvent.

It is interesting that workers in quite different fields have found the concept of charge transfer as enunciated by Mulliken to be useful in analysing their results. Thus X-ray crystallographers (Thewalt and Bugg, 1972) seem to have no difficulty in accepting the idea of charge transfer and it has proved very useful to them in analysing structures of molecular complexes. Similarly infrared spectroscopists have also used the concept of charge transfer complexing in analysing and describing the infrared spectra of molecular complexes (Yarwood, 1973).

The greatest difficulty in reading the literature on charge transfer complexing and interaction in biological systems is that only too frequently, the terms have been used to describe systems in which bright colours are observed without checking whether in fact these arose from charge transfer interaction or were the result of chemical interaction. Unfortunately the indiscriminate use of the terms charge transfer, charge transfer complexes, etc. in cases where there was no proof whatsoever that such interactions were present has made some scientists rather sceptical as to the validity of any of the literature.

Another point which has aroused much controversy is that charge transfer complexing can coexist with or precede chemical interactions, depending on the physical conditions, temperature, concentration, etc. The same system might under one set of conditions appear to exhibit only conventional chemical interactions, whereas under other conditions the charge transfer complexing might predominate. When this is not recognized, then different groups adopt diametrically opposed views each casting doubt on the expertise of the other if only by implication. Matters of opinion are reconcilable but apparent differences in results can only lead to the opinion that one at least is wrong.

Before discussing the significance of charge transfer in biology, we must first be in a position to state that charge transfer exists and to frame a set of criteria for recognizing it. Clearly the writer of this article has to start from the viewpoint that it does exist and that there is a recognizable set of criteria, otherwise the article could be encompassed in a single sentence.

It will be assumed that the description and arguments of Mulliken and those following his general approach are correct. It must be emphasized that the author does indeed hold these views and therefore no claims for impartiality can be made.

In the author's opinion the following criteria for charge transfer interaction must be observed in order for the description to be used. There must be a reversible interaction, which may vary considerably in magnitude, between molecules which can be described as electron acceptors and donors. The strength of interaction will lie within a range of less than 1 kcal mol^{-1} to about 30 kcal mol^{-1}, thus clearly differentiating these interactions from conventional chemical bonding. In certain cases the interaction will be characterized by a broad absorption band, the charge transfer transition, being a property of the complex as a whole rather than one individual member. This condition is rather controversial as certain authors believe that the term charge transfer complex can only be applied to a complex which exhibits a charge transfer band (Foster, 1969) and it is a description of the absorption of the complex rather than of the type of complex. However there are many reasons why the absorption is not observed. Firstly, it could occur in a region of intense local absorption. The extinction coefficients of charge transfer transitions are in many cases much weaker than those of molecular electronic absorptions (Briegleb, 1961). From the Mulliken treatment one can derive a relationship between the frequency of the charge transfer transition $h\nu$ and the ionization potential I_p and electron affinity E_A of the complex viz.

$$h\nu = I_p - E_A - \text{other terms}$$

an expression which has been well proven experimentally, so that the position of the charge transfer transition could occur in any region of the spectrum and could be easily missed if it lay outside the range of the spectrometer.

The equation just derived gives another method of deciding whether or not charge transfer complexing exists within a certain range of compounds. If some absorption band is observed and its position can be altered by an alteration of the ionization potential or electron affinity of one of the components than a *prima facie* case has been made out. With biological material the use of substituted molecules allow this to be done easily. Thus hydroxyl and methyl groups are electron donating and should increase the ionization potential and decrease the electron affinity of the parent compounds, whereas the converse should happen with electron-withdrawing groups such as chloro and nitro. The comparative effects of the different substituents can be obtained from the change in dipole moment or the behaviour of these groups substituted into benzene with conventional acceptors or donors.

The dissociation energies of charge transfer complexes can be derived from the Mulliken approach. This gives:

$$\Delta H_0 \propto I_P - E_A - \text{other terms.}$$

So that the binding energy should correlate in a similar manner with ionization potential and electron affinities. However even on the Mulliken approach charge transfer binding may represent only a small part of the overall binding, and in addition such factors as steric hindrance must be taken into account.

Although the original discovery of charge transfer complexes arose spectroscopically and much of the experimental work is carried out spectroscopically, other techniques might give useful criteria or even evidence as to the presence or role of charge transfer complexes in biological systems.

It is clear from the Mulliken theory that charge transfer forces will influence the structure of molecular complexes in two ways. From the principle of maximum superposition the shape will be determined. The lowest potential energy occurs when the two molecules in a bimolecular complex are orientated to give maximum possible overlap of the electron donor orbital and the electron acceptor orbital in the absence of other forces. This is illustrated by the structure of π-conjugated donors and acceptors which stack with the planes of the molecules parallel but slightly offset to each other. Over the last decade many studies have been made of the structures of complexes of biomolecules and charge transfer has been suggested at times to explain in part these structures.

Another result arising from X-ray crystallography is that in general, although there may be exceptions, the intermolecular distance between the partners of a charge transfer complex is less than, where only normal intercrystalline forces operate. Thus Wallwork (1961) has suggested that interatomic distances of less than 3.4 Å are characteristic of charge transfer bonded molecules whereas Van der Waals bonded molecules have slightly larger intermolecular distances.

Although there has been much less activity in the infrared spectroscopy of charge transfer complexes than visible or ultraviolet spectroscopy, there has been sufficient work done to be able to characterize, to some extent, the spectra of charge transfer complexes. Thus the author and his coworkers (Slifkin and Walmsley, 1969; Slifkin, 1970) have shown that quinone complexes are characterized by shifts of the carbonyl group of the quinone which are quite distinctive. In the extreme case where charge transfer is more or less complete in the ground state then one simply observes the infrared spectrum of the ionized donor and acceptor (Matsunaga, 1966).

There are other somewhat more esoteric techniques which have been or could be used in the study of biological molecules.

The Mulliken theory suggests that charge transfer complexes should be paramagnetic due to the displacement of charge from donor to acceptor. This has been proved experimentally although the results, many of them carried out four decades ago are rather confused, (Bhatnaga *et al.*, 1934; Sahney *et al.*, 1946). If as suggested by Dewar and Thompson (1966) there are locally excited states involved, and these have a physical reality, then these will certainly give rise to a diamagnetic contribution to the overall magnetism.

As diamagnetism is temperature sensitive unlike paramagnetism then a study of the magnetic properties vs. temperature will reveal not only the paramagnetic contribution but the diamagnetic contribution as well. It is possible that the somewhat varied results obtained by earlier workers in this field was due to the unsuspected presence of a diamagnetic contribution, in which case a true description of charge transfer complexing might have to include as well as bonding and non-bonding states, locally excited states, reconciling to some extent the views of Mulliken and Dewar.

Another technique which has been used in the study of organic charge transfer complexes is that of dipole moment measurement of the complexes (Kobinata and Nagakura, 1966). This might prove a diagnostic tool with biological systems if changes in dipole moments can be sufficiently characterized. From the Mulliken theory, transfer of charge does give rise to a change in dipole moment or in the cases of non-polar components, a dipole moment. Most biological molecules do already possess dipole moments and hence one would perforce have to look at changes in dipole moment in the systems.

Electrochemistry is being increasingly applied to the study of biological systems. One technique is to study the conductivity of a system as a function of the concentration of the different components. It has been suggested that in the case of chlorpromazines and related compounds that the results are explicable in terms of the formation of charge transfer complexes (Guttmann and Keyzer, 1966).

The form in which biological systems should be studied is the next point of discussion. Most of the work on organic charge transfer complexes is carried out in organic solvents. Clearly, except for special cases this cannot be done for biological molecules. If one wishes to work in solution, then buffered aqueous solvent of one kind or another will have to be used. The study of biological systems *in situ* represents all kinds of experimental problems and nearly all the discussions in this article will be confined to model systems of isolated purified biomolecules in fairly simple solvent systems. Unfortunately, different buffers appear to participate themselves in weak interactions between different biomolecules. The effect can completely inhibit any purported charge transfer interaction so that whilst in a common buffer such as phosphate buffer, one observes a charge transfer interaction, another common buffer such as citric buffer might completely inhibit it. As we shall also recount, the concentration of the biomolecules themselves can completely change the observed interactions as the concentrations are varied.

As an alternative to solutions, one can work in the solid state. In many cases one can precipitate from solution or even evaporate from solution, systems which might be charge transfer complexes. One can certainly carry out spectroscopic studies, magnetic studies, ESR studies, and so on with these solid samples. The question must then be answered how relevant are any of these studies to real systems. Many of the biomolecules to be studied are found in highly organized parts of the systems, for instance many of the components of the mitochondria are potentially charge transfer complexers. Their behaviour in solution does not seem relevant to their behaviour in the mitochondria neither of course does their behaviour in crystal. Probably one should work in water-based gels such as the polyacrylamides, although to the author's knowledge no such experiments have been carried out.

To summarize it is possible to set up various experimental criteria for charge transfer complexes. It is further possible to observe the effects of charge transfer complexing on model biological systems. It has not as yet proved possible to examine biological systems *in situ* so that any conclusions must necessarily be made by extrapolation from the model to the real case.

Mulliken in his many papers on charge transfer complexes has drawn up many categories of charge transfer complexes. For our purposes we need only consider a few general cases. In order for molecules to be charge donors they must possess easily ionizable electrons. There are two cases, π electrons in conjugated electron systems and localized lone-pair or n electrons such as occur in amino groups. The first case is well recognized in this field. The second case has been somewhat slow to find recognition possibly because much of the early work on organic charge transfer complexes was with benzenoid compounds. In addition there has been some discussion about the role of localized charge transfer in which the donation is not specifically referred to some electron but to some specific area of electronic charge in the donor molecule.

Electron acceptors can be classified into three main groups. The first group is where the donated electron goes into a π orbital. The second group is where the donated electron goes into a σ orbital. Thirdly, there is some suggestion that where there are localized regions of low charge in a molecule these can combine with localized regions of high charge in a donor to form localized charge transfer complexes.

The simple description of this latter case is no different than that of electrostatic binding. There is a difference however in that with electrostatic binding there is movement of the ions only. In the case of localized charge transfer complexes, there is localized movement of the electronic charge from the region of high charge to that of low charge.

The terms donor and acceptor are by no means mutually exclusive. Small molecules might be electron donors in complexes with more electropositive molecules than themselves and electron acceptors on interacting with more electronegative molecules. The very large molecules found in biological molecules could equally act simultaneously as electron acceptors and donors from different parts of the molecule. Another intriguing possibility which will be discussed further is the change from donor to acceptor with slight changes in pH.

From the foregoing categorizations, one might expect very many molecules occurring in biological systems to act as charge donors or acceptors.

There are very many biological amines which provided they existed in some part in the neutral amino form could be expected to form charge transfer complexes. Certainly any molecules with conjugated π electron systems such as the indoles could reasonably be expected to behave also as charge donors. Conversely, quinones, which are widely dispersed in biological systems could be expected also to act as acceptors.

Certain groups of molecules will be discussed in further detail.

6.3   Amino Acids, Proteins, and Analogous Molecules

The amino acids would not suggest themselves to a classical chemist as likely candidates for charge transfer interactions as at physiological pH's they are almost entirely zwitterionic, i.e. in the form $COO^-RCHNH_3^+$, and do not therefore possess an easily ionizable electron (Figure 6.1). However 'almost' is not the same as completely. Approximately 1 part per 1000 of the amino acids is in the neutral $-NH_2$ form at pH 7 so that there does exist the possibility of them behaving as charge donors via the lone-pair n electrons localized on the nitrogen of the amino moiety. There is considerable dispute about whether the amino acids form charge transfer complexes particularly with simple quinones or whether they react chemically to form monosubstituted or disubstituted quinones.

The case against charge transfer interactions occurring between aliphatic amino acids and simple quinones has been put most forcibly by Foster (1974), himself a very distinguished worker in the field of charge transfer complexes. Briefly, the case is put that primary aliphatic amines form disubstituted quinones or occasionally monosubstituted quinones and if amino acids interact in the neutral form they should interact as primary

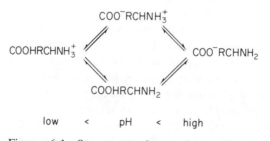

Figure 6.1   Structures of α-amino acids at
different pH

amines. Products have been isolated from solutions of amino acid–quinone mixtures which are believed to be the disubstituted quinones.

The evidence to the contrary is based on very similar experimental results. The addition of quinone to amino acid results in slow changes of the absorption spectra (Moxon and Slifkin, 1972b). Reversibility has been shown to exist by looking either at the development of these spectral changes as a function of temperature, or in those cases where an equilibrium is reached looking at the effect of temperature on the equilibrium position. If these changes are reversible with temperature then it can be safely assumed that some form of complex is formed. It is recognized that with certain amines and certainly at high temperature, greater than 35 °C, chemical interactions do take place. What is disputed is whether the initial process is the formation of weak complexes. The growth rates of the spectral changes have been analysed on the assumption that a 1 : 1 interaction of the form $A + B \rightleftharpoons AB$ takes place. Over a range of temperature the curves have all been found to be consistent with this analysis giving equilibrium constants, rate constants, and thermodynamic parameters which are consistent with each other. There are however some interesting points about the measured parameters. For instance the entropy changes are extraordinarily large indicating clearly that very strong solvation or other orientation effects must be taking place. The back rate constants are found to be much more dependent on pH and temperature than the forward rate constant. This suggests that solvation effects involve the complex much more than the separated components. The infrared spectra of solids evaporated from these systems are very similar to those of quinhydrone charge transfer complexes (Slifkin and Walmsley, 1970). However such spectra could perhaps arise from the direct bonding of the amino acids to the quinone ring, so that the infrared evidence is not conclusive. Although the interaction of primary amines with quinones at room temperature gives immediate chemical reactions without apparently any charge transfer complexing, Nogami and coworkers (1971) showed that by commencing at much lower temperatures, the first step preceding the chemical reaction was the formation of an amine–quinone charge transfer complex.

There is no inherent difficulty in suggesting that such chemical interactions can be preceded by charge transfer complexing, the problem is to show that this is indeed the case.

The concentrations used by Slifkin and coworkers in their experiments were unusually high, the amino acid concentration usually in excess of 10^{-2} M. Workers in this field would normally work at concentrations around 10^{-4} M or even less. Some recent work has been carried out by the writer to see whether the discrepancy between the results of the different groups arose from this difference, Figure 6.2.

In the first series of experiments the forward and back rate constants were

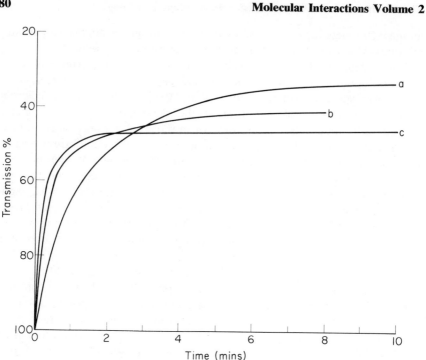

Figure 6.2   Change of transmission with time on mixing of 2×10^{-4} M p-benzoquinone and 7×10^{-2} M proline in (a) pH buffer only; (b) pH 7 buffer and 20% acetone; (c) pH 7 buffer and 40% acetone

measured in the presence of increasing DMSO concentrations. Lorentz (1974) had already shown that the proline–quinone interaction was a CT interaction in the presence of DMSO but not apparently in certain other solvents. The result summarized in Table 6.1 shows clearly that as the DMSO concentration is increased the forward and back rate constants increase as does the apparent equilibrium constant. Conversely the entropy change and the energy of dissociation apparently decrease with increasing DMSO concentration. Similar effects were found on looking at the proline–p-benzoquinone reaction with decreasing proline concentration (see Table 6.1). In addition to these effects the conversion to the chemical reaction went much quicker. Quite clearly the complex is only stabilized in a very polar medium such as produced by the presence of high concentration of the dipolar amino acid zwitterion or very specific interaction take place in which the zwitterion solvates the complex. Working at the lower concentration one observes the final chemical interaction and would be unable to characterize the intermediate charge transfer interaction.

Accepting therefore that a charge transfer complex can be stabilized in the presence of zwitterions or in a highly polar medium, it could be argued

Table 6.1   Charge-transfer parameters of proline–p-benzoquinone complexes

Parameters		Proline concentration	DMSO concentration
k_1 0.27 (1180) k_2 0.49 K_c 0.55 (2408)	$\Delta H^0 = -23.6$ $\Delta S^0 = -28.2$	4×10^{-1}	
k_1 0.22 (961.5) k_2 0.04 K_c 5.5 (24,038)	$\Delta H^0 = -17.5$ $\Delta S^0 = -5.3$	4×10^{-2}	
k_1 0.22 (961.5) k_2 0.005 K_c 44 (192,300)	$\Delta H^0 = -16.8$ $\Delta S^0 = +0.3$	4×10^{-3}	
k_1 0.22 (961.5) k_2 0.002 K_c 110 (480,750)	$\Delta H^0 = -16.3$ $\Delta S^0 = +4.9$	2×10^{-4}	
k_1 0.20 (874) k_2 0.0015 K_c 133.3 (582,667)		2×10^{-5}	
k_1 9.99 k_2 0.09 K_c 100	$\Delta H^0 = -7.8$ $\Delta S^0 = -17$	4×10^{-2}	0.1
k_1 10.83 k_2 0.103 K_c 105	$\Delta H^0 = -6.4$ $\Delta S^0 = -12$	4×10^{-2}	0.2
k_1 11.48 k_2 0.104 K_c 110	$\Delta H^0 = -5.3$ $\Delta S^0 = -8.4$	4×10^{-2}	0.3
k_1 19.32 k_2 0.139 K_c 126.84	$\Delta H^0 = -4.5$ $\Delta S^0 = -5.25$	4×10^{-2}	0.5
k_1 46.36 k_2 0.262 K_c 176	$\Delta H^0 = -3.4$ $\Delta S^0 = -1.15$	4×10^{-2}	0.7

k_1 in l. $mol^{-1} min^{-1}$; k_2 in min^{-1}; K_c in l. mol^{-1}; ΔH^0 in kcal mol^{-1}; ΔS^0 in entropy units.
Values are apparent, based on weights of dissolved proline.
Values in parentheses are corrected values allowing for the dissociation of proline.
DMSO dimethyl sulphoxide

that such an effect could not be too important. However bearing in mind that quinones can and do occur in highly organized regions as for instance in mitochrondria, such reactions could well occur. A point which will be discussed later.

Lorentz (1974) has turned the interaction of amino acids with p-benzoquinone into an analytical tool and has used it to determine the amount of amino nitrogen in serum and urine. Lorentz determines the amount of charge complex formed in the presence of organic solvents. This greatly increases the speed of interaction as compared to water alone. The method is claimed to combine high sensitivity with good reproducibility and is rapid, cheap, and simple. Lorentz (1976) has also considered whether the coloured entities obtained in his work are due to the presence of charge transfer complexes or to direct formation of mono- and disubstituted quinones. A study of the interactions of p-benzoquinone with proteins has led him to the conclusion that the coloured complexes formed are indeed charge transfer complexes with 1:1 stoichiometry and are quite different in colour to the mono- and disubstituted aminoquinones. However in common with other authors he finds that these substituted quinones are formed at a later stage of the interaction.

It seems clearly accepted that the aromatic amino acids form charge transfer complexes with acceptors but acting as conventional π-electron donors (Fujimori, 1959). In addition it has been suggested that the aromatic amino acids operate in a similar manner to the aliphatic amino acids as n- or lone-pair electron donors (Slifkin, 1971a).

X-ray structural analysis has been used by Ash *et al.* (1977) to determine the structure of the 1-methyl-3-carbamidopyridinium N-acetyl-L-tryptophanate charge transfer complex which they take as a model for charge transfer complexes between 1-substituted nicotinamide derivatives and protein.

Matsunaga (1973) has shown that the tryptophan picrate salt is a charge transfer complex formed between tryptophan as π donor and the picric acid ion as acceptor.

Interchain charge transfer complexing on polysarcosine chains having both terminal electron donor and acceptor groups in chloroform has been studied as a model for protein interactions (Sisido *et al.*, 1977). The characteristic charge transfer band was observed. The association constants and enthalpies of dissociation were measured. These were quite low, much lower for example than the value obtained for free amino acids with quinone acceptors. However the main and most interesting conclusion to be drawn is that although the charge transfer complexing forces were weak they nevertheless played an important part in determining the polymer conformation in solution.

Palumbo *et al.* (1977) have investigated charge transfer complexes of a poly(L-tryptophan) sequence with imidazolium HCl and poly(L-histidine HCl) using a variety of techniques including ultraviolet spectrophotometry, circular dichroism (CD), and fluorescence. The association constants of the charge transfer complexes between the indole and imidazolium groups were derived. A similar investigation was carried out (Palumbo, 1975) on the interaction of poly(L-histidine HCl) and indole. The characteristic CD spectrum of the mixture shows that charge transfer complexing takes place between the histidine and indole with an association constant of $2.2\,kcal\,mol^{-1}$.

The interaction of aromatic amino acids with nucleic acid bases has been established clearly by Hélène who used mainly fluorescence spectroscopy to establish the interaction. Slifkin (1971a, 1973) has produced evidence that the aliphatic amino acids also complex with various nucleic acid bases but in contrast to Montenay-Garestier and Hélène (1973) working at low temperature, this was carried out at room temperature on solids isolated by evaporation from mixed solutions.

During an investigation of substrate–protein interaction in cryptonase from *Bacillus alvei* Fenske and DeMoss (1975) found that holotryptophanase in the presence of a anthranilic acid displays a broad absorption band at 350 nm which they interpreted as being a charge transfer transition occurring between the anthranilate as acceptor and the indole portion of the pyridoxal-P as donor.

A study of fluorescence quenching in phenylalanine and model compounds (Tournon *et al.*, 1972) provides some evidence that an intramolecular charge transfer mechanism may be responsible for both the intersystem crossing, rate enhancement, and fluorescence quenching in these compounds. In these compounds, charge transfer proceeds from the phenyl group, i.e. π-electron donation. The belief in the existence of charge transfer in these systems was caused by using electron donating and accepting substituents on the model compounds and showing that the changes in luminescence properties mirrored their ionization potentials.

Studies of the binding of aromatic hydrocarbon carcinogens have been carried out by various groups (for a recent example see Ma *et al.*, 1977) such binding is generally ascribed to hydrophobic stacking forces. The carcinogens are good donors and potentially at least good acceptors (Foster, 1969; Slifkin, 1971a). However the association constants are rather larger than those between amino acids and good acceptors like the quinones. Whilst it is possible there is some stabilization of the binding by charge transfer complexing, there is as yet little proof of this and it must remain as a speculation.

There is little dissension about the ability of aromatic amino acids to behave as π-electron donors with suitable acceptors. The behaviour of aliphatic amino acids in acting as n donors is hotly disputed but in the author's opinion the evidence for formation of complexes with quinones is overwhelming, although only under certain conditions and the complexes can pass over to compounds. The conditions under which the complexes are stabilized, i.e. in regions of high polarity, may be more biologically relevant than those conditions in which the aliphatic amino acids and quinone interact rapidly and chemically.

Histamine undergoes an interesting set of reactions which may well have considerable biological interest. At pH 7 it forms complexes with quinones in a similar manner to the amino acids with the donated electron coming from a lone pair on the amino nitrogen (Moxon and Slifkin, 1972a). However Shinitzky and coworkers (1966) have demonstrated that histamine behaves as an acceptor to indoles at pH 6. At this pH the histamine is in fact protonated, thus blocking the side-chain lone-pair electron from which the donating electron came in the case of quinone complexes, and the imidazole ring now functions as the electron acceptor. The role of pH in mediating the charge transfer complexing properties of biological molecules has not been discussed by other authors but in the light of these experiments, pH changes, even very small ones, could have a marked regulatory role. Amines are very

Table 6.2   Charge transfer parameters of amino acid analogue complexes

Amino acid analogue complex	$K(\text{l. mol}^{-1})$	$\Delta H(\text{kcal mol}^{-1})$	$\Delta S(\text{e.u.})$	Temperature (K)	References
N-Acetyl-N',N'-dimethyl-p-phenylene-diamine-N-(3,5-dinitro-benzoyl)sarcosine dimethylamide	0.89	−1.40	−4.87	313	a
Interchain charge transfer complex on polysarcosine chain					
chain length = 6	0.167	−1.4	−8.3	313	a
chain length = 16	0.04	−0.43	−7.8	313	a
chain length = 21	0.25	−0.26	−8.2	313	a
Poly(L-tryptophan)–imidazolium	ca. 2				b
Poly(L-histidine HCl)–indole	2.2			295	c
Human serum albumin–benz(a)anthracene	1.56×10^5			298	d
Human serum albumin–benzo(a)pyrene	1.46×10^5			298	d

[a] Sisido *et al.* (1977) in chloroform.
[b] Palumbo *et al.* (1977) in 2,2,2-trifluoroethanol.
[c] Palumbo and Peggion (1975) in water.
[d] Ma *et al.* (1977) in pH 7 buffer.

common in biological systems and protonation–deprotonation effects have been very well studied. As has been shown the lone-pair electron on the amine nitrogen can take part in CT interaction but obviously only when the amine is free and not protonated. For amino acids the amount of unprotonated amine at pH 7 is very small (*ca.* 1 part in 1000) but as the pK_a's of these substances are in the region of 9–10, relatively small increases in pH will cause large increases in the amount of free amine group.

6.4   Nucleic Acid Bases

There has been much study into the charge transfer complexing properties of the nucleic acid bases. Much of this work has been reviewed recently by the author (Slifkin, 1973). Only some salient points will be dealt with here. There have been suggestions that complexes are formed between nucleic acid bases and the aromatic hydrocarbons, chloranil, acridine, aquacobalamin, haematoporphyrin, iodine, riboflavin, steroids, and amino acids. In all these cases the nucleic acid bases are presumed to be acting as the electron donor. Probably the most convincing evidence arises from correlation studies of association constants or some similar parameter with the ionization potential of the donor or the electron acceptor of the acceptor. In the section on flavins, the work of Hush and Cheung (1975) on purine–flavin complexes is described. In earlier years a lot of emphasis was put on the ability of the nucleic acid bases to solubilize various electron donors and acceptors but with a few exceptions (Slifkin, 1971a) attempts to correlate the solubilizing effect with various charge transfer parameters, failed. However this almost certainly reflects on the solubility experiments rather than the correlation. As pointed out for example by Hayman (1962) one needs to measure the difference between the solubility caused by charge transfer complexing and the solubizing effect of introducing the nucleic bases into solution, a very difficult exercise.

A considerable amount of work has been carried out on X-ray crystallography of various purine complexes. In many cases particularly those of 8-azaguanine, guanine, and guanosine the stacking of the molecules in the crystal is very reminiscent of the stacking in conventional organic charge transfer complexes which has led to the suggestion of MacIntyre (1965) that 8-azaguanine form a self charge transfer complex.

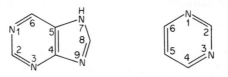

Figure   6.3 Chemical   formulae   of
purine and pyrimidine

Other evidence for the association between purines and pyrimidines comes from the work of Hélène and Michelson (1967) who from a study of the emission of various nucleic acid bases at 77 K concluded that association took place between the bases, the strength of interaction being purine/purine > purine/pyrimidine > pyrimidine/pyrimidine. It was suggested that charge transfer complexing had some part to play in the interaction. Montenay-Garestier and Hélène (1970) have also obtained evidence for the complexing between different protonated forms of the bases from emission spectra and luminescence titration studies. They concluded that charge transfer complexing between cytidine and its cation leads to stacked structures. The cytidine acts as the donor and the cation protonated at the $N_{(3)}$ position acts as the acceptor. In addition polycytidylic acid is double stranded at pH's close to the pK_a point, i.e. where neutral and protonated forms exist in equal concentration, which gives rise to the interesting speculation that it is charge transfer between the two forms which stabilizes the double-strand structure.

Several studies also suggest that the purines and pyrimidines can form weak complexes with amino acids. Evaporates from mixed purine–amino acid solutions are coloured and possess both emission and excitation spectra different from those of the components. Similar effects are observed in frozen solutions (Slifkin, 1971a). The infrared spectra of these evaporates show changes which are reminiscent of weak charge transfer complexes. Tryptophan–purine complexes exhibit interesting spectra in that the IR spectrum of the tryptophan in the complex is that of the neutral (non-zwitterionic form) suggesting that the complexing is an $n-\pi$ complex involving a lone-pair electron of the amino group of the amino acid, as donor.

Quite recently two conventional spectroscopic studies have been carried out on the interaction of nucleic acid bases with catechol and with epinephrine (Al-Obeidi and Borazan, 1976a, b). Changes in the absorption spectra were observed, and in the latter case presumed charge transfer transition as was observed. From these changes, equilibrium constants and other thermodynamic parameters were derived see Table 6.3. These were all in the range of values found for conventional charge transfer complexes. However in these studies it is assumed that the nucleic acids are the acceptors, instead of acting as donors, which is more usual.

Lung-Nan Lin and coworkers (1974) have studied the interaction of 9-cyclohexyladenine with the well-known acceptor iodine in organic solvents.

The systems behave in all respects like conventional charge transfer complexes. The crystalline structure of the complexes (van der Helm *et al.*, 1973; van der Helm, 1973) is explicable in terms of stabilization by charge transfer by an *n* electron from a ring nitrogen. The spectroscopic properties of the complex are regular, enabling association constants and dissociation enthalpies to be obtained. Similar results have been obtained using solubility methods. A comparison with the pyridine–iodine complex shows that cyclohexyladenine forms a similar but somewhat stronger complex than pyridine.

Pyrimidine dimers can be photomonomerized by both indoles and proteins (Chen *et al.*, 1976) and it appears that for the indoles this ability is related to the electronic properties. Those with high electron-donating ability are good photosensitizers in this respect, whilst those containing electron-withdrawing substituents are inactive. This has led to the postulated mechanism for the photosensitized splitting thus:

$$D \xrightarrow{h\nu} D$$

$$D + A\text{---}A \rightarrow (D \cdot A\text{---}A) \rightarrow (D^+\text{---}A^-\text{---}A)$$

$$(D^+\text{---}A^-\text{---}A) \rightarrow (D^+\text{---}A^- \cdot A) \rightarrow (D^+\text{---}A^-) + A$$

$$(D^+\text{---}A^-) \rightarrow D \cdot A \rightarrow D + A$$

where D is the donor or sensitizer and A—A is the dimer. It is seen that several different charge transfer complexes are included in this scheme. In the reactions charge transfer complexes have been observed between tryptophan and other indoles and a number of nucleic acid bases. It should be emphasized that the charge transfer complexes postulated in these reactions are formed in the excited states by excited molecules and are thus rather different from the normal charge transfer complexes.

A detailed study has been carried out by Morita (1974) on the charge transfer complexes formed between tryptophan and ATP or its analogues. Charge transfer transitions were seen and the author was able to derive full parameters of these 1:1 complexes. The stability of different nucleotide–tryptophan complexes varied in the order:

$$IMP > CMP > AMP > TMP > UMP.$$

No difference was observed between the complexing ability of AMP and ATP. One particularly interesting point to emerge from this study was that in the presence of ethylene glycol, although the stability of the complex increased as measured by the enthalpy of dissociation, the extinction coefficient of the complex decreased, thus making the complex more difficult to observe. This re-emphasizes the point made by the writer in many previous publications that the apparent lack of a charge transfer transition may not mean that a complex does not exist but merely that its absorbance is low or perhaps that the band is obscured by the absorption of the components of the system.

Sharples and Brown (1976) have studied the correlation of the base specificity of DNA-intercalating ligands with certain physicochemical

Table 6.3   Kinetic and thermodynamic parameters of purine and pyrimidine charge transfer complexes

Complex	K(l. mol^{-1})	ΔH(kcal mol^{-1})	ΔS(e.u.)	Temperature (K)	References
Adenine–catechol	1.59	−1.015	−2.6	291	a
Cytosine–catechol	1.00	−3.295	−11.3	291	a
Thymine–catechol	1.40	−2.541	−8.1	291	a
Uracil–catechol	0.51	−2.842	−11.0	291	a
Uridine–tryptophan	0.088			298	b
Uridine–tryptophan	0.17			298	c
ATP–tryptophan		−7.02	−17.7	296	d
AMP–tryptophan		−6.81	−18.5	296	d
IMP–tryptophan		−7.94	−21.3	296	d
CMP–tryptophan		−7.38	−23.8	296	d
TMP–tryptophan		−5.63	−15.3	296	d
UMP–tryptophan		−3.57	−8.87	296	d
ATP–nactryptophan-amide		−4.61	−10.5	296	d
AMP–nactryptophan-amide		−4.83	−11.5	296	d
AMP–tryptophan with ethylene glycol		−12.2	−38.6	296	d
9-Cyclohexyladenine–iodine		−9.15	21.6	296	e
Adenine–epinephrine	0.52	−3.76	−14.5	291	f
Thymine–epinephrine	0.73	−2.541	−9.3	291	f
Uracil–epinephrine	0.69	−6.936	−24.5	291	f
GMP–proflavin	9.3×10^5				g
GMP–acridine orange	14.0×10^5				g
GMP–pyronine G	18.7×10^5				g
GMP–thionine	12.7×10^5				g
GMP–methylene blue	22.8×10^5				g
GMP–toluidine blue	32.6×10^5				g
GMP–neutral red	3.1×10^5				g
GMP–phenosafranine	18×10^5				g
DNA–proflavin	8.8×10^4				h
DNA–acridine orange	2.0×10^5				h
DNA–pyronine G	1.3×10^4				h
DNA–thionine	3.7×10^4				h
DNA–methylene blue	1.2×10^4				h
DNA–toluidine blue	2.4×10^4				h
DNA–neutral red	6.6×10^4				h
DNA–phenosafranine	8.1×10^3				h

[a] Al-Obeidi and Borazan (1976a) in 0.1 N HCl.
[b] Chen et al. (1976) in H_2O.
[c] Dimicoli and Hélène (1971) in 1 M sodium cacodylate.
[d] Morita (1974) in pH 8 buffer.
[e] Lin et al. (1974) in CCl_4.
[f] Al-Obeidi and Borazan (1976b) in 0.1 N HCl.
[g] Sharples and Brown (1976) in pH 7 buffer. These are relative affinities of binding.
[h] Müller and Crowthers (1975) in pH 7. These are binding affinities.

parameters. In particular they have looked at bathochromic shifts of the ligands in the presence of GMP. From this data they have calculated affinity constants for GMP and correlated this with binding constants for calf thymus DNA. They interpret the binding with GMP as being due to the formation of charge transfer complexes. However from their correlation with DNA binding they feel that binding affinity of the ligands to DNA is due in some part to hydrophobic factors, although base specificity seems to be almost wholly determined by charge transfer affinity.

There seems little doubt that the nucleic acid can form charge transfer complexes in model systems acting both as donors and acceptors with the appropriate partners. The results for nucleic acids are summarized in Table 6.3.

6.5 Flavins

There is an enormous literature on flavin complexes (Slifkin, 1971a) with different and controversial views being expressed as to the role or even existence of charge transfer forces. The flavins would appear to be ideal candidates for both donors and acceptors in complexing. They exist in nine possible forms (Figure 6.4), these are three fully oxidized forms, the quinone in the normal, protonated, and deprotonated form depending on pH; the semiquinone which similarly exists in neutral, protonated, and deprotonated forms, and the hydroquinone also existing in the three forms. Quinones are usually good acceptors whereas hydroquinones are good donors. Presumably the semiquinone will have both moderate acceptor and donor properties. Complexes have been demonstrated or postulated as existing between flavins and indoles, phenols, purines, pyrimidines, aromatic hydrocarbons, coenzymes, and flavins, i.e. dimerization (Slifkin, 1971a). The evidence in general is based on spectral changes on complexing and also, more recently, on correlation data. A particularly good correlation has been shown to exist between the association constants of some purine–riboflavin complexes and the calculated highest occupied molecular orbitals of the purines (Slifkin,

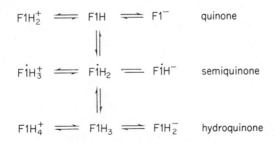

Figure 6.4   The nine states of flavin

1971b). Hush and Cheung (1975) have shown also that a very good correlation exists between the experimentally determined ionization potentials of the bases and the association constants with riboflavin, which they conclude is evidence of charge transfer complexing; moreover the average amplitude of the dative waveform in the ground state of these complexes is 0.3.

A very detailed study has been undertaken by Bruice and Yano (1975) on radical mechanisms for the reduction of carbonyl compounds, including such well-known charge transfer acceptors as p-benzoquinone, by 1,5-dihydro-5-methylflavin. It is shown that in general one-electron transfer takes place to the carbonyls to yield the oxidized flavin. Although the authors express the reaction in the general form

$$FlH_2 + \; \diagdown C=O \rightleftharpoons FlH^{\cdot} \; \diagdown C^{\cdot}(O^-)$$

the authors, specifically for the case of p-benzoquinone, propose the formation of an intramolecular complex with the reduced flavins together with intracomplex electron transfer which would appear to be a case of conventional charge transfer complexing, albeit expressed in slightly different terms. The equilibrium constant for the complex is of the order of 2×10^3 l. mol^{-1}. This value is rather high for most aqueous systems (Higuchi and Connors, 1965) although Yagi et al. (1968) have found a value of 2×10^4 l. mol^{-1} for the complex of cinnamate with flavoprotein and Yagi et al. (1965) have found equilibrium constants of this order for the interaction of various nitrophenols with FAD. It must also be pointed out that the flavin in this case is acting as the donor whereas most of the previous studies (see Slifkin, 1971a) concern the behaviour of flavins as acceptors.

Some quite different kind of evidence has been used to show the existence of charge transfer complexing between a flavin and a purine. Scarborough et al. (1976) have prepared a crystal of the lumiflavin–2,6-diamino-9-ethylpurine complex. Not only is this a deep-red colour, reminiscent of a charge transfer transition, but more important there are extensively overlapped flavin–flavin and flavin–adenine stacking interactions in which are contacts much closer than the van der Waals distances. This has been postulated by Prout and Wright (1968) as a property of charge transfer complexes. However the overall contribution to the binding from the charge transfer complexing is not believed to be predominant as other flavin–purine complexes do not exhibit such effects (Hamlin et al., 1965; Kyogoku et al., 1966). However, very similar effects are observed in other flavin complexes (Leijonmarck and Werner, 1970; Karlsson, 1972). It is suggested that charge transfer complexing might be important in causing minor reductions in the chemical activation energies thus leading to acceleration of the reaction. Flavin–flavin charge transfer interaction might be important in flavoenzymes which use multiple non-equivalent flavin coenzymes.

Similarly an intramolecular charge transfer complex in FAD could have a catalytic role in certain flavoenzyme-mediated reactions (Massey and Ghisla, 1976). These authors do not however believe, in common with Scarborough et al. (1976), that charge transfer complexes play a role in the binding of flavins to a substrate or a coenzyme.

It is interesting that those flavin–adenine complexes, which do not apparently exhibit charge transfer complexing, stack in an irregular manner (Voet and Rich, 1971; Weber et al., 1974), whereas the ones discussed stack regularly. Does the regular stacking arise from the presence of the charge transfer complexes or does the charge transfer complexing arise because the regular stacking brings donor and acceptor orbitals in sufficient proximity for charge transfer to take place? At the present moment there does not seem to be any answer to this question, but if the former were true then clearly charge transfer complexing plays a significant role in the structure of these flavin–adenine complexes.

Although our discussion has been concerned up to this point with π interaction there is also some evidence that localized charge transfer can operate with some flavins on complexing with purines. It has been shown that substitution of the 6-position in the purine ring or increasing alkylation of the amino group in the 6-position has a marked effect on the complexing of purines with riboflavin (Tsibris et al., 1965; Chassy and McCormick, 1965). The infrared spectrum of the riboflavin–9-ethyladenine complex shows band shift which could arise from localized charge transfer (Kyogoku and Yu, 1969). If this phenomenon were true then in considering the effect of charge transfer in structures of flavin, departures from the behaviour expected in terms of π donation only, might be explicable in terms of localized transfer between regions of particularly high and low electron density.

The solvent has been shown to have a very important influence on the complexing of flavins with various acceptors and donors (Harbury et al., 1959). In the examples quoted above we confined ourselves to complexes in aqueous media it being felt that these are more relevant to biochemical systems than interactions in organic solvents. In water, hydrogen bonding can occur with the flavin, and Harbury et al. (1959) have suggested that a hydrogen-bonded flavin is a better electron acceptor than the unbonded flavin, as hydrogen bonding to the isolloxazine ring of the flavin renders the ring more electropositive than before. In any speculation of the role of charge transfer complexing in these systems, the immediate environment must also be taken into account.

Porter et al. (1977) have looked at the structure of flavin–nicotinamide biscoenzyme and models for flavin–nicotinamide interactions using X-ray crystallography. Little evidence was found to suggest that charge transfer forces were involved in these structures in the crystalline form. This is quite

Table 6.4   Parameters of flavin complexes

Complex	$K(\text{l. mol}^{-1})$	$\Delta H(\text{kcal mol}^{-1})$	$\Delta S(\text{e.u.})$	Temperature (K)	References
FAD–L-tryptophan	22.3	−3.0	−3.9	298	a
FAD–D-tryptophan	20.5	−2.8	−3.5	298	a
7,8-Dimethyl–1,10-ethylene-isoalloxazine–triphenylphosphine	1.75×10^3			298	b
7,8-Dimethyl ethyoxy 9-methyl-isoalloxazine–triphenylphosphine	1.17×10^4			298	b
1,7,8,9-Tetramethyl-isoalloxazine–triphenylphosphine	7.85×10^3			298	b
7,8,9-Trimethyl-isoalloxazine–triphenylphosphine	3.28×10^2			298	b
FAD–sulphite	0.396			298	c
FMN–sulphite	0.52			298	c
6,7,8-Dimethyl-isoalloxazine–sulphite	0.625			298	c
10-Methyl-isoalloxazine–sulphite	6.58			298	c
3,7,10-Trimethyl-8-carboxy-isoalloxazine–sulphite	8.77			298	c
2-Thio-7,8,10-trimethyl-isoalloxazine–sulphite	26			298	c
3,10-Dimethyl-7,10-dichloro-isoalloxazine–sulphite	69			298	c
1,10-Dihydro-7,8-dimethyl-isoalloxazine–sulphite	7.57×10^3			298	c

[a] Takahashi and Maeda (1974) in pH 7 buffer.
[b] Müller (1972) in acetonitrile.
[c] Müller and Massey (1969) in pH 7 buffer.

contrary to the views put forward by Blankenhorn (1975) for similar systems. However Blankenhorn's work was with these molecules in solutions. It is more than likely that whereas charge transfer complexing could be operative in solution enabling face-to-face structures to be formed, in crystalline form other forces could align the components of these complexes in such a way that electron-donating and -accepting regions are far removed from each other. The crystal structures shown by Porter *et al.* (1977) are not conducive to the formation of charge transfer complexes.

Takahashi and Maeda (1974) have found different CT parameters between the complexes of FAD with L- and D-tryptophan showing the importance of steric factors in these interactions. From optical activities and polarographic data they find that FAD in water has a dimerization constant of $22.4 \, \mathrm{l. \, mol^{-1}}$.

Müller (1972) has prepared some flavin–phosphine derivatives. These form weak complexes whose spectra are not greatly different from those of other flavin complexes however the association constants are particularly large but not more so than amine–quinone complexes.

This parallels earlier work by Muller and Massey (1969) on the interaction of various flavins with sulphite. The association constants appear to correlate rather well with the oxidation–reduction potentials of the corresponding flavoquinones.

To summarize this section it does appear that not only has the ability of the flavins to form charge transfer complexes been amply demonstrated both as donors and acceptors but these interactions do have some significance in explaining various biological interactions involving the flavins.

6.6 Phenothiazines and Tranquillizers

There has been a lot of interest in the physicochemical properties of tranquillizer drugs and it has been suggested on several occasions that they are effective because of their low ionization potentials or their ability to form charge transfer complexes readily (Bloor *et al.*, 1970). There has been a lot of reported work on the charge transfer complexing of tranquillizers such as phenothiazine and chlorpromazine (Figure 6.5).

Recently Dwivedi *et al.* (1975) have examined charge transfer complexes of these compounds in some detail. Rate data and thermodynamic data have been evaluated for complexes between phenothiazine and a whole range of conventional electron acceptors. The results are typical of those found with normal charge transfer complexes except that in many cases the complexes gradually change to compounds. Thus the intensity of the charge transfer band varies as the enthalpy of dissociation and the position of the charge transfer also varies with the enthalpy of dissociation but in reverse order. Similar results were found by Barigand and coworkers (1970) in their study

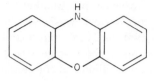

Phenothiazine

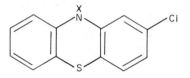

Figure 6.5 Chemical formulae
of phenothiazine and promazine

of the interaction of phenothiazine with chloranil. By comparing the behaviour of other aromatic and amine donors with the same acceptors, the
authors conclude that in the case of phenothiazine it behaves as a π-electron
donor rather than a lone-pair electron donor. The results for chlorpromazine are rather different, in most cases charge transfer bands are not
observed but instead the radical cation is observed. It is suggested that an
n-π complex might be formed initially in the interaction of chlorpromazine
from the amine side chain. The authors feel that the different behaviour of
the two closely related drugs may not be related to their tranquilizing ability
but the tranquilizer properties of these drugs may be related to their ability
to form radical cations.

Bhat (1976) has also studied the complexing of phenothiazine and *N*-
methylphenothiazine with some common acceptors and has produced association constants and enthalpies of dissociation. He finds that *N*-
methylphenothiazine is a weaker donor than the parent compound. Although one might suppose that the effect of adding the electron-donating
methyl group to phenothiazine would increase the charge transfer complexing ability the opposite appears to be the case, a result predicted by Bloor *et
al.* (1970). The explanation appears to be that the addition of the methyl
group to the compound, causes steric interactions resulting in increased
folding of the molecule and less interaction of *n* electrons with the π-
electron system of the molecule.

Guttmann and coworkers in a series of papers have studied the electrical
properties of chlorpromazine and other psychotropic compounds (Brau *et
al.*, 1972 and references therein). Unlike most of the studies reported herein
they have usually used conductivity titration. Of particular interest are the

studies they have carried out on complexes of these materials (Guttmann *et al.*, 1974). Conductivity titrations of some of these complexes point to some interaction which can be identified spectroscopically as being charge transfer complexing. Evidence is presented for complexing between chlorpromazine and acetylcholine, serotonin and 6-hydroxydopamine. It is suggested that *in vivo* interaction of chlorpromazine with 6-hydroxydopamine could involve the formation of a charge transfer complex on the surface of a biological membrane which acts as an electrode particularly at the postsynaptic membrane onto which chlorpromazine is specifically absorbed.

It is suggested that the therapeutic action of chlorpromazine might involve a fast charge transfer complex interaction with 6-hydroxydopamine. This complex then can act to prevent the accumulation of the free hydroxydopamine and the ensuing attack of the nerve endings.

Table 6.5   Phenothiazine and related compounds

Complex	K(l. mol^{-1})	ΔH(kcal mol^{-1})	ΔS(e.u.)	Temperature (K)	References
Phenothiazine–TCNQ	4.5	−4.2			[a]
Phenothiazine–TCNB	1.8	−1.8			[a]
Phenothiazine–TCNE	5.8	−5.6			[a]
N-Methylphenothiazine–TCNQ	1.5	−3.0			[a]
N-Methylphenothiazine–TCNB	1.1	−1.5			[a]
N-methylphenothiazine–TCNE	1.8	−1.8			[a]
Phenothiazine–naphthaquinone	0.4	−1		293	[b]
Phenothiazine–trinitrobenzene	1.6	−1.3		293	[b]
Phenothiazine–chloranil	3.5	−3.4		293	[b]
Phenothiazine–bromanil	3.3	−5.0		293	[b]
Phenothiazine–iodanil	2.7	−4.5		293	[b]
Atropine–iodine					
in chloroform	1920			293	[c]
in carbon tetrachloride	2300			293	[c]
Quinone–iodine (in chloroform)	2460			293	[c]
Brucine–iodine (in chloroform)	4100			293	[c]
N-Butylscopol- ammonium bromide– iodine	4430			293	[c]

[a] Bhat (1976) in CH_2Cl_2.
[b] Dwivedi *et al.* (1975) in CH_2Cl_2.
[c] Taha *et al.* (1974).

Crystallographic studies on ion-radical molecular complexes of phenothiazine have been reported by Singhbhandu and coworkers (1975). These complexes are of 1:1 stoichiometry and there is some evidence from the donor–acceptor distance of 3.36 Å that charge transfer interaction is important in the stabilizing of these structures.

Other drugs which appear to form strong charge transfer complexes are the alkaloids. Spectroscopic studies have been carried out by Taha and Gomaa (1976) on the charge transfer complexes formed with iodine in ethylene dichloride. The technique was utilized to analyse mixtures of weakly absorbing and strongly absorbing alkaloids.

6.7 Porphyrins

The porphyrins (Figure 6.6) have been the object of much study and speculation, however, unlike some of the other groups of biomolecules it seems to be generally agreed that they do form weak molecular complexes. Spectroscopic data has been presented for the interaction of porphyrins both with electrons donors such as indoles, purines, and amino acids and with electron acceptors such as trinitrobenzene and trinitrofluorenone in both organic and aqueous solvents (Slifkin, 1971a). Enthalpies of dissociation are usually in the range of 3–8 kcal mol^{-1}. The question of whether these are charge transfer complexes or not has been the subject of some discussion.

The spectra of the porphyrins are usually very strongly coloured so that charge transfer bands in the visible spectra would be difficult to observe. The general effect of complexing is observed in the shift of the absorption spectrum of the porphyrin. However the enthalpies of dissociation do correlate quite well with the presence of electron-withdrawing or electron-donating groups. Thus etioporhyrin with eight electron-donating groups on the pyrrole ring is a stronger complexer with trinitrobenzene than tetraphenylporphyrin with only four, which is in turn weaker than Zn tetraphenylporphyrin. One would expect that zinc in the form of Zn^{2+} would

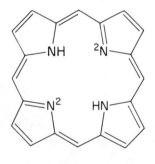

Porphyrin

Figure 6.6   Chemical formula of
porphyrin

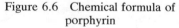

act as an electron-withdrawing group on the ring. A similar conclusion has been drawn by Mehdi *et al.* (1976) in a study of various porphyrins with some common acceptors in chloroform. An interesting point made by these authors is that the complexing properties of various metalloporphyrins are not greatly affected by the type of metal ion, although the absorbance of the complex varies very greatly. It would appear that the metal ion has only a small effect on the interaction of the ring with the acceptor even though there are obviously marked interactions between the porphyrin ring and the metal ion. Solvation effects are found to have an important influence on the stability constants of these complexes. Unfortunately there appears to be little data on the effect of solvent on the enthalpies of dissociation or perhaps more important on the entropy changes of complexes. Large changes in this latter parameter would give weight to the idea of Mehdi *et al.* (1976) that the interaction involves the solvent thus $AS + D.S_b \rightleftharpoons A.D.S._x + yS$ where S represent solvent. Such ideas have already been suggested to explain the somewhat contradictory conclusions put forward for amino acid–quinone interactions.

NMR studies of the interaction of metalloporphyrins also confirm the formation of 1:1 molecular complexes with various π-electron donors and acceptors (Fulton and LaMar, 1976). They confirm the conclusions of Mehdi *et al.* (1976) that the complexes arise from $\pi-\pi$ interaction and that the metal ion has little part to play in the interaction. They show clearly the ability of these porphyrins to act equally well with the appropriate partner as both donors and acceptors. However it is also demonstrated that these interactions could play some part in modulating the electronic structure of haem iron in such systems as haem–protein linkages.

The one problem of the missing charge transfer band has been answered in two ways. Certain authors have claimed to have observed such bands, although the majority have not. The most convincing study has observed a charge transfer band in an etioporphyrin complex at room temperature at 1400 nm (Siderov, 1974), a region not covered by most spectroscopic studies. This again reinforces the viewpoint that the failure to observe charge transfer bands in biological systems may be, that due to their low ionization potentials, they are to be found in the near infrared, not in the visible as is more common for organic charge transfer complexes, a view first put forward by the late E. D. Bergmann (1968).

Gasco (1976) has carried out a conventional spectroscopic study on the complexing of haematoporphyrin with pteridine derivatives over the pH range 7.5–10. In this case the pteridines are presumably acting as electron acceptors with the haematoporphyrin as the electron donor. The association constants cover the range 55–1240 and the extinction of the new absorption at 502 nm shows a slight variation of 9220–12,230. Although the author calls this the charge transfer band, no spectra are given and in view of the

Table 6.6 Kinetic and thermodynamic parameters of porphyrin charge transfer complexes

Complex	K(l. mol^{-1})	ΔH(kcal mol^{-1})	ΔS(e.u.)	Temperature (K)	References
p-CH$_3$-TPPCo– trinitrobenzene	17.5	−5.7	−14	298	a
p-CH$_3$-TPPCo– trinitrotoluene	3.3			298	a
p-CH$_3$-TPPCo– dinitrobenzene	2.9			298	a
p-CH$_3$-TPPCo– tetramethyl- phenylenediamine	10.2			298	a
Haematoporphyrin– folic acid					
pH 7.5	1240			298	b
pH 9	207			298	b
pH 10	65			298	b
Haematoporphyrin– methorexate					
pH 7.5	233			298	b
pH 9	165			298	b
pH 10	167			298	b
Haematoporphyrin– folinic acid (pH 7.5)	55			298	b
Dimethylprotoporphyrin– dinitrobenzene	14.6			298	c
Dimethylprotoporhphyrin– trinitrobenzene	118			298	c
Dimethylprotoporphyrin– trinitrofluorenone	1778			298	c
Dimethylprotoporphyrin– tetranitrofluorenone	1288			298	c
Dimethylprotoporphyrin– trinitrofluorenone					
in benzene	1318			298	c
in chloroform	1778			298	c
in 80% cyclohexane/ 20% benzene	25,118			298	c
in 80% cyclohexane/ 20% chloroform	42,657			298	c

[a] Fulton and LaMar (1976) in CDCl$_3$.
[b] Gasco (1976).
[c] Mehdi et al. (1976).

previous discussion, it is likely that this is merely a perturbation on the absorption of haematoporphyrin.

It is believed that these weak interactions which on the balance of the evidence appear to be charge transfer interactions are important in the respiratory transport chain, microsomal oxidations, and as mentioned haem protein.

6.8   Charge Transfer Interaction and Biological Regulation

Up to now we have dealt with various molecules and attempted to see what part they might play in charge transfer complexing. For many years different authors have been discussing the role or possible role of charge transfer complexing in biology. Most notable among these is Szent-Györgyi (1957, 1960, 1968, 1972, 1973, 1974, 1977) who for over two decades has been urging biochemists to see biochemistry not as the interaction of chemicals to produce other chemicals but in terms of physical and electronic changes and interactions. Whilst impossible to review all his manifold contribution to the field certain of his ideas will be presented here. Szent-Györgyi was apparently the first person to suggest that proteins could act as semiconductors. If so then clearly the role of charge transfer complexing becomes important. Just as with conventional semiconductors, the addition of charge donors and acceptors can produce much smaller energy gaps and activation energies, then presumably protein, biological charge donors and acceptors, such as the ones reviewed herein, could alter considerably the semiconducting properties of protein.

Szent-Györgyi has suggested that glyoxyl derivatives which are found in all living cells act as electron acceptors for proteins thus being responsible for converting cells from the anaeorobic state to the aerobic state.

It can be demonstrated that normally colourless protein becomes deeply coloured when incubated with aldehydes or glyoxyls and also produces ESR signals. This it is believed is evidence of the formation of a charge transfer complex with the protein as donor. The donated electron is postulated as being an n electron on the nitrogen of the amino group in the peptide bond.

It is stated by Szent-Györgyi that the interaction of proteins with carbonyls like glyoxal is not some rare isolated event but a prime mover in the main signs of life, motion, reflexes, and secretions.

Further work has shown that the charge transfer interaction can be catalysed both by sulphur and oxygen, charge transfer between amines and dicarbonyls by the SH group of glutathione, and charge transfer between amines and unsaturated aldehydes by molecular oxygen. Indeed it may be that molecular oxygen has not only the vital biological role as the final electron acceptor in oxidation but may be just as important is its function in catalysing charge transfer in living cells. Thus death by oxygen removal might be due to the cessation of the latter function as much as the former.

These ideas have been extended to explain the creation of cancerous cells. The cells become cancerous when they are in the proliferating anaerobic state. This is the state where there is no charge transfer as there are no carbonyl molecules present to limit proliferation in the cell. Some evidence for this is that cancerous cells are colourless, testifying to the lack of charge transfer in them whilst normal cells of the same organs are coloured. The addition of dicarbonyls to cancerous cells restores the deep colour of the corresponding normal cells. There is some evidence that dicarbonyls act as carcinostatic agents.

Szent-Györgyi (1974) believes that there are three different forms of charge transfer complex each with its own role to play in biology.

The first kind he calls charge transfer in the excited state which corresponds to charge transfer after absorption of light, i.e. the normal charge transfer reaction. The second kind he calls charge transfer in the ground state corresponds to spontaneous transfer in the ground state, i.e. a salt-like structure. This is the strong charge transfer complexing of organic chemistry in which even in the ground state substantial amounts of charge are donated (Mulliken and Person, 1969). He also believes that there is an intermediate situation where the molecules have similar donor and acceptor strength so that there are two charge transfers simultaneously. As a general rule good donor properties are accompanied by bad acceptor properties, although there are obvious exceptions like the porphyrins discussed herein. Certainly very large complex biological molecules could and almost certainly do possess regions with both good acceptor and good donor ability. Szent-Györgyi believes that such complexes would not be detected as they would exhibit neither ESR or CT transitions. This leads us into the realms of metaphysics rather than biophysics. Back donation is not unknown in organic chemistry, the effect appears to shift the charge transfer transition to higher energies, i.e. into the blue, where it might of course be obscured by locally excited transitions (Mulliken and Person, 1969).

Perhaps the most interesting idea put forward by Szent-Györgyi is that evolution came about because of charge transfer complexing. Proteins in the primordial soup were dielectric in nature, undifferentiated, semiliquid, and proliferated wildly without regulation. It was the introduction of charge acceptors into the primordial soup which enabled the proteins to act as semiconductors by the abstraction of electrons from the filled conduction bands, thus converting the proteins into semisolid regulating proteins.

In his most recent work, Szent-Györgyi (1977) has found that charge transfer reactions of methylglyoxal are catalysed by sulphydryl and this may play a central role in the electronic desaturation of proteins.

6.9 Concluding Remarks

In this rather brief survey, the possible role of charge transfer interactions in different biological situations has been examined. Clearly the attraction for these interactions apart from the novelty is that they represent an intermediate regime between conventional chemical interactions and casual interactions caused by diffusion. Furthermore they are specific in direction and unsaturated. In addition to all that they involve electronic charge with its connotation of modern technology. The speculations about charge transfer have ranged from very specific roles in isolated systems to the view that it is fundamental to the whole process of evolution and maintenance of life. One of the more interesting speculations has not been touched on in this review.

Reid (1957) has pointed out that the highly organized arrays occurring in the mitochondria consists of molecules with both electron donor and acceptor properties. They might operate by transferring from one adjacent molecule to the next in a zipper action. This would explain the rate of charge transport in the phosphorylating chain which is not apparently explicable in terms of chemical interaction, which is too fast or a diffusion mechanism, which is too slow.

The author does not feel that he has offered conclusive proof for the role of charge transfer interactions in biological systems. However ·a case has surely been made that firstly, charge transfer complexing is a real identifiable process and not merely a loose description of a phenomenon that could equally well if not better be described in other pre-existing terms, and furthermore that it does occur with biochemicals. Consequently, it is felt that charge transfer interactions are significant in biology and discussions of biological function must not overlook them. However, it is very easy to call any ill-understood system a charge transfer complex when simple experiments of the kind outlined would confirm their existence or otherwise. Unfortunately too many workers in the past have been content to call any new colour in their system charge transfer complexes without the slightest evidence that it was so, thus accounting for the somewhat jaundiced eye that many scientists have cast upon reports of charge transfer.

With the increasing amount of evidence being published on biological systems, it is to be hoped that a less partisan and more rational view will be taken of this subject.

6.10 References

Al-Obeidi, F. A. and Borazan, H. N. (1976a). *J. Pharm. Sci.*, **65**, 882.
Al-Obeidi, F. A. and Borazan, H. N. (1976b). *J. Pharm. Sci.*, **65**, 982.
Ash, R. P., Herriot, J. R., and Deranleau, D. A. (1977). *J. Amer. Chem. Soc.*, **99**, 4471.

Barigand, M., Orszach, J., and Tondeau, J. J. (1970). *Bull. Soc. Chim. Belg.*, **79,** 177.

Bergmann, E. D. (1968). *Molecular Associations in Biology* (Ed. Pullman, B.) Academic Press, New York.

Bhat, S. N. (1976). *Ind. J. Chem.*, **14A,** 791.

Bhatnaga, S. S., Verma, M. R., and Kapur, P. C. (1934). *Ind. J. Phys.*, **9,** 131.

Blankenhorn, G. (1975). *Eur. J. Biochem.*, **50,** 351.

Bloor, J. E., Gilson, B. R., Haas, R. J., and Zirkle, C. L. (1970). *J. Med. Chem.*, **13,** 922.

Brau, A., Farges, J. P., and Gutmann, F. (1972). *Electrochim. Acta*, **17,** 1803.

Briegleb, G. (1961). *Electronen-Donator-Acceptor-Complexe*, Springer-Verlag, Berlin.

Bruice, T. C. and Yano, V. (1975). *J. Amer. Chem. Soc.*, **97,** 5263.

Chassy, B. M. and McCormick, D. B. (1965). *Biochemistry*, **4,** 2612.

Chen, J., Huang, C. W., Hinman, L., Gordon, M. P., and Deranleau, D. A. (1976). *J. Theoret. Biol.*, **62,** 53.

Dewar, M. J. S. and Thompson, C. C., Jr. (1966). *Tetrahedron Suppl.*, **7,** 97.

Dimicoli, J. L. and Hélène, C. (1971). *Biochimie*, **53,** 331.

Dwivedi, P. C., Gurudath, K., Bhat, S. N., and Rao, C. N. R. (1975). *Spectrochim. Acta*, **31A,** 129.

Fenske, J. D. and DeMoss, R. D. (1975). *J. Biol. Chem.*, **250,** 7554.

Foster, R. (1969). *Organic Charge Transfer Complexes*, Academic Press, London.

Foster, R. (1974). In *Molecular Complexes* (Ed. Foster, R.), Vol. 2, Elek Science, London, p. 251.

Fujimori, E. (1959). *Proc. Natl. Acad. Sci. U.S.*, **455,** 133.

Fulton, G. P. and LaMar, G. N. (1976). *J. Amer. Chem. Soc.*, 2119.

Gasco, M. R. (1976). *Il Farmacia*, **31,** 308.

Guttmann, F. and Keyzer, H. (1966). *Electrochim. Acta*, **11,** 535.

Guttmann, F., Smith, L. C., and Slifkin, M. A. (1974). *The Phenothiazines and Structurally Related Drugs* (Ed. Forrest, J. S.) Raven, New York.

Hamlin, R., Lord, R. C., and Rich, A. (1965). *Science*, **148,** 1734.

Harbury, H. A., LaNue, K. F., Loach, P. A., and Amick, R. M. (1959). *Proc. Natl. Acad. Sci. U.S.*, **45,** 1708.

Hayman, H. J. G. (1962). *J. Chem. Phys.*, **37,** 2290.

Hélène, C. and Michelson, A. M. (1967). *Biochim. Biophys. Acta*, **142,** 12.

Higuchi, T. and Connors, K. A. (1965). *Adv. Anal. Chem. Instrum.*, **4,** 1117.

Hush, N. S. and Cheung, A. S. (1975). *Chem. Phys. Lett.*, **34,** 11.

Karlsson, R. (1972). *Acta Cryst.*, **B28,** 2358.

Kobinata, S. and Nagakura, S. (1966). *J. Amer. Chem. Soc.*, **88,** 3905.

Kyogoku, Y., Lord, R. C., and Rich, A. (1966). *Science*, **154,** 518.

Kyogoku, Y. and Yu, B. S. (1969). *Bull. Chem. Soc. Japan*, **42,** 1387.

Leijonmarck, M. and Werner, P. E. (1970). *Acta Chem. Scand.*, **24,** 2916.

Lin, L-N., Christian, S. D., and Childs, J. D. (1974). *J. Amer. Chem. Soc.*, **96,** 2727.

Lorentz, K. (1974). *Anal. Chem.*, **269,** 182.

Lorentz, K. (1976). *Experientia*, **32,** 1502.

Ma, J. K. H., Fu, P. P., and Luzzi, L. A. (1977). *J. Pharm. Sci.*, **66,** 209.

MacIntyre, W. M. (1965). *Science*, **147,** 507.

Massey, V. and Ghisla, S. (1976). *Ann. N.Y. Acad. Sci.*, **227,** 446.

Matsunaga, Y. (1966). *Nature*, **211,** 183.

Matsunaga, Y. (1973). *Bull. Chem. Soc. Japan*, **46,** 998.

Matsunaga, Y. and Suzuki, Y. (1973). *Bull. Chem. Soc. Japan*, **46,** 719.

Mehdi, S. H., Brisbin, D. A., and McBryde, W. A. E. (1976). *Biochim. Biophys. Acta*, **444,** 407.

Montenay-Garestier, T. and Hélène, C. (1970). *Biochemistry*, **9**, 2865.
Montenay-Garestier, T. and Hélène, C. (1973). *J. Agr. Food Chem.*, **21**, 11.
Morita, F. (1974). *Biochim. Biophys. Acta*, **343**, 674.
Moxon, G. H. and Slifkin, M. A. (1972a). *Biochim. Biophys. Acta*, **286**, 98.
Moxon, G. H. and Slifkin, M. A. (1972b). *J. Chem. Soc. Perkin II*, 1159.
Müller, F. (1972). *Z. Naturforsch.*, **276**, 1023.
Müller, W. and Crowthers, D. (1975). *Eur. J. Biochem.*, **54**, 267.
Müller, F. and Massey, V. (1969). *J. Biol. Chem.*, **244**, 4007.
Mulliken, R. S. (1950). *J. Amer. Chem. Soc.*, **72**, 600.
Mulliken, R. S. and Person, W. B. (1969). *Molecular Complexes*, Wiley–Interscience, New York.
Nogami, T., Yamaoka, T., Yoshihara, K., and Nagakura, S. (1971). *Bull. Chem. Soc. Japan*, **44**, 380.
Palumbo, M., Cosani, A., Terbojevich, M., and Peggion, E. (1977). *Biopolymers*, **16**, 109.
Palumbo, M. and Peggion, E. (1975). *Biopolymers*, **14**, 431.
Porter, D. J. T., Bright, H. J., and Voet, D. (1977). *Nature*, **269**, 213.
Prout, C. K. and Wright, J. D. (1968). *Angew. Chem. Intern. Ed. Engl.*, **7**, 659.
Reid, C. (1957). *Excited States in Chemistry and Biology*, Butterworths, London.
Sahney, R. C., Aggarwal, S. L., and Singh, M. (1946). *J. Ind. Chem. Soc.*, **23**, 335.
Scarborough, F. E., Shieh, H-S, and Voet, D. (1976). *Proc. Natl. Acad. Sci. U.S.*, **73**, 3807.
Sharples, O. and Brown, J. R. (1976). *Febs. Lett.*, **69**, 37.
Shinitzky, M., Katchalski, E., Grisaro, V., and Sharon, N. (1966). *Arch. Biochem. Biophys.*, **116**, 332.
Siderov, A. N. (1974). *Biofizika*, **19**, 45.
Singhbhandhu, A., Robinson, P. D., Fang, J. H., and Geiger, W. E. (1975). *Inorg. Chem.*, **14**, 318.
Sisido, M., Takagi, H., Imanishi, Y., and Higashimura, T. (1977). *Macromolecules*, **10**, 125.
Slifkin, M. A. (1970). *Chem. Phys. Lett.*, **7**, 195.
Slifkin, M. A. (1971a). *Charge Transfer Interaction of Biomolecules*, Academic Press, London.
Slifkin, M. A. (1971b). *Chem. Phys. Lett.*, **9**, 416.
Slifkin, M. A. (1973). *Charge Transfer Complexes of Purine and Pyrimidines* in *Physico-Chemical Properties of Nucleic Acids* (Ed. Duchesne, J.), Vol. I, Chap. 3, Academic Press, London.
Slifkin, M. A. and Walmsley, R. H. (1969). *Experientia*, **25**, 1930.
Slifkin, M. A. and Walmsley, R. H. (1970). *Spectrochim. Acta*, **26A**, 1237.
Szent-Györgyi, A. (1957). *Bioenergetics*, Academic Press, New York.
Szent-Györgyi, A. (1960). *Introduction to a Submolecular Biology*, Academic Press, New York.
Szent-Györgyi, A. (1968). *Bioelectronics*, Academic Press, New York.
Szent-Györgyi, A. (1972). *The Living State*, Academic Press, New York.
Szent-Györgyi, A. (1973). *Acta. Biochim. Biophys., Acad. Sci. Hung.*, **8**, 117.
Szent-Györgyi, A. (1974). *Life Sciences*, **15**, 863.
Szent-Györgyi, A. (1977). *Proc. Natl. Acad. Sci. U.S.*, **74**, 2844.
Taha, M., Ahmad, A. K. S., Gomaa, C. S., and El-Fatatry, H. (1974). *J. Pharm. Sci.*, **63**, 1853.
Taha, A. M. and Gomaa, C. S. (1976). *J. Pharm. Sci.*, **65**, 986.
Takahashi, F. and Maeda, H. (1974). *Bull. Chem. Soc. Japan*, **47**, 488.
Thewalt, U. and Bugg, C. E. (1972). *Acta Cryst.*, **B28**, 82.

Tournon, J., Kuntz, E., and El-Bayoumi, M. A. (1972). *Photochem. Photobiol.*, **16,** 425.

Tsibris, J. C. M., McCormick, D. B., and Wright, L. D. (1965). *Biochemistry,* **4,** 2612.

Van der Helm, D. (1973). *J. Cryst. Mole. Struct.,* **3,** 249.

Van der Helm, D., Christian, S. D., and Lin, L. N. (1973). *J. Amer. Chem. Soc.,* **95,** 2409.

Voet, D. and Rich, A. (1971). *Flavins and Flavoproteins* (Ed. Kamin, H.), Vol. 13, UPP., Baltimore.

Wallwork, S. C. (1961). *J. Chem. Sci.,* 494.

Weber, G., Tanaka, F., Okamoto, B. Y., and Drickamer, H. G. (1974). *Proc. Natl. Acad. Sci. U.S.,* **71,** 1264.

Yagi, K., Ozawa, T., Naoi, M., and Kotaki, A. (1968). *Flavins and Flavoproteins* (Ed. Yagi, K.), Vol. 237, University of Tokyo Press.

Yagi, K., Ozawa, T., and Okada, K. (1965). *Biochim. Biophys. Acta,* **35,** 102.

Yarwood, J. (1973). *Spectroscopy and Structure of Molecular Complexes,* Plenum Press, London.

Molecular Interactions, Vol. 2
Edited by H. Ratajczak and W. J. Orville-Thomas
© 1980 John Wiley & Sons Ltd

7 *Dipole moment, dielectric loss, and molecular interactions*

C. P. Smyth

7.1 Introduction

Dielectric constant measurements have been extensively used to obtain the electric dipole moments of molecules with the object of determining the geometrical structures of the molecules, the polarities of their bonds, and significant shifts of electric charge in them. The calculation of the molecular dipole moment from the measured dielectric constant of the bulk material depends upon the molecular interactions with their neighbours. A large part of the theoretical research in dielectrics has been concerned with the effects of these interactions as expressed in the effect of the internal field upon dielectric behaviour. It is obviously beyond the scope of a chapter such as this to describe the many partially or almost wholly successful treatments of this problem of the internal field. Similarly, measurement of the dielectric loss as a function of frequency gives a maximum at an angular frequency ($\omega = 2\pi f$), which is the the reciprocal of a time, the macroscopic dielectric relaxation time of the material. An approximate molecular or microscopic relaxation time can be calculated from the directly observed bulk or macroscopic relaxation time. As the value of the relaxation time is strongly influenced by molecular interactions, it may be used as quantitative or semiquantitative evidence of these interactions. It is the primary aim of this chapter to describe and, as far as possible, interpret the evidence of molecular interactions given by measurements of dielectric constant or permittivity and dielectric loss.

7.2 Dipole Moment and Dielectric Permittivity

An electric dipole is a pair of electric charges, equal in size but opposite in sign and very close together. The dipole moment is the product of one of the two charges and the distance between them. The system of positive and negative charges formed by a molecule may be treated as a single dipole, or as a system of dipoles lying in molecular axes, or bonds or groups, having a vector sum which is the molecular dipole moment. When an electric field acts upon a molecule, it shifts to form an induced dipole, usually much smaller than the permanent dipole due to structural asymmetry of the molecule. The order of magnitude of a molecular dipole moment is that of the product of an electronic charge, 4.80×10^{-10} electrostatic units, times an atomic radius, 10^{-8} cm, that is, 4.80×10^{-18} esu cm. Dipole moment values are expressed in 10^{-18} esu cm, frequently called a 'Debye' and abbreviated to 'D'.

The relation of the low-frequency or so-called static dielectric constant or permittivity ε_0 to dipole moment μ is given by the equation of Debye (1929):

$$P = \frac{(\varepsilon_0 - 1)}{(\varepsilon_0 + 2)} \frac{M}{d} = \frac{4\pi N}{3} \left(\alpha_0 + \frac{\mu^2}{3kT} \right) \tag{7.1}$$

in which P is the molar polarization, M is the molecular weight of the substance in question, d is the density, N is the Avogadro number, the number of molecules per mole, α_0 is the molecular polarizability, that is, the dipole moment induced in the molecule by unit electric field, k is the Boltzmann constant, and T is the absolute temperature. As an approximation, α_0 may be obtained from the expression

$$\frac{4\pi N}{3} \alpha_0 = \frac{(n_D^2 + 1)}{(n_D^2 + 2)} \frac{M}{d} \tag{7.2}$$

where n_D is the refractive index for the sodium D line. The molar polarization P may be written as the sum of three terms $P_E + P_A + P_M$, where P_M is the dipole orientation contribution, P_E the electronic polarization, and P_A the so-called atomic polarization. $P_E + P_A = 4\pi N \alpha_0/3$, the distortion polarization, is usually approximated by equation (7.2) and subtracted from P to obtain P_M for dipole moment calculation, but can also be obtained from gas measurements of P as a function of temperature (Smyth, 1955). P_A, the usually small difference between the distortion polarization and P_E, was thought to arise from movement of atomic nuclei relative to one another, which produced changes in dipole moment through changes in bond lengths and angles and through the bending and twisting of polar groups with consequent absorption in the infrared region. The absorption and reflection coefficients of several liquids were measured by Cartwright and Errera (1936) for radiation in the far infrared obtained by the method of Reststrahlen and used to calculate the refractive indices. These latter gave values of $P_E + P_A$ of the same magnitude as those obtained by the polarization–temperature method for gases (Smyth, 1955), showing the correctness of the attribution of P_A to infrared effects. Since the molecules orienting in the alternating electric field used in the measurements must be able to attain equilibrium with the field in order that equation (7.1) may hold, dipole moments are measured most accurately in the vapour state but more conveniently in dilute solutions in non-polar solvents to reduce the effects of the internal field in the liquid (Smyth, 1955, 1972; Hill *et al.*, 1969, Vaughan *et al.*, 1972). For liquids the Onsager equation (Onsager, 1936) is a better approximation that that of Debye since it makes possible the calculation of dipole moment from the dielectric permittivity of a pure polar liquid, sometimes with an error no greater than that in measurements on dilute solutions (Smyth, 1955). The equation may be written

$$\frac{(\varepsilon_0 - \varepsilon_\infty)(2\varepsilon_0 + \varepsilon_\infty)}{\varepsilon_0(\varepsilon_\infty + 2)^2} = \frac{4\pi Nd}{3M} \frac{\mu^2}{3kt} \tag{7.3}$$

where ε_∞ has been substituted (Smyth, 1955) for the square of Onsager's 'internal refraction index' n^2; ε_∞ is called the infinite frequency or optical dielectric constant.

The Debye equation (7.1) neglects the local directional forces exerted on the molecular dipoles by neighbouring molecules and the local field due to molecules within distances large compared to molecular dimensions but small compared to the thickness of the material. The Onsager equation (7.3) also neglects the local directional forces on the molecules due to their neighbours, but takes account of the local field due to molecules within distances large compared to molecular dimensions but small compared to the thickness of the material.

Kirkwood (1939) obtained a more general equation

$$\frac{(\varepsilon_0 - 1)(\varepsilon_0 + 2)}{9\varepsilon_0} \frac{M}{d} = \frac{4\pi N}{3}\left(\alpha_0 + \frac{g\mu^2}{3kT}\right) \tag{7.4}$$

in which a correlation parameter g is a measure of the local ordering in the material. Positive deviations of g from unity result when short-range hindering torques favour parallel orientation of the dipoles of neighbouring molecules, while negative deviations result from antiparallel orientation. If g is unity, it means that fixing the position of one dipole does not influence the positions of the others except through the long-range electrostatic forces. Using an approximately tetrahedral structure for liquid water, Kirkwood calculated a value of g which, inserted in equation (7.4), gave a permittivity value for water at 25 °C differing by only 0.4% from the observed value. However, at 83 °C, the permittivity calculated from μ and g was 12% higher than the observed and similar calculations for five alcohols showed discrepancies of 10–29%. Kirkwood noted that the Onsager calculation gave a permittivity of 31 for water at 25 °C as compared to 78.2 observed.

As previously mentioned, correct dipole moment values are obtained by application of the Debye equation to permittivity values measured for gases, while dipolar molecules in dilute solution in non-polar solvents are sufficiently far apart to yield moments differing by only a few per cent from the values for the vapour state. The application of the Onsager and Kirkwood (with g taken as unity) equations to pure dipolar liquids usually gives moment values differing by less than 10% from the gas values. If the gas value of μ is known, the value of g for the liquid state can be calculated from the permittivity and density of the pure liquid. Any large departure of a g value from unity is normally indicative of molecular association. The calculation of g from liquid structure is, at best, a laborious process leading to an approximate and not very informative result. The departure of the value of g from unity is an indication of the extent to which the molecular dipoles tend to orient parallel or antiparallel to their neighbours. Fröhlich (1949, 1958) and Cole (1957) later derived equations rather similar to those of Onsager and Kirkwood. These and other equations and their differing assumptions and derivations have been described and discussed in detail elsewhere (Smyth, 1955; Hill et al., 1969; Böttcher, 1973). The equations have been derived by treating the molecules as point dipoles with isotropic polarizability or embedded in a sphere of isotropic polarizable material. Later Scholte (1949) and others extended the Onsager treatment to ellipsoidal molecules. This extension to ellipsoidal molecules has been given by Böttcher (1973) in the form:

$$\mu^2 = \frac{3kTM(\varepsilon - \varepsilon_\infty)(2\varepsilon + 1)}{4\pi N d\varepsilon(2\varepsilon + \varepsilon_\infty)} \frac{[\varepsilon + (\varepsilon_\infty - \varepsilon)A_a^2]}{[\varepsilon + (1 - \varepsilon)A_a][1 + (\varepsilon - 1)A_a]^2} \qquad (7.5)$$

The number of molecules per cm^3 in the equation quoted by Böttcher has been converted to the Avogadro number as used in equation (7.1) by multiplying by the molar volume M/d. Factors A_a, A_b, or A_c depend on the form of the ellipsoid, the principal axes of which are $2a$, $2b$, and $2c$.

$$A_\alpha = \frac{abc}{2} \int_0^\infty \frac{ds}{(s + a^2)R} \quad (\alpha = a, b, c) \qquad (7.6)$$

with $R^2 = (s + a^2)(s + b^2)(s + b^2)(s + c^2)$. s is a variable. Extensive tabulations of A_α as a function of a, b, and c are available. For spherical molecules, $a = b = c$ and $A_\alpha = \frac{1}{3}$, and equation (7.5) reduces to the Onsager equation (7.3). Böttcher (1973) has calculated a large number of dipole moment values from the permittivities of the pure liquids by means of the Onsager equation (7.3) obtaining differences from the gas values of the same magnitude as those previously noted. Selecting from these substances those in whose molecules the dipole was directed along one of the axes, taken as the a axis, of the molecular ellipsoid and estimating the ratios of the axes, a, b, and c, from values of interatomic distances and atomic radii in the literature, he calculated moment values from the permittivities and densities of the pure liquids by means of equation (7.5), getting larger values than the Onsager values when $a > b$ and $a > c$ and smaller when $a < b$ and $a < c$. Usually, the values calculated with equation (7.5) were closer to the gas values than those calculated by the Onsager equation (7.3), but there were some exceptions. Naturally, associated liquids may give very different moment values from those measured for the gas, higher if association causes increased dipole orientation in the same direction, lower if it causes increased mutual opposition of the dipoles (this problem will be discussed later).

7.3 Solvent Effect

In the early determinations of molecular dipole moments by applying the Debye equation to solutions of dipolar molecules separated from one another by non-polar solvent molecules, it soon became apparent that the moment values obtained depended somewhat on the solvent used. The difference between the moment value measured in the vapour and that in solution is the so-called solvent effect. Raman and Krishan (1928) developed an equation for permittivity, which represented the molecule as both optically and electrically anisotropic, but it was open to criticism on theoretical grounds and contained too many unknown quantitites to be of practical value, although it contained the basic explanation of the solvent effect. Weigle (1933) took account of molecular anistropy in calculating the inductive effects of the polar solute molecules upon the non-polar solvent molecules and obtained values for the solvent effect of the same magnitude as the observed differences between the moments μ_0 for the gaseous state and μ_s for solution. Müller (1933–1937, Müller and Mortier, 1935) found that careful measurements for several substances in dilute solution could be approximately represented by an empirical equation

$$\frac{\mu_s}{\mu_0} = 1 - C(\varepsilon - 1)^2 \tag{7.7}$$

in which ε is the permittivity of the solvent and C is an empirical constant. The equation worked well in a number of cases for $C = 0.038$, but, in other cases, was inadequate. Frank (1935), Higasi (1936), Conner and Smyth (1943), and Ross and Sack (1950), among others, have developed equations involving the anisotropy of the solute molecule as given by the geometry of the molecular ellipsoid or the polarizabilities along the eltipsoidal axes. Other equations have been summarized and discussed elsewhere (Smith, 1948; Ross and Sack, 1950; Smyth, 1955). Three of these equations may be written very simply as follows:

Frank: $\quad \dfrac{\mu_s}{\mu_0} = 1 + A' \dfrac{\varepsilon - 1}{\varepsilon}$ (7.8)

Higasi: $\quad \dfrac{\mu_s}{\mu_0} = 1 + 3A \dfrac{\varepsilon - 1}{\varepsilon}$ (7.9)

Conner and Smyth: $\quad \dfrac{\mu_s}{\mu_0} = 1 + 0.43 A (\varepsilon - 1)$ (7.10)

A is calculable, usually having a negative sign, if the molecular dimensions or axial polarizabilities are known or capable of approximate estimation. The observed values of μ_s/μ_0 usually lie between 1.12 and 0.87. Values less than 1.0, corresponding to negative solvent effects, occur much more frequently than those greater than 1.0. Equations (7.9) and (7.10) appear to have been tested more extensively than the many other equations which have been proposed. Both equations give the correct magnitude and direction of the solvent effect and the values which they give usually differ from the observed by no more than the combined effects of the experimental errors in the observed μ_s/μ_0 and the uncertainties in the information used in calculating A. However, none of these equations takes strict account of all the possible factors involved.

A simplistic picture of solvent effect may be obtained (Smyth, 1955) by considering the effect of a dipole field upon its surroundings. If a dipole of moment μ is located at the origin of a system of polar coordinates pointing along the z axis, the field E due to it at a point r, θ, φ, where θ is the angle between r and the z axis and φ is the angle between the projection of r in the xy plane and the x axis, has the components

$$E_x = \frac{3\mu}{r^3} \cos \theta \sin \theta \cos \varphi \qquad (7.11a)$$

$$E_y = \frac{3\mu}{r^3} \cos \theta \sin \theta \sin \varphi \qquad (7.11b)$$

$$E_z = \frac{\mu}{r^3} (3 \cos^2 \theta - 1) \qquad (7.11c)$$

Let us suppose that μ is at the centre of a spherical solute molecule surrounded by six spherical solvent molecules of polarizability α symmetrically located around it so that the solute molecule is at the centre of a regular octahedron, with a solvent molecule at each of the six corners, at a distance r from μ, and with the solute dipole in an axis of symmetry connecting opposite corners of the octahedron. The field in the direction of the dipole axis is given by equation (7.11c), while the field perpendicular to this direction is shown by (7.11a) and (7.11b) to be zero. For the two solvent molecules at the ends of the symmetry axis containing the dipole, θ is $0°$ and $180°$ and the dipole moment m induced in each of the two is, therefore,

$$m = \frac{\mu}{r^3}(3\cos^2\theta - 1)\alpha = \frac{2\mu}{r^3}\alpha \qquad (7.12)$$

in the same direction as μ. For each of the four solvent molecules in the plane perpendicular to the dipole axis, θ is $90°$ and the moment m induced in each of the four is

$$m = \frac{-\mu}{r^3}\alpha \qquad (7.13)$$

acting in the opposite direction to that of μ. The sum of these four induced moments obviously cancels that of the two moments induced in the direction of μ, resulting in a zero solvent effect, as observed in the case of the nearly spherical molecule of t-butyl chloride. If the solute molecule is elongated so that the two solvent molecules in the line of the dipole axis are farther from the dipole, r in equation (7.12) is increased and m is decreased, resulting in a net negative value for the total induced moment, that is a negative solvent effect, as in chlorobenzene, where one solvent molecule is farther from μ_0 than the other. If the molecular axis in which the dipole lies is shortened, r is decreased and m in (7.12) is increased to give a positive solvent effect as in the case of the chloroform molecule, for which μ_s/μ_0 in benzene solution is 1.12. For the majority of the dipolar solute molecules which have been studied, a negative solvent effect has been observed.

The small differences between the observed gas values of the moment and the apparent values calculated by means of the Debye equation (7.1) from the measured permittivities and densities of dilute solutions of polar molecules in non-polar solvents may be calculated correctly as to magnitude and sign by means of the solvent effect equations which have been discussed or referred to in this section, although none of these equations takes strict account of all the factors involved. However, the approximately calculated electrostatic interactions between the molecules are consistent with the observed variations of the apparent dipole moments of the molecules in liquids. A prohibitively large number of parameters would be required to account exactly for the observed effects and the total probable error due to errors in the parameter values would often be as large as those in the oversimplified calculations which have been discussed.

7.4   Reaction Field of a Polarizable Point Dipole

For a polarizable point dipole, it can be shown (Böttcher, 1973) that the
reaction field R of a dipole in a pure dipolar liquid is approximately

$$R = \frac{4\pi dN}{3M} \frac{2(\varepsilon - 1)}{(2\varepsilon + n_D^2)} \frac{(n_D^2 + 2)}{3} \mu \qquad (7.14)$$

in which n_D is the refractive index of the liquid for the sodium D line. The
value calculated (Böttcher, 1973) for nitrobenzene at 20 °C is 43×10^6 volts/cm and that for chloroform is 9.6×10^6 volts/cm. Under the influence of the reaction field the dipole moment μ is increased by an induced
moment αR to give

$$\mu^* = \mu + \alpha R \qquad (7.15)$$

Substitution of α and R as given by equations (7.2) and (7.14) gives for the
ratio

$$\frac{\mu^*}{\mu} = \frac{(2\varepsilon + 1)}{(2\varepsilon + n_D^2)} \frac{(n_D^2 + 2)}{3} \qquad (7.16)$$

This equation gives values of μ^*/μ at 20 °C of 1.44 for pure liquid
nitrobenzene and 1.24 for pure liquid chloroform (Böttcher, 1973). In a
very dilute solution of nitrobenzen in benzene the reaction field of the
dipole is reduced to 16×10^6 volts/cm and the ratio μ^*/μ to 1.17.

7.5   Permanent Dipoles, Cohesion, Boiling Point, and Solubility

In a pure dipolar liquid the interaction energy of the dipole of one molecule
with all of the other molecules is $-\mu R$. For one mole of the liquid the part
of the energy due to the molecular dipoles (Böttcher, 1973) is

$$W = -\tfrac{1}{2}N\mu R \qquad (7.17)$$

Because each term $-\mu R$ is the interaction energy of one dipole with all of
the other molecules, the factor $\frac{1}{2}$ is inserted to prevent the counting of the
interaction energy between two particular molecules twice. W may be
obtained from (7.17) by using equation (7.14) to calculate an approximate
value of R. In the absence of molecular association the cohesion energy of
the liquid is due to W and the London–van der Waals interaction. Böttcher
(1973) has tabulated the values of λ, the heat of vapourization, and $-W$ at
the boiling point $T_b(\text{K})$ for a number of dipolar liquids. The change of
internal energy on vapourization as given by

$$\Delta U = \lambda - RT_b \qquad (7.18)$$

Table 7.1   Dipole and London–van der Waals interaction energies (kcal/mol) in dipolar liquids

Liquid	ε	μ	$-W$	λ	ΔU	$\Delta U + W$
Chloroform	4.23	1.0	0.18	6.98	6.31	6.13
Chlorobenzene	4.20	1.7	0.38	8.29	7.48	7.10
Nitrobenzene	15.6	4.2	3.2	10.05	9.08	5.9
Nitromethane	27.8	3.5	4.4	7.00	6.25	1.9
Acetonitrile	26.2	3.9	5.6	7.12	6.41	0.8

has also been listed for several liquids. R in (7.18) is the gas constant, not the reaction field as in (7.17). A few of these data are quoted in Table 7.1.

The approximate values in Table 7.1 give us the magnitudes of the dipole (W) and London–van der Waals ($\Delta U + W$) interaction energies. W depends on the square of the dipole moment (equations 7.14 and 7.17) and upon the permittivity, which depends in part upon the square of the dipole moment (equation 7.1). Consequently, the twofold increase in W from chloroform to chlorobenzene results from increase in dipole moment, the eightfold increase from chlorobenzene to nitrobenzene results from increase in both moment and permittivity, and the relatively smaller increases in W from nitrobenzene to nitromethane and acetonitrile result from decrease in molecular size and consequent increase in permittivity due to increase in the number of dipoles per unit volume. For chloroform and chlorobenzene, the dipole interaction energies W are much smaller than the London–van der Waals interaction energies, indeed smaller than the probable errors in the values of the latter. For nitrobenzene, the dipole interaction energy is half as large as the London–van der Waals energy, while, for nitromethane and acetonitrile, it is more than twice as large. Böttcher (1973) has compared the values of $\Delta U + W$ for six dipolar compounds with those of ΔU for approximately isomorphous compounds having small or zero dipole moments ($W = 0$) and, presumably, somewhat similar London–van der Waals energies. ΔU for the nearly or wholly non-polar compounds should be the London–van der Waals energy for these compounds and should be approximately the same as that calculated as $\Delta U + W$ for the corresponding dipolar compounds. For five out of six, it is slightly higher, the difference averaging 0.8 kcal/mol but, for the pair, acetonitrile, $CH_3C{\equiv}N$, and propyne, $CH_3C{\equiv}CH$, the difference is 4.3, the calculated London–van der Waals energy of acetonitrile being unexpectedly low in comparison with that of propyne. This anomaly may be significant, but is more probably the result of an unusually large accumulation of errors in the approximations involved in the calculations. According to that old approximation known as Trouton's law, $\lambda/T_b \approx 23$, increase in λ by dipole interaction energy should raise the boiling point correspondingly. Thus, the boiling point of acetonitrile is

105° higher than that of propyne, that of nitromethane 273° higher than that of ethane, and that of nitrobenzene 100° higher than that of toluene, Böttcher (1973) has concluded that, for p-dichlorobenzene ($\mu = 0$), the quadrupole contribution to the cohesion energy is of the same order of magnitude as W for o-dichlorobenzene ($\mu = 2.50$) and m-dichlorobenzene ($\mu = 1.70$), with the result that the boiling points T_b are 447, 455, and 445°K, respectively, in spite of the differences in overall molecular dipole moment. It would appear that the cohesive energy is determined by the number and size of the C—Cl dipole moments in the molecule rather than by the resultant moment of the molecule as a whole. This would seem to follow naturally from the nearness of the C—Cl dipoles to the molecular surfaces, which makes λ small in the expression for their field strength (equations 7.11a, 7.11b, 7.11c).

Analogous conclusions have been drawn from solubility and related studies on solutions of mono-, di-, and trisubstituted nitrobenzenes in benzene. Hildebrand and coworkers (Hildebrand and Carter, 1930; Hildebrand and Scott, 1950) have shown that, in spite of their different dipole moments, the dinitrobenzenes show the same deviation from ideal behaviour in benzene solution, while nitrobenzene shows less deviation, and 1,3,5-trinitrobenzene shows much more. He concluded that it was the number of nitro groups in the molecule and not their resultant moment which determined the degree of unlikeness to benzene. In an investigation of the solubilities of twelve tetra-, penta-, and hexa-chlorine- and methyl-substituted benzenes in benzene, Smyth and Lewis (1940) found that, although the occurrence of molecular rotation in the solid tended to reduce its solubility, the solubilities were approximately ideal and independent of the molecular dipole moments, which had values ranging from 0 to 3.01, but the solubilities and entropies of fusion tended to be low and the melting points high when the molecules possessed rotational freedom in the solid.

7.6 Dielectric Loss and Relaxation Time

At very high frequencies or high viscosities, dipolar molecules may not attain equilibrium with the applied field, energy is absorbed and the dielectric permittivity decreases. The ratio of the loss current to the charging current is then the loss tangent $\varepsilon''/\varepsilon'$, where ε'' is the dielectric loss and ε' is the measured dielectric permittivity. The relations of these quantitities to ε_0, ε_∞, and the angular frequency $\omega(2\pi$ frequency in c/s) are given by the equations of Debye (1929):

$$\varepsilon' = \varepsilon_\infty + \frac{\varepsilon_0 - \varepsilon_\infty}{1 + \omega^2\tau^2} \tag{7.19}$$

$$\varepsilon'' = \frac{(\varepsilon_0 - \varepsilon_\infty)\omega\tau}{1 + \omega^2\tau^2} \tag{7.20}$$

in which τ is the dielectric relaxation time. Dielectric relaxation is the exponential decay with time of the polarization in a dielectric when an externally applied field is removed; τ is the time in which the polarization is reduced to $1/e$ times its original value, where e is the natural logarithmic base. Equation (7.20) shows that ε'' approaches zero both for small and for large values of $\omega\tau$, while it reaches a maximum value ε_m when $\omega\tau = 1$. The corresponding value of the angular frequency ω_m is evidently the reciprocal of τ. This relationship provides an experimental definition of τ as $\tau = 1/\omega_m$. The various methods of using these equations to obtain τ are described in detail elsewhere (Smyth, 1955).

When two mutually independent relaxation processes exist in a system, having relaxation times τ_1 and τ_2, equations (7.19) and (7.20) become (Bergmann, 1957; Bergmann *et al.*, 1960),

$$\frac{\varepsilon' - \varepsilon_\infty}{\varepsilon_0 - \varepsilon_\infty} = C_1 \frac{1}{1 + (\omega\tau_1)^2} + C_2 \frac{1}{1 + (\omega\tau_2)^2} \tag{7.21}$$

$$\frac{\varepsilon''}{\varepsilon_0 - \varepsilon_\infty} = C_1 \frac{\omega\tau_1}{1 + (\omega\tau_1)^2} + C_2 \frac{\omega\tau_2}{(1 + \omega\tau_2)^2} \tag{7.22}$$

where C_1 and C_2 are the relative weights of each relaxation term and $C_1 + C_2 = 1$. These equations may be used when the loss involves dipole orientation by two independent relaxation processes.

Equations (7.19) and (7.20) were derived by separation of real and imaginary parts in the equation

$$\varepsilon^* = \varepsilon_\infty + \frac{\varepsilon_0 - \varepsilon_\infty}{1 + i\omega\tau} \tag{7.23}$$

in which ε^*, the complex dielectric constant, is defined as

$$\varepsilon^* = \varepsilon' - i\varepsilon'' \tag{7.24}$$

Cole and Cole (1941) combined equations (7.19) and (7.20) to obtain

$$\left[\varepsilon' - \left(\frac{\varepsilon_0 + \varepsilon_\infty}{2}\right)\right]^2 + \varepsilon''^2 = \left(\frac{\varepsilon_0 - \varepsilon_\infty}{2}\right)^2 \tag{7.25}$$

This equation represents a circle but, since all values must be positive, it gives a semicircle when ε'' is plotted as ordinate against ε' as abscissa, the diameter of the semicircle lying on the abscissa axis. When the measured dielectric constants or permittivities and losses of a material give such a semicircle, they conform to the Debye theory, and the material may be described as showing Debye behaviour. Many materials give a semicircular arc intersecting the abscissa axis at the values of ε_∞ and ε_0 and having one end of its diameter at ε_∞. For such materials, equation (7.23) was modified

(Cole and Cole, 1941) by the introduction of an empirical constant α to give

$$\varepsilon^* = \varepsilon_\infty + \frac{\varepsilon_0 - \varepsilon_\infty}{1 + (i\omega\tau_0)^{1-\alpha}} \qquad (7.26)$$

where τ_0 is the most probable relaxation time, the reciprocal of the angular frequency ω_m, at which ε'' has its maximum value, and α, an empirical constant with a value between 0 and 1, but commonly less than 0.1, is a measure of the distribution of the relaxation times. The centre of the circle of which the arc is a part lies below the abscissa axis and the diameter drawn through the centre from the ε_∞ point in the abscissa axis makes an angle $\alpha\pi/2$ with the latter. α and τ_0 may be obtained from the Cole–Cole plot (Smyth, 1955).

Instead of the usual symmetrical arc plot, a skewed-arc plot is sometimes obtained, representable by the Cole–Davidson equation (Davidson and Cole, 1951)

$$\varepsilon^* = \varepsilon_\infty + \frac{\varepsilon_0 - \varepsilon_\infty}{(1 + i\omega\tau_0)^\beta} \qquad (7.27)$$

in which β is an empirical constant with a value between 0 and 1 and τ_0 is a characteristic relaxation time. When $\alpha = 0$ in equation (7.26) or $\beta = 1$ in equation (7.27), the equation becomes identical with equation (7.23), which represents Debye behaviour (Hill *et al.*, 1969). When two distinct relaxation processes occur, equations (7.21) and (7.22) should be applicable, but unless the relaxation times τ_1 and τ_2 are considerably separated, the absorption regions for the two processes overlap to such an extent that the plot of the dielectric loss against log frequency may merely show a slightly flattened and widened maximum and the arc plot may appear symmetrical with an appreciable value of α, suggesting a distribution of relaxation times around a most probable value. If the intensities of the two absorption processes differ considerably, the ε''–log frequency curve will be unsymmetrical and the arc plot may appear to skewed.

7.7 Effect of Internal Field on Relaxation Time

The quantity $\tau = 1/\omega_m$ is a directly measurable bulk or macroscopic relaxation time which we will term τ_m. The relation of the relaxation process to molecular interactions should be more profitably discussed in terms of a microscopic or molecular relaxation time τ_μ. The Debye theory gives

$$\frac{\tau_m}{\tau_\mu} = \frac{\varepsilon_0 + 2}{\varepsilon_\infty + 2} \qquad (7.28)$$

Just as the Debye theory was inadequate for the calculation of the molecular dipole moment from the permittivity of a pure dipolar liquid (equation 7.1), this equation is inadequate to calculate the molecular relaxation time from the macroscopic relaxation time, the difficulty lying in the Lorentz field factor $(\varepsilon_0 + 2)/(\varepsilon_\infty + 2)$. Powles (1953) using an approximate *ad hoc* assumption based on the Onsager model obtained the ratio

$$\beta = \frac{\tau_m}{\tau_\mu} = \frac{3\varepsilon_0}{2\varepsilon_0 + \varepsilon_\infty} = 1 + \frac{\varepsilon_0 - \varepsilon_\infty}{2\varepsilon_0 + \varepsilon_\infty} \tag{7.29}$$

Later, Cole (1965) developed a sound generalization of the relaxation time ratio identical with that given by equation (7.29). It may be noted that, for ε_0 close to ε_∞, τ_m/τ_μ in both equations (7.28) and (7.29) is close to unity, justifying the use of the macroscopic relaxation time of a dilute solution of dipolar molecules in a non-polar solvent as a good approximation to the microscopic relaxation time of the dipolar molecules in the environment provided by the solvent. O'Dwyer and Sack (1952) developed an equation which may be written

$$\beta = \frac{\tau_m}{\tau_\mu} = 1 + x + \frac{3x^2}{(1 - 2x)(1 + x)} \tag{7.30}$$

where

$$x = \frac{\varepsilon_0 - \varepsilon_\infty}{2\varepsilon_0 + \varepsilon_\infty}$$

Omission of the third term on the right of this as an approximation would reduce it to equation (7.29). However, equation (7.30) and equation (7.28) require τ_m/τ_μ to approach infinity as ε_0 approaches infinity, while equation (7.29) evidently requires that τ_m/τ_μ approach 1.5.

The validity of these equations was examined by Miller and Smyth (1957) by assuming that the relaxation time of a weakly dipolar liquid or that of a dilute solution of strongly dipolar molecules in a non-polar liquid was the microscopic or molecular relaxation time τ_μ, which was to be compared with the directly measured macroscopic relaxation time of the pure dipolar liquid, τ_m. For example, the dipole moment, 0.4 D, of the toluene molecule is so small that at 20 °C, $\varepsilon_0 - \varepsilon_\infty = 0.11$ and equation (7.28) gives $\tau_m/\tau_\mu = 1.025$, (7.29) gives 1.016, and equation (7.30) gives 1.005, which means that the calculated microscopic or molecular relaxation time τ_μ is, in this case, indistinguishable from the directly measured macroscopic relaxation time τ_m, that is, $\tau_\mu = \tau_m$. As the molecule of 4-methylpyridine, or γ-picoline, is almost identical in size and shape with that of toluene, one may reasonably suppose that its molecular relaxation time should be the same as that of toluene except for the effects of difference in liquid viscosities and internal field. Since many liquids with unsymmetrical molecules had been found to show a rough proportionality between relaxation time and viscosity, the effect of viscosity was minimized by multiplying the relaxation time of one liquid by the inverse ratio of the two viscosities, η_1 and η_2, where subscript $_1$ refers to the dilute solution of the weakly dipolar liquid (1) and subscript $_2$ refers to the strongly dipolar liquid (2). The assumption then is that

$$\tau_{\mu 2} = \tau_{\mu 1}(\eta_2/\eta_1)$$

and, consequently, since, for toluene, $\tau_{\mu 1} = \tau_{m1}$,

$$\beta = \tau_{m2}/\tau_{\mu 2} = \tau_{m2}\eta_1/\tau_{m1}\eta_2 \tag{7.31}$$

Data at one to four temperatures for 26 pairs of liquids like toluene (1) and 4-methylpyridine (2) were used (Miller and Smyth, 1957) to calculate β by means of equation (7.31) and the results were plotted as ordinates against the corresponding values of $\varepsilon_0 - \varepsilon_\infty$ as abscissas. No single curve gave a good representation of all of the 54 points. All of the points lay below the curve given by equation (7.28) and four were practically on the curve given by equation (7.30), but none were above it. The best approximation to the points was given by the curve for equation (7.29). However, in spite of the fact that equation (7.29) limits the range of β values to 1.0–1.5, there were several values between 1.5 and 2.1, two between 0.9 and 1.0, and three in the range 2.5–5.4, the latter for molecules containing CN or NO_2 groups and consequently having exceptionally large dipole moments, which give rise to intense electric fields. The equations examined do not take into account the effects of differences in molecular shape or dipole location in the molecule. The assumption of proportionality of relaxation time to liquid viscosity is probably the most serious source of error in the estimated time ratios, as it is approximately correct only when the polar molecules are, at least, three times as large as the molecules surrounding them. Indeed, as no proportionality has been observed in the case of pure liquids consisting of approximately spherical molecules, the relaxation times of pairs of liquids consisting of such molecules are better left uncorrected for viscosity effects. It is evident that we cannot expect this method of evaluating the equations for the internal field effects upon relaxation times to be more than approximate, but it suffices to show that equation (7.29) is the best approximation of the three considered for the calculation of the molecular relaxation time from the macroscopic. This equation is, however, a rough but useful approximation, which appears to be inadequate in the cases of the two most intense dipole fields examined.

Some evidence of the effect of internal field on molecular relaxation time was given by binary mixtures of polar liquids (Forest and Smyth, 1965; Kilp et al., 1966). In an equimolar mixture of chlorobenzene and bromobenzene, the relaxation times calculated for the two components differed from those for the pure liquids by no more than the uncertainties in the values. When the much larger molecules of 1-chloronaphthalene and 1-bromonaphthalene were mixed with those of chlorobenze, the calculated relaxation times of the large molecules were considerably lowered and those of the smaller molecules raised by the change in liquid viscosity, the viscosity of the halogenated naphthalene being about 4 times that of the halogenated benzene, so that any effect of internal field upon relaxation time was masked by that of viscosity. When solutions in benzene were measured for comparison, the ratio of τ_μ obtained from measurements on the solutions in chlorobenzene to these values of τ_μ in benzene solution were 1.78 for chloronaphthalene and 1.51 for bromonaphthalene as compared to 1.22 and 1.21 calculated by means of equation (7.29). The actual effect of internal field was apparently larger than the calculated, but the discrepancy was in the general range of those previously discussed. These results indicated that, when the polar components were similar in molecular moment and shape, the dielectric behaviour of the mixtures could be accurately represented in terms of plausible values assigned to the relaxation times of the individual components, but the absence of large differences between the relaxation times reduced the significance of their interpretation. In the absence of an effect of internal field, one would expect the relationship

$$C_1/C_2 = n_1\mu_1^2/n_2\mu_2^2 \qquad (7.32)$$

where C_1 and C_2 are the relative weights of each relaxation term in equations (7.21) and (7.22), n_1 and n_2 are the numbers of molecules of the two components per cm^3, and μ_1 and μ_2 are their dipole moments. The ratio n_1/n_2 is given by the ratio of the mole fractions. For the halogenated benzenes and naphthalenes, the values of the dipole moments are so similar that the values C_1 and C_2 are indistinguishable from the mole fractions. However, in a second series of binary mixtures, in which one polar component had a much larger relaxation time than the other (Kilp et al., 1966), the dipole moments differed considerably from one another, and not only did the values obtained experimentally for C_1 and C_2 differ from those of the mole fractions, but the ratios C_1/C_2 were from 1.35 to 2.31 times those calculated by means of equation (7.32). The large discrepancy was attributed to the effect of internal field. Three mixtures of benzophenone with tetrahydrofuran containing, respectively, mole fractions 0.108, 0.235, and 0.492 of benzophenone gave C_1/C_2 ratios averaging 1.5 times those calculated by means of equation (7.22). When the three polar mixtures were diluted with benzene to give three series of dilute solutions containing less than 0.025 mole fraction of the polar mixtures, the disagreement between the observed and the calculated ratios C_1/C_2 disappeared, and the relaxation times of the two polar components were independent of their concentrations relative to each other at these high dilutions. These results would seem to confirm the correctness of the attribution of the discrepancies in the concentrated polar mixtures to the effect of internal field, which is greatly reduced or eliminated in the dilute solutions.

The binary mixtures in which the two polar components have very different relaxation times show that their dielectric constants and losses cannot be represented by the single arc of a Cole–Cole plot, but can be represented by a superposition of two distinct arcs, one for each of the two polar components. The fact that the dielectric behaviour of these binary mixtures of polar molecules can be represented in terms of the behaviours of the individual molecules argues against the existence of larger liquid regions such as those proposed by some earlier workers for binary systems which they investigated.

7.8 Rate Theory of Dielectric Relaxation and Viscosity

Eyring and coworkers (Glasstone *et al.*, 1941) developed a rate theory of dielectric relaxation on the assumption that the rotational orientation of the dipolar molecule required an activation energy sufficient to overcome the energy barrier separating two mean equilibrium positions. The number of times such a rotational process occurs per second is $1/\tau$ and τ is given by the expression:

$$\tau = \frac{\hbar}{kT} \exp\left(\frac{\Delta F^*}{RT}\right) = \frac{\hbar}{kT} \exp\left(\frac{-\Delta S^*}{R}\right) \exp\left(\frac{\Delta H^*}{RT}\right) \tag{7.33}$$

where $\hbar$ is the Planck constant, ΔF^* is the free energy of activation for the molecular orientation process, ΔS^* is the molar entropy of activation, and ΔH^* is the molar enthalpy of activation. Eyring has obtained an analogous expression for the viscosity η of a liquid:

$$\eta = \frac{N\hbar}{V} \exp\left(\frac{-\Delta S_\eta^*}{R}\right) \exp\left(\frac{\Delta H_\eta^*}{RT}\right) \tag{7.34}$$

where V is the molar volume of the liquid. Hill (Hill *et al.*, 1969) has pointed out that, since the process of viscous flow is not identical with that of dielectric relaxation, the values of the quantities involved in the two equations are not necessarily the same. They may, however, throw some light on the nature of the molecular interactions involved in the two processes.

7.9 Relaxation Time, Inner Friction, and Viscosity

In considering the rotational orientation of dipolar molecules in a field F, Debye (1929) wrote the torque tending to turn the molecule in the direction of the field as

$$M = -\mu F \sin \theta \qquad (7.35)$$

where θ is the angle between the dipole axis and the direction of F and the negative sign takes account of the fact that the torque is opposite in direction to that of increase in θ. A constant torque due to an inner frictional force resulting from impacts of the surrounding molecules and proportional to the angular velocity $d\theta/dt$ would just balance a constant impressed torque M so that

$$M = \zeta \frac{d\theta}{dt} \qquad (7.36)$$

where ζ is a constant measuring the inner friction. Debye (1929) treated the molecule as a sphere of radius a rotating in a continuous viscous medium of bulk viscosity η. For such a sphere,

$$\zeta = 8\pi\eta a^3 \qquad (7.37)$$

and, since

$$\tau = \frac{\zeta}{2kT} \qquad (7.38)$$

$$\tau_\mu = \frac{4\pi\eta a^3}{kT} \qquad (7.39)$$

Here the relaxation time is evidently τ_μ, which may be calculated from the measured τ_m by means of equation (7.29). In many cases equation (7.39) is found to give a reasonable order of magnitude for the quantities involved, but, if known values of η and a are inserted, τ_μ comes out too large, sometimes by a factor of 5–10, and, occasionally by factor values out of this range. There are, among others, two serious difficulties: (1) the molecules are rarely spherical, and (2) the medium is discontinuous.

Perrin (1934) and Budó et al. (1939) have treated the molecule as an ellipsoid with dipole moment components in its three principal axes. This has proved useful in the case of protein solutions (Oncley, 1943), where the solute molecule is so large that the discontinuities in the surrounding solvent are relatively unimportant.

Gierer and Wirtz (1953) (Hill et al. 1969) treated the rotating molecule as a sphere of radius a surrounded not by a continuous medium but by spherical layers of spherical molecules of radius a'. They obtained an expression for the inner friction constant

$$\zeta = \frac{8\pi\eta}{3} \left(\sum_{m=1}^{\infty} \frac{2a'}{R_m^4} \right)^{-1} \qquad (7.40)$$

where R_m, the radius of the spherical interface between the mth and

$(m+1)$th layer of molecules, is $a+2ma'$. If the rotating molecule is very large in comparison with those of the surrounding medium, that is, $a \gg a'$, the summation becomes equal to $1/(3a^3)$, and $\zeta = 8\pi\eta a^3$, which is identical with equation (7.37), used by Debye. When a' is of the same order of magnitude as a, a reasonable approximation gives

$$\sum_{m=1}^{\infty} \frac{2a'}{R_m^4} = \frac{1}{3a^3}\left(\frac{6a'}{a} + \frac{1}{(1+2a'/a)^3}\right) \tag{7.41}$$

When a' approaches a, the second term in parentheses in equation (7.41) becomes negligible in comparison with the first, and the summation becomes $2a'/a^4$, ζ becomes

$$\zeta = \frac{4\pi\eta a^4}{3a'} \quad \text{and} \quad \tau_\mu = \frac{2\pi\eta}{3kT}\frac{a^4}{a'} \tag{7.42}$$

For $a'=a$,

$$\zeta = \frac{4\pi\eta a^3}{3} \quad \text{and} \quad \tau_\mu = \frac{2\pi\eta a^3}{3kT} \tag{7.43}$$

Equation (7.43) gives for the special case $a'=a$, a value for τ_μ only $\frac{1}{6}$ as large as that given by (7.39) and much nearer to the magnitude commonly observed. In a pure dipolar liquid, where all of the molecules are the same, the Wirtz equation (7.43) would seem to be much preferable to the Debye equation (7.39). This treatment is a reminder of the fact that in an analysis of the effect of viscosity upon the dielectric relaxation of dipolar molecules in dilute solution in a non-polar solvent, it is not the macroscopic viscosity of the solution or that of the solvent which is important, but the interaction between the relatively small number of solute molecules and the solvent molecules which surround them.

Andrade (1934) assumed a model for liquids in which molecules of mass m vibrated with frequency f about temporary equilibrium positions, which changed relatively slowly with time and derived an expression for the viscosity

$$\eta = \frac{cfm}{3S} \tag{7.44}$$

in which S is the mean intermolecular distance, and c is the probability of a molecular collision being effective in the transfer of drift momentum. c may vary exponentially with temperature according to the equation

$$c = c_0 \exp(H/kT) \tag{7.45}$$

where H is an activation energy.

Hill and coworkers (1969) have applied this treatment to obtain the viscosity η of a mixture of liquids A and B in the expression

$$\eta S = X_A^2 \eta_A S_A + X_B^2 \eta_B S_B + 2 X_A X_B \eta_{AB} S_{AB} \qquad (7.46)$$

in which X_A and X_B are the mole fractions of A and B. η_A is the contribution to the viscosity arising from collisions of molecules of A with other A molecules, η_B is that arising from B–B molecular collisions, η_{AB} is termed the mutual viscosity of the mixture, and intermolecular distances S_A, etc, can be obtained from the relation $N S_A^3 = V_A = M_A/d_A$, where V_A is the molar volume, M_A the molecular weight, and d_A the density. S_{AB} need not be calculated separately from η_{AB}. Hill (Hill et al. 1969) has shown the considerable success of equation (7.46) in predicting the viscosities of mixtures, but has pointed out that the model used in its derivation cannot be expected to apply in the case of flexible, log-chain molecules or mixtures in which the molecules of A and B differ considerably in size.

Hill et al. (1969) have developed expressions for the microscopic relaxation times of the dipolar molecules of B

$$\tau_{\mu B} = (6 X_A \eta_{AB} S_{AB} K^2 + 3 A X_B \eta_B S_B K_B^2)/2kT = X_A \tau_{\mu AB} + X_B \tau_{\mu BB} \qquad (7.47)$$

where $\tau_{\mu AB}$ is the limiting relaxation time in the dilute solution, $\tau_{\mu BB}$ is the limiting relaxation time in the pure dipolar liquid, $K^2 = I_{AB} I_B (m_A + m_B)/m_A m_B (I_{AB} + I_B)$, $K_B^2 = I_B/m_B$, I_B is the moment of inertia of B about its centre of mass, and I_{AB} is the moment of inertia of molecule A about the centre of mass of B at the instant of collision. The factor A is a constant which, according to Hill, may plausibly be taken as $(3 - 2^{\frac{1}{2}}) = 1.59$.

In dilute solutions of the dipolar substance B in the non-polar liquid $A(X_A \gg X_B)$, the relaxation time depends on the mutual viscosity η_{AB} rather than the solution viscosity so that

$$\tau_{\mu B} = 3 \eta_{AB} S_{AB} K^2 / 2kT = \tau_{\mu AB} \qquad (7.48)$$

In the limiting case of a pure dipolar liquid ($X_B = 1$, $X_A = 0$) it follows from equation (7.47) that

$$\tau_{\mu B} = \tau_{\mu BB} = 3 A \eta_B S_B K_B^2 / 2kT \qquad (7.49)$$

This equation has the same form as the Debye equation (7.39), but the smaller factor by which η/T is multiplied, gives smaller values for τ, which agree better with the experimental results.

Hill (Hill et al. 1969) pointed out that, in the derivation of these equations, the molecules have been treated as rigid masses oscillating about positions of equilibrium, a model which should not necessarily apply to flexible long-chain molecules forming viscous liquids, or to mixtures of molecules differing considerably in size. There are other possible complicating factors, such as molecular shape, dipole location in the molecule, and intermolecular forces, the effects of which will be noted and discussed qualitatively in subsequent paragraphs.

Oncley (1943) measured the macroscopic relaxation time for dilute solutions of proteins in water and alcohols and found the values of the relaxation times to be of the magnitude to be expected from the values of molecular weight, asymmetry, and hydration determined by other methods. The differences between the microscopic and macroscopic relaxation times of these large molecules were small in spite of the large static dielectric constant values, which would tend to give a maximum difference. The relaxation times of the large protein molecules showed approximate proportionality to viscosity, presumably, because the solvent molecules surrounding those of the solute were so small in comparison with the large solute molecules as to give an approximation to the homogeneous fluid postulated by Debye. The dielectric method has proven useful in the investigation of very large molecules. Lack of proportionality between the relaxation times for small molecules and the viscosities was evident in the results of a number of early measurements. For example, twelve sets of measurements by various investigators on nitrobenzene in various solvents gave relaxation times from which equation (7.39) gave apparent radius values for the molecule varying from 0.25 to 2.4 Å.

A convenient way of examining equation (7.39) further is to calculate the hypothetical volume V_D of N molecules in a mole from the relation (Conner and Smyth, 1943)

$$V_D = \frac{4\pi Na^3}{3} = \frac{\tau RT}{3\eta} \qquad (7.50)$$

Values of V_D obtained for a variety of liquids are commonly only 10–20% of the directly measured molar volume $V = M/d$, where M is the molecular weight and d is the density, while if the liquid were a system of close-packed molecular spheres, the ratio of the actual volume of the spheres to the total volume of the liquid would be 0.74. The values of V_D for the straight-chain alkyl bromides (Hennelly et al., 1948) rise continuously from a value of 8.19 cm^3 at 25 °C for ethyl bromide to 13.84 cm^3 for n-hexyl bromide and then decrease to 8.55 cm^3 for n-hexadecyl bromide, a value less than 3% of M/d. Although the relaxation times increase with increasing chain lengths for these straight-chain molecules, it appears that orientation of the extended molecule by turning around its long axis and increased opportunity for orientation by twisting around the C—C bond, probably, cause the increase in relaxation time to fall farther and farther behind the increase in viscosity. The influence of at least one of these two factors is evidenced by the fact that the relaxation time 8.6×10^{-11} s at $25\,°C$ for α-bromonaphthalene, which has a somewhat flat, rigid molecule, is 2.5 times the value 3.4×10^{-11} for n-decyl bromide (Hennelly *et al.* 1948), which has the same number of carbon atoms arranged in a long flexible chain, while the viscosity, 4.52 cP, of α-bromonaphthalene is only 1.25 times the value 3.60 of n-decyl bromide.

The viscosities of a variety of fatty acid esters are slightly lower than those of alkyl bromides of approximately the same molecular length, but the relaxation times are only about half as large (McGeer *et al.* 1952). The rate of increase of the relaxation time with increase in viscosity is much less for the esters than for the alkyl bromides.

The results which have been discussed as typical indicate that the molecular or microscopic viscosity is much smaller than the directly measured or macroscopic viscosity which is available for calculating the molecular radius. This is not surprising when it is considered that the process of viscous flow used in measuring the macroscopic viscosity involves both translational and rotational motion of the molecules, while dipole orientation which determines the dielectric relaxation time primarily involves only molecular rotation, which may, however, necessitate some displacement and, hence, translational motion of the neighbouring molecules.

The molecules of methyl- and halogen-tetrasubstituted methanes have the shape of a tetrahedron with rounded corners and indented edges (Smyth, 1954). Their roughly spherical symmetry makes possible the orientation of their dipoles by molecular rotation in the crystal lattice. For such molecules in the pure liquid equation (7.39) should be at its best. For six of these molecules, substitution in equation (7.39) of measured values of η and values of τ_μ calculated from measured values of τ_m by means of equation (7.29) gives values for the apparent radius only 0.2–0.4 of the van der Waals radii of the molecules, in spite of the reduction of the discrepancies by the cube-root relationship. However, equation (7.43) gives values larger by a factor $6^{\frac{1}{3}} = 1.82$, which makes them 0.36–0.72 of the van der Waals radii, a somewhat improved agreement between calculated and observed. Another way of testing these equations is to insert correct values of the radii together with those of the microscopic relaxation times τ_μ and calculate the apparent or effective values of the viscosity, which may be called the microscopic viscosity η_μ. These calculated microscopic viscosities show discrepancies with the measured macroscopic viscosities analogous to those between the apparent volumes V_D calculated by means of equation (7.50) and the measured volumes. However, while the apparent volumes V_D are significant mainly in showing the extent of the deviation from the calculated behaviour, the values of the microscopic viscosity η_μ indicate that the resistance to rotation of the nearly spherical molecules is very much smaller than the resistance to translational motion, since the microscopic viscosities are only

0.008–0.06 of the macroscopic, which involve both kinds of motion. There appears to be little parallelism between the values of the microscopic and the macroscopic viscosities, which latter tend to increase with increasing molecular polarizability, but do not evidence an effect of molecular dipole moment, except, perhaps, in the case of 2,2-dinitropropane. For solutions of three methylchloromethanes in heptane and in a viscous paraffin oil, the relaxation times and, consequently, the microscopic viscosities, increased only 50% from heptane to Nujol, although the macroscopic viscosity increased 257 fold. In several of these tetrasubstituted methanes, the dielectric relaxation time is shortened by solidification of the liquid (Powles et al., 1953), showing that rotational orientation of these nearly spherical molecules may occur more readily in the crystal than in the liquid.

A comparison (Pitt and Smyth, 1959a) between the observed relaxation time and that calculated by the method of Hill (1943, 1944) for solutions of six dipolar substances in benzene and two in dioxan showed the calculated values for the benzene solutions to be three to nine times the observed, while those calculated for the dioxan solutions were 1.5–2.7 larger than the observed solution values. Although the use of mutual viscosity instead of solvent viscosity is, clearly, a logical step in the right direction, the relaxation times calculated with it for these solutions are much higher than those observed.

The relaxation times of camphor in solution in several different solvents having viscosities from 0.37 to 0.97 cP have been found to range from 5.8 to 10.7×10^{-12} s at 20 °C, not paralleling the viscosities. The camphor molecule is somewhat spherical in shape, showing rotational freedom over a wide range of temperature in the solid state. The relaxation time 17.5×10^{-12} s in Nujol solution at 20 °C is only slightly larger, although the viscosity of Nujol is 211 centipoises at 20 °C. This is in accordance with observations that nearly spherical molecules can rotate with little dependence on the viscosity of the medium. The long molecule of 4-bromobiphenyl in Nujol solution shows a relaxation time 148 times that of camphor at 20 °C because the orientation of its dipole by rotation involves the extensive displacemement of neighbouring molecules with consequent dependence on the viscosity of the liquid.

Application of the concept of mutual viscosity (Hill, 1954) to Nujol solutions shows its inadequacy for such data, which give too low and sometimes negative values for the mutual viscosity (Kalman and Smyth, 1960). Meakins (1958) concluded from measurements on a few molecules in different solvents that, when the solute molecule was at least three times as large as the solvent molecule, the solutions gave good agreement with Debye's simple equation, taking no account of molecular shape and using the macroscopic viscosity. It was found that a large flat porphyrazine molecule with its dipole in the long radius gave a relaxation time in benzene solution in rough agreement with that calculated by equation (7.39), the Debye equation, while a similarly shaped molecule with dipole in the short axis gave a relaxation time considerably longer than that calculated by equation (7.39). These results indicate that, although some parallelism has been found between the magnitudes of the relaxation time and the viscosity in pure liquids, the macroscopic viscosity has little meaning in the calculation of the absolute value of the relaxation time of a polar molecule unless it is surrounded by considerably smaller molecules.

It was found (Pitt and Smyth, 1959b) for two large porphyrazine molecules in benzene solution that equation (7.39) gave, at least, as good agreement with the measured relaxation times as did several previously proposed equations which took account of the departure of the molecule from sphericity. Since the volume swept out by an unsymmetrical molecule in rotational relaxation is dependent on the orientation of the dipole axis in the molecules, it is obvious that equation (7.39) cannot, at best, be expected to reproduce the exact value of the relaxation time of a non-spherical molecule.

The relation of molecular relaxation time to liquid viscosity and molecular size was investigated further by Nelson and Smyth (1964), who treated the molecule as a sphere of volume $V = 4\pi a^3/3$, and replaced a^3 in equation (7.39) by $3V/4\pi$ to obtain

$$\tau_\mu = \frac{3\eta V}{kT} \tag{7.51}$$

This differs from equation (7.50) only in that τ_μ specifies the microscopic or molecular relaxation time, $k = R/N$, and V is the volume of the individual molecule, which was calculated approximately from the atomic contributions to the van der Waals volumes of molecules compiled by Edward (1956). Two large molecules, bis(diphenylmethyl) ether and tetraphenylcyclopenta-dienone were measured in each of four solvents. The two solutions in which the volume of the solute molecule was only twice that of the solvent molecule averaged a relaxation time 0.65 of the calculated value, while the three solutions of bis(diphenylmethyl) ether in which the ratio of the volume of the solute molecule to that of the solvent molecule was about 4.0 averaged 0.88 of the calculated value and the three solutions of tetraphenyl-cyclopentadienone in which the volume ratio was about 4.3 averaged 1.12 of the calculated value. Except in the diphenylmethane solutions, the small discrepancies were rather less than might be expected from the differences in molecular shape involved. They were in strong contrast to the tremendous discrepancies in the cases of pure liquids consisting of small, nearly spherical molecules (Smyth, 1954; Miller and Smyth, 1956; Kalman and Smyth, 1960). The dropping of the relaxation times below the calculated values for the solutions in the large diphenylmethane molecules was small in comparison with the very large discrepancies observed in Nujol solutions, where the solvent molecules were much larger and the observed relaxation times were frequently as small as 0.1–0.01 and, occasionally, even 0.001 of the values to be expected from equation (7.51).

Calculation of the relaxation times for solutions of some 60 relatively rigid aromatic molecules for which measured values had been reported in the literature showed that, when the solute molecules were of about the same sizes as the sovent molecules, their observed relaxation times might be as small as 0.05 of the values calculated with equation (7.51), but as the solute to solvent size ratio increased, the observed value increased rapidly and approached much more closely to the calculated. When the ratios of observed to calculated relaxation times were plotted against the ratios of solute to solvent molecular volumes, the plot showed the extreme inadequacy of equation (7.51) and, hence, (7.39) when the volume of the polar solute molecule was smaller than that of the solvent molecule, the same size, or only a little larger. The relatively small differences between the observed and the calculated values of the relaxation time when the solute molecule has a volume three to five times that of the solvent molecule confirms the conclusion of Meakins (1958) and the subsequent observation of Pitt and Smyth (1959b) that the simple Debye equation (7.39) gives an approximate representation of the relation between dielectric relaxation time, viscosity of the medium, and volume of the relaxing molecule when the latter is, at least, three times as large as the volumes of the surrounding molecules. This is also in accord with the equations (7.40)–(7.43) of Gierer and Wirtz which cover a much wider range of size ratios.

It should be added that, although these many points for τ_{obs}/τ_{calc} against $V_{solute}/V_{solvent}$ may be very roughly represented by an empirical equation and a corresponding curve, the points should not lie on a single curve because the molecules to which they correspond have different shapes and different dipole orientations relative to the molecular axes. This is illustrated by the fact that paraldehyde, which is similar in viscosity and in molecular size and shape to 2,4,6-trimethylpyridine, has about twice as long a relaxation time, presumably, because its resultant dipole is perpendicular to the triangular slab form of the molecule, while, in 2,4,6-trimethylpyridine, it is in the plane of the slab (Miller and Smyth, 1956). A more extreme example is given by the fact that 4-iodobiphenyl in dilute solution in Nujol has a relaxation time about five times as large as that of 2-iodobiphenyl in Nujol (DiCarlo and Smyth, 1962), although the two molecules have virtually equal volumes and differ by only a moderate amount in length. It seems highly probable, however, that the relaxation times of rigid molecules at least three times as large as the rigid molecules surrounding them are roughly proportional to their volumes and to the viscosity of the liquid as required by equation (7.39), except for deviations caused by molecular shape and dipole orientation.

7.10 Individual Dipole–Dipole Interactions

In section 3 we considered the inductive effect of a molecular dipole upon neighbouring molecules in a liquid and upon the liquid as a whole as observed in the solvent effect and, in section 4, we considered the reaction field of a polarizable point dipole as preliminary to discussing, in section 5, the effects of the dipole field upon cohesion, boiling point, and solubility. It is convenient, at this point, to examine briefly the manner in which dipoles in adjacent molecules may orient relative to one another. In dipolar liquids and in solutions of dipolar molecules in non-polar solvents, particularly, in concentrated solutions, dipoles may line up head-to-tail, pointing in the same direction and adding their moments, or parallel to one another, pointing in opposite directions and cancelling one anothers' moments, that is, they are antiparallel. These are the two extreme orientations. Intermediate orientations usually occur because of the influence of molecular size and shape and dipole location in the molecule. Price (Hill *et al.* 1969) points out that the interaction energy of two dipoles from head-to-tail orientation is approximately $2\mu^2/r^3$ and for antiparallel orientation μ^2/r^3 and calculates an interaction energy of 74×10^{-15} erg/molecule for head-to-tail orientation when $\mu = 1$ D and $r = 3$ Å, as compared to $kT = 42 \times 10^{-15}$ erg/molecule at 300 °K. The larger the dipole moments and the smaller and more concentrated the dipolar molecules, the greater is the tendency toward one or the other extreme orientation, modified by the previously mentioned effects of molecular shape and dipole location in the molecule and, in solutions, reduced by dilution.

In early work on pure dipolar liquids, the fact that the Debye equation gave much lower dipole moments for the molecule in the pure liquid than in the vapour state or dilute solution in a non-polar solvent was sometimes attributed to a tendency to form antiparallel orientations of mutually opposing dipoles resulting in small or zero moments. However, the development of the Onsager equation largely eliminated or greatly reduced these differences except when definite molecular complex formation was known to occur. For example, acetic acid in the pure liquid state or even in moderately dilute solution shows dipole orientations in which the moments are largely cancelled by hydrogen bonding into complexes with dipoles in the antiparallel orientation, while hydrogen cyanide molecules in the pure liquid form head-to-tail orientations stabilized by hydrogen bonding into chain-like complexes with large moments, resulting in high dielectric permittivities (Coates and Coates, 1944): 206 at the melting point −13.3 °C, and

106 at the boiling point $25.7\,°C$. Cole (1954) has calculated a correlation factor $g = 4.04$ at $-13.3°\,C$ and 2.38 at $25.7°\,C$.

Occurrence of the antiparallel orientation provides a possible explanation of deviations from the non-linear dependence of polarization upon high electric field intensity predicted by Debye. Debye obtained an expression for the change of permittivity $\Delta\varepsilon$ produced by a field of intensity E acting on a liquid or gas,

$$\Delta\varepsilon = \frac{-4\pi N_1\mu^4 E^2}{45(kT)^3}A \tag{7.52}$$

in which N_1 is the number of dipolar molecules per cm^3 and $A = \{(\varepsilon + 2)^4\}/81$. Introduction of the Onsager equation by Thiébaut (1968) and Thiébaut *et al.* (1968) gave an expression of the same form, but with

$$A = \frac{\varepsilon^4(\varepsilon_\infty + 2)^4}{(2\varepsilon + \varepsilon_\infty)^2(2\varepsilon^2 + \varepsilon_\infty)^2} \tag{7.53}$$

The predicted decrease, $-\Delta\varepsilon$, in dielectric permittivity was found to occur in several simple liquids having molecules with moderate-sized dipole moments, such as ethyl ether, chloroform, and chlorobenzene (Kautsch, 1928), but A. and B. Piekara (1936) measured a large increase, $+\Delta\varepsilon$, for the much more polar nitrobenzene, and 1,2- and 1,3-nitrotoluene (Chelkowski, 1958) similarly showed large increases, $+\Delta\varepsilon$. Dilution of the highly polar liquid by a non-polar solvent, benzene, caused $\Delta\varepsilon$ to decrease rapidly to zero at a concentration of about 20% benzene in nitrobenzene, becoming negative for less concentrated solutions of nitrobenzene. Piekara explained this behaviour quantitatively as due to a transitory pairwise association of the highly polar molecules to give approximately antiparallel orientations of their dipoles and consequent lowering of the effective molecular dipole moments. The application of the strong electric fields used in the experiments tends to shift the molecular dipoles from the antiparallel orientation to one in the direction of the field, thus increasing their effective moments to such an extent that $\Delta\varepsilon$ is positive. The addition of the non-polar solvent reduces the amount of antiparallel dipole orientation and causes the less concentrated solutions to behave in the normal fashion with negative $\Delta\varepsilon$. Davies, Jones, and Gregson (Hill *et al.* 1969) have measured a solution of poly(γ-benzyl-L-glutamate) in dioxan at a field intensity $E = $ ca. $40\,kV/cm$. As the solute molecule has a huge dipole moment of about 2000 D, the observed $\Delta\varepsilon$ value, $-0.10 = \varepsilon$ (solvent) $- \varepsilon$(solution), was obtained, which means that dielectric saturation occurred.

Although it is an intramolecular rather than an intermolecular phenomenon, there would seem to be a rough parallelism between the proposed explanation of the nitrobenzene behaviour and the shift of rotational isomers, such as those of 1,2-dibromoethane, from the non-polar *trans* form to the polar *gauche* form in a strong field seemingly responsible for the $\Delta\varepsilon$ observed by Piekara and Chelkowski (Piekara *et al.*, 1937–58), However, dilution of the liquid evidently does not alter the proportion of *trans* form to the extent that it does the antiparallel orientation in nitrobenzene and, consequently, $\Delta\varepsilon$ retains a positive sign.

7.11 Hydrogen Bonding, Dipole Moment, and Relaxation Time

Among several books on the subject of the hydrogen bond, one may mention, as the most recent, the three-volume work edited by Schuster, Zundel, and Sandorfy (1976), which devotes an 89-page chapter, 'Dielectric Properties of Hydrogen-bonded Systems', to a survey of some of the work published in the period from about 1966 to 1973. Hydrogen bonding is most familiar in the interaction of the hydrogen atom in the O—H, N—H, and F—H groups with an unshared electron pair in another atom. Dipole–dipole forces contribute to the bonding energy and in much of the work on molecular association by hydrogen bonding, change of bond moment by association has, of necessity, been neglected. A comparison (Smyth, 1955) of dipole moments of potentially hydrogen-bonding molecules measured in benzene and in dioxan shows for n-butylamine and t-butylamine no increase in dipole moment from the solutions in benzene to those in dioxan, with whose oxygens they might be expected to form hydrogen bonds. However, hydrogen bonding to dioxan raises by 0.24, 0.37, and 0.28 the dipole moments of aniline, p-chloroaniline, and 2,4,6-tribromoaniline, in which charge shifts can occur through the ring structures. Dimethylaniline and 2,4,6-tribromodimethylaniline in which the hydrogen-bonding amino hydrogens are replaced by methyl groups show no significant difference in moment between the benzene and the dioxan solutions. The increase effected by hydrogen bonding seems to be greater the greater the initial polarity of the structure. p-Aminoacetophenone, p-aminobenzonitrile, p-nitroaniline, and 1-nitro-4-naphthylamine show rises of 0.50, 0.50, 0.57, and 0.59, while 2-nitro-6-naphthylamine shows a rise of 1.96 because of the greater charge separation in its polar structures.

Sobczyk, Engelhardt, and Bunzl (Schuster *et al.*, 1976) in their previously mentioned chapter define the polarity of the hydrogen bond as the excess dipole moment $\Delta\mu$ which is localized along the A—H bond in A—H $\cdots$ B. They tabulate the dipole moments of a number of phenols measured in solution in carbon tetrachloride, benzene, and dioxan, and list values of $\Delta\mu$, which, when corrected for charge transfer in the phenol derivative itself, lie between 0.33 and 0.60, with the exception of that for mesitol, 0.15. Malarski and Sobczyk (1969) have shown that even weak proton donors, such as the haloforms, phenylacetylene, diphenylamine, and triphenylcarbinol in solution in weak proton acceptors, such as benzene, show an increase of dipole moment from hydrogen interaction, which is seen in infrared spectra. It may be noted that the seven phenols measured in both carbon tetrachloride and benzene showed an average difference, $\mu_b - \mu_{ct} = 0.05$, a value too small to have much significance in itself, but consistent with the previous conclusion.

Measurements of dielectric relaxation time, proton magnetic resonance, and infrared absorption were combined in an investigation (Fong *et al.* 1966) of selected alkyl-substituted anilines and phenols. It had been found previously (Fong and Smyth, 1964) that, in 2,6-dimethylanisole, the two methyl groups hindered the rotation of the methoxy group, but, in this investigation of anilines and phenols, no such hindrance of rotation of the smaller OII and NH_2 groups was observed. Indeed, it appeared that *ortho*-alkyl groups shielded these groups from interaction with the solvent benzene molecules, which apparently occurred when the two methyl groups were in the *meta* instead of the *ortho* positions.

Dielectric relaxation times give evidence as to hydrogen bonding, but may be complicated by the effect of viscosity. Chloroform (Anthony and Smyth, 1964) has a relaxation time τ at 20 °C of 16.5×10^{-12} s in dilute solution in dioxan, 7.1 in benzene, 5.0 in carbon tetrachloride, 3.2 in cyclohexane, and 5.4 in the pure liquid. The value in the pure liquid is more than double the relaxation time 2.0×10^{-12} s for pure fluorotrichloromethane, the molecule of which is slightly larger, similar in shape, but lacking any hydrogen bonding. The values for chloroform show a large effect of hydrogen bonding diminishing to close to zero in cyclohexane. The use of the so-called reduced relaxation time τ/η to minimize the effects of the viscosity η changes the relative magnitudes somewhat, but does not alter the conclusions. It was pointed out that, if the relaxation time of a proton donor increases as the accepting ability of the solvent increases, other factors being equal, the order to be expected for the relaxation times of chloroform in the different solvents should be, just as observed, dioxan > benzene > carbon tetrachloride > cyclohexane, since this is the order found for the shifts in the OH stretching frequency of an alcohol in solution in these solvents. The work on chloroform has been extended to 1,2-dichloroethane in solution in several solvents at 20 °C (Chitoku and Higasi, 1967), to several haloethanes in cyclohexane and *p*-xylene at 25 °C (Crossley and Walker, 1968), and then to seven haloethanes in five solvents at 20 or 25 °C and 55 °C (Crossley and Smyth, 1969).

For all of the chloroethanes, except 1,1,2,2-tetrachloroethane, whose relaxation time in mesitylene is slightly longer than in dioxan, the relaxation times increase in the solvent order: cyclohexane < benzene < *p*-xylene < mesitylene < dioxan. Plots of the relaxation times of the solutes against the ionization potentials of the corresponding hydrocarbon solvents (Crossley and Smyth, 1969) proved to be linear, indicating solute–solvent interaction as the major factor in lengthening the relaxation times. The free energy of activation difference $\Delta\Delta F$ for molecular reorientation in two solvents was calculated from the equation

$$\frac{\tau_B}{\tau_c} = \exp\left(\frac{\Delta\Delta F}{RT}\right) \tag{7.54}$$

in which τ_B is the relaxation time in a solvent capable of acting as a proton acceptor (benzene, dioxan, etc.) and τ_c is that in cyclohexane. The values obtained for the free energy of activation differences between the reorientation process for 1,1,1-trichloroethane in cyclohexane, 140 cal/mol, and that for pentachloroethane in dioxan, 850 cal/mol, were of the magnitude to be expected and were comparable with similar values obtained from NMR measurements.

The abnormal dielectric behaviour of alcohols became evident many years ago (Smyth and Stoops, 1929) when maxima and minima were observed in the curves for their dielectric polarizations as functions of concentration in solution in non-polar solvents. An extensive literature has developed on the alcohol behaviour and a possibly more extensive one on water, liquid and solid. A brief treatment of some aspects of the alcohol behaviour will be given here because it relates to the rest of the chapter, but water behaviour is largely beyond the scope of this chapter.

Böttcher and his coworkers (1973), as well as several earlier investigators, treated alcohols as mixtures of multimers, whose dielectric behaviour as a function of temperature, pressure, and concentration could described in terms of the Kirkwood correlation factor g, which might have values approaching zero, as for 3-methyl-3-heptanol at 200 K, or approaching *ca.* 4, as for methanol, 1-butanol, and 1-heptanol. At high temperatures or very low concentrations, where the multimers are largely dissociated, g approaches 1.

Dielectric relaxation measurements provide a different approach to this complicated situation. The dielectric constant and loss (Garg and Smyth, 1965) curves for the straight-chain alcohols in the pure liquid state show a strong absorption in a frequency region (relaxation time, τ_1) too low to arise from rotational orientation of monomeric molecules, a weak, approximately measurable, but nonetheless reproducible absorption with a relaxation time τ_2 of the magnitude to be expected for the rotational orientation of monomeric molecules, and a third weak, but more accurately measurable absorption with a relaxation time τ_3 of the same size as that attributed to rotational orientation of the OH group in other compounds (Fong and Smyth, 1963). The range of the relaxation time values at 40 °C for the series of alcohols from n-propyl to n-dodecyl was $\tau_1 = 286–954 \times 10^{-12}$ s, $\tau_2 = 17.7–29.7 \times 10^{-12}$ s, and $\tau_3 = 1.80–3.87 \times 10^{-12}$ s. The range for the corresponding activation energies was $\Delta H_1 = 5.0–8.0$ kcal/mol, $\Delta H_2 = 1.5–$

3.1 kcal/mol, and $\Delta H_3 = 1.0-1.2 \pm 0.1$ kcal/mol. The sharpness of the low-frequency absorption region indicated that it did not arise from the rotational orientation of a variety of large multimeric molecules.

In dilute solutions (mole fraction 0.02–0.05) of methyl, n-butyl, and n-decyl alcohol in benzene, the low-frequency absorption region dominant in the pure alcohols does not appear (Johari and Smyth, 1969), as the solvent has broken up the liquid structure responsible for it, but two absorption regions are observed, one with $\tau_2 = 13-36 \times 10^{-12}$ s, presumably arising from rotational orientation of monomeric molecules, and one with $\tau_3 = 2-5 \times 10^{-12}$ s, attributable to OH group rotation. When dioxan is used as a solvent instead of benzene (Johari and Smyth, 1969), a single absorption region is observed with a distribution of relaxation times around a most probable relaxation time $\tau_0 = 3-9 \times 10^{-12}$ s. This behaviour is attributable to hydrogen bonding between solute and solvent, which is present in large excess, but the actual mechanism is obscured by the variety of structures possibly contributing to the result. Similar measurements (Glasser *et al.* 1972) on 1-butanol, 1-hexanol, 1-heptanol, and 1-decanol at concentrations from 0.03 to 0.3 mole fraction in n-heptane at 25 °C showed no low-frequency absorption region at concentrations up to mole fraction 0.1, but at 0.15 its contribution was 0.2–0.3 of the total and it seems probable that fragments of the structure which gives rise to the low-frequency absorption, 0.9 of the total in pure 1-heptanol, persist down to concentrations lower than 0.1. At the lower concentrations, the absorption occurs in two overlapping regions, one with relaxation time τ_2 ranging from 25 to 100×10^{-12} s, not far from that to be expected for rotational orientation of monomeric molecules and concentration dependent, the other with τ_3 about 3×10^{-12} s, almost independent of concentration and, presumably, the relaxation time for OH orientation by rotation around its C—O bond. Probably, the apparent increase in τ_2 with concentration is due to increase in association not fully taken care of in the analysis. In measurements (Crossley *et al.*, 1971) on six isomeric octyl alcohols at concentrations of 0.03–0.35 mole fraction in solution in n-heptane, those isomers in which a CH_3 group and the OH group were attached to the same carbon or to adjacent carbons did not show the low-frequency absorption region characterized by τ_1, but did show the other two, one with τ_2 of the magnitude commonly observed for rotational molecular relaxation and the other with $\tau_3 = ca.$ 3×10^{-12} s, previously attributed to relaxation of the OH group by rotation around its O—C bond. Presumably, the CH_3 group in these molecules was sufficiently close to the OH to shield it at least partially from the close approach of the OH groups in neighbouring molecules, thus combining with the solvent molecules to reduce hydrogen bonding between the alcohol molecules. For 5-methyl-3-heptanol, 4-octanol, and 2-octanol, in which the OH group is less shielded from the OH groups of neighbouring molecules, the low-frequency absorption amounts to about 0.2–0.4 of the total absorption at mole fraction = 0.33, and is probably present in smaller amounts at greater dilutions, where it is obscured by the other two absorption regions ($\tau_2 = 20$–60×10^{-12} s and $\tau_3 = ca.\ 3 \times 10^{-12}$ s).

A few mixtures of two alcohols, measured at relatively low frequencies have been reported (Denney and Cole, 1955; Bordewijk *et al.*, 1969) as showing a single relaxation region, but this has also been observed (Garg *et al.* 1966) for mixtures of two unassociated liquids, the loss curve being calculable from the concentration-weighted mean of the relaxation times of the two component liquids. Strong intermolecular forces may give rise to some departure from the predicted ideal curve, but only a very large difference in the two relaxation times produces two obvious maxima (Garg *et al.* 1966).

In conclusion, departure of the Kirkwood g values from a value of 1 usually gives positive evidence of departure of a liquid from normal behaviour and of the variation of the departure with temperature, pressure, and concentration, although it has been pointed out that a value of 1 for g for an alcohol does not necessarily prove the absence of hydrogen bonding because the effects of two kinds of multimers may cancel each other. It gives evidence of the probable presence of multimers of different sizes and dipole moments, but cannot give specific information as to the sizes of the multimers or of their moments. It may be a little more sophisticated than the old method of treating a liquid alcohol as a mixture of monomers, dimers, trimers, tetramers, etc., formulating a structure for each multimer, and calculating a dipole moment for each structure. The concentrations of multimers then provided as many adjustable parameters as might be needed to describe any dielectric behaviour. If applied to a system in which very few different multimers were possible and used with judgement, this approach was capable of yielding some conclusions, however speculative.

The dielectric relaxation method is necessarily somewhat approximate and the interpretation of some of its results is speculative, but it sheds some light upon mechanisms which other methods leave in darkness. Some analogy has been suggested between the mechanism of the low-frequency absorption (τ_1) in the hydrogen-bonded structures of the pure liquid alcohols and that proposed for the solid rotation phase of long-chain alcohols, in which proton transfer within sheets of hydrogen-bonded OH groups was suggested. The mechanism in the liquid alcohol could consist of the breaking of a hydrogen bond, which requires an activation energy close to the values $\Delta H = 5.0$–8.0 kcal/mol observed for the dielectric absorption (Garg and Smyth, 1965), followed by rotational orientation of the now partially free ROH, whose rotation time is dependent on its length, as is τ_1. The observation of this low-frequency absorption region in the more concentrated solutions of the alcohols indicates the existence of fragments of the strongly hydrogen-bonded structure of the pure liquid. The intermediate absorption region (τ_2) presumably corresponds to the rotational orientation of single molecules together with a few small fragments of the pure liquid structure or multimers. The high-frequency absorption appears to correspond to the rotational orientation of non-hydrogen-bonded OH groups.

7.12 Millimetre and Submillimetre Absorption

The dielectric loss or absorption observed in dipolar liquids at millimetre wavelengths usually consists almost but not quite entirely of dipole rotational absorption and, as such, has been used in the calculation of the relaxation times which have been discussed in previous sections. Non-polar liquids had been generally regarded as having no appreciable dielectric loss until Whiffen's (1950) careful measurements confirmed by those of Heston and Smyth (1950) showed, for several pure non-polar liquids, a loss detectable at 3.5 cm wavelength and increasing approximately linearly with frequency up to that corresponding to a wavelength of 8 mm. Well within their experimental error the points also lay on apparent Debye curves, which yielded apparent dipole moments of about 0.04–0.10 D and relaxation times of about 10^{-12} s for the molecules. The dipole moment values were of the magnitude estimated as induced during molecular collisions in the liquids and the relaxation times could be regarded as describing the decay of the induced moments. They were of the magnitude of the times between molecular collisions in the liquids and decreased with rising temperature as do these times. The millimetre wave measurements were subsequently extended down to 2 mm to obtain dielectric loss values (Garg *et al.* 1965) which continued to show the increase with frequency previously observed at the slightly lower frequencies.

Improved infrared methods soon made possible the extension of these measurements into the infrared. Chantry and coworkers (1967), using a Michelson interferometer, measured benzene, carbon tetrachloride, and carbon disulphide, and Garg *et al.* (1968) with the assistance of H. J. Labbé and using a grating spectrometer, measured these same liquids, as well as cyclohexane, tetrachloroethylene, and several mixtures. The two sets of measurements by different methods gave almost identical results. Dielectric measurements by Garg *et al.* (1965) were also carried out at wavelengths of 0.22, 1.25, and 3.22 cm or frequencies of 4.5, 0.80, and 0.31 cm^{-1} and found to give absorptions which lay well on the low-frequency extensions of the infrared curves drawn from 170 to 17 cm^{-1} and possessing broad maxima. The existence of a similar band in the same range of frequency for several different liquids suggested that this absorption was associated with the behaviour of the liquid phase rather than with that of specific molecules as such. Further indication was given by the fact that carbon tetrachloride vapour showed no absorption at frequencies below that due to the vibration at 315 cm^{-1} (Gabelnick and Strauss, 1967). Evidently the liquid absorption arose from intermolecular forces. Chantry *et al.* (1967) used the effect of pressure on the absorption bands to conclude that the absorption in the non-polar liquids was multipole-induced dipole absorption, and Garg *et al.* (1968) found that the apparent dipole moment values calculated from the absorption were in satisfactory agreement with those calculated as induced by the quadrupole or octapole moments of neighbouring molecules. Dielectric measurements by Dasgupta and Smyth (1974) upon several non-polar liquids and mixtures showed that the absorption coefficients were increased by the presence in the molecules of mutually opposing and compensating permanent dipoles and of π electrons. This established that, as would be expected, local dipole-induced dipole moments may also contribute to the far-infrared absorption. In a dipolar liquid the contribution would be small in comparison to that of the permanent molecular dipole moments. Bending of dipolar bonds or of bonds to dipolar groups, a source of atomic polarization (Coop and Sutton (1938); Smyth (1955)) is presumably a mechanism contributing to this absorption, as was proposed by Gabelnick and Strauss (1967) for carbon tetrachloride.

Before the development of Fourier spectroscopy, Poley (1955) found that the measured dielectric constants and losses of six monosubstituted benzenes fitted on Cole–Cole semicircles with zero distribution factors and values of ε_∞ consistently larger than the square of the infrared refractive indices n_i^2 given by the previously mentioned measurements of Cartwright and Errera (1936) at 70 cm^{-1}. The difference $(\varepsilon_\infty - n_i^2)$ was roughly proportional to the square of the molecular dipole moment. These differences, increasing from 0.18 for fluorobenzene, the least polar of the six molecules, to 1.64 for nitrobenzene, the most polar of the six were much larger than the small values of the loss ε'', which served to establish the existence of a loss or absorption region for non-polar liquids extending from the millimetre wave region into the very far-infrared region. Poley attributed these $(\varepsilon_\infty - n_i^2)$ differences to further absorption existing between the highest frequency 1.2 cm^{-1} of his dielectric measurements and the frequency 70 cm^{-1} at which n_i had been obtained and arising from oscillation of the dipolar molecules about locally and temporarily defined positions in the liquid. Hill (1963) developed this idea quantitatively by treating the absorption as arising from a partial orientation of a dipolar molecule librating in a cage formed by its neighbours and derived and approximate expression

$$\frac{\varepsilon_\infty - n_i^2}{\varepsilon_0 - n_i^2} = \frac{2kT}{I\omega_0^2} \tag{7.55}$$

in which ε_∞ is now the value terminating the Debye dispersion and assumed to come at a frequency below the start of the librational absorption, n_i is now the refractive index at the high-frequency end of the librational absorption, I is the moment of inertia of the librating molecule, and ω_0 is the frequency of the angular oscillation or libration of the caged molecule. Chantry and Gebbie (1965) and Leroy and Constant (1966) used inter- ferometric methods to observe absorption bands in the 50 cm^{-1} region for several dipolar liquids including chlorobenzene. These observed frequencies differed within a factor of about 2 from the resonance frequencies predicted by the approximate equation (7.55). The chlorobenzene absorption centred near 48 cm^{-1} coincides with the region of two lattice vibrations observed for the solid (Chantry and Gebbie, 1965). It would seem that these oscillations of the dipolar molecules could often be responsible for a large part of the atomic polarization P_A of dipolar molecules, which was found to depend in part upon the size of the dipole moments (Smyth, 1955) and attributed to charge displacement associated with infrared absorption.

In terms of the Hill treatment (Davies *et al.* 1968), the Debye dispersion $(\varepsilon_0 - \varepsilon_\infty)$ involves the major part of the reorientation of the dipolar molecule by cooperative movement with the immediately surrounding molecules, but the molecule can still librate in a local energy well or fluctuating molecular cage and the mean degree of reorientation by this process is responsible for the dispersion $(\varepsilon_\infty - n_i^2)$.

It should be mentioned at this point that, in the derivation of his equations for dielectric behaviour, Debye justifiably neglected the second term in the expression for the torque exerted on a dipolar molecule by a field F

$$-\mu F \sin \theta = \zeta \frac{d\theta}{dt} + I \frac{d^2\theta}{dt^2} \tag{7.56}$$

where ζ is the frictional force constant and I is the molecular moment of inertia. The second term is usually negligibly small at frequencies below about 10^{11} Hz. Rocard (1933) and Powles (1948) have treated the effects of inclusion of this term in the derivation of equation (7.20), and Davies' review (Hill *et al.* 1969) shows that its inclusion can account for a considera- ble fraction of the absorption in the region of the broad band in the far-infrared region for dipolar liquids, which have their maxima at frequen- cies above 33 cm^{-1}. The problem has been further treated by Steele (1963), Constant *et al.* (1965), and Leroy *et al.* (1967).

Kroon and van der Elsken (1967) have calculated the total intensity of the absorption of the regions $(\varepsilon_0 - \varepsilon_\infty)$ and $(\varepsilon_\infty - n_i^2)$ as

$$A = \frac{1}{N_1} \int_0^\infty \alpha(\bar{\nu}) \, d\bar{\nu} = \frac{\pi}{3c^2} \sum \mu_z^2 \left(\frac{1}{I_x} + \frac{1}{I_y} \right) \tag{7.57}$$

in which N_1 is the number of molecules per cm^3, $\tilde{\nu}$ is the frequency, α is the absorption coefficient, μ_z is the dipole moment along the polar axis, I_x and I_y are the moments of inertia of the molecule perpendicular to the polar axis, and c is the velocity of light. Values of A observed for a number of polar liquids were 70–90% of the calculated. The insertion into equation (7.57) of a dipole moment of the small size (0.03–0.12 D) of the collision-induced moments obtained by Garg *et al.* (1968) for several non-polar liquids leads to an absorption value of the same magnitude as that observed.

Davies (1970) in an excellent review has pointed out that the Debye relaxation, the molecular inertial effect, and the librational absorption all relate to a single process, for which representation can be obtained in terms of a correlation function $\varphi(t)$ for the time dependence of the angular orientation of the molecule (Cole, 1965; Gordon, 1965, 1966; Brot, 1967; Lassier and Brot, 1967, 1969). G. J. Davies and M. Evans (1976) have used generalized Langevin theory to describe far-infrared absorptions in non-dipolar liquids and, in a second paper Evans and Davies (1976), have studied the effect of pressure and temperature on the intermolecular mean-square torque in liquid carbon disulphide and carbon tetrachloride. G. J. Evans and M. W. Evans (1976) have derived a simple equation to describe the molecular rotational processes giving rise to the microwave and far-infrared absorption bands of dipolar molecules in the liquid phase, and have tested it with a series of liquids covering the extremes of molecular isotropy and anisotropy. G. J. Davies *et al.* (1977) used a specially developed polarizing interferometer for the range 2–200 cm^{-1} to obtain the rotational type absorptions of t-butyl chloride and some halobenzenes in the liquid state. The broad bands obtained had contributions from both permanent and induced temporary dipoles, the absorption due to the latter being negligibly small at frequencies below about 2 cm^{-1}, but gradually becoming predominant as the frequency increased. The authors found that their theoretical treatment was unable to reproduce the experimental results satisfactorily and referred to Deutch (1977), who, in an assessment of the theoretical and practical difficulties of studying molecular motion by means of dielectric measurements, concluded that, in the far infrared, differences in the predictions of the different models increased and the assumptions as to the effects of intramolecular motions and polarizabilities became more uncertain.

It would be beyond the limitations of this chapter to give a complete and meaningful account of the recent developments in the investigation of very far-infrared absorption in liquids, which merges into and, indeed, is virtually inseparable from dielectric absorption. The papers which have been discussed briefly or referred to give much more information and many additional references to work in these bordering and overlapping areas, in which molecular interactions are so important.

7.13 References

Andrade, E. N. da C. (1934). *Phil. Mag.*, **17**, 497.

Anthony, A. A. and Smyth, C. P. (1964). *J. Amer. Chem. Soc.*, **86**, 152.

Bergmann, K. (1957). *Doctoral Dissertation*, Freiburg, West Germany.

Bergmann, K., Roberti, D. M., and Smyth, C. P. (1960). *J. Phys. Chem.*, **64**, 665.

Bordewijk, P., Gransch, F., and Böttcher, C. J. F. (1969). *J. Phys. Chem.*, **73**, 3255.

Böttcher, C. J. F. (1973). *Theory of Electric Polarization*, 2nd edn., completely revised by O. C. Van Belle, P. Bordewijk, and A. Rip, Vol. I, Elsevier Scientific Publishing Company, Amsterdam.

Brot, C. (1967). *J. Phys. Radium*, **28**, 789.

Budó, A., Fisher, E., and Miyamoto, S. (1939). *Physik. Z.*, **40**, 337.

Cartwright, C. H. and Errera, J. (1936). *Proc. Roy. Soc. (London)*, **A154**, 138.

Chantry, G. W. and Gebbie, H. A. (1965). *Nature*, **208**, 378.

Chantry, G. W., Gebbie, H. A., Lassier, B., and Willie, G. (1967). *Nature*, **214**, 163.

Chelkowski, A. (1958). *J. Chem. Phys.*, **28**, 1249.

Chitoku, K. and Higasi, K. (1967). *Bull. Chem. Soc. Japan*, **40**, 773.

Coates, G. E. and Coates, J. E. (1944). *J. Chem. Soc.*, 77.

Cole, K. and Cole, R. H. (1941). *J. Chem. Phys.*, **9**, 341.

Cole, R. H. (1954). *J. Amer. Chem. Soc.*, **77**, 2012.

Cole, R. H. (1957). *J. Chem. Phys.*, **27**, 33.

Cole, R. H. (1965). *J. Chem. Phys.*, **42**, 637.

Conner, W. P., Clarke, R. P., and Smyth, C. P. (1942). *J. Amer. Chem. Soc.*, **64**, 1379.

Conner, W. P. and Smyth, C. P. (1943). *J. Amer. Chem. Soc.*, **65**, 382.

Constant, E., Leroy, Y., and Raczy, L. (1965). *Compt. Rend.*, **261**, 4687.

Coop, I. E. and Sutton, L. E. (1938). *J. Chem. Soc.*, 1269.

Crossley, J., Glasser, L., and Smyth, C. P. (1971). *J. Chem. Phys.*, **55**, 2197.

Crossley, J. and Smyth, C. P. (1969). *J. Amer. Chem. Soc.*, **91**, 2482.

Crossley, J. and Walker, S. (1968). *J. Chem. Phys.*, **48**, 4742.

Dasgupta, S. and Smyth, C. P. (1974). *J. Chem. Phys.*, **60**, 1746.

Davidson, D. W. and Cole, R. H. (1951). *J. Chem. Phys.*, **18**, 1417; **19**, 1484.

Davies, G. J. and Evans, M. (1976). *J. Chem. Soc. Faraday II*, **72**, 1194.

Davies, G. J., Evans, G. J., and Evans, M. (1977). *J. Chem. Soc. Faraday II*, **73**, 1071.

Davies, M. (1970). *Ann. Reports*, **67**, 65.

Davies, M., Pardoe, G. W. F., Chamberlain, J. E., and Gebbie, H. A. (1968). *Trans. Faraday Soc.*, **64**, 847.

Debye, P. (1929). *Polar Molecules*, Chemical Catalog Co., New York.

Denney, D. J. and Cole, R. H. (1955). *J. Chem. Phys.*, **23**, 1767.

Deutch, J. M. (1977). *Faraday Symp. Chem. Soc.*, **11**, 26.

DiCarlo, E. N. and Smyth, C. P. (1962). *J. Phys. Chem.*, **66**, 1105.

Edward, J. T. (1956). *Chem. Ind. (London)*, 774.

Evans, G. J. and Evans, M. W. (1976). *J. Chem. Soc. Faraday II*, **72**, 1169.

Evans, M. and Davies, G. J. (1976). *J. Chem. Soc. Faraday II*, **72**, 1206.

Fong, F. K., McTague, J. P., Garg, S. K., and Smyth, C. P. (1966). *J. Chem. Phys.*, **70**, 3567.

Fong, F. K. and Smyth, C. P. (1963). *J. Amer. Chem. Soc.*, **85**, 1565.

Fong, F. K. and Smyth, C. P. (1964). *J. Chem. Phys.*, **40**, 2404.

Forest, E. and Smyth, C. P. (1965). *J. Phys. Chem.*, **69**, 1302.

Frank, F. C. (1935). *Proc. Roy. Soc. (London)*, **152A**, 171.

Fröhlich, H. (1949, 1958) *Theory of Dielectrics*, Oxford University Press, London.

Gabelnick, H. S. and Strauss, H. L. (1967). *J. Chem. Phys.*, **46**, 396.
Garg, S. K., Kilp, H., and Smyth, C. P. (1965). *J. Chem. Phys.*, **43**, 2341.
Garg, S. K., Kilp, H., and Smyth, C. P. (1966). *J. Chem. Phys.*, **45**, 2799.
Garg, S. K. and Smyth, C. P. (1965). *J. Phys. Chem.*, **69**, 1294.
Garg, S. K., Bertie, J. E., Kilp, H., and Smyth, C. P. (1968). *J. Chem. Phys.*, **49**, 2551.
Gierer, A. and Wirtz, K. (1953). *Z. Naturforsch.*, **8a**, 532.
Glasser, L., Crossley, J., and Smyth, C. P. (1972). *J. Chem. Phys.*, **57**, 3977.
Glasstone, S., Laidler, K. J., and Eyring, H. (1941). *The Theory of Rate Processes*, McGraw-Hill, New York.
Gordon, R. G. (1965). *J. Chem. Phys.*, **43**, 1307; (1966) *J. Chem. Phys.*, **44**, 1803.
Hennelly, E. J., Heston, W. M., Jr., and Smyth, C. P. (1948). *J. Amer. Chem. Soc.*, **70**, 4102.
Heston, W. M., Jr. and Smyth, C. P. (1950). *J. Amer. Chem. Soc.*, **72**, 99.
Higasi, K. (1936). *Sci. Papers Inst. Phys. Chem. Research (Tokyo)*, **28**, 284.
Hildebrand, J. H. and Carter, J. M. (1930). *Proc. Natl. Acad. Sci. U.S.*, **16**, 285.
Hildebrand, J. H. and Scott, R. L. (1950). *The Solubility of Nonelectrolytes*, 3rd edn., Reinhold, New York.
Hill, N. E. (1943, 1944). *Proc. Phys. Soc. (London)*, **47**, 277; **48**, 76.
Hill, N. E. (1954). *Proc. Phys. Soc. (London)*, **67B**, 149.
Hill, N. E. (1963). *Proc. Phys. Soc. (London)*, **82**, 723.
Hill, N. E., Vaughan, W. E., Price, A. H., and Davies, M. (1969). *Dielectric Properties and Molecular Behavior*, Van Nostrand Reinhold, London.
Johari, G. P. and Smyth, C. P. (1969). *J. Amer. Chem. Soc.*, **91**, 6215.
Kalman, O. F. and Smyth, C. P. (1960). *J. Amer. Chem. Soc.*, **82**, 783.
Kautsch, F. (1928). *Physik. Z.*, **29**, 109.
Kilp, H., Garg, S. K., and Smyth, C. P. (1966). *J. Chem. Phys.*, **45**, 2799.
Kirkwood, J. G. (1939). *J. Chem. Phys.*, **7**, 911.
Kroon, S. G. and van der Elsken, J. (1967). *Chem. Phys. Lett.*, **1**, 285.
Lassier, B. and Brot, C. (1967). *Chem. Phys. Lett.*, **1**, 581; (1969) *Discuss. Faraday Soc.*, **48**, 39.
Leroy, Y. and Constant, E. (1966). *Compt. Rend.*, **262**, 1391.
Leroy, Y., Constant, E., and Desplanques, P. (1967). *J. Chim. Phys.*, **64**, 1499.
Malarski, Z. and Sobczyk, L. (1969). *Compt. Rend.*, **C269**, 874.
McGeer, P. L., Curtis, A. J., Rathmann, G. B., and Smyth, C. P. (1952). *J. Amer. Chem. Soc.*, **74**, 3541.
Meakins, R. J. (1958). *Trans. Faraday Soc.*, **54**, 1160.
Miller, R. C. and Smyth, C. P. (1956). *J. Chem. Phys.*, **24**, 814; *J. Phys. Chem.*, **60**, 1354.
Miller, R. C. and Smyth, C. P. (1957). *J. Amer. Chem. Soc.*, **79**, 3310.
Müller, F. H. (1933). *Physik. Z.*, **34**, 689.
Müller, F. H. (1934). *Physik. Z.*, **35**, 346; *Trans. Faraday Soc.*, **30**, 731.
Müller, F. H. (1937). *Physik. Z.*, **38**, 283.
Müller, F. H. and Mortier, P. (1935). *Physik. Z.*, **36**, 371.
Nelson, R. D., Jr. and Smyth, C. P. (1964). *J. Phys. Chem.*, **68**, 2704.
O'Dwyer, J. J. and Sack, R. A. (1952). *Australian J. Sci. Research*, **A5**, 647.
Oncley, J. L. (1943). In: *Proteins, Amino Acids and Peptides* (Eds. Cohn, E. J. and Edsall, J. T.), Reinhold, New York.
Onsager, L. (1936). *J. Amer. Chem. Soc.*, **58**, 1486.
Perrin, F. (1934). *J. Phys. Radium*, **5**, 497.
Piekara, A. (1937). *Acta Phys. Pol.*, **6**, 130, 150; (1939) *Proc. Roy. Soc. (London)*, **A172**, 360.

Piekara, A. and Chelkowski, A. (1956). *J. Chem. Phys.*, **25**, 795.
Piekara, A. and Kielich, S. (1958). *J. Chem. Phys.*, **29**, 1297.
Piekara, A. and Piekara, B. (1936). *Compt. Rend.*, **203**, 825.
Pitt, D. A. and Smyth, C. P. (1959a). *J. Amer. Chem. Soc.*, **81**, 783.
Pitt, D. A. and Smyth, C. P. (1959b). *J. Phys. Chem.*, **63**, 582.
Poley, J. P. (1955). *J. Appl. Sci. Res.*, **B4**, 337.
Powles, J. G. (1948). *Trans. Faraday Soc.*, **44**, 802.
Powles, J. G. (1953). *J. Chem. Phys.*, **21**, 633.
Powles, J. G., Williams, D. E., and Smyth, C. P. (1953). *J. Chem. Phys.*, **21**, 136.
Raman, C. V. and Krishnan, K. S. (1928). *Proc. Roy. Soc. (London)*, **117A**, 589.
Rocard, Y. (1933). *J. Phys. Radium*, **4**, 247.
Ross, I. G. and Sack, R. A. (1950). *Proc. Phys. Soc. (London)*, **36B**, 893.
Scholte, P. G. (1949). *Physica*, **15**, 437.
Schuster, P., Zundel, G., and Sandorfy, C. (1976). *The Hydrogen Bond*, Vol. III, Chap. 20, North-Holland, Amsterdam.
Smith, J. W. (1948). *Sci. Progr.*, **36**, 483.
Smyth, C. P. (1954). *J. Phys. Chem.*, **58**, 580.
Smyth, C. P. (1955). *Dielectric Behavior and Structure*, McGraw-Hill, New York.
Smyth, C. P. (1972). *Physical Methods of Chemistry* (Eds. Weissberger, A. and Rossiter, B. W.), Vol. I, Part IV, Chap. VI, Interscience, New York.
Smyth, C. P. and Lewis, G. L. (1940). *J. Amer. Chem. Soc.*, **62**, 949.
Smyth, C. P. and Stoops, W. N. (1929). *J. Amer. Chem. Soc.*, **51**, 3312, 3330.
Steele, W. A. (1963). *J. Chem. Phys.*, **38**, 2404, 2411.
Thiébaut, J. M. (1968). *Thesis*. Nancy.
Thiébaut, J. M., Weisbecker, A., and Ginet, C. (1968). *Compt. Rend.*, **267**, 66.
Vaughan, W., Smyth, C. P., and Powles, J. G. (1972). *Physical Methods of Chemistry* (Eds. Weissberger, A. and Rossiter, B. W.), Vol. I, Part IV, Chap. V, Interscience, New York.
Weigle, J. (1933). *Helv. Phys. Acta*, **6**, 68.
Whiffen, D. H. (1950). *Trans. Faraday Soc.*, **46**, 124.

Molecular Interactions, Volume 2
Edited by H. Ratajczah and W. J. Orville-Thomas
© 1980 John Wiley & Sons Ltd.

8 Nuclear quadrupole resonance studies of molecular complexes

L. Guibé and G. Jugie

8.1 General Introduction

Complexes are studied with help of many different techniques among which spectroscopic methods are of major importance as they give information at the microscopic level: transitions are observed between electronic or nuclear energy levels in the molecules forming the complex and thus are of fundamental importance for understanding the modification in the electron distribution during the formation of the complex. In this respect, nuclear quadrupole resonance spectroscopy (NQRS), the results of which depend upon the components of the electric field gradient (EFG) tensor at the site of the resonant nucleus caused by the surrounding electronic distribution, is particularly attractive.

It should be emphasized that NQRS gives information about the fundamental state of the molecule and valence orbitals. In this it differs from optical or UV spectroscopy which give data on the spacing of excited and fundamental electronic states. On the other hand, as quadrupolar nuclei act as local probes sensitive to the EFG at their site, NQRS gives local information differing from global parameters resulting from dipole moments or thermochemical measurements. NQRS appears as intrinsically different from other methods of chemical physics and is complementary in at least two ways: first it yields special information on the electronic structure which are otherwise unobtainable and, second, in the discussion of NQRS results (about the symmetry of the site of the resonant nuclei, bond angles in relation to the hybridization schemes, distances to neighbour atoms or molecules contributing to intermolecular interactions) IR and X-ray diffraction data can be used.

Since its discovery (Dehmelt and Krüger, 1950), NQRS has been widely extended to a great number of ionic, molecular, or complex and organic as well as inorganic compounds. As early as 1953, the ^{35}Cl NQR in two complexes, nitrobenzene·tin tetrachloride and diethyl ether·aluminium trichloride, was reported (Dehmelt, 1953). Now, the number of molecular complexes investigated by means of NQRS amounts to several hundred and the present chapter aims to review the results obtained in the field.

First, principles of NQRS and interpretation of experimental data in terms of chemical bonding are briefly recalled. Then, available results on molecular complexes are surveyed. Several reviews of complexes studied by NQRS have already been published among which the following can be cited: Maksyutin *et al.* (1970b), Weiss (1972, 1974), Semin *et al.* (1972), Ramakrishnan *et al.* (1977). Complexes are usually considered according to the nature of the orbitals of the electrons participating in the charge transfer and then classified as π–π, π–σ, n–σ, and σ–σ complexes. In this review, owing to the very great number of complexes so far investigated by NQRS, it seems convenient to adopt a systematic plan based on the scanning of the different columns of the periodical chart of the elements. It also appears useful to give a list of the molecular complexes studied by NQRS with corresponding literature references; to facilitate examination of the list and finding of complexes of a given molecule, each complex is cited twice, with the donor–acceptor (D–A) and acceptor–donor (A–D) formulations. Finally, ionic complexes which are extensively reviewed in a recent paper (Ramakrishnan *et al.* 1977) will not be considered here except for a few of them.

8.2 Nuclear Quadrupole Resonance Spectroscopy

8.2.1 *Introduction*

NQRS is now well known and many review articles and textbooks are available for reference (Das and Hahn, 1958; Lucken, 1969; Schempp and Bray, 1970; Smith, 1971; Semin *et al.*, 1972–1975; Chihara and Nakamura, 1972; Voronkov and Feshin, 1973; Grechishkin, 1973; Cousseau, 1976). Lists of nuclear quadrupole resonance frequencies and coupling constants can also be found (Biryukhov *et al.*, 1969; Semin, 1972–1975). Let us just recall that nuclear quadrupole resonance (NQR) frequencies are related to the quadrupole coupling constant (QCC), e^2Qq, between the electric quadrupole moment, eQ, of the resonant nucleus which acts as a probe inside the molecule, or the ion, under study, and the electric field gradient (EFG), eq, at the site of the nucleus. The EFG is produced by the electric charges surrounding the nucleus; they are the valence electrons in a molecule of a molecular compound and the neighbouring ions in a ionic crystal. In a molecular crystal the QCC, e^2Qq, is thus related to the distribution of the valence electrons of the resonant atom. Charges farther from the resonant nucleus also contribute to the EFG, but to a smaller extent and to a first, generally satisfying, approximation, they may be neglected. The relation between e^2Qq and the distribution of valence electrons is usually described by the Townes and Dailey's theory (Townes and Dailey, 1949; Dailey and Townes, 1955). Theoretical chemists study the molecules by approximate methods using LCAO–MO's and obtain the population of the valence orbitals on the resonant atom for comparison with experimental results or, using *ab initio* methods, calculate the components of the EFG directly.

NQRS is a sensitive method in the sense that the linewidth of the resonance is small (typically a few kilohertz for chlorine resonance at 30 MHz, corresponding to a relative value of $\sim 10^{-4}$). One may thus expect to be able to detect very small variations of the electronic distribution in molecules upon formation of a complex. Unfortunately, besides the valence electrons on the resonant atom, there are other contributions to the observed QCC and its variation upon complexation and their effects are quite unpredictable. They are (i) the contribution of the other electrons in the molecule which are also affected by the complexation, (ii) the effect of neighbouring molecules, (iii) the effect of molecular motion which averages the QCC of the resonant nucleus. All these effects limit the accuracy of the estimation of the influence of complexation on the distribution of valence electrons. Such strong limitations to the application of NQRS to the study of electronic distribution should be kept in mind when discussing experimental results. On the other hand, NQRS has still other potentialities resulting from the influence of temperature, pressure, and magnetic field on the QCC, as well as of temperature on the relaxation time, and they have been fruitfully employed in several circumstances for the study of molecular complexes. Since the first observation of NQR in complexes (Dehmelt, 1953), many studies have been developed and reported with interesting results worthy of a review.

8.2.2 *Basic principles*

NQRS consists in the observation of NQR spectra resulting from transitions between energy levels due to the coupling of the nuclear electric quadrupole moment with the electric field gradient at the site of the nucleus considered. The nuclear electric quadrupole moment is a tensor with components given by

$$Q_{ij} = \int x_i x_j \rho \, d\tau$$

where x_i and x_j are the ith and jth components of the nuclear element of volume $d\tau$, ρ is the electric charge density in units of the electronic charge e, and the integral is carried over the nuclear volume. In practice, this tensor can be completely determined by a single component Q along the direction of the nuclear spin; it also appears that only nuclei with spin $I \geqslant 1$ can have a

Table 8.1 Nuclear spin, quadrupole moment, and isotopic abundance of some nuclei studied in molecular complexes

Isotope	Spin	Quadr. mom.	Isot. abund. (%)
^{2}H(D)	1	2.77×10^{-3}	1.56×10^{-2}
^{14}N	1	2×10^{-2}	99.635
^{17}O	5/2	-4×10^{-3}	3.7×10^{-2}
^{25}Mg	5/2		10.05
^{27}Al	5/2	0.149	100
^{35}Cl	3/2	-7.97×10^{-2}	75.4
^{37}Cl	3/2	-6.21×10^{-2}	24.6
^{75}As	3/2	0.3	100
^{79}Br	3/2	0.33	50.57
^{81}Br	3/2	0.28	49.43
^{121}Sb	5/2	-0.8	57.25
^{123}Sb	7/2	-1	42.75
^{127}I	5/2	-0.75	100
^{209}Bi	9/2	-0.4	100

quadrupole moment. A list of such nuclei, commonly found in the study of molecular complexes, is given in Table 8.1. The electric field gradient (EFG) is also a tensor, with components

$$V_{ij} = \partial^2 V / \partial x_i \, \partial x_j$$

where V is the electric potential at the site of the nucleus considered. When referred to its principal set of axes, the EFGT is completely determined by two components which are chosen as

$$V_{zz} \quad \text{and} \quad \eta = (V_{xx} - V_{yy})/V_{zz}.$$

The Hamiltonian associated with the nuclear quadrupole coupling is

$$H_Q = \left[3I_z^2 - I^2 + \frac{\eta}{2}(I_x^2 + I_y^2) \right] e^2 Qq/4I(2I-1).$$

The expressions giving the value of the energy levels depend on the nuclear spin I. The most important ones are those for spin $I = 1(^{14}\text{N})$, $\frac{3}{2}(^{35-37}\text{Cl}$, $^{79-81}\text{Br}$, $^{75}\text{As})$, $\frac{5}{2}(^{121}\text{Sb}$, $^{127}\text{I})$, and $\frac{7}{2}(^{123}\text{Sb})$. They are given, for $I = 1$, $\frac{3}{2}$, and $\frac{5}{2}$, together with transition frequencies, in Table 8.2.

From these expressions it is clear that for $I = 1$ the observation of the two lines ν_+ and ν_- leads to $e^2 Qq = 3(\nu_+ + \nu_-)/2$ and $\eta = 3(\nu_+ - \nu_-)/(\nu_+ + \nu_-)$. Observation of $\nu_d = \nu_+ - \nu_-$ is useful to check the pairing of the lines ν_+ and ν_- when complicated spectra are obtained resulting from the presence of chemically equivalent nuclei located at unequivalent crystalline sites, or chemically unequivalent nuclei. For $I = \frac{3}{2}$ there is only one transition observed, $\nu = (1 + \eta^{2/3})^{1/2} e^2 Qq/2$, and it is not possible to get

Table 8.2 Nuclear quadrupole energy levels in an electric field gradient and transition frequencies

---

$I = 1$ $E_{\pm 1} = (1 \pm \eta)A$ $E_0 = -2\eta A$ with $A = e^2 Qq/4$

$\nu_+ = (3 + \eta)A/h$ $\nu_- = (3 - \eta)A/h$ $\nu_d = \nu_+ - \nu_- = 2\eta A/h$

$I = \frac{3}{2}$ $E_{\pm 3/2} = -E_{\pm 1/2} = \left(1 + \frac{\eta^2}{3}\right)^{1/2} e^2 Qq/4$

$\nu_{3/2-1/2} = \left(1 + \frac{\eta^2}{3}\right)^{1/2} e^2 Qq/2h$

$I = \frac{5}{2}$ $E_{\pm 5/2} = (10 + \frac{5}{9}\eta^2 + \frac{85}{2916}\eta^4 + \cdots)A$ with $A = e^2 Qq/40$

$E_{\pm 3/2} = (-2 + 3\eta^2 - \frac{23}{12}\eta^4 + \cdots)A$

$E_{\pm 1/2} = (-8 - \frac{32}{9}\eta^2 + \frac{1376}{729}\eta^4 + \cdots)A$

$\nu_{5/2-3/2} \cong 3(1 - 0.2037\eta^2)e^2 Qq/10h$

$\nu_{3/2-3/2} \cong 3(1 + 1.0926\eta^2)e^2 Qq/20h$

$I = 3, \frac{7}{2}$, and $\frac{9}{2}$: see Table II in 'Tables of NQR frequencies' (Biryukov *et al.*, 1969) or Table II.1 in 'Crystal field effects in NQR' (Weiss, 1972).

---

separate values for $e^2 Qq$ and η. For bromine and chlorine atoms which are singly bonded the asymmetry parameter η is often small, i.e. less than 10%, and it is possible to consider 2ν instead of $e^2 Qq$ for the discussion of the results. This assumption is not always valid because η is higher than 10% when there is conjugation of π electrons; η has a value of 25% in $COCl_2$, 21% in chloranil, 15% in hexachlorobenzene, and 11% in 1,3,5-trichlorobenzene, to be compared with 4% in hexachlorocyclohexane (Biryukov *et al.*, 1969). However, it is possible to obtain the value of η by Zeeman effect measurements (splitting of the resonance line on the application of a steady magnetic field to the sample) on single crystals and, though the procedure is somewhat tedious, it has been extensively used for molecular complexes (Okuda *et al.*, 1968, 1970, 1972a; Chihara and Nakamura, 1971; Yamada, 1977). The Zeeman effect can also be observed on polycrystalline samples and modification of the shape and width of the resonance line yields information on η but with much less accuracy than in the case of a single crystal and no data concerning the orientation of the EFGT is obtained (Morino and Toyama, 1961; Brooker and Creel, 1974; Smith and Tong, 1971a; Darville 1974).

For $I = \frac{5}{2}$ or $\frac{7}{2}$ two or three resonance lines are observed yielding both $e^2 Qq$ and η and even a third parameter ($I = \frac{7}{2}$) which may be the hexadecapole coupling term but its effect is usually small and its study is beyond the scope of the present paper.

The temperature and pressure dependence of the NQR frequencies and QCC also provide useful information about complexation. The corresponding theory is given in standard textbooks cited at the beginning of this section. Let us just recall that the temperature dependence is related to the molecular motion which averages the EFG seen by the nucleus. Within this simple scheme the QCC decreases with increasing temperature. Intermolecular bonding can reduce the amplitude of the thermal vibration which results in a smaller temperature coefficient. In practice, thermal excitation of some bonding orbitals can produce an increase of the QCC and a positive temperature coefficient is then observed (see Fichtner and Weiss, 1976). In a similar manner the effect of pressure on the QCC can be discussed and used to appreciate the importance of intermolecular bonding like in the case of the chloranil–hexamethylbenzene complex (Jugie and Smith, 1974).

NQRS which permits the determination of QCC from the frequency of the NQR lines is not the only method to obtain QCC. The nuclear quadrupole coupling can be superimposed on another coupling and estimated from the perturbation it causes to the other coupling. This is the case for NMR on a single crystal where the NMR line is split by the nuclear quadrupole interaction; the same holds for Mössbauer spectroscopy which is reviewed in Chapter 9. In microwave spectroscopy, QCC's are derived from the perturbed rotational spectra of molecules in the vapour phase; as the complexes reviewed in this chapter are formed in the condensed phase it is likely that microwave spectroscopy has not been used for their study.

8.2.3 *Experimental*

As spectrometers are described in the literature, only short indications are given in this paragraph. Spectrometers may be of different types, continuous wave (CW) marginal oscillator, superregenerative oscillator (SRO), pulse spectrometers, and double-resonance spectrometers. SRO spectrometers are easily built and operated, almost free of microphonics and very sensitive to small signals. Their major drawback is the presence of a spectrum of superregenerative sidebands which considerably complicate the NQR spectrum itself making the distinction between the components of the spectrum difficult. As a general rule, CW spectrometers are less sensitive to small signals than SRO's but very sensitive to microphonics and radiofrequency interferences and thus require great care for their construction and operation. However, for low frequencies (between 1 and 5 MHz) they perform quite satisfactorily and they have been extensively used for the detection of nitrogen resonances. As QCC's are characteristic of the samples studied the resonance condition is achieved by tuning the spectrometer and, for unknown resonances, the search is made by sweeping a broad frequency range. The strength or, more exactly, the signal-to-noise ratio of a given resonance line depends on the abundance of the resonant nucleus, of its relaxation time, and on the linewidth: some signals are easily observed on the screen of an oscilloscope and others are obtained on a strip chart recorder with a poor signal-to-noise ratio in spite of a long integrating time constant (10 or 20 s). As a consequence, the search for a new line may take a few minutes or several days.

Pulse spectrometers have the great advantage of permitting the observation of free induction nuclear signals and accumulation of the signals is rather easy. However their construction is more elaborated than that of CW or SRO spectrometers. Double-resonance techniques are useful for the observation of weak signals, especially those of nuclear species of low isotopic abundance or those occurring at low frequency (below 1 or 2 MHz). They have thus been proved efficient in extending the field of investigated nitrogen resonances towards low frequencies and in permitting the detection of the resonances of oxygen-17 and deuterium (see, for example, Goren, 1974; Schwartz and Ragle, 1974; Hsieh et al., 1977).

The size of the sample needed for NQR experiments depends on the type of spectrometer used and of the range of frequency scanned and is usually between 1 and 20 g. It is important that the sample, which is necessarily in the solid state (samples liquid at room temperature are solidified by cooling at liquid-nitrogen temperature), should be in a good crystallized state to reduce spurious broadening due to crystal distortion, otherwise the line intensity could be smaller than the sensitivity of the spectrometer. It may be mentioned that in some instances resonances have been observed in disordered crystals (Myasnikova et al., 1965) or in complexes of polymeric substances (Shoshtakovskii et al., 1973; Orlov et al., 1974).

8.2.4   Interpretation of quadrupole coupling constants in complexes

Interpretation of QCC's in molecules is based on the theory proposed by Townes and Dailey (Townes and Dailey, 1949; Dailey and Townes, 1955). The EFG at the emplacement of a quadrupolar nucleus is produced mainly by the electrons of the atom to which the nucleus belongs. In a first approximation, core electrons can be considered as disposed in complete shells of spherical symmetry and so do not contribute to the EFG. In fact the polarizability of the core electrons is taken into account by the Sternheimer antishielding factor (Lucken, 1969) but, as far as comparison of QCC's is considered with a view to chemical interpretation, instead of discussion of the absolute values in themselves, the polarization of core electrons can be ignored. Among valence electrons, those in s orbitals of spherical symmetry do not contribute to the EFG and the main contribution is that of the p electrons; the contribution of the d and f electrons is much less important because of their greater distance from the nucleus and of the small amount of d and f hybridization. Quantitative consideration of the contributions to the EFG leads to expressions of the following type valid for chlorine in organic chlorides

$$q_{zz}^{Cl} = [(1 - s^2 + d^2 - I - \pi) + I(s^2 + d^2)]q_{at}^{Cl} \qquad (8.1)$$

where s^2 and d^2 are the amounts of s and d hybridization of the bonding orbital on the chlorine atom, I the ionicity of the bond (the chlorine atom bears a negative charge), and $e^2 Qq_{at}^{cl}$ is the quadrupole coupling constant of a p electron on chlorine. Other expressions may be obtained when the populations of the different valence orbitals are explicitly considered; some results are discussed in terms of the populations N_x, N_y, and N_z of the three atomic p orbitals, the axes x, y, and z being usually chosen so that z lies along the direction of the bond when halogens are considered or along some symmetry direction of the system of several bonds when atoms like nitrogen or arsenic are considered; other results are more conveniently discussed in terms of the populations of the bonding and lone-pair orbitals.

From what precedes, interpretation of QCC's in terms of orbital populations may look quite straightforward. In fact NQRS gives, at best two parameters ($e^2 Qq$ and η), and in many cases ($I = \frac{3}{2}$) only an approximate value of $e^2 Qq$ is available, and the relation (8.1) above contains four parameters s, d, I, and π which cannot be derived from only one or two experimental parameters. Approximations are necessary which neglect d hybridization, π bonding in some cases and consider that s hybridization is small and remains constant for the series of compounds under study so that the variations of $e^2 Qq$ are directly related to the variations in ionicity or p-electron donation in case of hydrogen bonding.

In the case of nitrogen, with spin $I = 1$ the situation is more favourable as the experiment yields two parameters and the s–p hybridization can be deduced from bond angles when the molecular geometry is known. On the other hand, the atomic QCC, $e^2 Qq_{at}^{N}$ of a nitrogen-14 p electron is not known with accuracy as it is for chlorine, bromine, or iodine and only an approximate value of about 9 MHz is available, but it can be considered as reliable because it results from the discussion of a great number of experimental results.

As stated in the introduction of this section, there are, besides the above limitations to the discussion of experimental values of QCC's in terms of orbital populations, physical contributions altering the QCC so that the value obtained from experiment is different from that expected for hypothetic molecule, or complex unit, isolated and at rest. In the case of molecular complexes, there is an additional contribution to the modification of the QCC during the formation of the complex which is the change of hybridization resulting from a change in the geometry of the complexing molecules. A typical example is provided by complexes of MX_3 molecules where M is an element of Group IIIa or Group Vb of the periodical chart of the elements, and X a halogen or a methyl group. Upon complexation, the angles XMX are modified and the corresponding hybridization is altered, modifying the QCC even in the absence of any charge transfer: it is then difficult to discuss experimental data in an unequivocal way. In spite of all these difficulties a great number of results have been collected and consistent interpretative schemes proposed.

8.3 Electron Transfer in Molecular Complexes as Studied by NQRS

A point to be noted when dealing with charge transfer complexes as studied by NQRS is the great variety of situations found among these compounds: besides the simple donor–acceptor complexes like $H_3N \rightarrow BH_3$ or $Br_3Sb \rightarrow AlBr_3$, one finds complexes in which a given molecule like Br_3Sb can be donor ($Br_3Sb \rightarrow AlBr_3$) or acceptor ($C_6H_6 \rightarrow Br_3Sb$), the donating and accepting centres being different, i.e. Sb and Br, respectively, in the examples proposed; the presence of donating and accepting centres in the same molecule can lead to the formation of dimeric units like Al_2Br_6 in which bridging bromine atoms act as donors towards accepting aluminium atoms; intramolecular coordination is also possible and is found in *ortho*-halophenols (Korshunov *et al.*, 1971) or in some mercury or phosphorus compounds leading to the 'α effect' (Wulfsberg, 1975; Saatsazov *et al.*, 1975). This explains why the present review is based on a systematic exploration of the periodic chart of the elements for examination of the complexes containing quadrupolar nuclei found in each column of the chart though, at first sight, a classification according to the chemical nature of the complexes would have seemed more satisfying from a logical point of view.

Before reviewing the different complexes studied by NQRS, some of the problems encountered in the interpretation of the experimental results can be outlined briefly.

Are the shifts observed between the NQR frequencies in the complexed and the pure, uncomplexed starting materials due to charge transfer or to another cause? This question of basic importance should be considered with care in each case as its answer depends on several factors. When charge transfers are small, they are misrepresented by the frequency shifts in which the contribution of crystalline or solid-state effects, already cited in the preceding section and which are due to differences in the lattice environment of the molecules between the complex and the pure components (crystalline electric field, intermolecular interaction, thermal motion), may be as high as several hundreds of kilohertz (a value of 200 kHz is commonly considered in the case of chlorine resonance as a measure of these effects). When the observed shifts are stronger (one or several megahertz) they are not likely to be due to crystalline effects and they can be safely ascribed to electronic charge transfer effects. However, other contributions, such as change in hybridization, should also be considered. Change in hybridization accompanying the deformation of a planar $AlMe_3$ molecule into a pyramid (Patterson and Carnevale, 1973) formally produces a change in the QCC at the aluminium atom even if there is no variation of the population and ionicity of the bonding orbitals; the result is that both charge transfer and change in hybridization can be present and contribute to the shift of the NQR frequency. A more complex situation is found when strong inter-molecular interactions exist in the uncomplexed compounds: the breaking of these interactions upon complexation may result in a higher QCC in the complex, in contradiction with the usual simple scheme predicting a de-crease of the QCC in the complex; this is the case with mercury halide complexes, $HgBr_2 \cdot$ and $HgI_2 \cdot$dioxan (Brill, 1970). A further step in the discussion of experimental results can be reached if several quadrupolar nuclei are present in the complex, permitting the comparison of the shifts for each of them; structural data, usually from X-ray diffraction, are often useful. Finally, theoretical discussion may be helpful in appreciating the relative importance of the different contributions to the observed frequency shift (Bowmaker et al., 1973b).

How is an assignment made when several lines are detected in the complex? A first indication results from a comparison with the spectra of the uncomplexed materials: in the complex $2POCl_3 \cdot SnCl_4$, the resonances of the phosphorus oxychloride chlorine atoms are found at 30 MHz and those of $SnCl_4$ at 19–21 MHz, values to be compared to 29 and 24.1 for the respective pure compounds (Rogers and Ryan, 1968). Observing the lines corresponding to different isotopes of a given atom (^{35}Cl and ^{37}Cl, ^{79}Br and ^{81}Br, ^{63}Cu and ^{65}Cu, ^{121}Sb and ^{123}Sb, etc . . .), provided the strength of the lines permits their observation, is very helpful, as well as a comparison of the intensities of lines corresponding to chemically of crystallographically non-equivalent atoms. In some cases, the special configuration of the complex, like the cis or trans position of chlorine atoms in complexes of $SnCl_4$ can also be taken into account in the discussion. In short, the assignment is a matter of examining all the available data (special features of the NQR spectrum, X-ray, and IR data) which may help in developing a model consistent with the experimental results and common chemical evidence.

8.4   Complexes Containing Elements of Group I

Among elements Ia, the proton with spin $I = \frac{1}{2}$ has no quadrupole moment but its isotope ^2H, deuterium, has a quadrupole moment. The corresponding QCC is commonly of the order of one or a few hundred kilohertz; this is a low value but can be measured with double-resonance techniques. The deuterium quadrupole coupling constant has been determined, in particular, in some complexes of deuterochloroform with diethyl ether, acetone, and mesitylene (Ragle *et al.*, 1974), and in the complex of bromodichloromethane with diethyl ether-d_{10} (Schwartz and Ragle, 1974). The shift of the deuterium QCC in these complexes, except in the $CDCl_3$·mesitylene complex, is consistent with the model of a C—H $\cdots$ O hydrogen bond.

Considering the alkali metals, none of them appears to be involved in charge transfer complexes, at least as a centre of complexation. Among the complexes which contain alkali metals are those of NaCl with $AlCl_3$, giving the $AlCl_4^-$ anion (Evans and Lo, 1967a), and with other Group IIIa halides giving $GaCl_4^-$, $Ga_2Cl_7^-$, and $Al_2Br_7^-$ anions with K^+, Rb^+, and Cs^+ as well as Na^+ cations (Deeg and Weiss, 1975). On the other hand, several complexes in which copper and gold (elements Ib) are the centre of coordination are known and have been studied by NQRS. They are, particularly, interesting in that coordination numbers of 2 and 3 are shown by these metals.

Bowmaker *et al.* (1976b) have studied a series of dihalocuprate(I) complexes using infrared, Raman, and NQR spectroscopies and they reported halogen and copper NQR frequencies in three complexes, $(Bu_4N)CuCl_2$, $(Ph_4As)CuCl_2$, and $(Bu_4N)CuBr_2$. The temperature dependence of the resonance frequencies is quite normal. From the halogen QCC it appears that the Cu—Cl bond has a 20% covalent character. However, this value is an upper limit as electrical contribution from neighbouring charges can also add to the measured QCC. In an extension of this work, the NQR in $(Ph_3MeP)Cu_2I_3$ was investigated and lines due to ^{63}Cu, ^{65}Cu, and ^{127}I were detected (Bowmaker *et al.*, 1973a). The discussion of the experimental results is interesting: the high value of the resonance frequencies indicates a low symmetry at the copper ion site and favours a structural model with discrete anions over a polymeric and polydimensional structure found in other salts of Cu(II) such as $CsCuCl_3$. From the relative intensity of the two components of the copper NQR, it seems that the anion is $Cu_4I_6^{2-}$ rather than $Cu_2I_3^-$, which is consistent with IR results. Unfortunately, the weakness of the iodine lines does not allow a complete discussion of the data for this nucleus.

In another series of complexes, the copper ion is included in a cation instead of an anion and it is tricoordinated instead of bicoordinated. These complexes are $[Cu(Me_3PS)_3]ClO_4$ and $[Cu(Me_3)AsS)_3]ClO_4$ (Bowmaker *et al.*, 1975). Non-ionic systems, $(Me_3PSCuCl)_3$ and $(Me_3PSCuBr)_n$ are also reported. Only copper resonances were recorded in these complexes. The multiplicity of the lines in the spectra is consistent with the available structural data. A thorough discussion of the possible bonding schemes (sp^2 and $3d_z^2s$) and comparison with the experimental copper frequencies (in the range 25–29 MHz for the isotope ^{63}Cu), together with the influence of the donation from the ligand towards the copper atom, which lowers the NQR frequency, also leads to the conclusion that the *ds* scheme is the more satisfying.

Investigation of gold-containing complexes has been reported by Bowmaker and Whiting (1976). These complexes are bis(N,N-dibutyldithiocarbamato)gold(III) dichloroaurate(I), $(Bu_2NCS_2)_2AuAuCl_2$, compounds and tetraalkylammonium salts. Application of the Townes and Dailey theory gives an average ionic character of 68% for the Au—Cl bond in the $AuCl_2^-$ anion, to be compared with 82% in $CuCl_2^-$. These values are consistent with those expected from ionization potentials of gold and copper and d^{10}–$d^9 s$ promotion energies of the cations Au^+ and Cu^+. An interesting discussion results from the comparison of NQR frequencies of chlorine in $AuCl_2^-$ and $AuCl_4^-$; the frequency is smaller for $AuCl_2^-$ and this is explained by the fact that the charge on the central atom remains small and the charge transfer from chlorine to the gold atom needed to almost neutralize the charge on the central atom is accordingly smaller for $AuCl_4^-$ than for $AuCl_2^-$ which means that the Au—Cl bond is more ionic in $AuCl_2^-$ than in $AuCl_4^-$.

8.5 Complexes Containing Elements of Group II

8.5.1 *Magnesium-containing complexes*

Complexes of $MgBr_2$ and $RMgBr$ with ether and ether-like molecules have been studied (Kress and Guibé, 1977). The bromine resonances, between 30 and 50 MHz, were recorded at 77 K with a superregenerative spectrometer. In these compounds the magnesium atom is the acceptor and the oxygen atom of the ether molecule the donor. The measured frequencies of the bromine NQR can be related to the nature of the substituent R— in the complex of phenylmagnesium bromide with ethyl ether (1:2), the frequency, 33.53 MHz, is about 10% higher than in the corresponding complex of ethylmagnesium bromide, 30.55 MHz; such a trend is expected from the more electron-releasing character of the ethyl group as compared to the phenyl. They can also be related to the nature of the ligand: the frequency in the tetrahydrofuran complex of phenylmagnesium bromide, 31.14 MHz, is lower than in the corresponding diethyl ether complex and shows that THF is a stronger donor than DEE. In the diethyl ether complex of magnesium bromide the temperature dependence of the bromine NQR frequency shows the conformational phase transition associated to a change of the conformation of the diethyl ether molecule at about −40 °C (Guibé and Montabonel, 1978).

No NQR study of complexes containing the other Group IIa elements, calcium, strontium, or barium, seems to have been reported so far, except for those of resonances in some halide hydrates, $BaCl_2 \cdot 2H_2O$ (Graybeal and

Pathania, 1973), $BaI_2 \cdot 2H_2O$ (Volkov, 1973), in which the ^{135}Ba and ^{137}Ba resonances were investigated; but these compounds are not charge transfer complexes and, moreover, the water molecule in these hydrates is only considered as an electric dipole contributing to the EFG at the barium ion site and hydrogen bonding is discussed.

8.5.2 *Cadmium- and zinc-containing complexes*

Complexes of cadmium and zinc salts with pyridine have been investigated (Hsieh *et al.*, 1977: the pyridine nitrogen-14 NQR was detected at 77 K with a double-resonance method similar to that described by Edmonds (1974) and the change in the QCC, when compared to that of pure pyridine, is interpreted in terms of electron transfer and change in hybridization accompanying the complexation. Some of the salts studied by Hsieh are nitrates in which the resonance of the nitrato nitrogen shows a distortion of the EFG. In zinc thiocyanate, the resonance of the thiocyanato nitrogen shows that this atom is coordinated to the zinc ion.

8.5.3 *Mercury-containing complexes*

A great number of complexes containing mercury as acceptor have been studied. They can be put in two main groups according to the nature of the mercury-containing molecule: mercuric halide ($HgHal_2$ with Hal = Br, Cl, or I), and organomercuric halide (pentachlorocyclopentadienyl, trichloromethane, *trans*-2-ethylenemercuric chlorides ...). In the mercuric halide complexes, the resonance of the halogen is usually investigated though resonances of ^{201}Hg are reported in $HgCl_2 \cdot 2dioxan$ and $HgBr_2 \cdot dioxan$ complexes (Patterson *et al.*, 1973). These two complexes were considered because they contain $HgHal_2$ units which are almost completely isolated from each other with no strong intermolecular halogen–mercury interaction as observed in the pure mercury halides. The observation of the mercury resonance is of particular interest as this atom is the centre of complexation. It is shown that the variation of the mercury QCC upon complexation results from oxygen–mercury coordination and from an increase of the ionicity of the Hg—Hal bonds.

Besides the determination of the resonance frequencies, the temperature dependence of these frequencies has been studied in some complexes. Bayer's theory (see textbooks on NQR) applied to molecular compounds predicts a decrease of the NQR frequency with increasing temperature as a consequence of the averaging effect of thermal motion on the components of the EFG tensor. On the other hand, if the strength of a coordinative bond is lowered by thermal excitation, the contribution of the bond to the QCC of the resonant nucleus is reduced and the QCC increases (in the cases where the coordinative bond lowers the QCC as compared to the uncomplexed materials). In other words, the resultant temperature coefficient of the QCC is, in absolute value, smaller than in the uncomplexed material as is observed in $HgCl_2 \cdot 2dioxan$, or even of the opposite sign, i.e. positive, as in $HgCl_2 \cdot dioxan$, $HgCl_2 \cdot trioxan$, and $HgCl_2 \cdot trithiane$ (Fichtner and Weiss, 1976). An even more significant example is that of $HgCl_2 \cdot o$-chlorobenzaldehyde: the variation of the resonance frequency, between 77 and 300 K, is about 0.8 MHz for the ring chlorine atom and less than 0.1 for each of the two chlorine atoms bound to the mercury atom.

The donors which have been most widely investigated are oxygen- and sulphur-containing molecules (methyl alcohol, aldehydes, ketones, ether, di- and trioxans, and thianes). Only a few nitrogen-containing donors are reported: sym-collidine (Fichtner and Weiss, 1976), 2- and 4-chloro-pyridines (Ardjomande and Lucken, 1975), dimethylformamide (Maksyutin et al., 1970b), and tetramethylethylenediamine (Bryukhova et al., 1973). The donor molecule is either a monodentate ligand (dimethylsulphoxide, ether, methyl alcohol, tetrahydrofuran, tetrahydrothiophene, ...) or a bi- or polydentate ligand (dioxan, glymes, tetramethylethylenediamine, ...). Halogen atoms are also donor like in the case of pure mercuric halides where they contribute to intermolecular bonding: the mercury atom is surrounded by four halogen atoms pertaining to neighbouring molecules and situated in an equatorial plane (see, for example, Patterson et al., 1973), or in the case of pentachlorocyclopentadienylmercury complexes, $C_5Cl_5HgCl \cdot HgCl_2$ and $C_5Cl_5HgBr \cdot HgBr_2$ (Wulfsberg, 1975; Wulfsberg et al., 1975).

The dioxan complexes were studied with special interest, particularly the $1:1$ complex, the structure of which has been determined by X-ray (cf. Brill, 1970, ref. 2). Though these complexes are weak, the NQRS of chlorine and mercury in the $1:2$ complex (Patterson et al., 1973) has shown the electronic charge transfer accompanying the complexation. The crystal structure of some other complexes of $HgCl_2$ is also known (1,3,5-trithiane, tetrahydro-thiophene, 1,4-cyclohexanedione, and sym-collidine) but these complexes have been less extensively studied (Fichtner and Weiss, 1976) than the dioxan complexes.

Generally, the following features of the spectra are discussed: number of lines observed, which is usually consistent with structural data when they are available; relative strength of donors towards mercury gives new pieces of information when differences exist between thermochemical and IR data as is the case for mercury halide dioxanates (Brill, 1970) (the NQR order for decreasing interaction in these complexes is $HgCl_2 > HgBr_2 > HgI_2$. In some cases, the derivation of a classification of donors, or acceptors, from the frequency shifts observed upon complexation may be difficult because the shift contains a contribution from the cleaving of intermolecular bonding in the pure mercuric halide in addition to the contribution of the charge transfer resulting from the complexation (Brill, 1970; Bryukhova et al., 1973). In other cases, for a series of donors, a good correlation is found between the frequencies measured in the chloride complexes and those in the bromide complexes which are thus considered as isostructural with the chlorides (Wulfsberg, 1976). In a similar manner, the IR Hg—Cl stretching frequencies and the ^{35}Cl NQR frequencies correlate well over a range extending from 12 to almost 24 MHz for NQR and from 250 to 375 cm^{-1} for IR frequencies (Scaife, 1971; Wulfsberg, 1976). In such favourable circumstances it is possible to discuss experimental results in detail and obtain information on the basicity of donors and discover cases with an unusual coordination environment for the mercury atom (Wulfsberg, 1976). Another kind of correlation is also observed between the NQR chlorine frequency and the Hg—Cl distance (Scaife, 1971).

Specific structural results can be derived from the multiplicity and the spacing of the components in the NQR spectra of complexes. More precisely, $HgCl_2$ is known to form a tricoordinate $HgCl_3^-$ ion with Me_4N^+ as cation and an investigation was undertaken to find similar examples in complexes of mercuric chloride with organic molecules (Brill and Hughus, 1970). Though a great variety of structure and stoichiometry occurs in the chemistry of mercury halide complexes, discussion of the NQR results, when no X-ray data are available, suggests that a dimeric structure is the most probable in 1:1 complexes, where tricoordination was expected at the beginning, like complexes of THF, DMSO, BZP, ACP, resulting in tetracoordination of the mercury atom as shown in the figure

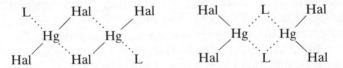

The dimeric structure at the left shows two bridging and two terminal halogen atoms while the structure at the right shows only terminal halogen atoms and two bridging ligands. It is clear that bridging and terminal halogen atoms have different electronic environment and it is experimentally observed that a bridging halogen has a QCC smaller than a terminal. Extension of the investigation to complexes of $HgBr_2$ with the preceding donors and also with other donors like phenoxathiin, methanol, pyridine-N-oxide (Brill and Hughus, 1971) leads to a similar conclusion for all the 1:1 complexes which present a splitting of the halogen resonance, ascribed to the difference in QCC between terminal and bridging atoms; the splittings as they result from the measurements of Brill (Brill and Hughus, 1971) are shown in Table 8.3; it is seen that they are smaller for complexes of composition 2:1, which have a structure resembling that of the pure halides.

Table 8.3 Splittings of the halogen resonance (MHz) in
$HgX_2 \cdot D$ and $HgX_2 \cdot 2D$ complexes as calculated from the frequencies reported by Brill (Brill and Hughus, 1971)

Donor	$HgCl_2$	$HgBr_2$	HgI_2
Pure HgX_2	0.200	0.95	1.87
THF	1.69	6.52	
BZP	0.943	2.91	
ACP	1.476	6.83	
2 Phenox.		1.27	
2 MeOH	0.153		
Py.—N—O	0.588	9.18	7.88
4-Pic.—N—O	1.401	9.78	7.40

The distinction between terminal and bridging atoms is quite obvious when a single dimeric unit is considered, as in the figure above, but it is not so clear cut in the actual crystal where intermolecular interactions exist between adjacent dimeric units, explaining that the splitting may vary to a rather large extent when similar complexes are compared to each other. In some complexes, the splittings are too small even to be ascribed to the presence of terminal and bridging atoms (Bryukhova et al., 1973). Moreover, it should be borne in mind that the actual crystalline structure, when known, is derived from distorted octahedra so that it is sometimes difficult to determine the coordination number of the mercury atom unambiguously; some neighbouring atoms are more or less close to the central atom and their distance is just less, or more, than the sum of the van der Waals radii so that it is difficult to conclude if they are coordinated or not.

In the case of iodine-127, of nuclear spin $I = \frac{5}{2}$, the asymmetry parameter η is easily obtained and gives new pieces of information for structural discussion. From a value of 48% in HgI_2, η drops to less than 5.5% in the complexes with dimethylthioethane and tetramethylethylenediamine showing that iodine is probably not in a bridging position in the complex (Bryukhova et al., 1973).

The study of organomercurials is interesting in several respects, in connection with charge transfer. Even in uncomplexed compounds intramolecular and intermolecular coordination should be considered to explain experimental results: the QCC is much smaller in the solid than in the gas phase and abnormally high splittings are observed in the NQR spectra. When considering a series of compounds, the change in the resonance frequency is better understood by introduction of intermolecular bonding than by simple consideration of electronic effect along the carbon chain (Bryukhova et al., 1974b). Complexes of pentachlorocyclopentadienylmercuric halides were studied in the course of a more general investigation of pentachlorocyclopentadienyl derivatives (Wulfsberg and West, 1971; Wulfsberg et al., 1973,

1975; Wulfsberg, 1975) and the corresponding data, together with those for trichloromethylmercury derivatives, constitute a contribution to the study of the 'α effect' in chlorinated organometallic compounds (Wulfsberg, 1975) and to its rationalization in terms of orbital interaction between the chlorine atom, the mercury atom, and the adjacent carbon atoms. It is interesting to examine the development of the argument showing the presence of the 'α effect' in the case of cyclopentadienylmercurials. A pentachlorocyclopenta-dienyl radical has five chlorine atoms, one allylic, Cl^5, and four vinylic falling into two groups, $(1, 4)$ and $(2, 3)$. Non-metallic derivatives of pentachloro-cyclopentadiene are first considered and, with all those for which the σ^* Taft constant of the substituent R is known, it is possible to derive a

correlation between the NQR frequency of the different sorts of chlorine atoms in the pentachlorocyclopentadienyl group and the Taft constant of the substituent in the form of expressions

$$\nu_5 = a_5 + b_5\sigma^* \tag{8.2}$$

$$\nu_{1,4} = a_{1,4} + b_{1,4}\sigma^* \tag{8.3}$$

$$\nu_{2,3} = a_{2,3} + b_{2,3}\sigma^* \tag{8.4}$$

The a_i and b_i are determined from the available experimental values ν_i and σ^*. Then, considering mercury derivatives, it is possible, from expressions (8.3) and (8.4), put into the reverse form

$$\sigma^* = (\nu_i - a_i)/b_i \tag{8.5}$$

to obtain the σ^* value for the mercury group and, from this value of σ^*, to calculate $\nu_{5\text{calc.}}$ from (8.2) and to compare it with the experimental value $\nu_{5\text{exp.}}$. The result of the comparison shows that $\nu_{5\text{exp.}}$ is higher than $\nu_{5\text{calc.}}$ and this constitutes the 'α effect'. Moreover, the values calculated from NQR experiments in cyclopentadienyl derivatives can be correlated with those resulting from a study of trichloromethylmercurials including complexes with mercury halides (Wulfsberg, 1975).

8.6 Complexes Containing Elements of Group III

Elements of column IIIa, boron, aluminium, gallium, and indium act as acceptors and a great number of complexes in which they are involved have been studied. On the other hand, no complex containing elements of Group IIIb seems to have been studied by NQR.

8.6.1 Boron-containing complexes

The chlorine resonance has been studied in complexes of BCl_3 with nitriles, ethers and thioethers, arsines, amines, phosphines, pyridines, and phosphorus oxychloride (Gilson and Hart, 1970; Kaplansky and Whitehead, 1970; Ardjomande and Lucken, 1971, 1975; Smith and Tong, 1971b; Jost, 1972; Dillon et al., 1976a, 1976b; Carter and Jugie, 1977). Other resonances have been observed in complexes of BH_3 with arsines (Carter and Jugie, 1977), BD_3 with trimethylamine (Merchang and Fung, 1969), and BF_4H with trimeric phosphonitrilic chloride (Whitehead, 1964).

Boron trichloride complexes range from weak to strong ones in which the boron atom has a regular tetrahedral environment of four identical ligands, BCl_4^-, as in $BCl_3 \cdot PCl_5$ (Jost, 1972), in tetramethylammonium tetrachloroborate (Ardjomande and Lucken, 1971), and in other tetrachloroborates (Dillon et al., 1976a). Several features of these complexes are worth of mention. The tetrahedral symmetry of the sp^3 hybridized BCl_4^- ion facilitates calculation of the orbital populations using Townes and Dailey's theory and it is possible to compare the populations obtained to that of the sp^2 planar BCl_3 molecule (Ardjomande and Lucken, 1971); the result is somewhat surprising as the charge on the boron atom appears to be higher in BCl_4^-. It can be shown that, if the simple assumptions used in the Townes and Dailey theory are revised to take into account some interaction of the back-coordination type between the p orbitals of the non-bonding electrons on the chlorine atoms, the difference between the charges in BCl_4^- and BCl_3 can be reduced to a value more acceptable according to common chemical evidence.

Another characteristic feature of the chlorine spectra in BCl_3 complexes investigated is that the frequency of the chlorine NQR does not vary much (all the resonances are found between about 21 and 22 MHz) with the donor considered (Ardjomande and Lucken, 1971), in contrast with the complexes of $GaCl_3$ (Tong, 1969). However, the order of the relative donor ability of the different ligands, as given by the shifts in the NQR frequency of chlorine, is similar to that found for $GaCl_3$ (Tong, 1969; Ardjomande and Lucken, 1971). The electron-attractive power of BCl_3 can be compared to that of other acceptors through the frequency shift of the chlorine resonance of the donor in several complexes with a common donor. The latter being $POCl_3$, the following order is obtained (Smith and Tong, 1971b)

$$BCl_3 > SbCl_5 > FeCl_3 > SnCl_4.$$

Though the NQR frequency of the chlorine does not vary much among the complexes of BCl_3, there is an exception with some nitrile complexes in which lines are found at 22 and also at 38 MHz as in $CCl_3CN \cdot BCl_3$ (the three lines of the chlorine atoms in the trichloroacetic group are found at a still higher frequency, ~41 MHz); this is explained by the transfer of a chlorine atom of the BCl_3 group to the nitrile carbon atom

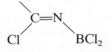

Such a structure is confirmed by the frequency of the bands observed in the IR spectrum and, also, by a molecular weight determination by cryoscopy in benzene (Ardjomande and Lucken, 1971).

In some cases, the rather small sensitivity of the chlorine NQR in BCl_3 to the nature of the donor molecule can be overcome by investigating the effect of complexation on the NQR of chlorinated donors such as chloropyridines and chlorobenzonitriles (Ardjomande and Lucken, 1975). Upon complexation the NQR frequency of the chlorine atoms in chloropyridines increases. However, the frequency shift between the complexed and uncomplexed molecules decreases as the distance of the chlorine atom from the nitrogen donor centre increases. The contribution of variation in double-bonding character and ionicity of the C—Cl bond can be discussed by reference to other complexes such as $ClCN \cdot MCl_5$ (M = Sb, Nb, or Ta), methylethers·BCl_3 and methyl thioethers·BCl_3. In particular it has been shown that whenever the C—Cl double bond exists in the donor, complexation increases it and that transfer of charge from the chlorine to the donor centre is of an order of 0.1 electron. These effects rapidly become smaller as the distance of the chlorine atom from the donor centre increases (2-, 3-, or 4-chloropyridines or -chlorobenzonitriles).

Other authors have tried to relate experimental results on BCl_3 to chemical bonding through theoretical calculations based on BEEM-π or CNDO methods (Kaplansky and Whitehead, 1970). Extension to complexes is less straightforward as the structure (especially bond angles) of the complexes is often unknown. However, it appears that complexation increases the ionic character of the B—Cl bond while π bonding is reduced and, as the corresponding effects on the QCC are of opposite sign, this explains why the shift in the BCl_3 NQR frequency is small. Calculations on amine·boron trichloride complexes, $RNH_2 \cdot BCl_3$, shows that the substitution of amino hydrogen atoms by methyl groups increases the ionic character of the B—Cl bonds to a maximum, after which further substitution in the chain has little effect.

Another extensive study of boron trichloride complexes has been made by Smith (Smith and Tong, 1971b) who considered a series of boron compounds of the type $(NH—BCl)_3$ or $RBCl_2$, which are outside the scope of the present article, together with complexes $Me_3N \cdot BCl_3$, $POCl_3 \cdot BCl_3$, $PhOMe \cdot BCl_3$, $Py. \cdot BCl_3$, $Et_2S \cdot BCl_3$, $PhNO_2 \cdot BCl_3$, $MeCN \cdot BCl_3$, $Et_2O \cdot BCl_3$ and $THF \cdot BCl_3$. Experimental results are discussed in terms of Townes and Dailey's theory with the help of the semi-empirical SGOBE method.

Though the boron NQR is not conveniently observable, its QCC was determined in pure BCl_3 by NMR in a polycrystalline sample (Casabella and Oja, 1969); the value obtained, 2.54 MHz, corresponds to a pure NQR frequency of 2.27 MHz. Some of the semi-empirical calculations mentioned above (Kaplansky and Whitehead, 1970) take this result into account when discussing the electronic structure of the uncomplexed BCl_3 molecule, but no result for complexes of BCl_3 seems to have been reported so far.

The temperature dependence of the chlorine NQR has been investigated in some BCl_3 complexes. In $Me_3N \cdot BCl_3$ the results correlate well with linewidth measurement of the proton NMR (Gilson and Hart, 1970); rotation of the methyl groups is observed as a broadening of the chlorine NQR line over a temperature range of about 40° centred at 110 K, and rotation of the BCl_3 groups makes the NQR line broaden and disappear at 250 K.

8.6.2 Aluminium, gallium-, and indium-containing complexes

A strong interest has recently been taken in these complexes. In the original paper of Dehmelt (1953), the frequency of the chlorine resonance in $AlCl_3 \cdot Et_2O$ is reported. A later study (Evans and Lo, 1967a) is concerned with $NaAlCl_4$ which is not exactly a molecular complex though a charge transfer results from the fixation of a Cl^- ion by an $AlCl_3$ molecule.

The first complexes of $AlBr_3$ studied have ether or amine-type ligands, $AlBr_3 \cdot OPh_2$, $- \cdot SMe_2$, and $- \cdot Py$. (Maksyutin et al., 1968). The effect of steric and electronic factors on the bromine NQR frequency during complexation is discussed and the case of the ether and thiother complexes of $AlBr_3$ is interesting. In these molecular complexes, the angle of the ether or thioether C—O—C (or C—S—C) plane with the Al—O (Al—S) direction is not the same in the different complexes, resulting in two distinct configurations. In one configuration two bromine atoms are more strongly affected by the ether molecule and, in the second configuration, only one bromine atom (which is in the plane of symmetry of the complex) is affected by the thioether molecule.

The range of the compounds studied is extended in other publications. Lobanova (Lobanova et al., 1975) presents the NQR frequencies in a series of aliphatic and aromatic ketone complexes of $AlBr_3$ and points out that the frequency shifts due to crystalline effects are important. Okuda (Okuda et al., 1975) reports on a Zeeman effect study of three complexes, $AlBr_3 \cdot SbBr_3$, $AlI_3 \cdot SbI_3$, and $AlBr_3 \cdot BiBr_3$; the conclusion is that the structure of the antimony-containing complexes is ionic, $SbX_2^+ \, AlX_4^-$, with weak bonds between the halogen atoms of the AlX_4^- anion and the SbX_2^+ cation. $AlBr_3 \cdot BiBr_3$ is different and it seems that a bridge structure exists between two Bi atoms (one of the two bromine NQR signals has a smaller e^2Qq and a higher η than the other).

On the other hand, Poleshchuk (Poleshchuk and Maksyutin, 1975b) considers that the ethane-like structure observed by electron diffraction in gaseous $SbBr_3 \cdot AlBr_3$ is preserved in the solid. In the case of $BiBr_3 \cdot AlBr_3$ he agrees with Okuda (Okuda *et al.*, 1975) about the bridging structure in Bi_2Br_6 units. It should be mentioned that Poleshchuk measurements were made at 77 K while those of Okuda were made at room temperature and that some phase change may occur between these temperatures or some polymorphism may exist depending upon the conditions of the preparation of the complex, which would explain the difference between the results of Okuda and Poleshchuk about the $SbX_3 \cdot AlX_3$ complexes. Perhaps a re-examination of these complexes would be useful to clarify this point.

Complexes of $AlBr_3$ with sulphones have been investigated (Guryanova *et al.*, 1974). In the complexes $R_2SO_2 \cdot AlBr_3$, only one oxygen atom of the sulphone group is coordinated to an $AlBr_3$ molecule and the two oxygen atoms of the sulphone group are coordinated to an aluminium bromide molecule in the $R_2SO_2 \cdot 2AlBr_3$ complexes. The lowering of the bromine NQR frequency in these complexes is ascribed to a charge transfer from the sulphone oxygen atoms towards the $AlBr_3$ molecules. A correlation is found between the average frequency shift of the bromine NQR in each complex and the enthalpy of formation of the complex.

Complexes of $AlCl_3$ with Me_3As and Et_3As as donors have been investigated through the ^{75}As NQR and its temperature dependence (Carter and Jugie, 1977). The ^{75}As NQR frequency is strongly modified by the complexation ($\nu_Q = 96.7$ MHz in pure Me_3As and 54.91 in the complex); the effect is still stronger with BCl_3 and $GaCl_3$ showing that, if geometrical factors can be neglected, $AlCl_3$ is a weaker acceptor than BCl_3 and $GaCl_3$ with respect to Me_3As.

Complexes of $AlMe_3$ and $InMe_3$ with trimethyl derivatives of arsenic, nitrogen, phosphorus, and antimony, have been investigated (Patterson and Carnevale, 1973). The complexation leads to a decrease of the QCC for both the metal and ligand atoms. The primary cause of the decrease is the transfer of electron from the lone pair of the ligand to the empty orbital of the metal atom. The electronic transfer is accompanied by a change in the molecular geometry. The metal trimethyl molecule which is planar when uncomplexed becomes slightly pyramidal upon complexation while the pyramid of the ligand is slightly flattened. The results are a lowering of the metal QCC and an increase of the ligand QCC. Closer examination of the experimental results and comparison with heats of complexation lead to a consideration of the contribution of $d-d$ or $p-d$ π bonding between the acceptor and donor molecules. Consideration of π bonding helps to rationalize the data of heats of complexation and QCC and its contribution though small can explain the deviations observed between the NQR and thermochemical data.

Deeg (Deeg and Weiss, 1976) has studied a series of aluminium and gallium halide ($AlBr_3$, $AlCl_3$, and $GaCl_3$) complexes with aromatic oxygenated ligands (benzophenones, benzoyl chlorides, and 4-chlorophenol). The formation of the complexes is observed through the shift of the halogen NQR frequency as well as that of the IR frequency of the carbonyl group in the case of benzophenones and benzoyl chlorides. However, from NQR measurements, benzoyl chlorides appear to be weaker donors than benzophenones whereas the shift of the carbonyl stretching frequency is smaller for benzophenones. In benzoyl chlorides complexes $L \cdot AlCl_3$ a correlation is found bètween the shift of the NQR frequency assigned to the chlorine atom linked to the carbonyl atom and the shift of the corresponding C—Cl stretching frequency. The charge transfer was calculated for some of the $GaCl_3$ complexes and found to vary linearly with the gallium NQR frequency. In the case of $AlCl_3$ complexes with benzoyl chlorides, NQR and IR results show that they are covalent and not ionic.

The complex $GaCl_3 \cdot POCl_3$ (Jost and Schneider, 1975) presents a phase transition which is probably related to the onset of hindered rotation of the PCl_3 group; the positive temperature coefficient of the ^{69}Ga NQR frequency above the transition at 200 K can be explained by an increase of the P—O—Ga bond angle increasing in turn the vibrational frequency.

Complexes considered in the preceding paragraphs are of the n–σ type; π–σ complexes with aluminium bromide are also known which have been investigated by NQR: benzene·$AlBr_3$ in which the benzene acts as a weak donor of π electrons towards dimeric Al_2Br_6 molecules (Okuda et al., 1972b). Ortho- and para-xylene are also weak donors forming 1:2 complexes (Volkov et al., 1976). On the other hand, mesitylene and durene are stronger donors and the corresponding complexes have the composition 1:1. It seems that benzophenone can form two complexes with $AlBr_3$, a 1:1 complex of the n–σ type and a 1:2 complex in which one of the $AlBr_3$ molecules is bound via a n–σ interaction and the second via a π–σ interaction. It is also worth mentioning that π donation takes place towards the halogen atom while n donation takes place towards the empty valence orbital of the aluminium atom.

Somewhat unusual is the study of polymeric compounds: signals corresponding to broad resonances were found in complexes of $AlBr_3$ with poly-4-vinylpyridine and CCl_3COOH with poly-N-vinylpyrrolidone (Orlov et al., 1974). In these complexes the degree of charge transfer is estimated by comparison with the resonance in similar complexes with non-polymerized ligands, respectively $AlBr_3 \cdot$4-picoline and $CCl_3COOH \cdot N$-isopropylpyrrolidone. As there is no significant difference between the resonance frequencies for similar complexes of both sorts it is concluded that they have almost the same strength and that no steric effect modifies the complex formation in the polymeric compounds.

8.7 Complexes Containing Elements of Group IV

A great number of complexes containing typical elements of Group IVa (carbon and tin) have been studied by NQRS but no results seem to have reported concerning complexes of germanium, silicon, or lead compounds. On the other hand, among Group IVb elements, titanium, zirconium, hafnium, and thorium, only titanium is found in a molecular complex studied by NQRS: $TiCl_4 \cdot POCl_3$ (Rogers and Ryan, 1968). The structure of this complex is bimolecular in the sense that the crystalline unit may be represented as $(TiCl_4 \cdot OPCl_3)_2$. There is an electronic charge transfer from the Lewis base $POCl_3$ towards the Lewis acid $TiCl_4$, during the formation of the complex, via a new metal–oxygen bond. The complex was studied with others and, from the shift of the NQR frequency of the chlorine atoms attached to the phosphorus atom in $POCl_3$, the following order was found for the strengths of the Lewis acids:

$$SbCl_5 > FeCl_3 > SnCl_4 > TiCl_4.$$

8.7.1 *Carbon-containing complexes*

Carbon is a typical element of column IV, with a valence shell of four electrons; when it is four times covalently bonded, as is usually the case, this atom cannot accommodate, or give, any electron and therefore cannot act as acceptor or donor. However, in unsaturated systems, π electrons can be transferred to σ orbitals or to π systems of acceptor molecules and form π–σ or π–π complexes. In the first case are found Menshutkin's complexes of the type $MX_3 \cdot Ar$, or $MX_3 \cdot 2Ar$, with M = Al, As, Sb, or Bi, X = Br or Cl, and Ar = aromatic molecule such as benzene, substituted benzenes or polycyclic aromatic hydrocarbons; these complexes have been extensively studied by NQRS (Biedenkapp and Weiss, 1964, 1968; Negita *et al.*, 1966; Okuda, 1971; Yusupov *et al.*, 1971; Grechishkin and Yusupov, 1973; Shoshtakovskii *et al.*, 1973; Ainbinder *et al.*, 1974; Volkov *et al.*, 1976). There are also complexes of benzenes with bromine (Cornil *et al.*, 1964; Hooper, 1964), halomethanes such as chloroform or carbon tetrachloride (Grechishkin and Kyuntsel, 1966a; Maksyutin *et al.*, 1970b; Weiss, 1972; Kyuntsel and Rosenberg, 1973; Grechishkin *et al.*, 1974), and perhalogenated molecules. In the second case, π–π complexes, a great number of complexes have been studied since the first observation of the NQR in the hexamethylbenzene·chloranil complex (Douglass, 1960); a review has been made by Maksyutin (Maksyutin et al., 1970b). The temperature and pressure dependence of the NQR in the HMB·chloranil complex has been useful to get a better insight of this complex and is cited in section 8.2.2.

The frequency shift accompanying the formation of the complexes is often small, making it difficult to separate the effect of a charge transfer from solid-state effects (Cornil et al., 1964; Hooper, 1964; Biedenkapp and Weiss, 1964'). A striking example is provided by the complexes of p-xylene with CBr_4 and CCl_4, and of benzene with Br_2 (Hooper, 1964): the halogen resonance in the p-xylene complexes is split into two components and, on the basis of the crystal structure of the complexes, the low-frequency component is assigned to a couple of halogen atoms of the CX_4 molecule which are close to the benzene ring while the other component is assigned to the halogen atoms which are the farthest from the benzene rings. The lower frequency is smaller than in pure CX_4 and this explained by a small charge transfer from the π system of the benzene ring towards the halogen atom. In the complex $C_6H_6·Br_2$, the bromine NQR frequency is higher than in pure bromine as if intermolecular interactions were stronger in pure bromine than in the complex. However, the observation of a well-resolved spectrum is evidence for the definite composition of the compound. Shoshtakovskii (Shoshtakovskii et al., 1973) has studied a series of complexes nSbCl₃·Ar with various compositions and he observed that when the concentration of $SbCl_3$ is increased above the stoichiometric value corresponding to a complex, lines of pure $SbCl_3$ are also seen together with those of the complex of higher n.

In many instances however, the effect of charge transfer is stronger than in the preceding complexes and it is possible to make a distinction between weak and strong complexes. In weak complexes of aluminium bromide with aromatic compounds, aluminium bromide remains in its dimeric form, Al_2Br_6, characterized by two sets of lines corresponding to the bridge and terminal atoms of Al_2Br_6. In stronger complexes the dimeric unit is broken into two monomer units, $AlBr_3$, and the resonance lines are spread on a frequency range smaller than the separation between the lines corresponding to the bridge and terminal atoms in the dimer. The strength of an aromatic donor is related to the number of methyl group substitutents on the ring; o- and p-xylene are weak donors which complex the dimer $\dot{A}l_2Br_6$, whereas durene and mesitylene are stronger donors forming 1:1 complexes with $AlBr_3$ (Volkov et al., 1976). The increase in strength of the donor follows a decrease in the ionization potential of the aromatic compound (Yusupov et al., 1971, Volkov et al., 1976). The order of magnitude of the frequency shift observed for the bromine NQR lines is 0.7 and 0.4% for the two xylenes (o- and p-, respectively) and 7.2 and 8.1% for the mesitylene and durene complexes of $AlBr_3$.

The complexes considered above are pure aromatic hydrocarbons without any heteroatoms such as nitrogen or oxygen which would act as n donors. In practice, aromatic compounds with heteroatoms can form complexes of both π–σ and n–σ types as occurs for the 1 : 2 complex of benzophenone with $AlBr_3$: one $AlBr_3$ molecule is bound to the ketonic oxygen atom (n–σ bond) and the second to the benzene ring (π–σ bond). On the other hand, in the case of the complexes of benzophenone with $SbCl_3$ (1 : 1 and 1 : 2), it is considered (Shoshtakovskii et al., 1973) that the interaction only takes place between the $SbCl_3$ molecule and the π electrons of the ring because no major change occurs in the chlorine NQR frequency when passing from the 1 : 1 to 1 : 2 complex. In a similar way, three complexes of p-terphenyl with $SbCl_3$ are observed (1 : 1, 1 : 2, 1 : 3). For the 1 : 3 complex it is assumed that each $SbCl_3$ molecule interacts with one of the three rings of the p-terphenyl molecule.

The localization of the interaction in a complex is a problem to which NQRS can bring useful pieces of information. In the case of the Al_2Br_6 complexes it is observed that complexation produces a shift of the NQR frequency of the terminal bromine atoms greater than that of the bridge atoms. From the importance of the ^{75}As frequency shift it appears that the arsenic atom is the centre of complexation in complexes of $AsBr_3$ (Yusupov et al., 1971).

Besides the determination of resonance frequencies, Zeeman effect investigations have been carried out on a few complexes, bringing further data (Okuda, 1971; Okuda et al., 1972b). The accuracy of Zeeman effect measurements is checked through a comparison with X-ray data when available. For $Al_2Br_6 \cdot C_6H_6$ it is found that the value of the bond angle Br— Al—Br (terminal bromine atoms) obtained by the Zeeman effect on NQR agrees with that from X-ray diffraction within 1 degree of angle. However, the agreement is not so good for the angle Al—Br—Al (bridging bromine atom) and an explanation is proposed which supposes some bending of the Br—Al bonds.

Another type of investigation in these complexes is the study of the temperature dependence of the NQR frequency ν_Q and relaxation times (Yusupov et al., 1971; Kyuntsel and Rosenberg, 1973; Ainbinder et al., 1975). They will not be considered in detail here as they do not seem to bring any special new insight into the charge transfer mechanism itself, though in some cases, like that of chloroform, the mobility of the —CCl_3 group, greater in the complex than in pure chloroform, gives an additional clue to the hydrogen bonding nature of the interaction in the complex (Kyuntsel and Rosenberg, 1973).

Mention should be made of complexes in which one component is a polymer. Results have been published concerning $SbCl_3 \cdot$polystyrene, $AlBr_3 \cdot$4-vinylpyridine, (Orlov *et al.*, 1974) and also $SbCl_3 \cdot$polyphenylacetylene (Shoshtakovskii *et al.*, 1973). The NQR spectrum in these complexes is compared to that of similar but unpolymerized compounds, $SbCl_3 \cdot$cumene and $AlBr_3 \cdot$4-methylpyridine. In the first one the QCC is greater in the polymeric compound than in the unpolymerized one whereas no difference is found for the second couple of compounds. It is inferred that the $SbCl_3 \cdot$polystyrene complex is weaker than $SbCl_3 \cdot$cumene and this may be due to some steric hindrance in the polymer: the size of the acceptor molecule prevents it from closely approaching the complex centre in the polystyrene units.

8.7.2 Tin-containing complexes

A great number of complexes containing tin halides or organic-substituted tin halides have been studied by NQRS. About 70 of them are listed in section 14. They were studied with a view to obtaining information about the Sn—Hal and the Sn—Lig bonds. Experimental data are considered in terms of the NQR frequency shift with respect to the uncomplexed compounds and assignment of the lines, especially when the donor also contains a chlorine atom and when the splitting has a conformational origin. Then, information about the electronic distribution is derived and the results may be summarized in three major groups:

(a) Characterization of the complex formation and determination of the centre of complexation: for example, in the case of $SnCl_4$ and $POCl_3$, a $1:2$ complex is known and if a $1:1$ mixture is used in preparing the complex, NQR lines of pure $SnCl_4$ are observed together with those of the $1:2$ complex (Kravchenko *et al.*, 1968). On the other hand, it is also reported that a mixture of $SnCl_4$ and $PSCl_3$ gives broad signals centred on the frequencies of the lines of the pure starting materials showing that no complex is formed. The centre of complexation may be identified by discussing the values of the shifts observed: in the $2:1$ complex $CH_3OCH_2Cl \cdot SnCl_4$ the charge transfer is made via the oxygen atom of the chloroether; in the complex $C_6H_5CH_2Cl \cdot SnCl_4$ it is quite probable that the chlorine atom of the donor is involved in the charge transfer (Maksyutin *et al.*, 1970c) whereas the two oxygen atoms are donors in the complex $C_6H_5NO_2 \cdot SnCl_4$. Most of the tin complexes studied appear as strong complexes in the sense that the NQR frequency shift between the NQR frequencies in the complex and in the uncomplexed materials is much higher than usually expected from crystal field effect.

(b) Influence of the complexation on the electronic distribution: a simple valence bond model can be used to explain the shift of the NQR frequency upon complexation. In $SNCl_4 \cdot 2POCl_3$, the $POCl_3$ molecules act as donors of electronic charge towards $SnCl_4$; as a result, the Sn—Cl bond becomes more ionic and the corresponding chlorine QCC and NQR frequency are lowered while the P—Cl bond becomes more covalent, increasing the QCC and NQR frequency of the chlorine atoms in $POCl_3$ (Biedenkapp and Weiss, 1964).

(c) Structural arrangement of the atoms about the central tin atom. The effect of the conformation on the QCC and NQR frequencies can be explained theoretically, but detailed models, using MO's instead of valence bond schemes are required for this purpose (Kravchenko *et al.*, 1968).

Among the complexes studied are those with $SnHal_4$ as acceptor, the donor being mostly phosphroc chloride or derivatives (Biedenkapp and Weiss, 1964; Kravchenko *et al.*, 1968; Rogers and Ryan 1968; Safin *et al.*, 1975; Andreeva *et al.*, 1975, 1976) though some other donors have also been studied (Dehmelt, 1953; Biedenkapp and Weiss, 1964; Kravchenko *et al.*, 1968, Maksyutin *et al.*, 1970c, 1973; Petrosyan *et al.*, 1973; Poleschuk *et al.*, 1974b), and those with methyltin as acceptor, the donor being oxygen- or nitrogen-containing compounds (Petrosyan *et al.*, 1973).

Andreeva (Andreeva *et al.*, 1975) and Safin (Safin *et al.*, 1975) report results on several octahedral complexes of tin tetrachloride with organophosporus compounds of composition $1:2$. Safin's preliminary paper shows the relation with other studies by NMR and NQR spectroscopies in connection with the *cis* or *trans* structure of the complexes. Andreeva carried out an investigation on a series of about 35 complexes. The structure, *cis* or *trans* of the complexes as derived from the splittings observed in the NQR spectra is given and a correlation between the frequency shift of the chlorine NQR in the donor molecule and the NQR frequency of the chlorine in the $SnCl_4$ acceptor molecule is shown for a series of *trans* complexes. The temperature dependence of the chlorine NQR frequency is also given for a couple of complexes. Further study of that kind of complex is found in Andreeva's paper (Andreeva *et al.*, 1976) concerned with amido-phosphorus derivatives as donor. Chlorine and nitrogen resonances were observed and the structure of these complexes was found to be *trans*; the donor character is stronger for $\geqq$P decreasing in the order $\geqq$P: $> \geqq$P:S $> \geqq$P:O. In the latter, the oxygen atom is the complexation centre. The asymmetry of the electric field gradient at the site of the nitrogen atom is reduced upon complexation. However, the ionicity of the N—P bond appears to be only slightly sensitive to a change of the substituent in the donor in the series reported.

Tin complexes with donors other than phosphorus derivatives have been investigated for a basic study of the conformation in $2L \cdot SnCl_4$ complexes (Kravchenko et al., 1968) together with $2POCl_3 \cdot SnCl_4$. In such complexes tin is hexacoordinated with an octahedral structure; two configurations, *trans* and *cis* are also possible. In the figure below, the scheme cis_2, at the right of the figure, differs from the scheme *cis*, in the middle, by the angle of viewing which now shows clearly the two kinds of chlorine atoms, axial and equatorial. From a simple MO model, it is shown that the average frequency shift

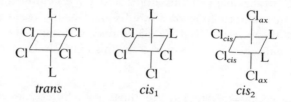

upon complexation, for the chlorine atoms in $SnCl_4$, is not the same for the *cis* and *trans* structure and the NQR frequency of an axial atom is smaller than that of an equatorial one (Maksyutin et al., 1973). In the same paper, NQR results are compared to those of IR and γ-ray spectroscopies and agreement is found between them and a relation appears between the axial–equatorial NQR frequency splitting and the average NQR frequency shift. It is also shown that the difference between the *cis* and *trans* frequency shift is maximum at lower charge transfer.

A difference in behaviour of *cis* and *trans* complexes is also observed when the temperature of the sample is varied. Of the two compounds reported (Safin et al., 1975; Andreeva et al., 1975), $[(C_2H_5)_2P(O)-SC_2H_5)]_2 \cdot SnCl_4$ and $[(C_4H_9O)_2P(O)(SCH_3)]_2 \cdot SnCl_4$, the first one is *trans* and its spectrum presents a doublet of small separation (0.09 MHz). When the temperature is increased from 77 to 280 K the frequency of these two lines decreases almost linearly with the same slope for the two lines. On the contrary, the second compound has a *cis* structure and its spectrum presents two components separated by about 1.2 MHz and while the low-frequency component has a linear temperature dependence, the high-frequency one has a strongly curved aspect which clearly demonstrates the unequivalence of the two sites of the resonant nuclei.

As to whether *cis* or *trans* structures are the more probable for a given complex, one statement has been made that *cis* structure is favoured for ligands which do not cause steric hindrance, otherwise *trans* complexes are formed (Safin et al., 1975). An illustration of this rule is found with the complexes $2C_2H_5OH \cdot SnCl_4$ and $2(C_2H_5)_2O \cdot SnCl_4$, the former being *cis* and the latter *trans* (Kravchenko et al., 1968; Maksyutin et al., 1973). Another argument is that for ligands of similar size, the *trans* structure is favoured by a higher donor capability of the ligand (Andreeva et al., 1975). Anyway, this point is not a simple one because the conformation may differ in the solid phase and in solution as is found for $2(C_2H_5)_2O \cdot SnCl_4$ (Kravchenko et al., 1968).

Calculations show that upon complexation the charge is mainly transferred to the chlorine atoms, the charge increase on tin being small (Poleschuk et al., 1974b).

Complexes of methyltin halides (Petrosyan et al., 1973) present this difference with tin tetrahalide complexes that whereas the latter are mostly octahedrally hexacoordinated, trimethyltin bromide complexes are pentacoordinated with a 1:1 composition when the ligand is monodentate; pyridine·trimethyltin bromide has a trigonal bipyramidal structure. From the observed shifts of the bromine NQR frequency in these compounds, the ligands studied have been arranged in a series of increasing donor activity with respect to $(CH_3)_3SnBr$. Dialkyltin halides, $(CH_3)_2SnBr_2$ and $(CH_3)_2SnCl_2$, from octahedral complexes of which the structure should be one of five possible configurations; to determine which, help is found in examining the splitting of the components in the spectra and from data given by Mössbauer spectroscopy. Some monomethyltin trihalide complexes have also been studied; the results suggest (i) that the uncomplexed molecules CH_3SnBr_3 and CH_3SnCl_3 are in dimeric state and (ii) that the structure of the complexes depends on the nature of the ligand. A few complexes of tin tetrabromide and tin tetraiodide are also reported: the shift of the halogen NQR frequency is, compared to the corresponding tin tetrachloride complexes, stronger.

Besides complexes of tin, intramolecular coordination between a tin (or silicon, or germanium) atom is also an example of charge transfer which has been investigated by NQRS. In a series of compounds $RR'R''MCH_2Cl$, the frequency of the chlorine resonance is higher in compounds with M = Si, Ge, or Sn, than in the compound with M = C, which is surprising in view of the higher electronegativity of the carbon atom. This anomaly was explained by the contribution of an intramolecular electronic interaction (α effect) between the unshared pair of the chlorine atom and a vacant d orbital of the atom M, as shown in the scheme below (Voronkov, 1971):

Theoretical developments on this 'α effect' were presented later by the same authors (Feshin and Voronkov, 1973).

8.8 Complexes Containing Elements of Group V

Column V is exceptionally important due to the number of complexes in which elements of the column participate as centre of complexation, and almost all the elements of the column are found in complexes. With five electrons, the valence shell of elements of column V is more than half-filled, as far as s and p subshells are concerned, and can only accomodate three more electrons to form three covalent bonds as in compounds of the AB_3 type. The lone pair on the A atom is then available for donation (NMe_3, PMe_3, $AsMe_3$, $SbMe_3$ compounds and like). In the case of As, Sb, and Bi, d orbitals are available for higher hybridization and their compounds can also be acceptor whether they are trivalent or pentavalent. Owing to the great variety of complexes containing elements of group V, it is difficult to classify them in a satisfactory manner and a simple way for a review is to consider each element of the column in turn.

8.8.1 Nitrogen-containing complexes

Molecules with a nitrogen atom are basically n donors apart from a few exceptions like aniline which may also be a π donor or tetracyanoethylene which is a strong acceptor. In most of the complexes studied by NQR, the nitrogen atom is either nitrilo or amino (amino in a broad sense including aliphatic, aromatic, and heterocyclic amines like trimethylamine, aniline, or pyridine).

8.8.1.1 *Complexes of nitriles* Nitriles complexed to a great variety of acceptors have been investigated though all the acceptors contain halogen atoms.

Complexes of nitriles with pentahalogenides. The first one reported is $CH_3CN \cdot SbCl_5$ (Schneider and Di Lorenzo, 1967); NQR permits one to obtain clues about the structure: two possible structures could be expected; one is octahedral with four equatorial chlorine atoms and one axial; the second is a 2:2 ionic structure with a $SbCl_6^-$ anion and a $SbCl_4(CH_3CN)_2^+$ cation. The relative intensities of the two chlorine lines observed strongly suggest the first structure which is also supported by the fact that the chlorine frequencies are somewhat different from those measured in a compound with $SbCl_6^-$ ions. Other complexes of nitriles with $SbCl_5$ were studied later (Poleschuk *et al.*, 1974a) and an estimation of the charge transfer, within the frame of the Townes and Dailey's theory, leads to a value of ~0.3 e, the charge being transferred to the chlorine atoms of $SbCl_5$ with even a decrease of the charge on the antimony atom respective to the uncomplexed molecule. In some cases, the NQRS is observed but no definite conclusion can be drawn about the structure or the charge transfer as is the case for $NbF_5 \cdot CH_2ClCN$ or $NbF_5 \cdot C_5H_5N$ (Fuggle *et al.*, 1974). It may be pointed out that only Nb resonances were observed whereas both Sb and Cl resonances were detected in the preceding complexes, the Cl resonances giving useful information.

A special nitrile is ClCN and its complex with $SbCl_5$ has been studied by Burgard and Lucken (1972) and its complexes with $NbCl_5$ and $TaCl_5$ by Burgard and MacCordick (1972). X-ray diffraction measurements on $SbCl_5 \cdot NCCH_3$ have confirmed the structure proposed by Schneider and Di Lorenzo (1967, cited above). Complexes of CH_3CN with $NbCl_5$ and $TaCl_5$ were also studied by NQR (Burgard and MacCordick, 1972). In the complexes of ClCN it is quite certain that the complexation takes place through the nitrogen atom and only a slight effect is observed on the chlorine resonance. Unfortunately, the nitrogen resonance which would be useful for further discussion has not been reported so far. Comparison of the ionic character of the Cl—Sb bond in several complexes shows that CH_3CN is a better donor than ClCN though the difference is not very great. On the same basis, $POCl_3$ appears to be a donor of the same strength. An estimated value of the charge transfer in the $ClCN \cdot SbCl_5$ complex is 0.25 e.

Complexes of nitriles with trihalogenides. Complexes of nitriles with BCl_3 and $GaCl_3$ have been studied. NQR spectra of $CH_3CN \cdot GaCl_3$ and $PhCN \cdot GaCl_3$ were recorded in the frame of a general study of $GaCl_3$ complexes showing the correlation between the ^{69}Ga and ^{35}Cl NQR frequencies (Tong, 1969); no special attention is brought to the complexes with nitriles but it is seen that the corresponding points on the plot showing the above correlation are close to that of $POCl_3$ which is consistent with the statement about the donor strengths of acetonitrile and $POCl_3$ cited in the preceding paragraph.

Complexes of nitriles with BCl_3 have been more extensively studied. Several papers are concerned with BCl_3 complexes with no special attention for nitrile donors though some comments are given about the number of components in the NQR spectrum in connection with the crystal structure (Smith and Tong, 1971b; Jost, 1972). The detailed study by Kaplansky and Whitehead (1970) deals with the resonances observed in five nitriles; the interpretation of NQR results is made in terms of the Townes and Dailey theory and, to avoid some arbitrariness inherent to the method, calculations using CNDO, SAVE–CNDO and BEEM-π techniques were also made. In particular it is shown that the addition of methyl groups in the donor increases the ionic character of the chlorine atom in the BCl_3 accepting group. In other words, this means that the methyl groups increase the donor ability of the nitriles. A similar conclusion was obtained in a study of complexes of nitriles with halogens (see below). Ardjomande and Lucken (1971) report on complexes of acetonitrile and chloroacetonitriles with BCl_3 as studied by Graybeal (1958) and discuss the molecular structure of these complexes as was explained in the paragraph on complexes of boron. Finally, Ardjomande and Lucken (1975) have also studied a series of complexes of chlorobenzonitriles with BCl_3 in the course of a more general study with a view to determining the effect of complexation on the donor molecule. In a general way, an increase of the NQR frequency of the substituent chlorine is observed on complexation which is explained by the diminution of the electron density on the donor.

Complexes of nitriles with halogens. Negita *et al.* (1973) report the ^{14}N and halogen quadrupole coupling constants measured in the complexes $2CH_3CN \cdot Cl_2$ and $2CH_3CN \cdot Br_2$ and the corresponding change in the charge on the nitrogen and chlorine atoms with respect to the uncomplexed compounds. From the decrease in the ^{14}N QCC it is concluded that the charge transfer takes place from the nitrogen atom towards the halogen; the complexes are considered as weak ones as far as the change in QCC and the calculated charge transfer (0.03 e) are concerned. The increase of electronic charge on the bromine atom is smaller than the decrease on the nitrogen which can be explained if the acceptor orbital on the bromine atom is *d* hybridized. No phase change is visible in a study of the temperature dependence of the NQR frequency in the bromine complex. In a later investigation it is said that the composition of the complex is 1:1 instead of 2:1 (Dresvyankin 1975); despite this difference, the results of this new investigation are quite similar and the average charge transfer, for a series of six nitriles including benzonitrile·Cl_2 and 2-benzonitrile·Cl_2, is between 0.02 and 0.03 e. The chlorine complex of adiponitrile (a dinitrile) has a polymeric structure suggested by the presence of only one resonance line, explaining the high melting point of the complex. Calculation of the charge transfer (Dresvyankin *et al.*, 1975) shows that most of the charge is transferred to the terminal chlorine atom; the calculation is made with a three-centre de-localized σ orbital model.

Tetracyanoethylene (TCNE), as mentioned at the beginning of the nitrile section, deserves a special mention as it is a polynitrile which is a strong π acceptor. Its complex with naphthalene has been studied (Onda *et al.*, 1973) by recording the ^{14}N NQR spectrum. In the complex, a small increase of both e^2Qq and η with respect to pure TCNE is observed and though the interpretation of this effect is not quite obvious in terms of the simple Townes and Dailey's theory it appears that the complex ion is a weak one.

Though complex ions with a metal as central atom are outside the scope of this chapter, the case of some complex ions of palladium, nickel, and zinc, containing the thiocyanate ion, NCS^-, should be mentioned since, in a recent publication (Cheng *et al.*, 1977), it has been shown that NQR permits the distinction between *S-* and *N*-coordinated thiocyanate ions. The nitrogen-14 QCC is, indeed, much lower ($\sim 1\,MHz$) in complex ions with *N*-coordinated $[NCS]^-$ group than in complex ions with *S*-coordinated $[SCN]^-$ group ($\sim 3.2\,MHz$).

8.8.1.2 *Complexes of amines* The variety of complexes with amines is at least as great as that of complexes with nitriles.

Complexes of aliphatic amines and amines with saturated ring. Several complexes of amines with chloroform have been studied in which both ^{14}N and ^{35}Cl resonances were recorded; the estimated charge transfer is 0.05 e (Lucas and Guibé, 1970). In complexes with BCl_3 (Kaplansky and Whitehead, 1970) only the chlorine resonance is observed and, as BCl_3 is only very slightly sensitive to charge transfer effect, not much information can be gained from NQR measurements, except for molecular motion from the temperature dependence of the resonance frequency and linewidth in the case of trimethylamine·BCl_3 (Gilson and Hart, 1970).

Complexes of aromatic amines. A series of complexes of aniline and haloanilines with 2,4,6-trihalophenols have been investigated by both infrared and NQR spectroscopies (Korshunov *et al.*, 1971). Their composition is either 1:1 or 1:2. The experimental results seem to indicate a contribution of the $-NH_3^+$ structure and suggest the presence of different hydrogen bonds either intra- or intermolecular as well as weak $\pi-\pi$ interaction between aniline and phenol molecules. Consideration of IR spectroscopy results helps to formulate a description of possible hydrogen-bonded structures.

Complexes of heterocyclic amines. The heterocyclic amines studied by NQR in their complexes are mainly pyridine and its derivatives. Different sorts of acceptor have been used. In complexes with inorganic acceptors such as BCl_3 (Jost, 1972), IBr, ICl (Bowmaker and Hacobian, 1969), NbF_5 (Fuggle *et al.*, 1974), $SbCl_5$ (Poleshchuk *et al.*, 1974a), $HgCl_2$, $SnCl_2$, $BiCl_3$, BCl_3, $SbCl_5$, HCl, $HClO_4$, and H_2SO_4 (Ardjomande and Lucken, 1975), it is the resonance of quadrupolar nuclei in the acceptor which was observed whereas halogen resonance in the donor was observed in some complexes of chloropyridines with organic acceptors like trinitrophenol and CH_3I (Ardjomande and Lucken, 1975). In complexes of $ZnBr_2$, $ZnCl_2$, ZnI_2, $CdCl_2$, $Zn(NO_3)_2$, and $Zn(NCS)_2$ with two pyridine molecules (Hsieh *et al.*, 1977), and of pyridine, 4-methylpyridine, 3,5-dimethylpyridine with $CHCl_3$ (Lucas and Guibé, 1970), the ^{14}N resonance was detected giving useful data for the estimation of the charge transfer which is small for the $CHCl_3$ complexes (0.05 e) and higher for the complexes with inorganic acceptors (~ 0.2 e for Zn salts and ~ 0.6 for the $HPyNO_3$ compound). The experimental results for the complexes of pyridine with methanol and pyrrole were analysed in terms of hydrogen bonding between the two molecules forming the complex (Ha and O'Konski, 1970).

The donor properties of the nitrogen atom through its lone pair are also important in biological systems like the metalloporphyrins where it would be useful to have some experimental data about the electronic structure in the vicinity of the central metal atom. The NQR of the pyrrolic nitrogen atom was investigated in chlorophyll-a and in magnesium phthalocyanin (Lumpkin, 1975). The QCCs are slightly different in these two molecules and this can be related to a difference' in the molecular structure. The magnesium resonance was also recorded but discussion of the QCC is difficult because the actual state of coordination of the magnesium atom is not exactly known.

8.8.2 *Phosphorus-containing complexes*

Only a few phosphorus-containing complexes in which the phosphorus atom acts as donor have been reported as studied by NQRS. They are compounds of trivalent phosphorus of the type R_3P with R = Cl, Me, or Et; in complexes of pentavelent phosphorus, of the type $R'R''R'''PO$ or $R'R''R'''PS$, the oxygen or sulphur atom is the donor (see, for example, Andreeva *et al.*, 1976).

Complexes of trimethylphosphorus with trimethylaluminium, -gallium, and -indium, have been investigated by several authors (Patterson and Carnevale, 1973; Kasakov *et al.*, 1977). Phosphorus seems to be a stronger donor than N, As, or Sb in complexes with $InMe_3$ or $GaMe_3$ but the order is not exactly the same for complexes with $AlMe_3$: nitrogen then appears as a better donor than P, As, and Sb.

The NQR frequencies of chlorine in the complex $Ph_3P \cdot BCl_3$ are not very significant as it is known that the chlorine frequencies in BCl_3 complexes do not vary much in a series of complexes (Ardjomande and Lucken, 1971). Complexes of trivalent phosphorus with $SnCl_4$ have also been investigated in $Et_2ClP \cdot SnCl_4$ (2:1); the NQR frequency of the chlorine in the stannic group (16.705 MHz) is smaller than the values of 17, 18, or 19 MHz found for complexes of the type $R'R''R'''PO \cdot SnCl_4$ (2:1) (Andreeva *et al.*, 1975); the same trend is observed for hexamethylphosphorus triamide, $(Me_2N)_3P$, complexed with $SnCl_4$ ($\nu'_{av} = 18.53$ MHz). The increase in the frequency of the chlorine NQR in Et_2ClP on complexation with $SnCl_4$ is also an indication of a strong electronic transfer from the phosphorus towards the $SnCl_4$ molecule.

Comparison of the resonance frequencies of the chlorine atoms of $GaCl_3$ in complexes with Me_3P and Et_3P (Carter *et al.*, 1976) shows that phosphorus is a stronger donor, as compared to oxygen or sulphur in ethers or thioethers, and that the resonance frequency is close to that found in $GaCl_4^-$.

Intramolecular dative bonds have also been investigated in some halomethylated phosphine oxides. In only one compound of the series studied, $(CH_2Br)_3P$, are the splittings of the bromine resonances large enough to be ascribed to some intramolecular effect of the α type (Saatsazov *et al.*, 1975).

8.8.3   *Arsenic- and antimony-containing complexes*

There is some similarity between the complexes of these two elements: their trihalides are acceptors of π electrons from aromatic molecules in Menshutkin's complexes and their trimethyl derivatives are donors towards trimethyl derivatives of elements of Group III. On the other hand, antimony pentachloride acts as an acceptor to form octahedral complexes with many n donors.

NQR in Menshutkin's complexes has been extensively studied since its first observation (Grechishkin and Kyuntsel, 1963a; Biedenkapp and Weiss, 1968 and references therein). Resonance frequencies of different isotopes (^{35}Cl, ^{75}As, ^{81}Br, ^{121}Sb, and ^{123}Sb) (Grechishkin and Kyuntsel, 1964b; Grechishkin and Yusupov, 1970b, Anferov, 1971) have been reported as well as the temperature dependence of the resonance frequency (Grechishkin and Kyuntsel, 1964b; Okuda *et al.*, 1972a) and relaxation times (Grechishkin and Gordeev, 1966a; Gordeev *et al.*, 1971) and the Zeeman effect in single crystals (Okuda *et al.*, 1970, 1972a; Okuda, 1971). The size of the frequency shift between the complexed and pure compounds depends strongly upon the complex considered. In the case of $AsCl_3$ complexes it appears that the ^{75}As NQR frequency is smaller in the complexes than in pure $AsCl_3$ and this is explained by the fact that the As atom is the acceptor centre in the complex (Grechishkin and Yusupov, 1970a). In the case of the 1:1 complex $AsBr_3 \cdot 1,3,5\text{-Me}_3\text{Ph}$, the decrease in the ^{75}As frequency is as high as 17% (Grechishkin and Yusupov, 1973). Moreover, the amplitude of the frequency shift of the ^{75}As NQR is linearly related to the ionization potential of the aromatic molecules (Grechishkin and Yusupov, 1970b; Yusupov *et al.*, 1971).

If complexes of $SbCl_3$ (Shoshtakovskii, 1973) are considered, it is seen that the antimony QCC in the complexes is higher than in pure $SbCl_3$ and that there is no close linear relation between the QCC and the ionization potential of the donor molecule, though a general linear trend can be discerned which can be analysed and represented by the relation $e^2Qq = 350 + 3.83I$ MHz, where I is the ionization potential in eV. It is possible that, owing to the bigger size of the antimony atom as compared to that of arsenic, the change in the hybridization in the complexed $SbCl_3$ molecule is higher than in the complexed $AsCl_3$, but as Shoshtakovskii points out, it is difficult to make a distinction between steric and electronic effects for weak complexes. Another argument present by Shoshtakovskii is that higher d hybridization on the antimony atom in the complex can explain the observed higher QCC relative to uncomplexed $SbCl_3$; this high d hybridization being almost constant in a series of complexes, the QCC decreases with increasing donor ability of the aromatic molecule.

An interesting result of the study of the Zeeman effect in the complex naphthalene·2SbCl$_3$ (Okuda, 1971) is the determination of the asymmetry parameter η and QCC of the three different chlorine atoms in the complex. The maximum splitting observed between the three QCCs (2.98 MHz) is much higher than the commonly accepted value of 0.2–0.3 MHz for crystalline effects and this is also observed in the complex ethylbenzene·SbCl$_3$ (Okuda et al., 1972a,b) and is attributed to the trans effect which favours charge transfer towards the chlorine atom in the trans position with respect to the aromatic donor molecule.

Complexes of SbCl$_3$ and SbBr$_3$ with aniline have also been investigated (Gordeev and Kyuntsel, 1974) and present a strong shift of the Sb QCC ($\sim 25\%$) which is due to n donation from the lone pair of the nitrogen atom in aniline. The 1:1 complex aniline·SbBr$_3$ also presents an unexpectedly low asymmetry parameter of the antimony resonance. A calculation, based on a point charge and tangential sphere model, shows that this low value of η may be explained by the cancellation of opposite effects (Brill, 1974). Similarly, complexes of SbBr$_3$ and SbCl$_3$ with anisole (Kyuntsel and Rozenberg, 1971) and of SbCl$_3$ with C$_6$H$_5$Cl (Grechishkin and Yusupov, 1970a) are also n–σ complexes studied by NQR.

The study of the relaxation time and temperature dependence of the resonance frequencies and relaxation time was developed in several Menshutkin's complexes to obtain rotational frequencies and average lifetime of rotation oscillation (Grechishkin, 1966; Yusupov et al., 1971). The strong temperature dependence of the frequency in some complexes was considered as of interest for possible thermometric applications (Grechishkin and Kyuntsel, 1963a). Other authors have also reported on the temperature dependence of the frequency and relaxation time (Biedenkapp and Weiss, 1964; Okuda, 1971; Kyuntsel 1971).

Whereas arsenic and antimony appear as acceptors in the complexes of AsHal$_3$ and SbHal$_3$, these atoms act as donors in the complexes of AsMe$_3$, AsEt$_3$, and SbMe$_3$ with trimethyl derivatives and trichlorides of Group III metals like aluminium, gallium, and indium. In the complexes of Me$_3$As and Et$_3$As with GaCl$_3$ (Carter et al., 1976) the chlorine NQR frequency is almost as low as it is in GaCl$_4^-$, showing that the Ga—Cl bond has a pronounced ionic character which, in turn, indicates that the donor power of the arsenic derivatives considered is strong. Complexes of Me$_3$As and Me$_3$Sb (as well as of Me$_3$N and Me$_3$P) with trimethyl derivatives of Group III metals have been extensively studied by Patterson and Carnevale (1973), as mentioned in the paragraph on aluminium and indium (section 8.6.2; p. 363). Let us just recall that the change in the QCC results (i) from charge transfer across the As—Al bond and (ii) from modification of the geometry of the complexed molecules (change in the Me—As—Me angle and in the Me—As bond length). In other publications (Dewar et $al.$, 1973; Kasakov et $al.$, 1977), complexes of Me_3As and Me_3Sb with trimethyl derivatives of Group III metals are also considered but the attention is focused on the NQR of the metal.

Another series of complexes to be reviewed is that of the pentacoordinated antimony compounds. Several complexes of $SbCl_5$ have been investigated in which Sb is the acceptor and tends to adopt an octahedral sp^3d^2 hybridization. The first complexes of this type reported are those with CH_3CN and $POCl_3$ (Schneider and Di Lorenzo, 1967; Rogers and Ryan, 1968); the results for the CH_3CN complex (1:1 composition with C_{4v} symmetry) were discussed in the paragraph on CH_3CN (section 8.8.1, p. 372). In the complex $Cl_3PO \cdot SbCl_5$, the oxygen atom of Cl_3PO is the donor and the multiplicity of the NQR spectra is in accordance with X-ray diffraction data. Comparison of the NQR results for $Cl_3PO \cdot SbCl_5$ with those for other acceptors leads to the following order of the Lewis acid strengths: $SbCl_5 > FeCl_3 > SnCl_4 > TiCl_4$. Burgard and Lucken (1972) have studied the complex $ClCN \cdot SbCl_5$ observing the chlorine-35 resonances (in the 41–42 MHz region for ClCN and in the 27–29 MHz region for $SbCl_5$) and the ^{121}Sb (1 line) and ^{123}Sb (2 lines) resonances. They found, in a comparison with the $CH_3CN \cdot SbCl_5$ complex that the structure of the $ClCN \cdot SbCl_5$ complex is probably of the C_s type rather than C_{4v}. From the small variation in the ^{35}Cl NQR frequency, the ionicity of the Sb—Cl bound appears to be higher in the complex with CH_3CN than in the complex with ClCN, and from the frequency of the antimony isotopes, a value of 0.25 e is found for the charge transfer.

In the complex $(CH_2Cl)_2O \cdot SbCl_5$ the chlorine NQR spectrum shows that the lone pair of the oxygen atom is involved in the formation of the complex; if the chlorine atoms were contributing, two different chlorine NQR frequencies would be detected, which is not the case (Makridin et $al.$, 1973). Moreover, the charge transferred to the chlorine atoms of $SbCl_5$, through the antimony atom, is quite close to that in two other complexes, $Cl_3PO \cdot SbCl_5$ (Rogers and Ryan, 1968) and $CH_3CN \cdot SbCl_5$ (Schneider and Di Lorenzo, 1967).

In 1974, Poleshchuk (Poleshchuk et $al.$, 1974a) reported on an extensive series of $SbCl_5$ complexes with CH_3CN, C_6H_5CN, C_7H_9N (quinoline), NH_3, $(C_2H_5)_2O$, and $(CH_2)_4O_2$. The L—Sb bond appears to be strong in these complexes and a 0.3 e value is estimated for the charge transfer. Discussion of the results is difficult on a quantitative basis because the sign of the QCC of the antimony nuclei is not known. A later investigation by Mössbauer spectroscopy shows that the EFG component along the bond direction in the complex is positive (Sychev et $al.$, 1975b); it then appears that all the charge is transferred from the ligand to the chlorine atoms and, moreover, some charge is withdrawn from the Sb atom by the chlorine atoms resulting in an electron deficiency on the antimony (Poleshchuk and Maksyutin, 1975a). Another point of interest is the linear relation found between the binding energy E of the $3d_{5/2}$ electrons of the Sb atom in the complex and the change Δa_{Cl} in the electron density on the chlorine atom and, finally, the charge transfer is estimated to be 0.3 e for the $CH_3CN \cdot SbCl_5$ complex. The theoretical calculations (Sychev *et al.*, 1975a) on $SbCl_5$ complexes show a decrease in the s orbital population of the Sb atom, in agreement with the isomer shift measured in the Mössbauer experiment and also a decrease in the population of the p_z orbital in agreement with the discussion of the NQR results. In complexes with 2- and 3-chloropyridines (Ardjomande and Lucken 1975) $SbCl_5$ appears to be a weaker acceptor than $BiCl_3$ or $HgCl_2$. The antimony halides which have been most studied are $SbCl_3$, $SbCl_5$, and $SbBr_3$, but some antimony resonances have been observed in complexes of SbF_3 with some inorganic salts like KCl, K_2SO_4, and Cs_2SO_4 (Bryukhova *et al.*, 1974a).

8.8.4 Bismuth-containing complexes

Complexes of $BiCl_3$ with acetone, acetonitrile, anisole, and diethyl ether have been studied by NQRS (Maksyutin *et al.*, 1970a). These complexes appear to be relatively weak and the observed QCC depends strongly on the geometrical structure of the complexes. The spectra obtained are discussed in relation to the structure of solid $BiCl_3$ which contains bridged halogen bonds; the bridges are broken upon complexation. The weakness of the anisole complex is shown by the presence of the resonance of uncomplexed $BiCl_3$ in the sample, explained by some breaking of the complex prepared. In the case of the $BiCl_3 \cdot DEE$ complex, the strong difference between the values of the asymmetry parameter for the two sites observed is ascribed to the presence of two rotational isomers of the DEE molecule. It also appears that in complexes of $BiCl_3$ the contribution of d hybridization to the QCC is so important that it is difficult to obtain significant information about the charge transfer (Maksyutin *et al.*, 1970a). Later, complexes of $BiCl_3$ with 2- and 4-chloropyridines were examined in the course of a study of the effect of complexation on the QCC of the chlorine substituent in different chloropyridines (Ardjomande and Lucken, 1975). $BiCl_3$ is also an acceptor whose strength compares with that of $SnCl_4$, trinitrophenol, or HCl, but is weaker than $HgCl_2$ and BCl_3 from the shift observed in the chlorine NQR frequency in 2-chloropyridine.

8.8.5 Complexes containing elements Vb (Niobium and tantalum)

Complexes of $NbCl_5$ and $TaCl_5$ with ClCN have been investigated (Burgard and MacCordick, 1972) by recording the chlorine resonance in the coordinated nitrile. The effect of the complexation on the frequency of the NQR of this chlorine atom is small and the change observed in the resonance frequency upon complexation is of the order of the crystalline effects.

Fuggle (Fuggle *et al.*, 1974) has studied complexes of NbF_5 with XeF_2 and organic bases. NQR results are consistent with a pseudo-octahedral structure for the complexes with CH_2ClCN and Me_2O whereas nothing is said about the structure of the $NbF_5 \cdot$pyridine complex which is known to be ionic in solution.

8.9 Complexes Containing Elements of Group VI

Among elements of column VI, oxygen is the most frequently found as donor centre in molecules containing elements of this column. Some sulphur-containing donors have also been considered and their effect on the NQR frequencies compared to that of the oxygen-containing parent compound. Only a few selenium- and tellurium-containing compounds have been examined which are selenium- and telluriumdimethyl and -diethyl complexes with galliumtrimethyl as acceptor. There is not much difference between the selenium and tellurium compounds but the dimethyul derivatives appear more efficient than the diethyl compounds in the lowering the gallium NQR frequency (Kasakov *et al.*, 1977).

At first examination, the results found in the available references do not seem concerned with the electronic properties and donor ability of the molecules considered. It should be emphasized that (i) in many complexes studied the properties of the acceptor, mainly structural, are considered, (ii) valuable information about the different donor molecules can be obtained only through an investigation of a series of complexes with the same acceptor and, in many cases, only limited series were studied, (iii) when a series is investigated, the discussion should take several factors into account besides charge transfer, such as change in the importance of the *s* hybridization and the geometry of the complex, and, finally (iv) in contrast to donor atoms such as As, Sb, or N, the abundant isotopes of oxygen or sulphur have no quadrupole moment which would be helpful for obtaining information about the electronic distribution on the donor centre.

Despite these unfavourable circumstances, many valuable results have been obtained which concern the electronic distribution and donor activity of oxygen- and sulphur-containing molecules as well as the acceptor activity of accepting molecules. In 1968, Rogers and Ryan gave the classification $SbCl_5 > FeCl_3 > SnCl_4 > TiCl_4$, for the accepting activity of these molecules from the study of their complexes with $OPCl_3$ as donor. On the other hand, the study of the NQR of antimony and chlorine nuclei in $SbCl_5$ complexed with different donors makes it possible to compare the activities of the donors. From the chlorine resonances reported by Poleshchuk (Poleshchuk *et al.*, 1974a; Poleshchuk and Maksyutin, 1975a) the following order of increasing donor effect is obtained:

$$POCl_3 < THF \lessapprox ROH \lessapprox Diox. < DMF < HMPT < DEE.$$

This conclusion should not be taken too strictly as no provision is made for possible steric effect or change in hybridization of the Sb—Cl bonds. However, comparison of other classifications mentioned below in this section leads to the order

$$POCl_3 < THF < DMF < HMPT$$

which is consistent with the preceding one.

Brunette *et al.* (1977) have considered a different series including mostly inorganic donors and have compared the results from NQR measurements with those of thermochemical and IR studies:

m.p.: $VOCl_3 < SOCl_2 < CH_3COCl < SeOCl_2 < POCl_3$

IR: $VOCl_3 < SOCl_2 < POCl_3 < SeOCl_2$

NQR: $VOCl_3 < SOCl_2 < C_6H_5COCl < POCl_3 < SeOCl_2$

It can be seen that the orders found from different methods are quite similar except for the relative positions of $POCl_3$ and $SeOCl_2$.

An example of a study of an extended series of complexes with oxygen-containing donors is that by Petrosyan *et al.* (1973) who considered complexes of Me_3SnBr and related compounds. In an NQR study of complexes of Me_3SnBr, he found the following order for the increasing donor activity of the ligands:

$$DEE < Diox. < Acetone < DME < THF \ll DMF < HMPT < Py. < DMSO$$
$$< TMED$$

Classifications of that kind were previously reported by Tong (1969) and Smith and Tong (1971b) as obtained from the chlorine NQR in $GaCl_3$ complexes:

$$PhNO_2 < PhOMe < POCl_3 < DEE < THF < Et_2S$$

and in BCl_3 complexes:

$$PhNO_2 < Et_2S < PhOMe < POCl_3 < THF \simeq DEE.$$

These two series, though they differ from each other by the position of Et_2S, are similar but can hardly be compared to that of Petrosyan as DEE and THF are the only donors common to the series. A more recent study of $GaCl_3$ complexes by Carter *et al.* (1976) confirms that Et_2S is a stronger donor than Et_2O while, on the other hand, a study of complexes of $AlMe_3$ (Dewar *et al.*, 1973) shows that THF reduces the QCC of ^{27}Al more than THT, which is in qualitative agreement with the fact that ethers are stronger bases than sulphides. A similar result is reported for Me_2O and Me_2S

though in this case the difference is smaller than for THF and THT. From a study of complexes of $MgBr_2$ (Kress and Guibé, 1977), THF appears to be a stronger donor than Et_2O.

From a study of chlorine and nitrogen resonances in complexes of $SnCl_4$ with phosphorus derivatives (Andreeva *et al.*, 1975, 1976), the strengths of the complexes as derived from the shift of the frequency of the $SnCl_4$ chlorine atoms and of the chlorine atoms bound to the phosphorus, is in the order

$$P: > P{=}S \approx P{=}O.$$

In complexes of mercury halides, sulphur-containing donors (trithiane and tetrahydrothiophene) are stronger donors than the respective oxygen-containing molecules (Fichtner and Weiss, 1976). Other investigations of mercury halide complexes lead to the series:

$$DME < TMED < DMTE \quad \text{(Bryukhova } et \ al., \ 1973\text{)}$$

and

$$DME < Diox. < CH_3OH < Py.{-}N{-}Oxide \quad \text{(Wulfsberg, 1976).}$$

To complete this section, it is worthwhile mentioning the great variety oxygen-containing donor molecules investigated: alcohols, phenols and halophenols, benzaldehyde and derivatives, ketones (aliphatic, cyclic, and aromatic), alkoyl chlorides, quinone and chloranil, nitro derivatives (nitrobenzenes, picryl chloride), as well as inorganic molecules ($POCl_3$ and derivatives).

Results concerning the structure of the complexes, such as the *cis–trans* isomerism in the case of tin halide complexes, are not considered here as they are surveyed in the respective acceptor paragraphs. However, there is the case of the complexes of pyridine-N-oxide derivatives with copper chloride and bromide (Frausto da Silva and Wooton, 1970; Frausto da Silva *et al.*, 1971) which are not considered in the section dealing with Group I. These complexes were investigated with the aim of checking the usual ideas about terminal and bridging halogen atoms in metal complexes by a study of the halogen NQR. As a general result, the halogen resonances occur at frequencies higher than in the pure copper halide, which is known to have only bridging halogen atoms, and it is concluded that only terminal halogen atoms are found in the complexes, which leads one to question some of the previous proposals, made on the basis of X-ray and IR data, about the bridging character of the halogen atoms.

8.10 Complexes Containing Elements of Group VII

In this section, molecular complexes in which the halogen atom is the centre of complexation will be reviewed as well as complexes involving molecules in which the halogen atom(s), though not participating directly to the complex bond, can be considered as important contributor(s) to the electron-attractive property of the acceptor. In other words, we shall consider complexes of the pure halogens, interhalogens, and related trihalogen and hydrogen dichloride ions, and halogenated molecules like the halomethanes.

8.10.1 Complexes of pure halogens

NQR results have been reported for complexes of pure halogens accepting π electrons from benzene (Cornil et al., 1964; Hooper, 1964) or substituted benzenes (Cornil et al., 1964), or n electrons from ethers, ketones (Cornil et al., 1964), nitriles (Negita et al., 1973; Grechishkin et al., 1974; Dresvyankin, 1975), and amines (Bowmaker and Hacobian, 1969). Complexes of amines with interhalogens have also been investigated (Bowmaker and Hacobian, 1969; Fleming and Hanna, 1971; Bowmaker, 1976). In the case of the bromine complex of benzene (Cornil et al., 1964; Hooper, 1964) the upward shift of the bromine NQR frequency, as compared to pure bromine, is very small and in the direction opposite to that expected for a charge transfer from the benzene π system towards the bromine σ orbitals. A later study (Kadaba et al., 1971) confirms the results obtained by Cornil, which differ slightly from those of Hooper for the value of the NQR frequency of bromine in the complex, a difference which may be explained if the sample used by Hooper was in a crystalline form different from that of the samples used by Cornil and Kadaba. Kadaba has also detected a phase transition at 60 K in this complex and, from the influence of the temperature on the proton relaxation time, he found that the motion of the benzene ring is quite similar to that in pure benzene; on the other hand, the bromine motion appears to have an unusually large amplitude. In the bromine complexes of halobenzenes (fluoro-, chloro-, and bromobenzene) the frequency shift of the bromine resonance with respect to pure bromine is also very small amounting to 1% (Cornil et al., 1964). The chlorine complex of benzene has also been investigated and the temperature dependence of the NQR frequency and relaxation time is reported (Gordeev et al., 1970); a phase transition occurs at 140 K and, at 80 K, the chlorine frequency is about 1% higher than in pure chlorine. In this compound and in the complex dioxan·chlorine, the potential barriers, as calculated from the chlorine NQR relaxation time, are compared to those obtained from the proton NMR. An overall result of the NQR study of these complexes is that the electronic effect of the charge transfer, as seen by NQR, is very small and some authors consider them as van der Waals complexes (Weiss, 1972).

In bromine complexes of n donors (ethers, ketones, and nitriles), the effect of the charge transfer on the NQR frequency is also small and even in the opposite direction, like in the diethyl ether·bromine complex (Cornil et al., 1964), but in most compounds it is downwards and, in the case of the acetonitrile·bromine complex in which the nitrogen resonance is also observed, the charge transfer is estimated to be 0.03 e (Negita *et al.*, 1973) from the shift of the nitrogen resonances; that the effect on the bromine resonance is smller than expected is explained by the *d* nature of the accepting orbitals on the bromine molecule (see also section 8.8.1.1 p. 374). A series of chlorine complexes of nitriles has been studied and the charge transfer also estimated to be 0.03 e (Dresvyankin, 1975; Dresvyankin *et al.*, 1975).

The charge transfer is much higher in halogen complexes of amines which will be considered together with the interhalogen complexes. Interhalogen complexes have been extensively studied by Bowmaker (Bowmaker and Hacobian, 1969) who considered a series of IBr and ICl complexes of pyridine and pyridine derivatives. From the measured halogen NQR frequencies, after proper assignment to quadrupolar nuclei in the complexes considered, the electronic charges on the resonant nuclei, compared to the uncomplexed molecules, are calculated and show a depletion on the nitrogen and iodine atoms, and an increase on the end halogen, Br or Cl. The results are consistent with the accepted gradation in acceptor strengths: $Br_2 < IBr < ICl$. Discussion of the NQR data in terms of a charge transfer model shows that a ionic structure $NX^+ X^-$, which does not involve a transfer of charge from the donor N to the acceptor XX, should also be considered. Later, Fleming (Fleming and Hanna, 1971) again considered ICl complexes of pyridine and 4-methylpyridine and could observe both iodine and chlorine resonances and calculated the electronic charges on the three atoms, N, I and Cl, resulting from the complex formation, to be 0.26, 0.35, and -0.61 e, respectively. Electrostatic forces in the no-band wave function, $\psi(D) \cdot \psi(A)$, are considered to play an important role in the stabilization of the complex and explain an induced dipole moment in ICl. More recently, Bowmaker, 1976 taking advantage of the possibilities afforded by double-resonance techniques, obtained the nitrogen QCC in the 3,5-dichloropyridine·ICl complex and gave an estimate of 0.3–0.4 e for the magnitude of the charge transfer, a value consistent with that obtained from the nitrogen QCC in a similar complex 3,5-dibromopyridine·Br_2. Complexes of amines with iodine have also been studied (Brüggemann *et al.*, 1972) together with compounds of ionic structure, [Py.—I—Py.]$^+$ I_3^-, and the NQR data are discussed with the help of simple MO calculations which show that the iodine atom bound to the nitrogen atom of the amine bears a small positive charge.

8.10.2 *Polyhalides*

Polyhalides ions were studied as early as 1955 by Kojima (Kojima *et al.*, 1955b) who observed the iodine resonances in Me_4NI_3 and Me_4NI_5. Extensive investigation of this kind of compound followed: Cornwell (Cornwell and Yamasaki, 1957) and Yamasaki (Yamasaki and Cornwell, 1959) studied the chlorine and iodine resonances in a series of complexes $MI \cdot Cl_2$ of composition $1:1$ or $1:2$ with $M=Na$, K, Rb, or Cs; composition $1:1$ corresponds to the linear ion ICl_2^- and $1:2$ to the planar ICl_4^- ion. Bowmaker (Bowmaker and Hacobian, 1968) considered a series containing mostly triiodide ions: NH_4I_3, CsI_3, Me_4NI_3, Et_4NI_3, $(n-Bu)_4NI_3$ and a few others Cs_2I_8, NH_4IBr_2; the triiodide ion I_3^- can be found in both a symmetrical and an unsymmetrical form, depending upon the size of the cation: I_3^- is unsymmetrical in NH_4I_3 and symmetrical in Me_4NI_3 (a similar feature is also observed in the hydrogen dichloride ion (see below, p. 374)). Bowmaker also mentions the similarity existing between trihalide ions and halogen complexes of amines, which is shown by the bond lengths, IR, and Raman vibrational data. Experimental data are discussed in terms of a charge transfer model and atomic charges are calculated showing that the charge of the ion is concentrated on the terminal iodine atoms. Later, the spectra of several polyiodides were investigated in detail and the temperature dependence examined (Harada et al., 1974). In a well-documented article, the question of NQR in polyiodides has been reviewed and new results presented (Nakamura and Kubo, 1975). Positive I_2Cl^+ and I_3^+ ions are also found in complexes of ICl with $AlCl_3$, $2ICl \cdot AlCl_3$ and $ICl \cdot I_2 \cdot AlCl_3$ (Merryman et al., 1972). It is worth mentioning that the nature of the anion $AlCl_4^-$ was identified by NQRS which permitted the distinction from another and, a priori, possible anion, $Al_2Cl_7^-$. This was done, in particular, by a comparison with the spectrum obtained in $Na^+AlCl_4^-$. The frequency of the central iodine atom in the ICl_2^+ cation is much higher than in I_3^+ and related ions and a progression towards high frequencies, observed in the series I_3^+, I_2Cl^+ and ICl_2^+, is in agreement with the expected increase of charge on the central iodine atom. Compounds containing other trihalide ions, Br_3^-, $BrCl_2^-$, Cl_3^- as well as ICl_2^-, have been investigated in Ph_4AsCl_3, Et_4NCl_3, Me_4NBrCl_2, Me_4NICl_2 (Riedel and Willett, 1975), $CsBr_3$, $(Me_3NH)_2Br_4$, PBr_7, and NH_4Br_3 (Breneman and Willett, 1967). In connection with polyhalogen ions, a mention can be made of the iodobenzene dichloride. This compound was first investigated by Evans (Evans and Lo, 1967c) who mainly considered the structure of the three-halogen group. Moreover, in the crystal, there is some intermolecular coordination between the iodine atom of one molecule and a chlorine atom of a second molecule and between the iodine atom of the second molecule and a chlorine atom of the first molecule, resulting in a kind of dimeric structure. The relation between the intermolecular coordination and the NQR frequencies has been investigated (Nesmeyanov et al., 1976) in a series of substituted iodobenzene dichlorides. As a result, the intermolecular coordination, or the QCC, does not show any dependence on the nature of the substituent and it seems that steric effects are predominant.

Another kind of triatomic ion studied by NQR is the hydrogen dichloride ion HCl_2^-. The first investigations of this ions were those of Evans and Lo (1966, 1967b) and Haas and Welsh (1967). Later, the problem was further investigated by Smith (1974). The study of this ion, which is found in solid HCl and in dihydrochloride salts and belongs to a general class of hydrogen-bonded species, is of interest in connection with an understanding of the hydrogen bond (Evans and Lo, 1966). One problem concerns the ion structure which may be either non-centrosymmetric (I) or centrosymmetric (II). The $ClHCl^-$ ion in tetramethylammonium hydrogen dichloride is unsymmetrical from IR vibrational spectroscopic data and the two chorine atoms are thus uniequivalent. However, only one chlorine NQR could be detected in this salt (Evans and Lo, 1966). But this was not considered to be too surprising because the frequency expected for a symmetrical structure on the basis of rather simple arguments is much smaller (~ 13 MHz) than the frequency (~ 20 MHz) of the resonance observed in this salt. A theoretical calculation (Haas and Welsh, 1967) also supports this conclusion and it is thought that the resonance of the second chlorine atom is at a frequency too low to be detected with the spectrometer used for the study. Since then a chlorine resonance at ~ 12 MHz in tetraethylammonium hydrogen dichloride has been observed (Evans and Lo, 1967b) showing that the HCl_2^- ion is symmetrical there, which agrees with IR spectroscopic data. Later, Smith (1974) provided new information about the effect of deuteration of the ion (stronger for ion of type I) on the NQR frequency and the temperature dependence of the frequency and relaxation time of the chlorine resonance, thus providing new features with which to characterize hydrogen dichloride ions I and II. Another example of a non-symmetrical HCl_2^- ion studied by NQR is the triethylamine dihydrochlororide studied by chlorine NQR and proton NMR (Cousseau *et al.*, 1973). Discussing the relation between the proton chemical shift and the chlorine NQR frequency, the authors give some tentative arguments about the conditions in which the ion is to be found in a symmetrical or unsymmetrical form.

8.10.3 · *Carbon chloride complexes* (CCl_4, $CHCl_3$)

Carbon tetrahalides bear this similarity to pure halogens in that the complexing atom is an halogen. Reports are found of a NQR study of CCl_4 and CBr_4 complexes of aromatic compounds: p-xylene (Hooper, 1964), benzene, o- and p-xylene, durene, mesitylene, 2,6-lutidine, and 2,4,6-collidine (Gilson and O'Konski, 1968). As in some of the complexes reviewed above, the bromine NQR frequency shift is small and of the size of the 'environmental shift' due to intermolecular and motional effects; but it is worth mentioning the fact, quoted by Gilson and O'Konski (1968), that if the effect of charge transfer on the bromine resonance is small enough to be indistinguishable from intermolecular contributions, charge transfer has been observed by UV spectroscopy in similar carbon tetrachloride complexes. One can also recall the comment about the smallness of the bromine NQR shift in the $2CH_3CN \cdot Br_2$ complex, saying that the d accepting orbitals on the bromine atom have a smaller QCC than the p-orbitals usually considered in bond formation. Further investigation of CCl_4 complexes of aromatic molecules has been reported in Wakefield's thesis (1970). Among the halomethane complexes, those of chloroform have been the most widely studied by NQR. These complexes are also very weak and the amount of charge transfer as derived from the shift of both the chlorine and nitrogen resonances in chloroform complexes of amines is of the order of 0.05 e (Lucas and Guibé, 1970). Complexes of nitriles have been investigated (Dresvyankin et al., 1975); complexes of benzaldehyde, aromatic hydrocarbons, ethers, acetophenone have also been studied (Grechishkin and Kyunstel, 1966a; Bennett and Hooper, 1967; Anferov et al., 1973; Grechishkin et al., 1974) and the relation between the NQR frequency shift and the ionization potential of the donor molecule has been considered (Anferov et al., 1973). The formation of the chloroform complexes of amines is thought to take place through the formation of a $N: \cdots H—C$ hydrogen bond (Lucas and Guibé, 1970). However, considering the analogy with complexes of CCl_4 it is considered that in some complexes of $CHCl_3$ a charge transfer involving the chlorine atoms plays a major role (Grechishkin et al., 1973). Relaxation by orientational motion of the $—CCl_3$ group has been investigated in some $CHCl_3$ complexes (Kyuntsel and Rozenberg, 1973). Complexes of $CHCl_3$ and $CHBrCl_2$ have been investigated by considering the deuterium resonance of the d species using double-resonance techniques (Ragle et al., 1974; Schwartz and Ragle, 1974).

Halogens can be found as donors in some dimeric molecules like Al_2Br_6 in which the resonance of both the terminal and bridging bromine atoms can be observed (Das and Hahn, 1958) or participating in weak hydrogen bonds in complexes of chlorophenols (Korshunov et al., 1971).

8.11   Complexes Containing Elements of Group 0

Complexes of xenon difluoride, $XeF_2 \cdot NbF_5$ and $XeF_2 \cdot 2NbF_5$, in which the fluorine atom of the xenon difluoride acts as donor towards the niobium atom of NbF_5, have been studied by NQRS of ^{93}Nb (Fuggle et al., 1974). ^{93}Nb having a spin $I = 9/2$ gives several NQR lines from the frequency of which both e^2Qq and η can be derived. As there was some discussion about the formulation of these complexes (no X-ray data on their structure is available) NQRS was employed to obtain information about the electronic distribution in the vicinity of the niobium nuclei. From the QCC observed in pure NbF_5 and in the two complexes considered, and with the help of the crystal structure of NbF_5, as known from X-ray diffraction analysis, it was possible to show that NQR data strongly favour the *cis* conformation of the $XeF_2 \cdot 2NbF_5$ complex (similar to that found by X-ray in $XeF_2 \cdot 2SbF_5$) over the *trans* conformation. In addition, it was shown that the effect of the bridging fluorine atom between FXe— and —NbF$_5$ on the QCC of the niobium nucleus was different from that of the bridging fluorine atom between the NbF$_5$ molecules in the complex. The QCC of ^{93}Nb in XeF_2NbF_5 is much smaller (34 MHz) than that of the terminal ^{93}Nb^1 (110 MHz) in $XeF_2 \cdot 2NbF_5$; in the latter, the QCC of the niobium atom, ^{93}Nb^2 bound to the XeF_2 molecule is higher (74 MHz) than in $XeF_2 \cdot NbF_5$ (34 MHz) owing to the contribution of the second NbF$_5$ molecule to its EFG.

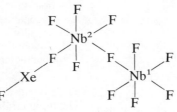

8.12 Chemical Index

Acetone 380, 382
Acetonitrile 373, 380
— ·bromine 385
Acetophenone (ACP) 357, 358, 388
Adiponitrile 374
Alcohol
 Methyl 356
Alcohols 383
Aldehydes 356
Alkoli metals 353
Alkoyl chlorides 383
Allylic (-chlorine atom) 359
Aluminium 359, 362, 378
— atom 359, 352, 364
— bromide 363, 364, 366
— halides 364
— trichloride 378
 Trimethyl- 376, 378, 379
Amidophosphorus derivatives 369
Amine·boron trichloride complexes 361
Amines 360, 375, 384, 385, 386, 388
 Aliphatic, aromatic, heterocyclic 372, 375
Amine-type ligands 362
Amino hydrogen atom 361
Amino nitrogen atom 372
Aniline 372, 375, 378

— ·SbBr₃ 378
Anisole 378, 380
Antimony 363, 377, 378, 380
— atom 372, 377, 379
— compounds 379
— halides 380
— nuclei 379, 381
— pentachloride 377
Pentocoordinated—compounds **379**
— QCC 377, 379
— resonance 379, 380
 Trimethyl- 363, 377
Aromatic compounds 366
Aromatic hydrocarbons 365, 367
— molecules 377, 378, 388
Arsenic 350, 363, 367, 377, 378
— NQR frequency 363, 377
 Trimethyl- 377
Arsines 360

Barium 354
— halide hydrate 354
Benzaldehyde 383, 388
 Chloro-(o-) 356
Benzene 361, 364, 365, 366, 384
— ·AlBr₃ 354
 Bromo-, chloro-, fluoro- 384

Hexachloro- 347
   Substituted 365, 384
   1,2,5-Trichloro- 347
Benzonitrile·Cl 374
   2-·Cl₂ 374
   Chloro-s 361, 373
Benzophenone 364, 367
Benzoyl chloride 364
Biological systems 376
Bismuth 380.
   — atom 363
Bis(N,N-dibutyldithiocarbamato)gold(III)
   dichloroaurate(I) 354
Boron 359, 360, 361, 373
   — atom 360
   — complexes 361
   — NQR 362
   — trichloride 360, 361
Bromides 357
Bromine 350, 365, 367, 374, 384, 385
   — atom 347, 351, 362, 367, 374, 388
   — frequency (and shift of-) 354, 362,
     363, 366, 371, 384, 387, 388
   — molecule 385
   — resonance 354, 366, 376, 385
Bromodichloromethane-diethyl   ether-
   d_{10} 353
BZP (Benzophenone) 357

Cadmium 355
Calcium 354
Carbon 358, 365
   — atom 359, 361, 371
   — tetrachloride 365, 388
   — tetrahalides 387
Chloranil 347, 383
   — ·hexamethylbenzene 348
Chlorides 350, 357, 387
Chlorine 350, 354, 361, 365, 369, 374,
   376, 380, 384
   — atom(s) 347, 350, 352, 356, 359,
     360, 361, 365, 368, 369, 370,
     371, 372, 374, 376, 378, 379,
     380, 381, 383, 386, 387, 388
   — complexes 374, 375, 376
   — frequency 357, 360, 364, 369, 372,
     376, 378, 379, 381, 384
   — nucleus 381
   — resonance (line, spectrum) 351,
     356, 360, 361, 362, 367, 369,
     371, 372, 373, 375, 379, 380,
     381, 382, 383, 385, 386, 387, 388

Chloroacetonitriles 373
Chlorobenzaldehyde (o-) 356
Chlorobenzonitriles 361, 373
Chloroether (dimethyl-) 368
Chloroform 365, 367, 375, 388
Chlorophenol(s) 364, 388
Chlorophyll-o 376
Chloropyridines 356, 361, 375, 380
Collidine (sym) 356, 387
Copper 353, 354
   — atom 354
   — bromide, chloride 383
   — halides 383
   — ion 353
   — resonance 353
Cumene 368
1,4-Cyclohexanedione 356
Cyclopentadienylmercurials 359

DEE (Diethyl ether) 380, 382
Deuterium 349, 352, 388
Deuterochloroform·acetone
   ·diethylether, ·mesitylene 353
Dialkyl halides 371
Diethyl ether 354, 380
   — ·aluminium trichloride 344
   BiCl₃·- 380
   — ·bromine 384
Dihalocuprates 353
Dihydrochloride salts 387
Dimethylformamide 356
   — sulphoxide 356
   — thioethane 358
Dioxan 356, (dioxanates) 382, 383
   — ·chlorine 384
   — ·HgCl₂ 355, 356
Dithiane 356
DME (Dimethoxyethane) 382, 383
DMF (Dimethylformamide) 382
DMSO (Dimethylsulphoxide) 382
DMTE (Dimethylthioethane) 383
Durene 364, 366, 367, 387

Ether(s) 354, 356, 360, 362, 376, 382,
   384, 388 (See also diethylether)
   Methyl-·BCl₃ 361
   Methylthio-·BCl₃ 361
Ether-type ligands 362
Ethyl,-group 354
   — magnesium bromide 354
Ethylbenzene·SbCl₃ 378

Fluorine atom 388, 389

Gallium 359, 362
— halide 364
— NQR frequency 364, 381
— trichloride 378
Trimethyl- 376, 378, 381
Germanium 365, 371
Glyme(s) 356
Gold 353, 354

Hafnium 365
Halides 353, 357
Haloanilines 375
Halobenzenes 384
Halogenated molecules 384
Halogen atom 351, 356, 363, 364, 366, 372, 374, 383
— bond 380
Bridging-atom 357, 383
— complexes 385, 386
End-atom 357, 383, 385
— frequencies 364, 371, 385
— resonances 355, 357, 358, 366, 375, 383
Halogens 350, 373, 384, 387, 388
Halomethanes 365, 384, 388
Halophenols 383
Ortho- 351
Hexachlorobenzene 347
Hexachlorocyclohexane 347
Hexamethylbenzene·chloranil 366
Hexamethylenephosphorus triamide 376
HMPT (Hexamethylphosphorus tri-amide) 382
Hydrogen dichloride ion 384, 386, 387

Indium 359, 362, 378
— trichloride 378
Trimethyl- 376, 378
Interhalogen 384
— complexes 385
Iodine 350, 358, 385
— atom 385, 386
— resonance 385, 386
Iodobenzene dichloride 386
Substituted-s 386

Ketone complexes of AlBr₃ 362
Ketones 356, 383, 384

Lead 365
Lutidine (2,6-) 387

Magnesium atom resonance 376
— bromide 354
Ethylmagnesium bromide 354
Phenylmagnesium bromide 354
— phthalocyanin 376
Menshutkin's complexes 365, 377
Mercuric chloride 357
— halides 355, 356, 357, 359, 383
— complexes 352, 355
Mercury 355, 356
— atom 356, 357, 358, 359
— compounds (and derivatives) 351, 359
Pentachlorocyclopentadienylmercury chloride 355
Pentachlorocyclopentadienylmercury halides 358
— resonance 355
Trans-2-chloroethylenemercury chloride 355
Trichloromethylmercury chloride 355, 359
Mesitylene 364, 366, 367, 387
Metalloporphyrins 376
Methanol 357, 375 (see also 'Alcohol, Methyl-')
Methyl ether·BCl₃ complexes 361
— group(s) 351, 361, 362, 366, 373
4-methylpyridine 368, 375, 385
— thioether·BCl₃ complexes 361
— tin 369
— methyl tin halides 371
Monomethyltin trihalides 371

Naphthalene 374
— ·2SbCl₃ 378
Nickel 374
Niobium atom, nucleus 380, 388, 389
Nitrates 355
Nitriles 360, 372, 373, 374, 375, 381, 384, 385, 388
— complexes 361, 372, 373, 374
Nitrobenzene 383
— ·tin tetrachloride 344
Nitro derivatives 383
Nitrogen 350, 361, 363, 367, 372, 376
— atom 361, 369, 373, 374, 375, 378, 385
— containing compounds 369
— containing donors 356
Nitrato- 355
Nitrilo- 372

— QCC 374, 385
— resonances 348, 349, 355, 369, 373, 383, 385, 388
— trimethyl 363

Organic chlorides 350
Organomercurials 355, 358, 359
Organometallic compounds 359
Organophosphorus compounds 369
Oxygen 349, 367, 376, 381, 382
— atom 354, 363, 367, 368, 369, 376, 379
— containing compounds 356, 369, 381, 382, 383

Palladium 374
Pentachlorocyclopentadiene 359
Pentachlorocyclopentadienyl derivatives 358 (See also 'mercury')
Pentachlorocyclopentadienyl group 359
— mercurials 359
— radical 359
Pentahalogenides 372
Perhalogenated molecules 365
Phenol(s) 375, 383
Chloro- 364, 388 (See also 'chlorophenols')
Phenoxathiin 358
Phenyl group 354
Phosphines 360
— oxides 376
Phosphonitrilic chloride trimeric 360
Phosphoric chloride 369
Phosphorus (-atoms,-compounds) 351, 363, 365, 370, 376, 383
Amido- derivatives 369
— oxychloride 352, 360
Picoline (4-) 365
— N—O 358
Picryl chloride 383
Polyhalide ions 385
Polyhalogen ions 386
Polyiodides 386
Polynitrile 374
Polystyrene 368
Polyvinylacetylene 368
Poly-4-vinylpyridine 364, 368
Poly-N-vinylpyrrolidone 364
Proton 352, 362, 384, 387
Pyridine 355, 360, 372, 375, 381, 382, 385

— derivatives 385
3,5-Dibromo-·Br2 385
3,5-Dichloro- 385
3,5-Dimethyl- 375
4-Methyl- 368, 375, 385
— molecule 375
— ·trimethyltin bromide 371
Pyridine—N—O 357, 358, 383
Pyrrole 375
Pyrrolic (nitrogen atom) 376

Quinone 383

Selenium, diethyl- and dimethyl- 381
Silicon 365, 371
Stannic group 376
Strontium 354
Sulphides 382
Sulphone group 363
Sulphur (and -containing molecules) 356, 381, 383
— atom 376

Tellurium, diethyl- and dimethyl- 381
Terphenyl (p-) 367
Tetraalkylammonium salts 354
Tetrachloroborates 360
Tetracyanoacetylene 372, 374
Tetraethylammonium hydrogen chloride 387
Tetrahydrofuran 354, 356
Tetrahydrothiophene 356, 383
Tetramethylammonium hydrogen chloride 387
Tetramethylammonium tetrachloroborate 360
Tetramethylethylenediamine 356, 358
THF (Tetrahydrofuran) 354, 357, 358, 362, 382, 383
Thiocyanate 355, 374
— nitrogen atom 355
Thioethers 360, 362, 376
Thorium 365
THT (Tetrahydrothiophene) 382, 383
Tin 365, 369, 370, 371
— complexes 368, 370, 371
— halides 368, 383
Methyl- 369
— halides 371
Tin tetrabromide 371
— tetrachloride 369, 371
— tetrahalides 371

Titanium 365
TMED   (Tetramethylethylenediamime)
    382, 383
Triatomic ions 387
Trichloroacetic group 361
Trichlorobenzene 347
Trichloromethylmercuric   chloride   355,
    359
Triethylamine dihydrochloride 387
Trihalides 377, 386
Trihalogenides 373
Trihalogen ions 384
Trihalophenols (2,4,6-) 375
Triiodide ion 386
Trimethylamine 360, 372
    — ·BCl₃ 375

Trimethyl derivatives 363, 377, 378
Trimethyltin bromide 371
Trinitrophenol 375, 380
Trioxan 356
Trithiane 356, 383

4,Vinyl pyridine 368
Vinylic (chlorine atom) 359

Xenon difluoride 388
Xylene (p-) 366, 387
Xylenes (o- and p-) 364, 366, 367, 387

Zinc 355, 374
    — thiocyanate 355
Zirconium 365

8.13   Formulae Index

AB_3 372
Al 365
^{27}Al 346, 382
$AlBr_3$ 362, 363, 364, 366, 367
    — ·$BiBr_3$ 362, 363
    — ·4-Methylpyridine 368
    — ·OPh_2 362
    — ·4-Picoline 365
    — ·Poly-4-vinylpyridine 364, 368
    — ·Pyridine 362
    — ·$SbBr_3$ 362, 363
    — ·SMe_2 362
$AlCl_3$ 362, 363, 364, 386
    — ·Et_2O 362
$AlCl_4^-$ 353, 386
$AlI_3.SbI_3$ 362
$AlMe_3$ 352, 363, 382
AlX_4^- 363
Al_2Br_6 351, 364, 366, 367, 388
    — ·C_6H_6 367
$Al_2Br_7^-$ 353
$Al_2Cl_7^-$ 386
Ar 365
As 365, 372, 376, 377, 381
^{75}As 346, 363, 367, 377
$AsBr_3$ 367
    — ·1,3,5-Me_3Ph 377
$AsCl_3$ 377
$AsEt_3$ 378
$AsHal_3$ 378
$AsMe_3$ 372, 378

Au^+ 354
$AuCl_2^-$ 354
$AuCl_4^-$ 354

BCl_3 360, 361, 362, 363, 373, 375, 376,
    380
    — ·PCl_5 360
BCl_4^- 360
BD_3 360
BF_4H 360
BH_3 360
^{135,137}Ba 355
$BaCl_2·2H_2O$ 354
$BaI_2·2H_2O$ 355
Bi 363, 365, 372
^{209}Bi 346
$BiBr_3·AlBr_3$ 363
$BiCl_3$ 375, 380
    — ·Acetone, acetonitrile, anisole, 2-,
        4-chloropyridine, diethyl ether 380
Bi_2Br_6 363
Br 355, 365, 385
^{79,81}Br 346, 352, 377
$BrCl_2^-$ 386
Br_2 366, 385
Br_3^- 386
BrSb 351
    — ·$AlBr_3$ 351
Bu_4NCuBr_2 353
Bu_4NCuCl_2 353
Bu_4NI_3 386

C 371
CBr$_4$ 366, 387
— CCl$_3$ 367, 388
CCl$_3$CN·BCl$_3$ 361
CCl$_3$COOH·N-Isopropylpyrrolidone 365
— ·Poly-N-vinylpyrrolidone 364
CCl$_4$ 366, 387, 388
CDCl$_3$·Mesitylene 353
CHBrCl$_2$ 388
CHCl$_3$ 375, 388
(CH$_2$Br)$_3$P 376
CH$_2$ClCN 381
(CH$_2$Cl)$_2$O·SbCl$_5$ 379
(CH$_2$)$_4$O$_2$ 379
CH$_3$CN 373, 379
2-·Br$_2$ 374, 388
2-·Cl$_2$ 374
— ·GaCl$_3$ 373
— ·NbCl$_5$ 373
— ·SbCl$_5$ 372, 379, 380
— ·TaCl$_5$ 373
CH$_3$COCl 382
CH$_3$I 375
CH$_3$OCH$_2$Cl·SnCl$_4$ 368
CH$_3$OH 383
CH$_3$SnBr$_3$ 371
CH$_3$SnCl$_3$ 371
(CH$_3$)$_2$SnBr$_2$ 371
(CH$_3$)$_2$SnCl$_2$ 371
(CH$_3$)$_3$SnBr 371
COCl$_2$ 347
CX$_4$ 366
(C$_2$H$_5$)$_2$O 379
[(C$_2$H$_5$)$_2$O]$_2$·SnCl$_4$ 370, 371
(C$_2$H$_5$OH)$_2$·SnCl$_4$ 370
[(C$_2$H$_5$)$_2$P(O)(SC$_2$H$_5$)]$_2$·SnCl$_4$ 370
[(C$_4$H$_9$O)$_2$P(O)(SCH$_3$)]$_2$·SnCl$_4$ 370
C$_6$H$_5$CH$_2$Cl·SnCl$_4$ 368
C$_6$H$_5$CN 379
C$_6$H$_5$COCl 382
C$_6$H$_5$NO$_2$·SnCl$_4$ 368
C$_6$H$_6$·Br$_2$ 366
— ·Br$_3$Sb 351
C$_7$H$_9$N 379
CdCl$_2$ 375
Cl 355, 365, 372, 376, 385
[35]Cl 343, 346, 352, 373, 375, 377
[37]Cl 346, 352
Cl$^-$ 362
ClCN 373, 379, 380

— ·MCl$_5$ 361
— ·NbCl$_5$ 373
— ·SbCl$_5$ 373, 379
— ·TaCl$_5$ 373
ClHCl$^-$ 387
Cl$_2$ 374
Cl$_3^-$ 386
Cl$_3$PO·SbCl$_5$ 379
Cs 386
Cs$^+$ 353
CsBr$_3$ 386
CsCuCl$_3$ 353
CsICl$_2$ 386
CsI$_3$ 386
Cs$_2$I$_8$ 386
Cs$_2$SO$_4$ 380
[63,65]Cu 352, 353
Cu$^+$ 354
CuCl$_2$ 354
[Cu(Me$_3$AsS)$_3$]ClO$_4$ 353
[Cu(Me$_3$PS)$_3$]ClO$_4$ 353
Cu$_3$I$_3^-$ 353
Cu$_4$I$_6^{2-}$ 353

Et 376
Et$_2$ClP 376
— ·SnCl$_4$ 376
Et$_2$O 382, 383
— ·BCl$_3$ 362
Et$_2$S 382
— ·BCl$_3$ 362, 382
Et$_3$As 363
— ·GaCl$_3$ 378
Et$_3$P·GaCl$_3$ 376
Et$_4$I$_3$ 386
Et$_4$NCl$_3$ 386

FeCl$_3$ 360, 365, 379, 381

[69]Ga 364, 373
GaCl$_3$ 360, 363, 364, 373, 376, 378
— ·POCl$_3$ 364
GaCl$_4^-$ 353, 376, 378
GaMe$_3$ 376
Ga$_2$Cl$_7^-$ 353
Ge 371

HCl 375, 380, 387
HCl$_2^-$ 387
HClO$_4$ 375
(HNBCl)$_3$ 361
HPyNO$_3$ 375

H_2SO_4 375
$H_3N \cdot BH_3$ 351
^2H 346, 352
^{201}Hg 355
$HgBr_2$ 356, 357, 358
— ·Dioxan 352, 355
$HgCl_2$ 356, 357, 358, 375, 380
— ·Dioxan, trioxan, trithiane 356
— ·2Dioxan 355, 356
— o-chlorobenzaldehyde 356
$HgCl_3^-$ 357
$HgHal_2$ 355
HgI_2 356, 358
HgX_2 358
— ·Dioxan 352

I 355, 385
^{127}I 346, 353
IBr 375, 385
ICl 375, 385, 386
2-·$AlCl_3$ 386
— ·$I_2 \cdot AlCl_3$ 386
ICl_2^- 386
ICl_4^- 386
I_2Cl^+ 386
I_3^- 386
I_3^+ 386
$InMe_3$ 363, 376

K 386
K^+ 353
KCl 380
$KI \cdot Cl_2$ 386
K_2SO_4 380

$L \cdot AlCl_3$ 364
$2L \cdot SnCl_4$ 370

$MI \cdot Cl_2$ 386
MX_3 351
— ·Ar 365
— ·2Ar 365
Me 376
$MeCN \cdot BCl_3$ 362
MeOH 358
$(Me_2N)_3P \cdot SnCl_4$ 376
Me_2O 381, 382
Me_2S 382
Me_3As 363, 378, 379
— ·$GaCl_3$ 378
Me_3N 378

— ·BCl_3 362
$(Me_3NH)_2Br_4$ 386
Me_3P 378
— ·$GaCl_3$ 376
$(Me_3PSCuBr)_n$ 353
$(Me_3PSCuCl)_3$ 353
Me_3Sb 378, 379
Me_3SnBr 382
Me_4N^+ 357
Me_4NBrCl_2 386
Me_4NICl_2 386
Me_4NI_3 385, 386
Me_4NI_5 385
^{25}Mg 346
$MgBr_2$ 354, 383

N 376, 381, 385
^{14}N 346, 374, 375
$(NCS)^-$ 374
NH_3 379
—NH_3^+ 375
NH_4Br_3 386
NH_4IBr_2 386
NH_4I_3 386
NMe_3 372
Na 386
Na^+ 353
$NaAlCl_4$ 362
$Na^+AlCl_4^-$ 386
$NaCl \cdot AlCl_3$ 353
$NaICl_2$ 386
Nb 372
^{93}Nb 388, 389
Nb 372
^{93}Nb 388, 389
$NbCl_5$ 373, 380
NbF_5 375, 381, 388, 389
— ·CH_2ClCN 372
— ·C_5H_5N 372, 381
— ·XeF_2 388, 389
2-·XeF_2 388, 389

^{17}O 346
$OPCl_3$ 381

P 376
PBr_7 386
PCl_3 364
PMe_3 372
$POCl_3$ 365, 368, 369, 373, 379, 382, 383

— ·BCl_3 362
2-·$SnCl_4$ 352, 370
$PSCl_3$ 368
$PhCN·GaCl_3$ 373
$PhNO_2·BCl_3$ 362, 382
— ·$GaCl_3$ 382
$PhOMe·BCl_3$ 362, 382
— ·$GaCl_3$ 382
$(Ph_3MeP)Cu_2I_3$ 353
$Ph_3P·BCl_3$ 376
Ph_4AsCl_3 386
$Ph_4AsCuCl_2$ 353
$Py·BCl_3$ 362
$(Py—I—Py)^+I_3^-$ 385

$RBCl_2$ 361
$RMgBr$ 354
$RNH_2·BCl_3$ 361
ROH 382
$RR'R''MCH_2Cl$ 371
$RR'R''P(O)$ 376
— ·$SnCl_4$ 376
$RR'R''P(S)$ 376
$R_2SO_2·AlBr_3$ 363
— ·$2AlBr_3$ 363
R_3P 376
Rb 386
Rb^+ 353
$RbI·Cl_2$ 386

$(SCN)^-$ 374
$SOCl_2$ 382
Sb 365, 372, 376, 378, 379, 380, 381
$^{121,123}Sb$ 346, 352, 377, 379
$SbBr_3$ 378, 380
— ·$AlBr_3$ 363
— ·Aniline, anisole 378
$SbCl_3$ 366, 367, 377, 378, 380
$SbCl_3·$Aniline, anisole 378
— ·C_6H_5Cl 378

— ·Cumene 368
— ·Polyphenylacetylene 368
— ·Polystyrene 368
n-·Ar 366
$SbCl_4(CH_3CN)_2^+$ 372
$SbCl_5$ 360, 365, 372, 373, 375, 379, 380, 381
— ·CH_3CN 373, 379, 380
— ·C_6H_5CN, C_7H_9N (quinoline), NH_3, $(C_2H_5)_2O$, $(CH_2)_4O_2$ 379, 380
$SbCl_6^-$ 372
SbF_3 380
— ·KCl, K_2SO_4, Cs_2SO_4 380
$SbHal_3$ 378
$SbMe_3$ 372, 378
SbX_2^+ 363
SbX_2^+ AlX_4^- 362
$SbX_3·AlX_3$ 363
$SeOCl_2$ 382
Si 371
Sn 371
$SnCl_2$ 375
$SnCl_4$ 352, 360, 365, 368, 369, 370, 376, 379, 380, 381, 383
— ·$2POCl_3$ 352, 369, 370
$SnHal_4$ 369

$TaCl_5$ 373, 380
$TiCl_4$ 365, 379, 381
— ·$OPCl_3$ 365

$VOCl_3$ 382

XeF_2 389
$XeF_2·NbF_5$ 381, 388, 389
— ·$2NbF_5$ 388, 389
— ·$2SbF_5$ 389

Zn, —Br_2, —Cl_2, —I_2, —$(NO_3)_2$, —$(NCS)_2$ 375

8.14 Nuclear Quadrupole Resonance Investigation of Molecular Complexes: List of Complexes

The numbers given refer to the references in Section 8.16

$AlBr_3·AlBr_3$ [Dimer]   R: 52
  — ·$BiBr_3$ R: 144, 153
  — ·Br_3Sb R: 144, 153
  — ·C_2H_6S [Me_2S]   R: 122
  — ·C_3H_6O [Me_2CO]   R: 116
  — ·$C_4H_6O_2S$ [$EtSO_2C\equiv CH$] $\langle 2:1 \rangle$ R: 95

—·— $\left[\bigcirc SO_2 \right]$ ⟨2:1⟩   R: 95

—·C_4H_8O [Butanone]   R: 116

—·$C_4H_{10}O$ [Et_2O]   R: 122

—·$C_4H_{10}O_2S$[Et_2SO_2] ⟨1:1; 2:1⟩   R: 95

—·C_5H_5N [Py.]   R: 122

—·$C_5H_{10}O$ [2-Pentanone]   R: 116

—·— [3-Me-2-Butanone]   R: 116

—·C_6H_6 [Ph] ⟨2:1⟩   R: 141

—·C_6H_7N [4-MePy.]   R: 146

—·$C_6H_{10}O$ [Cyclohexanone]   R: 116

—·$C_6H_{10}O_2S$ [$C_4H_9SO_2C{\equiv}CH$] ⟨2:1⟩   R: 95

—·$C_6H_{12}O$ [3,3-Me_2-butanone]   R: 116

—·C_7H_7N [4-Vinylpyridine] ⟨1:n⟩   R: 146

—·C_8H_8O [ACP] ⟨1:1⟩   R: 116

—·— ⟨2:1⟩   R: 181

—·C_8H_{10} [1,2-Me_2Ph] ⟨2:1⟩   R: 181

—·— [1,4-Me_2Ph] ⟨2:1⟩   R: 181

—·$C_8H_{10}O$ [EtOPh]   R: 122

—·$C_8H_{18}O_2S$ [(iso-Bu)$_2SO_2$] ⟨1:1⟩   R: 95

—·C_9H_{12} [1,3,5-Me_3Ph] ⟨1:1⟩   R: 181

—·$C_9H_{12}O_2S$ [$PhSO_2Pr$] ⟨2:1⟩   R: 95

—·$C_{10}H_{14}$ [1,2,4,5-Me_4Ph]   R: 181

—·$C_{12}H_{10}O$ [Ph_2O]   R: 122

—·$C_{12}H_{10}O_2S$ [Ph_2SO_2] ⟨1:1, 2:1⟩   R: 95

—·$C_{13}H_8Cl_2O$[4,4'-Cl_2BZP]   R: 48

—·$C_{13}H_9BrO$ [4-BrBZP]   R: 48

—·$C_{13}H_9ClO$ [4-ClBZP]   R: 48

—·$C_{13}H_{10}O$ [BZP] ⟨1:1⟩   R: 48, 116

—·— ⟨2:1⟩   R: 181

—·$C_{14}H_{12}O$ [4-MeBZP]   R: 48

—·$C_{14}H_{12}O_2$ [4-MeOBZP]   R: 48

—·$C_{14}H_{14}O_2S$ [(p-MePh)$_2SO_2$] ⟨1:1, 2:1⟩   R: 95

—·$C_{15}H_{14}O$ [4,4'-Me_2BZP]   R: 116

—·$C_{15}H_{14}O_3$ [4,4'-(MeO)$_2$BZP]   R: 48

—·$C_{20}H_{16}O$ [2,4'-Ph_2ACP] ⟨1:1⟩   R: 116

$AlCl_3$· CH_3Cl_4P [MePCl$_4$]   R: 53

—·$C_2H_5Cl_4P$ [P(Et)Cl$_4$]   R: 53

—·C_2H_6O [Me_2O]   R: Junp

—·C_2H_6S [Me_2S]   R: Junp

—·C_3H_9As [Me_3As]   R: 36

—·C_3H_9N [Me_3N]   R: Junp

—·C_3H_9P [Me_3P]   R: Junp

—·$C_4H_{10}O$ [Et_2O]   R: 50, Junp

—·$C_4H_{10}S$ [Et_2S]   R: Junp

—·$C_6H_{15}As$ [Et_3As]   R: 36

—·$C_6H_{15}N$ [Et_3N]   R: Junp

—·$C_6H_{15}P$ [Et_3P]   R: Junp

—·$C_7H_3Cl_3O$ [3,4-Cl_2BZCl]   R: 48

—·$C_7H_4Cl_2O$ [2-ClBZCl]   R: 48

—·— [4-ClBZCl]	R: 48
—·C$_7$H$_5$ClO [BZCl]	R: 48
—·C$_8$H$_7$ClO [4-MeBZCl]	R: 48
—·C$_{13}$H$_{10}$O [BZP]	R: 48
—·C$_{14}$H$_{12}$O$_2$ [4-MeOBZP]	R: 48
—·C$_{18}$H$_{15}$Cl$_2$P [Ph$_3$PCl$_2$]	R: 53
—·ClI ⟨1:2⟩	R: 130
—·ClI·I$_2$	R: 130
—·Cl$_3$I	R: 59

AlI$_3$·I$_3$Sb ⟨1:1⟩	R: 144

AsBr$_3$·C$_8$H$_{10}$ [1,4-Me$_2$Ph] ⟨2:1⟩	R: 90
—·C$_8$H$_{10}$ [PhEt] ⟨1:1⟩	R: 72, 90, 93
—·— ⟨2:1⟩	R: 72, 93
—·C$_9$H$_{12}$ [1,3,5-Me$_3$Ph] ⟨1:1⟩	R: 72, 195
—·C$_{10}$H$_8$ [Naphth.] ⟨1:1⟩	R: 72, 195
—·C$_{13}$H$_{12}$ [Ph$_2$CH$_2$] ⟨2:1⟩	R: 72, 90

AsCl$_3$·C$_3$H$_9$N [Me$_3$N]	R: 36, Junp
—·C$_6$H$_5$Cl [PhCl] ⟨2:1⟩	R: 91, 85, 86, 89, 169
—·C$_6$H$_6$ [Ph] ⟨2:1⟩	R: 5, 11, 90, 92, 142, 195
—·C$_7$H$_8$ [PhMe]	R: 5, 90, 92, 195
—·C$_8$H$_{10}$ [1,2-Me$_2$Ph] ⟨1:1⟩	R: 90, 91
—·— ⟨2:1⟩	R: 5, 90, 91
—·— [1,3-Me$_2$Ph] ⟨1:1⟩	R: 5, 90, 91
—·— ⟨2:1⟩	R: 5, 90, 91, 92
—·— [1,4-Me$_2$Ph] ⟨2:1⟩	R: 5, 11, 90, 92, 195
—·— [EtPh] ⟨1:1⟩	R: 5, 11, 90, 91, 195
—·— ⟨2:1⟩	R: 5, 11, 90, 91
—·C$_9$H$_{12}$ [1,3,5-Me$_3$Ph] ⟨1:1⟩	R: 5, 90, 91
—·— ⟨2:1⟩	R: 5, 91
—·C$_{10}$H$_8$ [Naphth.] ⟨2:1⟩	R: 5, 11, 90, 92, 195
—·C$_{12}$H$_{10}$ [PhPh] ⟨1:1⟩	R: 91
—·— ⟨2:1⟩	R: 5, 90, 91, 195
—·C$_{13}$H$_{12}$ [Ph$_2$CH$_2$] ⟨2:1⟩	R: 5, 11, 90, 92, 195
—·Cl$_3$OP [OPCl$_3$]	R: 12

AsI$_3$·S$_8$ ⟨1:3⟩	R: 138, 160

AuCl·C$_8$H$_{20}$ClN [Et$_4$NCl] ⟨1:1⟩	R: 19
—·C$_9$H$_{18}$NS$_2$ [bdtc] ⟨2:2⟩	R: 19
—·C$_{16}$H$_{36}$ClN [(n-Bu)$_4$NCl] ⟨1:1⟩	R: 19

BCl$_3$·CClN	R: 6
—·C$_2$Cl$_3$N [Cl$_3$CN]	R: 7
—·C$_2$HCl$_2$N [CHCl$_2$CN]	R: 7
—·C$_2$H$_2$ClN [CH$_2$ClCN]	R: 7
—·C$_2$H$_3$N [MeCN]	R: 7, 102, 106, 175
—·C$_2$H$_5$ClO [MeOCH$_2$Cl]	R: 6
—·C$_2$H$_5$ClS [MeSCH$_2$Cl]	R: 6
—·C$_2$H$_6$Cl$_3$P [Me$_2$PCl$_3$]	R: 53
—·C$_2$H$_6$O [Me$_2$O]	R: Junp

—·C_2H_6S [Me_2S]	R: Junp
—·C_3H_3N [CH_2=CHCN]	R: 106
—·C_3H_5N [$MeCH_2CN$]	R: 106
—·C_3H_9As [Me_3As]	R: 36
—·$C_3H_9AsCl_2$ [Me_3AsCl_2]	R: 54
—·C_3H_9N [Me_3N]	R: 7, 69, 102, 106, 175, Junp
—·C_3H_9P [Me_3P]	R: Junp
—·C_4H_7N [Me_2CHCN]	R: 106
—·C_4H_8O [THF]	R: 7, 175
—·$C_4H_{10}Cl_3P$ [Et_2PCl_3]	R: 53
—·$C_4H_{10}O$ [Et_2O]	R: 175, Junp
—·$C_4H_{10}S$ [Et_2S]	R: 175, Junp
—·$C_4H_{12}ClN$ [Me_4NCl]	R: 7
—·$C_5H_3Cl_2N$ [2,6-Cl_2Py.]	R: 6
—·— [3,5-Cl_2Py.]	R: 6, 7
—·C_5H_4ClN [2-ClPy.]	R: 6
—·— [3-ClPy.]	R: 6
—·— [4-ClPy.]	R: 6, 7
—·C_5H_5N [Py.]	R: 7, 102, 175
—·C_5H_9N [C_4H_9CN]	R: 106
—·$C_6H_5AsCl_4$ [$AsPhCl_4$]	R: 54
—·$C_6H_5PCl_4$ [$PPhCl_4$]	R: 53
—·$C_6H_5NO_2$ [$PhNO_2$]	R: 7, 175
—·C_6H_7N [4-MePy.]	R: 7
—·$C_6H_{15}As$ [Et_3As]	R: 36
—·$C_6H_{15}Cl_2P$ [Et_3PCl_2]	R: 53
—·$C_6H_{15}N$ [Et_3N]	R: 106, Junp
—·$C_6H_{15}P$ [Et_3P]	R: Junp
—·$C_7H_3Cl_2N$ [2,6-Cl_2PhCN]	R: 6
—·C_7H_4ClN [2-ClPhCN]	R: 6
—·— [3-ClPhCN]	R: 6
—·— [4-ClPhCN]	R: 6, 7
—·C_7H_8O [PhOMe]	R: 175
—·C_7H_8S [PhSMe]	R: 7
—·$C_{12}H_{10}AsCl_3$ [$AsPh_2Cl_3$]	R: 54
—·$C_{12}H_{10}PCl_3$ [PPh_2Cl_3]	R: 53
—·$C_{13}H_{10}O$ [BZP]	R: 7
—·$C_{18}H_{15}AsCl_2$ [$AsPh_3Cl_2$]	R: 54
—·$C_{18}H_{15}PCl_2$ [PPh_3Cl_2]	R: 53
—·$C_{18}H_{15}P$ [Ph_3P]	R: 7, 102
—·Cl_3OP [$OPCl_3$]	R: 102, 175
—·Cl_5P	R: 102
BD_3·C_3H_9N [Me_3N]	R: 129
BF_4H·$(Cl_2NP)_3$	R: 187
BH_3·$C_6H_{15}As$ [Et_3As]	R: 36
$BiBr_3$·$AlBr_3$	R: 144, 153
$BiCl_3$·C_2H_3N [MeCN] ⟨1:1⟩	R: 124
—·C_3H_6O [$MeCOCH_3$] ⟨1:1⟩	R: 124

—·C$_4$H$_{10}$O [Et$_2$O] ⟨1:1⟩   R: 124
—·C$_5$H$_4$ClN [2-ClPy.]   R: 6
—·— [4-ClPy.]   R: 6
—·C$_7$H$_8$O [PhOMe] ⟨1:1⟩   R: 124

BrClCu·C$_5$H$_5$NO [Py.—O] ⟨1:2⟩   R: 66

BrClO$_4$·C$_5$H$_5$N [Py.]   R: 101

BrCs·Br$_2$ R: 21

BrCu·C$_3$H$_9$PS [Me$_3$PS] ⟨n:n⟩   R: 20
—·C$_{16}$H$_{36}$BrN [(n-Bu)$_4$NBr] ⟨1:1⟩   R: 16

BrH·C$_2$H$_6$BrN [BrCH$_2$CH$_2$NH$_2$]   R: 100

BrI·C$_5$H$_3$Br$_2$N [3,5-Br$_2$Py.]   R: 17
—·C$_5$H$_4$BrN [2-BrPy.]   R: 17
—·C$_5$H$_4$ClN [2-ClPy.]   R: 17
—·— [4-ClPy.]   R: 17
—·C$_5$H$_5$N [Py.]   R: 17
—·C$_6$H$_7$N [4-MePy.]   R: 17
—·C$_6$H$_{12}$N$_4$ [HMT]   R: 17

BrLi·4H$_2$O·C$_6$H$_{12}$N$_4$ [HMT]   R: 79

Br$_2$·BrCs   R: 21
—·BrH$_4$N   R: 21
—·C$_2$H$_3$N [MeCN] ⟨1:2⟩   R: 79, 135
—·C$_3$H$_6$O [Me$_2$CO]   R: 41
—·C$_3$H$_{10}$BrN [(Me)$_3$NHBr]   R: 21
—·C$_4$H$_8$O$_2$ [Diox.]   R: 41
—·C$_4$H$_{10}$O [Et$_2$O]   R: 41
—·C$_4$H$_{12}$IN [Me$_4$NI]   R: 18
—·C$_5$H$_3$Br$_2$N [3,5-Br$_2$Py.]   R: 17
—·C$_5$H$_4$BrN [2-BrPy.]   R: 17
—·C$_5$H$_4$ClN [2-ClPy.]   R: 17
—·C$_6$H$_5$Br [PhBr]   R: 41, 158
—·C$_6$H$_5$Cl [PhCl]   R: 41
—·C$_6$H$_5$F [PhF]   R: 41
—·C$_6$H$_6$ [Ph]   R: 41, 99, 105
—·C$_8$H$_{10}$ [1,4-Me$_2$Ph]   R: 99
—·C$_8$H$_{20}$IN [Et$_4$NI]   R: 18
—·H$_4$IN [NH$_4$I]   R: 18

BrH$_4$N·Br$_2$ R: 21

Br$_2$Cd·C$_5$H$_5$N [Py.] ⟨1:2⟩   R: 169

Br$_2$Cu·C$_5$H$_5$NO [Py.—O] ⟨1:1⟩   R: 66, 67
—·— [Py.—O] ⟨2:1⟩   R: 66
—·C$_6$H$_7$NO [4-Pic.—O] ⟨1:1⟩   R: 66

Br$_2$Hg·C$_2$H$_6$OS [DMSO] ⟨1:2⟩   R: 125
—·C$_3$H$_7$NO [DMF] ⟨1:1⟩   R: 125
—·C$_4$H$_8$O [THF]   R: 24, 190
—·C$_4$H$_8$O$_2$ [Diox.] ⟨1:1⟩   R: 23, 148, 190

—·— ⟨1:2⟩ — R: 148, 190
—·C₄H₁₀O₂ [gly.] ⟨1:1⟩ — R: 30, 190
—·C₄H₁₀S₂ [MeS(CH₂)₂SMe] — R: 30
—·C₅BrCl₅Hg [C₅Cl₅HgBr] — R: 188, 189
—·C₅H₅NO [Py.—O] — R: 24, 190
—·C₆H₇NO [4-MePy.—O] — R: 24, 190
—·C₆H₁₄O₃ [Digly.] — R: 190
—·C₆H₁₆N₂ [Me₂N(CH₂)₂NMe₂] — R: 30
—·C₈H₈O [ACP] — R: 24, 190
—·C₈H₁₈O₄ [Trigly.] — R: 190
—·C₁₀H₂₂O₅ [Tetragly.] — R: 190
—·C₁₂H₂₄O₆ [18-Crown-6-ether] — R: 190
—·C₁₂H₂₆O₃ [Diglybuty.] — R: 190
—·C₁₃H₁₀O [BZP] — R: 24, 190
—·C₁₄H₃₀O₇ [Hexagly.] — R: 190

Br₂Mg·C₄H₁₀O [Et₂O] ⟨1:1⟩ — R: 113
—·— ⟨1:2⟩ — R: 113
—·— ⟨1:3⟩ — R: 113

Br₂Zn·C₅H₅N [Py.] — R: 101

Br₃Sb·AlBr₃ — R: 144, 153
—·C₆H₆ [Ph] ⟨2:1⟩ — R: 83, 85, 86, 87, 89, 134, 143

—·C₆H₇N [PhNH₂] ⟨1:3⟩ — R: 73
—·C₇H₈ [PhMe] — R: 85, 86, 87, 89
—·C₇H₈O [PhOMe] ⟨1:1⟩ — R: 83, 85, 86, 87, 89, 115

—·C₈H₁₀ [1,2-Me₂Ph] — R: 81, 85, 86, 87, 89, 90, 134

—·— [1,3-Me₂Ph] — R: 85, 86, 87, 89
—·— [1,4-Me₂Ph] ⟨1:1⟩ — R: 85, 89, 138
—·— ⟨2:1⟩ — R: 90
—·— [EtPh] — R: 85, 87, 89, 90, 134
—·C₈H₁₀O [PhOEt] — R: 85, 86, 87, 89
—·C₉H₁₂ [1,3,5-Me₃Ph] ⟨2:1⟩ — R: 85, 86, 87, 89
—·C₁₀H₈ [Naphth.] ⟨2:1⟩ — R: 81, 85, 86, 87, 89
—·C₁₂H₁₀ [PhPh] ⟨2:1⟩ — R: 81, 85, 86, 87, 89
—·C₁₃H₁₂ [PhCH₂Ph] ⟨2:1⟩ — R: 81, 83, 85, 90
—·C₁₄H₁₀ [Phenanthrene] ⟨2:1⟩ — R: 134

Br₄Sn·CH₃Cl₂OP [MePOCl₂] ⟨1:2⟩ — R: 3
—·C₂H₆OS [DMSO] ⟨1:2⟩ — R: 151
—·C₄H₈O [THF] ⟨1:2⟩ — R: 151
—·C₄H₈O₂ [Diox.] ⟨1:1⟩ — R: 151
—·C₄H₁₀O [Et₂O] ⟨1:2⟩ — R: 151
—·C₄H₁₀O₂ [gly.] ⟨1:1⟩ — R: 151
—·C₅H₅N [Py.] ⟨1:2⟩ — R: 151
—·C₆H₁₈N₃P [HMPT] ⟨1:2⟩ — R: 151

CBrCl₂D·C₄D₁₀O [Et2O-d₁₀] — R: 168

CBrCl₃Hg [CCl₃HgBr]·C₄H₈O₂ [Dioox.] — R: 32

CBr$_4$·C$_6$H$_6$ [Ph] ⟨1:1⟩   R: 70
  ·C$_7$H$_9$N [2,6-Me$_2$Py.]   R: 70
  —·C$_8$H$_{10}$ [1,2-Me$_2$Ph]   R: 70
  —·— [1,4-Me$_2$Ph]   R: 70, 99
  —·C$_8$H$_{11}$N [s-Me$_3$Py.]   R: 70
  —·C$_9$H$_{12}$ [1,3,5-Me$_3$Ph]   R: 70
  —·C$_{10}$H$_{14}$ [Durene]   R: 70

CCIN [ClCN]·BCl$_3$ R: 6
  —·Cl$_4$Sn   R: 33
  —·Cl$_5$Nb   R: 34
  —·Cl$_5$Sb   R: 33, 34, 152, 176
  —·Cl$_5$Ta   R: 34

CCl$_3$D·C$_3$H$_6$O [MeCOMe]   R: 156
  —·C$_4$H$_{10}$O [Et$_2$O]   R: 156
  —·C$_9$H$_{12}$ [1,3,5-Me$_3$Ph]   R: 156

CCl$_4$·C$_3$H$_6$O [MeCOMe] ⟨1:1⟩   R: 84
  —·C$_4$H$_{10}$O [Et$_2$O] ⟨1:1, 1:2⟩   R: 84
  —·C$_5$H$_5$N [Py.] ⟨2:1⟩   R: 184
  —·C$_6$H$_6$ [Ph] ⟨1:1⟩   R: 84, 184
  —·— ⟨2:1⟩   R: 84, 184
  —·— ⟨3:1⟩   R: 84
  —·C$_8$H$_{10}$ [1,4-Me$_2$Ph]   R: 99
  —·C$_8$H$_{11}$N [s-Me$_3$Py.] ⟨1:1⟩   R:. 184
  —·C$_9$H$_{12}$ [1,2,4-Me$_3$Ph] ⟨1:1⟩   R: 184
  —·— [1,3,5-Me$_3$Ph] ⟨1:1⟩   R: 184
  —·C$_{10}$H$_{14}$ [1,2,4,5-Me$_4$Ph]   R: 184
  —·C$_{18}$H$_{15}$Cl$_2$P [Ph$_3$PCl$_2$] ⟨1:2⟩   R: 53

CCl$_4$Hg [CCl$_3$HgCl]·C$_4$H$_{10}$O$_2$ [gly.]   R: 32

CHCl$_3$·CH$_4$O [MeOH] ⟨1:1⟩   R: 84, 114
  —·C$_2$H$_3$N [MeCN] ⟨1:1⟩   R: 4, 57
  —·C$_2$H$_8$N$_2$ [ETDA]   R: 10
  —·C$_3$H$_4$BrN [β-Br(CH$_2$)$_2$CN] ⟨1:1⟩   R: 57
  —·C$_3$H$_5$N [MeCH$_2$CN] ⟨1:1⟩   R: 57
  —·C$_3$H$_6$O [MeCOMe] ⟨1:1⟩   R: 10, 84, 114
  —·C$_3$H$_9$N [Me$_3$N]   R: 149
  —·C$_4$H$_6$O$_2$ [(MeCO)$_2$] ⟨1:1⟩   R: 4
  —·C$_4$H$_6$O$_3$ [(MeCO)$_2$O] ⟨1:1⟩   R: 4, 79
  —·C$_4$H$_8$O$_2$ [Diox.] ⟨2:1⟩   R: 114
  —·C$_4$H$_{10}$N$_2$ [Piperazine]   R: 117
  —·C$_4$H$_{10}$O [Et$_2$O] ⟨2:1⟩   R: 84, 114
  —·— ⟨1:1⟩   R: 84, 114
  —·— ⟨1:2⟩   R: 84
  —·— ⟨1:3⟩   R: 84, 114
  —·C$_4$H$_{11}$N [DEA]   R: 10, 117
  —·C$_5$H$_5$N [Py.] ⟨1:1⟩   R: 10, 96, 114, 117
  —·C$_5$H$_{11}$N [Piperidine]   R: 10, 117
  —·C$_6$H$_6$ [Ph]   R: 10
  —·C$_6$H$_7$N [4-MePy.]   R: 117

—·$C_6H_8N_2$ [NC(CH$_2$)$_4$CN] ⟨2:1⟩	R: 57
—·$C_6H_{10}O_4$ [(COOC$_2$H$_5$)$_2$] ⟨1:1⟩	R: 4
—·$C_6H_{11}N$ [Me(CH$_2$)$_4$CN] ⟨1:1⟩	R: 57
—·$C_6H_{12}N_2$ [TED]	R: 117
—·$C_6H_{15}N$ [Et$_3$N] ⟨1:1⟩	R: 10, 114, 117
—·C_7H_5N [PhCN] ⟨1:1, 2:1⟩	R: 57, 79
—·C_7H_6O [BZA] ⟨2:1⟩	R: 79
—·C_7H_9N [3,5-Me$_2$Py.]	R: 117
—·C_8H_8O [ACP] ⟨1:1⟩	R: 4, 79
—·— ⟨2:1⟩	R: 79
—·C_8H_{10} [1,2-Me$_2$Ph]	R: 10, 84
—·— [1,3-Me$_2$Ph]	R: 10, 84
—·— [1,4-Me$_2$Ph]	R: 10, 79
—·C_9H_{12} [1,3,5-Me$_3$Ph] ⟨1:1⟩	R: 10, 84, 114
—·— [1,2,4-Me$_3$Ph] ⟨2:1⟩	R: 79
—·$C_{10}H_{14}$ [1,2,4,5-Me$_4$Ph]	R: 10
CHI$_3$·S$_8$ ⟨1:3⟩	R: 160
CH$_2$Cl$_3$OP [(ClCH$_2$)P(O)Cl$_2$]·Cl$_4$Sn	R: 3
CH$_3$Br$_3$Sn·C$_4$H$_8$O [THF] ⟨1:2⟩	R: 151
—·C$_4$H$_8$O$_2$ [Diox.] ⟨2:1⟩	R: 151
CH$_3$ClHg [MeHgCl]·MeHgI	R: 31
CH$_3$Cl$_2$OP [MeP(O)Cl$_2$]·Br$_4$Sn ⟨2:1⟩	R: 3
—·Cl$_4$Sn ⟨2:1⟩	R: 3
CH$_3$Cl$_2$O$_2$P [MeOP(O)Cl$_2$]·Cl$_4$Sn ⟨2:1⟩	R: 3
CH$_3$Cl$_3$Sn·C$_4$H$_8$O [THF] ⟨1:2⟩	R: 151
—·C$_4$H$_8$O$_2$ [Diox.] ⟨1:2⟩	R: 151
—·C$_4$H$_{10}$O [Et$_2$O] ⟨1:2⟩	R: 151
CH$_3$Cl$_4$P [MePCl$_4$]·AlCl$_3$	R: 53
—·ClI	R: 53
CH$_3$HgI·CH$_3$ClHg [MeHgCl]	R: 31
CH$_3$I·C$_5$H$_4$ClN [3-ClPy.]	R: 6
CH$_3$NO$_2$·Cl$_3$Ga	R: 178
—·Cl$_4$Sn	R: 112
CH$_4$O [MeOH]·CHCl$_3$	R: 84, 114
—·C$_5$H$_5$N [Py.]	R: 96
—·Cl$_2$Hg ⟨2:1⟩	R: 24, 190
—·Cl$_5$Sb	R: 152
C$_2$Cl$_3$N [CCl$_3$CN]·BCl$_3$	R: 7
C$_2$Cl$_4$Hg [CCl$_2$=CClHgCl]·C$_2$H$_6$OS [DMSO]	R: 169
—·C$_3$H$_7$NO [DMF]	R: 169
C$_2$Cl$_6$Hg [(CCl$_3$)$_2$Hg]·C$_4$H$_8$O [THF]	R: 32
—·C$_4$H$_{10}$O$_2$ [DME]	R: 32

C₂HCl₂N [CHCl₂CN]·BCl₃ — R: 7

C₂HCl₃O₂ [CCl₃COOH]·C₆H₉NO [*N*-Vinylpyrrolidone] — R: 146
—·C₇H₁₃NO [*N*-iso-Pr-pyrrolidone] — R: 146

C₂H₂ClN·BCl₃ [CH₂ClCN] — R: 7
—·F₅Nb — R: 68

C₂H₂Cl₂Hg [CHCl=CHHgCl]·C₄H₈O [THF] — R: 169

C₂H₃N [MeCN]·BCl₃ — R: 7, 102, 106, 175
—·BiCl₃ — R: 124
—·Br₂ ⟨2:1⟩ — R: 79, 135
—·CHCl₃ — R: 4, 57
—·Cl₂ ⟨1:1⟩ — R: 56, 57
—·— ⟨2:1⟩ — R: 135
—·Cl₃Ga — R: 178
—·Cl₅Sb — R: 152, 154, 167

C₂H₄Cl₂O [(CH₂Cl)₂O]·Cl₅Sb — R: 120, 152, 153

C₂H₅AlCl₂ [Dimer] — R: 52

C₂H₅BrMg [EtMgBr]·C₄H₁₀O [Et₂O] ⟨1:2⟩ — R: 113
—·C₆H₁₄O [(iso-Pr)₂O] ⟨1:2⟩ — R: 113

C₂H₅ClO [MeCOCH₂Cl]·BCl₃ — R: 6
—·Cl₄Sn ⟨2:1⟩ — R: 123, 126
—·Cl₅Sb — R: 152

C₂H₅ClS [MeSCH₂Cl]·BCl₃ — R: 6

C₂H₅Cl₂OP [EtP(O)Cl₂]·Cl₄Sn — R: 3

C₂H₅Cl₄P [EtPCl₄]·AlCl₃ — R: 53

C₂H₆AlBr [Dimer] — R: 52

C₂H₆AlCl [Dimer] — R: 52

C₂H₆AsCl₃ [AsMe₂Cl₃]·ClI — R: 54

C₂H₆BrN [BrCH₂CH₂NH₂]·BrH — R: 100

C₂H₆Br₂Sn [Me₂SnBr₂]·C₂H₆OS [DMSO] ⟨1:2⟩ — R: 151
—·C₃H₆O [Me₂CO] ⟨1:2⟩ — R: 151
—·C₃H₇NO [DMF] ⟨1:2⟩ — R: 151
—·C₄H₈O [THF] ⟨1:2⟩ — R: 151
—·C₄H₈O₂ [Diox.] ⟨1:1⟩ — R: 151
—·C₄H₁₀O₂ [DME] ⟨1:1⟩ — R: 151
—·C₅H₅N [Py.] ⟨1:2⟩ — R: 151
—·C₆H₁₈N₃OP [HMPT] ⟨1:2⟩ — R: 151

C₂H₆ClOP [Me₂P(O)Cl]·Cl₄Sn ⟨2:1⟩ — R: 3

C₂H₆ClPS [Me₂P(S)Cl]·Cl₄Sn ⟨2:1⟩ — R: 3

C₂H₆Cl₂NOP [Me₂NP(O)Cl₂]·Cl₄Sn ⟨2:1⟩ — R: 3

$C_2H_6Cl_2Sn$ [Me_2SnCl_2]·C_3H_6O [Me_2CO] $\langle 1:2 \rangle$ R: 151
  —·$C_4H_8O_2$ [Diox.] $\langle 1:1 \rangle$ R: 151
  —·$C_4H_{10}O_2$ [DME] $\langle 1:1 \rangle$ R: 151

$C_2H_6Cl_3P$ [Me_2PCl_3]·BCl_3 R: 53
  —·ClI   R: 3, 53
  —·Cl_5Sb R: 53

C_2H_6O [Me_2O]·$AlCl_3$ R: Junp
  —·BCl_3 R: Junp
  —·C_3H_9Al [Me_3Al]   R: 52, 107
  —·Cl_3Ga R: Junp
  —·F_5Nb R: 68

C_2H_6OS [DMSO]·Br_2Hg R: 125
  —·Br_4Sn R: 151
  —·C_2Cl_4Hg [$Cl_2C{=}CClHgCl$]   R: 169
  —·Cl_2Hg R: 25, 190
  —·$C_2H_6Br_2Sn$ [Me_2BrSn] $\langle 2:1 \rangle$ R: 151
  —·C_3H_9BrSn [Me_3BrSn] $\langle 1:1 \rangle$ R: 151
  —·C_5H_5NO [Py.—O].ClCu   R: 66

C_2H_6S [Me_2S]·$AlBr_3$ R: 122
  —·$AlCl_3$ R: Junp
  —·BCl_3 R: Junp
  —·C_3H_9Al [Me_3Al]   R: 52, 107
  —·C_3H_9Ga [Me_3Ga]   R: 107
  —·Cl_3Ga R: 35, 178

C_2H_6Se [Me_2Se]·C_3H_9Ga [Me_3Ga]   R: 107

C_2H_6Te [Me_2Te]·C_3H_9Ga [Me_3Ga]   R: 107

$C_2H_7O_3P$ [$(MeO)_2P(O)H$]·Cl_4Sn $\langle 2:1 \rangle$ R: 3, 163

C_2H_8IN [Me_2NH_2I]·I_2 R: 98, 133

$C_2H_8N_2$ [ETDA]·$CHCl_3$ R: 10

$C_2K_2O_4$·F_3Sb $\langle 1:1 \rangle$ R: 29

$C_2N_2S_2Zn$ [$(NCS)_2Zn$]·C_5H_5N [Py.] $\langle 1:2 \rangle$ R: 101

C_3H_2ClNO [$CH_2ClCOCN$]·Cl_5Sb R: 152

C_3H_3N [$CH_2{=}CHCN$]·BCl_3 R: 106

C_3H_4BrN [CH_2BrCH_2CN]·$CHCl_3$ $\langle 1:1 \rangle$ R: 57
  —·[CH_2BrCH_2CN]·Cl_2 $\langle 1:1 \rangle$ R: 56, 57
  —·— $\langle 2:1 \rangle$ R: 56, 57

C_3H_5N [CH_3CH_2CN]·BCl_3 R: 106
  —·$CHCl_3$ $\langle 1:1 \rangle$ R: 57
  —·Cl_2 R: 56, 57

C_3H_6O [MeCOMe]·$AlBr_3$ R: 116
  —·$BiCl_3$ $\langle 1:1 \rangle$ R: 124
  —·Br_2 R: 41

—·CCl$_4$ ⟨1:1⟩	R: 84
—·CDCl$_3$	R: 156
—·CHCl$_3$ ⟨1:1⟩	R: 10, 84, 114
—·C$_2$H$_6$Br$_2$Sn [Me$_2$Br$_2$Sn] ⟨2:1⟩	R: 151
—·C$_2$H$_6$Cl$_2$Sn [Me$_2$Cl$_2$Sn] ⟨2:1⟩	R: 151
—·C$_3$H$_9$BrSn [Me$_3$BrSn] ⟨1:1⟩	R: 151
C$_3$D$_6$O·CDCl$_3$	R: 156
C$_3$H$_6$O$_3$ [Triox.]·Cl$_2$Hg	R: 64
C$_3$H$_6$S$_3$ [Trithiane]·Cl$_2$Hg	R: 64
C$_3$H$_7$NO [DMF]·Br$_2$Hg	R: 125
—·C$_2$Cl$_4$Hg [CCl$_2$=CClHgCl]	R: 169
—·C$_2$H$_6$Br$_2$Sn [Me$_2$Br$_2$Sn] ⟨2:1⟩	R: 151
—·C$_3$H$_9$BrSn [Me$_3$BrSn] ⟨1:1⟩	R: 151
—·Cl$_2$Hg	R: 190
—·Cl$_5$Sb	R: 28, 152
C$_3$H$_8$O [C$_3$H$_7$OH]·Cl$_5$Sb	R: 152
C$_3$H$_9$Al [Me$_3$Al]·C$_2$H$_6$O [Me$_2$O]	R: 52, 107
—·C$_2$H$_6$S [Me$_2$S]	R: 52, 107
—·C$_3$H$_9$Al [Dimer]	R: 52
—·C$_3$H$_9$As [Me$_3$As]	R: 52, 107, 147
—·C$_3$H$_9$N [Me$_3$N]	R: 52, 107, 147
—·C$_3$H$_9$P [Me$_3$P]	R: 52, 107, 147
—·C$_3$H$_9$Sb [Me$_3$Sb]	R: 52, 107, 147
—·C$_4$H$_8$O [THF]	R: 52, 107
—·C$_4$H$_8$O$_2$ [Diox.] ⟨2:1⟩	R: 52, 107
—·C$_4$H$_8$S [THT]	R: 52, 107
C$_3$H$_9$AlO·C$_3$H$_9$AlO [Me$_3$AlO—Trimer]	R: 52
C$_3$H$_9$AlS·C$_3$H$_9$AlS [Me$_3$AlS—Polymer]	R: 52
C$_3$H$_9$As [Me$_3$As]·AlCl$_3$	R: 36
—·BCl$_3$	R: 36
—·C$_3$H$_9$Al [Me$_3$Al]	R: 52, 107, 147
—·C$_3$H$_9$Ga [Me$_3$Ga]	R: 107
—·C$_3$H$_9$In [Me$_3$In]	R: 107, 147
—·ClCuO$_4$ [CuClO$_4$] ⟨3:1⟩	R: 20
—·Cl$_3$Ga	R: 35, 36
C$_3$H$_9$AsCl$_2$ [Me$_3$AsCl$_2$]·BCl$_3$	R: 54
C$_3$H$_9$AsS [Me$_3$AsS]·CuCl ⟨n:n⟩	R: 20
—·ClCuO$_4$ [CuClO$_4$] ⟨3:1⟩	R: 20
C$_3$H$_9$BrSn [Me$_3$SnBr]·C$_2$H$_6$OS [DMSO] ⟨1:1⟩	R: 151
—·C$_3$H$_6$O [Me$_2$CO] ⟨1:1⟩	R: 151
—·C$_3$H$_7$NO [DMF] ⟨1:1⟩	R: 151
—·C$_4$H$_8$O [THF] ⟨1:1⟩	R: 151
—·C$_4$H$_8$O$_2$ [Diox.] ⟨2:1⟩	R: 151
—·C$_4$H$_{10}$O [Et$_2$O] ⟨1:1⟩	R: 151
—·C$_4$H$_{10}$O$_2$ [DME] ⟨1:1⟩	R: 151

—·C$_5$H$_5$N [Py.] $\langle 1:1 \rangle$ R: 151
—·C$_6$H$_{16}$N$_2$ [TMED] $\langle 2:1 \rangle$ R: 151
—·C$_6$H$_{18}$N$_3$OP [HMPT] $\langle 1:1 \rangle$ R: 151

C$_3$H$_9$Cl$_2$P [Me$_3$PCl$_2$]·BCl$_3$ R: 53
—·ClI   R: 53

C$_3$H$_9$Ga [Me$_3$Ga]·C$_2$H$_6$S [Me$_2$S]   R: 107
—·C$_2$H$_6$Se [Me$_2$Se]   R: 107
—·C$_2$H$_6$Te [Me$_2$Te]   R: 107
—·C$_3$H$_9$As [Me$_3$As]   R: 107
—·C$_3$H$_9$N [Me$_3$N]   R: 107
—·C$_3$H$_9$P [Me$_3$P]   R: 107
—·C$_3$H$_9$Sb [Me$_3$Sb]   R: 107
—·C$_4$H$_{10}$O [Et$_2$O]   R: 107
—·C$_4$H$_{10}$S [Et$_2$S]   R: 107
—·C$_4$H$_{10}$Se [Et$_2$Se]   R: 107
—·C$_4$H$_{10}$Te [Et$_2$Te]   R: 107

C$_3$H$_9$In [Me$_3$In]·C$_3$H$_9$As [Me$_3$As]   R: 107, 147
—·C$_3$H$_9$N [Me$_3$N]   R: 107, 147
—.C$_3$H$_9$P [Me$_3$P]   R: 107, 147
—·C$_3$H$_9$Sb [Me$_3$Sb]   R: 107, 147
—·C$_4$H$_{10}$O [Et$_2$O]   R: 107

C$_3$H$_9$N [Me$_3$N]·AlCl$_3$ R: Junp
—·AsCl$_3$ R: 36, Junp
—·BCl$_3$ R: 7, 69, 102, 106,
                                                              175, Junp

—·BD$_3$ R: 129
—·CHCl$_3$ R: 149
—·C$_3$H$_9$Al [Me$_3$Al]   R: 52, 107, 147
—·C$_3$H$_9$Ga [Me$_3$Ga]   R: 107˙
—·C$_3$H$_9$In [Me$_3$In]   R: 107, 147
—·Cl$_3$Ga   R: 107, Junp
—·I$_2$ R: 17

C$_3$H$_9$O$_2$P [Me$_2$P(O)OMe]·Cl$_4$Sn $\langle 2:1 \rangle$ R: 3, 163
— [Et(MeO)P(O)H]·Cl$_4$Sn $\langle 2:1 \rangle$ R: 3, 163

C$_3$H$_9$P [Me$_3$P]·AlCl$_3$ R: Junp
—·BCl$_3$ R: Junp
—·C$_3$H$_9$Al [Me$_3$Al]   R: 52, 107, 147
—·C$_3$H$_9$Ga [Me$_3$Ga]   R: 107
—·C$_3$H$_9$In [Me$_3$In]   R: 107, 147
—·Cl$_3$Ga   R: 35

C$_3$H$_9$PS [Me$_3$PS]·BrCu $\langle n:n \rangle$ R: 20
—·ClCu $\langle 3:3 \rangle$ R: 20
—·ClCuO$_4$ [CuClO$_4$] $\langle 3:1 \rangle$ R: 20
—·CuI $\langle n:n \rangle$ R: 20

C$_3$H$_9$Sb [Me$_3$Sb]·C$_3$H$_9$Al [Me$_3$Al]   R: 52, 107, 147
—·C$_3$H$_9$Ga [Me$_3$Ga]   R: 107
—·C$_3$H$_9$In [Me$_3$In]   R: 107, 147

$C_3H_{10}BrN$ [$(CH_3)_3NHBr$]·Br_2 — R: 21

$C_3H_{13}B_{10}BrHg$ [$MeC(B_{10}H_{10})CHgBr$] (*C*-Methyl-*C*-mercury-*o*-borane)·$C_{10}H_8N_2$ [4,4′-bipy.] — R: 136, 169

C_4H_5N [Pyrrole]·C_5H_5N [Py.] $\langle 1:1 \rangle$ — R: 96

$C_4H_6O_2$ [$(MeCO)_2$]·$CHCl_3$ $\langle 1:1 \rangle$ — R: 4, 79

$C_4H_6O_2S$ [$EtSO_2C \equiv CH$]·$AlBr_3$ $\langle 1:2 \rangle$ — R: 95

$- \left[\bigcirc SO_2 \right]$·$AlBr_3$ $\langle 1:2 \rangle$ — R: 95

$C_4H_6O_3$ [$(MeCO)_2O$]·$CHCl_3$ $\langle 1:1 \rangle$ — R: 4

C_4H_7N [$(Me)_2CHCN$]·BCl_3 — R: 106

C_4H_8O [Butanone]·$AlBr_3$ — R: 116

C_4H_8O [THF]·BCl_3 — R: 7, 175
— ·Br_2Hg — R: 24, 190
— ·Br_4Sn $\langle 2:1 \rangle$ — R: 151
— ·CH_3Br_3Sn [$MeSnBr_3$] $\langle 2:1 \rangle$ — R: 151
— ·CH_3Cl_3Sn [$MeSnCl_3$] $\langle 2:1 \rangle$ — R: 151
— ·C_2Cl_6Hg [$(CCl_3)_2Hg$] — R: 32
— ·$C_2H_2Cl_2Hg$ [*Trans*-$CHCl{=}CHHgCl$] — R: 169
— ·$C_2H_6Br_2Sn$ [Me_2SnBr_2] $\langle 2:1 \rangle$ — R: 151
— ·C_3H_9Al [Me_3Al] — R: 52, 107
— ·C_3H_9BrSn [Me_3SnBr] — R: 151
— ·C_5Cl_6Hg [C_5Cl_5HgCl] $\langle 1:2 \rangle$ — R: 188, 189
— ·C_6H_5BrMg [$PhMgBr$] $\langle 2:1 \rangle$ — R: 113
— ·$C_{10}Cl_{10}Hg$ [$(C_5Cl_5)_2Hg$] $\langle 2:1 \rangle$ — R: 188, 189
— ·Cl_2Hg — R: 24, 25, 64, 190
— ·Cl_3Ga — R: 178
— ·Cl_5Sb — R: 152

$C_4H_8O_2$ [*p*-Diox.]·Br_2 — R: 41
— ·Br_2Hg $\langle 1:1 \rangle$, $\langle 2:1 \rangle$ — R: 23, 148, 190
— ·— $\langle 2:1 \rangle$ — R: 148, 190
— ·Br_4Sn $\langle 1:1 \rangle$ — R: 151
— ·$CBrCl_3Hg$ [CCl_3HgBr] — R: 32
— ·CBr_4 — R: 70
— ·$CHCl_3$ $\langle 1:2 \rangle$ — R: 114
— ·CH_3Br_3Sn [$MeSnBr_3$] $\langle 1:2 \rangle$ — R: 151
— ·CH_3Cl_3Sn [$MeSnCl_3$] $\langle 1:2 \rangle$ — R: 151
— ·$C_2H_6Br_2Sn$ [Me_2SnBr_2] $\langle 1:1 \rangle$ — R: 151
— ·$C_2H_6Cl_2Sn$ [Me_2SnCl_2] $\langle 1:1 \rangle$ — R: 151
— ·C_3H_9Al [Me_3Al] $\langle 1:2 \rangle$ — R: 52, 107
— ·C_3H_9BrSn [Me_3SnBr] $\langle 1:2 \rangle$ — R: 151
— ·Cl_2 $\langle 1:1 \rangle$ — R: 71
— ·Cl_2Hg $\langle 1:1, 2:1 \rangle$ — R: 12, 23, 25, 64, 148, 165, 190

— ·— $\langle 2:1 \rangle$ — R: 64, 148, 190
— ·Cl_5Sb — R: 152, 154
— ·HgI_2 — R: 23, 148, 190

C$_4$H$_8$S [THT]·C$_3$H$_9$Al [Me$_3$Al]	R: 52, 107
—·Cl$_2$Hg	R: 64
C$_4$H$_9$AlCl$_2$ [iso-BuAlCl$_2$] ⟨Dimer⟩	R: 52
C$_4$H$_9$BrMg[t-BuMgBr]·C$_4$H$_{10}$O [Et$_2$O] ⟨1:n⟩	R: 113
C$_4$H$_{10}$AlBr [Et$_2$AlBr] ⟨Dimer⟩	R: 52
C$_4$H$_{10}$AlCl [Et$_2$AlCl] ⟨Dimer⟩	R: 52
C$_4$H$_{10}$AlI [Et$_2$AlI] ⟨Dimer⟩	R: 52
C$_4$H$_{10}$ClOP [Et$_2$P(O)Cl]·Cl$_4$Sn ⟨2:1⟩	R: 3
C$_4$H$_{10}$ClP [Et$_2$PCl]·Cl$_4$Sn ⟨2:1⟩	R: 3
C$_4$H$_{10}$Cl$_3$P [Et$_2$PCl$_3$]·BCl$_3$	R: 53
C$_4$H$_{10}$N$_2$ [Piperazine]·CHCl$_3$	R: 117
C$_4$H$_{10}$O [Et$_2$O]·AlBr$_3$	R: 122
—·AlCl$_3$	R: 50, Junp
—·BCl$_3$ ⟨1:1⟩	R: 175, Junp
—·BiCl$_3$ ⟨1:1⟩	R: 124
—·Br$_2$	R: 41
—·Br$_2$Mg ⟨1:1, 2:1, 3:1⟩	R: 113
—·Br$_4$Sn ⟨2:1⟩	R: 151
—·CCl$_4$ ⟨1:1, 2:1⟩	R: 84
—·CHCl$_3$ ⟨1:2, 1:1, 3:1⟩	R: 84, 114
—·— ⟨2:1⟩	R: 84
—·CDCl$_3$	R: 156
—·CHCl$_3$Sn [MeSnCl$_3$] ⟨2:1⟩	R: 151
—·C$_2$H$_5$BrMg [EtMgBr] ⟨2:1⟩	R: 113
—·C$_3$H$_9$BrSn [Me$_3$SnBr] ⟨1:1⟩	R: 151
—·C$_3$H$_9$Ga [Me$_3$Ga]	R: 107
—·C$_3$H$_9$In [Me$_3$In]	R: 107
—·C$_4$H$_9$BrMg [t-BuMgBr] ⟨n:1⟩	R: 113
—·C$_6$H$_5$BrMg [PhMgBr] ⟨2:1⟩	R: 113
—·C$_6$H$_{15}$Ga [Et$_3$Ga]	R: 107
—·Cl$_2$Hg	R: 125
—·Cl$_3$Ga	R: 35, 178
—·Cl$_3$GeH	R: 112
—·Cl$_4$Sn ⟨2:1⟩	R: 12, 112, 123, 155
—·Cl$_5$Sb	R: 152, 154
C$_4$D$_{10}$O [Et$_2$O-d$_{10}$]·CBrCl$_2$D	R: 168
—·CCl$_3$D	R: 156
C$_4$H$_{10}$O$_2$ [Gly.]·Br$_2$Hg	R: 30, 190
—·Br$_4$Sn ⟨1:1⟩	R: 151
—·CCl$_4$Hg [CCl$_3$HgCl]	R: 32
—·C$_2$Cl$_6$Hg [(CCl$_3$)$_2$Hg]	R: 32
—·C$_2$H$_6$Br$_2$Sn [Me$_2$SnBr$_2$] ⟨1:1⟩	R: 151
—·C$_2$H$_6$Cl$_2$Sn [Me$_2$SnCl$_2$] ⟨1:1⟩	R: 151
—·C$_3$H$_9$BrSn [Me$_3$SnBr] ⟨1:1⟩	R: 151

—·C$_5$BrCl$_5$Hg [C$_5$Cl$_5$HgBr]   R: 188, 189, 192
—·C$_5$Cl$_6$Hg [C$_5$Cl$_5$HgCl] ⟨3:5⟩   R: 188, 189
—·C$_{10}$Cl$_{10}$Hg [(C$_5$Cl$_5$)$_2$Hg]   R: 188, 189
—·Cl$_2$Hg   R: 30, 190
—·Cl$_4$Sn ⟨1:1⟩   R: 112, 123
—·HgI$_2$ R: 30

C$_4$H$_{10}$O$_2$S [Et$_2$SO$_2$]·AlBr$_3$ R: 95

C$_4$H$_{10}$S [Et$_2$S]·AlCl$_3$ R: Junp
—·BCl$_3$ R: 175, Junp
—·Br$_2$Hg   R: 30
—·C$_3$H$_9$Ga [Me$_3$Ga]   R: 107
—·Cl$_2$Hg   R: 30
—·Cl$_3$Ga   R: 35, 178

C$_4$H$_{10}$S$_2$ [DMTE]·Br$_2$Hg   R: 30
—·Cl$_2$Hg   R: 30
—·HgI$_2$ R: 30

C$_4$H$_{10}$Se [Et$_2$Se]·C$_3$H$_9$Ga [Me$_3$Ga]   R: 107·

C$_4$H$_{10}$Te [Et$_2$Te]·C$_3$H$_9$Ga [Me$_3$Ga]   R: 107

C$_4$H$_{11}$ClNPS [Me$_2$NP(S)ClEt]·Cl$_4$Sn ⟨2:1⟩   R: 2, 3

C$_4$H$_{11}$N [D.E.A.]·CHCl$_3$ R: 10, 117

C$_4$H$_{12}$AlN [AlMe$_4$N] ⟨Dimer⟩   R: 52

C$_4$H$_{12}$BrN [Me$_4$NBr]·Cl$_2$ R: 159

C$_4$H$_{12}$ClN [Me$_4$NCl]·BCl$_3$ R: 7
—·Cl$_2$ R: 159
—·ClD   R: 97
—·HCl   R: 60, 62, 97

C$_4$H$_{12}$IN [Me$_4$NI]·Br$_2$ R: 18
—·Cl$_2$ R: 159
—·I$_2$ ⟨1:1⟩   R: 18, 98, 110, 133
—·— ⟨1:2⟩   R: 110

C$_4$H$_{12}$NO$_3$P [Me$_2$NP(O)(OMe)$_2$]·Cl$_4$Sn ⟨2:1⟩   R: 3

C$_4$H$_{12}$NPS [Me$_2$NP(S)Me$_2$]·Cl$_4$Sn ⟨2:1⟩   R: 2, 3

C$_5$BrCl$_5$Hg [C$_5$Cl$_5$HgBr]·Br$_2$Hg   R: 188, 189
—·C$_4$H$_{10}$O$_2$ [Gly.]   R: 188, 189, 192
—·C$_6$H$_{14}$O$_3$ [Digly.]   R: 188, 189
—·C$_8$H$_{18}$O$_4$ [Trigly.]   R: 188, 189

C$_5$Cl$_5$Tl·C$_7$H$_8$ [MePh] ⟨1:2⟩   R: 191

C$_5$Cl$_6$Hg [C$_5$Cl$_5$HgCl]·C$_4$H$_8$O [THF] ⟨2:1⟩   R: 188, 189
—·C$_4$H$_{10}$O$_2$ [Gly.]   R: 188, 189
—·C$_6$H$_{14}$O$_3$ [Digly.]   R: 188, 189, 192
—·C$_8$H$_{18}$O$_4$ [Trigly.]   R: 188, 189
—·Cl$_2$Hg   R: 188, 189, 190, 191

$C_5H_3Br_2N$ [3,5-Br$_2$Py.]·BrI — R: 17
  —·Br$_2$ — R: 17
  —·ClI — R: 17
  —·HClO$_4$ — R: 17

$C_5H_3Cl_2N$ [2,6-Cl$_2$Py.]·BCl$_3$ — R: 6
  — [3,5-Cl$_2$Py.]·BCl$_3$ — R: 6, 7
  —·ClI — R: 14

C_5H_4BrN [2-BrPy.]·BrI — R: 17
  —·Br$_2$ — R: 17
  —·ClI — R: 17
  —·HClO$_4$ — R: 17

C_5H_4ClN [2-ClPy.]·BCl$_3$ — R: 6
  —·BiCl$_3$ — R: 6
  —·BrI — R: 17
  —·Br$_2$ — R: 17
  —·C$_6$H$_3$N$_3$O$_7$ [(NO$_2$)$_3$Ph$_2$OH] — R: 6
  —·C$_7$H$_8$O$_3$S [4-MePhSO$_3$H] — R: 6
  —·ClI — R: 17
  —·Cl$_2$Hg — R: 6
  —·Cl$_2$Sn — R: 6
  —·Cl$_5$Sb — R: 6, 152
  —·HCl — R: 6
  —·HClO$_4$ — R: 6

C_5H_4ClN [3-ClPy.]·BCl$_3$ — R: 6
  —·CH$_3$I — R: 6
  —·C$_6$H$_3$N$_3$O$_7$ [(NO$_2$)$_3$PhOH] — R: 6
  —·C$_7$H$_8$O$_3$S [4-MePhSO$_3$H] — R: 6
  —·Cl$_2$Sn — R: 6
  —·HCl — R: 6
  —·HClO$_4$ — R: 6
  —·H$_2$SO$_4$ — R: 6

C_5H_4ClN [4-ClPy.]·BCl$_3$ — R: 6, 7
  —·BiCl$_3$ — R: 6
  —·BrI — R: 17
  —·C$_6$H$_3$N$_3$O$_7$ [(NO$_2$)$_3$PhOH] — R: 6
  —·ClI — R: 17
  —·Cl$_2$Hg — R: 6
  —·HCl — R: 6
  —·HClO$_4$ — R: 6
  —·H$_2$SO$_4$ — R: 6

C_5H_5N [Py.]·AlBr$_3$ — R: 122
  —·BCl$_3$ — R: 7, 102, 175
  —·BrClO$_4$ ⟨2:1⟩ — R: 101
  —·BrI — R: 17
  —·Br$_2$Cd — R: 169
  —·Br$_2$Zn ⟨2:1⟩ — R: 101
  —·Br$_4$Sn ⟨2:1⟩ — R: 151
  —·CCl$_4$ ⟨1:1⟩ — R: 184

—·$CHCl_3$ ⟨1:1⟩ R: 10, 96, 114, 117
—·CH_4O [MeOH] R: 96
—·$C_2H_6Br_2Sn$ [Me_2SnBr_2] ⟨2:1⟩ R: 151
—·$C_2N_2S_2Zn$ [$(NCS)_2Zn$] R: 101
—·C_3H_9BrSn [Me_3SnBr] ⟨1:1⟩ R: 151
—·C_4H_5N [Pyrrole] ⟨1:1⟩ R: 96
—·$C_6H_{15}Al$ [Et_3Al] R: 150
—·$C_9H_{21}Al$ [Pr_3Al] R: 150
—·$C_{18}H_{39}Al$ [Hex_3Al] R: 150
—·$CdCl_2$ ⟨2:1⟩ R: 101
—·ClI ⟨1:1⟩ R: 17, 65
—·$ClIO_4$ ⟨2:1⟩ R: 101
—·Cl_2Zn ⟨2:1⟩ R: 101
—·Cl_3Ga R: 178
—·Cl_5Sb R: 152, 154
—·F_5Nb R: 68
—·HNO_3 R: 101
—·I_2 ⟨1:1⟩ R: 27
—·I_2Zn ⟨2:1⟩ R: 101
—·I_4Sn ⟨2:1⟩ R: 151
—·N_2O_6Zn [$Zn(NO_3)_2$] ⟨2:1, 3:1⟩ R: 101

C_5H_5NO [Py.—O]·BrClCu ⟨2:1⟩ R: 66
—·Br_2Cu ⟨1:1⟩ R: 66, 67
—·— ⟨2:1⟩ R: 66
—·Br_2Hg R: 24, 190
—·Cl_2Cu ⟨1:1, 2:1⟩ R: 66, 67
—·Cl_2Cu·DMSO R: 66
—·Cl_2Hg R: 24, 165, 190
—·HgI_2 R: 24

C_5H_9N [*n*-BuCN]·BCl_3 R: 106
—·Cl_2 R: 56, 57

C_5H_{10} [2-Pentanone]·$AlBr_3$ R: 116

$C_5H_{10}O$ [3-Me_2-Butanone]·$AlBr_3$ R: 116

$C_5H_{11}N$ [Piperidine]·$CHCl_3$ R: 10, 117

$C_5H_{13}O_3PS$ [$(Et)_2P(O)(SMe)$]·Cl_4Sn ⟨2:1⟩ R: 3, 163

$C_5H_{15}N_2OP$ {[$(Me)_2N]_2P(O)Me$}·Cl_4Sn ⟨2:1⟩ R: 2, 3

$C_5H_{15}N_2O_2P$ [$(Me_2N)_2P(O)(OMe)$]·Cl_4Sn ⟨2:1⟩ R: 2, 3

C_6ClF_5 [$ClPhF_5$]·C_6H_5Cl R: 125, 186
—·C_6H_6 R: 170
—·C_7H_7N [$PhNH_2$] R: 170
—·$C_6H_{15}N$ [Et_3N] R: 170
—·C_7H_8 [MePh] R: 170, 186
—·C_9H_{12} [1,3,5-Me_3Ph] R: 170

$C_6Cl_3F_3$ [1,3,5-Cl_3-2,4,6-F_3Ph]·C_9H_{12} [1,3,5-Me_3Ph] R: 170
—·$C_{12}H_{18}$ [HMB] R: 170

$C_6Cl_4O_2$ [Chloranil]·$C_6H_8N_2$ [1,4-Ph(NH$_2$)$_2$]   R: 40
  —·$C_8H_{11}N_2$ [PhN(Me)$_2$]   R: 40
  —·C_9H_7NO [8-Hydroxyquinoline]   R: 40
  —·C_9H_{12} [1,3,5-Me$_3$Ph]   R: 84
  —·$C_{10}H_{10}N_2$ [1,5-(NH$_2$)$_2$Naphth.]   R: 40
  —·$C_{10}H_{14}$ [Durene]   R: 40
  —·$C_{12}H_{18}$ [HMB]   R: 40, 55, 104
  —·$C_{16}H_{10}$ [Pyrene]   R: 40
  —·$C_{16}H_{20}N_2$ [Tetramethylbenzidine]   R: 40
  —·$C_{18}H_{12}CuN_2O_2$ [Copper(II)-bis-
      8-hydroxyquinolinate]   R: 40
  —·$C_{20}H_{12}$ [Perylene]   R: 40

$C_6H_2ClN_3O_6$ [Picryl chloride]·C_6H_5Br [PhBr]   R: 121
  —·C_6H_5Cl [PhCl]   R: 121
  —·C_6H_5F [PhF]   R: 121
  —·C_6H_5I [PhI]   R: 121
  —·$C_6H_5NO_2$ [PhNO$_2$]   R: 121
  —·C_6H_6 R: 121
  —·C_6H_6O [PhOH]   R: 121
  —·C_7H_7Br [p-BrPhMe]   R: 121
  —·C_7H_7Cl [p-ClPhMe]   R: 121
  —·C_7H_7F [p-FPhMe]   R: 121
  —·C_7H_7I [p-IPhMe]   R: 121
  —·C_7H_8 [PhMe]   R: 121
  —·C_7H_8O [p-MePhOH]   R: 121
  —·C_8H_{10} [p-Ph(Me)$_2$]   R: 121
  —·C_9H_{12} [1,3,5-Me$_3$Ph]   R: 121
  —·$C_{10}H_8$ [Naphth.]   R: 108, 121
  —·$C_{10}H_{14}$ [Durene]   R: 121
  —·$C_{11}H_{15}$ [Ph(Me)$_5$]   R: 121
  —·$C_{14}H_{10}$ [Anthracene]   R: 121
  —·— [Phenantrene]   R: 121

$C_6H_3BrCl_2O$ [2,6-Cl$_2$-4-BrPhOH]·C_6H_6BrN
    [1,4-BrPhNH$_2$] ⟨2:1⟩   R: 111
  —·C_6H_6ClN [1,4-ClPhNH$_2$] ⟨2:1⟩   R: 111
  —·C_6H_6IN [1,4-IPhNH$_2$] ⟨2:1⟩   R: 111
  —·C_6H_7N [PhNH$_2$] ⟨1:1⟩   R: 111

$C_6H_3Br_2ClO$ [2,6-Br$_2$-4-ClPhOH]
    ·C_6H_6ClN [1,4-ClPhNH$_2$] ⟨1:1⟩   R: 111
  —·C_6H_7N [PhNH$_2$] ⟨1:1⟩   R: 111

$C_6H_3Br_3O$ [2,4,6-Br$_3$PhOH]·C_6H_6ClN [1,4-ClPhNH$_2$] ⟨1:1⟩ R: 111
  —·C_6H_6IN [1,4-IPhNH$_2$] ⟨1:1⟩   R: 111
  —·C_6H_7N [PhNH$_2$] ⟨1:1⟩   R: 111

$C_6H_3ClN_2O_4S$ [2,4-(NO$_2$)$_2$PhSCl]·Cl$_5$Sb   R: 9

$C_6H_3Cl_3O$ [2,4,6-Cl$_3$PhOH]·C_6H_6BrN [1,4-BrPhNH$_2$] ⟨1:1⟩ R: 111
  —·C_6H_6ClN [1,4-ClPhNH$_2$] ⟨2:1⟩   R: 111

—·C$_6$H$_6$IN [1,4-IPhNH$_2$] ⟨1:1⟩   R: 111
—·C$_6$H$_7$N [PhNH$_2$] ⟨1:1⟩   R: 111

C$_6$H$_3$N$_3$O$_7$ [(NO$_2$)$_3$PhOH]·C$_5$H$_4$ClN [2-ClPy.]   R: 6
—·C$_5$H$_4$ClN [3-ClPy.]   R: 6
—·C$_5$H$_4$ClN [4-ClPy.]   R: 6

C$_6$H$_4$BrCl$_2$OP [p-BrPhP(O)Cl$_2$]·Cl$_4$Sn ⟨2:1⟩   R: 196

C$_6$H$_4$ClNO$_2$ [4-ClPhNO$_2$]·Cl$_4$Sn ⟨1:1⟩   R: 112

C$_6$H$_4$Cl$_3$OP [p-ClPhP(O)Cl$_2$]·Cl$_4$Sn ⟨2:1⟩   R: 196

C$_6$H$_4$N$_2$O$_4$ [1,3-(NO$_2$)$_2$Ph]·Cl$_3$Sb   R: 85, 169

C$_6$H$_4$O$_2$ [BZQ]·Cl$_2$Hg   R: 25, 190

C$_6$H$_5$AlCl$_2$ [PhAlCl$_2$] ⟨Dimer⟩   R: 52

C$_6$H$_5$AsCl$_4$ [PhAsCl$_4$]·BCl$_3$ R: 54

C$_6$H$_5$Br [BrPh]·Br$_2$ R: 41, 158
—·C$_6$H$_2$ClN$_3$O$_6$ [Picryl chloride]   R: 121

C$_6$H$_5$BrMg [PhMgBr]·C$_4$H$_8$O [THF] ⟨1:2⟩   R: 113
—·C$_4$H$_{10}$O [Et$_2$O] ⟨2:1⟩   R: 113

C$_6$H$_5$Cl [PhCl]·AsCl$_3$ ⟨1:2⟩   R: 85, 86, 89, 91, 169
—·Br$_2$ R: 41
—·Cl$_3$Sb   R: 85, 86, 89
—·C$_6$ClF$_5$ R: 125, 186
—·C$_6$H$_2$ClN$_3$O$_6$ [Picryl chloride]   R: 121
—·C$_9$H$_{12}$ [1,3,5-Me$_3$Ph] ⟨1:1⟩   R: 84

C$_6$H$_5$Cl$_2$OP [PhP(O)Cl$_2$]·Cl$_4$Sn ⟨2:1⟩   R: 3, 196

C$_6$H$_5$Cl$_4$P [PhPCl$_4$]·BCl$_3$ R: 53
—·Cl$_5$P   R: 53
—·Cl$_5$Sb   R: 53

C$_6$H$_5$F [FPh]·Br$_2$ R: 41
—·C$_6$H$_2$ClN$_3$O$_6$ [Picryl chloride]   R: 121

C$_6$H$_5$I [IPh]·C$_6$H$_2$ClN$_3$O$_6$ [Picryl chloride]   R: 121
—·Cl$_2$ R: 61, 137

C$_6$H$_5$NO$_2$ [PhNO$_2$]·BCl$_3$ R: 7, 175
—·C$_6$H$_2$ClN$_3$O$_6$ [Picryl chloride]   R: 121
—·Cl$_3$Ga   R: 178
—·Cl$_4$Sn ⟨1:1⟩   R: 12, 50, 112, 123

C$_6$H$_6$·Al$_2$Br$_6$ R: 141
—·AsCl$_3$ ⟨1:2⟩   R: 5, 11, 90, 92, 142,
                                                                           195
—·Br$_2$ R: 41, 99, 105
—·Br$_3$Sb ⟨1:2⟩   R: 83, 85, 86, 87, 89,
                                                                           134, 143
—·CBr$_4$ R: 70

—·CCl$_4$ ⟨1:1, 1:2⟩	R: 84, 184
—·— ⟨1:3⟩	R: 84
—·CHCl$_3$	R: 10
—·C$_5$ClF$_5$	R: 170
—·C$_6$H$_2$ClN$_3$O$_6$ [Picryl chloride]	R: 121
—·Cl$_2$	R: 71, 82
—·Cl$_3$Sb ⟨1:1⟩	R: 171
—·— ⟨1:2⟩	R: 11, 12, 80, 81, 85, 86, 87, 88, 89, 90, 142, 171
C$_6$H$_6$BrN [4-BrPhNH$_2$]·C$_6$H$_3$BrCl$_2$O	
[2,6-Cl$_2$-4-BrPhOH] ⟨1:2⟩	R: 111
—·C$_6$H$_3$Cl$_3$O [2,4,6-Cl$_3$PhOH] ⟨1:2⟩	R: 111
—·HCl	R: 125
C$_6$H$_6$ClN [4-ClPhNH$_2$]·C$_6$H$_3$BrCl$_2$O	
[2,6-Cl$_2$-4-BrPhOH] ⟨1:2⟩	R: 111
—·C$_6$H$_3$Br$_2$ClO [2,6-Br$_2$-4-ClPhOH] ⟨1:1⟩	R: 111
—·C$_6$H$_3$Br$_3$O [2,4,6-Br$_3$PhOH] ⟨1:1⟩	R: 111
—·C$_6$H$_3$Cl$_3$O [2,4,6-Cl$_3$PhOH] ⟨1:1⟩	R: 111
—·HBr	R: 125
—·HCl	R: 125
C$_6$H$_6$IN [4-IPhNH$_2$]·C$_6$H$_3$BrCl$_2$O	
[2,6-Cl$_2$-4-BrPhOH] ⟨1:2⟩	R: 111
—·C$_6$H$_3$Br$_3$O [2,4,6-Br$_3$PhOH] ⟨1:1⟩	R: 111
—·C$_6$H$_3$Cl$_3$O [2,4,6-Cl$_3$PhOH] ⟨1:1⟩	R: 111
—·HCl	R: 125
C$_6$H$_6$O [PhOH]·C$_6$H$_{12}$N$_4$ [HMT]	R: 127, 128
—·C$_6$H$_2$ClN$_3$O$_6$ [Picryl chloride]	R: 121
—·Cl$_3$Sb ⟨1:2⟩	R: 81, 85, 86, 87, 89, 169
C$_6$H$_6$O$_2$ [2-HOPhOH]·C$_6$H$_{12}$N$_4$ [HMT]	R: 127
—·[3-HOPhOH]·C$_6$H$_{12}$N$_4$ [HMT]	R: 127
C$_6$H$_7$N [4-MePy.]·AlBr$_3$	R: 146
—·BCl$_3$	R: 7
—·BrI	R: 17
—·CHCl$_3$	R: 117
—·ClI ⟨1:1⟩	R: 17, 65
—·I$_2$	R: 27
C$_6$H$_7$N [PhNH$_2$]·Br$_3$Sb	R: 73
—·C$_6$Cl$_5$F	R: 170
– ·C$_6$H$_3$BrCl$_2$O [2,6-Cl$_2$-4-BrPhOH] ⟨1:1⟩	R: 111
—·C$_6$H$_3$Br$_2$ClO [2,6-Br$_2$-4-ClPhOH] ⟨1:1⟩	R: 111
—·C$_6$H$_3$Br$_3$O [2,4,6-Br$_3$PhOH] ⟨1:1⟩	R: 111
—·C$_6$H$_3$Cl$_3$O [2,4,6-Cl$_3$PhOH] ⟨1:1⟩	R: 111
—·Cl$_3$Sb ⟨1:1⟩	R: 22, 73
—·— ⟨2:1⟩	R: 73
—·— ⟨3:1⟩	R: 73

C_6H_7NO [4-MePy.—O]·Br_2Cu $\langle 1:1 \rangle$ R: 66
  —·Br_2Hg R: 24, 190
  —·Cl_2Cu $\langle 1:1 \rangle$ R: 66
  —·Cl_2Hg R: 24, 190
  —·HgI_2 R: 24

$C_6H_8N_2$ [$NC(CH_2)_4CN$]·$CHCl_3$ $\langle 1:2 \rangle$ R: 57
  —·Cl_2 R: 56, 57

$C_6H_8N_2$ [1,4-$(NH_2)_2Ph$]·$C_6Cl_4O_2$ [Chloranil]   R: 40

$C_6H_8N_3OP$ [$(Me_2N)_3PO$].Cl_5Sb R: 152

$C_6H_8O_2$ [1,4-CHDN]$HgCl_2$ R: 64, 148, 190

C_6H_9NO [N-Vinylpyrrolidone]·$C_2HCl_3O_2$
  [CCl_3COOH] $\langle n:n \rangle$ R: 146

$C_6H_{10}O$ [Cyclohexanone]·$AlBr_3$ R: 116

$C_6H_{10}O_2S$ [$C_4H_9SO_2C{\equiv}CH$]·$AlBr_3$ $\langle 1:2 \rangle$ R: 95

$C_6H_{10}O_4$ [$(COOC_2H_5)_2$]·$CHCl_3$ $\langle 1:1 \rangle$ R: 4

$C_6H_{11}N$ [$Me(CH_2)_4CN$]·$CHCl_3$ R: 57
  —·Cl_2 R: 56, 57

$C_6H_{12}N_2$ [TED]·$CHCl_3$ R: 117

$C_6H_{12}N_4$ [HMT]·BrI   R: 17
  —·C_6H_6O [PhOH] $\langle 1:3 \rangle$ R: 127, 128
  —·$C_6H_6O_2$ [2-HOPhOH]   R: 127
  —·— [3-HOPhOH]   R: 127
  —·C_7H_8O [3-MePhOH] $\langle 1:3 \rangle$ R: 127
  —·D_2O $\langle 1:6 \rangle$ R: 127
  —·ClI   R: 17
  —·H_2O R: 127
  —·$4H_2O$·BrLi   R: 79
  —·$4H_2O$·ClLi   R: 79
  —·$4H_2O$·ILi   R: 79
  —·$8H_2O$·$CaCl_2O_8$ [$Ca(ClO_4)_2$]   R: 8
  —·I_2 R: 27

$C_6H_{12}O$ [3,3-Me_2-butanone]·$AlBr_3$ R: 116

$C_6H_{14}AlCl$ [$(n$-Pr$)_2AlCl$] $\langle$Dimer$\rangle$ R: 52

$C_6H_{14}O$ [$(iso$-Pr$)_2O$]·C_2H_5BrMg R: 113

$C_6H_{14}O_2$ [$MeO(CH_2)_4OMe$]·Cl_4Sn $\langle 1:1 \rangle$ R: 112, 123

$C_6H_{14}O_3$ [Digly.]·Br_2Hg R: 190
  —·C_5BrCl_5Hg [C_5Cl_5HgBr]   R: 188, 189
  —·C_5Cl_6Hg [C_5Cl_5HgCl]   R: 188, 189, 192
  —·Cl_2Hg R: 190
  —·HgI_2 R: 190

$C_6H_{15}Al$ [Et_3Al] $\langle$Dimer$\rangle$ R: 52
  —·C_5H_5N [Py.]   R: 150

$C_6H_{15}AlO$ [Et$_3$AlO] ⟨Dimer⟩   R: 52

$C_6H_{15}As$ [Et$_3$As]·AlCl$_3$ R: 36
— ·BCl$_3$ R: 36
— ·BH$_3$ R: 36
— ·C$_6$H$_{15}$Ga [Et$_3$Ga]   R: 107
— ·Cl$_3$Ga   R: 35

$C_6H_{15}Cl_2P$ [Et$_3$PCl$_2$]·BCl$_3$ R: 53

$C_6H_{15}Ga$ [Et$_3$Ga]·C$_4$H$_{10}$O [Et$_2$O]   R: 107
— ·C$_6$H$_{15}$As [Et$_3$As]   R: 107

$C_6H_{15}N$ [Et$_3$N]·AlCl$_3$ R: Junp
— ·BCl$_3$ R: 106, Junp
— ·CHCl$_3$ ⟨1:1⟩   R: 10, 114, 117
— ·C$_6$ClF$_5$ R: 170
— ·HCl ⟨1:2⟩   R: 44

$C_6H_{15}OPS$ [Et$_2$P(O)SEt]·Cl$_4$Sn ⟨2:1⟩   R: 3, 163

$C_6H_{15}P$ [Et$_3$P]·AlCl$_3$ R: Junp
— ·BCl$_3$ R: Junp
— ·Cl$_3$Ga   R: 35

$C_6H_{15}PS_2$ [Et$_2$P(S)SEt]·Cl$_4$Sn ⟨2:1⟩   R: 3, 163

$C_6H_{16}IN$ [(n-Pr)$_2$NH$_2$I]·I$_2$ R: 98, 133

$C_6H_{16}N_2$ [Me$_2$N(CH$_2$)$_2$NMe$_2$]·Br$_2$Hg   R: 30
— ·C$_3$H$_9$BrSn [Me$_3$SnBr]   R: 151
— ·Cl$_2$Hg   R: 30
— ·HgI$_2$ R: 30

$C_6H_{17}N_2OP$ [(Me$_2$N)$_2$P(O)Et]·Cl$_4$Sn ⟨2:1⟩   R: 3

$C_6H_{17}N_2PS$ [(Me$_2$N)$_2$P(S)Et]·Cl$_4$Sn ⟨2:1⟩   R: 3

$C_6H_{18}N_3OP$ [(Me$_2$N)$_3$P(O)]·Br$_4$Sn ⟨2:1⟩   R: 151
— ·C$_2$H$_6$Br$_2$Sn [Me$_2$SnBr$_2$] ⟨2:1⟩   R: 151
— ·C$_3$H$_9$BrSn [Me$_3$SnBr] ⟨1:1⟩   R: 151
— ·Cl$_4$Sn ⟨2:1⟩   R: 2, 3
— ·Cl$_5$Sb   R: 152
— ·I$_4$Sn ⟨2:1⟩   R: 151
$C_6H_{18}N_3P$ [(Me$_2$N)$_3$P]·Cl$_4$Sn ⟨2:1⟩   R: 2, 3

C_6N_4 [TCNE]·C$_{10}$H$_8$ [Naphth.]   R: 145
$C_7H_3Cl_2N$ [2,6-Cl$_2$PhCN]·BCl$_3$ R: 6

$C_7H_3Cl_3O$ [3,4-Cl$_2$BzCl]·AlCl$_3$ R: 48

C_7H_4ClN [2-ClPhCN]·BCl$_3$ R: 6
— [3-ClPhCN]·BCl$_3$ R: 6
— [4-ClPhCN]·BCl$_3$ R: 6

$C_7H_4Cl_2O$ [2-ClBZCl]·AlCl$_3$ R: 48
— [4-ClBZCl]·AlCl$_3$ R: 48

C_7H_5ClO [PhCOCl]·AlCl$_3$ R: 48
— ·Cl$_4$Sn $\langle 1:1 \rangle$ R: 126
— ·Cl$_5$Sb   R: 6, 28

C_7H_5ClO [2-ClBZA]·Cl$_2$Hg $\langle 1:1 \rangle$ R: 64
— [3-ClBZA]·Cl$_2$Hg $\langle 2:1 \rangle$ R: 64
— [4-ClBZA]·Cl$_2$Hg $\langle 3:1 \rangle$ R: 64

C_7H_5N [PhCN]·CHCl$_3$ $\langle 1:1, 1:2 \rangle$ R: 57, 79
— ·Cl$_2$ $\langle 1:1, 2:1 \rangle$ R: 56, 57
— ·Cl$_3$Ga   R: 178
— ·Cl$_5$Sb   R: 152, 154

C_7H_6O [BZA]·CHCl$_3$ R: 79
— ·Cl$_2$Hg   R: 64

C_7H_7Br [p-BrPhMe]·C$_6$H$_2$ClN$_3$O$_6$ [Picryl chloride]   R: 121

C_7H_7Cl [PhCH$_2$Cl]·Cl$_4$Sn $\langle 2:1 \rangle$ R: 123, 126
— ·C$_6$H$_2$ClN$_3$O$_6$ [Picryl chloride]   R: 121

$C_7H_7Cl_2OP$ [p-MePhP(O)Cl$_2$]·Cl$_4$Sn $\langle 2:1 \rangle$ R: 196

$C_7H_7Cl_2O_2P$ [p-MeOPhP(O)Cl$_2$]·Cl$_4$Sn $\langle 2:1 \rangle$ R: 196

C_7H_7F [PhCH$_2$F]·C$_6$H$_2$ClN$_3$O$_6$ [Picryl chloride]   R: 121

C_7H_7I [PhCH$_2$I]·C$_6$H$_2$ClN$_3$O$_6$ [Picryl chloride]   R: 121

C_7H_7N [4-Vinylpyridine]·AlBr$_3$ $\langle n:1 \rangle$ R: 146

C_7H_8 [MePh]·AsCl$_3$ $\langle 1:2 \rangle$ R: 85, 90, 92, 195
— ·Br$_3$Sb   R: 86, 87, 89
— ·C$_5$H$_5$Tl $\langle 2:1 \rangle$ R: 191
— ·C$_6$ClF$_5$ R: 170
— ·C$_6$H$_2$ClN$_3$O$_6$ [Picryl chloride]   R: 121
— ·Cl$_3$Sb $\langle 1:1, 1:2 \rangle$ R: 85, 86, 87, 89, 90

C_7H_8O [3-MePhOH]·C$_6$H$_{12}$N$_4$ [HMT]   R: 127
— [4-MePhOH]·C$_6$H$_2$ClN$_3$O$_6$ [Picryl chloride]   R: 121

C_7H_8O [PhOMe]·BCl$_3$ R: 175
— ·BiCl$_3$ R: 124
— ·Br$_3$Sb   R: 81, 83, 85, 86, 87,
                                                   89, 115
— ·Cl$_3$Ga   R: 178
— ·Cl$_3$Sb $\langle 1:1, 1:2 \rangle$ R: 77, 81, 85, 86, 87,
                                                   89, 115
— ·— $\langle 1:1 \rangle$ R: 87
— ·— $\langle 1:2 \rangle$ R: 77

$C_7H_8O_3S$ [4-MePhSO$_3$]·C$_5$H$_4$ClN [2-ClPy.]   R: 6
— ·C$_5$H$_4$ClN [3-ClPy.]   R: 6

C_7H_8S [PhSMe]·BCl$_3$ R: 7

C_7H_9N [3,5-Me$_2$Py.]·CHCl$_3$ R: 117
— ·[2,6-Me$_2$Py.]·CBr$_4$ R: 70

C_7H_9NO [2,6-Me_2Py.—O].Cl_2Cu R: 66

$C_7H_{10}O_2$ [DMP]·Cl_2Hg R: 25, 190

$C_7H_{13}NO$ [N-(*iso*-Pr)pyrrolidone]·$C_2HCl_3O_2$ [CCl_3COOH]   R: 146

$C_7H_{17}OPS$ [(*iso*-Pr)$_2$P(O)SMe]·Cl_4Sn ⟨2:1⟩   R: 3, 163

$C_7H_{17}O_3PS$ [(*iso*-C_3H_7O)$_2$P(O)SMe]·Cl_4Sn ⟨2:1⟩   R: 3, 163

C_8H_6 [PhC≡CH]·Cl_3Sb ⟨$n:n$⟩   R: 171

C_8H_7ClO [4-ClPhCOMe]·Cl_5Sb R: 6

C_8H_7ClO [4-MePhCOCl]·$AlCl_3$ R: 48

C_8H_8 [Styrene]·Cl_3Sb ⟨$n:n$⟩   R: 148, 171

C_8H_8O [ACP]·$AlBr_3$ ⟨1:1⟩   R: 116
  —·— ⟨1:2⟩   R: 181
  —·Br_2Hg R: 24, 190
  —·$CHCl_3$ ⟨1:1⟩   R: 4, 79
  —·— ⟨1:2⟩   R: 79
  —·Cl_2Hg R: 24, 25, 190
  —·Cl_3Sb R: 81, 85, 86, 89, 169

C_8H_{10} [1,2-Me_2Ph]·$AlBr_3$ ⟨1:2⟩   R: 181
  —·$AsCl_3$ ⟨1:1⟩   R: 90, 91
  —· ⟨1:2⟩   R: 5, 90, 91
  —·Br_3Sb ⟨1:1⟩   R: 81, 85, 86, 87, 89, 90
  —·CBr_4 R: 70
  —·$CHCl_3$ R: 10, 84
  —·Cl_3Sb ⟨1:1⟩   R: 87, 90, 169
  —·— ⟨1:2⟩   R: 81, 85, 86, 87, 89, 90, 169

C_8H_{10} [1,3-Me_2Ph]·$AsCl_3$ ⟨1:1, 1:2⟩   R: 5, 90, 91
  —·Br_3Sb ⟨1:1⟩   R: 85, 86, 87, 89, 169
  —·$CHCl_3$ R: 10, 84
  —·Cl_3Sb ⟨1:1⟩   R: 87, 90
  —·— ⟨1:2⟩   R: 87

C_8H_{10} [1,4-Me_2Ph]·$AlBr_3$ ⟨1:1⟩   R: 181
  —·$AsBr_3$ ⟨1:2⟩   R: 90
  —·$AsCl_3$ ⟨1:2⟩   R: 5, 11, 90, 92, 195
  —·Br_2 R: 99
  —·Br_3Sb ⟨1:1, 1:2⟩   R: 81, 85, 89
  —·CBr_4 R: 70, 99
  —·CCl_4 R: 99
  —·$CHCl_3$ R: 10, 79
  —·$C_6H_2ClN_3O_6$ [Picryl chloride]   R: 121
  —·Cl_3Sb ⟨1:1⟩   R: 87, 169
  —·— ⟨1:2⟩   R: 11, 81, 85, 86, 87, 89, 90, 169

C_8H_{10} [EtPh]·AsBr$_3$ ⟨1:1, 1:2⟩   R: 72, 90, 93
—·— ⟨1:1⟩   R: 90
—·AsCl$_3$ ⟨1:1⟩   R: 5, 11, 90, 91, 195
—·— ⟨1:2⟩   R: 5, 11, 90, 91
—·Br$_3$Sb ⟨1:1⟩   R: 81, 85, 86, 87, 89,
    134, 169
—·Cl$_3$Sb ⟨1:1⟩   R: 11, 80, 81, 85, 86,
    87, 88, 89, 90, 140,
    169, 171
—·— ⟨1:2⟩   R: 11, 85, 86, 87, 89,
    90, 169

$C_8H_{10}O$ [EtOPh]·AlBr$_3$ R: 122
—·Br$_3$Sb ⟨1:1⟩   R: 85, 86, 87, 89, 169
—·Cl$_3$Sb ⟨1:1⟩   R: 80, 87

$C_8H_{11}Al$ [PhAlMe$_2$] ⟨Dimer⟩   R: 52

$C_8H_{11}N$ [2,4,6-Me$_3$Py.]·CBr$_4$ R: 70
—·CCl$_4$ R: 184
—·Cl$_2$Hg   R: 64

$C_8H_{11}N$ [PhN(Me)$_2$]·C$_6$Cl$_4$O$_2$ [Chloranil]   R: 40

$C_8H_{15}B_{10}BrHg$ [PhC(B$_{10}$H$_{10}$)CBrHg]·C$_{10}$H$_8$N$_2$ [biPy.]   R: 136

$C_8H_{18}AlCl$ [(iso-Bu)$_2$AlCl] ⟨Dimer⟩   R: 52

$C_8H_{18}O_2S$ [(iso-C$_4$H$_9$)$_2$SO$_2$]·AlBr$_3$ ⟨1:1⟩   R: 95

$C_8H_{18}O_4$ [Trigly.]·Br$_2$Hg   R: 190
—·C$_5$BrCl$_5$Hg [C$_5$Cl$_5$HgBr]   R: 188, 189
—·C$_5$Cl$_6$Hg [C$_5$Cl$_5$HgCl]   R: 188, 189
—·Cl$_2$Hg   R: 190
—·HgI$_2$ R: 190

$C_8H_{20}ClN$ [Et$_4$NCl]·AuCl ⟨1:1⟩   R: 19
—·Cl$_2$ R: 159
—·HCl   R: 60

$C_8H_{20}IN$ [Et$_4$NI]·Br$_2$ R: 18
—·I$_2$ R: 18, 98, 133

C_9H_7N [Quinoline]·Cl$_5$Sb   R: 152, 154

C_9H_7NO [8-Hydroxyquinoline]·C$_6$Cl$_4$O$_2$ [Chloranil]   R: 40

$C_9H_{11}Al$ [Me$_2$AlC≡CPh] ⟨Dimer⟩   R: 52

C_9H_{12} [1,2,4-Me$_3$Ph]·CCl$_4$ R: 184
—·CHCl$_3$ R: 79

C_9H_{12} [1,3,5-Me$_3$Ph]·AlBr$_3$ ⟨1:1⟩   R: 181
—·AsBr$_3$ ⟨1:1⟩   R: 72, 195
—·AsCl$_3$ ⟨1:1⟩   R: 5, 90, 91
—·— ⟨1:2⟩   R: 5, 91
—·Br$_3$Sb ⟨1:2⟩   R: 85, 86, 87, 89

—·CBr$_4$	R: 70
—·CCl$_4$	R: 184
—·CHCl$_3$ $\langle 1:1 \rangle$	R: 10, 84, 114
—·CDCl$_3$	R: 156
—·C$_6$ClF$_5$	R: 170
—·C$_6$Cl$_3$F$_3$	R: 170
—·C$_6$Cl$_4$O$_2$ $\langle 1:1 \rangle$	R: 84
—·C$_6$H$_2$ClN$_3$O$_6$ [Picryl chloride]	R: 121
—·C$_6$H$_5$Cl $\langle 1:1 \rangle$	R: 84
—·Cl$_3$Sb	R: 85, 87

C$_9$H$_{12}$ [Cumene]·Cl$_3$Sb	R: 146, 171
C$_9$H$_{12}$O$_2$S [PhSO$_2$Pr]·AlBr$_3$ $\langle 1:2 \rangle$	R: 95
C$_9$H$_{13}$O$_2$P [Ph(iso-Pr)P(O)SMe]·Cl$_4$Sn	R: 3, 163
C$_9$H$_{18}$NS$_2$ [bdtc]·AuCl $\langle 2:2 \rangle$	R: 19
C$_9$H$_{21}$Al [Pr$_3$Al]·C$_5$H$_5$N [Py.]	R: 150
C$_9$H$_{21}$O$_3$PS (cis-[(n-BuO)$_2$P(O)SMe])·Cl$_4$Sn $\langle 2:1 \rangle$	R: 3, 163

C$_{10}$Cl$_{10}$Hg [C$_5$Cl$_5$HgC$_5$Cl$_5$]·C$_4$H$_8$O [THF]	R: 188, 189
—·C$_4$H$_{10}$O$_2$ [Gly.] $\langle 3:5 \rangle$	R: 188, 189

C$_{10}$H$_8$ [Naphth] AsBr$_3$	R: 72, 195
—·AsCl$_3$ $\langle 1:2 \rangle$	R: 5, 11, 90, 92, 195
—·Br$_3$Sb $\langle 1:2 \rangle$	R: 81, 85, 86, 87, 89, 169
—·C$_6$N$_4$ [TCNE]	R: 145
—·C$_6$H$_2$ClN$_3$O$_6$ [Picryl chloride]	R: 108, 121
—·Cl$_3$Sb $\langle 1:1 \rangle$	R: 73, 171
—·Cl$_3$Sb $\langle 1:2 \rangle$	R: 1, 11, 12, 77, 80, 81, 85, 86, 87, 89, 90, 139, 169, 171

C$_{10}$H$_8$N$_2$ [biPy.]·C$_3$H$_{13}$B$_{10}$BrHg [MeC(B$_{10}$H$_{10}$)CHgBr]	R: 136
—·C$_8$H$_{15}$B$_{10}$BrHg [PhC(B$_{10}$H$_{10}$)CHgBr]	R: 136
—·Cl$_3$Ga	R: 102

C$_{10}$H$_{10}$N$_2$ [Diaminonaphthalene]·C$_6$Cl$_4$O$_2$ [Chloranil]	R: 40

C$_{10}$H$_{14}$ [durene]·AlBr$_3$ $\langle 1:1 \rangle$	R: 181
—·CBr$_4$	R: 70
—·CCl$_4$	R: 184
—·CHCl$_3$	R: 10
—·C$_6$Cl$_4$O$_2$ [Chloranil]	R: 40
—·C$_6$H$_2$ClN$_3$O$_6$ [Picryl chloride]	R: 121

C$_{10}$H$_{17}$N$_2$OP [(Me$_2$N)$_2$P(O)Ph]·Cl$_4$Sn $\langle 2:1 \rangle$	R: 2, 3
C$_{10}$H$_{17}$N$_2$PS [(Me$_2$N)$_2$P(S)Ph]·Cl$_4$Sn $\langle 2:1 \rangle$	R: 2, 3

C$_{10}$H$_{22}$O$_5$ [Tetragly.]·Br$_2$Hg	R: 190
—·Cl$_2$Hg	R: 190
—·HgI$_2$	R: 190

$C_{11}H_{15}$ [Me$_5$Ph]·C$_6$H$_2$ClN$_3$O$_6$ [Picryl chloride]   R: 121

$C_{12}H_8N_2$ [Phenazine]·I$_2$ R: 27

$C_{12}H_{10}$ [PhPh]·AsCl$_3$ ⟨1:1⟩   R: 91
—·— ⟨1:2⟩   R: 5, 90, 91, 195
—·Br$_3$Sb ⟨1:2⟩   R: 81, 85, 86, 87, 89, 169
—·Cl$_3$Sb ⟨1:2⟩   R: 12, 85, 86, 87, 88, 89, 90, 169

$C_{12}H_{10}AsCl_3$ [Ph$_2$AsCl$_3$]·BCl$_3$ R: 54

$C_{12}H_{10}Cl_3P$ [Ph$_2$PCl$_3$]·BCl$_3$ R: 53
—·Cl$_5$P   R: 53
—·Cl$_5$Sb   R: 53

$C_{12}H_{10}O$ [PhOPh]·AlBr$_3$ R: 121
—·Cl$_3$Ga   R: 178

$C_{12}H_{10}O_2S$ [Ph$_2$SO$_2$]·AlBr$_3$ ⟨1:1, 1:2⟩   R: 95

$C_{12}H_{18}$ [HMB]·C$_6$Cl$_3$F$_3$ [1,3,5-Cl$_3$-2,4,6-F$_3$Ph]   R: 170
—·C$_6$Cl$_4$O$_2$ [Chloranil]   R: 40, 55, 104

$C_{12}H_{18}IN$ [(n-Pr)$_4$NI]·I$_2$ R: 98, 133

$C_{12}H_{24}O_6$ [18-Crown-6-ether]·Br$_2$Hg   R: 190
—·Cl$_2$Hg   R: 190
—·HgI$_2$ R: 190

$C_{12}H_{26}O_3$ [Diglybu.]·Br$_2$Hg   R: 190
—·Cl$_2$Hg   R: 190
—·HgI$_2$ R: 190

$C_{12}H_{27}Al$ [(iso-Bu)$_3$Al] ⟨Dimer⟩   R: 52

$C_{12}H_{27}OP$ [(n-Hex)$_2$P(O)H]·Cl$_4$Sn ⟨2:1⟩   R: 3, 163

$C_{13}H_8Cl_2O$ [4,4'-Cl$_2$BZP]·AlBr$_3$ R: 48

$C_{13}H_9BrO$ [4-BrBZP]·AlBr$_3$ R: 48

$C_{13}H_9ClO$ [4-ClBZP]·AlBr$_3$ R: 48
—·Cl$_3$Ga   R: 48

$C_{13}H_9N$ [Acridine]·I$_2$ ⟨1:1⟩   R: 27

$C_{13}H_{10}O$ [BZP]·AlBr$_3$ ⟨1:1⟩   R: 48, 116
—·— ⟨1:2⟩   R: 181
—·AlCl$_3$ ⟨1:1⟩   R: 48
—·BCl$_3$ R: 7
—·Br$_2$Hg   R: 24, 190
—·Cl$_2$Hg   R: 24, 25, 190
—·Cl$_3$Ga ⟨1:1⟩   R: 48
—·Cl$_3$Sb ⟨1:1, 1:2⟩   R: 171

$C_{13}H_{12}$ [PhCH$_2$Ph]·AsBr$_3$ ⟨1:2⟩   R: 72, 90, 195
—·AsCl$_3$ ⟨1:2⟩   R: 5, 11, 90, 92, 195

—·Br$_3$Sb ⟨1:2⟩ R: 81, 83, 85, 90, 169

—·Cl$_3$Sb ⟨1:2⟩ R: 11, 77, 80, 83, 90

C$_{13}$H$_{13}$OP [Ph$_2$P(O)Me]·Cl$_4$Sn ⟨2:1⟩ R: 3

C$_{14}$H$_{10}$ [Anthracene]·C$_6$H$_2$ClN$_3$O$_6$ [Picryl chloride] R: 121
—·Cl$_3$Sb ⟨1:1, 1:2, 1:3⟩ R: 171

C$_{14}$H$_{10}$ [Phenanthrene]·Br$_3$Sb ⟨1:2⟩ R: 134
—·C$_6$H$_2$ClN$_3$O$_6$ [Picryl chloride] R: 121
—·Cl$_3$Sb ⟨1:1, 1:2⟩ R: 171

C$_{14}$H$_{12}$O [4-MeBZP]·AlBr$_3$ R: 48

C$_{14}$H$_{12}$O$_2$ [4-MeOBZP]·AlBr$_3$ R: 48
—·AlCl$_3$ R: 48
—·Cl$_3$Ga R: 48

C$_{14}$H$_{14}$O$_2$S [(p-MePh)$_2$SO$_2$]·AlBr$_3$ ⟨1:1, 1:2⟩ R: 95

C$_{14}$H$_{16}$NOP [Me$_2$NP(O)Ph$_2$]·Cl$_4$Sn ⟨2:1⟩ R: 2, 3

C$_{14}$H$_{30}$O$_7$ [Hexagly.]·Br$_2$Hg R: 190
—·Cl$_2$Hg R: 190
—·HgI$_2$ R: 190

C$_{15}$H$_{14}$O [4,4'-Me$_2$BZP]·AlBr$_3$ R: 116

C$_{15}$H$_{14}$O$_3$ [4,4'-(MeO)$_2$BZP]·AlBr$_3$ R: 48

C$_{16}$H$_{10}$ [Pyrene]·C$_6$Cl$_4$O$_2$ [Chloranil] R: 40
—·Cl$_3$Sb R: 171

C$_{16}$H$_{20}$N$_2$ [Tetramethylbenzidine]·C$_6$Cl$_4$O$_2$ [Chloranil] R: 40

C$_{16}$H$_{36}$BrN [(n-Bu)$_4$NBr]·BrCu R: 16

C$_{16}$H$_{36}$ClN [(n-Bu)$_4$NCl]·AuCl ⟨1:1⟩ R: 19
—·ClCu R: 16

C$_{16}$H$_{36}$IN [(n-Bu)$_4$NI]·I$_2$ R: 18, 98, 133

C$_{18}$H$_{12}$ [Chrysene]·Cl$_3$Sb ⟨1:4⟩ R: 171

C$_{18}$H$_{12}$CuN$_2$O$_2$ [Copper(ii)-bis-8-hydroxyquino-
        linate]·C$_6$Cl$_4$O$_2$ [Chloranil] R: 40

C$_{18}$H$_{15}$ [p-Terphenyl]·Cl$_3$Sb ⟨1:1, 1:2, 1:3⟩ R: 171

C$_{18}$H$_{15}$Al [AlPh$_3$]·BCl$_3$ R: 7, 102

C$_{18}$H$_{15}$AsCl$_2$ [Ph$_3$AsCl$_2$]·BCl$_3$ R: 54

C$_{18}$H$_{15}$AsO [Ph$_3$AsO]·Cl$_2$Hg ⟨2:1⟩ R: 165

C$_{18}$H$_{15}$Cl$_2$P [Ph$_3$PCl$_2$]·AlCl$_3$ R: 53
—·BCl$_3$ R: 53
—·CCl$_4$ ⟨2:1⟩ R: 53
—·Cl$_5$P R: 53
—·Cl$_5$Sb R: 53

$C_{18}H_{15}P$ [Ph$_3$P]·BCl$_3$	R: 7, 102
$C_{18}H_{39}Al$ [Hex$_3$Al]·C$_5$H$_5$N [Py.]	R: 150
$C_{19}H_{18}IP$ [Ph$_3$P(Me)I]·CuI $\langle 1:2 \rangle$	R: 15
$C_{20}H_{12}$ [Perylene]·C$_6$Cl$_4$O$_2$ (Chloranil)	R: 40
$C_{20}H_{16}O$ [2,4'-Ph$_2$ACP]·AlBr$_3$	R: 116
$C_{24}H_{20}AsCl$ [Ph$_4$AsCl]·ClCu	R: 16
—·Cl$_2$	R: 159
$C_{32}H_{16}MgN_4$ [Magnesium phthalocyanine]	R: 119
$C_{55}H_{72}MgN_4O_5$ [Chlorophyll-a]	R: 119
$CaCl_2O_8$ [Ca(ClO$_4$)$_2$]·8H$_2$O·2C$_6$H$_{12}$N$_4$ [HMT]	R: 8
$CdCl_2$·C$_5$H$_5$N [Py.] $\langle 1:2 \rangle$	R: 101
$ClCu$·C$_3$H$_9$AsS [M$_3$AsS] $\langle n:n \rangle$	R: 20
—·C$_3$H$_9$PS [Me$_3$PS] $\langle 3:3 \rangle$	R: 20
—·C$_{16}$H$_{36}$ClN [Bu$_4$NCl] $\langle 1:1 \rangle$	R: 16
—·C$_{24}$H$_{20}$AsCl [Ph$_4$AsCl] $\langle 1:1 \rangle$	R: 16
$ClCuO_4$ [CuClO$_4$]·C$_3$H$_9$AsS [Me$_3$AsS] $\langle 1:3 \rangle$	R: 20
—·C$_3$H$_9$PS [Me$_3$PS] $\langle 1:3 \rangle$	R: 20
ClD·C$_4$H$_{12}$ClN [Me$_4$NCl]	R: 97
ClI [ICl]·AlCl$_3$ $\langle 2:1 \rangle$	R: 130
—·I$_2$·AlCl$_3$ $\langle 1:1:1 \rangle$	R: 130
—·CH$_3$Cl$_4$P [MePCl$_4$]	R: 53
—·C$_2$H$_6$AsCl$_3$ [Me$_2$AsCl$_3$]	R: 54
—·C$_2$H$_6$Cl$_3$P [Me$_2$PCl$_3$]	R: 3, 53
—·C$_3$H$_9$Cl$_2$P [Me$_3$PCl$_2$]	R: 53
—·C$_5$H$_3$Br$_2$N [3,5-Br$_2$Py.]	R: 17
—·C$_5$H$_3$Cl$_2$N [3,5-Cl$_2$Py.]	R: 14
—·C$_5$H$_4$BrN [2-BrPy.]	R: 17
—·C$_5$H$_4$ClN [2-ClPy.]	R: 17
—·— [4-ClPy.]	R: 17
—·C$_5$H$_5$N [Py.]	R: 17, 65
—·C$_6$H$_7$N [4-MePy.]	R: 17, 65
—·C$_6$H$_{12}$N$_4$ [HMT]	R: 17
—·Cl$_2$ $\langle 2:2 \rangle$	R: 42, 59, 131
$ClIO_4$·C$_5$H$_5$N [Py.] $\langle 1:2 \rangle$	R: 101
ClK·F$_3$Sb $\langle 1:1 \rangle$	R: 29
$ClLi$·4H$_2$O·C$_6$H$_{12}$N$_4$ [HMT]	R: 79
Cl_2·C$_2$H$_3$N [MeCN] $\langle 1:1 \rangle$	R: 56, 57
—·— $\langle 1:2 \rangle$	R: 135
—·C$_3$H$_4$BrN [CH$_2$BrCH$_2$CN] $\langle 1:1 \rangle$	R: 57
—·— $\langle 1:2 \rangle$	R: 56, 57
—·C$_3$H$_5$N [CH$_3$CH$_2$CN]	R: 56, 57
—·C$_4$H$_8$O$_2$ [Diox.]	R: 71

—·C$_4$H$_{12}$BrN [Me$_4$NBr]	R: 159
—·C$_4$H$_{12}$ClN [Me$_4$NCl]	R: 159
—·C$_4$H$_{12}$IN [Me$_4$NI]	R: 159
—·C$_5$H$_9$N [Me(CH$_2$)$_3$CN]	R: 56, 57
—·C$_6$H$_5$I [PhI]	R: 61, 137
—·C$_6$H$_6$	R: 71, 82
—·C$_6$H$_8$N$_2$ [NC(CH$_2$)$_4$CN]	R: 56, 57
—·C$_6$H$_{11}$N [Me(CH$_2$)$_4$CN]	R: 56, 57
—·C$_7$H$_5$N [PhCN] ⟨1:1, 1:2⟩	R: 56, 57
—·C$_8$H$_{20}$ClN [Et$_4$NCl]	R: 159
—·C$_{24}$H$_{20}$AsCl [Ph$_4$AsCl]	R: 159
—·ClI	R: 42. 59, 131
—·CsI ⟨1:1, 2:1⟩	R: 42, 194
—·IK ⟨1:1, 2:1⟩ and H$_2$O	R: 42, 194
—·INa ⟨2:1⟩·2H$_2$O	R: 42, 194
—·IRb ⟨1:1, 2:1⟩	R: 42, 194
Cl$_2$Cu·C$_2$H$_6$OS [DMSO]·C$_5$H$_5$NO [Py.—O]	R: 66
—·C$_5$H$_5$NO [Py.—O] ⟨1:1, 1:2⟩	R: 66, 67
—·C$_6$H$_7$NO [4-Pic.—O] ⟨1:1⟩	R: 66
—·C$_7$H$_9$NO [2,6-Me$_2$Py.—O] ⟨1:1⟩	R: 66
Cl$_2$Hg·CH$_4$O [MeOH] ⟨1:2⟩	R: 24, 190
—·C$_2$H$_6$OS [DMSO]	R: 25, 190
—·C$_3$H$_6$O$_3$ [Triox.]	R: 64
—·C$_3$H$_6$S$_3$ [Trithiane]	R: 64
—·C$_3$H$_7$NO [DMF]	R: 190
—·C$_4$H$_8$O [THF]	R: 24, 25, 64, 190
—·C$_4$H$_8$O$_2$ [Diox.] ⟨1:1⟩	R: 12, 23, 25, 64, 148, 165, 190
—·— ⟨1:2⟩	R: 64, 148, 190
—·C$_4$H$_8$S [THT]	R: 64
—·C$_4$H$_{10}$O [Et$_2$O]	R: 125
—·C$_4$H$_{10}$O$_2$ [Gly.]	R: 30, 190
—·C$_4$H$_{10}$S$_2$ [DMTE]	R: 30
—·C$_5$Cl$_6$Hg [C$_5$Cl$_5$HgCl]	R: 188, 189, 190, 191
—·C$_5$H$_4$ClN [2-ClPy.]	R: 6
—·— [4-ClPy.]	R: 6
—·C$_5$H$_5$NO [Py.—O]	R: 24, 165, 190
—·C$_6$H$_4$O$_2$ [BZQ]	R: 25, 190
—·C$_6$H$_7$NO [4-Pic.—O]	R: 24, 190
—·C$_6$H$_8$O$_2$ [CHDN]	R: 64, 148, 190
—·C$_6$H$_{14}$O$_3$ [Digly.]	R: 190
—·C$_6$H$_{16}$N$_2$ [TMED]	R: 30
—·C$_7$H$_5$ClO [2-ClBZA]	R: 64
—·— [3-ClBZA] ⟨1:2⟩	R: 64
—·— [4-ClBZA] ⟨3:2⟩	R: 64
—·C$_7$H$_6$O [BZA]	R: 64
—·C$_7$H$_{10}$O$_2$ [4-DMP]	R: 25, 190
—·C$_8$H$_8$O [ACP]	R: 24, 25, 190
—·C$_8$H$_{11}$N [2,4,6-Me$_3$Py.]	R: 64
—·C$_8$H$_{18}$O$_4$ [Trigly.]	R: 190

—·$C_{10}H_{22}O_5$ [Tetragly.]   R: 190
—·$C_{12}H_{24}O_6$ [18-Crown-6-ether]   R: 190
—·$C_{12}H_{26}O_3$ [Diglybu.]   R: 190
—·$C_{13}H_{10}O$ [BZP]   R: 24, 25, 190
—·$C_{14}H_{30}O_7$ [Hexagly.]   R: 190
—·$C_{18}H_{15}AsO$ [Ph_3AsO] ⟨1:1⟩   R: 165

$Cl_2OS·Cl_5Sb$ R: 28

$Cl_2OSe·Cl_5Sb$ R: 28

$Cl_2Sn·C_5H_4ClN$ [2-ClPy.]   R: 6
—·— [3-ClPy.]   R: 6

$Cl_2Zn·C_5H_5N$ [Py.] ⟨1:2⟩   R: 101

$Cl_3Fe·Cl_3OP$ R: 161

$Cl_3Ga·CH_3NO_2$ R: 178
—·C_2H_3N [MeCN]   R: 178
—·C_2H_6O [Me_2O]   R: Junp
—·C_2H_6S [Me_2S]   R: 35, 178
—·C_3H_9As [Me_3As]   R: 35, 36
—·C_3H_9N [Me_3N]   R: 107
—·C_3H_9P [Me_3P]   R: 35
—·C_4H_8O [THF]   R: 178
—·$C_4H_{10}O$ [Et_2O]   R: 35, 178
—·$C_4H_{10}S$ [Et_2S]   R: 35, 178
—·C_5H_5N [Py.]   R: 178
—·$C_6H_5NO_2$ [$PhNO_2$]   R: 178
—·$C_6H_{15}As$ [Et_3As]   R: 35
—·$C_6H_{15}N$ [Et_3N]   R: Junp
—·$C_6H_{15}P$ [Et_3P]   R: 35
—·C_7H_5N [PhCN]   R: 178
—·C_7H_8O [PhOMe]   R: 178
—·$C_{10}H_8N_2$ [biPy.] ⟨2:2⟩   R: 102
—·$C_{12}H_{10}O$ [Ph_2O]   R: 178
—·$C_{13}H_9ClO$ [4-ClBZP]   R: 48
—·$C_{13}H_{10}O$ [BZP]   R: 48
—·$C_{14}H_{12}O_2$ [4-MeOBZP]   R: 48
—·Cl_3OP [$OPCl_3$]   R: 102, 103, 178
—·Cl_4Te R: 143
—·Cl_5P R: 102

$Cl_3GeH·C_4H_{10}$ [Et_2O] ⟨1:2⟩   R: 112

$Cl_3I·AlCl_3$ R: 59

$Cl_3OP·AsCl_3$ R: 12
—BCl_3 R: 102, 175
—Cl_3Fe R: 161
—Cl_3Ga R: 102, 103, 178
—Cl_4Sn ⟨2:1⟩   R: 3, 12, 112, 123, 126, 161
—Cl_4Ti R: 161

—·Cl$_5$Nb   R: 169

—·Cl$_5$Sb   R: 28, 102, 152, 153, 154, 161, 167

—·Cl$_5$Ta   R: 169

Cl$_3$OV·Cl$_5$Sb   R: 28

Cl$_3$Sb·C$_6$H$_4$N$_2$O$_4$ [1,3-(NO$_2$)$_2$Ph]   R: 85, 169

—·C$_6$H$_5$Cl [PhCl]   R: 85, 86, 89, 169

—·C$_6$H$_6$ [Ph] ⟨1:1⟩   R: 171

—·— ⟨2:1⟩   R: 11, 12, 80, 81, 85, 86, 87, 89, 90, 142, 171

—·C$_6$H$_6$O [PhOH] ⟨2:1⟩   R: 81, 85, 86, 89, 169

—·C$_6$H$_7$N [PhNH$_2$] ⟨1:1⟩   R: 22, 73

—·— ⟨1:2⟩   R: 73

—·— ⟨1:3⟩   R: 73

—·C$_7$H$_8$ [PhMe] ⟨1:1, 2:1⟩   R: 85, 86, 87, 89, 90

—·C$_7$H$_8$O [PhOMe] ⟨1:1, 2:1⟩   R: 81, 85, 86, 87, 89, 115

—·— ⟨2:1⟩   R: 77

—·C$_8$H$_6$ [PhC≡CH] ⟨n:n⟩   R: 171

—·C$_8$H$_8$ [Styrene] ⟨n:n⟩   R: 146, 171

—·C$_8$H$_8$O [ACP]   R: 81, 85, 86, 87, 89

—·C$_8$H$_{10}$ [PhEt] ⟨1:1⟩   R: 11, 80, 81, 85, 86, 87, 88, 89, 90, 140, 169, 171

—·— ⟨2:1⟩   R: 11, 85, 86, 87, 89, 90, 169

—·C$_8$H$_{10}$ [1,2-Me$_2$Ph] ⟨1:1⟩   R: 87, 90, 169

—·— ⟨2:1⟩   R: 81, 85, 86, 87, 89, 90, 169

—·— [1,3-Me$_2$Ph] ⟨1:1⟩   R: 87, 90

—·— ⟨2:1⟩   R: 87

—·— [1,4-Me$_2$Ph] ⟨1:1⟩   R: 87, 169

—·— ⟨2:1⟩   R: 11, 81, 85, 86, 87, 89, 90, 169

—·C$_8$H$_{10}$O [PhOEt]   R: 80, 87

—·C$_9$H$_{12}$ [1,3,5-Me$_3$Ph]   R: 85, 87

—·C$_9$H$_{12}$ [iso-PrPh] ⟨1:1⟩   R: 146, 171

—·C$_{10}$H$_8$ ⟨1:1⟩   R: 73, 171

—·C$_{10}$H$_8$ ⟨2:1⟩   R: 1, 11, 12, 77, 80, 81, 85, 86, 87, 89, 90, 139, 169, 171

—·C$_{12}$H$_{10}$ [PhPh] ⟨2:1⟩   R: 12, 85, 86, 87, 88, 89, 90, 169

—·C$_{13}$H$_{10}$O [PhCOPh] ⟨1:1, 2:1⟩   R: 171

—·C$_{13}$H$_{12}$ [PhCH$_2$Ph] ⟨2:1⟩   R: 11, 77, 80, 83, 90

—·C$_{14}$H$_{10}$ [Anthracene] ⟨1:1, 2:1, 3:1⟩   R: 171

—·— [Phenanthrene] ⟨1:1, 2:1⟩   R: 171

—·C$_{16}$H$_{10}$ [Pyrene] ⟨2:1⟩   R: 171

—·C$_{18}$H$_{12}$ [Chrysene] ⟨4:1⟩   R: 171

—·C$_{18}$H$_{15}$ [p-Terphenyl] ⟨1:1, 2:1, 3:1⟩   R: 171

$Cl_4Sn \cdot CClN$ [ClCN]   R: 33

—$\cdot CH_2Cl_3OP$ [ClCH$_2$P(O)Cl$_2$] ⟨1:2⟩   R: 3

—$\cdot CH_3Cl_2OP$ [MeP(O)Cl$_2$] ⟨1:2⟩   R: 3

—$\cdot CH_3Cl_2O_2P$ [MeOP(O)Cl$_2$] ⟨1:2⟩   R: 3

—$\cdot CH_3NO_2$ R: 112

—$\cdot C_2H_5ClO$ [MeOCH$_2$Cl] ⟨1:2⟩   R: 123, 126

—$\cdot C_2H_5Cl_2OP$ [EtP(O)Cl$_2$] ⟨1:2⟩   R: 3

—$\cdot C_2H_6ClOP$ [Me$_2$P(O)Cl] ⟨1:2⟩   R: 3

—$\cdot C_2H_6ClPS$ [Me$_2$P(S)Cl] ⟨1:2⟩   R: 3

—$\cdot C_2H_6Cl_2NOP$ [Me$_2$NP(O)Cl$_2$] ⟨1:2⟩   R: 3

—$\cdot C_2H_6O$ [EtOH] ⟨1:2⟩   R: 112, 123, 155

—$\cdot C_2H_7O_3P$ [[(MeO)$_2$P(O)H] ⟨1:2⟩   R: 3, 163

—$\cdot C_3H_9O_2P$ [(Et)(MeO)P(O)H] ⟨1:2⟩   R: 3, 163

—$\cdot C_3H_9O_2P$ [Me$_2$P(O)OMe] ⟨1:2⟩   R: 3, 163

—$\cdot C_4H_{10}ClOP$ [Et$_2$P(O)Cl] ⟨1:2⟩   R: 3

—$\cdot C_4H_{10}ClP$ [Et$_2$PCl] ⟨1:2, 1:1⟩   R: 3

—$\cdot C_4H_{10}O$ [Et$_2$O] ⟨1:2⟩   R: 12, 112, 123, 155

—$\cdot C_4H_{10}O_2$ [Gly.] ⟨1:1⟩   R: 112, 123

—$\cdot C_4H_{11}ClNPS$ [Me$_2$NP(S)ClEt] ⟨1:2⟩   R: 2, 3

—$\cdot C_4H_{12}NO_3P$ [Me$_2$NP(O)(OMe)$_2$] ⟨1:2⟩   R: 3

—$\cdot C_4H_{12}NPS$ [Me$_2$NP(S)Me$_2$] ⟨1:2⟩   R: 2, 3

—$\cdot C_5H_{13}O_3PS$ [(EtO)$_2$P(O)SMe] ⟨1:2⟩   R: 3, 163

—$\cdot C_5H_{15}N_2OP$ [(Me$_2$N)$_2$P(O)Me] ⟨1:2⟩   R: 2, 3

—$\cdot C_5H_{15}N_2O_2P$ [(Me$_2$N)$_2$P(O)OMe] ⟨1:2⟩   R: 2, 3

—$\cdot C_6H_4BrCl_2OP$ [p-BrPhP(O)Cl$_2$] ⟨1:2⟩   R: 196

—$\cdot C_6H_4ClNO_2$ [4-ClPhNO$_2$] ⟨1:1⟩   R: 112

—$\cdot C_6H_4Cl_3OP$ [p-ClPhP(O)Cl$_2$] ⟨1:2⟩   R: 196

—$\cdot C_6H_5Cl_2OP$ [PhP(O)Cl$_2$] ⟨1:2⟩   R: 3, 196

—$\cdot C_6H_5NO_2$ [PhNO$_2$] ⟨1:1⟩   R: 12, 50, 112, 123

—$\cdot C_6H_{14}O_2$ [MeO(CH$_2$)$_4$OMe] ⟨1:1⟩   R: 112, 123

—$\cdot C_6H_{15}OPS$ [trans-Et$_2$P(O)SEt] ⟨1:2⟩   R: 3, 163

—$\cdot C_6H_{15}PS_2$ [Et$_2$P(S)SEt] ⟨1:2⟩   R: 3, 163

—$\cdot C_6H_{17}N_2OP$ [(Me$_2$N)$_2$P(O)Et] ⟨1:2⟩   R: 3

—$\cdot C_6H_{17}N_2PS$ [(Me$_2$N)$_2$P(S)Et] ⟨1:2⟩   R: 3

—$\cdot C_6H_{18}N_3OP$ [(Me$_2$N)$_3$P(O)] ⟨1:2⟩   R: 2, 3

—$\cdot C_6H_{18}N_3P$ [(Me$_2$N)$_3$P] ⟨1:2⟩   R: 2, 3

—$\cdot C_7H_5ClO$ [PhCOCl] ⟨1:1⟩   R: 126

—$\cdot C_7H_7Cl$ [PhCH$_2$Cl] ⟨1:2⟩   R: 123, 126

—$\cdot C_7H_7Cl_2OP$ [p-MePhP(O)Cl$_2$] ⟨1:2⟩   R: 196

—$\cdot C_7H_7Cl_2O_2P$ [p-MeOPhP(O)Cl$_2$] ⟨1:2⟩   R: 196

—$\cdot C_7H_{17}OPS$ [(iso-Pr)$_2$P(O)SMe] ⟨1:2⟩   R: 3, 163

—$\cdot C_7H_{17}O_3PS$ [(iso-PrO)$_2$P(O)SMe] ⟨1:2⟩   R: 3, 163

—$\cdot C_9H_{13}O_2P$ [(Ph)(iso-Pr)P(O)H] ⟨1:2⟩   R: 3, 163

—$\cdot C_9H_{21}O_3PS$ {cis-[(n-BuO)$_2$P(O)SMe]} ⟨1:2⟩   R: 3, 163

—$\cdot C_{10}H_{17}N_2OP$ [(Me$_2$N)$_2$P(O)Ph] ⟨1:2⟩   R: 2, 3

—$\cdot C_{10}H_{17}N_2PS$ [(Me$_2$N)$_2$P(S)Ph] ⟨1:2⟩   R: 2, 3

—$\cdot C_{12}H_{27}OP$ [(n-Hex)$_2$P(O)H] ⟨1:2⟩   R: 3, 163

—$\cdot C_{13}H_{13}OP$ [Ph$_2$P(O)Me] ⟨1:2⟩   R: 3

—$\cdot C_{14}H_{16}NOP$ [(Me$_2$N)P(O)Ph$_2$] ⟨1:2⟩   R: 2, 3

—$\cdot Cl_3OP$ [OPCl$_3$] ⟨1:2⟩   R: 3, 12, 112, 123, 126, 161

$Cl_4Te \cdot Cl_3Ga$	R: 143
$Cl_4Ti \cdot Cl_3OP$	R: 161
$Cl_5Nb \cdot CClN$ [ClCN]	R: 34
—$\cdot Cl_3OP$ [OPCl$_3$]	R: 169
$Cl_5P \cdot BCl_3$	R: 102
—$\cdot C_6H_5Cl_4P$ [PhPCl$_4$]	R: 53
—$\cdot C_{12}H_{10}Cl_3P$ [Ph$_2$PCl$_3$]	R: 53
—$\cdot C_{18}H_{15}Cl_2P$ [Ph$_3$PCl$_2$]	R: 53
—$\cdot Cl_3Ga$	R: 102
$Cl_5Sb \cdot CClN$ [ClCN] $\langle 1:1 \rangle$	R: 3, 34, 152, 176
—$\cdot CH_4O$ [MeOH]	R: 152
—$\cdot C_2H_3N$ [MeCN] $\langle 1:1 \rangle$	R: 152, 154, 167, 176
—$\cdot C_2H_4Cl_2O$ [(CH$_2$Cl)$_2$O]	R: 120, 152, 153
—$\cdot C_2H_5ClO$ [CH$_2$ClOMe]	R: 152
—$\cdot C_2H_6Cl_3P$ [Me$_2$PCl$_3$]	R: 53
—$\cdot C_2H_6O$ [EtOH]	R: 152
—$\cdot C_3H_2ClNO$ [CH$_2$ClCOCN]	R: 152
—$\cdot C_3H_7NO$ [DMF]	R: 28, 152
—$\cdot C_3H_8O$ [C$_3$H$_7$OH]	R: 152
—$\cdot C_4H_8O$ [THF]	R: 152
—$\cdot C_4H_8O_2$ [Diox.]	R: 152, 154
—$\cdot C_4H_{10}O$ [DEE]	R: 152, 154
—$\cdot C_5H_4ClN$ [2-ClPy.]	R: 6, 152
—— [3-ClPy.]	R: 6
—$\cdot C_5H_5N$ [Py.]	R: 152, 154
—$\cdot C_6H_3ClN_2O_4S$ [2,4-(NO$_2$)$_2$PhSCl]	R: 9
—$\cdot C_6H_5Cl_4P$ [PhPCl$_4$]	R: 53
—$\cdot C_6H_8N_3OP$ [(Me$_2$N)$_3$P(O)]	R: 152
—$\cdot C_6H_{18}N_3OP$ [HMPT]	R: 152
—$\cdot C_7H_5ClO$ [PhCOCl]	R: 6, 28
—$\cdot C_7H_5N$ [PhCN]	R: 152, 154
—$\cdot C_8H_7ClO$ [4-ClPhCOMe]	R: 6
—$\cdot C_9H_7N$ [Quinoline]	R: 152, 154
—$\cdot C_{12}H_{10}Cl_3P$ [Ph$_2$PCl$_3$]	R: 53
—$\cdot C_{18}H_{15}Cl_2P$ [Ph$_3$PCl$_2$]	R: 53
—$\cdot Cl_2OS$	R: 28
—$\cdot Cl_2OSe$	R: 28
—$\cdot Cl_3OP$ [OPCl$_3$]	R: 28, 102, 152, 153, 154, 161, 167
—$\cdot Cl_3OV$	R: 28
—$\cdot H_3N$	R: 152, 154
—$\cdot H_3PO_4$	R: 152
$Cl_5Ta \cdot CClN$ [ClCN] $\langle 1:1 \rangle$	R: 34
—$\cdot Cl_3OP$ [OPCl$_3$]	R: 169
$Cl_6N_3P_3$ [(Cl$_2$NP)$_3$]$\cdot$HBF$_4$	R: 187
—$\cdot$HClO$_4$	R: 187

CsI·Cl$_2$ ⟨1:1, 1:2⟩   R: 42, 194
— ·I$_2$ ⟨1:1, 2:3⟩   R: 18, 98, 133, 164

Cs$_2$O$_4$S·F$_3$Sb ⟨3:4⟩   R: 29

CuCl·C$_3$H$_9$AsS [Me$_3$AsS]   R: 20

CuI·C$_3$H$_9$PS [Me$_3$PS] ⟨n:n⟩   R: 20
— ·C$_{19}$H$_{18}$IP [Ph$_3$MePI] ⟨2:1⟩   R: 15

FK·F$_3$Sb ⟨1:1, 2:1⟩   R: 29

F$_2$Xe·F$_5$Nb ⟨1:1, 1:2⟩   R: 68

F$_3$Sb·C$_2$K$_2$O$_4$ ⟨1:1⟩   R: 29
— ·ClK ⟨1:1⟩   R: 29
— ·Cs$_2$O$_4$S ⟨4:3⟩   R: 29
— ·FK ⟨1:1, 1:2⟩   R: 29
— ·K$_2$SO$_4$ ⟨1:1, 2:1⟩   R: 29

F$_5$Nb·C$_2$H$_2$ClN [CH$_2$ClCN]   R: 68
— ·C$_2$H$_6$O [Me$_2$O]   R: 68
— ·C$_5$H$_5$N [Py.] ⟨1:2⟩   R: 68
— ·F$_2$Xe ⟨1:1, 2:1⟩   R: 68

HBF$_4$·Cl$_6$N$_3$P$_3$ [(Cl$_2$NP)$_3$]   R: 187

HBr·C$_2$H$_6$BrN [BrCH$_2$CH$_2$NH$_2$]   R: 100
— ·C$_6$H$_6$ClN [p-ClPhNH$_2$]   R: 125

HCl·C$_4$HI$_2$ClN [Me$_4$NCl]   R: 60, 62, 97
— ·C$_5$H$_4$ClN [2-ClPy.]   R: 6
— ·— [3-ClPy.]   R: 6
— ·— [4-ClPy.]   R: 6
— ·C$_6$H$_6$BrN [p-BrPhNH$_2$]   R: 125
— ·C$_6$H$_6$ClN [p-ClPhNH$_2$]   R: 125
— ·C$_6$H$_6$IN [p-IPhNH$_2$]   R: 125
— ·C$_6$H$_{15}$N [Et$_3$N] ⟨2:1⟩   R: 44
— ·C$_8$H$_{20}$ClN [Et$_4$NCl]   R: 60

HClO$_4$·C$_5$H$_3$Br$_2$N [3,5-Br$_2$Py.]   R: 17
— ·C$_5$H$_4$BrN [2-BrPy.]   R: 17
— ·C$_5$H$_4$ClN [2-ClPy.]   R: 6
— ·— [3-ClPy.]   R: 6
— ·— [4-ClPy.]   R: 6
— ·Cl$_6$N$_3$P$_3$ [(Cl$_2$NP)$_3$]   R: 187

HNO$_3$·C$_5$H$_5$N [Py.] ⟨1:1⟩   R: 101

H$_2$SO$_4$·C$_5$H$_4$ClN [2-ClPy.]   R: 6
— ·— [3-ClPy.]   R: 6
— ·— [4-ClPy.]   R: 6

H$_3$N·Cl$_5$Sb   R: 152, 154

H$_3$PO$_4$·Cl$_5$Sb   R: 152

H_4IN [NH_4I]·Br_2 R: 18
  —·I_2 R: 18, 109, 110, 133, 164

HgI_2·$C_4H_8O_2$ [Diox.]   R: 23, 148, 190
  —·$C_4H_{10}O_2$ [Gly.]   R: 30, 190
  —·$C_4H_{10}S_2$ [MeS(CH_2)$_2$SMe]   R: 30
  —·C_5H_5NO [Py.—O]   R: 24
  —·C_6H_7NO [4-Pic.—O]   R: 24
  —·$C_6H_{14}O_3$ [Digly.]   R: 190
  —·$C_6H_{16}N_2$ [$Me_2N(CH_2)_2NMe_2$]   R: 30
  —·$C_8H_{18}O_4$ [Trigly.]   R: 190
  —·$C_{10}H_{22}O_5$ [Tetragly.]   R: 190
  —·$C_{12}H_{24}O_6$ [18-Crown-6-ether]   R: 190
  —·$C_{12}H_{26}O_3$ [Diglybu.]   R: 190
  —·$C_{14}H_{30}O_7$ [Hexagly.]   R: 190

IK·Cl_2 ⟨1:1, 1:2⟩ and ·H_2O R: 42, 194

ILi·$4H_2O$·$C_6H_{12}N_4$ [HMT]   R: 79

INa·Cl_2 ⟨1:2⟩ and ·$2H_2O$ R: 42, 194

IRb·Cl_2 ⟨1:1, 1:2⟩   R: 42, 194
  —·I_2 R: 133, 164

ITl·I_2 R: 98, 133

I_2·C_2H_8IN [Me_2NH_2I]   R: 98, 133
  —·C_3H_9N [Me_3N]   R: 17
  —·$C_4H_{12}IN$ [Me_4NI] ⟨1:1, 2:1⟩   R: 18, 98, 110, 133
  —·C_5H_5N [Py.] ⟨1:1⟩   R: 27
  —·C_6H_7N [4-Pic.] ⟨1:1⟩   R: 27
  —·$C_6H_{12}N_4$ [HMP]   R: 27
  —·$C_6H_{16}IN$ [(n-Pr)$_2NH_2I$]   R: 98, 133
  —·$C_8H_{20}IN$ [Et_4NI]   R: 18, 98, 133
  —·$C_{12}H_8N_2$ [Phenazine] ⟨1:1⟩   R: 27
  —·$C_{12}H_{18}IN$ [(n-Pr)$_4NI$]   R: 98, 133
  —·$C_{13}H_9N$ [Acridine] ⟨1:1⟩   R: 27
  —·$C_{16}H_{36}IN$ [(n-Bu)$_4NI$]   R: 18, 98, 133
  —·CsI ⟨1:1⟩   R: 18, 133, 164
  —·— ⟨3:2⟩   R: 18, 98, 133
  —·H_4IN [NH_4I]   R: 18, 109, 110, 133, 164

  —·IRb   R: 133, 164
  —·ITl   R: 98, 133

I_2Zn·C_5H_5N [Py.]   R: 101

I_3Sb·$AlBr_3$ R: 144
  —·AlI_3 R: 144
  —·S_8 ⟨1:3⟩   R: 138, 160

I_4Sn·S_8 ⟨1:2, 1:4⟩   R: 138
  —·—   R: 151

—·C_5H_5N [Py.] $\langle 1:2 \rangle$ R: 151
—·$C_6H_{18}N_3OP$ [HMPT]   R: 151

K_2O_4S [K_2SO_4]·F_3Sb $\langle 1:1,\ 1:2 \rangle$ R: 29

N_2O_6Zn [(NO_3)$_2$Zn]·C_5H_5N [Py.]   R: 101

8.15   List of Abbreviations

ACP = acetophenone
bdtc = bis(N,N-dibutyldithiocarbamato)gold(III) dichloroaurate(I),
         [(Bu_2NCS_2)$_2$Au]$AuCl_2$
  Bu = butyl group, —(CH_2)$_3CH_3$
 BZA = benzaldehyde
BZCl = benzoyl chloride
 BZP = benzophenone
 BZQ = benzoquinone
CHDN = cyclohexanedione
 DEA = diethylamine
 DEE = diethyl ether
Digly. = diglyme, CH_3—(OCH_2CH_2)$_2$—OCH_3
Diglybu. = diglybutyl, (CH_3)$_3$C(OCH_2CH_2)$_2$OC(CH_3)$_3$
Diox. = dioxan (usually 1,4-dioxan)
 DME = dimethoxyethane = glyme, see below
 DMF = dimethylformamide $HCON(CH_3)_2$
 DMP = 2,6-dimethyl-γ-pyrone
DMSO = dimethylsulphoxide (CH_3)$_2$SO
DMTE = dimethylthioethane, $CH_3SCH_2CH_2SCH_3$
   Et = ethyl group, —CH_2CH_3
ETDA = ethylenediamine
  gly. = glyme = $CH_3OCH_2CH_2OCH_3$
 HMB = hexamethylbenzene
HMPT = hexamethylphosphorus triamide
 HMT = hexamethylenetetramine
  Me = methyl group, —Ch_3
 Mes. = mesitylene
Naphth = naphthalene
   Ph = phenyl group, —C_6H_{5-n}, n being the number of substituents
  Pic. = picoline = methylpyridine
Pic.—O = picoline-oxide
   Pr = propyl group —(CH_2)$_2CH_3$
  Py. = pyridine, $C_5H_{5-n}N$, n being the number of substituents
Py.—O = pyridine-oxide
TCNE = tetracyanoethylene
 TED = triethylenediamine
 THF = tetrahydrofuran

THT = tetrahydrothiophene
TMED = tetramethylethylenediamine
Triox. = trioxan $(CH_2O)_3$

8.16 References

References are presented in alphabetical order according to the name of the first author (and, if necessary, of the second and following authors). When several publications with the same first author have been published the same year, the year number is followed by an identification letter, a, b, c, . . . For a given author (or set of authors) the most recent references are given first. In Russian references, numbers between square brackets refers to the page in the English translation of the journal considered. The numbers preceding the references are used in the list of complexes.

1. Ainbinder, N. E., Manzhura, Yu. I., and Kyuntsel, I. A. (1974) [1975]. *Fiz. Tverd. Tela*, **16**, 3518 [2288].
2. Andreeva, A. I., Kuramshin, I. Ya., Muratova, A. A., Osokin, D. Ya., Pudovik, A. N., and Safin, I. A. (1976). *Koord. Khim.*, **2**, 683.
3. Andreeva, A. I., Kuramshin, I. Ya., Muratova, A. A., Osokin, D. Ya., Pudovik, A. N., and Safin, I. A. (1975). *Izv. Akad. Nauk SSSR, Ser. Fiz.*, **39**, 2590 [119].
4. Anferov, V. P., Grechishkin, V. S., and Yusupov, M. Z. (1973). *Zh. Fiz. Khim.*, **47**, 1267 [713].
5. Anferov, V. P., Versinov, V. S., Grechishkin, V. S., and Yusupov, M. Z. (1971). *Zh. Strukt. Khim.*, **12**, 924.
6. Ardjomande, S. and Lucken, E. A. C. (1975). *J. Chem. Soc. Perkin II*, 453.
7. Ardjomande, S. and Lucken, E. A. C. (1971). *Helv. Chim. Acta*, **54**, 176.
8. Azizov, E. O. and Luganskii, Yu. M. (1975). *Izv. Akad. Nauk SSSR, Ser. Fiz.*, **39**, 2627 [151].
9. Babushkina, T. A. and Kalinin, M. I. (1969). *Izv. Adad. Nauk SSSR, Ser. Khim.*, 157.
10. Bennett, R. A. and Hooper, H. O. (1967). *J. Chem. Phys.*, **47**, 4855.
11. Biedenkapp, D. and Weiss, A. (1968). *Z. Naturforsch.*, **23b**, 174.
12. Biedenkapp, D. and Weiss, A. (1964). *Z. Naturforsch.*, **19a**, 1518.
13. Biryukov, I. P., Voronkov, M. G., and Safin, I. A. (1969). *Tables of Nuclear Quadrupole Resonance Frequencies*, IPST, Jerusalem.
14. Bowmaker, G. A. (1976). *J. Chem. Soc. Faraday II*, **72**, 1964.
15. Bowmaker, G. A., Brockliss, L. D., Earp, C. D., and Whiting, R. (1973a). *Austr. J. Chem.*, **26**, 2593.
16. Bowmaker, G. A., Brockliss, L. D., and Whiting, R. (1973b). *Austr. J. Chem.*, **26**, 29.
17. Bowmaker, G. A. and Hacobian, S. (1969). *Austr. J. Chem.*, **22**, 2047.
18. Bowmaker, G. A. and Hacobian, S. (1968). *Austr. J. Chem.*, **21**, 551.
19. Bowmaker, G. A. and Whiting, R. (1976). *Austr. J. Chem.*, **29**, 1407.
20. Bowmaker, G. A., Whiting, R., Ainscough, E. W., and Brodie, A. M. (1975). *Austr. J. Chem.*, **28**, 1431.
21. Breneman, G. L. and Willett, R. D. (1967). *J. Phys. Chem.*, **71**, 3684.
22. Brill, T. B. (1974). *J. Magn. Res.*, **15**, 395.
23. Brill, T. B. (1970). *J. Inorg. Nucl. Chem.*, **32**, 1869.
24. Brill, T. B. and Hughus, Z. Z. (1971). *J. Inorg. Nucl. Chem.*, **33**, 371.

25. Brill, T. B. and Hughus, Z. Z. (1970). *Inorg. Chem.*, **9**, 984.
26. Brooker, H. R. and Creel, J. B. (1974). *J. Chem. Phys.*, **61**, 3658.
27. Brüggemann, R., Reiter, F., and Voitländer, J. (1972). *Z. Naturforsch.*, **27a**, 1525.
28. Brunette, J. P., Burgard, M., Leroy, M. J. F., and Lucken, E. A. C. (1977). *J. Mol. Struct.*, **36**, 269.
29. Bryukhova, E. V., Egorov, V. A., Alymov, I. M., and Semin, G. K. (1974a). *Izv. Akad. Nauk SSSr, Ser. Khim.*, No. 8, 1919 [1849].
30. Bryukhova, E. V., Erdyneev, N. S., and Prokofev, A. K. (1973). *Izv. Akad. Nauk SSSR, Ser. Khim.*, No. 8, 1895 [1846].
31. Bryukhova, E. V., Prokofev, A. K., Melnikova, T. Ya., and Semin, G. K. (1974b). *Izv. Akad. Nauk SSSr, Ser. Khim.*, No. 2, 477 [448].
32. Bryukhova, E. V., Velichko, F. K., and Semin, G. K. (1969). *Izv. Akad. Nauk SSSR, Ser. Khim.*, No. 4, 960 [880].
33. Burgard, M. and Lucken, E. A. C. (1972). *J. Mol. Struct.*, **14**, 397.
34. Burgard, M. and MacCordick, J. M. (1972). *Inorg. Nucl. Chem. Lett.* **8**, 185.
35. Carter, J. C., Haran, R., and Jugie, G. (1976). *C. R. Acad. Sci. (Paris)*, **282C**, 623.
36. Carter, J. C. and Jugie, G. (1977). 4th ISNQRS, Abstr., p. 34, Osaka, Sept. 13–16.
37. Casabella, P. A. and Oja, T. (1969). *J. Chem. Phys.*, **50**, 4814.
38. Cheng, C. P., Brown, T. L., Fultz, W. C., and Burmeister, J. L. (1977). *J. Chem. Soc. Chem. Commun.* 599.
39. Chihara, H. and Nakamura, N. (1972). Nuclear quadrupole resonance spectroscopy in MTP, *International Review of Science, Physical Chemistry Series, Magnetic Resonance*, (Ed. McDowell, C. A.), Vol. IV, Butterworths, University Park Press.
40. Chihara, H. and Nakamura, N. (1971). *Bull. Chem. Soc. Japan*, **44**, 2676.
41. Cornil, P., Duchesne, J., Read, M., and Cahay, R. (1964). *Bull. Acad. Roy. Belg.*, **L**, 235.
42. Cornwell, C. D. and Yamasaki, R. S. (1957). *J. Chem. Phys.*, **27**, 1060.
43. Cousseau, J. (1976). *L'Actualité Chimique (France)*, No. 7, 15, Sept.; No. 8, 16, Oct.
44. Cousseau, J., Gouin, L., Jones, L. V., Jugie, G., and Smith, J. A. S. (1973). *J. Chem. Soc. Faraday II*, **69**, 1821.
45. Dailey, B. P. and Townes, C. H. (1955). *J. Chem. Phys.*, **23**, 118.
46. Darville, J., Gerard, A., and Calende, M. T. (1974). *J. Magn. Res.*, **16**, 205.
47. Das, T. P. and Hahn, E. L. (1958). *Nuclear Quadrupole Resonance Spectroscopy*, Suppl. I to *Solid State Physics* (Eds. Seitz, F. and Turnbull D.), Academic Press, New York.
48. Deeg, T. and Weiss, A. (1976). *Ber. Bunsenges. Phys. Chem.*, **80**, 2.
49. Deeg, T. and Weiss, A. (1975). *Ber. Bunsenges. Phys. Chem.*, **79**, 497.
50. Dehmelt, H. G. (1953). *J. Chem. Phys.*, **21**, 380.
51. Dehmelt, H. G. and Krüger, H. (1950). *Naturwiss.*, **37**, 111.
52. Dewar, M. J. S., Patterson, D. B., and Simpson, W. I. (1973). *J. Chem. Soc. Dalton*, 2381.
53. Dillon, K. B., Lynch, R. J., Reeve, R. E., and Waddington, T. C. (1976a). *J. Chem. Soc. Dalton*, **13**, 1243.
54. Dillon, K. B., Lynch, R. J., and Waddington, T. C. (1976b). *J. Chem. Soc. Dalton*, **15**, 1478.
55. Douglass, D. C. (1960). *J. Chem. Phys.*, **32**, 1882.
56. Dresvyankin, B. V. (1975). *Zh. Strukt. Khim.*, **16**, 478 [444].
57. Dresvyankin, B. V., Grechishkin, V. S., and Lunegov, V. I. (1975). *Izv. Adak. Nauk SSSR, Ser. Fiz.*, **39**, 2584 [114].
58. Edmonds, D. T. (1974). *Pure Appl. Chem.*, **40**, 193.
59. Evans, J. C. and Lo, G. Y-S. (1967a). *Inorg. Chem.*, **6**, 836.

60. Evans, J. C. and Lo, G. Y-S. (1976b). *J. Phys. Chem.*, **71**, 3697.
61. Evans, J. C. and Lo, G. Y-S. (1967c). *J. Phys. Chem.*, **71**, 2730.
62. Evans, J. C. and Lo, G. Y-S. (1966). *J. Phys. Chem.*, **70**, 2702.
63. Feshin, V. P. and Voronkov, M. G. (1973). *Dokl. Acad. Nauk SSSR*, **209**, 400 [262].
64. Fichtner, W. and Weiss, A. (1976). *Z. Naturforsch.*, **31b**, 1626.
65. Fleming, H. C. and Hanna, M. W. (1971). *J. Amer. Chem. Soc.*, **93**, 5030.
66. Frausto da Silva, J. J. R., Vila Boas, L. F., and Wooton, R. (1971). *J. Inorg. Nucl. Chem.*, **33**, 2029.
67. Frausto da Silva, J. J. R. and Wooton, R. (1970). *J. Chem. Soc. Chem. Commun.*, 403.
68. Fuggle, J. C., Tong, D. A., Sharp, D. W. A., Winfield, J. M. and Holloway, J. H. (1974). *J. Chem. Soc. Dalton*, 205.
69. Gilson, D. F. R. and Hart, R. M. (1970). *Canad. J. Chem.* **48**, 1976.
70. Gilson, D. F. R. and O'Konski, C. T. (1968). *J. Chem. Phys.*, **48**, 2767.
71. Gordeev, A. D., Grechishkin, V. S., Kyuntsel, I. A., and Rozenberg, Yu. I. (1970). *Zh. Strukt. Khim.*, **11**, 773.
72. Gordeev, A. D., Grechishkin, V. S., and Yusupov, M. Z. (1971). *Zh. Strukt. Khim.*, **12**, 725.
73. Gordeev, A. D. and Kyuntsel, I. A. (1974). *Zh. Strukt. Khim.*, 935.
74. Goren, S. D. (1974). *J. Chem. Phys.*, **60**, 1892.
75. Graybeal, J. D. (1958). *J. Phys. Chem.*, **62**, 483.
76. Graybeal, J. D., Pathania, M., and Ing, S. D. (1973). *J. Magn. Res.*, **9**, 27.
77. Grechishkin, V. S. (1966). *Opt. Spektrosk.*, **21**, 517 [289].
78. Grechishkin, V. S. (1973). *Nuclear Quadrupole Interactions in Solids* (in Russian), Nauka, Moscow.
79. Grechishkin, V. S., Anferov, V. P., Dresvyankin, B. V., and Yusupov, M. Z. (1974). *Zh. Fiz. Khim.*, **48**, 1580 [931].
80. Grechishkin, V. S. and Gordeev, A. D. (1966a). *Zh. Strukt. Khim.*, **7**, 205.
81. Grechishkin, V. S. and Gordeev, A. D. (1966b). *Tr. ENI Permskom. Gos. Univ.*, **12** (4), 29.
82. Grechishkin, V. S., Gordeev, A. D., and Galichevskii, Yu. A. (1969). *Zh. Strukt. Khim.*, **10**, 743.
83. Grechishkin, V. S., Gushkin, S. I., and Shishkin, V. A. (1970). *Zh. Strukt. Khim.*, **11**, 145.
84. Grechishkin, V. S. and Kyuntsel, I. A. (1966a). *Zh. Strukt. Khim.*, **7**, 119.
85. Grechishkin, V. S. and Kyuntsel, I. A. (1966b). *Tr. ENI Permskom. Gos. Univ.*, **12**, (1), 9.
86. Grechishkin, V. S. and Kyuntsel, I. A. (1964a). *Opt. Spektrosk.*, **16**, 161.
87. Grechishkin, V. S. and Kyuntsel, I. A. (1964b). *Zh. Strukt. Khim.*, **5**, 53.
88. Grechishkin, V. S. and Kyuntsel, I. A. (1963a). *Fiz. Tverd. Tela*, **5**, 948. [694].
89. Grechishkin, V. S. and Kyuntsel, I. A. (1963b). *Opt. Spektrosk.*, **15**, 832.
90. Grechishkin, V. S. and Yusupov, M. Z. (1973). *Zh. Strukt. Khim.*, **14**, 1028.
91. Grechishkin, V. S. and Yusupov, M. Z. (1970a). *Opt. Spektrosk.*, **29**, 804 [428].
92. 92. Grechishkin, V. S. and Yusupov, M. Z. (1970b). *Zh. Fiz. Khim.*, **44**, 2933 [1673].
93. Grechishkin, V. S. and Yusupov, M. Z. (1970c). *Proc. Congress Ampere*, 16th, 1033.
94. Guibé, L. and Montabonel, M.-C. (1978). *J. Magn. Res.*, **31**, 419.
95. Guryanova, E. N., Puchkova, V. V. and Volkov, A. F. (1974). *Zh. Strukt. Khim.*, **15**, 439.
96. Ha, T. K. and O'Konski, C. T. (1970). *Z. Naturforsch.*, **25a**, 1509.
97. Haas, T. E. and Welsh, S. M. (1967). *J. Phys. Chem.*, **71**, 3363.
98. Harada, H., Nakamura, D., and Kubo, M. (1974). *J. Magn. Res.*, **13**, 56.
99. Hooper, H. O. (1964). *J. Chem. Phys.*, **41**, 599.

100. Hooper, H. O. and Bray, P. J. (1960). *J. Chem. Phys.*, **33**, 334.
101. Hsieh, Y. N., Rubenacker, G. V., Cheng, C. P., and Brill, T. B. (1977). *J. Phys. Chem.*, **99**, 1384.
102. Jost, J. W. (1972). *Diss. Abstr. Int. B*, **32**, 5130.
103. Jost, J. W. and Schneider, R. F. (1975). *J. Phys. Chem. Solids*, **36**, 349.
104. Jugie, G. and Smith, J. A. S. (1974). *J. Chem. Soc. Faraday II*, **71**, 608.
105. Kadaba, P. K., O'Reilly, D. E., Peterson, E. M., and Scheie, C. E. (1971). *J. Chem. Phys.*, **55**, 5289.
106. Kaplansky, M. and Whitehead, M. A. (1970). *Canad. J. Chem.*, **48**, 697.
107. Kasakov, V. P., Bryukhova, E. V., Bregadze, V. I., Golubinskaya, L. M., and Semin, G. K. (1977). *Teor. Eksp. Khim.*, **13**, 55.
108. Kitagorodskii, A. I. and Fedin, E. I. (1961). *Zh. Strukt. Khim.*, **2**, 216.
109. Kojima, S., Shimauchi, A., Hagiwara, S., and Abe, Y. (1955a). *J. Phys. Soc. Japan*, **10**, 930.
110. Kojima, S., Tsukada, K., Ogawa, S., and Shimauchi, A. (1955b). *J. Chem. Phys.*, **23**, 1963.
111. Korshunov, A. V., Babushkina, T. A., Shabanov, V. F., Volkov, V. E., and Semin, G. K. (1971). *Opt. Spectrosc.*, **30**, 476.
112. Kravchenko, E. A., Maksyutin, Yu. K., Guryanova, E. N., and Semin, G. K. (1968). *Izv. Akad. Nauk SSSR, Ser. Khim.*, **6**, 1271 [1200].
113. Kress, J. and Guibé, L. (1977). *Canad. J. Chem.*, **55**, 1515.
114. Kyuntsel, I. A. and Rozenberg, Yu. I. (1973). *Opt. Spectrosc.*, **34**, 341.
115. Kyuntsel, I. A. and Rozenberg, Yu. I. (1971). *Teor. Eksp. Khim.*, **7**, 565.
116. Lobanova, L. A., Guryanova, E. N., Volkov, A. F., and Shifrina, R. R. (1975). *Zh. Obshch. Khim.*, **45**, 1857 [1819].
117. Lucas, J.-P. and Guibé, L. (1970). *Mol. Phys.*, **19**, 85.
118. Lucken, E. A. C. (1969). *Nuclear Quadrupole Coupling Constants*, Academic Press, London.
119. Lumpkin, O. (1975). *J. Chem. Phys.*, **62**, 3281.
120. Makridin, V. P., Bryukhova, E. V., and Semin, G. K. (1973). *Izv. Akad. Nauk SSSR, Ser. Khim.*, No. 6, 1414 [1378].
121. Maksyutin, Yu. A., Babushkina, T. A., Guryanova, E. N., and Semin, G. K., (1969). *Theor. Chim. Acta*, **14**, 48.
122. Maksyutin, Yu. K., Bryukhova, E. V., Semin, G. K., and Guryanova, E. N. (1968). *Izv. Akad. Nauk SSSR, Ser. Khim.*, No. 11, 2658 [2528].
123. Maksyutin, Yu. K., Guryanova, E. N., Kravchenko, E. A., and Semin, G. K. (1973). *J. Chem. Soc. Chem. Commun.*, 429.
124. Maksyutin, Yu. K., Guryanova, E. N., and Semin, G. K. (1970a). *Izv. Akad. Nauk SSSR, Ser. Khim.*, No. 7, 1632 [1534].
125. Maksyutin, Yu. K., Guryanova, E. N., and Semin, G. K. (1970b). *Usp. Khim.*, **39**, 727 [334].
126. Maksyutin, Yu, K., Makridin, V. P., Guryanova, E. N., and Semin, G. K. (1970c). *Izv. Akad. Nauk SSSR, Ser. Khim.*, No. 7, 1634 [1537].
127. Marino, R. A. (1974). *Advances in NQR* (Ed. Smith, J. A. S.), Vol. 1, Heyden, p. 391.
128. Marino, R. A. (1972). *J. Chem. Phys.*, **57**, 4560.
129. Merchant, S. Z. and Fung, B. M. (1969). *J. Chem. Phys.*, **50**, 2265L.
130. Merryman, D. J., Edwards, P. A., Corbett, J. D., and McCarley, R. E. (1972). *J. Chem. Soc. Chem. Commun.*, 779.
131. Morino, Y. and Toyama, M. (1961). *J. Chem. Phys.*, **35**, 1289.
132. Myasnikova, R. M., Robas, V. I., and Semin, G. K. (1965). *Zh. Strukt. Khim.*, No. 6, 474.
133. Nakamura, D. and Kubo, M. (1975). *Advances in NQR* (Ed. Smith, J. A. S.),

Vol. 2, Heyden, p. 117.

134. Negita, H., Okuda, T., and Kashima, M. (1966). *J. Chem. Phys.*, **45**, 1076.
135. Negita, H., Shibata, K., Furukawa, Y., and Yamada, K. (1973). *Bull. Chem. Soc. Japan*, **46**, 2662.
136. Nesmeyanov, A. N., Okhlobystin, O. Yu., Bryukhova, E. V., Bregadze, V. I., Kravtsov, D. N., Faingor, B. A., Golovchenko, L. S., and Semin, G. K. (1969). *Izv. Akad. Nauk SSSR, Ser. Khim.*, No. 9, 1928 [1785].
137. Nesmeyanov, A. N., Semin, G. K., Saatsazov, V. V., Burbelo, V. M., Lisichkina, I. N., Bryukhova, E. V., Khotsyanova, T. L., and Tolstaya, T. P. (1976). *Dokl. Akad. Nauk SSSR*, **231**, 396 [1072].
138. Ogawa, S. (1958). *J. Phys. Soc. Japan*, **13**, 618.
139. Okuda, T. (1971). *J. Sci. Hiroshima Univ.*, **A35**, 213.
140. Okuda, T., Furukawa, Y., and Negita, H. (1972a). *Bull. Chem. Soc. Japan*, **45**, 2940.
141. Okuda, T., Furukawa, Y., and Negita, H. (1972b). *Bull. Chem. Soc. Japan*, **45**, 2245.
142. Okuda, T., Nakao, A., Shiroyama, M., and Negita, H. (1968). *Bull. Chem. Soc. Japan*, **41**, 61.
143. Okuda, T., Terao, H., Ege, O., and Negita, H. (1970). *Bull. Chem. Soc. Japan*, **43**, 2398.
144. Okuda, T., Yamada, K., Ishihara, H., and Negita, H. (1975). *Chem. Lett.*, 785.
145. Onda, S., Ikeda, R., Nakamura, D., and Kubo, M. (1973). *Bull. Chem. Soc. Japan*, **46**, 2678.
146. Orlov, I. G., Koshelev, K. K., Poleshchuk, O. Kh., and Maksyutin, Yu. K. (1974). *Izv. Akad. Nauk SSSR, Ser. Khim.*, No. 8, 1915 [1843].
147. Patterson, D. B. and Carnevale, A. (1973). *J. Chem. Phys.*, **59**, 6464.
148. Patterson, D. B., Peterson, G. E., and Carnevale, A. (1973). *Inorg. Chem.*, **12**, 1282.
149. Péneau, A. (1975). *Thesis*, No. 1405, Orsay.
150. Petrakis, L. and Swift, H. (1968). *J. Phys. Chem.*, **72**, 546.
151. Petrosyan, V. S., Yashina, N. S., Reutov, O. A., Bryukhova, E. V., and Semin, G. K. (1973). *J. Organometall. Chem.*, **52**, 321.
152. Poleshchuk, O. Kh. and Maksyutin, Yu. K. (1975a). *Izv. Akad. Nauk SSSR, Ser. Fiz.*, **39**, 2579 [110].
153. Poleshchuk, O. Kh. and Maksyutin, Yu. K. (1975b). *Teor. Eksp. Khim.*, **11**, 406 [345].
154. Poleshchuk, O. Kh., Maksyutin, Yu. K., and Orlov, I. G. (1974a). *Izv. Akad. Nauk SSSR, Ser. Khim.*, No. 1, 109 [101].
155. Poleshchuk, O. Kh., Maksyutin, Yu. K., and Orlov, I. G. (1974b). *Izv. Akad. Nauk SSSR, Ser. Khim.*, No. 1, 234 [229].
156. Ragle, J. L., Minott, G., and Mokarram, M. (1974). *J. Chem. Phys.*, **60**, 3184.
157. Ramakrishnan, L., Soundararajan, S., Sastry, V. S. S., and Ramakrishna, J. (1977). *Coord. Chem. Rev.*, **22**, 123.
158. Read, M., Cahay, R., Cornill, P., and Duchesne, J. (1963). *C. R. Acad. Sci. (Paris)*, **257**, 1778.
159. Riedel, E. F. and Willett, R. D. (1975). *J. Amer. Chem. Soc.*, **97**, 701.
160. Robinson, H. G., Dehmelt, H. G., and Gordy, W. (1954). *J. Chem. Phys.*, **22**, 511.
161. Rogers, M. T. and Ryan, J. A. (1968). *J. Phys. Chem.*, **72**, 1340.
162. Saatsazov, V. V., Khotsyanova, T. L., and Kuznetsov, S. I. (1975). *Izv. Akad. Nauk SSSR, Ser. Khim.*, No. 4, 925 [839].
163. Safin, I. A., Kuramshin, I. Ya., Muratova, A. A., Pudovik, A. N., and Plekhov, V. P. (1975). *Zh. Obshch. Khim.*, **45**, 2400 [2359].
164. Sasane, A., Nakamura, D., and Kubo, M. (1967). *J. Phys. Chem.*, **71**, 3249.

165. Scaife, D. E. (1971). *Austr. J. Chem.*, **24,** 1753.
166. Schempp, E. and Bray, P. J. (1970). *Physical Chemistry, an Advanced Treatise* (Ed. Henderson, D.) Vol. IV, Chap. 11, Academic Press, New York.
167. Schneider, R. F. and Di Lorenzo, J. V. (1967). *J. Chem. Phys.*, **47,** 2343.
168. Schwartz, D. and Ragle, J. L. (1974). *J. Chem. Phys.*, **61,** 429.
169. Semin, G. K., Babushkina, T. A., and Yakobson, G. G. (1972). *Nuclear Quadrupole Resonance in Chemistry*, Khimia, Leningrad; IPST, John Wiley, 1975.
170. Semin, G. K., Robas, V. L., Shteingartz, V. D., and Yakobson, G. G. (1965). *Zh. Strukt. Khim.*, **6,** 160.
171. Shoshtakovskii, M. F., Poleshchuk, O. Kh., Maksyutin, Yu. K., and Orlov, I. G. (1973). *Izv. Akad. Nauk SSSR, Ser. Khim.*, No. 1, 15 [13].
172. Smith, J. A. S. (1971). *Chem. Educ.*, **48,** 39, 177, A147, A243.
173. Smith, J. A. S. (1974). *Advances in NQR* (Ed. Smith, J. A. S.) Vol. 1, Heyden, London, p. 115.
174. Smith, J. A. S. and Tong, D. A. (1971a). *J. Chem. Soc. A*, 173.
175. Smith, J. A. S. and Tong, D. A. (1971b). *J. Chem. Soc. A*, 178.
176. Sychev, O. F., Poleshchuk, O. Kh., and Maksyutin, Yu. K. (1975a). *Izv. Akad. Nauk SSSR, Ser. Fiz.*, **39,** 2675 [196].
177. Sychev, O. F., Poleshchuk, O. Kh., and Maksyutin, Yu. K. (1975b). *Izv. Akad. Nauk SSSR, Ser. Fiz.*, **39,** 2671 [192].
178. Tong, D. A. (1969). *J. Chem. Soc. Chem. Commun.*, 790.
179. Townes, C. H. and Dailey, B. P. (1949). *J. Chem. Phys.*, **17,** 782.
180. Volkov, A. F. (1973). *J. Magn. Res.*, **11,** 73.
181. Volkov, A. F., Romm, I. P., Guryanova, E. N., and Kocheshkov, K. A. (1976). *Izv. Akad. Nauk SSSR, Ser. Khim.*, No. 6, 1365 [1310].
182. Voronkov, M. G. and Feshin, V. P. (1973). *Determination of Organic Structure by Physical Methods* (Eds. Nachod, F. C. and Zuckermann, J. H.), Vol. 5, Academic Press, p. 169.
183. Voronkov, M. G., Feshin, V. P., Mironov, V. F., Mikhailyants, S. A., and Gar, T. K. (1971). *Zh. Obsch. Khim.*, **41,** 221 [2237].
184. Wakefield, S. L. (1970). *Diss. Abstr. Int. B*, **30,** 5004.
185. Weiss, A. (1974). *Advances in NQR* (Ed. Smith, J. A. S.), Vol. 1, Heyden, London, p. 1.
186. Weiss, A. (1972). *Topics in Current Chemistry*, Vol. 30, Springer, Berlin, p. 49.
187. Whitehead, M. A. (1964). *Canad. J. Chem.*, **42,** 1212.
188. Wulfsberg, G. A. (1975). *J. Organometall. Chem.*, **86,** 321.
189. Wulfsberg, G. A., West, R., and Rao, V. N. M. (1975). *J. Organometall. Chem.*, **86,** 303.
190. Wulfsberg, G. A. (1976). *Inorg. Chem.*, **15,** 1791.
191. Wulfsberg, G. A. and West, R. (1971). *J. Amer. Chem. Soc.*, **93,** 4085.
192. Wulfsberg, G. A., West, R., and Mallikarjuna Rao, V. N., (1973). *J. Amer. Chem. Soc.*, **95,** 8658.
193. Yamada, K. (1977). *J. Sci. Hiroshima Univ.*, **A41,** 77.
194. Yamasaki, R. S. and Cornwell, C. D. (1959). *J. Chem. Phys.*, **30,** 1265.
195. Yusupov, M. Z., Grechishkin, V. S., Kosulin, A. T., Anferov, V. P., and Versilov, V. S. (1971). *Org. Magn. Res.*, **3,** 515.
196. Zakirov, D. U., Kuramshin, I. Ya., Safin, I. A., Pudovik, A. N., and Zhelonkina, L. A. (1977). *Zh. Obshch Khim.*, **47,** [1661] 1522.
Jugie, G., unpublished results; J   unp.

Molecular Interactions, Volume 2
Edited by H. Ratajczah and W. J. Orville-Thomas
© 1981 John Wiley & Sons Ltd.

9 Mössbauer effect studies of molecular complexes

M. Pasternak and T. Sonnino

9.1 Introduction

It would be superfluous at this time to enumerate the developments in the physical sciences which followed the discovery of the Mössbauer effect in 1958 (Mössbauer, 1958a, b), an achievement which was recognized by the award of the Nobel Prize in 1961 (Mössbauer, 1961). A new spectroscopy was born and its applications extended to almost every field of experimental investigations in physics, chemistry, and biology.

Conferences devoted to applications of the Mössbauer effect (proceedings, 1976) and developments in its methodology (Gruverman, 1965) are held annually; an annual Index is devoted to the literature concerning this effect (Stevens and Stevens, 1965). Many books on this specific topic have been published since 1962, (see for example Wertheim, 1964; May, 1971), and there are also numerous publications on specific topics and applications (see for example Goldanskii and Herber, 1968; Greenwood and Gibb, 1971).

In this review, we consider only topics relevant to molecular interactions. Hence, after a brief summary of the physical phenomena which are implicated in the Mössbauer effect (section 2), and their relation to experimentally measured parameters (section 3), we shall discuss their use in deriving chemical information (section 4). The emphasis will be on a specific application—the study of the molecular complexes of iodine—where from the point of view of molecular interactions, the results are of great interest. Specific experiments will be described and interpreted according to the theories of molecular interactions. The iodine complexes are presented as a model, as they provide clear data, whose interpretation can contribute to the understanding of the chemical interaction in molecular complexes, due to a fortunate combination of circumstances.

The iodine atom participates directly in the interaction, rather than indirectly through other atoms in the molecule. Moreover interpretation of the experimental results is derived in terms of chemical parameters, and finally the theory for iodine molecular complexes is well established.

For the other systems where data are abundant as in iron and tin (Gutlich, 1975) the interpretation is less simple and these cases will only be considered briefly.

9.2 The Mössbauer Effect or Nuclear Gamma Resonance

The principles of energy and linear momentum conservation impose a recoil energy E_R on a free atom after a gamma ray with energy E_γ has been emitted by its radioactive nucleus. Similarly, an absorbing nucleus in its ground state will recoil as result of an impingent gamma ray. Due to the very narrow energy width of the absorption cross-section $\sigma(E)$ at the resonant energy E_0, no nuclear gamma resonance can be observed in free atoms under normal conditions. A solid, however, may provide suitable channels, for absorbing the thermal spike due to the recoil energy E_R. Providing this thermal energy is of the order of the solid's phonons' energy, a finite probability f, exists whereby a measurable fraction of gamma rays may be emitted and absorbed with the full transition energy, thus producing resonance absorption or scattering. Another way of expressing the requirement for resonance is that the energy of the thermal vibrations induced by the gamma ray emission (or absorption), in the crystal atoms, is within the range of the phonons' energy.

To satisfy the conditions of nuclear gamma resonance, these general rules must be obeyed:

1. The recoil energy E_R must be small:

$$E_R = E_\gamma/2Mc^2 \tag{9.1}$$

   where M is the free atom mass and c is the velocity of light.
2. The solids, where the emitting and absorbing atoms are embedded, must be 'stiff', that is the crystals should be characterized by a high Debye temperature θ_D. It can then be shown (Frauenfelder, 1962) that for very low temperatures ($T \cong 0$), the recoilless fraction f can be expressed as:

$$f = \exp\left(-E_R/k_B\theta_D\right) \tag{9.2}$$

   where k_B is the Boltzmann factor.

The exact treatment of the problem of the recoilless fraction is derived by the most general expression:

$$f = \exp\left(-E_R/k^2\langle x^2\rangle\right) \tag{9.3}$$

where now k is the wave number of the radiation, and $\langle x^2\rangle$ is the mean-square displacement of the emitting or absorbing nucleus in the solid.

The simplest theory dealing with atomic motions in solids is the Debye theory, that implies harmonic vibrations near the equilibrium point. The dynamics of the atoms in the crystal are described then by a single parameter the Debye temperature θ_D, that characterizes the distribution of vibrational motion, or the phonons' spectrum.

Using the Debye model, we can derive an expression relating the recoilless fraction to the absolute temperature T of the solids investigation as:

$$f = \exp\left\{-\frac{3E_R}{2k_B\theta_D}\left[1+4\left(\frac{T}{\theta_D}\right)^2\int_0^{\theta_D/T}\frac{x}{e^x-1}\,dx\right]\right\} \tag{9.4}$$

The formula (9.4) has two limits for high and low temperatures in respect to the Debye temperature:

$$f = \exp\left(-3E_R/2k_B\theta_D\right) \quad \text{for} \quad T \ll \theta_D \tag{9.5}$$

or

$$f = \exp\left(-E_R T/k_B\theta_D^2\right) \quad \text{for} \quad T > \theta_D \tag{9.6}$$

Debye temperatures for solids, as measured or calculated, are generally available from literature, and are used to estimate f.

Using equation (9.6) it is possible to derive the Debye temperature as measured by Mössbauer experiments, when the solids under study are investigated at different temperatures. From the values of the Debye temperatures, or the values of f, information is obtained on the dynamics of the atoms in the solid, and hence on the forces that bind it in the crystal.

In general, the lower the Debye temperature, the higher the recoilless fraction and the measured effect. In other words a high effect indicates a strong binding of the atom in the molecule or in the crystal.

Nuclear theory predicts the energy profile of the radiation emitted or absorbed, according to the formula of Breit–Wigner. The absorption cross-section $\sigma(E)$ is given by

$$\sigma(E) = \sigma_0 \frac{\Gamma^2}{\Gamma^2 + 4(E-E_0)^2} \tag{9.7}$$

Where σ_0 is the absorption cross-section at zero recoil, i.e. when the gamma ray has the full energy of the nuclear transition E_0. Γ is the linewidth of the radiation (gamma ray) dictated by the indetermination principle. This profile is called Lorentzian.

Experimentally a Mössbauer effect spectrum is an absorption spectrum, and the line(s) has a Lorentzian lineshape having a width of 2Γ at half of its maximum resonance intensity, according to (9.7). It is found that the experimental linewidth is generally broader than 2Γ due to a series of causes, especially the finite thickness of the samples, sources and absorbers, small hyperfine interactions, and chemical non-homogeneity.

In brief a Mössbauer effect experiment consists of the following:

1. A solid source of gamma rays, generally a crystal of cubic structure.
2. The absorber of the gamma rays, the sample to be studied.
3. A detector of the gamma rays, that selects only those of appropriate energy.
4. An apparatus for moving the source (absorber) with a relative velocity with respect to the stationary absorber (source), and then modulating the gamma ray energy via the Doppler Effect.

The gamma ray absorption or scattering is then scanned as function of the relative velocity v, and the abosrption spectrum is recorded.

Source and absorber are generally at low temperatures to increase the recoilless emission and absorption fractions f. Maximum absorption is obtained at σ_0 where absorption cross-section is maximum, equation (9.7); and the spectrum is represented by an inverted peak, generally near to zero relative velocity. Hyperfine interactions will shift the peak and split it in a series of absorption dips.

9.3 Hyperfine Interactions and Mössbauer Parameters

The atomic nuclei have mass, that is identified by their mass number and electric charges that determine the chemical identity of the element. Nuclei may be in different energy states, energy levels, that are characterized by the nuclear quantum numbers. The lowest energy level is the ground state, the others are excited states. Atomic nuclei interact differently with the electromagnetic field, external or that produced by their electrons, according to the energy state.

This interaction is described by an interaction Hamiltonian H, and has different effects depending on the quantum numbers of the nucleus and the structure of the field. Using the standard multipole expansion, it is possible to derive an expression for the consequences of this interaction as they are reflected in the Mössbauer effect spectra:

—shift in the energy levels,
—removal of the degeneracy of the nuclear levels inducing magnetic dipole splitting, or electric quadrupole splitting.

The formal expression for the Hamiltonian is:

$$H = H(e, 0) + H(m, 1) + H(1, 2) + \cdots \tag{9.8}$$

where the first term corresponds to the electric monopole interaction, between the nuclear electric charge distribution and the static electric field;

the second term to the magnetic dipole interaction between the nuclear magentic moments and magnetic fields; and the third term to the electric quadrupole interactions as determined by the nuclear quadrupole moments and the gradient of the electric field; higher-term interactions are scarcely observed in Mössbauer experiments.

The results of these interactions as observed in Mössbauer spectra are discussed below.

9.3.1 The monopole interaction and the isomer shift

A nucleus with a charge Ze and an alectrostatic radius R, interacts with the electrons' electric field at the nuclear region ($r = 0$), via the Coulomb potential. As a result the energy levels will be shifted by an amount $\delta(E)$ equal to:

$$\delta(E) = (2\pi/5)Ze^2R^2 |\psi(0)|^2 \tag{9.9}$$

where $|\psi(0)|^2$ is the electron density at the nuclear region, as determined by the electronic wavefunction ψ. Practically only s electrons have a finite density in the nuclear region, $r = 0$.

Since the Mössbauer effect is related to nuclear transitions from an excited state to the ground one, the difference in their relative shifts is of relevance ΔE:

$$\Delta E = \delta(E)_e - \delta(E)_g = (2\pi/5)Ze^2(R_e^2 - R_g^2) |\psi(0)|^2 \tag{9.10}$$

Experimentally, the difference between the emission (source, S) and absorption (absorber, A) energies is measured, and if their electron densities $|\psi(0)|^2$ are different, due to their different chemical composition, the observed quantity is:

$$\delta = (4\pi/5)Ze^2R^2(\Delta R/R)[|\psi(0)|_A^2 - |\psi(0)|_S^2] \tag{9.11}$$

The difference in nuclear radius ΔR, between the excited and the ground state is a nuclear property, and for a specific nucleus $\Delta R/R$ will determine the trend of the isomer shifts. If it is positive, an increase in electron density at the nuclear region in the absorber with respect to a fixed source, will lead to an increase of δ.

For ^{129}I the Mössbauer isotope that will be of most interest in this review, $\Delta R/R$ is positive; the same is true for ^{119}Sn, whereas in ^{57}Fe, for example, it is negative, leading to opposite results. The shifts are experimentally observed as the maximum resonant absorption, is not observed at zero relative velocity of the source with respect to the absorber, but the maximum dip is shifted to a velocity v:

$$v = (c/E_\gamma)(4\pi/5)Ze^2R^2(\Delta R/R)[|\psi(0)|_A^2 - |\psi(0)|_S^2] \tag{9.12}$$

where c is the velocity of light and E_γ is the energy of the gamma ray.

In the actual experiments a standard source, with a single narrow line is used and the properties of the absorbers are investigated. The values of isomer shifts are then always reported as relative to the chosen standard source, and indicate the relative velocity at maximum absorption. Under these conditions, then the isomer shift is a linear function of $|\psi(0)|^2_A$, the electron density at the nucleus of the Mössbauer atom imbedded in absorbers of different chemical composition:

$$v = \text{const.} \left[|\psi(0)|^2_A - C\right] \tag{9.13}$$

where the two constants, are derived from experiment. The relative velocities are generally reported in mm s^{-1}.

The electron density at the nuclear region is due mainly to s electrons, but chemical interest lies in changes in the valence shell structure which has two general influences on the electron density: first directly, due to valence s electrons and secondly indirect, via shielding of s electrons by p, d, valence orbital electrons: this causes the s-electron cloud to vary and results in a different value of $|\psi(0)|^2_A$ that is measured experimentally by the isomer shift.

9.3.2 *The electric quadrupole interaction*

Nuclear levels having a spin $I \geq 1$ possess an electric quadrupole moment which interacts with the electric field gradient produced by the non-spherical charge distribution. This interaction partially removes the nuclear level degeneracy splitting these energy levels into $(2I+1)/2$ states.

Thus for example, we observe two absorption lines in nuclei having a ground-state spin $I_g = \frac{1}{2}$, and an excited state spin $I_e = \frac{3}{2}$, due to the splitting of the excited-state level into $m = \pm\frac{3}{2}$ and $m = \pm\frac{1}{2}$ states, and consequently two transition lines. The general interaction Hamiltonian has the form:

$$H = \frac{e^2 q Q}{4I(2I-1)} \left[3I_z^2 - I^2 + \eta(I_+^2 - I_-^2)/2\right] \tag{9.14}$$

where eQ is the nuclear quadrupole moment and eq is the electric field gradient; the I are the components of the nuclear spin operator and η is the asymmetry parameter of the electric field gradient reflecting its deviation from axial (z) symmetry; ($0 \leq \eta \leq 1$). For the simple case of $I = \frac{3}{2}$ solution of equation (9.14) gives two energy levels split by an energy value:

$$\Delta E_Q = \frac{e^2 q Q}{4I(2I-1)} \left[3m^2 - I(I+1)\right](1 + \eta^2/3)^{1/2} \tag{9.15}$$

ΔE_Q, called the quadrupole splitting energy, is the measured quantity, i.e. the separation between the two absorption lines and is given in Mc s^{-1}, the unit used in NQR spectroscopy.

For nuclei whose spin is higher than 1 the number of levels increases and hence the number of transitions; in general then $(2I_e + 1)(2I_g + 1)/4$ lines are observed if the excited state has spin I_e and the ground state I_g. For ^{129}I, where $I_g = \frac{7}{2}$ and $I_e = \frac{5}{2}$ a maximum of 12 transitions should be observed, but if $\eta \cong 0$ only the transitions with $\Delta m = 0$; 1 are allowed producing 8 absorption lines. The sign of $e^2 q Q$ may be obtained by experimental results in cases where the nuclear spin is high, as for ^{129}I. When $I = \frac{3}{2}$ and only two lines are observed as in the cases of ^{57}Fe or ^{119}Sn the sign of this quantity cannot be derived from a simple experiment and it must be deduced from experiments with single oriented crystals, or when external magnetic fields are applied.

9.4 Chemical Information from Isomer Shift and Quadrupole Splitting Data

We shall consider here only information pertaining to iodine compounds. Correlations of isomer shift and quadrupole splitting with the electron configurations of iron, tin, and other elements, based on Mössbauer effect experiments have been described elsewhere. (Wertheim, 1964; Goldanskii and Herber, 1968; May, 1971; Greenwood and Gibb, 1971; Gutlich, 1975).

The standard source for iodine experiments is Zn^{129}Te; a cubic semiconductor. ^{129}Te decays to the excited state of ^{129}I, and a single relatively narrow line is obtained. Experiments are usually performed at low temperatures, e.g. 100 K and below.

Due to the simple outer shell configuration of iodine, the correlation of isomer shift and quadrupole splitting data with the electronic structure of the chemical compounds under study has been carried out with relative success (Review, 1974).

The outer shell configuration of atomic iodine is $5s^2 5p^5$ namely one p hole (h_p) in the full shell of Xe, equivalent to the ion I$^-$ ($5s^2 5p^6$). This full shell has been taken as reference for isomer shift and quadrupole splitting. Systematic studies of many iodine compounds, especially interhalogens have shown the trend when a chemical bond is formed primarily by p electrons. In fact, the results have demonstrated that there exists a linear correlation between the isomer shift δ and p and s holes, when data are referred to the standard ZnTe source.

$$\delta(\text{mm s}^{-1}) = -(9 \pm 1)h_s + (1.5 \pm 0.1)h_p - (0.54 \pm 0.02) \qquad (9.16)$$

This shows that s holes have an opposite and higher effect than p holes, and as $\Delta R/R$ is positive the higher the value of δ the more p holes there are in the electronic structure of the iodine atom in the sample. A few cases were found, particularly among the high oxidation states of iodine, where $s–p$ hybridization occurs; the isomer shift is then negative and large (Pasternak and Sonnino, 1968).

The number of p holes is correlated to the total population N of p electrons in the shell, considering the directional orbitals N_x, N_y, N_z then:

$$h_p = 6 - (N_x + N_y + N_z) \tag{9.17}$$

To correlate the quadrupole coupling data with the p electrons distribution in the different directions of the bond, that determines the electric field gradient at the nucleus, U_p is defined as:

$$U_p = eq_{zz}(\text{mol.})/eq_{zz}(\text{atom.}) \tag{9.18}$$

i.e. the ratio between the molecular value and the atomic value of the electric field gradient.

The classical theory of nuclear quadrupole resonance (NQR) gives:

$$U_p = N_{pz} - (N_{px} + N_{py})/2. \tag{9.19}$$

Thus for an axially symmetric bond, using the experimentally derived values of the quadrupole coupling, it is possible to deduce the p-electron population in the z direction (σ bond), and x, y directions (π bond).

In the case where q is not axially symmetric, the use of the experimental value of η, allows the solution for the p-electron populations in all three directions:

$$\eta = (N_{px} - N_{py})/N_{pz} \tag{9.20}$$

The immediate result is that η gives information on the symmetry of the bond.

9.5 Review of Selected Experimental Data

The Mössbauer effect can be observed only in solids, whereas the most interesting charge transfer complexes are found in liquid solutions. In 1963 it was suggested (Jaccarino and Wertheim, 1961) that liquids could be studied if transformed in fast-frozen solutions, obtained, for example, by their immersion in liquid nitrogen.

The I_2–benzene complex was the first molecular complex to be studied in this way in 1968 (Bukshpan et al., 1968), with the aim of obtaining information on the molecular parameters of iodine. Iodine was also studied, its molecule being isolated in inert gases and solvents. The results summarized in Table 9.1 confirm the basic assumption that the complex structure is preserved in the fast cooling process.

Further studies to obtain information on I_2 and iodine–halogen complexes with different organic solvents (Winter et al., 1969; Bukshpan et al., 1975), and solid complexes with strong polar molecules (Ichika et al., 1971) produced interesting results. The parameters deduced experimentally from Mössbauer spectroscopy, in solid solutions of inert solvents such as CCl_4 and

Table 9.1   Experimental data for I_2 in simple systems. ^{129}I is the Mössbauer atom (Bukshpan et al., 1975)

System	Isomer shift $(mm\ s^{-1})$	Quadrupole coupling* $(Mc\ s^{-1})$	Asymmetry parameter
I_2/solid Ar	0.93 ± 0.05	-2230 ± 20	0
I_2/inert solvents	0.95 ± 0.05	-2268 ± 20	0
I_2/benzene	0.77 ± 0.03	-2380 ± 25	0

* The quadrupole coupling data are reported in $Mc\ s^{-1}$, the resonance energy of ^{127}I in NQR.

hexane, are identical to those obtained when the I_2 molecule is trapped in a solid matrix of inert gas.

These results confirm the first investigation (Benesi and Hildebrand, 1949) of the spectroscopic properties of these solutions: they indicate clearly that the I_2 molecule is almost free having negligible interaction with the solid solvent, hence the parameters derived well represent the isolated state of the I_2 molecule.

The situation is different when frozen solutions of I_2 are prepared with strong polar solvents or with other solvents that are known to produce charge transfer complexes. In the case of the I_2–benzene complex, the data led to the conclusion that there is a charge transfer from benzene to the I_2 molecule, in accord with the theory of charge transfer complexes. It is possible to estimate the amount of charge transfered using the semi-empirical formulae (9.16–9.20). Employing the experimental values for isomer shift, quadrupole coupling, and asymmetry parameter it is possible to derive quantitative results for the s and p holes from the $5s^2 5p^6$ (I^-) configuration, and hence the net charge on, the iodide atoms, as well as the symmetry of the bond. In the particular case of the I_2–benzene complex the result $\eta = 0$ indicates a symmetrical arrangement as proposed by Aano (1959). Experimental data on other complexes are summarized in Table 9.2. Note that the Mössbauer spectra reveal the presence of two different states of iodine in ethyl ether and amino complexes, that are denoted as I' and I''. The ^{129}I Mössbauer effect in charge complexes of iodine has also been studied (Sakai et al., 1980).

Interpretation of these Mössbauer effect experiments is based on the theoretical work of Mulliken and his followers (Mulliken, 1950; Mulliken and Person, 1969). Without repeating the arguments previously presented (Bukshpan et al., 1975) we recall that there are two basic mechanisms proposed to explain charge transfer complex formation:

(a) Transfer of charge from the donor (D) molecule to the acceptor (A) molecule. This mechanism is described by a wavefunction for the ground

Table 9.2   Experimental data for ^{129}I in charge transfer complexes (Bukshpan *et al.*, 1968, 1975; Winter *et al.*, 1969; Ichika *et al.*, 1971)

Complex	Isomer shift (mm s^{-1})	Quadrupole coupling (Mc s^{-1})	Asymmetry parameter
Inert solvents			
Argon	0.93±0.05	−2230±20	0
Organic solvents	0.95±0.05	−2268±20	0
Cyclohexane	0.93±0.05	−2170±20	0.19
π donors			
Benzene	0.77±0.03	−2380±25	0
Toluene	0.93±0.03	−2240±20	0.16
p-Xylene	0.99±0.02	−2150±20	0.18
Oxygen compounds			
Methanol	0.80±0.05	−2320±20	0.07
Acetone	0.83±0.05	−2350±20	0.01
Ethyl ether I′	1.07±0.03	−2460±10	
Ethyl ether I″	0.75±0.03	−2280±15	
Amines			
Pyridine I′	0.28±0.05	−2600±15	
Acridine I′	1.64±0.02	−2840±20	0.13
Acridine I″	0.29±0.02	−1310±20	0.16
HMTA I′	1.51±0.02	−2580±20	0.19
HMTA I″	0.28±0.02	−1270±20	0.06
Phenazine	0.93±0.02	−2220±20	0.06

state of the complex of the type:

$$\psi_N = a\psi_0(A, D) + b\psi_1(A^-D^+)$$

in which b^2 is an approximate measure of the relative weight of the structure (A^-B^+), called the dative structure.

(b) The effects of pure inductive electrostatic forces, as proposed by Hanna (1968).

From our previous derivations (Bukshpan *et al.*, 1975), we conclude that both mechanisms play an important role and are responsible for complex formation. Calculating the amount of charge transferred to the I_2 molecule, we obtain the results presented in Table 9.3.

Considering the data of Table 9.2 for the two different iodine states of ether and amino complexes, there appear to be iodine atoms having positive and negative net charges. They can be identified as $I^{(+\delta)}$ and $I^{(-\delta)}$ where δ is this fractional charge that can also be calculated and whose values are given in Table 9.4.

Table 9.3   Charge transferred to the I_2
molecule in different complexes

Solvent	Charge transferred (electrons)
Hexane	0
Benzene	0.26
Methanol	0.22
Ethyl ether	0.07
Pyridine	0.16
HMTA	0.08
Acridine	0
Phenazine	0.18

The I^+ atom is the one that is bounded to the complexant organic molecule, as is clearly shown by the Mössbauer effect parameters, and by the fact that the pyridine–iodine–halogen complexes reveal only the spectrum characteristic of positive iodine (Winter et al., 1969).

In those studies the derivation of electronic charge states was relatively simple, the Mössbauer atom being that one directly involved in the charge transfer mechanism. In more complicated systems, the results are less clear but indicate interesting and potentially fruitful future lines of research.

Studies of CsI_3 and triiodide complexes of benzamide and amylose (Erlich and Kaplan, 1969) demonstrate the existence of non-equivalent iodine atoms: the terminal iodine atoms in these complexes being negatively charged, the central one positive.

Fluorine complexes of the type $IF_6^+AsF_6^-$ (Bukshpan et al., 1969) and more recently KrF_2MF_5 (M = As, Sb) (Holloway et al., 1977), have been studied, but in these cases the Mössbauer atom is surrounded by fluorine atoms, which are directly involved in the charge transfer.

Lastly, Mössbauer spectroscopy is being used to try to elucidate the structure of complexes of biological interest where charge transfer may be relevant.

Table 9.4   Partial charge on iodine atoms in I_2–
solvent complexes, when charge separation was
observed

Solvent	Partial charge (electrons)	
	I^+	I^-
Ethyl ether	0.07	0.14
Pyridine	0.29	0.45
HMTA	0.37	0.45
Acridine	0.45	0.45

9.6 References

Aano, S. (1959). *Progr. Theor. Phys. (Kyoto)*, **22**, 313.

Benesi, H. A. and Hildebrand, J. H. (1949). *J. Amer. Chem. Soc.*, **71**, 2703.

Bukshpan, S., Goldstein, C., and Sonnino, T. (1968). *J. Chem. Phys.*, **49**, 5477.

Bukshpan, S., Goldstein, C., Sonnino, T., May, L., and Pasternak, M. (1975). *J. Chem. Phys.*, **62**, 2606.

Bukshpan, S., Soriano, J., and Shamir, J. (1969). *Chem. Phys. Lett.*, **4**, 241.

Erlich, B. S. and Kaplan, M. (1969). *J. Chem. Phys.*, **51**, 603.

Frauenfelder, H. (1962). *The Mössbauer Effect*, W. A. Benjamin Inc., New York.

Goldanskii, V. I. and Herber, R. H. (Eds.) (1968). *Chemical Applications of Mössbauer Spectroscopy*, Academic Press, New York.

Greenwood, N. N. and Gibb, T. C. (1971). *Mossbauer Spectroscopy*, Chapman and Hall, London.

Gruverman, I. J. (Ed.). (1965). *Mössbauer Effect Methodology* published annually since 1965, by Plenum Press, New York.

Gutlich, P. (1975). In *Mössbauer Spectroscopy* (Ed. Gonser, U.), Springer–Verlag, Berlin, Heidelberg.

Hanna, M. W. (1968). *J. Amer. Chem. Soc.*, **90**, 285.

Holloway, J. H., Schrobilgen, G. J., Bukshpan, S., Hilbrants, W., and de Waard, H. (1977). *J. Chem. Phys.*, **66**, 2627.

Ichika, S., Sakai, H., Negita, H., and Morda, Y. (1971). *J. Chem. Phys.*, **54**, 1627.

Jaccarino, V. and Wertheim, G. K. (1961). In *The Proceedings of the Second International Conference on M.E.* (Eds. Compton, D. M. J. and Shoen, A. H.), John Wiley, 1962.

May, L. (Ed.). (1970). *An Introduction to Mössbauer Spectroscopy*, Plenum Press, New York.

Mössbauer, R. L. (1958a). *Z. Phyzik.*, **151**, 124.

Mössbauer, R. L. (1958b). *Naturwiss.* **45**, 548.

Mössbauer, R. L. (1962). *Le Prix Nobel en 1961*, Nobel Foundation, Stockholm.

Mulliken, R. S. (1950). *J. Amer. Chem. Soc.*, **72**, 600.

Mulliken, R. S. and Person, W. B. (1969). *Molecular Complexes*, Wiley–Interscience, New York.

Pasternak, M. and Sonnino, T. (1968). *J. Chem. Phys.*, **48**, 1997.

Sakai, H., Maeda, Y., Ichika, S., and Negita, H. (1980). *J. Chem. Phys.*, **72**, 6192.

See, for example, *The Proceedings of the International Conference on the Applications of the Mössbauer Effect*, Corfu, Greece, 1976, *Journal de Physique*, C 6, (1976). A review of ^{129}I results and interpretations may be found in the 1974 *Data Index* (ref. 5) p. 447.

Stevens, J. and Stevens, V. (1965). *Mössbauer Effect Data Index*, published annually since 1965, IFI, Plenum Press, New York.

Wertheim, G. K. (1964). *Mössbauer Effect. Principles and Applications*, Academic Press, New York.

Winter, C. I., Hill, J., Bledsoe, W., Shenoy, G. K., and Ruby, S. L. (1969). *J. Chem. Phys.*, **50**, 3872.

Molecular Interactions, **Volume 2**
Edited by H. Ratajczak and W. J. Orville-Thomas
© 1981 John Wiley & Sons Ltd.

10 Electrical conductivity of solid molecular complexes

K. Pigoń and H. Chojnacki

10.1 Introduction

The solid charge transfer and hydrogen-bonded systems were in the latest decades extensively studied for their crystal structure and electrical properties. There exists a number of essential reasons for the interest paid to these classes of compounds. The investigation of the crystal structures adds much to the understanding of the nature and properties of the intermolecular interactions, especially weak ones which do not result in formation of definite molecular entities in the dissolved or gaseous state. The spectroscopic properties of CT crystals offer an insight into the problem of molecular excitations and energy transfer in molecular solids. The discovery of astonishingly high electrical conductivity in many CT crystals started a new, dynamically developing chapter of solid-state physics and offered hope for new, spectacular applications.

It would hardly be possible to overestimate the perspectives of these studies in life science. From the very beginning of the investigation into the electrical properties of organic solids, these implications were stressed by Szent-Györgi (1941) and a great deal of work, especially on hydrogen-bonded solids has been done along this line ever since.

The molecular complexes and ion-radical salts are nowadays perhaps the most widely studied class of organic solids.

10.2 Crystal Structures of Molecular Complexes

10.2.1 *Phase diagrams of systems with* CT *interaction and stoichiometry of* CT *complexes*

Although the characteristic features of the phase diagrams of systems with CT interaction had been noted in the early stages of the study of molecular complexes and a large quantity of experimental material had been collected (cf. e.g. Landolt–Börnstein Tables, 1956), further interest in this topic has been aroused recently and new data have been published.

In a typical binary donor–acceptor system a congruently melting equimolar CT complex forms simple eutectics with each of the components. It has been tacitly assumed that there is no mutual solubility within the solid phases and only rarely is information on possible polymorphism in a CT complex available. The 1:1 stoichiometry is also usually found for CT complexes precipitated from solutions of the components; only a few examples of a non-equimolar composition are known, except for the radical salts and complexes of TCNQ with 1:2 donor–acceptor ratio (cf. Gutmann and Lyons, 1967, Table 8.5). As a rule there is only one complex formed from a given donor–acceptor pair, regardless of the donor–acceptor ratio in the solution, exceptions being some iodine complexes (Kommandeur and Hall, 1961; Uchida and Akamatu, 1961; Mainthia et al., 1964) and 2,3-dihalogeno-5,6-dicyano-p-benzoquinone complexes (Matsunaga, 1965a). Other examples are the perylene–TCNQ and hexamethylbenzene–TCNE systems, where DA as well as D_3A (or D_2A resp.) complexes may be obtained from solution, depending on the conditions of precipitation (Vincent et al., 1974; Truong and Bandrauk, 1976).

After differential scanning calorimetry (DSC) had been introduced as a routine tool in thermal analysis of binary donor–acceptor systems, their phase diagrams turned out to be far more complicated. As an example the phase diagram of the pyrene–picryl chloride system is shown in Figure 10.1 (Bando and Matsunaga, 1976).

The existence of as many as five compounds has been claimed in this case viz.: DA, D_4A_3, D_2A, D_3A, and D_4A, all but the equimolar one melting incongruently. Solid–solid phase transitions have been observed by DSC and X-ray analysis in DA, D_2A, and D_3A compounds.

The reexamination of the phase diagram of the pyrene-picryl chloride system revealed the existence of two viz. DA and D_3A_2 complexes only (Krajewska, Wasilewska to be published).

The formation of incongruently melting non-equimolar compounds has been proved for a number of other systems, e.g. pyrene–pyromellitic dianhydride (PMDA) (Herbstein and Snyman 1969), perylene–PMDA (Boeyens and Herbstein 1965), anthracene–2,4,7-trinitrofluorenone (TNF) and pyrene–TNF (Krajewska and Pigoń, 1978), o-bromoaniline–picric acid (Komorowski et al., 1976). The phenomenon is presumably quite common, at least among certain donors and/or acceptors.

The phase diagram of the o-bromoaniline–picric acid system shown in Figure 10.2 reveals some interesting features.

The D_2A complex of picric acid and o-bromoaniline is of the 'charge-and-proton-transfer' type (CPT), i.e. the picric acid molecule acts as an electron acceptor (Lewis acid) and a proton donor (Brønsted acid) simultaneously,

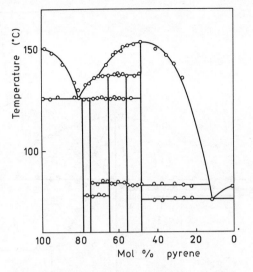

Figure 10.1 Phase diagram of the pyrene–picryl chloride system (Bando and Matsunaga, 1976). The endothermic peaks at 77, 81, and 82–83 °C have been ascribed to solid–solid transitions in 3:1, 1:2, and 2:1 compounds respectively. The peritectic points of 2:1 and 3:1 compounds (136 °C) are too close to each other to be separated by DSC. Similarly, the eutectic point and peritectic points of the 4:1 and 3:1 compounds at (126 °C) are unresolved

one bromoaniline molecule being a Lewis base and the other a Brønsted base. This interesting class of compounds was first described by Kofler in 1944, and has since been studied in detail by Matsunaga and Saito (Matsunaga and Saito, 1972; Saito and Matsunaga, 1973).

The interactions mentioned above are also acting in the 1:1 compound in this system, and the phase transition occurring at 375 K is due to a change in the bonding character. The low-temperature modification is a salt (ionic pair) with proton transferred to the bromoaniline molecule, whereas in the high-temperature phase CT interaction between neutral molecules prevails. The transition is accompanied by a change in colour from yellow (salt) to red (CT complex), as well as by appropriate changes in the IR spectrum. This 'complex isomerism' (Hertel, 1926) has been studied extensively by Briegleb (Briegleb and Delle, 1960a,b) and Matsunaga et al. (1975).

The phase diagram shown in Figure 10.2 reveals also a considerable solubility of the components in the 1:1 compound and a change in the phase transition temperature due to non-stoichiometry.

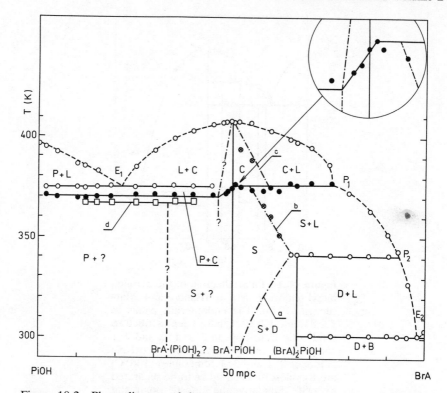

Figure 10.2   Phase diagram of the *o*-bromoaniline–picric acid system. [Reproduced by permission of Gordon and Breach Science Publishers from Komorowski *et al.*, 1976.] Solid–solid salt–CT complex transition points are given by dots, composition of solid terminal solutions by – · · – ·

Solid solutions of two CT complexes of different donors with the same acceptor are also known. Lower (1969) studied phase equilibria within the pseudo-binary system: anthracene–1,3,5-trinitrobenzene (TNB)–phenanthrene–TNB, and found the existence of two sets of solid solutions with the general formula: [(anthracene)$_x$(phenanthrene)$_{1-x}$]–TNB. Those with $0 \leqslant x \leqslant 0.6$ crystallized with a phenanthrene–TNB structure, those with $0.8 \leqslant x \leqslant 1.0$ with an anthracene–TNB structure; for $0.6 < x < 0.8$ a miscibility gap was found. More complicated phase relations have been established in the anthracene–phenanthrene–picric acid system (Koizumi and Matsunaga, 1974). In the system anthracene–phenanthrene–1,2,4,5-tetracyanobenzene, however, donor components form a continuous series of mixed CT complexes with acceptor (Wright *et al.*, 1976). The structural and spectral properties of such ternary CT complexes seem to be of considerable interest.

10.2.2   *Crystal structures of* CT *complexes*

The majority of charge transfer complexes studied with respect to their electric, photoelectric, and magnetic properties belong to the $\pi-\pi$ type, the more important exception being the $n-\pi$ complexes of iodine with aromatic hydrocarbons only.

A characteristic feature of the $\pi-\pi$ complex crystal structure is the presence of stacks of nearly parallel donor and acceptor molecules, which alternate within a stack in a sequence corresponding to the complex stoichiometry (e.g. ..ADAD.. or ..ADDADD..). In this respect CT complexes differ significantly from the highly conducting ion-radical salts (as TTF–TCNQ) where components form separate stacks in the crystal lattice. However, some modifications of the usual alternating stacks arrangement are also known. The crystals of the 8-hydroxyquinoline–1,3,5-trinitrobenzene (TNB) 1:1 complex is built up from hydrogen-bonded 8-hydroxyquinoline dimers alternating within the stacks with pairs of isolated centrosymmetrically related TNB molecules (Castellano and Prout, 1971). In some complexes with non-equimolar stoichiometry, e.g. $(pyrene)_2$–$(picryl\ bromide)_3$, molecules of the excess component occupy interstitial positions between the mixed stacks (Herbstein and Kaftory, 1975). In the o-phenanthroline–TCNQ complex D and A molecules are arranged as separate entities rather than in the form of continuous stacks (Goldberg and Shmueli, 1977).

Interplanar distances between adjacent donor and acceptor molecules within a stack are somewhat shorter (3.2–3.3 Å) than the van der Waals contact distance (3.4 Å in aromatic hydrocarbons), but the effect is due to dipolar or hydrogen bonding interactions rather than to the CT interactions.

According the Mulliken's 'Overlap and Orientation Principle' (Mulliken, 1956) the mutual orientation of donor and acceptor molecules should provide the most effective overlap of the orbitals involved in CT interaction. Prout and Wright (1968) and Mayoh and Prout (1972) examined the problem in detail and concluded that the relative orientation of donor and acceptor is the result of a compromise between the requirements of the best CT stabilization and the dipole-dipole or hydrogen bonding interaction, and only when these stronger interactions show little sensitivity to orientation is the maximum orbital overlap realized. The authors compared the value of CT stabilization energy in the real crystal structures with its maximum value corresponding to the 'best' orientation. In 19 of the 29 CT complexes examined, 80–100% of the maximum stabilization energy was attained, in others, where the requirements of the dipolar or hydrogen bonding interaction were competitive with the orbital overlap principle, smaller CT stabilization was found or none at all.

The condition of maximum CT interaction does not usually correspond to the centre-on-centre donor–acceptor orientation (see Figure 10.3a). When now and then such an arrangement would be favourable with regard to the energy, it is precluded by the crystal packing conditions, unless the size and shape of donor and acceptor molecules are similar, as in the anthracene–TCNQ complex (see Figure 10.3b). In this case molecular planes are strictly perpendicular to the stacking axis (see Figure 10.4c). Two other alignments of donor–acceptor stacks commonly found in CT crystals are those shown in Figures 10.4a and 10.4b. The angle between the normal to the molecular plane and the stacking axis amounts here to 5–30°.

The interactions between donor and acceptor molecules within the stack are often quite insensitive to tiny in-plane rotation of molecules, a fact which accounts for the rather large amplitudes of librational motions as well as for the considerable degree of disorder found in the majority of CT crystals (Shmueli and Goldberg, 1973). In some structures there are two, or more, non-equivalent positions of the donor or/and acceptor molecule corresponding to the minima of potential energy of equal or nearly equal depth, separated by a potential barrier. The positions are occupied in proportions depending on the relative depths of the potential wells, e.g. in 0.5 to 0.5 ratio in the naphthalene–1,2,4,5-tetracyanobenzene (TCNB) complex (Kumakura et al., 1967), or in 0.93 to 0.07 ratio in pyrene-d_8–TCNE complexes (Larsen et al., 1975) (see Figure 10.5).

When the potential barrier between the minima is fairly low molecules will reorientate between the corresponding positions more or less freely, especially at higher temperatures. This type of dynamical disorder was studied by the NMR technique (temperature dependence of linewidth) by several authors. Alexandre and Rigny (1972) found that in the azulene–TNB complex both donor and acceptor molecules perform rotational jumps with characteristic frequencies $\omega_i = 1.1 \times 10^{15} \exp(-E_i/kT)$, where E_i amounts to 0.55 and 0.45–0.49 eV for TNB and the azulene molecule respectively. Similar motions have been found by the authors in complexes of TNB with other donors viz. naphthalene, indole, and benzothiophene. A much lower barrier, of the order of kT, separates two donor positions (about 12° apart) in the o-phenanthroline–TCNQ complex. Thus almost free rotation of donor molecules could be expected here (Goldberg and Shmueli, 1977).

In the low-temperature phase of the pyrene–PMDA complex pyrene molecules occupy two different sites and the stacks are arranged after the pattern .. D_1AD_2A .. Above an order–disorder transition temperature (near 200 K) a single 'average' pyrene molecule only is found in the stack, which results in halving of the c-axis length (Herbstein and Snyman, 1969). NMR observations lead to the conclusion that the disorder is of a dynamic nature and that at room temperature pyrene molecules move not only between two sites but by complete reorientation in their molecular planes (Fyfe, 1974).

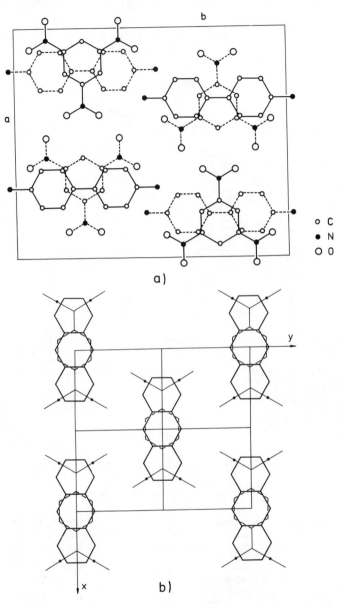

o C
• N
O O

a)

b)

Figure 10.3 (a) Projection of the molecular arrangement onto the (001) plane for the benzidine–TNB solvent-free molecular complex (Tachikawa *et al.*, 1974). (b) Projection of the molecular arrangement onto the common molecular plane for the anthracene–TCNQ molecular complex (Williams and Wallwork 1968). [(a) and (b) reproduced by permission of the International Union of Crystallography]

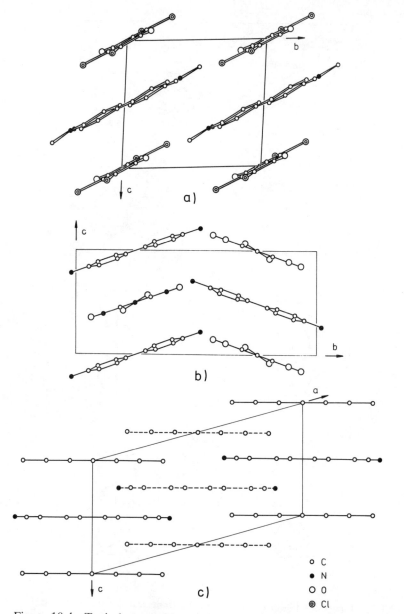

Figure 10.4 Typical arrangements of donor–acceptor stacks in crystal structures of CT complexes. (a) N,N,N',N'-Tetramethylbenzidine–chloranil (Yakushi et al., 1973); (b) benzidine–TNB (Tachikawa et al., 1974); (c) anthracene–TCNQ (Williams and Wallwork 1968). [(a), (b), and (c) reproduced by permission of the International Union of Crystallography]

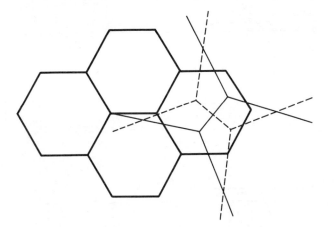

Figure 10.5   Two possible arrangements of the acceptor molecule in the structure of the pyrene-d_8–TCNE complex at 105 K. The solid-drawn TCNE molecule occupies the more populated position. [Reproduced by permission of the International Union of Crystallography from Larsen *et al.*, 1975]

Other types of structural disorder have also been reported for CT complexes. Kobayashi (1973) found in the carbazole–TCNQ complex a disordered domain structure as a result of the existence of two alternative orientations of carbazole molecules with respect to TCNQ. These domains are assumed to be of the opposite polarity and a strong electrostatic interaction between them would account for the singular features of the complex (e.g. for its low solubility in comparison to similar TCNQ complexes). The same author reported (1974) the existence of a sinusoidal modification of crystal structure of the phenothiazine–TCNQ complex, due to periodic distortion of the unit cell by weak hydrogen bonds between donor and acceptor molecules of adjacent stacks.

A large degree of disorder seems to be a characteristic feature of CT crystals and it would be interesting to discuss its influence on the electrical properties of these materials.

10.2.3   Crystal structures of hydrogen-bonded crystals

When only one hydrogen atom is able to form a hydrogen bond, crystal structures are relatively simple. Imidazole which is a constituent of histamine and histidine may be representative of this type of conducting solid. Its unit cell of $P2_1/c$ symmetry comprises four molecules (Martinez-Carrera, 1966). The imidazole rings, linked by hydrogen bonds of 2.86 Å in length, form infinite chains extended along the c axis (Figures 10.6 and 10.7) and

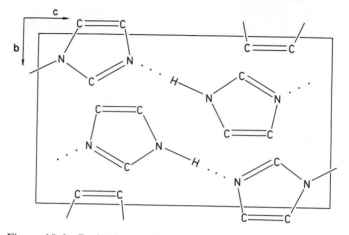

Figure 10.6   Projection of the imidazole unit cell on the (100) plane. Infinite chains are extended along the *c* axis. [Reproduced by permission of the International Union of Crystallography from Martinez-Carrera, 1966]

two perpendicular cleavage planes are observed. The twisting angle between two adjacent molecules within the chain is 61.5°. The same value for the hydrogen bond distance was found by neutron determinations (Craven *et al.*, 1977) and by X-ray studies (Martinez-Carrera, 1966) and no significant changes in the lattice parameters were detected. It is concluded from a neutron scattering study that possible proton transfers between two sites in

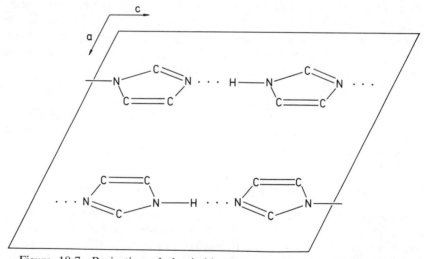

Figure 10.7   Projection of the imidazole unit cell on the (010) plane. [Reproduced by permission of the International Union of Crystallography from Martinez-Carrera, 1966]

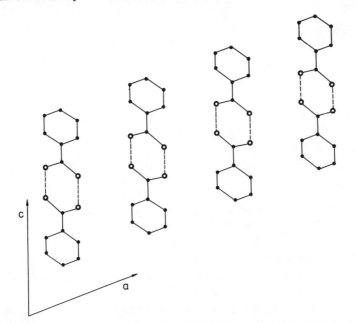

Figure 10.8 Projection of dimers in the benzoic acid crystal on the (010) plane. [Reproduced by permission of the International Union of Crystallography from Sim *et al.*, 1955]

the hydrogen bonds have frequencies below $3 \times 10^8 \, s^{-1}$ (Palacios-Gomes *et al.*, 1977).

When two hydrogen bonding atoms per molecule are present then the possibilities for different structures are much more extensive. For the benzoic acid crystal infinite chains (Figure 10.8) are built of cyclic dimers linked to each other by two identical O—H · · · O hydrogen bonds 2.64 Å in length (Sim *et al.*, 1955). All other nearest neighbouring distances are of the order of 3 Å, and correspond to normal van der Waals interactions. The X-ray results (Sim *et al.*, 1955) suggest that in the benzene carboxylic acids ionic structure (**1**) makes a considerable contribution as well as structures (**2**) and (**3**)

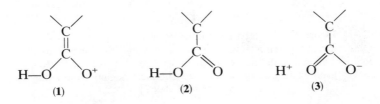

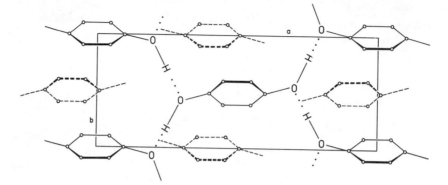

Figure 10.9   Projection of molecules of γ-hydroquinone on the (001) plane
(Kitaigorodsky, 1949)

The γ-crystal modification of hydroquinone is an example of a structure with two hydrogen bonds where a two-dimensional network is formed (Kitaigorodsky, 1949) (Figure 10.9).

Within the $P2_1/a$ unit cell of this modification there are four molecules and only one cleavage plane is found. However, in the rhombohedral ($C\bar{3}$) α-crystal modification of hydroquinone, molecules are bridged to each other (see Figure 10.10) by hydrogen bonds 2.75 Å in length, forming a three-dimensional network without a cleavage plane (Palin and Powell, 1947). Large intermolecular distances enable the inclusion of water, SO_2, H_2S,

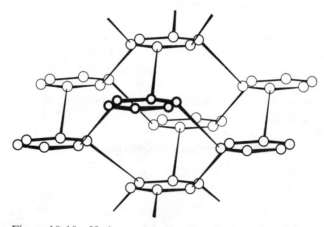

Figure 10.10   Hydrogen bonding in α-hydroquinone. The hexagons denote the hydrogen bonds and the longer lines connecting different hexagons are the O ··· O axes. [Reproduced by permission of the Chemical Society from Palin and Powell, 1947]

HCl, CH_3OH, and other small molecules where the hydrogen bonding effect seems to be a decisive parameter in forming clathrates. Libration motions or free rotation of included molecules are often observed.

Quinhydrone is a system where charge transfer interactions play an important role and hydrogen bonds seem to affect the crystal structure. In monoclinic α-quinhydrone different molecules are connected by hydrogen bonds formed between the hydroxyl and the carbonyl groups resulting in an infinite molecular chain (Figure 10.11) (Sakurai, 1968). In β-quinhydrone, hydroquinone, and quinone molecules are linked by hydrogen bonds to form zigzag molecular chains packed side by side by the charge transfer interactions to form a molecular sheet (Figure 10.12) (Sakurai, 1965). The structure of the molecular sheet is similar to that in the monoclinic form, although the remarkable deformation of the hydroquinone molecule, reported for monoclinic quinhydrone, does not exist in the triclinic modification. Interaction of the two kinds of molecules of the hydrogen bonding type prevails and charge transfer interactions are not as strong as in other charge transfer complexes (Fukushima and Sakurada, 1976).

More details about the crystal structures of hydrogen-bonded solids can be found in excellent books by Hamilton and Ibers (1968) and by Kitaigorodsky (1973).

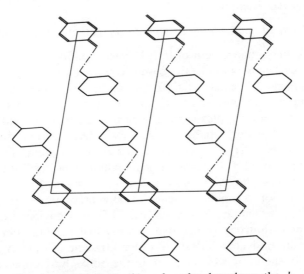

Figure 10.11 Projection of molecules along the *b* axis in the crystal of α-quinhydrone. Hydroquinone and quinone amolecules are linked by O—H $\cdots$ O hydrogen bonds to form a zigzag molecular chain along the (120) direction. [Reproduced by permission of the International Union of Crystallography from Sakurai, 1968]

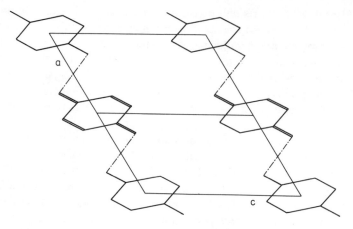

Figure 10.12 Projection of molecules along the b axis in the crystal of β-quinhydrone. [Reproduced by permission of the International Union of Crystallography from Sakurai, 1965]

10.3 Survey of Experimental Results on Electrical Properties of Molecular Complexes

10.3.1 *Electrical conductivity and photoconductivity in* CT *complexes*

The remarkable enhancement of the electrical conductivity of charge transfer complexes as compared to the parent components was established more than 20 years ago (Akamatu *et al.*, 1954) and it was mainly that observation which stimulated extensive studies on this class of compounds. A vast number of different complexes have been tested for their conductivity ever since, the most widely studied were those derived from the acceptors containing nitro or cyano groups (1,3,5-trinitrobenzene (TNB), 2,4,7-trinitrofluorenone (TNF), tetracyanoquinodimethane (TCNQ)), carbonyl groups (quinones, haloquinones, pyromellitic dianhydride (PMDA)) or iodine, the donor component being usually aromatic or heteroaromatic hydrocarbons, or diamines (e.g. benzidine (BZ), *p*-phenylenediamine (PDA) or its tetramethylnitrogen derivative (TMPD)) or sulphur-containing compounds (tetrathiofulvalene (TTF), tetrathiotetracene (TTT)). A comprehensive review of the early work is given in the books by Gutmann and Lyons (1967) and by Meier (1974). There is no need to complete the list, as this would be beyond the scope of this article; some general remarks will be given instead.

The specific conductivity values of CT complexes cover a range of more than 12 orders of magnitude (we have excluded the best conducting complexes being in fact ion-radical salts, which we will not deal with here), and it is reasonable to divide the CT complexes into two classes: the well or moderately conducting with room temperature conductivity between 10^{-6} and $10^{-1} \, \text{ohm}^{-1} \, \text{m}^{-1}$ and poorly conducting ones ($10^{-13} < \sigma_{300} < 10^{-6} \, \text{ohm}^{-1} \, \text{m}^{-1}$). The classification, although arbitrary as to the position of the borderline, is conterminous with the concept of strong (dative) and weak (non-dative) complexes, based on the degree of charge transfer between the donor and acceptor molecules (Kuroda *et al.*, 1962c; Matsunaga, 1965a). Strong CT complexes are formed by acceptors of high electron affinity (iodine, TCNQ, halo- or halocyanoquinones) and donors of fairly low ionization potential (phenazine, phenothiazine, TMPD, TTT). They are paramagnetic and their IR spectra differ distinctly from the superposition of the spectra of their components. Although the conductivity and paramagnetism of strong CT complexes seem to be mutually related, e.g. they change in the same direction on doping, no direct relationship could be found either between the value of specific conductance and the concentration of unpaired spins, or between the energies of thermal activation for the two effects. The typical values of the activation energy of conductivity in well-conducting CT complexes (evaluated from the relationship $\sigma = \sigma_0 \exp(-\varepsilon_a / kT)$) amount to 0.2–0.6 eV; activation energies of unpaired spins concentration (ε_s) are of 0–0.15 eV only.

It is difficult to account for this behaviour because we are still lacking models of generation and transport of charge carriers in organic solids as well as an accepted opinion on the nature of the paramagnetic centres in CT complexes.

Some authors assume the paramagnetic centres in moderately or even in weakly conducting complexes to be either triplet states originating from ionic ground singlet states, or doublets formed from neutral singlets; in both instances the intensity of the ESR signal varies with temperature as $T^{-1} \exp(-\varepsilon_s / kT)$ approximately. However, electrons in the excited states do not contribute to the conduction, thus an extra energy ($\varepsilon_a - \varepsilon_s$) is necessary to separate charge carriers and make them mobile (Ottenberg *et al.*, 1963; Bhat and Rao, 1969). On the other hand, Munnoch and Wright (1976) when assuming intrinsic conductivity in poorly conducting complexes (e.g. hexamethylbenzene–TCNQ or perylene–chloranil) ascribe the observed ESR signal to antiferromagnetic interactions between spins in clusters of ionic 'impurities' inevitably existing in such compounds. These 'impurities' are presumably $M^+(\text{acceptor})^-$ entities formed spontaneously in concentration up to 10^{20} per mole, due to charge transfer between intrinsic or extrinsic donors and the intrinsic acceptor. According to this model there is no relation between spins and free electrons densities nor between ε_a and ε_s. Unlike the former authors, Carnochan and Pethig (1976), who studied dielectric dispersion in the perylene–chloranil complex, interpreted the conduction process in terms of activated hopping of electrons between localized trap sites. Here the activation energy represents the depth of the traps and bears no reference to the 'energy gap' of an intrinsic semiconductor.

Kuroda and coworkers (1962a, b, c) have established a linear relationship between the 'energy gap' ε_a' taken as double the value of activation energy ε_a and the position of the maximum of the charge transfer band evaluated from the solution spectrum

$$\varepsilon_a' = h\nu_{CT(\text{solution})} + a$$

a is constant for a given series of complexes and of the order of -0.5 to $+0.5$ eV. The relationship holds for a series of poorly conducting complexes formed by a given acceptor (e.g. TNB or TCNE) and a number of chemically related donors (e.g. aromatic hydrocarbons or amines).

To explain this relationship, as well as photoconduction phenomena in CT complexes, an energy level diagram shown in Figure 10.13 has been proposed by the authors (Akamatu and Kuroda, 1963). Absorption in a charge transfer band leads to an excitonic state in which pairs of carriers are localized on neighbouring molecules. The lowest conduction state is characterized by the presence of widely separated donor and acceptor ions which

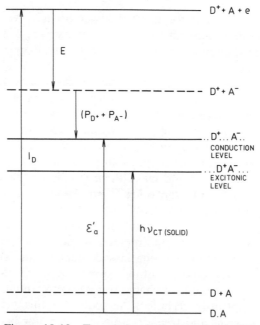

Figure 10.13 Energy level diagram for a CT complex. [Reproduced by permission of the American Institute of Physics from Akamatu and Kuroda, 1963]

do not interact electrostatically. A charge carrier attached to an acceptor (or donor) site can thus move freely, or almost freely, to the neighbouring neutral acceptor (or donor) taking part in the conduction process. The energy of formation of such a separated pair of ions is related to the ionization potential of the donor molecule I_D and the electron affinity of the acceptor molecule E_A by the following relationship:

$$\varepsilon_a' = I_D - E_A - (P_{D^+} + P_{A^-})$$

where $(P_{D^+} + P_{A^-})$ is the sum of the polarization energies of the lattice by D^+ and A^- ions. It has been estimated in a similar scheme by Munnoch and Wright (1976) to be 3.0–3.5 eV approximately.

On the other hand, the energy of CT excitation is related to I_D and E_A by the formula

$$h\nu_{CT(solution)} = I_D - E_A - E_{Coul} - E_{res}$$

where E_{Coul} and E_{res} are the Coulombic and resonance energies respectively. The authors claim the difference

$$h\nu_{CT(solution)} - \varepsilon_a' = (P_{D^+} + P_{A^-}) - E_{Coul} - E_{res}$$

to be approximately constant in a selected series of complexes.

One has to be aware, however, of rather poor reproducibility of activation energy values for badly defined, polycrystalline samples. The substantial role of injection and trapping of carriers in the conduction of weakly conducting materials has been repeatedly stressed and a remarkable influence of the degree of crystallinity on the values of σ and ε_a has also been observed. Thus there is some arbitrariness in selecting ε_a' values, and the relationship between ε_a' and $h\nu_{CT}$ can be quite accidental. Moreover, it is an open question, whether the carrier recombination is a bimolecular or monomolecular process. In the latter case the 'energy gap' value should be equal to ε_a rather than to $\varepsilon_a' = 2\varepsilon_a$ as was assumed by Kuroda *et al.*

The energy level diagram shown in Figure 10.13 is compatible both with the band and the hopping model of carrier transport. The latter seems to be more realistic in view of general concepts of electrical conduction in organic solids as well as of experimental evidence. Carnochan and Pethig (1976) have found a dielectric absorption with a relaxation time amounting to 8 s at 294 K and depending exponentially on $1/T$ in the perylene–chloranil complex. An unrealistically large value of the dipole moment has to be assumed to explain the results within the framework of the Debye relaxation mechanism, and an interpretation in terms of an activated hopping electron model was considered by the authors as most likely.

Additional supporting evidence for charge carrier hopping in CT complexes is provided by the character of the temperature dependence of carrier mobility (cf. next paragraph) and the observation of a frequency dependence of the a.c. photoconductivity (Schoenes and Kanazawa, 1976). While the conduction in a band is a frequency-independent process, hopping among localized states yields a frequency-dependent a.c. photoconduction. This is the case for the poly(N-vinylcarbazole)–TNF complex, where a $\sigma_{ph} \sim \omega^n$ relationship has been found by the authors, exponent n increasing from 0.25 at 300 K to 0.8 at 100 K. The latter value corresponds to single hops between pairs of sites, the former to multiple hops the probability of which increases with increasing temperature.

The energy level diagram in Figure 10.13 indicates that the excitation in the CT band does not necessarily result in photoconduction. Indeed, the principal peak in the photoconductivity action spectrum corresponds usually to the donor absorption band and not to CT absorption. Photocurrent carriers are generated either by the bimolecular recombination of singlet excitons or by their interaction with defects (especially near the surface). In some cases, however, there is a second peak corresponding to the CT absorption. This has been found in defective pyrene–TCNE crystals, by Akamatu and Kuroda (1963), who assumed thermal activation of the CT excitonic state to the conduction state (cf. Figure 10.13), and by Kramarenko et al. (1977) who explain the effect in terms of bimolecular recombination of CT excitons. According to Haarer and Karl (1973) and Möhwald and Sackmann (1974) excitation in the CT band results in formation of mobile triplet excitons of donors. Frankevich et al. (1977) have proposed an energy level diagram, shown in Figure 10.14, to explain the

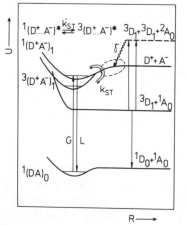

Figure 10.14 Energy level diagram proposed by Frankevich et al. (1977) to explain carrier generation via CT excitation. [Reproduced by permission of Akademie-Verlag, Berlin from Frankevich et al., 1977]

influence of the magnetic field on the photoconduction of the anthracene–dimethylpyromellitimide complex crystals.

The first excited singlet CT state of the complex $^1(D^+A^-)_1$ produced by the absorption of $h\nu_{CT}$ quanta decays either by fluorescence emission or by two alternative radiationless processes, *viz.* dissociation on defects and intersystem transition. The latter leads to an excited triplet CT state $^3(D^+A^-)_1$ equivalent to the anthracene triplet exciton $^3D_1$. The annihilation of two strongly excited (singlet or triplet) CT states $^{1,3}(D^+ \cdots A^-)^*$ (presumably ion-radicals spaced at small distances in their mutual Coulomb field), being intermediate in the $^3D_1-^3D_1$ annihilation, is the second charge carrier generation process operating here. The external magnetic field influences the photoconductivity due to the change of rate constants of the intersystem crossing and $^3D_1-^3D_1$ annihilation.

Important information about the mechanism of conduction in CT complexes could be deduced from a study of the anisotropy of conductivity, if reliable data were not so difficult to obtain. Contrary to the highly conducting ion-radical salts where several measurements of the directional dependence of conductivity have been reported, only a few results have been published for CT complexes (Kronick and Labes, 1961; Kokado *et al.*, 1964). High anisotropy values given by Kronick and Labes for 1,6-diaminopyrene–chloranil and 1,5-diaminonaphthalene–chloranil ($1:10^2$ or even $1:10^3$) do not confirm the conductivity and mobility data found for other complexes and are presumably due to contact resistance in these low-conducting materials.

An attempt to link electric properties of CT complexes with their crystal structure has been given by Vincent and Wright (1974), who compared resistivity, activation energy, and quantum yield for photoconductivity for a series of 24 structurally related complexes of aromatic hydrocarbons with TCNQ and some other acceptors. While the dark conductivity revealed extrinsic behaviour, presumably due to the presence of $M^+(acceptor)^-$ entities, where M^+ is an impurity ion such as sodium, photoconductivity showed intrinsic character. The authors discussed the quantum yield for photoconduction in terms of the kinetics of competing processes by which the excited charge transfer state can dissociate or decay and indicated two structural conditions for the high photoresponse in CT complexes: (i) π orbitals of adjacent donor molecules or adjacent acceptor molecules (but not both) should overlap appreciably (which is a prerequisite of the effective separation of the ion-pair and of free migration of the charge carriers through the lattice), and (ii) π orbitals of neighbouring donor and acceptor molecules involved in the CT transition should not overlap (to minimize the decay of excited charge transfer state to the ground state). Inspection of the crystal structures of the complexes studied proved the conditions to be fulfilled for chrysene–TCNQ and tetrabenzonaphthalene–TCNQ complexes in which the highest photoconductivity was also found.

An interesting, yet still open problem is the role of impurities and excess of one of the components in conductivity of CT complexes. The latter effect has been studied in several complexes of iodine, though the results are inconclusive (cf. Gutmann and Lyons, 1967, Table 8.5). Recently Doi and coworkers (1977) have pointed out the rather complicated stoichiometry of some iodine complexes. They postulate the formula $(\text{phenothiazine}_2)^+\text{I}_5^-$ for the stoichiometric phenothiazine–iodine complex and ascribe the phenothiazine$_2(\text{I}_2)_3$ composition, given for this complex hitherto, to the presence of $\frac{1}{2}$ mol of excess I_2 included within the lattice. The excess iodine results, according to the authors, in an increase in conductivity and decrease of the activation energy.

Similar effects caused by an excess of donor have been reported for the trimethylamine–chloranil complex (Lorenz and Pigoń, 1972). An attempt by the authors to explain the increase of conductivity in the system by doping is reviewed in section 3.3.

An increase in conductivity, sometimes by as much as 5 orders of magnitude, has been observed as a result of inclusion of solvent molecules (e.g. CH_2Cl_2) into crystals of the benzidine–TCNQ complex (Ohmasa et al., 1971; Takahashi et al., 1976). The effect, however, is due to a change in the crystal structure and cannot be viewed as doping.

Summarizing, in spite of numerous papers published every year on conductivity of CT complexes there is still much ambiguity and controversy concerning the major questions, viz. the generation and transport of current carriers in these materials, and the relationship between their chemical composition and electrical properties.

10.3.2 Current carrier mobility

Much more direct insight into the mechanism of the transport processes should be offered by measurements of the mobility of the current carriers. However, surprisingly few mobility data for CT complexes have been published hitherto (see Table 10.1).

There are probably two reasons for this situation; the difficulty in obtaining sufficiently large and sufficiently perfect single crystals and the relatively weak photoresponse of many CT complexes. In contrast to the ion-radical TCNQ salts for which some successful Hall mobility measurements have been reported (Farges et al., 1970; Ong and Ports, 1977; Copper et al., 1976, 1977), the drift mobility of photogenerated carriers in weak CT complexes has been almost exclusively measured by the transit time (Kepler, 1960) or initial photocurrent (Rose, 1955) techniques. Both of them require fairly large photocurrent pulses unaffected by carrier trapping—a condition not easy to fulfil. Thus at present reliable results have been obtained for three CT complexes only, viz. anthracene–TNB (Zboiński, 1974, 1976),

Table 10.1 Mobility of current carriers (m^2 V^{-1} s^{-1}) in single crystals of CT complexes at room temperature unless otherwise stated

Complex	$\mu_e \times 10^4$	Temperature dependence	$\mu_h \times 10^4$	Temperature dependence	Method	References
Pyrene–TCNE	$\cong 0.03$		20–30	Activated $E_a = 0.5$ eV	Initial photocurrent	Tobin and Spicer (1965)
Pyrene–TCNE	$\perp bc$ 0.15		0.05–0.08		Transit time	Kuroda et al. (1963)
Pyrene–TCNE	$\dfrac{\mu\|c}{\mu\perp bc} \cong 9$ $\dfrac{\mu\|c}{\mu\|b} \cong 8$ $\theta = 0°$		$\dfrac{\mu\|c}{\mu\perp bc} \cong$ $\dfrac{\mu\|c}{\mu\|b} \cong 5$		Analysis of pulse shape	Kaino (1974, 1975, 1976)
Pyrene–TCNE	$\cong 0.01$				?	Kronick and Labes (1967)
N-Isopropyl-carbazole-picryl chloride	0.12	$\sim T^0$	0.33	Activated $E_a = 0.1$ eV	Transit time	Sharp (1967)
Anthracene–TNB*	μ_1 0.0052 μ_2 0.0059 μ_3 0.0289 $\theta = 0°$	$\sim T^1$	μ_1 0.0313 μ_2 0.0945 μ_3 0.1387 $\theta = 33°$	$\sim T^1$ very weak	Transit time	Zboiński (1974, 1976)
Anthracene–PMDA	(001) 0.15 ± 0.01	$t > 80\,°C \sim T^0$ $t < 80\,°C$ activated $E_a = 0.22$ eV $\sim T^{-2}$	(001) 0.012 160 °C	Activated $E_a = 0.58$ eV	Transit time	Karl and Ziegler (1975)
Phenanthrene–PMDA	(010) 0.02 120 °C ∥ stack 0.011 axis ⊥ stack 0.005 axis	Activated $E_a = 0.11$ eV Activated $E_a = 0.14$ eV	(010) 0.025 160 °C		Transit time	Möhwald et al. (1975)
Poly(N-vinyl-carbazole)—I$_2$	0.3				Hall mobility	Hermann and Rembaum (1967)

* μ_1, μ_2, μ_3—principal components of the mobility tensor, $\mu_2 \| b$, θ—angle between tensor axes and crystallographic a', c axes.

anthracene–PMDA (Karl and Ziegler, 1975), and phenanthrene–PMDA (Möhwald *et al.*, 1975, 1976; Haarer and Möhwald, 1975).

There is only one reference to Hall mobility measurements in CT complexes. Hermann and Rembaum (1967) found for a nearly stoichiometric sample of poly(N-vinylcarbazole)–iodine $1:1$ complex, a negative Hall voltage corresponding to a mobility value of $0.3 \times 10^{-4} \, m^2 \, V^{-1} \, s^{-1}$. Owing to the extreme experimental difficulties, as well as the sophisticated theory of the Hall effect in molecular crystals, the result—rather high for a heavily disordered material—cannot be considered as adequate.

Apart from the value of the carrier mobility two other pieces of information are sought for in-drift mobility measurements, *viz.* the directional and the temperature dependence of the mobility. As the influence of the molecular arrangement and of the temperature on carrier transport are more easily available from theory than exact values of the mobility itself, observation of these two effects seems to be especially relevant.

The exponential dependence of the mobility on reciprocal temperature found in some cases (see Figure 10.15), especially at lower temperatures, is due to the thermal release of carriers from shallow traps. The activation energy represents the depth of a trapping level provided that only a single level is responsible for the trapping process.

In purer or more perfect, samples or at higher temperatures, true microscopic mobility is measured. It is the order of $10^{-7} - 10^{-5} \, m^2 \, V^{-1} \, s^{-1}$; the values typical for hopping rather than for coherent transport of carriers. The temperature dependence of the microscopic mobility is different for each of

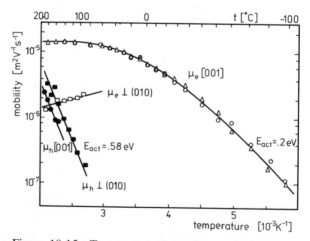

Figure 10.15 Temperature dependence of mobility of electrons (circles) and holes (dots) in single crystals of the anthracene–PMDA complex. [Reproduced by permission of North-Holland Publishing Company from Karl and Ziegler, 1975]

the complexes studied and even different in various directions of carrier motion in a given complex. It is difficult to deduce an unambiguous mobility–temperature relationship based on measurements taken within a rather narrow temperature range, especially when more sophisticated formulae are to be considered. None the less it seems that the band theory with its $\mu \sim T^{-n}$ ($1 \leqslant n \leqslant 2$) relationship (Friedman, 1964, 1965) does not account well for the observed temperature dependence. On the other hand, in Holstein's small polaron model (Holstein, 1959; Emin and Holstein, 1969) mobility is a decreasing function of temperature at low temperatures, and an increasing function at higher ones, a relationship more adequate to the actual situation in CT complexes, at least as far as is known at present. In fact both decrease, increase, and virtual independence of mobility on temperature have been observed. In this respect there exists a remarkable difference between CT complexes and aromatic or heteroaromatic hydrocarbons for which a $\mu \sim T^{-n}$ (with $n \cong 1$–3 relationship is commonly found (cf. the data compiled by Schein, 1977). It would be of interest to examine whether the considerable structural disorder inherent in many CT crystals might possibly account for the difference. Much work on the problem is still to be done, however, and precise measurements of mobility in various CT complexes within a broad temperature range are necessary before any definite conclusion can be formulated.

More conclusive results have been obtained by observation of the directional dependence of the mobility. In the crystal structure of all of the complexes listed in Table 10.1 stacks of alternate donor and acceptor molecules extend along the crystallographic c axis, and this is also the direction of the highest mobility of electrons. The anisotropy is not as pronounced as in conducting ion-radical salts of TCNQ, the reality of the effect, however, is beyond any doubt. It is worth noticing that in the anthracene–TNB complex transport of holes occurs preferentially in the direction of the shortest interstack distances rather than along the stack axis Zboiński, 1976). Unfortunately, there are no data on microscopic mobility of holes in other CT complexes to corroborate the conclusion on different mechanisms of transport of electrons and holes in these materials.

Several attempts have been made to deduce the mobility of carriers from the value of the Seebeck coefficient and/or from the preexponential factor in the conductivity–temperature relationship. The results are rather dubious and often contradictory as the employed models have been elaborated for covalent semiconductors and are hardly adequate for the situation in molecular solids.

10.3.3 *Ionic and electronic conductivity*

In some hydrogen-bonded systems, particularly in biologically related structures, speculations have been made about the possible participation of hydrogen bridges in charge transfer processes. The essential role of hydrogen bonding in the electrical conductivity could therefore be observed for systems involving mobile or easily splittable protons. In ice crystals it is established that in d.c. conductivity the rate-determining step involves translational jumps of protons from hydronium ions towards neighbouring water molecules (Eigen *et al.*, 1964). The bulk electrical conductivity in ice is 1.5×10^{-8} ohm^{-1} m^{-1} at 267 K with an activation energy 0.37 eV. On the other hand, spectroscopic investigations of the electronic structure of ordinary and heavy ice of hexagonal structure suggest (Shibaguchi *et al.*, 1977) that ice is essentially a kind of molecular crystal. Hydrogen-bonded crystals other than ice of the ferroelectric type, e.g. KH_2PO_4 have also prompted much discussion and investigation of the role of proton mobility in these systems (O'Keeffe and Perrino 1967). The electrical conductivity for these cases has been demonstrated to be protonic, mainly from coulometric determinations of ionic transport numbers (Schmidt, 1962; Murphy, 1965), as well as from other experimental information.

Another group of molecular systems suggested are hydrogen-bonded solids involving π electrons free to move within the molecular fragments. Except for polypeptides it has been found that chains or networks molecular crystals shows relatively large conductivities. Imidazole single crystals have a highly anisotropic conductivity and along the crystallographic direction where hydrogen-bonded chains extend throughout the crystal the protonic mechanism seems to be decisive (Kawada *et al.*, 1970; McGhie *et al.*, 1970). In this respect the electrical conductivity of imidazole bears a resemblance to protonic conduction in the ice crystal. In the *a* direction where no extended hydrogen bond exists the conduction mechanism was suggested to be electronic and extrinsic (Kawada *et al.*, 1970). However, along the *c* axis the conductivity was explained by considering that the deformation of the crystal lattice enabled an imidazole molecule to rotate around an axis in the plane of the molecules and perpendicular to the hydrogen-bonded chain (Figure 10.16). The cooperative transport of protons through the crystal should occur with the reorientation of the imidazole rings as the rate-determining step. However, the crystal activation energy for rotation, 1.7 eV, used to expand the lattice around the rotating molecule might be too high to account for. On the other hand, the cooperative rotation of imidazole rings within the infinite chains should lead to lowering of the potential barrier for this motion.

Protonic conductivity in molecular solids may also be accounted for by assuming proton defect formation. The migration of a proton defect is not very strongly temperature dependent and the energy of its formation is of the order of 0.5–1.0 eV (Jacobs and Lam Wee, 1972). This model of conductance mechanism was able to explain the electrical properties of organic acids in the solid state (Pollock and Ubbelohde, 1956). In any case

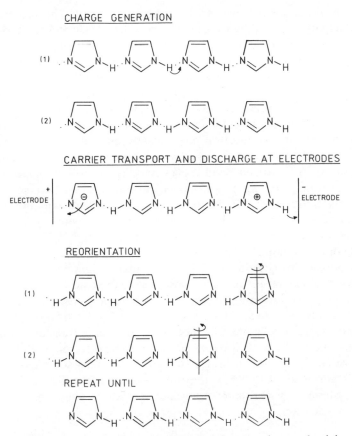

Figure 10.16 Possible mechanism of protonic conductivity along the *c* axis in an imidazole crystal. [Reproduced by permission of the American Institute of Physics from Kawada *et al.*, 1970]

the conductivity mechanisms should involve proton transfer at least as an intermediate step. Proton injection from electrodes cannot be excluded in hydrogen-bonded molecular solids (Thomas *et al.*, 1971; Weber and Flanagan, 1975).

In many hydrogen-bonded systems their electrical conductivity may be due to water traces adsorbed on internal surfaces or externally. This effect is mainly observed when polycrystalline samples are used and therefore more reliable results are those for single crystals. In some cases the effect of molecular structure on conductivity was found (Gravatt and Gross, 1967), being dependent on the size of the conjugated system and the structural parameters affecting the degree of inter-molecular interaction. It seems that in many hydrogen-bonded solids involving π electrons the electronic mechanism of conductivity cannot be entirely excluded. The magnitude of the conductivity is generally dependent upon the physical characteristics of the sample, such as purity and also crystallinity of the material. This would be the only common feature for hydrogen-bonded and charge transfer solids as the role of hydrogen bridges is still a moot point. It was found that hydrogen bonding within a crystal lattice strongly disperses some molecular levels leaving others relatively unaffected (Middlemiss and Santry, 1974a, b). The level of intermolecular charge transfer is low and the necessary condition for transfer of electron density between molecules in a crystal is that they should be crystallographically independent (Middlemiss and Santry, 1974a, b). A hydrogen bond itself may play the role of waveguide for electrons or it may be a kind of a trap for current carriers. Delocalization of π electrons across hydrogen bonds may also play a role in the electron transfer processes (Pullman, 1964). Proton migration within the hydrogen bonds, shown experimentally (Palacios-Gomes *et al.*, 1977), distributed randomly onto two equilibrium sites was found to lower the potential barrier for electron transfer (Chojnacki, 1969). Essential reduction in the potential barrier for proton transfer has been observed in the case of the cooperative proton motion (Weiner and Askar, 1970). In such a case the barrier was mainly the activation energy for conductivity and the activation energy for proton motion was negligible. The mechanism of charge transfer through the hydrogen bond presumably involves coupling of the electron with the proton motion within the hydrogen bridge. Model calculations show (Sabin *et al.*, 1969; Fischer *et al.*, 1969) that the proton–phonon coupling in hydrogen-bonded systems is also rather strong and the lattice dynamics need more theoretical attention.

10.4 Possible Mechanism of the Electrical Conductivity

10.4.1 *Problem of interaction in the solid state*

Experimental results showing a higher electrical conductivity both for charge transfer and hydrogen-bonded complexes than for the isolated components is unquestionably evidence for the attractive forces responsible for intermolecular bonding. The same conclusion can be drawn by observation of the spectral properties of the solid charge transfer as well as hydrogen-bonded systems. Within the first group of molecular complexes only direct interactions should occur, whereas in the second class indirect hydrogen bond interactions seem to be predominant.

According to the valence bond concept of Mulliken's theory (Mulliken and Person, 1969) the ground-state wave function of a charge transfer complex between a donor molecule D and an acceptor molecule A may be expressed as

$$\Phi_0 = a\Phi(D, A) + b\Phi(D^+ - A^-).$$

The real complex structure is therefore superposition of a no-bond structure and a dative resonance structure corresponding to an ionic and covalent bond. Usually it is regarded as sufficient for a first approximation to take as the dative structure merely the lowest charge transfer state where an electron has been transferred from the highest occupied orbital of the donor to the acceptor. It seems, however, that this approximation is not satisfactory for many complexes and other charge transfer states also make significant contributions to the bonding energy. As all charge transfer states may interact with the no-bond state, the wave function of the ground-state complex would be (Kuroda *et al.*, 1966)

$$\Phi_0 = a\Phi^0(D, A) + \sum_i \sum_j b_{ij}\Phi_{ij}^1(D^+ - A^-) + \sum_k \sum_l c_{kl}\Phi_{kl}^2(D^- - A^+).$$

Except for a hybrid of the ground no-bond state $\Phi^0(D, A)$, charge transfer $\Phi_{ij}^1(D^+ - A^-)$, and $\Phi_{kl}^2(D^- - A^+)$, the local excited states $\Phi(D^*, A)$ and $\Phi(D, A^*)$ should be included in the total wave function of the complex (Dewar and Thompson, 1966). Estimations of Kuroda *et al.* (1967) at the π-electronic level show that charge transfer interactions play an important role in the interaction scheme. More advanced *ab initio* calculations, however, point out that electrostatic interactions are the predominant term in the stabilization of weak charge transfer complexes (Lathan and Morokuma, 1975). A satisfactory agreement of the van der Waals interactions with the experimental energy of formation in the gas phase has been found (Mantione, 1969) for stacked structures of TCNE with different donors (Table 10.2). In this case, resembling typical lattice energy calculations in molecular crystals, the electrostatic interactions E_{el} were found within the point charge approximation, polarization energy E_{pol} was estimated by the charge distribution of the complementary molecule of the complex, whereas the dispersion and repulsion energy terms were calculated by the Kitaigorodsky

Table 10.2 Calculated interaction energies (kJ mol^{-1}) for equilibrium distance 3.50 Å of molecular complexes with different donors. [Reproduced by permission of Springer-Verlag from Mantione, 1969]

Donor	E_{el}	E_{pol}	E_{disp}	E_{rep}	E_{tot}	E_{exp}
p-Xylene	−11.30	−5.82	−42.68	21.88	−37.86	−31±2
o-Xylene	−11.24	−5.98	−44.77	24.77	−37.24	−33±2
Mesitylene	−11.52	−6.32	−48.16	24.98	−40.96	−38±2
Durene	−12.50	−6.74	−54.10	30.08	−43.47	−42±3

(1961) expressions

$$E_{disp} = -C_1 \left(\frac{r_0}{r}\right)^6$$

$$E_{rep} = C_2 \exp\left(-\alpha \frac{r}{r_0}\right)$$

respectively. The r_0 denotes distances for each pair of atoms, C_1 and C_2 are constant parameters.

The calculations show that charge transfer systems, at least those forming weak complexes, do not differ in kind from other van der Waals interactions. Furthermore, the results point out that the contribution of these interactions is expected to play the dominant role even in the stabilization energy of the $\pi-\pi$ molecular complexes.

The term hydrogen bond is given to the system of two bonds $X—H \cdots Y$, i.e. proton donor $X—H$ and its acceptor $H \cdots Y$, where X and Y are two electronegative atoms. The donor bond is usually shorter and hence normally stronger than the acceptor bond. As a rule hydrogen bonds connect atoms of higher electronegativity than hydrogen and in some cases the intermolecular interactions are so weak that chemists may consider that there is no hydrogen bonding (Kollman and Allen, 1972).

In some papers (Statz and Lippert, 1969) electrostatic models of hydrogen bonding were preferred. However, if this were completely adequate every hydrogen bond could be described as an entirely ionic effect. In fact, the hydrogen bonding effect seems to be only partially ionic, and valence bond theory (Coulson, 1957) gives a better interpretation of this kind of interaction. On the other hand, several authors (Mulliken, 1964; Bratož, 1967; Toyoda, 1969; Ratajczak, 1972; Ratajczak and Orville-Thomas, 1975) have emphasized the importance of the charge transfer mechanism in hydrogen bonding interactions.

In molecular crystals the experimental evidence for hydrogen bonding is the fact that the sum of the van der Waals radii of H and Y is greater than the non-bonded $H \cdots Y$ interatomic distance (Hamilton and Ibers, 1968). The geometrical criterion for hydrogen bond existence in crystals is that the following relationship is fulfilled (Zefirov, 1976)

$$d_{H \cdots Y} = 2.15\sqrt{R_Y} - 0.35 \, [\text{Å}]$$

where R_Y is the van der Waals radius of the Y atom. Valence bond analysis (Brown, 1976) leads to the conclusion that strong $O \cdots O$ hydrogen bonds shorter than 2.7 Å involve strains and are linear while the weaker ones have an extra degree of freedom and are generally bent. Crystal packing as well as an energetic preference (Kroon *et al.*, 1975) often lead to the formation of bent and hence, weaker hydrogen bonds.

For conformational analysis or crystal lattice calculations and structure predictions semi-empirical atom–atom or intermolecular potential functions have also been used. The Stockmayer potential (Stockmayer, 1941)

$$V_{ij}(r) = 4\varepsilon \left[\left(\frac{\sigma}{r_{ij}} \right)^{12} - \left(\frac{\sigma}{r_{ij}} \right)^6 \right] + \frac{1}{r_{ij}^3} \left[(\boldsymbol{\mu}_i \boldsymbol{\mu}_j) - \frac{3(\boldsymbol{\mu}_i r_{ij})(\boldsymbol{\mu}_j r_{ij})}{r_{ij}^2} \right]$$

was successfully used for dimethylglyoxime (Giglio, 1969) or imidazole (Lehmann, 1974) crystal structure predictions. In some cases hydrogen atom positions as well as thermal motion parameters could be determined (Poltev and Sukhorukov, 1968). It seems that the empirical intermolecular potentials should be improved by including induction as well as charge transfer terms as contributing to the stability of the hydrogen bonds (Momany, 1976).

The interaction mechanism in molecular charge transfer and hydrogen-bonded systems is complicated by many subtle interrelated effects and different terms of the same order of magnitude should be included in the total energy. The common feature of the two groups of molecular complexes seems to be the concentration of the charge density within the bonding space. The same effect was found for interactions between the oxygen and halogen atoms of quinoic molecules where their nature is essentially charge transfer resembling those of hydrogen-bonded systems (Gaultier et al., 1971).

Recently the importance of charge transfer contribution in molecular complexes has been questioned (Hanna, 1968) and it has been suggested, at least for isolated systems, that simple electrostatic effects are of predominant importance for hydrogen-bonded complexes. This supposition was corroborated by the *ab initio* calculations where for weak complexes and at long X—H $\cdots$ Y distances electrostatic interactions are very important (Lathan and Morokuma, 1975). Charge transfer interactions are of considerable importance mostly at short X—H $\cdots$ Y distances. However, in many cases the fundamental similarity between charge transfer complexes and hydrogen bond interactions is clearly demonstrated (Dumas and Gomel, 1975). Therefore, Mulliken's electronic model (Mulliken and Person, 1969) and Hush's intervalence theory (Hush, 1967) are limiting cases of the unified molecular charge transfer model invented by Chiu (1976). It covers all ranges of molecular interactions including electronic Franck–Condon factors and vibrational structural changes. It resembles also some assumptions used in the polaron study of an excess electron moving among an aggregate of molecules in the molecular crystal lattice (Siebrand, 1964).

10.4.2   *Band structure considerations*

For typical organic semiconductors, like the anthracene crystal, Le Blanc was the first to calculate the structure of one-electron bands for excess hole and electron within the right-binding scheme (Le Blanc, 1961). The mobility tensors evaluated from the band structures exhibit the anisotropy similar to that observed experimentally and the bandwidths were found to be of the order of magnitude required by the uncertainty principle (Fröhlich and Sewell, 1959).

Within the tight-binding formulation of the molecular orbital theory crystal orbitals are constructed as a linear combination of one-electron molecular wave functions

$$\Phi(r) = \sum_n \sum_s \sum_p c_{n,s}^p \Phi_s^p(r - r_n - r_s)$$

where $\Phi_s^p(r - r_{\bar{n}} - r_s)$ is the pth molecular orbital of the crystal site. The r_n vector denotes the centre of the unit cell while the r_s vector is directed from the centre of the cell to the geometric centre of the sth molecule. For the Hamiltonian H the eigenvalue problem results in a set of linear equations

$$\sum_n \sum_s \sum_p c_{n,s}^p H_{st}^{kl}(n, m) = E \sum_n \sum_s \sum_p c_{n,s}^p S_{st}^{kl}(n, m)$$

with

$$H_{st}^{kl}(n, m) = \int \Phi_t^k(r - r_m - r_t) H \Phi_s^l(r - r_n - r_s) \, dV$$

and

$$S_{st}^{kl}(n, m) = \int \Phi_t^k(r - r_m - r_t) \Phi_s^l(r - r_n - r_s) \, dV$$

Furthermore, using the Bloch theorem one can reduce the secular equation to the form

$$\sum_p \sum_s \left\{ \sum_n [H_{st}^{kl}(n, m) - E S_{st}^{kl}] \exp(ikr_{\bar{n}}) \right\} c_s = 0$$

for any value of l, t, and m.

Ab initio calculations within the tight-binding model, owing to their complexity and the numerical problems involved, are still very scarce and except for model calculations for one-dimensional systems (André and Leroy, 1970; Duke and O'Leary, 1973) no systematic studies of conducting molecular complexes have been done so far. Three-dimensional calculations at an all-valence level are also not available for this class of charge transfer or hydrogen-bonded complexes. The bandwidths calculated by the extended Hückel approach for imidazole chains (Figure 10.17) are of the order of magnitude of 0.004–4.25 eV (Chojnacki, 1975), however, the large bandgap of 5.91 eV excludes it as an intrinsic semiconductor. This seems to be confirmed by the fact that there is no agreement between the electronic

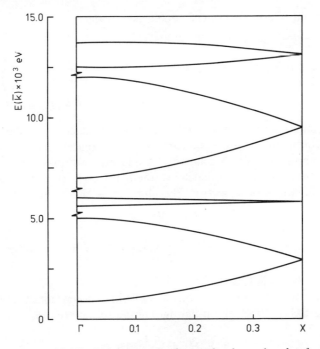

Figure 10.17 Lowest conduction and valence bands of imidazole chain calculated by the extended Hückel method. [Reproduced by permission of Société Scientifique de Bruxelles from Chojnacki, 1975]

absorption threshold and the relevant thermal activation energy for conductivity. The results found by INDO and MINDO methods for polyglicyne (Beveridge *et al.*, 1972), where the bandgaps are of the order of 8.51–9.53 eV, exclude the possibility of intrinsic conductivity for this system also. The direction of the polypeptide backbone was found to be more favourable for electrical conduction than hydrogen bonds (Suhai, 1974). The electronic structure both of imidazole (Chojnacki, 1975), and polyglicyne (Fujita and Imamura, 1970) reflects distinctly the effect of hydrogen bond formation. In all cases, however, comparison of quantum mechanical calculations with the corresponding experimental information is almost impossible.

Molecular complexes seem to be amenable to treatment within the π-electronic approximation when the secular equation takes the form

$$\sum_s \left\{ \sum_n [H_{st}(\boldsymbol{n}, \boldsymbol{m}) - ES_{st}(\boldsymbol{n}, \boldsymbol{m})] \exp{(i\boldsymbol{kr_n})} \right\} c_s = 0.$$

Neglecting overlap in the case of a crystal with four molecules in the unit cell we obtain (Chojnacki and Gołębiewski, 1969)

$$
\begin{pmatrix} E_{us} \\ E_{gs} \\ E_{ua} \\ E_{ga} \end{pmatrix} = \sum_n \exp(ikr_n) \begin{pmatrix} 1 & 1 & 1 & 1 \\ 1 & 1 & -1 & -1 \\ 1 & -1 & 1 & -1 \\ 1 & -1 & -1 & 1 \end{pmatrix} \begin{pmatrix} H_{11}(n,0) \\ H_{21}(n,0) \\ H_{31}(n,0) \\ H_{41}(n,0) \end{pmatrix}
$$

where we take $t = 1$ and $m = 0$. The results of the calculations depend on the approximations for the intermolecular transfer integrals $H_{11}(n,0)$, $H_{21}(n,0)$, $H_{31}(n,0)$, and $H_{41}(n,0)$. A relatively simple method for their evaluation at the π-electronic level is to calculate them in terms of Slater-type atomic orbitals assuming a one-electron Hamiltonian of the form

$$ H = -\tfrac{1}{2}\nabla^2 + V(r) $$

where

$$ V(r) = \sum_n \sum_s V_s(r - r_n - r_s) $$

determines the crystal field.

According to the Goeppert-Mayer and Sklar approximation (Goeppert-Mayer and Sklar, 1938) the potential energy terms are

$$ V(H) = -\frac{e^{-2r}}{r}(1+r) $$

for hydrogen atoms, and

$$ V(C) = -\frac{e^{-2\beta}}{r}(4 + 6\beta + 4\beta^2 + \tfrac{4}{3}\beta^3) $$

for carbon atoms, where $\beta = \alpha r$. Besides

$$ V(N_1) = -\frac{e^{-2\beta}}{r}(5 + \tfrac{15}{2}\beta + 5\beta^2 + \tfrac{5}{3}\beta^3) + \tfrac{3}{2}(3\cos^2\Theta - 1)R(r,\beta) $$

for nitrogen atoms of pyrrolic type, where Θ is an angle between the axis of a $2p\pi$ orbital and the vector r directed from this atom towards the reference point. Furthermore, for nitrogen atoms of pyridine type

$$ V(N_2) = -\frac{e^{-2\beta}}{r}(5 + \tfrac{15}{2}\beta + 5\beta^2 + \tfrac{5}{3}\beta^3) + Q(r,\beta)\cos\Theta + R(r,\beta)(3\cos^2\Theta - 1), $$

where Θ is an angle between the axis of the sp-type lone pair and the vector r. Here

$$ Q(r,\beta) = \frac{5\sqrt{6}}{9\beta r}[1 - e^{-2\beta}(1 + 2\beta + 2\beta^2 + \tfrac{6}{5}\beta^3 + \tfrac{2}{5}\beta^4)] $$

$$ R(r,\beta) = \frac{1}{\beta r}[1 - e^{-2\beta}(1 + 2\beta + 2\beta^2 + \tfrac{4}{3}\beta^3 + \tfrac{2}{3}\beta^4 + \tfrac{2}{9}\beta^5)]. $$

The three-centre integrals may also be of importance, especially when Hartree-Fock wavefunctions are used instead of Slater type. It was shown that those terms change the energy by about 25% (Katz *et al.*, 1963). The advantage of the π-electronic approach is, however, its relative simplicity enabling band structure calculations even for complicated molecular systems. The results show (Chojnacki and Jodkowski, 1978) that there are only small differences in the bandwidths for charge transfer complexes (Figure 10.18), hydrogen-bonded systems (Figure 10.19), and the anthracene crystal (Figure 10.20).

Within the simplified model, assuming a constant isotropic free path, the components of the mobility tensor should be equal to

$$\mu_{ij} = \frac{e\lambda}{kT} \langle v_i v_j \mid \boldsymbol{v}(\boldsymbol{k}) \rangle.$$

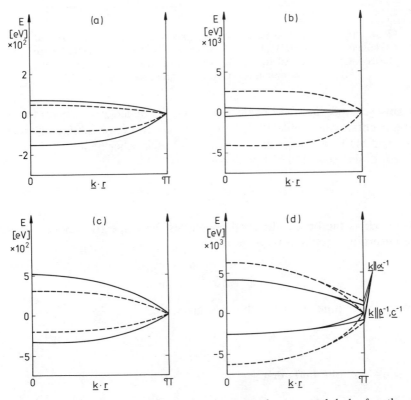

Figure 10.18 Band structure of an excess electron and hole for the anthracene–trinitrobenzene complex. (a) Hole band at 298 K; (b) electron band at 298 K; (c) hole band at 173 K; (d) electron band at 173 K calculated with the Hartree–Fock potential (solid lines) or Hartree potential (dashed lines) at the π-electronic level (Chojnacki and Jodkowski, 1978)

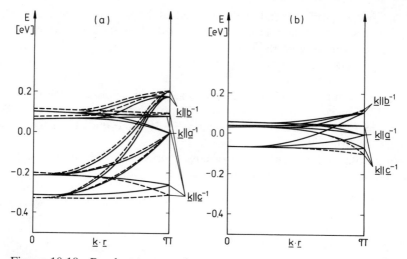

Figure 10.19 Band structure of an excess electron and hole for the imidazole crystal. (a) Hole band; (b) electron band calculated with the Hartree–Fock potential for 123 K (solid lines) and 293 K (dashed lines) at the π-electronic level (Chojnacki and Jodkowski, 1978)

In this scheme the reasonable agreement of calculated anisotropy with experimental data (Table 10.3) seems to point to the possibility of applying the band model to molecular complexes. On the other hand, if the mobility (Fröhlich and Sewell, 1959) is expressed as

$$\mu = \frac{e}{kT} \frac{\Delta E^2 a^2}{\hbar^2}$$

where ΔE is the bandwidth, and a denotes the respective lattice constant, the uncertainty relation requires that

$$\Delta E \geqslant \frac{\hbar}{\tau}.$$

Thus a lower limit on the mobility is given by

$$\mu \geqslant \frac{ea^2 \Delta E}{\hbar kT}.$$

With $a = 3$ Å one finds that

$$\mu \geqslant 10^{-5} \frac{\Delta E}{kT} [m^2 \ V^{-1} \ s^{-1}].$$

Hence, mobilities of the order of $1.0 \times 10^{-4} \ m^2 \ V^{-1} \ s^{-1}$ are on the borderline of the band description (Friedman, 1964). As mobilities of weak charge

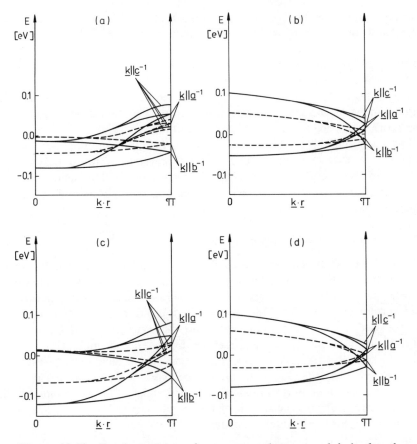

Figure 10.20 Band structure of an excess electron and hole for the anthracene crystal. (a) Hole band at 290 K; (b) electron band at 290 K; (c) hole band at 95 K; (d) electron band at 95 K calculated with the Hartree–Fock potential (solid lines) or Hartree potential (dashed lines) at the π-electronic level (Chojnacki and Jodkowski, 1978)

Table 10.3 Calculated (Chojnacki and Jodkowski, 1978) and experimental (Zboiński, 1976) anisotropy of current carrier mobilities for the anthracene–trinitrobenzene complex at room temperature. $\mu_2 = \mu_{bb}$ whereas Θ is an angle between the a' axis and the tensor component μ_1

	Theoretical		Experimental	
Mobility ratio	hole	electron	hole	electron
μ_2/μ_1	2.93	0.92	3.02	1.14
μ_3/μ_1	4.15	5.23	4.43	5.56
Θ	28.2°	11.3°	32.5°	−4.5°

transfer complexes are of the order of $1.0 \times 10^{-6}\,\text{m}^2\,\text{V}^{-1}\,\text{s}^{-1}$ (Zboiński, 1976) and those of hydrogen-bonded crystals presumably should be of the same order, their description by means of the band model may be in violation of the uncertainty principle. Furthermore, it was shown that the tight-binding model at the π-electronic level does not consider the role of hydrogen bonding satisfactorily (Chojnacki, 1975). In this case the essential improvement of the current carrier anisotropy was found when $2p\pi$ hydrogen orbitals were taken into account (Chojnacki, 1970).

The most difficult problem within the band approximation is a proper treatment of the scattering of excess charge carriers by the intermolecular motions of the constituent molecules (Friedman, 1965). This seems to be an essential effect not only at the π-electronic approximation which is itself a separate problem.

10.4.3 Tunnelling and hopping model

Another line of attack on the conductivity problem in molecular solids pictures the electrons and holes as jumping from one crystal site to another. As Eley et al. (1959) first suggested, this could occur by tunnelling through the intermolecular barrier. This mechanism does not require a charge carrier to be raised above the ground state and therefore the model cannot be rejected on the basis of the temperature dependence of mobilities.

In general, the transfer process may occur when current carriers come over the potential barrier or quantum mechanically tunnel across it. Thus the electrical conductivity should be proportional to the probability of the barrier crossing P and the length of the single step in electron transport l. For the same hitting frequency of the barrier we obtain

$$\sigma \sim \{\exp\left(-\varepsilon_a/kT\right) + [1 - \exp\left(-\varepsilon_a/kT\right)](D^+ - D^-)\} \cdot 1$$

where D^+ and D^- stand for transmission coefficients in the direction of the electric field and opposite to it. The effect of an electric field is to lower the barrier height for electrons moving along the field and raise it for electrons moving in the opposite direction. Therefore, the potential energy of an electron travelling along the x axis is represented by

$$V(x) = V_0(x) \mp eFx$$

where $V_0(x)$ denotes the potential energy in the absence of the electric field, F is the field strength due to the potential drop across the barrier, whereas the respective signs correspond to the electron motion along or against the applied field. According to the JWKB approximation the amplitude of the transmitted wave is related to the barrier shape by the formula

$$\ln \frac{1}{A^2} = \frac{(2m)^{1/2}}{\hbar} \int_{x_1}^{x_2} [V(x) - E]^{1/2}\,dx$$

where m is the mass of the tunnelling particle, l is the barrier width, and E denotes the mean kinetic energy of the particle. Taking into account the potential drop across the barrier the transmission coefficient for the parabolic potential can be expressed by

$$D^{\pm} = \exp\left\{-\frac{\pi}{4\hbar}(2m)^{1/2}l[V_0(x)-E-0.5eFx]^{1/2}\right.$$
$$\left. \times\left[1+\left(\frac{V_0(x)-E-0.5eFx}{V_0(x)-E+0.5eFx}\right)^{1/2}\right]\right\}.$$

The product Pl permitted interpretation of the effect of ambient amine on the electrical resistance of the p-chloranil–trimethylamine charge transfer complex (Figure 10.21). It was shown (Chojnacki *et al.*, 1973) that the theoretical dependences were very similar to those found experimentally, corroborating the assumed combined tunnelling–hopping model. In this case the anticipated field strength characteristics agreed reasonably well with the experimental values also. Despite the obvious oversimplifications of the tunnelling–hopping model it seems to be useful for understanding the doping effects observed within this group of molecular complexes.

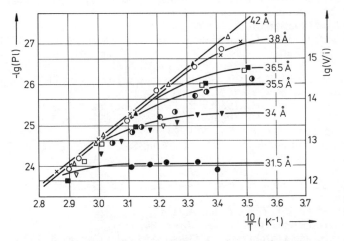

Figure 10.21 The effect of ambient amine on the apparent electrical resistance of a thin layer of the p-chloranil–trimethylamine complex (right-hand scale). × First run, $(T\uparrow)$, 10^{-1} Tr; ■$(T\uparrow)$, □$(T\downarrow)$, 40 Tr; ◑ $(T\uparrow)$, ◐ $(T\downarrow)$, 100 Tr; ▼ $(T\uparrow)$, ▽ $(T\downarrow)$, 270 Tr; ●$(T\uparrow)$, 675 Tr; ○ last run, $(T\downarrow)$, 10^{-1} Tr; ▲ $(T\uparrow)$, △ $(T\downarrow)$ electric resistance of a pellet at 10^{-3} Tr (scale lowered by three logarithm units). The solid lines denote the values of $-\log(Pl)$ for several widths of the rectangular barrier and $1\,\text{kV cm}^{-1}$ (left-hand scale). [Reproduced by permission of Akademie-Verlag, Berlin from Chojnacki *et al.*, 1973]

Apart from the question of the temperature dependence of current mobilities quantitative calculations show that the tunnelling model can account reasonably well for electron mobilities in the anthracene crystal (Keller and Rast, 1962). However, the bands may not be narrow enough to obtain adequate relations for the transfer probabilities (Glaeser and Berry, 1966). The assumption that tunnelling through the intermolecular barrier is the time-limiting process leads to good quantitative predictions regarding the excess electron and hole mobilities. As tunnelling–hopping probabilities depend upon vibrational states, a thermal activation effect would be expected too.

The tunnelling effect is expected to be important in many processes related to chemical reactions occurring in the solid state. The concept of long-range electron tunnelling has been tested experimentally permitting explanation of electron transfer results (Zamaraev and Khairutdinov, 1974).

10.5   References

Akamatu, H., Inokuchi, H., and Matsunaga, Y. (1954). *Nature*, **173**, 168.
Akamatu, H. and Kuroda, H. (1963). *J. Chem. Phys.*, **39**, 3364.
Alexandre, M. and Rigny, P. (1972). *Molecular Crystals and Liquid Crystals*, **17**, 19.
André, J. M. and Leroy, G. (1970). *Chem. Phys. Lett.*, **5**, 71.
Bando, M. and Matsunaga, Y. (1976). *Bull. Chem. Soc. Japan*, **49**, 3345.
Beveridge, D. L., Jano, I., and Ladik, J. (1972). *J. Chem. Phys.*, **56**, 4744.
Bhat, S. N. and Rao, C. N. R. (1969). *Canad. J. Chem.*, **47**, 3899.
Boeyens, J. C. A. and Herbstein, F. H. (1965). *J. Phys. Chem.*, **69**, 2153.
Bratož, S. (1967). *Advances in Quantum Chemistry* (Ed. Löwdin, P. O.), Vol. III, Academic Press, New York, N.Y. p. 209.
Briegleb, G. and Delle, H. (1960a). *Z. Elektrochem.*, **64**, 347.
Briegleb, G. and Delle, H. (1960b). *Z. Phys. Chem.* (*NF*), **24**, 359.
Brown, I. D. (1976). *Acta Cryst.*, **32A**, 24.
Carnochan, P. and Pethig, R. (1976). *J. Chem. Soc. Faraday I*, **72**, 2355.
Castellano, E. and Prout, C. K. (1971). *J. Chem. Soc. A*, 550.
Chiu, Y. N. (1976). *J. Phys. Chem.*, **80**, 992.
Chojnacki, H. (1969). *Electrical Conductivity and Hydrogen Bonding*, Thesis, Wroclaw.
Chojnacki, H. (1970). *Theor. Chim. Acta*, **17**, 244.
Chojnacki, H. (1975). *Ann. Soc. Sci. Brux.*, **89**, 165.
Chojnacki, H. and Gołebiewski, A. (1969). *Molecular Crystals*, **5**, 317.
Chojnacki, H. and Jodkowski, J. T. (1978). Unpublished results.
Chojnacki, H., Lorenz, K., and Pigoń, K. (1973). *Phys. Stat. Sol.* (*a*), **20**, 211.
Cooper, J. R., Miljak, M., Delplanqué, G., Jérome, D., Weger, M., Fabre, J. M., and Giral, L. (1977). *J. Phys.* (*Paris*), **38**, 1097.
Cooper, J. R., Weger, M., Delplanqué, G., Jérome, D. and Bechgaard, K. (1976). *J. Phys.* (*Paris*), **37**, L-349.
Coulson, C. A. (1957). *Research*, **10**, 149.
Craven, B. M., McMullan, R. K., Bell, J. D., and Freeman, H. C. (1977). *Acta Cryst.* **33B**, 2585.
Dewar, M. J. S. and Thompson, C. C., Jr. (1966). *Tetrahedron, Suppl.*, No. 7, 97.
Doi, S., Inabe, T., and Matsunaga, Y. (1977). *Bull. Chem. Soc. Japan*, **50**, 837.

Duke, B. J. and O'Leary, B. (1973). *Chem. Phys. Lett.*, **20,** 459.

Dumas, J. M. and Gomel, M. (1975). *J. Chim. Phys., Phys.-Chim. Biol.*, **72,** 1185.

Eigen, M., De Maeyer, L., and Spatz, H. Ch. (1964). *Berichte der Bunsen Gesellschaft*, **68,** 19.

Eley, D. D., Inokuchi, H., and Willis, M. R. (1959). *Discuss. Faraday Soc.*, **28,** 54.

Emin, D. and Holstein, T. (1969). *Ann. Phys.*, **53,** 439.

Farges, J. P., Brau, A., Vasilescu, D., Dupuis, P., and Néel, J. (1970). *Phys. Stat. Sol.*, **37,** 745.

Fischer, S. F., Hofacker, G. L., and Sabin, J. R. (1969). *Physics of Condensed Matter*, **8,** 268.

Frankevich, E. L., Tribel, M. M., Sokolik, I. A., and Kotov, B. V. (1977). *Phys. Stat. Sol. (a)*, **40,** 655.

Friedman, L. (1964). *Phys. Rev.*, **133A,** 1668.

Friedman, L. (1965). *Phys. Rev.*, **140A,** 1649.

Fröhlich, H. and Sewell, G. L. (1959). *Proc. Phys. Soc. (London)* **74,** 643.

Fujita, H. and Imamura, A. (1970). *J. Chem. Phys.*, **53,** 4555.

Fukushima, K. and Sakurada, M. (1976). *J. Phys. Chem.*, **80,** 1367.

Fyfe, C. A. (1974). *J. Chem. Soc. Faraday II*, **70,** 1633.

Gaultier, J., Hauw, Ch., and Schvoerer, M. (1971). *Acta Cryst.*, **27B,** 2199.

Giglio, E. (1969). *Nature*, **222,** 339.

Glaeser, R. M. and Berry, R. S. (1966). *J. Chem. Phys.*, **44,** 3797.

Goeppert-Mayer, M. and Sklar, A. L. (1938). *J. Chem. Phys.*, **6,** 645.

Goldberg, I. and Schmueli, U. (1977). *Acta Cryst.*, **33B,** 2189.

Gravatt, C. C. and Gross, P. M. (1967). *J. Chem. Phys.*, **46,** 413.

Gutmann, F. and Lyons, L. E. (1967). *Organic Semiconductors*, John Wiley, New York.

Haarer, D. and Karl, N. (1973). *Chem. Phys. Lett.*, **21,** 49.

Haarer, D. and Möhwald, H. (1975). *Phys. Rev. Let.*, **34,** 1447.

Hamilton, W. C. and Ibers, J. (1968). *Hydrogen Bonding in Solids*, W. A. Benjamin Inc., New York–Amsterdam.

Hanna, M. W. (1968). *J. Amer. Chem. Soc.*, **90,** 285.

Herbstein, F. H. and Kaftory, M. (1975). *Acta Cryst.*, **31B,** 45.

Herbstein, F. H. and Snyman, J. A. (1969). *Phil. Trans. Roy. Soc. (London)*, **264A,** 635.

Hermann, A. M. and Rembaum, A. (1967). *J. Polymer Sci.*, **17C,** 107.

Hertel, E. (1926). *Liebigs Ann.*, **451,** 179.

Holstein, T. (1959). *Ann. Phys.*, **8,** 343.

Hush, N. S. (1967). *Progr. Inorg. Chem.*, **8,** 391.

Jacobs, P. W. M. and Lam Wee, N. (1972). *J. Phys. Chem. Solids*, **33,** 2031.

Kaino, H. (1974). *J. Phys. Soc. Japan*, **36,** 1500.

Kaino, H. (1975). *J. Phys. Soc. Japan*, **39,** 708.

Kaino, H. (1976). *J. Phys. Soc. Japan*, **41,** 570.

Karl, N. and Ziegler, J. (1975). *Chem. Phys. Lett.*, **32,** 438.

Katz, J. L., Rice, S. A., Choi, S., and Jortner, J. (1963). *J. Chem. Phys.*, **39,** 1683.

Kawada, A., McGhie, A. R., and Labes, M. M. (1970). *J. Chem. Phys.*, **52,** 3121.

Keller, R. A. and Rast, H. E., Jr. (1962). *J. Chem. Phys.*, **36,** 2640.

Kepler, R. G. (1960). *Phys. Rev.*, **119,** 1226.

Kitaigorodsky, A. I. (1949). *Izv. Akad. Nauk SSSR, Ser. Khim.*, 263.

Kitaigorodsky, A. I. (1961). *Tetrahedron*, **14,** 230.

Kitaigorodsky, A. I. (1973). *Molecular Crystals and Molecules*, Academic Press, New York–London.

Kobayashi, H. (1973). *Bull. Chem. Soc. Japan*, **46**, 2675.

Kobayashi, H. (1974). *Acta Cryst.*, **30B**, 1010.

Kofler, A. (1944). *Z. Elektrochem.*, **50**, 200.

Koizumi, S. and Matsunaga, Y. (1974). *Bull. Chem. Soc. Japan*, **47**, 9.

Kokado, H., Hasegawa, K., and Schneider, W. G. (1964). *Canad. J. Chem.*, **42**, 1084.

Kollman, A. and Allen, L. C. (1972). *Chem. Rev.*, **72**, 283.

Kommandeur, J. and Hall, F. R. (1961). *J. Chem. Phys.*, **34**, 129.

Komorowski, L., Krajewska, A., and Pigoń, K. (1976). *Molecular Crystals and Liquid Crystals*, **36**, 337.

Krajewska, A. and Pigoń, K. (1978) *Sci. Papers. Inst. Org. Phys. Chem.*, Wroclaw Techn. Univ. No. 16, 199.

Krajewska, A. and Wasilewska, K., to be published.

Kramarenko, N. L., Naboikin, J. V., and Paivin, V. S. (1977). *Ukr. Fiz. Zh.*, **22**, 1673.

Kronick, P. L. and Labes, M. M. (1961). *J. Chem. Phys.*, **35**, 2016.

Kronick, P. L. and Labes, M. M. (1967), after M. M. Labes, *J. Polymer Sci.*, **17C**, 95.

Kroon, J., Kanters, J. A., Van Duijneveldt-van de Rijdt, J. G. C. M., Van Duijneveldt, F. B., and Vliegenthart, J. A. (1975). *J. Mol. Struct.*, **24**, 109.

Kumakura, S., Iwasaki, F., and Saito, Y. (1967). *Bull. Chem. Soc. Japan*, **40**, 1926.

Kuroda, H., Amano, T., Ikemoto, I., and Akamatu, H. (1967). *J. Amer. Chem. Soc.*, **89**, 6056.

Kuroda, H., Ikemoto, I., and Akamatu, H. (1966). *Bull. Chem. Soc. Japan*, **39**, 1842.

Kuroda, H., Kobayashi, M., Kinoshita, M., and Takemoto, S. (1962a). *J. Chem. Phys.*, **36**, 457.

Kuroda, H., Yoshihara, K., and Akamatu, H. (1962b). *Bull. Chem. Soc. Japan*, **35**, 1604.

Kuroda, H., Yoshihara, K., and Akamatu, H. (1963). *Bull. Chem. Soc. Japan*, **36**, 1365.

Kuroda, H., Yoshihara, K., Kinoshita, M., and Akamatu, H. (1962c). *Proc. Intern. Symp. Mol. Struct. Spectr.*, *Tokyo*, D 101.

Landolt-Börnstein (1956). *Zahlenwerte und Funktionen aus Physik, Chemie, Astronomie, Geophysik und Technik.*, Vol. II.2, Vol. II.3, 6th edn., Springer–Verlag, Berlin.

Larsen, F. K., Little, R. G., and Coppens, P. (1975). *Acta Cryst.*, **31B**, 430.

Lathan, W. A. and Morokuma, K. (1975). *J. Amer. Chem. Soc.*, **97**, 3615.

Le Blanc, O. H., Jr. (1961). *J. Chem. Phys.*, **35**, 1275.

Lehmann, M. S. (1974). *Acta Cryst.*, **30A**, 713.

Lorenz, K. and Pigoń, K. (1972). *Molecular Crystals and Liquid Crystals*, **16**, 189.

Lower, S. K. (1969). *Molecular Crystals and Liquid Crystals*, **5**, 363.

Mainthia, S. B., Kronick, P. L., and Labes, M. M. (1964). *J. Chem. Phys.*, **41**, 2206.

Mantione, M. J. (1969). *Theor. Chim. Acta*, **15**, 141.

Martinez-Carrera, S. (1966). *Acta Cryst.*, **20**, 783.

Matsunaga, Y. (1965a). *J. Chem. Phys.*, **42**, 1982.

Matsunaga, Y. (1965b). *Nature*, **205**, 72.

Matsunaga, Y., Osawa, E., and Osawa, R. (1975). *Bull. Chem. Soc. Japan*, **48**, 37.

Matsunaga, Y. and Saito, G. (1972). *Bull. Chem. Soc. Japan*, **45**, 963.

Mayoh, B. and Prout, C. K. (1972). *J. Chem. Soc.*, *Faraday II*, **68**, 1072.

McGhie, A. R., Blum, H., and Labes, M. M. (1970). *J. Chem. Phys.*, **52**, 6141.

Meier, H. (1974). *Organic Semiconductors*, Verlag Chemie, Weinheim.

Middlemiss, K. M. and Santry, D. P. (1974a). *J. Chem. Phys.*, **61**, 5400.

Middlemiss, K. M. and Santry, D. P. (1974b). *Chem. Phys. Lett.*, **28**, 140.

Möhwald, H. and Haarer, D. (1976). *Molecular Crystals and Liquid Crystals*, **32**, 215.

Möhwald, H., Haarer, D., and Castro, G. (1975). *Chem. Phys. Lett.*, **32**, 433.

Möhwald, H. and Sackmann, E. (1974). *Solid State Commun.*, **15**, 445.

Momany, F. A. (1976). *Environmental Effects on Molecular Structure and Properties* (Ed. Pullman, B.), D. Reidel Publishing Co., Dordrecht, Holland, p. 437.

Mulliken, R. S. (1956). *Rec. Trav. Chim. Pays-Bas*, **75**, 845.

Mulliken, R. S. (1964). *J. Chim. Phys., Phys.-Chim. Biol.*, **61**, 20.

Mulliken, R. S. and Person, W. B. (1969). *Molecular Complexes*, Wiley–Interscience, New York, N.Y.

Munnoch, P. J. and Wright, J. D. (1976). *J. Chem. Soc., Faraday I*, **72**, 1981.

Murphy, E. J. (1965). *Ann. N.Y. Acad. Sci.*, **118**, 727.

Ohmasa, M., Kinoshita, M., and Akamatu, H. (1971). *Bull. Chem. Soc. Japan*, **44**, 395.

O'Keeffe, M. and Perrino, C. T. (1967). *J. Phys. Chem. Solids*, **28**, 211.

Ong, N. P. and Ports, A. M. (1977). *Phys. Rev.*, **15B**, 1782.

Ottenberg, A., Hoffman, C. J., and Osiecki, J. (1963). *J. Chem. Phys.*, **38**, 1898.

Palacios-Gomes, J., Grimm, H., and Stiller, H. (1977). *Ber. Bunsenges. Phys. Chem.*, **81**, 286.

Palin, D. E. and Powell, H. M. (1947). *J. Chem. Soc.*, 208.

Pollock McC., J. and Ubbelohde, A. R. (1956). *Trans. Faraday Soc.*, **52**, 1112.

Poltev, W. I. and Sukhonikov, B. I. (1968). *Zh. Strukt. Khim.*, **9**, 298.

Prout, C. K. and Wright, J. D. (1968). *Angew. Chem.*, **80**, 688.

Pullman, A. (1964). *Quantum Aspects of Polypeptides and Polynucleotides* (Ed. M. Weissbluth), Interscience Publishers, New York, N.Y., *Biopolymers Symposia*, No. 1, p. 29.

Ratajczak, H. (1972). *J. Phys. Chem.*, **76**, 3000.

Ratajczak, H. and Orville-Thomas, W. J. (1975). *J. Mol. Struct.*, **26**, 387.

Rose, A. (1955). *Phys. Rev.*, **97**, 1538.

Sabin, J. R., Fischer, S. F., and Hofacker, G. L. (1969). *Int. J. Quantum Chem.*, **3**, 257.

Saito, G. and Matsunaga, Y. (1973). *Bull. Chem. Soc. Japan*, **46**, 714.

Sakurai, T. (1965). *Acta Cryst.*, **19**, 320.

Sakurai, T. (1968). *Acta Cryst.*, **24B**, 403.

Schein, L. B. (1977). *Phys. Rev.*, **15B**, 1024.

Schmidt, V. H. (1962). *Bull. Amer. Phys. Soc.*, **7**, 440.

Schoenes, J. and Kanazawa, K. K. (1976). *Phys. Rev. Lett.*, **37**, 1698.

Sharp, I. H. (1967). *J. Phys. Chem.*, **71**, 2587.

Shibaguchi, T., Onuki, H., and Onaka, R. (1977). *J. Phys. Soc. Japan*, **42**, 152.

Shmueli, U. and Goldberg, I. (1973). *Acta Cryst.*, **29B**, 2466.

Siebrand, W. (1964). *J. Chem. Phys.*, **40**, 2223.

Sim, G. A., Robertson, J. M., and Goodwin, T. H. (1955). *Acta Cryst.*, **8**, 157.

Statz, G. and Lippert, E. (1969). *Physics of Ice, Proceedings of the International Symposium on Physics of Ice*, Munich, Germany, September 9–14 (1968), Plenum Press, New York, N.Y., p. 152.

Stockmayer, W. H. (1941). *J. Chem. Phys.*, **9**, 398.

Suhai, S. (1974). *Theor. Chim. Acta*, **34**, 157.

Szent-György, A. (1941). *Nature*, **148**, 197.

Tachikawa, N., Yakushi, K., and Kuroda, H. (1974). *Acta Cryst.*, **30B**, 2770.

Takahashi, N., Yakushi, K., Ishii, K., and Kuroda, H. (1976). *Bull. Chem. Soc. Japan*, **49**, 182.

Thomas, J. M., Evans, J. R. N., and Lewis, T. J. (1971). *Discuss. Faraday Soc.*, **51**, 73.

Tobin, M. C. and Spicer, D. P. (1965). *J. Chem. Phys.*, **42**, 3652.

Toyoda, K. (1969). *Bull. Chem. Soc. Japan*, **42**, 1767.

Truong, K. D. and Bandrauk, A. D. (1976). *Chem. Phys. Lett.*, **44**, 232.

Uchida, T. and Akamatu, H. (1961). *Bull. Chem. Soc. Japan*, **34**, 1015.

Vincent, Verra, M., and Wright, J. D. (1974). *J. Chem. Soc., Faraday I*, **70**, 58.

Weber, K. E. and Flanagan, T. B. (1975). *Solid State Commun.*, **16**, 23.

Weiner, J. H. and Askar, A. (1970). *Nature*, **226**, 842.

Williams, R. M. and Wallwork, S. C. (1968). *Acta Cryst.*, **24B**, 168.

Wright, J. D., Ohta, T., and Kuroda, H. (1976). *Bull. Chem. Soc. Japan*, **49**, 2961.

Yakushi, K., Ikemoto, I., and Kuroda, H. (1973). *Acta Cryst.*, **29B**, 2640.

Zamaraev, K. I. and Khairutdinov, R. F. (1974). *Chem. Phys.*, **4**, 181.

Zboiński, Z. (1974). *Sci. Papers Inst. Org. Phys. Chem.*, Wroclaw Techn. Univ., No. 7, 224.

Zboiński, Z. (1976a). *Molecular Crystals and Liquid Crystals*, **32**, 219.

Zboiński, Z. (1976b). *Phys. Stat. Sol. (b)*, **74**, 561.

Zefirov, Y. V. (1976). *Zh. Obshch. Khim.*, **46**, 2636.

Molecular Interactions, Volume 2
Edited by H. Ratajczah and W. J. Orville-Thomas
© 1981 John Wiley & Sons Ltd.

11 Photoelectron spectroscopic studies of molecular complexes

I. H. Hillier

11.1 Introduction

Photoelectron (p.e.) spectroscopy provides an experimental method of measuring the whole manifold of molecular ionization energies (i.e.) arising from both the valence and the more deeply bound core electrons. The i.e.s of the valence electrons are usually studied using HeI or HeII ionizing radiation (ultraviolet photoelectron spectroscopy (UV p.e.)), whilst that of the core electrons are measured using X-ray photons giving rise to the technique of ESCA (Electron Spectroscopy for Chemical Analysis). These two techniques are fully described in standard works (Siegbahn *et al.*, 1969; Turner *et al.*, 1970; Eland, 1974).

In this review we shall be concerned with the use of these techniques to study the electronic structure and bonding in intermolecular complexes. In such complexes there is usually electronic charge migration from the donor molecule to the acceptor molecule as a result of which the i.e.s of the complex may differ from those of the uncomplexed molecules. Such changes in the i.e.s may provide definite information on the nature of the inter-molecular interactions. To see how the experimental i.e.s may yield this information we must first outline the theoretical interpretation of molecular i.e.s.

11.2 Theoretical Background

Within the orbital picture, the wavefunction for a $2n$ electron system is written in terms of n doubly occupied spin orbitals

$$\Psi = |\phi_1 \bar{\phi}_1 \phi_2 \bar{\phi}_2 \cdots \phi_n \bar{\phi}_n|$$

which are eigenfunctions of the Fock operator F (McWeeny and Sutcliffe, 1969).

Thus

$$F\phi = \varepsilon\phi,$$

where ε is the corresponding orbital eigenvalue. In molecular orbital (MO) calculations, the one-electron orbitals are expanded in terms of basis functions (χ) which are usually taken to represent atomic orbitals

$$\phi = \Sigma c_i \chi_i$$

giving rise to the Linear Combination of Atomic Orbitals (LCAO) method. When all the integrals necessary in such a calculation are evaluated exactly the calculation is termed *ab initio*. When further approximations are introduced to reduce the amount of numerical computation a variety of semi-empirical methods have evolved, the one most used to interpret i.e.s being the Complete Neglect of Differential Overlap (CNDO) and Intermediate Neglect of Differential Overlap (INDO) methods (Pople and Beveridge, 1970).

If the MOs are taken to be unchanged in form upon molecular ionization then the eigenvalues of the Fock operator, ε, are equal to the negative of the corresponding i.e. Thus if E_0 is the energy of the unionized molecule and E_i^+ is the energy of the ion formed by the removal of an electron from orbital ϕ_i, then

$$E_i^+ - E_0 = -\varepsilon_i$$

Thus within this approximation (Koopmans' theorem) (Koopmans, 1933) the negative of the eigenvalues of the Fock operator may be directly equated to the i.e.s measured by p.e. spectroscopy. This approximation neglects the orbital relaxation occurring upon ionization. However, the latter may be taken into account by performing a direct self-consistent field calculation on the molecular ionic state (ΔSCF calculation). In addition to the consideration of orbital relaxation effects it is often necessary to include correlation energy differences between the ground and ionic states to achieve agreement between the calculated and experimental i.e.s of better than 0.5 eV.

In donor–acceptor complexes the orbital of the donor molecule participating in the complex formation is frequently non-bonding. Thus on complex formation its character changes to that of a bonding orbital. For this reason its i.e. increases in the complex from its value in the free donor molecule. Associated with such bond formation there is charge transfer from the donor to the acceptor molecule which can result in a stabilization of other orbitals in the donor molecule, and a destabilization of orbitals in the acceptor molecule, both being due to electrostatic effects. For these reasons there may be a change in the i.e. of a number of orbitals of both the donor and acceptor molecules on formation of the intermolecular complex. These changes will be discussed more fully when we consider specific examples of the interpretation of the p.e. spectrum of donor–acceptor complexes.

11.2.1 Interpretation of core ionization energies

We turn now to the interpretation of core level p.e. spectra. Upon ionization of a core electron which is localized near a particular atomic nucleus the 'effective' nuclear charge experienced by the valence electrons localized on the same atom is increased by nearly unity. There is thus a large degree of valence electron reorganization occurring upon core electron ionization so that Koopmans' theorem is here a grosser approximation than in the case of valence electron ionization. It is the change in the core electron i.e. (chemical shift) upon change of atomic environment that contains information on the valence electron distribution. Such chemical shifts are commonly an order of magnitude *less* than the energy lowering produced by the orbital relaxation that occurs upon ionization, so that, at first sight it may appear that the detailed interpretation of valence electron distribution in terms of measured chemical shifts, will be unduly complicated. However, it has been found that in many simple systems such orbital relaxation is largely atomic in origin so that it does not contribute to the measured chemical shift. These shifts may be interpreted using Koopmans' theorem coupled with the core electron eigenvalues resulting from an all electron *ab initio* calculation when it is generally found that good agreement between theory and experiment results (Schwartz *et al.*, 1972). A rather more empirical interpretation utilizes the observation that the core electron binding energy depends upon the electrostatic potential at the atomic centre from which core electron ionization occurs. This potential may be partitioned into a term arising from the latter atom and a term from the other atoms in the molecule. This leads to the so-called potential model, where the chemical shift, ΔE_A is given as

$$\Delta E_A = kq_A + V + l \qquad (11.1)$$

where k and l are parameters determined by experimental data on a series of compounds, q_A is the charge on the atom considered and

$$V = \sum_{B \neq A} q_B / R_{AB} \qquad (11.2)$$

is the electrostatic potential at atom A, q_B being the charge on atom B and R_{AB} the distance between atoms A and B (Gelius, *et al.*, 1970). When the ESCA shifts are measured in the solid state, it may be necessary to include in a calculation of V, contributions from other atoms or molecules in the lattice. Equation (11.1) may often be simplified since the distance from one atom to neighbouring atoms in a molecule is often approximately constant, leading to

$$\Delta E_A = k' q_A \qquad (11.3)$$

where k' is a new parameter determined from experiment. By use of this potential model it is thus possible to relate the measured chemical shift to the molecular charge distribution via equation (11.1) or (11.3) so that ESCA should be a particularly fruitful technique for the study of the charge distribution in charge transfer complexes. We now discuss the use that has been made of ESCA to study donor–acceptor complexes.

11.3 ESCA Studies of Donor–Acceptor Complexes

11.3.1 *Pyridine–iodomonochloride*

In order to gain information on the electronic structure of donor–acceptor complexes from ESCA measurements it is most desirable to use gas-phase measurements to avoid solid-state contributions to the measured chemical shifts. Unfortunately the majority of such measurements have been carried out on solid samples for which the experimental technique is considerably simpler than for gaseous materials. However, one of the most well-known charge transfer complexes, that involving pyridine and iodomonochloride, has been studied in the gas phase by ESCA, and we will here describe the results obtained in some detail (Mostad *et al.*, 1973). In Figure 11.1 we reproduce the iodine $3d$ lines in the ESCA spectrum of this complex obtained at two different temperatures. From the changes in peak profile with temperature it is clear that the peak to low binding energy corresponds to complexed ICl, whilst that to higher binding energy corresponds to free ICl resulting from dissociation of the complex. There is thus a decrease in the I $3d$ i.e. on complex formation of $0.8\,\text{eV}$ which is a result of electron transfer from the pyridine donor molecule to the ICl acceptor molecule. Conversely the N $1s$ i.e. of pyridine is shifted to $0.9\,\text{eV}$ higher, i.e. on complex formation due to electron loss from the nitrogen atom. With the aid of equation (11.3), with k' taken to be $14\,\text{eV/unit}$ charge, a charge of $q = +0.1$ on the nitrogen atom in the complex was obtained. The degree of charge transfer obtained from the ESCA shift may be correlated with the usual description of the charge transfer complex in terms of a no-bond ($\Psi(\text{D, A})$) and a charge transfer configuration ($\Psi(\text{D}^+, \text{A}^-)$), where the wavefunction (Ψ) for the complex is given as

$$\Psi = a\Psi(\text{D, A}) + b\Psi(\text{D}^+, \text{A}^-) \qquad (11.4)$$

For comparison purposes, it may be noted that a value of b^2 of 0.16 has been estimated for this complex (Mulliken and Person, 1970). Investigation of the Cl $2p$ lines in the complex revealed only one component with no

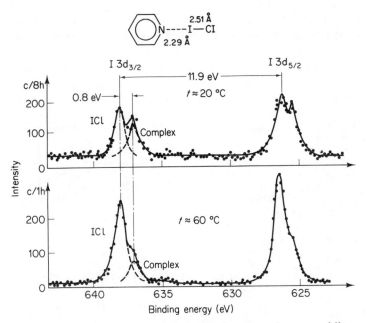

Figure 11.1 The nitrogen $1s$ electron line from pyridine-iodomonochloride and pyridine. The chemical shift between the two components is 0.9 eV.

broadening, so that it was concluded that the chemical shift on complex formation was less than 0.3 eV, supporting the idea that the charge is transferred to an empty orbital localized on the iodine atom. In passing it is worthy of note that since both the complex and its dissociation products can be seen in the ESCA spectrum of the complex, it may be possible to determine the equilibrium constant as a function of temperature leading to an evaluation of the heat of formation of the complex.

11.3.2   Chloranil–donor adducts

Ng and Hercules (1975) have studied a number of chloranil–donor adducts by ESCA. Since their data were obtained in the solid state, correction for crystal potential effects are necessary to correlate their ESCA chemical shifts with the calculated charge distribution. Their results show that ESCA may be used to elucidate the nature of the intermolecular interaction. Thus for the adduct between chloranil and HMB (hexamethylbenzene), no chemical shifts were observed for any of the core level peaks, the ESCA spectrum of the adduct being identical with that of a 1:1 powder mixture of the components. There is thus little charge transfer on complex formation, the ground state of the complex being predominantly composed of the no-bond configuration. On the other hand, the adduct between chloranil and TMPD (*N,N,N',N'*-tetramethyl-*p*-phenylenediamine) is an example of a product from charge transfer complexes involving relatively strong donors and strong acceptors. Actual charge transfer from donor to acceptor has been detected in solution by means of ultraviolet absorption spectroscopy, by the mechanism shown below (Foster, 1969) Pott and Kommandeur (1967) have

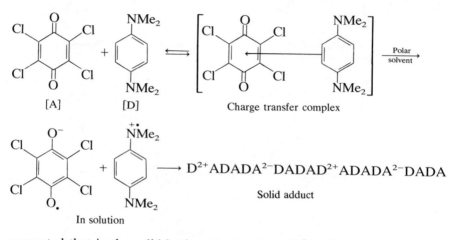

$D^{2+}ADADA^{2-}DADAD^{2+}ADADA^{2-}DADA$

Solid adduct

suggested that in the solid lattice, divalent ions (D^{2+}, A^{2-}) are formed as a result of disproportionation of monovalent ions, leading to the proposal that the crystal lattice of the complex is composed of chains of donors (D) and acceptors (A) in the following way

$$\ldots\ldots D^{2+}ADADA^{2-}DADAD^{2+}ADADA^{2-}DADAD^{2+}\ldots\ldots$$

in which one out of every five donors (or acceptors) will be a divalent ion. In the ESCA spectrum of this adduct, two N 1s peaks are observed at 400.0 and 402.1 eV with an intensity ratio of 6 : 1. The weaker peak to higher i.e. will correspond to the nitrogen atom having a positive charge. Thus these ESCA data suggest that one out of every seven donors or acceptors will be divalent, which is somewhat different from the conclusions of Pott and Kommandeur (1967) which were obtained from a consideration of the contraction of the chains caused by the presence of divalent ions. The O 1s i.e. data on this adduct parallels the nitrogen results. Thus there are two peaks at 531.8 and 533.1 eV with an intensity ratio of 1 : 5.3 suggesting that 1 out of every 6.3 atoms of oxygen in the solid adduct is negatively charged. Ng and Hercules (1975) suggest that more reliance should be placed upon the N 1s data and have demonstrated that ESCA is a means of accurately measuring the concentration of divalent ions in the lattice.

The interaction between a donor and acceptor molecule does not always lead to a molecular complex, for the reaction may proceed further. Ng and

Hercules (1975) have shown that ESCA data for the complex between chloranil and triphenylphosphine are consistent with a 1:1 addition product having the suggested structure

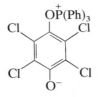

The work of these authors has thus shown that ESCA is a powerful tool in the study of a variety of different electron donor–acceptor systems.

11.3.3 Determination of charge transfer in TTF–TCNQ

The ability of ESCA measurements to distinguish between charged and uncharged donor and acceptor molecules in the crystal lattice, previously illustrated for the adduct between chloranil and TMPD, has been used to study the charge transfer in organic conductors. Such studies have highlighted the problem of unambiguously interpreting the measurements due to crystal field effects. This problem is well illustrated by considering the case of TTF–TCNQ (tetrathiofulvalenium–tetracyanoquinodimethane) which is the best organic conductor known (Ferraris $et\ al.$, 1973), and consists of parallel linear chains of separately stacked TTF cations and TCNQ anions. Both core and valence p.e. spectra of this system have been studied (Grobman $et\ al.$, 1974). Comparison of the UV p.e. spectrum of TTF–TCNQ with that of $(TCNQ)^-$ suggested that most of the photoemission (probably $\geqslant\frac{2}{3}$) is due to $(TTF)^+$ and $(TCNQ)^-$. Such a model is supported by the ESCA N $1s$ core level spectra of the complex, where there are at least two main N $1s$ peaks. If these are interpreted as representing different chemical states of nitrogen in TCNQ then the N $1s$ spectrum is consistent with both neutral $(TCNQ)$ and $(TCNQ)^-$ molecules in the solid, with $(TCNQ)^-$ dominating. Thus both valence and core p.e. spectra were considered to confirm the presence of mostly charged species in TTF–TCNQ. However, such an interpretation of the ESCA spectrum implies, at first sight that the atomic self-charge (equation 11.1) dominates the measured chemical shift and that the electrostatic potential (V, equation 11.2), which includes a Madelung term, has little effect on the chemical shift. However, it has been proposed that the electrostatic Madelung potential, which is different at the two crystallographically inequivalent nitrogen atoms (N_1, N_2) can shift the binding energy of the N $1s$ core level in N_1 with respect to N_2 (see Figure 11.2) (Metzger and Bloch, 1975; Epstein $et\ al.$, 1975). Thus a CNDO/2

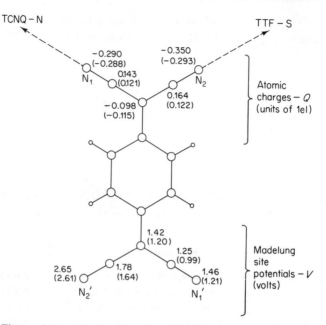

Figure 11.2 Atomic charges, atomic site potentials, for TCNQ⁻ in TTF⁺–TCNQ⁻ are shown using (a) the self-consistent crystal potential (bare numbers) or (b) quantities obtained from the CNDO calculation for isolated molecules (numbers in parentheses), before charge self-consistency is achieved.

calculation predicts essentially the same atomic charge (-0.29 e) on the two inequivalent nitrogen atoms but the Madelung potential is greater at N_2, which has positive charged TTF sulphurs near it, than at N_1. This leads to a predicted intramolecular chemical shift of ~ 1.3 eV so that the two N 1s peaks in the ESCA spectrum of the complex may be attributed to the different *crystalline* environment of the two nitrogen atoms. However, such an argument ignores the fact that the (TCNQ)⁻ atomic charges were obtained for the free molecular ion and that such atomic charges may be modified due to the Madelung potential. A self-consistent field calculation of the (TCNQ)⁻ molecular ion including the Madelung potential resulted in charge flow from the nitrogen (N_1) which is near other negatively charged nitrogens in the crystal and towards the nitrogen (N_2) which has positively charged TTF sulphurs near it (see Figure 11.2) (Grobman and Silverman, 1976). The charge difference between the inequivalent nitrogen atoms then becomes ~ 0.06 e and results in an intramolecular N 1s shift (proportional to Δq) that is larger than the splitting caused by the Madelung site potentials. The net result of these two effects is to yield an estimated intramolecular

N $1s$ chemical shift of ~-0.5 eV. The important result is that the valence charge redistribution occurring due to the lattice environment provides a mechanism for the reduction of the i.e. splittings at crystallographically inequivalent nitrogen atoms. This calculation thus provides support for the contention that the two peaks in the TTF–TCNQ N $1s$ core level spectrum are due to $(TCNQ)^-$ and $TCNQ^0$. The ratio of the areas of the two components is ~ 0.64 corresponding to ~ 0.6 electrons/formula unit of charge transfer. Thus the N $1s$ ESCA spectrum of TTF–TCNQ is strong evidence for partial charge transfer in the crystal with between 1.3 and 2 $TCNQ^-$ molecules for every $TCNQ^0$ molecule, and also highlights the need for careful inclusion of crystal effects in the calculation of intramolecular chemical shifts.

11.3.4 *Donor–acceptor complexes involving* BF_3

The examples we have discussed so far have illustrated that ESCA measurements are extremely useful in understanding the gross changes in electronic charge that occur upon formation of an intermolecular complex. However, these examples have been discussed only in terms of semiquantitative models and approximate molecular orbital calculations. Core binding energies of a number of nitrogen bases and their adducts with BF_3 have been reported which have been interpreted using both *ab initio* and semi-empirical MO calculations (Barber *et al.*, 1973). Such a comparison between calculated and experimental i.e.s yields a better understanding of the origin of the observed chemical shifts. In Table 11.1 we list the experimental $1s$ i.e.s of the carbon, nitrogen, boron, and fluorine atoms for a variety of such complexes and for the free donor molecules, together with the $1s$ i.e.s calculated, for the smaller systems, by means of *ab initio* calculations and Koopmans' theorem. It can be seen that there is a generally good agreement between the calculated and experimental chemical shifts, although the absolute values of the i.e.s are calculated to be too large due to the neglect of orbital relaxation effects. In Table 11.2 we show the calculated orbital populations and formal atomic charges obtained from the *ab initio* wavefunctions which allow an understanding of the origin of the observed chemical shifts. For all the bases the N $1s$ i.e. is increased by 2–3 eV on complex formation, and smaller, though still significant changes in the boron and F $1s$ i.e.s of the acceptor BF_3 group occur when the base is varied. In addition to the N $1s$ shift of the base, the core i.e. of the other atoms of the base may also be affected by formation of the B—N bond, in particular there is an increase in the intramolecular C $1s$ shift of CH_3CN and $C_2H_5NH_2$ compared with the values for the free bases. This can be seen in Figure 11.3 where the C $1s$ region of the ESCA spectrum of CH_3CN is compared with that of CH_3CNBF_3.

Table 11.1   Experimental and calculated 1s ionization potentials (eV)

	Carbon		Nitrogen		Boron		Fluorine	
	expt.	calc.	expt.	calc.	expt.	calc.	expt.	calc.
NH_3				422.5				
NH_3BF_3			402.1	427.6	195.1	210.8	686.8	714.5
$C_2H_5NH_2$ (a)*	285.8	306.8	399.1	422.8				
	285.2	305.8						
$C_2H_5NH_2BF_3$ (b)	287.0	308.6	401.6	427.3	194.8	210.5	686.8	714.4
	285.6	307.3						
CH_3CN (c)	287.4	308.4	399.6	424.9				
	286.5	307.8						
CH_3CNBF_3 (d)	289.3	311.8	402.0	428.0	195.7	211.1	687.2	714.9
	287.5	309.4						
C_5H_5N (e)	285.7		399.5					
$C_5H_5NBF_3$ (f)	286.3		401.6		194.5		685.8	
$2,6\text{-}(CH_3)_2C_5H_3N$ (g)	285.4		399.2					
$2,6\text{-}(CH_3)_2\text{-}$ $C_5H_3NBF_3$ (h)	284.9		401.4		194.3		686.1	

* These letters provide a code for the points in Figure 11.4.

The population analyses (Table 11.2) show that, as expected, there is a loss of σ-electron density (mainly $2p\sigma$) from the nitrogen atom on formation of the B—N bond. In addition, an increase in the N $2p\pi$ populations occurs to counterbalance this effect leading to a small net increase in the nitrogen atomic population. There is a large positive charge on the boron atom of the complexes, although not as large as that of free BF_3 due to an increase in the B $2s$ and B $2p\sigma$ orbital populations. The trends in the experimental B $1s$ i.e.s (Table 11.1) follow those of the formal atomic charges, the value for CH_3CNBF_3 being greater than those of NH_3BF_3 and $C_2H_5NH_2BF_3$. The very small differences between the F $1s$ i.e. parallel the trend in the boron values, and also correlates with the changes in the fluorine atom charges, although the latter are very small. The small F $1s$ shifts may thus be explained in terms of the fluorine and boron atomic charges (equation 11.1). The quite large changes in the N $1s$ i.e. of the base on formation of the donor–acceptor complex, are mainly due to the large positive charge on the neighbouring boron atom (the second term in equation 11.1), whilst the change in B $1s$ i.e. between the complexes reflects the formal boron charge (the first term in equation 11.1). It can be seen from the experimental data (Table 11.1, Figure 11.3) that the core i.e.s of the atoms of the base, other than the nitrogen atom, are also changed on complex formation. Thus in the BF_3 complexes of both CH_3CN and $C_2H_5NH_2$ an increased separation of the C $1s$ i.e.s is observed compared with the values in the free bases. For

Table 11.2 Calculated orbital populations from *ab initio* wavefunctions

	Nitrogen				Carbon				Boron				Fluorine
	$2s$	$2p(\sigma)$	$2p(\pi)$	charge	$2s$	$2p(\sigma)$	$2p(\pi)$	charge	$2s$	$2p(\sigma)$	$2p(\pi)$	charge	charge
NH_3	1.68	1.82	2.28	-0.78									
NH_3BF_3	1.67	1.51	2.62	-0.80					0.54	0.44	0.86	1.16	-0.50
$C_2H_5NH_2$	1.67	3.96		-0.62	1.41	3.21		-0.62*					
					1.35	2.96		-0.31					
$C_2H_5NH_2BF_3$	1.66	4.01		-0.67	1.41	3.22		-0.63*	0.54	0.45	0.85	1.16	-0.50
					1.36	2.96		-0.32					
CH_3CN	1.86	1.42	2.07	-0.35	1.42	0.86	2.37	-0.64*					
					1.07	0.79	1.97	0.17					
CH_3CNBF_3	1.74	1.33	2.34	-0.41	1.42	0.80	2.43	-0.66*	0.45	0.41	0.89	1.25	-0.49
					1.09	0.80	1.74	0.37					
BF_3									0.40	0.40	0.88	1.32	-0.44

*These values refer to the —CH_3 carbon atom.

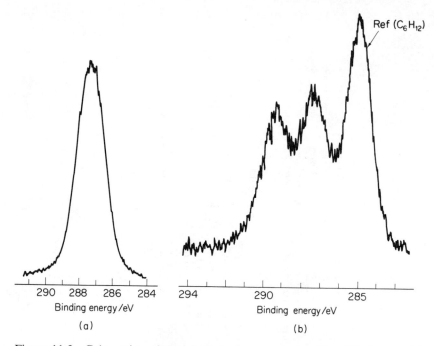

Figure 11.3 C 1s region of the photoelectron spectrum of (a) CH₃CN and (b) CH₃CNBF₃, the latter containing C_6H_{12} as calibrant.

CH_3CN, a calculation at the equilibrium molecular geometry predicts the —CH_3 C 1s i.e. to be the larger of the two. However, in the BF_3 complex the *ab initio* calculation gives the —CN C 1s i.e. to be larger than that of the —CH_3 group by 2.4 eV, to be compared with the experimental splitting of 1.9 eV (Figure 11.3). Thus with this assignment, the C 1s i.e. of the —CH_3 group has increased by 0.1 eV, and that of the —CN group has increased by 2.8 eV. These trends are reflected in the calculated charge distributions (Table 11.2). On complex formation, the charge on the —CH_3 carbon atom remains essentially unaltered, whilst that of the —CN carbon atom is increased by ~0.2 e, due to a loss of C $2p\pi$ electron density, the σ density remaining practically unchanged. In both $C_2H_5NH_2$ and its BF_3 complex, the *ab initio* calculations predict that the —CH_3 i.e. is the smaller of the two. With this assignment, the C 1s i.e. of the —CH_3 group is increased by 0.4 eV and that of the —CH_2— group by 1.2 eV on complex formation. The calculated carbon atomic charges do not vary significantly between the free and complexed base (Table 11.2), so that the changes in the C 1s i.e. values may be ascribed mainly to the influence of the positively charged boron atom and to the loss of electron density from the hydrogen atoms of the base. In the case of pyridine, lutidine and their BF_3 complexes, no resolution

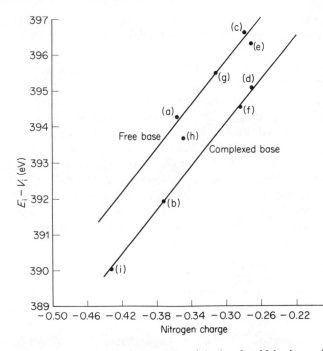

Figure 11.4 Point charge potential plot for N 1s i.e. using atomic charges from INDO wavefunctions (see Table 11.1 for lettering code).

of the C 1s i.e. occurs, although an increase of the full width at half maximum from 2.3 eV for the bases to 2.7 eV for their complexes suggests that an increase in the C 1s i.e. splittings has occurred on complex formation. The experimental i.e.s may also be interpreted using equation (11.1), by plotting $(E - V)$ vs. q when a straight line will be obtained if this semi-empirical relationship is valid. Figure 11.4 shows such a plot for the N 1s i.e. of both the free and complexed bases, the formal atomic charges used to compute V being obtained by an INDO calculation on these systems. A separation into two distinct lines, one for the free bases and one for the complexed bases, is evident indicating that the N 1s i.e. in the complexed base has been overestimated compared with that for the free base. For the carbon and boron 1s i.e.s the potential model gives satisfactory linear plots.

Thus the change in charge distribution on formation of the N—B bonds, calculated to be entirely of σ character, is found to be reflected in the change in the core i.e. of the atoms both of the base, and of the electron acceptor. The B 1s i.e. reflects the degree of σ donation and hence the strength of the B—N bond. Indeed a very good correlation exists between the experimental heats of formation of the complexes and the B $1s$ i.e. Thus the heats of formation of $C_5H_5NBF_3$ (50.6 kcal mol^{-1}) (Van der Meulen and Heller, 1932) and $C_2H_5NH_2BF_3$ are greater than that of CH_3CNBF_3 (26.5 kcal mol^{-1}) (Laubengayer and Sears, 1945), which is in line with B $1s$ i.e. values, that of CH_3CNBF_3 being the largest (195.7 eV), and that of $C_5H_5NBF_3$ the smallest (194.5 eV) of those studied. Both the *ab initio* and INDO wavefunctions agree with this order of stability.

The work described so far has illustrated the value of core i.e.s in understanding the details of the change in electron density that occurs on formation of a donor–acceptor complex. Associated with such a charge migration there will be a change in the energies of the valence MOs which may be observed experimentally by UV p.e. and interpreted by MO calculations. Unfortunately such studies are few in number at present, but the borane complexes with Lewis bases have been studied extensively by Lloyd and coworkers (Lloyd and Lynaugh, 1970, 1972).

11.4 Valence Photoelectron Spectra

11.4.1 *Complexes of borane with Lewis bases*

To illustrate the use of UV p.e. in the understanding of the bonding in donor–acceptor complexes we shall consider the two simple systems NH_3BH_3 and BH_3CO. In NH_3, the highest filled orbital is the $3a_1$ having mainly nitrogen lone-pair character, the next MO is the $1e$ which has N—H bonding character (Figure 11.5). On formation of the complex NH_3BH_3, the NH_3 $3a_1$ MO becomes delocalized and gives rise to the N—B σ bond and is thereby stabilized by ~ 3 eV (this orbital becomes the $4a_1$ in the complex; Figure 11.5). The $1e$ orbital is also stabilized experimentally by ~ 2 eV. These changes in the valence i.e.s are satisfactorily interpreted by means of Koopmans' theorem using *ab initio* wavefunctions. The situation is similar for borane complexes of ammonia, mono-, di-, and trimethylamine where the delocalization of the donor lone pair to the boron atom results in a stabilization of all but the lone-pair orbitals by about 1 eV which is most readily interpreted as an electrostatic effect from increasing the positive charge at the nitrogen atom. The donor lone pair is stabilized by about 3.0 eV and it is suggested that though some of this is due to the increased charge on the nitrogen atom, most of the stabilization comes from delocalization (Lloyd and Lynaugh, 1972).

In the case of carbonyl-borane (BH_3CO) the interpretation of the change in the valence i.e.s on complex formation is less straightforward. Thus both the 5σ carbon lone pair and the 1π (C—O bonding) orbital of CO are very little changed in energy on complex formation when the i.e.s of CO and BH_3CO are compared (Figure 11.5), and the 4σ MO is indeed destabilized

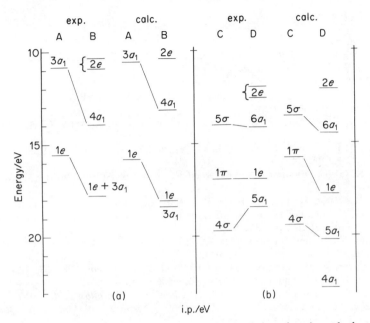

Figure 11.5 Comparison of experimental i.e. (exp.) and the eigenvalues from *ab initio* calculations (calc.) for borane complexes and for the free bases. Eigenvalues have been multiplied by 0.92, and are taken from Armstrong and Perkins (1969). The i.e. for NH_3 and CO are from Turner *et al.* (1970). A, NH_3; B, NH_3BH_3; C, CO; D, BH_3CO.

by about 1 eV. These experimental observations are in conflict with the results of an accurate *ab initio* calculation on BH_3CO near the Hartree–Fock limit (Ermler *et al.*, 1976). This calculation predicts, if Koopmans' theorem is valid, that the 5σ orbital is stabilized by 0.2 eV, the 1π by 1.5 eV and that the energy of the 4σ is essentially unchanged. Such a disagreement with the experimental data probably arises from the neglect of correlation effects in the calculation of the i.e.s, and requires further study.

11.5 Conclusions

Although the technique of photoelectron spectroscopy has only been exploited to a rather limited extent in the study of donor–acceptor complexes the work carried out to date has demonstrated its usefulness in this area. The changes in both the core and valence i.e.s on complex formation are sensitive monitors of the change in molecular charge distribution accompanying bond formation. With the aid of both semi-empirical and *ab initio* molecular orbital calculations the experimental i.e.s may be directly interpreted in terms of such charge migration and delocalization.

11.6 References

Armstrong, D. R. and Perkins, P. G. (1969). *J. Chem. Soc. A*, 1064.

Barber, M., Connor J. A., Guest, M. F., Hillier, I. H., Schwarz, M., and Stacey, M. (1973). *J. Chem. Soc., Faraday II*, **69**, 551.

Eland, J. H. D. (1974). *Photoelectron Spectroscopy*, Butterworths.

Epstein, A. J., Lipari, N. O., Sandman, D. J., and Nielsen, P. (1975). *Bull. Amer. Phys. Soc.*, **20**, 465.

Ermler, W. C., Glasser, F. D., and Kern, C. W. (1976). *J. Amer. Chem. Soc.*, **98**, 3799.

Ferraris, J. P., Cowan, D. O., Walatka, V., and Pealstein, J. H. (1973). *J. Amer. Chem. Soc.*, **95**, 948.

Foster, R. (1969). *Organic Charge-Transfer Complexes*, Academic Press, New York, N.Y.

Gelius, U., Roos, B., and Siegbahn, P. (1970). *Chem. Phys. Lett.*, **4**, 471.

Grobman, W. D., Pollak, R. A., Eastman, D. E., Maas, E. T., and Scott, B. A. (1974). *Phys. Rev. Lett.*, **32**, 534.

Grobman, W. D. and Silverman, B. D. (1976). *Solid State Commun.*, **19**, 319.

Koopmans, T. (1933). *Physica*, **1**, 104.

Laubengayer, A. W. and Sears D. S. (1945). *J. Amer. Chem. Soc.*, **67**, 164.

Lloyd, D. R. and Lynaugh, N. (1970). *Chem. Commun.*, 1545.

Lloyd, D. R. and Lynaugh, N. (1972). *J. Chem. Soc., Faraday II*, **68**, 947.

McWeeny, R. and Sutcliffe, B. T., (1969). *Methods of Molecular Quantum Mechanics*, Academic Press.

Metzger, R. M. and Bloch, A. N. (1975). *Bull. Amer. Phys. Soc.*, **20**, 415.

Van der Meulen, P. A. and Heller, H. A. (1932). *J. Amer. Chem. Soc.*, **54**, 4404.

Mostad, A., Svensson, S., Nilsson, R., Basilier, E., Gelius, U., Nordling, C., and Siegbahn, K. (1973). *Chem. Phys. Lett.*, **23**, 157.

Mulliken, R. S. and Person, W. B. (1970). *Molecular Complexes*, John Wiley, New York.

Ng, K. T. and Hercules, D. M. (1975). *J. Amer. Chem. Soc.*, **97**, 4168.

Pople, J. A. and Beveridge, D. L. (1970). *Approximate Molecular Orbital Theory*, McGraw-Hill.

Pott, G. T. and Kommandeur, J. (1967). *Mol. Phys.*, **13**, 373.

Schwartz, M. E., Switalski, J. D., and Stronski, R. E. (1972). In *Electron Spectroscopy* (Ed. Shirley, D. A.); North-Holland, Amsterdam.

Siegbahn, K., Nordling, C., Johansson, G., Hedman, J., Heden, P. F., Hamrin, K., Gelius, U., Bergmark, T., Werme, L. O., Manne, R., and Baer, Y. (1969). *ESCA Applied to Free Molecules*, North-Holland, Amsterdam.

Turner, D. W., Baker, C., Baker, A. D., and Brundle, C. R. (1970). *Molecular Photoelectron Spectroscopy*, John Wiley.

Molecular Interactions, Volume 2
Edited by H. Ratajczak and W. J. Orville-Thomas
© 1981 John Wiley & Sons Ltd.

12 Properties of molecular complexes in the electronic excited states

N. Mataga

12.1 Introduction

Studies of electronic structures and dynamic behaviour of molecular complexes in the excited electronic state are quite interesting and important from the viewpoint of elucidating the mechanisms of photochemical reactions. They are closely related not only to the mechanisms of photochemical reactions but also to the mechanisms of chemical reactions in general, since one of the most important elementary processes of chemical reactions is the change of electronic distribution, followed by the change of nuclear configurations in the course of the reaction, and photoexcitation of molecular complexes induces such changes.

Thus the molecular complex in the excited state may be regarded as a typical model system for elementary processes of chemical reactions. The circumstance seems to be quite analogous also in the case of molecular interactions occurring in the encounter collision in the excited electronic state. Molecular complexes or aggregates which are unstable (dissociative) in the ground state but are stable under electronic excitation are called 'exciplex'. The main path of formation of the exciplex is the encounter collision between excited- and ground-state partners. This sort of aggregate formed between identical molecules is called an 'excimer' and that formed between unlike molecules a 'heteroexcimer'. Many aromatic hydrocarbons form typical excimers, and typical heteroexcimers are formed between aromatic hydrocarbon and amine as well as between aromatic hydrocarbon and cyano-substituted aromatic hydrocarbon.

In the present article, we include in 'exciplex' not only the typical excimers and heteroexcimers but also the excited state of electron donor–acceptor (EDA) complexes stable in the ground state, hydrogen bonding complexes in the excited state, as well as some solute–solvent complexes formed due to strong electrostatic interactions in the excited state.

For the studies of electronic structures of these excited systems and their various relaxation processes such as solvation, hydrogen atom and proton transfers, ionic dissociation into solvated radical ions, and other photochemical as well as photophysical processes, the use of pulsed laser techniques seems to be most suitable, because of the rapidity of these processes. Namely, the picosecond (ps, $1 \text{ ps} = 10^{-12} \text{ s}$) and nanosecond (ns, $1 \text{ ns} = 10^{-9} \text{ s}$) time-resolved transient absorption and fluorescence measurements by means of laser flash spectroscopy are providing much insight into the nature of the molecular interactions in the excited electronic state.

This chapter is not intended to provide a comprehensive review of studies of structures and dynamic behaviour of the above exciplex systems. There are some more or less detailed recent review articles and books relevant to exciplex phenomena (Förster, 1969, 1975; Birks, 1970, 1975a, b; Mataga and Kubota, 1970; Ottolenghi, 1973; Klöpffer, 1973; Nagakura, 1975; Lippert, 1975; Beens and Weller, 1975; Davidson, 1975; Mataga and Ottolenghi, 1979). In this article, we consider some examples of investigations by means of laser as well as by conventional flash photolysis and luminescence measurements, and examine fundamental principles underlying the behaviour of these systems.

12.2 Interaction of Excited Polar Systems with Polar Solvent Molecules

We are examining in this article various dynamic processes following the photoexcitation of complex organic molecules and molecular complexes in solution. The energy levels, electronic structures, as well as dynamic behaviour of these systems in the excited electronic state are dependent upon the interaction with the surrounding solvent molecules. There are various solute–solvent interactions relevant to the dynamic behaviour of excited solute systems including electrostatic ones such as dipole–dipole and dipole–polarization interactions, as well as more specific interactions of hydrogen bonding and charge transfer. In this section we are concerned mainly with the electrostatic interactions, where both the excited solute system and solvent molecules are polar.

As it is well known, light absorption and emission in solution will occur in accordance with the Franck–Condon principle including not only the nuclear configurations of the solute system but also the configurations of surrounding solvent molecules, since the duration of a radiative electronic transition is so short ($\sim 10^{-15} \text{ s}$) that during such a transition the configurations of all nuclei remain almost unchanged.

The intramolecular vibrational relaxation time τ_{vib} is $ca.$ 10^{-13} s in its order of magnitude which does not seem to depend very much upon the temperature and the viscosity of the environment. However, the relaxation time τ_R of solute–solvent dipole–dipole interaction depends considerably upon the temperature and viscosity of the environment. Therefore, although the vibrational relaxation occurs very rapidly immediately after the light absorption, there may be various cases depending upon the values of τ_R compared with the lifetime τ_e of the excited state relaxed with respect to intramolecular vibration. We are mainly concerned in this section with the lowest excited singlet state (S_1), τ_e of which is *ca.* 10^2 ns–1 ns in its order of magnitude.

12.2.1   *Molecular emission process and solvent relaxation*

Immediately after the light absorption transition from the ground state (S_0) to the lowest excited singlet state (S_1), the solute–solvent reorientation relaxation starts. This dynamic behaviour originates from the fact that the electronic structure of the solute in S_1 differs from that in S_0. Accordingly, the equilibrium configuration of the surrounding solvent dipoles in S_0 is not the most stable one for S_1, leading to the reorientation relaxation from the non-equilibrium excited state to the relaxed equilibrium excited state as indicated schematically in Figure 12.1.

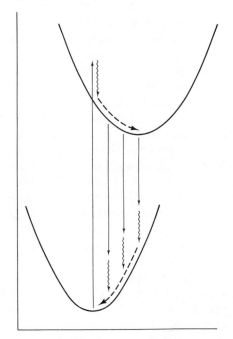

Configuration coordinate

Figure 12.1   Molecular absorption and emission as well as related relaxation processes. ξ : vibrational relaxation, $\rightarrow$ --→ : reorientation relaxation

If $\tau_R \ll \tau_e$, the equilibrium excited state with respect to the reorientation relaxation will be arrived at during the excited-state lifetime and the fluorescence will be emitted almost exclusively from the equilibrium excited state. The ground state immediately after the emission and vibrational relaxation is a non-equilibrium state, from which the reorientation relaxation to the equilibrium state occurs. The energy difference between non-equilibrium state immediately after intramolecular vibrational relaxation and the equilibrium state may be called the orientation destabilization energy. The sum of the orientation destabilization energies for the ground and excited state determines the Stokes' red shift due to the solute–solvent dipole–dipole interactions. The larger the difference of the electronic structures between S_0 and S_1 and the larger the polarity of the solvent, the greater the values of the Stokes' shift. A simplified equation for this Stokes' shift $(\tilde{\nu}_a - \tilde{\nu}_e)$ (equation 12.1) was first given by Lippert and Mataga (Lippert, 1955, 1957; Mataga et al., 1955, 1956).

$$hc(\tilde{\nu}_a - \tilde{\nu}_e) = \frac{2(\mu_e - \mu_g)^2}{a^3}\left[\frac{\varepsilon - 1}{2\varepsilon + 1} - \frac{n^2 - 1}{2n^2 + 1}\right] \tag{12.1}$$

where μ_e and μ_g are dipole moments of the solute in the excited and ground state, respectively, ε and n dielectric constant and refractive index of the solvent, and a is the cavity radius in Onsager's theory of reaction field.

If $\tau_R \gg \tau_e$, reorientation relaxation is not possible during the lifetime of the excited state. Such a condition may be realized in highly viscous, low-temperature rigid solutions, where the emission from the state non-relaxed with respect to the solute–solvent reorientation will be observed.

The condition $\tau_R \approx \tau_e$ may be realized when the excited-state lifetime is very short or τ_R becomes somewhat longer by temperature lowering as well as by raising viscosity compared with that in ordinary solution at room temperature. In this case, since the reorientation relaxation process competes with the emission, one can observe the spectra emitted during orientational relaxation from the non-relaxed to the equilibrium excited state.

Although accurate values of τ_R for various systems are not known, one may presume that the value of the dielectric relaxation time τ_D of a solvent may be close to the τ_R value. The τ_D values of the ordinary solvents which are not so viscous at room temperature are 1–10 ps (1 ps $= 10^{-12}$ s) in their orders of magnitude, which are much smaller than τ_e of S_1, as assumed for the derivation of equation (12.1). However, τ_R is a microscopic quantity which might be considerably affected by a short-range electrostatic interaction of the solvent dipole with large solute dipole in the excited state, while the τ_D value is a more or less macroscopic quantity. Nevertheless, if an accurate value of τ_R is lacking, τ_D will give an approximation to it.

As described above, by using low temperature or otherwise viscous solutions, the condition $\tau_R \approx \tau_e$ seems to be realized and, in this case, one can study the relaxation process by means of nanosecond time-resolved spectroscopy. However, there might be some temperature-dependent quenching process due to solute–solvent interaction other than the reorientation relaxation. Therefore, for a more direct measurement of the reorientation relaxation process, we need picosecond time-resolved spectroscopy. Such a measurement seems to be very important not only from the viewpoint of the elucidation of the dynamic behaviour of excited molecules, but also for the elucidation of the nature of the liquid state.

Recent development of picosecond laser spectroscopy has made possible such a direct measurement. Although the application of picosecond laser methods to this problem is rather rare at present, the mechanism of this relaxation process may be elucidated through the picosecond studies upon various systems.

12.2.2 Solvent dependence of electronic and geometrical structures of excited polar systems

The effect of the polar solvent upon the behaviour of the excited polar solute molecule is not limited to the stabilization of the solute fluorescent state with rigid dipole due to the reorientation of the surrounding solvent dipoles. From a theoretical point of view, electronic and geometrical structures of molecular composite systems such as intra- and intermolecular exciplex systems and excited EDA complexes stable in the ground state seem to be liable to change due to the interaction with polar solvent molecules, from the reasoning described below.

Generally speaking, the Schrödinger equation and the Hamiltonian of the solute–solvent system may be given by:

$$\mathcal{H}\Psi = E\Psi,$$
$$\mathcal{H} = \mathcal{H}_0 + \mathcal{H}' \tag{12.2}$$

where $\mathcal{H}_0$ is the Hamiltonian of the solute in the absence of the solvent and $\mathcal{H}'$ represents the solute–solvent interactions. $\mathcal{H}'$ depends upon the change distribution of solute, i.e. upon the wavefunction Ψ. However, Ψ depends not only upon $\mathcal{H}_0$ but also upon $\mathcal{H}'$. Thus the electronic structure of the solute is changed by the interaction with solvents and one must make a SCF-type calculation in order to solve equation (12.2).

In the case of the inter- and intramolecular charge transfer (CT) systems, the wavefunction may be given, for example, by,

$$\Psi = C_1\Psi(A^-D^+) + C_2\Psi(A^*D) + C_3\Psi(AD) \tag{12.3}$$

where the first term represents the CT configuration, the second is the locally excited configuration, and the last term is the ground configuration.

Ordinarily, the contribution of $\Psi(A^-D^+)$ is larger in the excited state than in the ground state. It will be further increased by the interaction of the solute with polar solvent molecules. Namely, the solvation of the CT system by polar solvents will oppose the electronic delocalization between A and D since the stabilization energy due to the solvation will increase with increase of the charge separation in the system, i.e. with its increasing dipole moment. Therefore, it might be possible that strong solvation induces a change of geometrical structure of the solute in the excited equilibrium state, leading to a decrease of the electronic delocalization interaction between A and D.

Since the solute–solvent interaction is essentially 'molecular' interaction between solute molecule and surrounding solvent molecules, for a theoretical treatment of this problem, one must calculate the interaction energy for each case by using some appropriate wavefunction like that of equation (12.3), assuming some specified form of $\mathscr{H}'$ for several interacting molecules. However, such a calculation seems to be rather difficult in practice. This is a rather old problem in the theoretical treatment of molecular interactions in liquid solutions.

Assuming the dielectric continuum model, $\mathscr{H}'$ may be given by using Onsager's reaction field,

$$\mathscr{H}' = -\mathbf{\mu}_{op}\mathbf{F} \tag{12.4}$$

where $\mathbf{\mu}_{op}$ is the dipole operator, and $\mathbf{F}$ is Onsager's reaction field. By means of this simplified model and by using the wavefunction of equation (12.3), the energies and wavefunctions of a composite system in solution may be given by the following procedure (Beens and Weller, 1975; Masaki et al., 1976).

The reaction field is given by;

$$\mathbf{F} = \mathbf{\mu} f_\varepsilon / a^3, \qquad f_\varepsilon = 2(\varepsilon - 1)/(2\varepsilon + 1) \tag{12.5}$$

where a is the cavity radius and $\mathbf{\mu}$ is given by:

$$\mathbf{\mu} = \langle \Psi | \mathbf{\mu}_{op} | \Psi \rangle \simeq C_1^2 \langle \Psi(A^-D^+) | \mathbf{\mu}_{op} | \Psi(A^-D^+) \rangle = C_1^2 \mathbf{\mu}_0 \tag{12.6}$$

The matrix elements of the Hamiltonian are given as follows:

$$\begin{aligned}
\langle \Psi(A^-D^+) | \mathscr{H} | \Psi(A^-D^+) \rangle &= \langle \Psi(A^-D^+) | \mathscr{H}_0 | \Psi(A^-D^+) \rangle \\
&\quad + \langle \Psi(A^-D^+) | \mathscr{H}' | \Psi(A^-D^+) \rangle \\
&= E_c - \langle \Psi(A^-D^+) | \mathbf{\mu}_{op}\mathbf{F} | \Psi(A^-D^+) \rangle \\
&= E_c - C_1^2 \left(\frac{\mu_0^2}{a^3} \right) f_\varepsilon
\end{aligned} \tag{12.7}$$

$$\left.\begin{array}{l} \langle \Psi(A^-D^+) \, | \, \mathcal{H} \, | \, \Psi(A^*D) \rangle \simeq \langle \Psi(A^-D^+) \, | \, \mathcal{H}_0 | \, \Psi(A^*D) \rangle = \alpha \\[4pt] \langle \Psi(A^-D^+) \, | \, \mathcal{H} \, | \, (AD) \rangle \simeq \langle \Psi(A^-D^+) \, | \, \mathcal{H}_0 | \, \Psi(AD) \rangle = \beta \\[4pt] \langle \Psi(A^*D) \, | \, \mathcal{H} \, | \, \Psi(A^*D) \rangle \simeq \langle \Psi(A^*D) \, | \, \mathcal{H}_0 | \, \Psi(A^*D) \rangle = E_e \\[4pt] \langle \Psi(AD) \, | \, \mathcal{H} \, | \, \Psi(AD) \rangle \simeq \langle \Psi(AD) \, | \, \mathcal{H}_0 | \, \Psi(AD) \rangle = E_g \end{array}\right\} \quad (12.8)$$

In terms of the matrix elements of equations (12.7) and (12.8), the secular equations can be written as:

$$\left.\begin{array}{l} C_1\left(E_c - C_1^2\left(\dfrac{\mu_0^2}{a^3}\right)f_\varepsilon - E\right) + C_2\alpha + C_3\beta = 0 \\[10pt] C_1\alpha + C_2(E_e - E) = 0, \qquad C_1\beta + C_3(E_g - E) = 0 \end{array}\right\} \quad (12.9)$$

Equation (12.9) is non-linear in the sense that the energy of $\Psi(A^-D^+)$ contains C_1^2. However, in this case, the solution can be obtained without iteration by solving equation (12.10).

$$\left\{ E_c - E - \frac{\alpha^2}{(E_e - E)} + \frac{\beta^2}{E} \right\}\left\{ 1 + \frac{\alpha^2}{(E_e - E)^2} + \frac{\beta^2}{E^2} \right\} = \left(\frac{\mu_0^2}{a^3}\right)f_\varepsilon \quad (12.10)$$

In order to obtain the total energy of the system, one must add to the energy obtained from equation (12.10) the polarization energy of the solvent, $E_{pol} = (\tfrac{1}{2})\boldsymbol{\mu}\boldsymbol{F}$. Then, the energies of the lowest excited equilibrium state (E_e^{eq}) and the Franck–Condon ground state (E_g^{FC}) will be given by equations (12.11) and (12.12) respectively, taking into consideration the orientation destabilization energy in the ground state.

$$E_e^{eq} = E_1 + \tfrac{1}{2}\boldsymbol{\mu}_e^{eq}\boldsymbol{F}_e^{eq} = E_1 + \frac{1}{2}\left(\frac{\mu_e^{eq^2}}{a^3}\right)f_\varepsilon = E_1 + \tfrac{1}{2}C_1^4\left(\frac{\mu_0^2}{a^3}\right)f_\varepsilon \quad (12.11)$$

$$E_g^{FC} = E_0 - \left(\frac{\boldsymbol{\mu}_g^{FC}\boldsymbol{\mu}_e^{eq}}{a^3}\right)(f_\varepsilon - f_n) + \frac{1}{2}\left(\frac{\mu_e^{eq^2}}{a^3}\right)(f_\varepsilon - f_n) - \frac{1}{2}\left(\frac{\mu_g^{FC^2}}{a^3}\right)f_n$$

$$= E_0 - C_1'^2 C_1^2\left(\frac{\mu_0^2}{a^3}\right)(f_\varepsilon - f_n) + \tfrac{1}{2}C_1^4\left(\frac{\mu_0^2}{a^3}\right)(f_\varepsilon - f_n) - \tfrac{1}{2}C_1'^4\left(\frac{\mu_0^2}{a^3}\right)f_{n'} \quad (12.12)$$

$$f_n = 2(n^2 - 1)/(2n^2 + 1)$$

where n is the refractive index of the solvent, $\boldsymbol{\mu}_e^{eq}$ is the solute dipole moment in the excited equilibrium state, $\boldsymbol{\mu}_g^{FC}$ is the dipole moment in the Franck–Condon ground state with respect to solvation, E_0 and E_1 are respectively the lowest and the second lowest eigenvalues obtained from equation (12.10), and C_1 and C_1' represent respectively, the coefficients of $\Psi(A^-D^+)$ in the wavefunctions for the lowest excited state and the ground state. Thus the solvent dependence of the fluorescence spectrum due to the solvation of the solute dipole including the solvation-induced electronic structure change of the solute molecule may be given by, $h\nu_e = E_e^{eq} - E_g^{FC}$.

Now, let us examine a simplified case where the matrix elements α and β due to the electronic delocalization interaction between A and D are negligibly small and $E_1 > E_e$ though they are fairly close to each other. In a non-polar solvent, the energy of the relaxed polar state,

$$E_p = E_1 + \frac{1}{2}\left(\frac{\mu_e^{eq^2}}{a^3}\right)f_\varepsilon = E_c - \frac{1}{2}\left(\frac{\mu_e^{eq^2}}{a^3}\right)f_{\varepsilon'}$$

is higher than E_e and the Franck–Condon destabilization energy due to the solvation can be neglected. With increase of the solvent polarity, E_p becomes lower and level inversion between E_p and E_e occurs at a value of the solvent polarity parameter. At the inversion point, there arises the orientation destabilization energy,

$$\delta E_g^{FC} = \frac{1}{2}\left(\frac{\mu_e^{eq^2}}{a^3}\right)(f_\varepsilon - f_n).$$

In solvents more polar than the inversion point the frequency of the fluorescence from the polar excited state may be given by:

$$h\nu_e = (E_c - E_g) - \left(\frac{\mu_e^{eq^2}}{a^3}\right)(f_\varepsilon - \tfrac{1}{2}f_n) \tag{12.13}$$

Since the polar state becomes lower than the non-polar state only after the solvent reorientation, emissions both from polar state and non-polar state may be observed simultaneously in solutions of appropriate polarity. These circumstances are indicated schematically in Figure 12.2. If α as well as β are not so small, the level inversion may not be so sharp. In any case, this sort of solvent-indiced charge transfer in the excited state may be

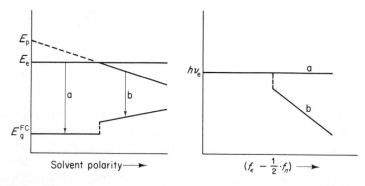

Figure 12.2 Schematic representation of dependences of energy levels and emission frequencies of excited charge transfer systems upon solvent polarity. (See text for the notations.) a. Fluorescence from locally excited state. b. Fluorescence from charge transfer state

closely related to the mechanism of photochemical electron transfer in polar solutions.

We examine now some actual examples of solvent-induced charge transfer in excited composite systems. Solvent-induced electronic structural change can be observed in the case of some typical intramolecular exciplex systems such as p-$(CH_3)_2NC_6H_4$—$(CH_2)_n$—(9-anthryl) (abbreviated as A_n) and p-$(CH_3)_2NC_6H_4$—$(CH_2)_n$—(1-pyrenyl) (abbreviated as P_n) (Masaki $et\ al.$, 1976), where anthryl and pyrenyl groups are electron acceptors in the excited state.

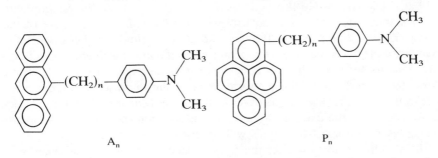

A_n $\qquad\qquad\qquad\qquad\qquad\qquad\qquad$ P_n

In the case of A_1, A_2, and P_1, fluorescence spectral measurements in various solvents of different polarity show clearly that the behaviour of these systems in the excited state is very close to the case indicated in Figure 12.2. Thus in the case of these systems, new fluorescence from the polar state (the CT fluorescence) appears at the value of the solvent polarity parameter $(f_\varepsilon - (\frac{1}{2})f_n) \approx 0.3$–$0.4$, where its red shift with the solvent polarity increase is especially large and the red shift of the CT fluorescence with further increase of the solvent polarity can be reproduced by equation (12.13) just as indicated in Figure 12.2.

Thus in these systems, intramolecular charge transfer is induced by the interaction with polar solvents in the excited state, for example A^*—CH_2—$D \rightarrow (A^-$—CH_2—$D^+)^*_{\text{solv.}}$.

In the case of A_3 and P_3, the fluorescence band of the CT type can be observed not only in polar solvents but also in non-polar solvents. In these cases, it is possible to form sandwich-type exciplexes where the stabilization of the CT state due to the coulombic force between the pair may be larger than the non-sandwich-type exciplexes. Thus the formation of a CT state in the excited state of these systems involves not only solvation but also a change in the geometrial structure of the solute, as indicated by equation (12.14).

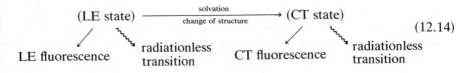

$$\text{(LE state)} \xrightarrow[\text{change of structure}]{\text{solvation}} \text{(CT state)} \qquad\qquad (12.14)$$

LE fluorescence $\qquad$ radiationless transition $\qquad$ CT fluorescence $\qquad$ radiationless transition where LE state represents the locally excited state A^*—$(CH_2)_n$—D. The dynamic aspect of the formation of a CT state from an LE state will be discussed in detail later.

In the case of the sandwich-type exciplexes of A_3 and P_3, the overlap of the wavefunctions of the partners may be larger than in the case of A_1 and P_1, resulting in somewhat larger values of the matrix elements α and β. If this is the case, the electronic structure of the sandwich-type exciplex will change with change of the solvent polarity according to equations (12.3)–(12.10). The increase of the exciplex dipole moment in the fluorescent state with increase of the solvent polarity will be reflected in a curvature of the $h\nu_e \sim (f_\varepsilon - (\frac{1}{2})f_n)$ plot due to the increase of $\mu_e^{eq^2}$. However, because of the scatter of the observed points due to more specific solute–solvent interactions, it seems to be rather difficult to confirm such a solvent-induced electronic structural change from the measurements of the fluorescence frequencies. Nevertheless, it is possible to observe the solvent-induced electronic structural change of the exciplex by means of absorption spectral measurements using laser flash spectroscopy as will be discussed later.

It is well known now that the fluorescence of p-(N,N-dimethylamino)-

benzonitrile (DMABN), $N\equiv C$—⬡—$N\begin{smallmatrix}CH_3\\CH_3\end{smallmatrix}$, shows a quite remarkable

dependence upon solvent polarity (Lippert *et al.*, 1962; Lippert, 1975). Namely, it exhibits dual fluorescence bands in polar solvents, F_a at the longer wavelength side (*ca.* 400–500 nm) and F_b at the shorter wavelength side (*ca.* 360 nm).

In non-polar solvents only the F_b band can be observed, and the F_a band shows a large red shift with increase of the solvent polarity, while the F_b band does not show such a red shift. Therefore, the excited state responsible for the F_a band has a strong intramolecular CT character while that for the F_b band does not. Although the latter state is lower than the former in non-polar solvents, inversion of the two states seems to occur in polar solvents during the excited-state lifetime, owing to the solvation. Thus the circumstance is quite analogous to the above-described case of intramolecular exciplex systems.

Measurement of fluorescence rise and decay processes of DMABN in viscous solution at low temperature where the condition $\tau_R \simeq \tau_e$ may be realized was made by exciting with a nanosecond laser pulse (Nakashima *et al.*, 1973a). According to this measurement, the F_b band exists immediately after excitation, while it takes about 7 ns for the F_a band to arrive at its peak intensity. Picosecond laser spectroscopic experiment at room temperature also indicates that the rise process of the F_a band is somewhat delayed although the F_b band exists immediately after excitation (Struve and Rentzepis, 1974a, b). These results are in agreement with the above argument that the intramolecular CT state is formed during the excited state lifetime.

It has been proposed furthermore that the relaxation process of excited DMABN involves not only the solvation of the CT state but also a change of its geometrical structure (Rotkiewicz et al., 1975, 1976). Namely, the dimethylamino group is perpendicular to the benzene plane in the F_a fluorescent state while it is coplanar with the benzene ring in the F_b fluorescent state. In the perpendicular conformation, charge transfer from a lone-pair orbital of the dimethylamino group to a vacant π orbital of the —C≡N group results in a highly polar state because of the lack of delocalization interaction. The F_b fluorescent state with planar conformation is less polar due to delocalization over the entire conjugated π-electron system. This interpretation has been confirmed by the observation that a compound **A** which is forced into a perpendicular conformation shows only the F_a band, while a compound **B** forced to take planar conformation exhibits only the F_b emission band.

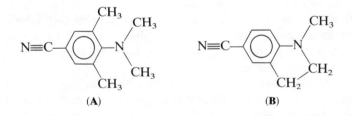

(A)   **(B)**

Thus the observed results of the excited state of DMABN show rather clearly that the solvent-induced electronic structural change results in a molecular conformation change sacrificing the π-electronic delocalization and strengthening the solvation.

The examples discussed so far are those systems which have two distinctly different groups, electron donor and acceptor, in a molecule. However, solvent-induced electronic structural change in the excited state can arise even in some systems composed of two identical groups. A well-known example is 9,9'-bianthryl which shows dual fluorescence in polar solvents, and the band at the longer-wavelength side shows a large red shift with increase of the solvent polarity, just as in the case of the intramolecular exciplex systems and DMABN (Lippert, 1975; Schneider and Lippert, 1968). The red shift of the fluorescence in polar solvents indicates that the relaxed fluorescent state has a dipole moment due to the intramolecular charge transfer from one anthracene nucleus to another, caused by interaction with polar solvent molecules in the excited state. If this interpretation is correct, the above result indicates a phenomenon of solvent-induced 'broken symmetry', as pointed out by the present authors (see Nakashima *et al.*, 1976), i.e. lowering of the symmetry of the solute molecule due to the interaction with polar solvent molecules, resulting in a charge transfer state even if the molecule is composed of two identical chromophores. A direct confirmation of intramolecular charge transfer in the excited state of this molecule was made by measuring the $S_n \leftarrow S_1$ absorption spectra in polar solvents by means of laser flash spectroscopy and demonstrating that the spectra are similar to the superposition of those of the anthracene anion and cation (Nakashima *et al.*, 1976).

Furthermore, it has been demonstrated that the polarity of the CT fluorescence state increases with increase of the solvent polarity, since the CT fluorescence lifetime, τ_f, increases and its yield, η_f, decreases with increase of the solvent polarity. The latter fact indicates the decrease of the radiative transition probability, $k_f = \eta_f/\tau_f$, with increase of the solvent polarity, which can be ascribed to the solvent-induced change in the electronic structure to a more polar one (Nakashima *et al.*, 1976).

A theoretical consideration of this system analogous to that given for the intramolecular heteroexcimer-type A—D system is possible. Just as in the case of the heteroexcimer-type A—D system, the electronic delocalization interaction between two groups is in competition with the stabilization by solvation which favours the charge separation by CT. If the delocalization interaction is strong, the formation of the polar state by solvation is very difficult (Beens and Weller, 1969).

In relation to above results for 9,9′-bianthryl, we discuss here further examples of solvent-induced polarization phenomena in the excited state of composite systems with identical halves, where, however, the solvent-induced polarization seems to be realized only transiently in the course of formation of an intramolecular excimer state with no dipole moment.

A remarkable example is [2.2] (1,3)pyrenophane(*meta*-pyrenophane) (mePy) (Hayashi *et al.*, 1977a).

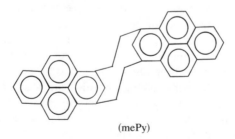

(mePy)

Although the absorption spectrum of mePy is a little broad and red shifted compared with that of pyrene, the nature of its fluorescent state is rather similar to that of pyrene since the fluorescence band shows a vibrational structure ($\lambda_{max} = 390$ nm) and fluorescence lifetime is fairly long ($\tau_f = 150$ ns) in methylcyclohexane. In polar solvents, however, mePy shows a broad structureless emission band at longer wavelength along with pyrene monomer-like emission at 390 nm.

This result is rather similar to the above-described cases of intramolecular heteroexcimer systems, DMABN and 9,9'-bianthryl. However, the broad emission band ($\lambda_{max} = 475$ nm) does not show any red shift with increase of the solvent polarity. Therefore, the 475 nm band may be ascribed to the intramolecular excimer state. Since the yield of the monomer-like emission decreases and that of the broad emission increases with increase of the solvent polarity, the intramolecular excimer state may be formed through the intramolecular CT state.

The *anti* → *syn* isomerization of mePy seems to be made easy by the formation of the CT state due to the interaction with polar solvents because of the transannular interaction in the CT state including coulombic attraction force. Estimation of the energy of the CT state by using oxidation and reduction potentials of pyrene appears to support this argument (Hayashi *et al.*, 1977a). When two pyrene rings approach each other in a CT state, the electronic delocalization interaction will increase and finally the non-polar excimer state will be realized.

'Solvent-induced polarization' phenomena analogous to the above example can be observed also in the case of some dianthrylethanes (Hayashi *et al.*, 1977b). Both 1,2-di(1-anthryl)ethane and 1,2-di(9-anthryl)ethane show intramolecular excimer fluorescence spectra at 460 nm in addition to the fluorescence from the locally excited state of anthracene. The ratio of excimer fluorescence yield to LE fluorescence yield increases with increasing solvent polarity. Thus the intermediate CT state produced in polar solvents seems to enhance the excimer formation.

In the case of 9,9'-bianthryl, the solvent-induced polarization is maintained during the excited-state lifetime because of its geometrical structure which inhibits any appreciable overlap of the wavefunctions between two anthracene rings. However, mutual approach of two anthracene rings, resulting in the overlap of wavefunctions, is possible due to the internal rotations, around the CH_2—CH_2 bond in the case of dianthrylethanes. This structural change in the CT state, during the excited-state lifetime, is the essential difference from the case of 9,9'-bianthryl.

Solvent-induced change of electronic structures as well as geometrical structures in the excited state is not limited to the above-discussed systems, but seems to be a more or less general phenomenon and has a close connection with the mechanism of photochemical electron transfer reactions in solution.

12.2.3   *Kinetics of solvation processes*

As has been discussed in the previous sections, if the condition that the solvent reorientation relaxation time τ_R is close to the fluorescence lifetime τ_e is realized, one can observe in principle the spectra emitted in the course of the reorientation relaxation. Thus if one can observe 'time-resolved' emission spectra, the emission spectra will show the time-dependent red shift. Furthermore, luminescence rise and decay curves observed at various wavelengths in the luminescence band will be different from each other. A model for such a luminescence process was first proposed by Bakhshiev (Bakhshiev *et al.*, 1966; Bakhshiev, 1972), and the first clear-cut observation of the time-dependent spectral shift was made by Ware *et al.* (1971) for the case of 4-aminophthalimide (**a**) in 1-propanol at low temperature, in the nanosecond time range.

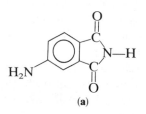

(a)

In Bakhshiev's treatment, it is assumed that the surrounding solvent is a dielectric continuum and the luminescence spectrum, which depends upon the length t during which the molecule is in the excited state, shifts as a whole without changing its shape. The total luminescence spectrum is a superposition of elementary spectra emitted in the time interval dt. According to this model, the time-dependent luminescence spectrum may be given by,

$$I(\tilde{\nu}, t) = Const \cdot \exp(-t/\tau_e) \cdot I(\tilde{\nu}_m(t) - \tilde{\nu}) \qquad (12.15)$$

where $I(\tilde{\nu}_m(t) - \tilde{\nu})$ represents the quantum intensity distribution in the elementary spectrum for the instant t, normalized to unity over the area, and $\tilde{\nu}_m(t)$ is the wavenumber of the maximum.

It has been shown by Ware *et al.* that the results of the measurements of the time-dependent red shift on 4-aminophthalimide in 1-propanol at low temperature can be reproduced approximately by equation (12.15) (Ware *et al.*, 1971). The red shift observed at -70 °C, during delay times ranging from a few nanoseconds through a few 10 nanoseconds, amounts to *ca.* 40 nm (from $\lambda_{max} = 460$ –500 nm). Since the λ_{max} measured by the stationary method in toluene is 410 nm and that in 1-propanol is 510 nm, the result of time-resolved fluorescence studies indicates the existence of a relaxation process which is completed during a delay time shorter than 1 ns. The rapid relaxation process may be due to some short-range solute–solvent interactions which are not taken into consideration in the case of the above continuum model.

A kinetic model of the solvation process, assuming solvation by a finite number of solvent molecules surrounding the excited solute in a cage, was given by Rapp *et al.* (1971). One by one reorientation of solvent dipoles of the cage was explicitly taken into consideration in this model, assuming that: (1) every solute molecule is surrounded by m polar solvent molecules; (2) every reorientation of solvent dipole of the cage results in a decrease of the wavenumber maximum $\tilde{\nu}_0$ by $(\Delta\tilde{\nu}_s/m)$ where $\Delta\tilde{\nu}_s$ represents maximum possible shift; (3) the total transition probability for spontaneous emission $n_s = (\eta_f/k_f)$ and the fluorescence yield η_f are independent of the solvent relaxation state of the system; (4) the relaxation process of the solvent dipoles is exponential with rate constant $k = 1/\tau_R$.

With these assumptions, the time-dependent emission spectrum corresponding to equation (12.15) was given by equation (12.16) for the system where n solvent dipoles have reoriented.

$$I(\tilde{\nu}, t) = \frac{Nn_s}{\kappa\Delta\tilde{\nu}} \sum_{n=0}^{m} \binom{m}{n} f\left(\tilde{\nu}, \tilde{\nu}_0 - n\frac{\Delta\tilde{\nu}_s}{m}\right)$$

$$\times \int_0^t \exp\left[-\left(mk + \frac{n_s}{\eta_f}\right)(t-t')\right][\exp\{k(t-t')\}-1]^n n_a(t')\, dt' \quad (12.16)$$

where N is the number density of the solute molecule, $n_a(t)$ is the probability for excitation of the solute, f represents the intensity distribution function (normalized to unity) of non-relaxed fluorescence band with $fwhm = 2\Delta\tilde{\nu}$ and with maximum at $\tilde{\nu}_0 - n(\Delta\tilde{\nu}_s/m)$, and κ is a constant.

Assuming appropriate parameter values, it was demonstrated that the luminescence signal in the small wavenumber region show a rising part extending to times where the exciting pulse has already negative slope or has disappeared, and the luminescence decay near $\tilde{\nu}_0$ is faster and it is quite slow near $(\tilde{\nu}_0 - \Delta\tilde{\nu}_s)$. Moreover, it can be shown that in the limiting case of $m \to \infty$ equation (12.16) becomes essentially the same as Bakhshiev's equation.

Although one by one solvation in a cage is considered in the above treatment, it is still not concerned with specific solute–solvent interactions but is concerned with the universal solvent-excited solute interactions. Of course, the universal interaction or continuum model is nothing but an approximation which explains gross features of the problem. More or less specific interactions as revealed by time-resolved fluorescence studies seem to be observed in several cases in alcohol solutions, which might be ascribed to hydrogen bonding interactions (DeToma *et al.*, 1976; DeToma and Brand, 1977). For example, time-resolved fluorescence spectra of 2-anilino-naphthalene (**b**) in cyclohexane containing ethanol can be explained as due

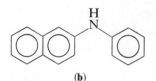

(b)

to the stoichiometric complex formation of excited 2-anilinonaphthalene with ethanol when the concentration of ethanol is low. When the concentration of ethanol is high, the complex formation in the excited state seems to be followed by the reorientation of the polar solvent dipoles in the cage (DeToma and Brand, 1977). That one should divide the solvent effects upon fluorescence spectra into short-range specific interactions and long-range dipolar interactions in the case of associating solvents such as alcohols, was pointed out previously by the present authors (Mataga et al., 1956).

Now, in the above-described examples, the condition $\tau_R \simeq \tau_e$ favourable for the observation of time-resolved fluorescence was realized by controlling the τ_R value by lowering the temperature and/or enhancing the viscosity of the solvent. However, instead of slowing down the solvent relaxation process, it may be possible to shorten τ_e by enhancing the induced emission probability by exciting the system with a strong laser pulse.

The system 2-amino-7-nitrofluorene (ANF) in 1,2-dichlorobenzene solution was found to show laser action when excited with the second harmonic of a Q-switched ruby laser (Gronau et al., 1972). By changing the exciting pulse intensity, the dye laser output power was strengthened 200 times, which led to a blue shift of the dye laser wavelength by 100 Å. Analogously, an increase of ANF concentration by a factor of 10 led to a blue shift of about 100 Å. Increase of concentration also indicates high density excitation leading to the enhancement of induced emission probability.

The observed blue shift means that, because of the enhancement of the induced emission, the τ_e value becomes comparable with the τ_R value which seems to be about 10 ps for this system. A similar phenomenon has been observed in the case of tetralin solution of p-(9'-anthryl)-N,N-dimethyl-aniline (ADMA) (c) (Nakashima et al., 1973b).

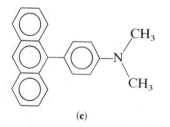

(c)

By exciting with a Q-switched ruby laser, the solution showed a blue shift of ca. 30 Å with increase of the concentration from 2×10^{-4} M to 1×10^{-2} M.

Although we are observing the phenomenon in the picosecond region by using the nanosecond laser pulse in the above measurements, it is necessary to use a picosecond laser method for a complete elucidation of the relaxation process, as is pointed out in section 12.2.1.

Direct picosecond time-resolved measurements have been made of the fluorescence shift of ANF in 2-propanol (Hallidy and Topp, 1977). Fluorescence rise and decay processes measured at different wavelengths (from 487 nm to 707 nm) were found to vary remarkably. The results seem to be explained approximately assuming that the relaxation process from the Franck–Condon state with respect to orientation (F) involves the reorientation process with rate constant k_r to a relaxed state (R) and consecutive deactivation of R with rate constant k_d.

The F $\rightarrow$ R orientation relaxation time k_r^{-1} measured at the shortest wavelength and at room temperature was shorter than 10 ps, which is somewhat shorter than the dielectric relaxation time of 2-propanol at room temperature assigned to the reorientation (25 ps). Thus the microscopic reorientational relaxation time seems to be somewhat different from the dielectric relaxation time. The deactivation process k_d needs activation energy amounting to 5.5 kcal/mol, and involves not only the radiative transition from the R state but also radiationless processes due to hydrogen bonding and hydrogen bond breaking processes. Thus it is essential to take into consideration the hydrogen bonding interactions in the case of associating solvents.

12.3   Mechanisms of Intermolecular Charge Transfer Processes in the Excited Electronic State

As discussed in section 2, the charge transfer process in the excited electronic state depends upon the interaction of the CT system with polar solvent molecules as well as the mutual geometry of the electron donor and acceptor groups. Studies of the behaviour of such intramolecular exciplex systems as P_n and A_n, in various solvents of different polarities, by means of nanosecond and picosecond laser spectroscopy will contribute greatly to the elucidation of the mechanism of intermolecular charge transfer in the excited state. Results of such studies will be discussed in detail in sections 4 and 5. In this section, the behaviour of another model system for the elucidation of the photochemical charge transfer mechanism, will be discussed at first. Namely, the charge transfer processes in the excited state of heteroexcimer-forming systems aggregated at low temperature and in a highly viscous or rigid state and some related cases will be considered. Next, possible mechanisms of photochemical charge transfer will be discussed in analogy to the mechanisms of intermolecular electronic excitation transfer.

12.3.1 Excitation of ground-state aggregates—change of structure due to intracomplex relaxation

Ordinarily, exciplexes are formed by encounter collisions in fluid solutions containing donors and acceptors dissolved in solvents. In this case, as will be discussed in section 4, direct observation of the formation of a CT fluorescent state from an encounter collisional state, seems to be rather difficult in the case of an intermolecular exciplex even if one uses the picosecond laser method.

In the case of aromatic hydrocarbon–amine two-component systems, an exciplex seems to be formed in fluid solution immediately after the excitation. However, the exciplex formation is suppressed more or less in a rigid microcrystalline state at 77 K (Mataga and Ezumi, 1967). For example, fluorescence spectra of anthracene and perylene in N,N-dimethylaniline (DMA) at room temperature are due exclusively to heteroexcimers, while at 77 K they are quite different from those at room temperature and rather close to the spectra in solvents which do not form any exciplex with these hydrocarbons. The fluorescence spectra of pyrene in DMA at 77 K are a little broad compared with the spectra of anthracene and perylene being a superposition of pyrene monomer fluorescence and a broad band due to weak heteroexcimers. Thus in the case of pyrene, heteroexcimer seems to be formed to some extent even in the rigid crystalline state of DMA. However, the wavelength of the heteroexcimer fluorescence is considerably shorter than that at room temperature.

The above results indicate clearly that rearrangements or reorientations of donor and acceptor pairs as well as the surrounding solvent molecules are necessary for the formation of the heteroexcimer, and are almost inhibited in a rigid crystalline state at 77 K. In the case of the pyrene–DMA system, only specific ground-state aggregate pairs of appropriate geometrical configuration seem to lead to the fluorescent heteroexcimer state upon excitation, other geometries leading to monomer fluorescence, since no reorientational motion during the excited-state lifetime seems to be possible.

The ground-state interactions between donor–acceptor pairs forming heteroexcimers upon excitation can be studied in a more detail in the cyclohexane matrix which seems to have adjacent sites with orientations favourable for the formation of an exciplex (Mataga et al., 1966). Thus perylene–DMA, pyrene–DMA, and pyrene–DEA (N,N-diethylaniline) systems show heteroexcimer fluorescence in a cyclohexane matrix at low temperatures. The absorption band (^{1}La band) of these systems is broadened and red shifted compared with the monomer band, and moreover, it has been confirmed that the excitation spectrum of the heteroexcimer fluorescence agrees with the red-shifted ^{1}La band (Okada and Mataga, 1976). Thus it is evident that only specific pairs interacting weakly in the ground state can form a heteroexcimer upon excitation. Since the absorption spectra of the specific pairs are not the CT absorption but are due to the transition to the slightly perturbed ^{1}La state, some structural change should occur, probably a change of the distance between the pair, before the exciplex fluorescence transition.

In the case of weak EDA complexes stable in the ground state, the CT absorption band will be observed. However, much stronger CT in the excited state than in the ground state may be followed by some structural change analogous to the above case of exciplex formation in a rigid matrix.

The formation process of the CT state in the above systems together with that for the exciplex in a non-polar solvent are indicated schematically in Figure 12.3 in terms of reaction coordinate diagrams. The Franck–Condon excited state of the heteroexcimer-forming pair in the rigid matrix, $(A \cdots D)^*$, might have a slight CT character, and the CT character of the

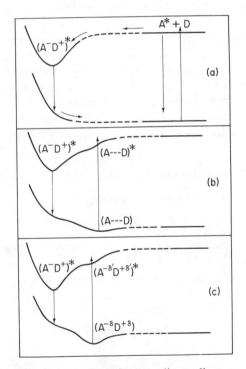

Figure 12.3   Reaction coordinate diagrams for charge transfer complex formation in the excited state. a. A heteroexcimer in a non-polar solvent. b. A heteroexcimer in a rigid matrix. c. Excitation of a ground-state EDA complex

Franck–Condon excited state of the EDA complex may be much larger than that in the ground state.

Now, in all examples in Figure 12.3, excited-state complexes with donors and acceptors interacting rather strongly at short distances in specified mutual orientations are formed. However, as discussed in section 2.2, for solvent-induced charge transfer in the intramolecular heteroexcimer systems, A—$(CH_2)_n$—D, close contact between A and D with specified mutual orientation does not seem to be necessary for the charge transfer, A*—$(CH_2)_n$—D → A⁻—$(CH_2)_n$—D⁺, in polar solvents. In the case of heteroexcimer formation in a rigid state of the hydrocarbon–amine two-component system, no reorientations of the surrounding amine molecules stabilizing the CT state are possible during the excited-state lifetime, resulting in the heteroexcimer formation of only specific pairs with favourable orientations.

The above arguments suggest that the excited-state intermolecular charge transfer in considerably polar solvents is possible at relatively long intermolecular distances and in loose mutual orientations of the pair compared with the case of a rigid state and non-polar solutions of two-component systems. Of course, the electronic interaction responsible for the charge transfer between A* and D becomes smaller with increase of the intermolecular distance. This means that the charge transfer process becomes slower with increase of the distance. It may be possible to distinguish several cases of charge transfer process depending upon the rate of charge transfer and of the rates of vibrational as well as solvation relaxations of the CT state (Mataga and Tanimoto, 1969; Mataga and Kubota, 1970).

12.3.2 Possible different types of charge transfer processes in the excited state

Let us consider the wavefunction of equation (12.17)

$$\Psi = C_1\Phi_i + C_2\Phi_f \tag{12.17}$$

where Φ_i represents the locally excited (LE) state A* · · · D or A · · · D*, and Φ_f is the CT state A⁻ · · · D⁺. At $t = 0$, the system is in the LE state, Φ_i. In the course of time, Φ_f appears and its probability density reaches a maximum value,

$$P_m = \frac{4\beta^2}{4\beta^2 + \Gamma^2} = \frac{1}{\{1 + (\Gamma^2/4\beta^2)\}} \tag{12.18}$$

when

$$t = t_m = \frac{h}{4\beta\{1 + (\Gamma^2/4\beta^2)\}^{1/2}} \tag{12.19}$$

where β is the matrix element of the electronic interaction between Φ_i and Φ_f, and Γ is the energy difference between these configurations. We define the transfer rate n as the maximum probability density (expectation value) divided by t_m.

$$n = \frac{4\beta\{1 + \Gamma^2/4\beta^2)\}^{1/2}}{h} \qquad (12.20)$$

In the case of resonance,

$$n = \frac{4\beta}{h}. \qquad (12.21)$$

The transfer rate n is not the ordinary transition probability but represents the frequency with which an electron is oscillating between two molecules. If the transfer is much faster than the periods of molecular vibrations, the electronic state of the system can be actually represented by equation (12.17) and we call this case strong interaction. Contrary to this, if the electronic interaction is very weak and the transfer is much slower than the vibrational relaxation, two molecules can be regarded as almost completely isolated with approximately no interaction between them. This case is called very weak interaction. If the transfer is comparable with the vibrational relaxation, it is called the weak interaction. These are the same classification as proposed by Förster (1965) for the intermolecular electronic excitation transfer.

The strong interaction case probably corresponds to the formation of exciplexes or the excitation of ground-state complexes as discussed in section 3.1. In the case of charge transfer phenomena in polar solutions, however, one must take into account the relaxation due to the solvation of the CT state by polar solvent, in addition to the vibrational relaxation. Thus even in the case of the excitation of the ground-state EDA complex, the intracomplex relaxation may be followed by reorientation of the surrounding polar solvent molecules, which makes the interaction between the halves of the complex smaller, leading finally to ionic dissociation in the case of strongly polar solutions. Although the interaction between the partners in the complex is weaker in the weak interaction case, the circumstance may be essentially similar to the above case of strong interaction.

In the limit of very weak interaction, one can define the ordinary transition probability in the form,

$$w = \frac{2\pi}{\hbar}|\beta|^2 \rho_f \qquad (12.22)$$

where ρ_f is the density level in the Φ_f state. After the vibrational relaxation following the CT transition, the CT state may be further stabilized by reorientation of surrounding polar solvent molecules, as indicated schematically in Figure 12.4.

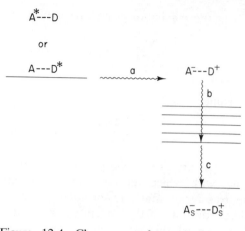

Figure 12.4 Charge transfer and related processes in the case of very weak interaction. a. Charge transfer. b. Vibrational relaxation. c. Solvation process

If the charge transfer is not accompanied by vibrational relaxation, but only solvation relaxation is associated with it, we must compare the rate of charge transfer with the rate of solvent reorientation. The rate of reorientation relaxation seems to be approximately 10^{12}–$10^{11}\,\mathrm{s}^{-1}$ in the case of an ordinary solvent at room temperature. If we put $n = 10^{12}$–$10^{11}\,\mathrm{s}^{-1}$ for charge transfer, $|\beta| = 10^{-3}$–10^{-4} eV. Although these $|\beta|$ values correspond to the case of very weak interaction when compared with the rate of vibrational relaxation, we are discussing the weak interaction case if compared with reorientation relaxation.

Since the rate of reorientation may be comparable with the rate of collision from the positions near the encounter, it may not be very clear whether it is meaningful or not to consider the case of very weak interaction compared with the reorientation relaxation in fluid solutions. Thus molecules will make encounter collision leading to strong interaction before the charge transfer due to very weak interaction.

However, the photochemical charge transfer due to the very weak interaction mechanism might become important when the molecular motion is slowed down by temperature lowering in fluid solutions or the donor–acceptor pair separated from each other is trapped in micelles or membranes containing some water molecules near the pair which stabilize the produced CT state by reorientation relaxations. These arguments suggest the importance of the very weak interaction mechanism of charge transfer in biological systems where the relevant molecules seem to be trapped in membrane systems (Yomosa, 1973; Hopfield, 1974, 1977; Jortner, 1976).

12.4 Kinetics of Exciplex Formation and Photochemical Charge Transfer Processes

In this section, some actual examples of exciplex formation and decomposition processes as well as charge transfer reactions in polar solvents will be discussed.

12.4.1 Kinetic fluorescence studies of exciplex formation

In the case of typical intermolecular exciplex formation in non-polar solvents as indicated in Figure 12.3a, the process is diffusion limited and can be expressed as,

$$A^* + D \underset{k_2}{\overset{k_1}{\rightleftharpoons}} (A^-D^+)^* \xrightarrow{k_3} \text{Prod.}$$

$$
\begin{array}{ccc}
k_f \nearrow & \searrow k_i & \quad k_f' \swarrow \quad \searrow k_i' \\
A + h\nu_f & A, {}^3A & A + D + h\nu_f' \\
& & A + D, {}^3A + D
\end{array}
$$

(12.23)

In polar solvents, however, the possibility arises that the solvated ion-radical formation occurs directly from the state immediately after the charge transfer, before solvent relaxation, in addition to the formation from the relaxed exciplex state.

$$A^* + D \underset{k_2}{\overset{k_c}{\rightleftharpoons}} (A^-D^+)^*$$

$$
\begin{array}{cccccc}
\swarrow k_f & \searrow k_i & \searrow k_a & \swarrow k_3 & \downarrow k_f' & \searrow k_i' \\
A + h\nu_f & A, {}^3A & \text{Prod. including} & A + D + h\nu_f' & A + D, {}^3A + D \\
& & \text{ion-radicals}
\end{array}
$$

(12.24)

From this reaction scheme, the time dependence of fluorescence intensities may be given as follows assuming an excitation pulse short relative to the decay of excited species in equation (12.24).

$$I(t) = \{[A^*]_0 k_f/(\beta - \alpha)\}\{(\beta - \mu) \exp(-\alpha t) + (\mu - \alpha) \exp(-\beta t)\}$$

$$I'(t) = \{[A^*]_0 k_c[D]/(\beta - \alpha)\}\{\exp(-\alpha t) - \exp(-\beta t)\}$$

$$\left.\begin{array}{c}\alpha\\\beta\end{array}\right\} = \tfrac{1}{2}[(\mu + \nu) \mp \{(\nu - \mu)^2 + 4k_c k_2[D]\}^{1/2}]$$

(12.25)

$$\mu = k_f + k_i + k_1[D], \qquad k_1 = k_a + k_c$$

$$\nu = k_f' + k_i' + k_2 + k_3$$

Some work by means of nanosecond pulse excitation has been done upon typical heteroexcimers of hydrocarbon–amine systems such as perylene–DMA (Ware and Richter, 1968), pyrene–DMA (Yoshihara *et al.*, 1971), pyrene–DEA (Nakashima *et al.*, 1973c), pyrene–TBA (tri-n-butylamine) (Nakashima *et al.*, 1973c), and anthracene–DMA (Hui and Ware, 1976). Ware and coworkers, especially, have made detailed studies by means of nanosecond pulse excitation not only upon hydrocarbon–amine systems but also upon hydrocarbon–olefin systems (Ware *et al.*, 1974a, b; O'Connor and Ware, 1976).

In the case of such typical heteroexcimers as hydrocarbon–amine systems, the formation process is a rapid diffusion-controlled reaction. In such a case, the rate constant k_c for heteroexcimer formation cannot be regarded as time independent in any accurate analysis of the fluorescence decay and rise curves. The time-dependent rate constant $k_c(t)$ can be expressed by equation (12.26) (Noyes, 1961).

$$k_c(t) = \frac{4\pi RD}{1 + \frac{4\pi RD}{k}} \left[1 + \frac{k}{4\pi RD} \exp(x^2) \operatorname{erfc}(x) \right]$$

(12.26)

$$x = \left(1 + \frac{k}{4\pi RD}\right)\left(\frac{\sqrt{Dt}}{R}\right)$$

where R is the molecular diameter of the reacting molecules, D is the sum of the diffusion coefficients of the reactants, and k is an elementary time-independent bimolecular rate constant. In the case of typical excimers and heteroexcimers, coupled differential equations based on the reaction scheme of equation (12.25) including the time-dependent rate constant $k_c(t)$ must be solved. It does not appear possible to accomplish this in a closed form but the solution can be obtained numerically as has been done for the anthracene–DMA system in a non-polar solvent (Hui and Ware, 1976).

The origin of the transient effect can be viewed in the following way (Noyes, 1961). At $t = 0$, the reactants A* and D are randomly distributed, but as time proceeds those distributions in which an A* is near a D are preferentially depleted since there is a higher probability for reaction than for those distributions where A* and D are far apart. This reaction produces a spatially non-uniform distribution of molecules, which leads to a flux of molecules from the more concentrated regions of the liquid. Since the distribution of molecules is changing with time, the rate constant for the reaction is also changing with time. In any way, because of its rapidity, is transient behaviour can be observed more directly by means of the picosecond laser method.

Although the picosecond or even the subpicosecond laser method may be necessary for direct observation of the exciplex formation process from the configuration near encounter complex, the circumstance seems to be somewhat different in the case of the intramolecular heteroexcimer systems such as P_n and A_n discussed in section 2.2. For example, in the case of P_3, it was found that the intramolecular heteroexcimer formation process itself is affected considerably by the presence of the methylene chain between the electron donor and the acceptor groups (Mataga et al., 1976; Okada et al., 1977).

The intramolecular heteroexcimer formation of P_3 was studied by means of fluorescence rise and decay curve measurements using a pulsed nitrogen gas laser for excitation (Mataga et al., 1976; Okada et al., 1977). In decalin solution, the time dependence of fluorescence intensities can be well reproduced by equation (12.25), with $\alpha^{-1} = 90$ ns and $\beta^{-1} = 7.5$ ns at room temperature. Since the two groups in P_3 seems to be within the encounter distance, this result for the intramolecular system shows that the intramolecular heteroexcimer formation takes quite a long time compared with the intermolecular case. This is due to the effect of hindered rotation about the CH_2—CH_2 bonds which need to take the sandwich form and which require activation energies of 3–4 kcal/mol. Temperature lowering makes the rise time even longer. For example, $\beta^{-1} = 25$ ns at 263 K and $\beta^{-1} = 58$ ns at 223 K.

The rise time of P_3 heteroexcimer fluorescence at room temperature is shorter than 2.5 ns (the rise time of the exciting nitrogen laser pulse) both in 1-pentanol and 2-butanol, although these solvents are more viscous than decalin. This result indicates the importance of the solvent polarity in the charge transfer process as has been discussed in section 2.2. Thus the intramolecular charge transfer of P_3 may be induced by interaction with polar solvents in the excited state without sandwich structure being taken. (Picosecond laser photolysis studies supporting this argument are described in section 4.2.)

12.4.2 Transient absorption studies by means of picosecond laser photolysis

Direct observations of the heteroexcimer formation in non-polar solvent as well as the solvated ion-radical formation in strongly polar solvent, including the transient effects, by means of the picosecond laser method were made for the anthracene–DEA system, by exciting anthracene with the second harmonic of a picosecond pulse from a mode-locked ruby laser and by monitoring the heteroexcimer or anthracene anion absorption with the fundamental pulse (Chuang and Eisenthal, 1975).

The same kinetic behaviour was assumed for both hexane and acetonitrile solutions.

$$A^* + D \xrightarrow{\ k_c(t)C\ } (A^-D^+)^* \text{ or } A_S^- + D_S^+$$

$$\left. \begin{array}{c} k_A \swarrow \\ A \end{array} \right. \qquad\qquad \left. \begin{array}{c} \downarrow k_E \\ A + D \end{array} \right. \qquad\qquad (12.27)$$

where C is the concentration of D, $k_c(t)$ is given by equation (12.26), and the back reaction from the heteroexcimer is neglected. In view of the fact that k_A and k_E are negligibly small in comparison with $k_c(t)C$ in the actual experimental conditions ($C = 0.2 \sim 8$ M), the concentration of the heteroexcimer or the solvated ion-radicals, $n_p(t)$, may be given by,

$$n_p(t) = n_A(0)\left[1 - \exp\left\{-C\int_0^t k_c(t')\,dt'\right\}\right] \qquad (12.28)$$

where $n_A(0)$ is the concentration of A* at $t = 0$.

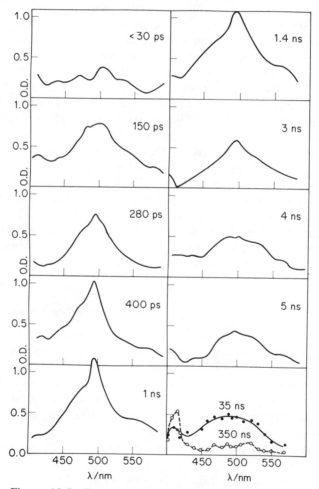

Figure 12.5 Transient absorption spectra of P_3 in 2-propanol at various delay times. [Taken from: Migita, M. *et al.*, 1978 by permission of North Holland Publishing Company]

Observed rise curves both for heteroexcimers and ion-radicals can be reproduced by using equation (12.26) for the time-dependent rate function with $k \approx 10^{11} \, \text{M}^{-1}\text{s}^{-1}$ and $R \approx 8 \, \text{Å}$. Thus the importance of solvent polarity in the charge transfer process as discussed in section 4.1 for intramolecular heteroexcimer systems was not recognized in the picosecond studies upon the intermolecular system.

The behaviour of excited P_n in polar solvents has been examined with the picosecond laser method by measuring the transient absorption spectra (Migita et al., 1978). Transient absorption spectra of P_3 in 2-propanol, excited by the second harmonic of a picosecond pulse from a mode-locked ruby laser and monitored at various delay times with picosecond continuum generated by collimating the fundamental pulse into polyphosphoric acid, are given in Figure 12.5.

The spectrum at the delay time smaller than 30 ps may be due to $S_n \leftarrow S_1$ transition of the pyrene group. With increase of the delay time, the spectra become more and more similar to the superposition of the absorption bands of the pyrene anion and the DMA cation. The sharp ion-like band observed at 400 ps to 1 ns delay times, however, changes to a broad band at longer delay times and coincides with that observed at 35 ns delay times by means of the nanosecond laser photolysis method (Hinatsu et al., 1978).

The spectral change indicated in Figure 12.5 may be explained in terms of the mechanism described at the end of section 4.1. Namely, the excited P_3 in 2-propanol may form a loose heteroexcimer, at first, due to a little approach of two moieties from the stretched form, which is followed by a structural change to the sandwich type as indicated in equation (12.29). The sharp ion-like band may be ascribed to the loose heteroexcimer and the broad band to the sandwich heteroexcimer.

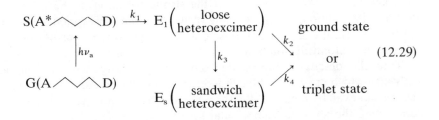

It may be possible that some fraction of P_3 in the ground state takes sandwich conformation which leads to the sandwich heteroexcimer immediately after excitation. However, since the spectrum at the earliest delay time in Figure 12.5 can be assigned to the $S_n \leftarrow S_1$ transition in the pyrene group, the ground-state sandwich species is neglected in equation (12.29), as a first approximation.

According to the reaction scheme of equation (12.29), the time dependence of each species may be given by equation (12.30).

$$[S] = [S]_0 \exp[-k_1 t]$$
$$[E_1] = A\{\exp[-(k_2 + k_3)t] - \exp[-k_1 t]\} \qquad (12.30)$$
$$[E_s] = B \exp[-k_4 t] + C \exp[-k_1 t] - D \exp[-(k_2 + k_3)t]$$

where A, B, C, and D are time independent. The observed rise and decay curves can be approximately reproduced by the superposition of these equations with rate constants of $k_1^{-1} \approx 400$ ps, $(k_2 + k_3)^{-1} \approx 1$ ns, and $k_4^{-1} \approx 100$ ns.

It should be noted here that the rise of the ion-like bands of P_2 in 2-propanol is considerably faster than that of P_3, which indicates that more extensive structural change is necessary in the case of P_3 compared with P_2 for the charge transfer to occur. Moreover, the ion-like bands of P_2 do not show the change of band shape to the broad one, the spectra observed at delay times of a few hundred picoseconds being similar to the spectra observed at the delay time of 25 ns, which may be ascribed to the fact that P_2 cannot take the sandwich structure.

In acetonitrile solution, the rise of the ion-like bands of P_3 is considerably faster than in 2-propanol solution, which might be ascribed to the stronger stabilization of the charge transfer state in acetonitrile and, accordingly, smaller structural change necessary for the charge transfer. The transient spectra of P_3 in acetonitrile are more sharp and more similar to the spectra of free ion-radicals compared with the spectra in 2-propanol. This result may be due to the fact that the electronic and geometrical structures of this heteroexcimer depend significantly upon the interaction with polar solvent (Mataga et al., 1967). Because of the strong solvation of the CT state in acetonitrile solution, the broadening of the transient absorption spectra due to the structural change from loose one to sandwich type was not observed clearly in acetonitrile solution.

Thus the details of the excited-state charge transfer process in polar solvents, which are rather difficult to elucidate in the case of an intermolecular system even by means of the picosecond laser method, can be clarified by employing the above intramolecular systems.

12.4.3 *Relation between the fluorescence quenching rate constant and the free energy change of the charge transfer processes*

As discussed in previous sections, the charge transfer process in the excited state is intimately connected with various relaxation processes including donor and acceptor pair as well as the surrounding solvent molecules.

In the case of heteroexcimer formation in slightly or moderately polar solvents, the heteroexcimer may be stabilized considerably by solvation. The stabilization process of a typical heteroexcimer due to the solvent reorientations at room temperature has not yet been measured directly by means of the picosecond laser method. However, as discussed in section 2.3, measurement of the time-dependent spectral shift due to the solvent reorientation process can be made in the nanosecond region by using highly viscous solution at low temperatures.

Such a nanosecond time-resolved fluorescence study was made for the heteroexcimer formation process in the case of the pyrene–tri-n-butylamine (TBA) system at 156 K, where TBA is in a highly viscous supercooled liquid state (Nakashima *et al.*, 1972). It was shown by this study that some of the heteroexcimers are formed immediately after the excitation and for these heteroexcimers the solvent (TBA) reorientation is not yet accomplished.

This heteroexcimer formation seems to occur for some specific donor–acceptor pairs which already have a favourable geometric configuration in the ground state, just as in the case of heteroexcimer formation in the cyclohexane matrix, discussed in section 3.1. Solvent relaxation occurs subsequently resulting in the time-dependent red shift of the emission.

The time-dependent red shift is accompanied by a simultaneous increase in intensity, which may be attributed to a rearrangement of solute-solvent configurations which lacked the initial geometry required for the immediate formation of heteroexcimers. These time-dependent changes occurred during the first 20 ns of delay times, after which the heteroexcimer emission decayed without changing its position.

In a strongly polar solvent such as acetonitrile, no heteroexcimer emission can be observed in an intermolecular system, although one can recognize very weak emission in an intramolecular system. This complete quenching of the intermolecular heteroexcimer emission is due to the rapid ionic dissociation of strongly solvated ions produced by charge transfer between fluorescer and quencher. Thus the fluorescene quenching process of heteroexcimer systems in strongly polar solvent is one of the most typical examples of solvent relaxation following excited-state charge transfer.

There may arise various cases depending upon the extent of the relaxations. Namely, both intramolecular vibrational relaxation and solvent relaxation may be involved, or solvent relaxation alone may operate to stabilize the ions produced. It may be possible also that a considerable amount of activation energy is necessary for producing ion-radicals.

The detailed mechanisms of the charge transfer process for these cases might be different from each other depending upon the extent of the possible relaxations accompanying the charge transfer, in view of the fact that in the case of P_3 somewhat long-range charge transfer is possible in polar solvents (see section 4.2), where solvation relaxation is possible, compared with the case of a non-polar solution where stabilization by solvation is not possible.

At any rate, a systematic experimental study of the fluorescence quenching in strongly polar solvents has revealed a relation between the stationary-state bimolecular quenching rate constant k_q and the free energy difference ΔG^0 between relaxed states in encounter before and after charge transfer (Rehm and Weller, 1969; 1970), $A^* \cdots D$ or $A \cdots D^* \rightarrow A_s^- \cdots D_s^+$.

Assuming the reaction scheme of equation (12.31), the stationary-state method gives equation (12.32) for the over-all rate constant (Marcus, 1956).

$$
\begin{array}{c}
{}^1A^* + D \\
\text{or} \\
A + {}^1D^*
\end{array}
\underset{k_{-1}}{\overset{k_1}{\rightleftharpoons}}
\begin{array}{c}
{}^1A^* \cdots D \\
\text{or} \\
A \cdots {}^1D^*
\end{array}
\underset{k_{-2}}{\overset{k_2}{\rightleftharpoons}}
{}^2A_s^- \cdots {}^2D_s^+ \overset{k_3}{\rightarrow}
\tag{12.31}
$$

$$
k_q = \frac{k_1}{1 + \dfrac{k_{-1}}{k_2} + \dfrac{k_{-1}}{k_3}\dfrac{k_{-2}}{k_2}}
\tag{12.32}
$$

Putting $k_2 = k_{20}\exp(-\Delta G^*/RT)$ and $(k_{-2}/k_2) = K^{-1} = \exp(\Delta G^0/RT)$, one obtains equation (12.33).

$$
k_q = \frac{k_1}{1 + \left(\dfrac{k_{-1}}{k_{20}}\right)\exp\left(\dfrac{\Delta G^*}{RT}\right) + \left(\dfrac{k_{-1}}{k_2}\right)\exp\left(\dfrac{\Delta G^0}{RT}\right)}
\tag{12.33}
$$

where ΔG^* is the activation free energy which is necessary to obtain the nuclear configurations of the donor–acceptor pair as well as the surrounding polar solvent molecules, where the free energies of the pair before and after the charge transfer are equal.

Since the charge transfer itself is very rapid compared with the nuclear motions (Franck–Condon principle), we are concerned here with non-equilibrium polarization systems. The rearrangement of the nuclear configurations to the state with activation free energy ΔG^* is necessary in order to conserve energy in the charge transfer process. ΔG^0 can be calculated by means of equation (12.34).

$$
\Delta G^0 = E_{ox} - E_{red} - \frac{e^2}{\varepsilon a} - \Delta E_{00}
\tag{12.34}
$$

where E_{ox} and E_{red} are the oxidation potential of donor and reduction potential of acceptor, respectively, $(e^2/\varepsilon a)$ is the coulomb interaction energy between A^- and D^+ at distance a in solvent of dielectric constant ε, and ΔE_{00} is the energy difference between the fluorescent and the ground state.

The results of measurements for many donor-acceptor pairs in acetonitrile indicate that when $\Delta G^0 \leqslant -10$ kcal/mol the reaction is diffusion-controlled with $k_q \simeq 10^{10} \, \text{M}^{-1} \, \text{sec}^{-1}$ and $\Delta G^* \sim 0$. However, when $\Delta G^0 > -10 \, \text{kcal/mol}$, $\Delta G^* > 0$. The following empirical relation was proposed for the relation between ΔG^* and ΔG^0 (Rehm and Weller, 1969).

$$\Delta G^* = (\Delta G^0/2) + \{(\Delta G^0/2)^2 + (\Delta G_0^*)^2\}^{1/2} \qquad (12.35)$$

where ΔG_0^* is the value of ΔG^* when $\Delta G^0 = 0$ and is equal to 2.4 kcal/mol.

The term k_3 in equation (12.31) will involve the process to the ground state $A \cdots D$ or the generation of the component triplet state. In many cases, since ΔG° between $A_s^- \cdots D_s^+$ and $A \cdots D$ is large, ΔG^* may be negligible for the deactivation to the ground state, according to equation (12.35). Thus deactivation to the ground state might be overwhelming in the process k_3. However, it is actually observed that in some cases the deactivation to the ground state is much slower than the dissociation to the free ions (see section 6).

The above results are all concerned with charge transfer in the fluorescent state. Similar studies upon the charge transfer reaction in the triplet state are rather scarce. Kramer and coworkers made detailed measurements upon the charge transfer reaction of the triplet states of thionine and lumiflavin with electron donors such as aromatic hydrocarbons and methoxy- and/or dimethylamino-substituted benzenes in methanol solution (Vogelmann *et al.*, 1976). According to their results, approximately the same relation between k_q and ΔG^0 as in the above case of the fluorescence quenching reaction holds also in the triplet-state quenching reaction. In this case, because of the difference of the spin multiplicity, the deactivation from the radical-pair state produced by charge transfer to the ground state is slow in general, resulting in almost complete dissociation of the radical-pair.

Another investigation of the charge transfer reaction in the triplet state has been made recently (Kikuchi *et al.*, 1977). According to this study, however, the reaction does not occur for a positive ΔG_0 value in contrast to the fact that one can observe charge transfer even for a positive ΔG_0 value in the case of fluorescence quenching. Thus a final conclusion about the k_q vs. ΔG_0 relation in the case of the charge transfer reaction in the triplet state has not yet been obtained.

12.5 Structures of Exciplexes

We have discussed in 2.2, 3.1, 4.1, and 4.2. some features of the electronic and geometrical structures of exciplexes and the excited state of weak EDA complexes.

It has been argued that, in relatively non-polar environments, geometrical restriction to the aromatic hydrocarbon–amine exciplexes is rather severe, i.e. the donor and acceptor must interact rather strongly at short distances in specified mutual orientations. However, the structure of the exciplex is rather sensitive to the solvent polarity, and, in moderately polar solvents, a loose-structure exciplex without close contact between donor and acceptor with specified mutual orientation seems to be possible. The circumstance seems to be similar also in the excited state of weak EDA complexes.

In this section, electronic and geometrical structures of excited charge transfer complexes and their solvent dependences as studied by transient absorption as well as luminescence spectral measurements will be discussed.

12.5.1 *Electronic and geometrical structures of some EDA complexes in the fluorescent and phosphorescent states*

Although quite numerous investigations have been made on the EDA complexes stable in the ground state by means of ground-state absorption spectral measurements, studies by fluorescence spectral measurements are rather scarce, except at a low-temperature and in a rigid state, because the fluorescence yields of EDA complexes in solution at room temperature are extremely small. This means that their excited-state (S_1) lifetimes are extremely short, which makes a study of their excited state, even by means of the nanosecond laser method, rather difficult. The development of the picosecond laser method will contribute to the elucidation of the excited state of those EDA complexes, but such studies are quite scarce at present.

However, some detailed studies were possible for TCNB (1,2,4,5-tetra-cyanobenzene)–aromatic hydrocarbon complexes since their fluorescence yields are considerable even in solution at room temperature when the solvent dielectric constant is not large (see for review, Nagakura, 1975).

The results of the measurements of fluorescence spectra, fluorescence yields, and fluorescence decay times of TCNB complexes at various temperatures and in various solvents, clearly indicated a large difference of energies as well as structures between the Franck–Condon excited state and the equilibrium excited state (Mataga and Murata, 1969). For example, it has been observed for TCNB–toluene as well as TCNB–benzene systems that the fluorescence decay times at room temperature are more than 100 ns, which is much longer than the radiative lifetimes ($\sim$50 ns) calculated from the intensity of the CT absorption band, thus indicating a considerable change in the structure of the complex during the excited-state lifetime. The molecular rearrangement including the intracomplex as well as the surrounding solvent relaxations produces an extraordinarily large Stokes shift of *ca.* 12,000 cm^{-1} at room temperature for the above systems.

The Stokes shifts of the CT fluorescence of stronger TCNB complexes than those above are a little smaller. Moreover, the smaller the ionization potential of the donor aromatic hydrocarbon, the smaller is the Stokes shift. This fact shows that, for the stronger complexes, the extent of the molecular rearrangements in the excited state is smaller (Mataga and Murata, 1969). This result seems to be quite reasonable.

The relaxation process leading to the large Stokes shift of the CT fluorescence of the TCNB–toluene system was observed directly by means of time-resolved fluorescence spectral studies in 1–100 ns regions in a highly viscous state at low temperature (147 K) (Egawa et al., 1971), just as in the case of the pyrene–TBA system at 156 K.

The wavefunctions for an EDA complex in the excited state (Ψ_e) and ground state (Ψ_g) may be written in general as,

$$\Psi_e \approx \sum_i a_i \Phi_i(A^* \cdot D) + \sum_j a_j \Phi_j(A \cdot D^*) + \sum_k a_k \Phi_k(A^- \cdot D^+)$$

$$+ \sum_l a_l \Phi_l(A^+ \cdot D^-) + a_m \Phi_m(A \cdot D) \tag{12.36}$$

$$\Psi_g \approx \sum_i b_i \Phi_i(A^* \cdot D) + \sum_j b_j \Phi_j(A \cdot D^*) + \sum_k b_k \Phi_k(A^- \cdot D^+)$$

$$+ \sum_l b_l \Phi_l(A^+ \cdot D^-) + b_m \Phi_m(A \cdot D) \tag{12.37}$$

For a weak complex, $|b_m| \gg \sum_i |b_i|, \sum_j |b_j|, \sum_k |b_k|, \sum_l |b_l|$ while $\sum_i |a_i|$ or $\sum_j |a_j| \gg |a_m|$ and $\sum_k |a_k|$ or $\sum_l |a_l| \gg |a_m|$. It should be noted that a's and b's are functions of nuclear configurations including the complex and the surrounding solvent molecules just as in the case calculated in section 2.2. The above-described relaxation will involve the change of the coefficients of Φ's, i.e. the increase of a_k as indicated in Figure 12.3c.

The electronic structures of excited CT systems can be observed more directly by means of absorption spectral measurements with the laser photolysis method. The absorption spectra of the relaxed fluorescent state of the TCNB–toluene system at room temperature were quite similar to those of the TCNB anion-radical (Potashnik and Ottolenghi, 1970; Masuhara and Mataga, 1970).

Furthermore, it has been shown that the $S_n \leftarrow S_1$ spectra of the systems containing TCNB in liquid donor solvents, such as benzene, toluene, mesitylene, at room temperature can be reproduced by superposition of the bands of the acceptor anion and the donor dimer cation (Tsujino et al., 1973). This result was explained as due to the exciplex formation of the type,

$$(A^- \cdot D^+)^* + D \to (A^- \cdot D_2^+)^* \tag{12.38}$$

Therefore, the extensive relaxation process of the excited TCNB–toluene system as proved by the time-resolved fluorescence studies may be not only due to the solvent reorientation relaxation but also due to the 1:2 exciplex formation.

The observed spectra of TCNB complexes in rigid matrices where the $1:2$ exciplex formation is inhibited can be reproduced approximately by the superposition of the absorption bands of the TCNB anion and donor cation (Masuhara *et al.*, 1973). Thus the fluorescent state of TCNB complexes is quite polar, being almost 'contact ion-pair' (*ca.* 100% CT character).

Theoretical calculations have shown that the TCNB–toluene complex has an asymmetrical structure analogous to the one indicated in Figure 12.6a in the ground equilibrium as well as the excited Franck–Condon state where the intermolecular overlap integral is considerable (Iwata *et al.*, 1966). According to this calculation, locally excited and CT configurations contribute almost equally to S_1 of the TCNB–toluene complex, which cannot explain the observed absorption spectra of the excited state.

It has been shown theoretically that in order to be compatible with the observed $S_n \leftarrow S_1$ spectra, one must assume the symmetrical overlapping structure of Figure 12.6b (Masuhara and Mataga, 1972). A similar conclusion has been derived also from the interpretation of the fluorescence radiative transition probability on the basis of the electronic structure of the S_1 state (Kobayashi *et al.*, 1971).

Thus the geometrical structure as well as the CT degree of the $1:1$ complex seem to change greatly during the excited-state lifetime even in rigid matrices. The structural change might be achieved by rotations of molecular planes as indicated in Figure 12.6a. Furthermore, the $S_n \leftarrow S_1$ spectra of the TCNB–toluene complex in polymethylmethacrylate measured at 4.2 K by the nanosecond laser photolysis method were also close to the absorption bands of ions (Masuhara and Mataga, 1973).

Therefore, even at 4.2 K, the structural change in the excited state occurs easily within a few nanoseconds. However, there is a possibility that the structure of the complex in the ground equilibrium state in rigid matrices or in solutions is not necessarily equal to that of Figure 12.6a but rather close

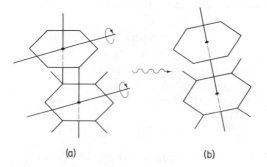

(a)   (b)

Figure 12.6   Geometrical structures of TCNB
complexes in the excited Franck–Condon state
(a) and excited equilibrium state (b)

to that of Figure 12.6b. If this is the case, an almost pure CT state can be realized without any large structural change.

The phosphorescent state of the TCNB complexes has been studied in detail by means of ESR and $T_n \leftarrow T_1$ spectral measurements made by Nagakura and coworkers (Nagakura, 1975). According to detailed ESR studies, the CT character in the phosphorescent state seems to be considerably smaller than that in the fluorescent state (Hayashi et al., 1969). In accordance with this result, the $T_n \leftarrow T_1$ spectra are rather different from the superposition of the absorption bands of the TCNB anion and the donor cation, although a band around $23 \times 10^3 \, cm^{-1}$ of the TCNB–hexamethyl-benzene complex whose T_1 state is essentially CT in nature ($\sim 80\%$) can be assigned approximately to excitation within the A^- anion (Matsumoto et al., 1972, 1973).

It should be noted here, that, by a comparison of observed $T_n \leftarrow T_1$ spectra with theoretical calculations based on different geometries, the conformation of the TCNB complexes in the phosphorescent state was concluded to be different from that in the fluorescent state and to be similar to that in the ground state.

12.5.2 Structures of exciplexes as revealed from the studies upon intramolecular complex systems

As discussed in section 5.1, the electronic structure of TCNB complexes in the fluorescent state is quite polar, being almost 'contact ion-pair', and the geometrical structure of the fluorescent state is of the symmetrical overlapping sandwich type. These conclusions have been derived from the $S_n \leftarrow S_1$ spectral measurements and theoretical considerations. The very polar electronic structure of the S_1 state can be deduced also from the measurements of the solvent polarity effect upon fluorescence wavenumber (Egawa et al., 1971). Quite similar studies of solvent effects upon fluorescence wavenumbers have been made for some typical heteroexcimers, the results of which show a very polar electronic structure for those heteroexcimers also (Mataga and Kubota, 1970). It has been shown that absorption spectra of some typical heteroexcimers are rather close to the superposition of ion bands (Fujiwara et al., 1977; Mataga and Ottolenghi, 1979).

However, it is rather difficult to elucidate geometrical structures of exciplexes formed by dynamic processes in the excited state. For this purpose, a study of model systems where the groups forming the exciplex are connected by methylene chains may be very useful.

As described in section 2.2 in the discussion of solvent-induced charge transfer in the excited state of the intramolecular heteroexcimer systems, $A—(CH_2)_n—D$ (P_n, A_n), the so-called $n = 3$ rule is not valid and there seem to be no strong geometrical restrictions for the formation of these intramolecular heteroexcimers.

The fluorescence of A_1, A_2, and P_1 in non-polar solvents is due to the excited state localized in the anthracene or pyrene part, while heteroexcimer fluorescence can be observed in the case of P_2, P_3, and A_3 even in non-polar solvents. This result indicates stronger coulombic interaction between two moieties in the CT state in the case of P_2, P_3, and A_3. P_3 and A_3 can take the sandwich configuration, which results in the larger coulombic interaction than for the other systems.

Although A_1, A_2, and P_1 show only the fluorescence from the anthracene or pyrene part in non-polar solvents, their heteroexcimer fluorescence can be observed in polar solvents. Therefore, the plane-parallel sandwich structure, which is possible only when $n = 3$, is not necessary for the heteroexcimer formation of these systems (Masaki et al., 1976).

The fact that the geometrical restriction for the heteroexcimer formation is not very critical was observed also for the intramolecular heteroexcimer of naphthalene and aliphatic amine combined with methylene chain (Chandross and Thomas, 1971).

The above results indicate that the overlap between the electron clouds of the halves in the typical heteroexcimers is not important for the determination of the geometry of the exciplex, and the bonding is essentially due to the electrostatic coulombic attraction between the ions. However, the circumstance is quite different in the case of 'homo' excimer state formation.

The $n = 3$ rule for the 'homo' excimer formation, confirmed in the case of diphenylalkanes (Hirayama, 1965) and dinaphthylalkanes (Chandross and Dempster, 1970), is not strictly valid also for 'homo' excimers since the excimer formation has been confirmed in the case of various anthracenophanes (Hayashi et al., 1976), 1,2-dianthrylethanes (Hayashi et al., 1977b), mePy(meta pyrenophane) (Hayashi et al., 1977a), and dipyrenylalkanes of various chain lengths (Zachariasse and Kühnle, 1976).

However, comparison of excimer fluorescence spectra of anthracenophanes and 1,2-dianthrylethanes shows clearly that the excimer fluorescence shows a larger red shift with increase of the overlap between the two anthracene rings (Hayashi et al., 1976). Thus the overlap plays an important role in the stabilization of the excimer state and in the destabilization of the Franck–Condon ground state.

As described in section 2.2, in the case of 1,2-dianthrylethanes and MePy in polar solvents, the excimer state has lower energy than the solvent-induced ion-pair state. Thus for the realization of excimer state, the overlap between the electron clouds of the halves is of crucial importance, resulting in a change of the structure to the overlapping type. Thus the 'homo' excimer interaction is essentially quantum mechanical just as in the case of homopolar bonding.

12.5.3 Solvent effects upon structures of exciplexes

In the early stage of fluorescence studies upon exciplexes, it was recognized that, upon increasing solvent polarity, both fluorescence quantum yield and decay time of typical heteroexcimers such as pyrene–DMA and anthracene–DEA decrease, and fluorescence yield decreases more rapidly with increasing solvent polarity than does the corresponding fluorescence decay time (Mataga et al., 1967; Knibbe et al., 1967).

Analogous results to the above were observed also for some EDA complexes stable in the ground state (Mataga and Murata, 1969; Prochorow and Siegoczyński, 1969; Prochorow and Bernard, 1974; Craig and Rodgers, 1976).

One possible interpretation of the above results is to assume solvent-induced change of electronic and geometrical structure of the heteroexcimer or the excited EDA complex (Mataga et al., 1967; Mataga and Murata, 1969; Prochorow, 1973). Namely, the CT fluorescence yield (η_f) may be expressed by,

$$\eta_f = \eta_e k_f/(k_f + k_d) = \eta_e k_f \tau_e \tag{12.39}$$

where η_e is the yield of the relaxed fluorescent-state formation from the encounter collisional state of the heteroexcimer system or the Franck–Condon excited state of the EDA complex, k_f and k_d are radiative and non-radiative rate constants from the fluorescent state, respectively.

According to the above interpretation, k_f and k_d will change depending upon solvent polarity, i.e. k_f will decrease and k_d will increase with increase of the solvent polarity. The electronic structure of the excited complex will become more polar with increasing polarity of the solvent according to the mechanism discussed in section 2.2. Assuming the wavefunction of equation (12.3), the transition moment of the CT fluorescence is mainly determined by the transition moment integral, $\langle \Psi(AD)|\boldsymbol{\mu}_{op}|\Psi(A^*D)\rangle$. Therefore, the k_f value depends on C_2^2 which decreases with increase of the solvent polarity. The strong solvation may induce a change of geometrical structure of the complex leading to a further decrease of the electronic delocalization interaction between A and D. The k_d value, however, will increase with solvent polarity because the energy gap between relevant electronic states become smaller in a more polar solvent. Moreover, with further increase of the solvent polarity, ionic dissociation of the excited complex will contribute to k_d.

There is another interpretation for the above result of the solvent effect upon CT fluorescence. Namely, it can be explained by assuming competition between non-fluorescent ion-pair and relaxed fluorescent-state formation in the encounter complex or in the excited Franck–Condon state (Knibbe et al., 1967; Masuhara et al., 1971; Shimada et al., 1973). With increase of the solvent polarity, η_e decreases, resulting in a larger decrease of η_f than τ_e.

Since the nanosecond laser photolysis and transient photoconductivity measurements on some typical heteroexcimer systems indicated the existence of ionic dissociation before relaxed heteroexcimer formation (Taniguchi and Mataga, 1972), both mechanisms seem to be possible in the case of the heteroexcimer systems. Discussions concerning the structures of transient complexes relevant to ionic dissociation in considerably or strongly polar solvent will be given also in section 6.

Now, the structures of exciplexes and effects of polar solvent upon them can be elucidated further by absorption spectral measurements, which are complementary to fluorescence measurements, and will give more direct information.

12.5.3.1 *Pyrene–DEA intermolecular heteroexcimer* The absorption spectra of a pyrene–DEA heteroexcimer measured by means of laser flash spectroscopy in hexane, toluene, and dimethoxyethane solutions (Fujiwara *et al.*, 1977) show clearly that the spectra depend rather strongly upon the solvent polarity.

Compared with the spectra in acetonitrile solution, which are due to dissociated ions, those of heteroexcimer in the above-described solutions are considerably broader although their peak wavelengths are almost the same as that of acetonitrile solution. The latter result indicates that the observed spectra of the heteroexcimer are mainly due to transitions of the type: $(A^-D^+)^* \rightarrow (A^{-*}D^+)^*$ and $(A^-D^+)^* \rightarrow (A^-D^{+*})^*$, i.e. the transitions approximately localized in each ion.

Thus the circumstance is somewhat analogous to the case of $S_n \leftarrow S_1$ spectra of TCNB complexes, which can be reproduced by the superposition of absorption bands of the TCNB anion and the donor cation. Nevertheless, the spectra of the pyrene–DEA heteroexcimer in non-polar solvents are much broader than the superposition of the bands of the ion-radicals.

The absorption spectra of the heteroexcimer become more and more similar to those of the separate ions as solvent polarity increases. Plausible explanations for this may be: (a) It can be ascribed to the solvent-induced changes in the electronic structure of the heteroexcimer (Mataga *et al.*, 1967; Orbach and Ottolenghi, 1975a). Namely, with increase of the solvent polarity, contribution of the CT configuration to the heteroexcimer wavefunction increases, leading to a decrease in intensities of back CT transitions. According to this interpretation, the broadening of the spectra in non-polar solvents can be ascribed to the effect of superposition of weak back CT bands upon strong ion bands. (b) An alternative interpretation may be to assume solvent-induced change of the geometry of the heteroexcimer which has the structure of the contact ion-pair even in non-polar solvents. The ion bands are broadened in non-polar solvents due to the strong field of gegen ion. In polar solvents, the acceptor–donor distance in the heteroex-cimer increases due to solvation, diminishing the effect of the field due to gegen ion.

12.5.3.2 Intramolecular heteroexcimers, P_n

The Transient behaviour and structures of P_3 and P_2 heteroexcimers in 2-propanol and acetonitrile studied by means of picosecond laser photolysis have been discussed already in section 4.2. Here, we make a comparison of the structures of inter-molecular pyrene–DMA heteroexcimers with those of P_n heteroexcimers in 2-propanol.

The spectra of the intermolecular heteroexcimers in 2-propanol as well as in 1-pentanol can be reproduced by super-position of the bands of the pyrene anion and the DMA cation. They are also similar to the spectra of P_1 and P_2 in 2-propanol, but they are much different from the quite broad spectra of the P_3 heteroexcimer in non-polar solvents (Hinatsu et al., 1978).

A possible explanation for the above difference between the spectra of the P_3 heteroexcimer and the intermolecular one may be to assume that a geminate ion-pair plays an important role in the intermolecular heteroex-cimer system. Namely, the heteroexcimers in those alcohols are produced by recombination of geminate ion-pairs or they are in equilibrium with gemi-nate ion-pairs (Goodall et al., 1974; Orbach and Ottolenghi, 1975a,b; Schulten et al., 1976). If the geminate ion-pair is the main chemical species which exists during the observed heteroexcimer fluorescence lifetime, the spectra should be reproduced by the superposition of ions bands.

However, according to the investigations of magnetic field modulation of the pyrene triplet and ion-radical concentrations, a heteroexcimer predomi-nates in 2-propanol instead of a geminate ion-pair (Michel-Beyerle et al., 1976).

Therefore, one should conclude that, although the relaxed structure of the P_3 heteroexcimer in 2-propanol is of the sandwich type, the pyrene–DMA intermolecular heteroexcimer has a looser structure owing to the more extensive solvation compared with the P_3 heteroexcimer.

12.5.3.3 Cyano-substituted layered cyclophanes

As a model for $1:2$ ex-ciplexes of TCNB and methyl-substituted benzenes, fluorescence and $S_n \leftarrow S_1$ absorption spectra of cyano-substituted cyclophanes indicated in Figure 12.7 were studied (Yoshida et al., 1976; Masuhara et al., 1977b).

All these compounds show the broad fluorescence band of the exciplex type. The red shifts of the fluorescence when the solvent is changed from cyclohexane to acetonitrile are ca. 2000 for **1** and 3300 cm^{-1} for **3** as well as **4**, while it is only 400 cm^{-1} for **2**. The solvent shifts of **3** and **4** are similar to those of the TCNB–toluene $1:2$ exciplex providing further evidence for the structure, $A^-(DD)^+$, of the exciplex.

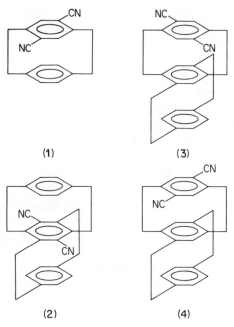

(1)   (3)

(2)   (4)

Figure   12.7   Structures   of   cyanosub-
stituted layered cyclophanes

Plots of fluorescence maxima against $(f_\varepsilon - \frac{1}{2}f_n)$ according to equation
(12.13) for **1**, **3**, and **4** show curvature especially in the case of **3** and **4**,
indicating solvent-induced electronic change to a more polar structure due
to the interaction with solvent. In accordance with this result, $S_n \leftarrow S_1$
spectra of **4** were confirmed to show a marked dependence upon solvent
polarity.

According to the theoretical consideration described in section 2.2, it is
very difficult for the solvent-induced electronic structural change to occur if
the matrix elements α and β of equation (12.8) are not small. However,
cyclophanes may be deemed to consist of strongly interacting electronic
systems and considered to be one molecule, the π-electronic system of
which cannot be divided into two groups as in the case of intramolecular
heteroexcimer systems, A—$(CH_2)_n$—D. Nevertheless, the solvent-induced
electronic structure change seems to occur.

12.6   Ionic Photodissociation and Related Processes

As discussed in previous sections, charge transfer in the excited state and
exciplex formation processes are affected considerably by the solvent polarity. Moreover, not only the exciplex formation processes but also the

electronic and geometrical structures of exciplexes depend upon solvent polarity.

In the medium of considerable polarity, dissociation to solvated ion-radicals becomes important as a non-radiative deactivation path of heteroexcimer systems as well as excited EDA complexes because of their strongly polar nature (Masuhara *et al.*, 1971; Taniguchi *et al.*, 1972; Taniguchi and Mataga, 1972). Details of the mechanism of this process are very important as a basis for elucidating various processes of conversion of light energy into chemical energy in chemistry and biology.

Some discussions have been given already in section 4 concerning the charge transfer processes in polar solvents. Especially, in the case of photochemical charge transfer reactions in strongly polar solvents, a systematic relation has been established between the quenching rate constant k_q and the free energy change ΔG^0 of the the charge transfer, $A^* \cdots D$ or $A \cdots D^* \rightarrow A_S^- \cdots D_S^+$. However, for elucidation of the ionic photodissociation mechanism, measurement of its quantum yield is of crucial importance, although such measurements are rather scarce.

Quantitative measurement of the yield has become possible by employing the laser photolysis method. According to some results obtained until now, the ionic photodissociation yield is affected not only by solvent polarity, but also depends strongly upon the chemical properties of electron donor and acceptor as well as the strength of CT interaction between them and spin multiplicity of the CT state (Masuhara *et al.*, 1975; Hino *et al.*, 1976b).

12.6.1 *Effects of strength of* CT *interaction, spin multiplicity, and solvent polarity upon the dissociation yield*

The results of measurements when excited pyrene is quenched by various electron donors or acceptors such as amines, nitriles, esters, and anhydrides in strongly polar solvents such as acetonitrile or acetone show clearly that there is no systematic relation between the ionic dissociation yield and ΔG^0, but the values of the yield depend strongly upon the chemical nature of the quencher. The values of yield can be grouped roughly into four, i.e. amines, nitriles, esters, and anhydrides. Thus the encounter collisional charge transfer between the fluorescer and quencher is not always followed by ionic dissocation but may be followed by other non-radiative processes as indicated schematically in Figure 12.8, the extent of which depends on the chemical nature of the fluorescer–quencher pair.

The nature of the CT state in Figure 12.8 is not very clear at present, but it may be a kind of exciplex produced immediately after charge transfer. There should be some interaction between the pair in the CT state so that the chemical property of the donor and the acceptor is reflected in the dynamic behaviour of the pair.

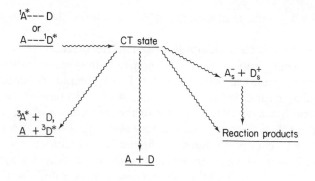

Figure 12.8 Dynamic behaviour following the encounter collisional charge transfer between the fluorescer and quencher

Although there is no systematic relation between the dissociation yield and ΔG^0 in general, one can recognize a relation between the strength of CT interaction and the dissocation yield in the case of a series of donor or acceptor of the same kind.

For example, when pyrene fluorescence is quenched by p-dicyanobenzene, TCNB, and tetracyanoethylene in acetonitrile, the relative ionic dissociation yields are 1.0, 0.5, and 0.3, respectively, decreasing with increase of the electron affinity of the quencher. Since increase of triplet pyrene was not recognized with the decrease of the dissociation yield, and also, no photochemical product was detected, the process competing with the ionic dissociation is the direct degradation to the ground state which increases with increase of the CT interaction, i.e. with decrease of the energy gap between CT and ground state.

A quite similar result has been observed also in the case of TCNB complexes with methyl-substituted benzenes in acetonitrile. Relative quantum yields of ionic dissociation for benzene, toluene, mesitylene, and hexamethylbenzene donors are 1.0, 1.0, 0.5, and 0.1, respectively.

Spin multiplicity of the CT state also affects the dissociation yield profoundly in some cases. The aromatic hydrocarbon–amine systems such as pyrene–DMA or DEA in acetonitrile show ca. 80% ionic dissociation yield from the S_1 state. However, in the case of the bacteriopheophytin (Bph)–p-benzoquinone (Q) system in acetone–methanol mixture (Holten *et al.*, 1976) as well as the chlorophyll-a–2,6-dimethylbenzoquinone system in ethanol (Huppert *et al.*, 1976), no ionic dissociation from the S_1 state was observed. These results are again remarkable examples of the dependence of the ionic dissociation yield upon the chemical properties of the donor–acceptor pair.

In the case of the Bph–Q system, only when $^3Bph^*$ was produced by means of external heavy atom effect adding high concentration (8 M) of methyl iodide, was ionic dissociation observed due to the reaction, $^3Bph^* + Q \rightarrow Bph_s^+ + Q_s^-$. Charge transfer in the S_1 state seems to be followed by rapid degradation to the ground state as indicated in Figure 12.8, while such rapid degradation may be difficult in the triplet CT state because of the difference of spin multiplicity.

In the reaction centre of bacterial photosynthesis, very rapid photoinduced charge transfer in the excited singlet state and very efficient charge separation seem to be occurring according to picosecond laser photolysis studies (Rockley *et al.*, 1975; Kaufmann *et al.*, 1975). Some ingenious device which prevents the rapid degradation to the ground state should be present in the actual photosynthetic system.

In relation to the mechanisms of photoinduced cationic polymerization of α-methylstyrene (α-MST) in the presence of TCNB or pyromellitic dianhydride (PMDA), ionic photodissociations of α-MST–TCNB and α-MST–PMDA complexes in methylene chloride have been investigated by transient absorption and transient photoconductivity measurements by the nanosecond laser photolysis method (Irie *et al.*, 1974).

In the case of the α-MST–TCNB system, the solvated ion-pair produced from the excited singlet complex, $^1(^2A_S^- \cdots {}^2D_S^+)$, which has a lifetime of *ca.* 100 ns, is deactivated to the ground state without dissociation into the free ions. The free ions, which can be detected by transient photoconductivity measurement, are produced slowly, probably from the triplet ion-pair formed from the triplet complex. In the case of the α-MST–PMDA system, ion-pair formation from the singlet excited state is not observed, but only the slow ionic dissociation, which is ascribed to the triplet state, can be observed. The deactivation of the singlet excited state due to the radiationless transitions to triplet as well as ground state of this system may be much faster than that of the TCNB complex. Thus the ionic dissociation from the triplet state plays an important role also in these systems just as in the case of the chlorophyll–quinone or Bph–quinone systems.

Although the singlet ion-pair of the α-MST–TCNB system has a rather long lifetime, it does not show the dissociation into free ions. This long lifetime of the singlet ion-pair is very different from the case of the exciplex systems of pyrene–DMA or anthracene–DMA (Taniguchi and Mataga, 1972). A similar result has been observed more clearly, however, in the case of the pyrene–p-DCNB (p-dicyanobenzene) exciplex system (Hino *et al.*, 1976b). The ion-pair formed by charge transfer in the S_1 state has a lifetime of *ca.* 100 ns in dichloromethane as well as 1,2-dichloroethane. Moreover, the singlet ion-pair of this system shows a rather efficient dissociation into free ions. The yield as well as the rate of dissociation increase with increase of the solvent polarity.

The reason for the different behaviour of pyrene–DMA and pyrene–p-DCNB systems in the ionic dissociation process is not very clear at the present stage of investigation.

The ionic photodissociation yield of the CT systems is affected strongly also by the polarity of the environment. Roughly speaking, the dissociation yield increases with increase of the solvent polarity. Although the dissociation yield depends strongly upon the chemical properties of the electron donor–acceptor pair, there exists a systematic relation for a donor–acceptor pair between the yield and the solvent polarity.

For example, the ionic photodissociation yield, ϕ_i, of the pyrene–DMA system or the TCNB–toluene system in various solvents can be reproduced by equation (12.40), where ε is the solvent dielectric constant and A and B are constants (Masuhara et al., 1975).

$$\log (1/\phi_i) = (A/\varepsilon) + B \qquad (12.40)$$

This equation can be derived from Onsager's ion recombination model modified by Mozumder (1968, 1969). Namely, it is assumed that a fraction $\alpha(\leqslant 1)$ of the dissociative state is converted into the geminate ion-pair with distance r_0. The geminate ion-pair undergoes further diffusional separation in competition with recombination. According to this model, ϕ_i is given by,

$$\phi_i = \alpha \exp (-r_c/r_0), \qquad r_c = e^2/\varepsilon kT \qquad (12.41)$$

Therefore, $\log (1/\phi_i)$ is proportional to $(1/\varepsilon)$.

There is another treatment of this problem, which can reproduce the observed results. It is based upon the Horiuchi–Polanyi relation, i.e. the proportionality relation between the activation free energy of ionic dissociation and the free energy difference between the initial and final state (Masuhara et al., 1975). The initial state is assumed here to be a CT state non-relaxed with respect to solvation and in the final state the ions are completely solvated, the free energy of which is assumed to be given by the Born's formula which contains the inverse of the solvent dielectric constant.

The CT state in Figure 12.8 involves various states from the non-relaxed one immediately after the charge transfer through the geminate ion-pair state, according to the above model.

12.6.2   Local triplet formation from ion-pairs and relevant states

It is well known that charge transfer quenching of fluorescence leads to efficient production of triplet states in some cases. In the case of heteroexcimer formation in non-polar solvents, local triplet states may be formed by intersystem crossing (ISC) in the heteroexcimer. There have been some controversies concerning the 'fast' ISC process preceding the generation of the relaxed heteroexcimer state (Mataga and Ottolenghi, 1979). However, it has been confirmed by means of picosecond and nanosecond laser photolysis studies that fast ISC from the non-relaxed heteroexcimer state is not important in the case of typical heteroexcimer systems such as anthracene–DEA (Nishimura *et al.*, 1977; Fujiwara *et al.*, 1977).

However, when the photodissociation to solvated ion-radicals is predominant in polar solvent, the local triplet formation in such typical heteroexcimer systems seems to occur through both slow and fast mechanisms. The slow mechanism in this case is due to the second-order homogeneous recombination process of radical-ions separated completely.

$$^2A_S^- + {}^2D_S^+ \rightarrow {}^3A^* + D \text{ or } A + {}^3D^* \\ \searrow A + D \tag{12.42}$$

The delay time from the exciting laser pulse where this process occurs is typically of $\sim 1\,\mu s$ in ordinary polar solvents. Because of the randomly aligned electron spins of radical-ions, the homogeneous recombination will lead approximately to 75% triplet and 25% singlet.

The triplet generation due to the fast mechanism occurs within a few nanoseconds, which cannot be ascribed to the homogeneous recombination of radical-ions (Orbach and Ottolenghi, 1975b). Therefore, the geminate recombination must be invoked in order to explain this fast process. Namely, following the charge transfer, the geminate ion-pairs which have singlet spin multiplicity are formed. Triplet ion-pairs must be produced from the singlet ion-pairs within a few nanoseconds, leading, finally, to the formation of a local triplet as indicated in equation (12.43).

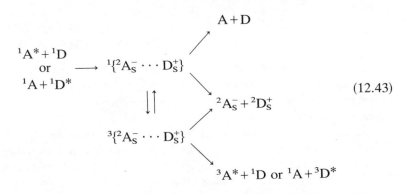

$$\tag{12.43}$$

The formation of the triplet geminate pairs from the singlet pairs within a few nanoseconds seems to be possible by means of the hyperfine coupling between unpaired electron spins and nuclear spins of the radical pairs, which induces transitions between degenerate spin states S_0, T_0, T_1, T_{-1} of the radical pairs (Brocklehurst, 1974).

The hyperfine coupling is modified by an external magnetic field, affecting the transition probabilities between the spin states. By analysing the magnetic field modulation of the local triplet concentrations in nanosecond laser photolysis experiments, the validity of the above model of the hyperfine coupling in the geminate ion-radical pairs has been confirmed for the pyrene–DEA system in methanol (Michel-Beyerle et al., 1976) and the pyrene–3,5-dimethoxy-N,N-dimethylaniline system in methanol (Schulten et al., 1976).

Thus the mechanism of the 'fast' triplet formation in the case of typical aromatic hydrocarbon–amine heteroexcimer systems in strongly polar solvent has been elucidated. In the pyrene–p-DCNB system, although the mode of ionic photodissociation is somewhat different from the case of hydrocarbon–amine systems, 'fast' generation of pyrene triplet has been observed (Hino et al., 1976a,b). Thus the triplet already exists at 50 ns after the nanosecond laser oscillation, before the dissociation of the solvated ion-pair. Although no detailed studies including magnetic field effect are available for this system, triplet generation from the geminate ion-pair within a few nanoseconds might be possible.

Although triplet generation from the non-relaxed charge transfer state or from the encounter complex is not plausible in the above systems, there are many cases in which the fast triplet generation mechanism is not clear or generation from the non-relaxed state is suggested. For example, in the case of oxonine (OxH^+) and allylthiourea (ATU) in strongly polar solvents, 'fast' generation of oxonine triplet was observed (Bonneau and Joussot-Dubien, 1976; Vogelmann and Kramer, 1976).

OxH^+:
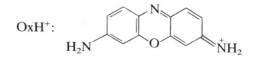

Since the energy of the radical pair, $OxH \cdots ATU^+$, seems to be too low to give back the dye triplet by recombination, the possibility of ISC from the non-relaxed charge transfer state or from the encounter complex should be taken into account.

Because of the initial absorbance present immediately after the nanosecond laser pulse, ISC of the TCNB complex at low temperatures from the non-relaxed charge transfer state was suggested (Masuhara et al., 1972). However, there is a possibility of accidental cancellation between absorbances of triplet and relaxed excited singlet states, and a final conclusion has not been reached.

12.6.3   Ionic photodissociation in micellar solutions

The photoinduced charge transfer and ionic dissociation processes of exciplex systems discussed up to now all depend rather critically upon the nature of the solvent. In view of this result in homogeneous organic solvents, studies of heteroexcimer systems in such inhomogeneous systems of biological importance as membranes and micelles seem quite interesting and important from the following standpoints.

1. One can compare the behaviour of heteroexcimer systems near the micelle–water or membrane–water interfaces with their accumulated results in homogeneous organic solvents, which will provide a basis for the elucidation of photoprimary processes in biological membrane systems.
2. Heteroexcimers will be useful as a new type of probe system to clarify the structures and functions of micelles and membrane systems. Namely, they will be sensitive not only to the rigidity of the membrane but also to its polarity because of their strongly polar electronic structure.

Photoprimary processes in micellar solutions, such as the quenching of the fluorescence of probe molecules trapped in micelles by water-soluble quenchers and electron ejection from the excited molecule in micelle to water, as well as the reaction of hydrated electrons with the molecule trapped in the micelle, have been studied extensively by Thomas, Grätzel, and others by means of laser photolysis and pulse radiolysis (See for review, Grätzel and Thomas, 1976; Thomas, 1977). However, studies upon heteroexcimers and excited charge transfer complexes in micellar solutions are scarce. Here a brief discussion will be given of the behaviours of typical heteroexcimer systems in micellar solutions studied recently (Masuhara *et al.*, 1977a, Waka *et al.*, 1978, 1979 Katušin-Ražem *et al.*, 1978).

Fluorescence of 1-pyrene sulphonic acid (PyS) in aqueous solution is quenched by *p-N,N*-dimethylaniline sulphonate (DMAS), satisfying the Stern–Volmer (S.-V.) equation with $k_q = 3.5 \times 10^9 \, M^{-1} s^{-1}$ at room temperature. However, in CTAC (cetyltrimethylammonium chloride, $C_{16}H_{33}N^+(CH_3)_3Cl^-$) and DTAC (dodecyltrimethylammonium chloride, $C_{12}H_{25}N^+(CH_3)_3Cl^-$) solutions, quenching of the PyS fluorescence by DMAS cannot be expressed by the S.-V. equation, but much stronger quenching is observed. Presumably, negatively charged fluorescers and quenchers are 'concentrated' on the positively charged micelle surfaces enhancing the quenching.

It has been confirmed that pyrene fluorescence is quenched strongly by added DMA in both CTAC and SDS (sodium dodecyl sulphate, $C_{12}H_{25}SO_4^-Na^+$) micellar solutions and no heteroexcimer emission is observed. Much stronger quenching than that expressed by the simple S.-V. equation was observed in both solutions. Strong quenching was observed also for the pyrene–DMAS system in cationic micellar solutions. The above results suggest that the quenching of pyrene fluorescence due to the charge transfer from DMA takes place in the region of relatively high polarity near

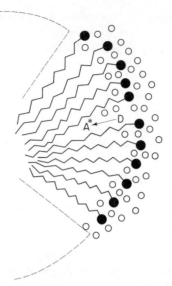

Figure 12.9   A schematic representation of photochemical charge transfer near the micelle surface. ●: Ionic end group of surfactant. ○: Water

the micellar surface as indicated in Figure 12.9. Moreover, strong quenching similar to the above case of DMA was observed also in the case of pyrene–DCNB (dicyanobenzene) systems in both cationic and anionic micellar solutions (Waka *et al.*, 1979).

The observed relative fluorescence intensity of the pyrene–DMA system in SDS solution can be expressed approximately by,

$$I/I_0 \approx \exp\left(-\alpha[Q]\right),$$

where α is a constant and $[Q]$ is the concentration of DMA. This result can be derived from a model proposed recently (Rodgers and Wheeler, 1978, Maestri *et al.*, 1978, Infelta, 1979, Henglein and Scheerer, 1978). It is based on the following assumptions.

(i) Fluorescers and quenchers are both associated with micellar aggregates.
(ii) Their distribution obeys Poisson statistics.
(iii) The quenching process cannot be described by a bimolecular rate constant, but by a first-order rate constant. The first-order rate constant is expressed by $k_m = mk_1$ for quenching in a micelle containing m quencher molecules.

From these assumptions (I/I_0) can be obtained as follows,

$$I/I_0 = \sum_{m=0}^{\infty} \frac{\exp(-[Q]/[Mic])}{m!} \left(\frac{[Q]}{[Mic]}\right)^m \left(1 + \frac{mk_1}{k_0}\right)^{-1}$$

where [Mic] is the concentration of micelles and k_0 is the rate constant of the decay of the fluorescent state in the absence of quencher.

When (k_1/k_0) is very large, which seems to be the case for the pyrene–DMA–SDS system, the above I/I_0 reduces to,

$$I/I_0 \approx \exp(-[Q]/[Mic]).$$

However, more detailed examination of the experimental results shows that the observed results cannot be fitted to these equations especially in the case of pyrene–DMA in cationic micellar solution and pyrene–DCNB in both kinds of micellar solutions (Waka et al., 1978b). It seems to be necessary to take into consideration the bimolecular quenching due to quenchers from the aqueous phase, in addition to the intramicellar first-order quenching process. On the basis of this model and assuming Poisson statistics for the distribution of quencher molecules in the micellar phase, the following equation can be derived for (I/I_0).

$$\frac{I}{I_0} = \sum_{m=0}^{\infty} \frac{\exp(-[Q]_M/[Mic])}{m!} \left(\frac{[Q]_M}{[Mic]}\right)^m \left(1 + \frac{mk_1}{k_0} + \frac{k}{k_0}[Q]_A\right)^{-1},$$

where $[Q]_M$ and $[Q]_A$ are quencher concentrations in micellar and aqueous phase, respectively, and k is the bimolecular quenching rate constant. For large values of (k_1/k_0), this equation can be reduced to,

$$I/I_0 \approx \exp(-[Q]_M/[Mic]) \left(1 + \frac{k}{k_0}[Q]_A\right)^{-1}.$$

These equations can explain the observed results satisfactorily in general.

The charge transfer mechanism of the fluorescence quenching in the PyS–DMAS, pyrene–DMAS, and pyrene–DMA systems in micellar solutions has been confirmed by means of the nanosecond laser photolysis method (Waka et al., 1978, Katušin-Ražem et al., 1978). Relative yields of ion-radicals of the pyrene–DMA system produced by charge transfer in micellar solutions measured immediately after the excitation are 1.1, 0.7, and 0.4 in CTAC, Brij 35 (polyoxyethylene dodecyl alcohol ether, a neutral surfactant, $C_{12}H_{25}(OC_2H_4)_nOH$), and SDS, respectively, taking the value in acetonitrile solution as a standard. This result might be explained as follows.

The intramicellar position of DMA may be near the surface in all kinds of micellar systems or it comes from the aqueous phase during the lifetime of the fluorescer molecule, while pyrene may be placed at a relatively inner side of the micelle compared with DMA. DMA cations formed by photochemical electron transfer in cationic micelles may be driven away outside the surface because of the instability due to the same plus charge, while the pyrene anion will be held inside the micelle, leading to the inhibition of recombination of radical-ions. In accordance with this interpretation, pyrene anions in CTAC do not show any decay even at the delay times where the ions in acetonitrile solution show a considerable decay due to recombination. In anionic micelles, pyrene anion may be driven away outside the micelle. However, it may not be as easy as in the case of the DMA cation in cationic micelles because of the relatively inner position of the pyrene anion compared with the DMA cation. Actually, rapid decay of ions probably due to recombination was observed. The small yield of ions in SDS even immediately after the excitation by a nanosecond laser pulse may be due to rapid recombination, details of which may be elucidated by means of picosecond laser spectroscopy.

Separation of produced ion-radicals through a micelle–water interface in the case of photoinduced charge transfer has been demonstrated also for the chlorophyll-a–duroquinone system in SDS micellar solution (Wolff and Grätzel, 1977). Namely, the recovery of the ground-state bleaching of chlorophyll-a in SDS micelle studied by means of microsecond flash photolysis became much slower in the presence of duroquinone, which seems to indicate the ejection of the duroquinone anion produced by light-induced charge transfer from chlorophyll-a outside the micelle, making recombination very difficult. Charge separation at the micelle–water interface was observed by means of laser photolysis also in the case of charge transfer from Fe^{2+}, adsorbed on the surface of SDS micelle, to triplet duroquinone in the micelle (Scheerer and Grätzel, 1977).

The behaviour of intramolecular heteroexcimer systems, P_n, in the excited state was examined also in micellar solutions (Masuhara et al., 1977a, Katušin-Ražem et al., 1978). Although fluorescence quenching was observed, the heteroexcimer emission was hardly detected. Quite efficient fluorescence quenching was observed in the case of DTAC solutions. The quenching may be ascribed to photoinduced intramolecular charge transfer at the micelle–water interface, which was confirmed by the nanosecond laser photolysis method. Since A^- and D^+ ions are combined by methylene chain, the ejection of D^+ or A^- outside the micelle is not possible and the two ions are kept within a short distance. This circumstance will result in a rapid decay of produced ions due to back charge transfer. Actually, it was confirmed that the lifetime of the ion-pair state is very short.

12.7   Proton Transfer and Related Processes

In previous sections, only the electron donor–acceptor interactions in the excited electronic state are discussed. In this section, proton donor–acceptor interactions and related processes such as hydrogen bonding interactions and charge tranfer followed by proton transfer resulting in eventual hydrogen atom transfer in some exciplex systems, will be discussed briefly.

12.7.1 Charge transfer followed by proton transfer in exciplex systems

The typical heteroexcimers of aromatic hydrocarbon–tertiary amine systems are rather strongly fluorescent in non-polar solvents. If the heteroexcimer state is rapidly deactivated by some efficient radiationless processes such as ionic dissocation or deactivation to the ground state, one can scarcely observe the heteroexcimer emission and the lifetime of the heteroexcimer state may be very short.

In some heteroexcimer systems, deactivation occurs frequently due to more specific chemical reactions (see for a review of this problem, Davidson, 1975). For example, heteroexcimers formed by primary or secondary amines with aromatic hydrocarbons are almost non-fluorescent or only very weakly fluorescent even in non-polar solvents (Mataga et al., 1966; Mataga and Ezumi, 1967; Mataga, 1975; Okada et al., 1976). This result indicates a fairly efficient deactivation process competing with fluorescence in the case of these heteroexcimer systems.

A possible deactivation path seems to be charge transfer followed by transfer of the $\overset{\frown}{>}$N—H proton, resulting ultimately in hydrogen atom transfer (Mataga, 1975) as indicated in equation (12.44), in accordance with the fact that aromatic hydrocarbons are reduced photochemically in the presence of primary and secondary amines (Davidson, 1975).

$$A^* + H\!-\!N\!\!\begin{array}{c}{\scriptstyle R}\\[-2pt]{\scriptstyle R'}\end{array} \rightarrow (\dot{A}^- \cdots H\!-\!\dot{N}^+\!\!\begin{array}{c}{\scriptstyle R}\\[-2pt]{\scriptstyle R'}\end{array})^* \rightarrow \dot{A}\!-\!H \cdots \dot{N}\!\!\begin{array}{c}{\scriptstyle R}\\[-2pt]{\scriptstyle R'}\end{array}$$

$$\rightarrow \dot{A}\!-\!H + \dot{N}\!\!\begin{array}{c}{\scriptstyle R}\\[-2pt]{\scriptstyle R'}\end{array} \quad (12.44)$$

The formation of radicals $\dot{A}$—H and $\dot{N}\!\!\begin{array}{c}{\scriptstyle R}\\[-2pt]{\scriptstyle R'}\end{array}$ in the case of pyrene and primary as well as secondary amines in hexane solutions have been confirmed by means of nanosecond and microsecond flash photolysis measurements (Mataga, 1975; Okada et al., 1976). However, complete measurement of the formation of a heteroexcimer and neutral radical pairs from it by means of the nanosecond laser method was rather difficult.

Measurements upon the pyrene–N-ethylaniline (NEA)–hexane system have been made by the picosecond laser method (Mataga et al., 1978). Some results are indicated in Figure 12.10. The spectra at 100 ps delay time are somewhat broad compared with the spectra at 200 ps delay time, which is due to the superposition of the $S_n \leftarrow S_1$ bands of free pyrene on the absorption bands of the heteroexcimer. The spectra observed at 200 ps

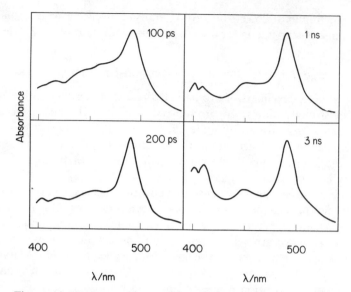

Figure 12.10   Transient absorption spectra of the pyrene–N-ethylaniline–hexane system at several delay times. Normalized at the peak of the pyrene anion-like band. Unpublished result by Nishimura, T., Karaki, I., Okada, T., and Mataga, N.

delay time are approximately due to the heteroexcimer. An interesting point concerning the absorption spectra of this heteroexcimer is the fact that the spectra are quite close to the superposition of the absorption bands of the pyrene anion and the amine cation, but are different from those of the pyrene–DEA heteroexcimer in hexane.

As discussed in section 5.3, the spectra of the pyrene–DEA heteroexcimer in hexane are much broader than the superposition of the bands of the ion-radicals. This result might be ascribed to the contributions of weak back CT transitions of the type: $(A^-D^+)^* \to (A^*D)$ and $(A^-D^+)^* \to (AD^*)$, to the spectra mainly due to the transitions localized in each ion: $(A^-D^+)^* \to (A^{-*}D^+)^*$ and $(A^-D^+)^* \to (A^-D^{+*})^*$.

The sharp ion-like bands of the pyrene–NEA heteroexcimer in hexane suggest that it has a more loose geometrical structure and more polar electronic structure than the pyrene–DEA heteroexcimer in hexane. Moreover, the strongly polar electronic structure of the pyrene–NEA heteroexcimer may be favourable for proton transfer from D^+ to A^-.

The spectra at 1 ns and 3 ns delay times show characteristic bands at 405 nm and 412 nm, respectively. They show growing-in the region of a few nanoseconds in accordance with the decay of the heteroexcimer. The 405 nm band is due to the 1-hydro-1-pyrenyl radical, $\dot{A}$—H (Okada *et al.*,

1976), and the 412 nm band is due to the pyrene triplet. Thus the heteroexcimer is formed at first, and the $\dot{A}$—H radical as well as the pyrene triplet state are formed via the heteroexcimer.

The mechanisms of intersystem crossing (ISC) or local triplet formation in excited charge transfer systems were discussed in section 6.2. The possibility of 'fast' ISC preceding the formation of the relaxed heteroexcimer state has been excluded in the case of typical heteroexcimer systems such as anthracene–DEA (Nishimura *et al.*, 1977) and pyrene–DEA (Fujiwara *et al.*, 1977); the triplet generation in the pyrene–NEA system in hexane is much faster.

In polar solvents, even the typical heteroexcimer-forming systems such as pyrene–DEA show fast triplet generation within a few nanoseconds, which is due to the geminate recombination of the produced radical ion-pairs (Michel-Beyerle *et al.*, 1976; Schulten *et al.*, 1976). The fast triplet generation in the pyrene–NEA–hexane system suggests reaction mechanisms analogous to the geminate recombination in polar solvents. Namely, a possible mechanism of triplet generation may be the geminate recombination of radicals produced by proton transfer as indicated in equation (12.45).

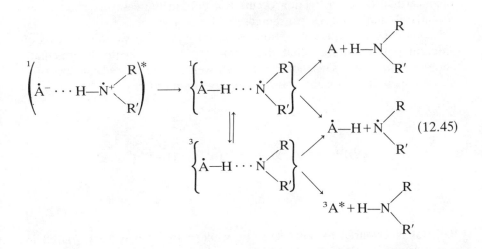

$$(12.45)$$

In this reaction mechanism, dissociation of the geminate pairs and formation of the triplet geminate pair from the singlet one, as well as the formation of the pyrene triplet state from the triplet geminate pair, should be faster than proton transfer in the heteroexcimer, since $\dot{A}$—H and $^3A^*$ appear almost simultaneously.

Another mechanism might be the fast formation of triplet heteroexcimer leading to rapid production of pyrene triplet state as shown in equation (12.46).

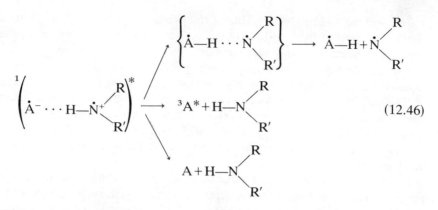

$$(12.46)$$

Which one of these mechanisms is actually operating is not very clear at the present stage of investigation.

There are many other systems where the mechanism of charge transfer followed by proton transfer in the exciplex system is assumed for the photochemical reaction in solution (Davidson, 1975). A well-known example is the triplet benzophenone quenched by amines. In this system, photochemical hydrogen abstraction occurs not only in the case of primary and secondary amines but also in the case of tertiary amines.

Nanosecond laser photolysis studies on this system in various solvents of different polarity suggest that the reaction proceeds via the intermediate heteroexcimer state since ion-radicals as well as neutral radicals are produced depending upon the solvent polarity (Arimitsu *et al.*, 1975).

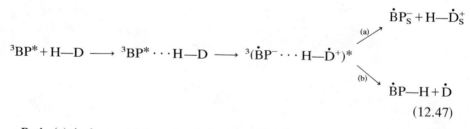

$$(12.47)$$

Path (a) is favoured in polar solvents while path (b) may be favoured in non-polar solvents. However, direct proof of the existence of the intermediate heteroexcimer by means of the picosecond laser method has not yet been reported.

In relation to charge transfer and proton transfer processes in the excited state, we discuss here briefly proton-induced quenching of fluorescence of some intramolecular charge transfer systems.

Fluorescence of naphthylamines in polar solvents is quenched in the presence of protons (Förster, 1972; Schulmann and Liedke, 1973). Although intermediates where a proton is interacting with naphthylamines were assumed for the quenching reaction, details of the reaction mechanism were not clear.

Measurements of fluorescence lifetimes show that the fluorescence of naphthylamine protonated on the amino nitrogen is not quenched by proton, only the fluorescence of neutral naphthylamine being significantly quenched (Tsutsumi and Shizuka, 1977). This result suggests that the quenching is caused by interaction between excited naphthylamine and a proton other than that on the amino nitrogen.

The extent of the intramolecular charge transfer from the amino group to the naphthalene ring of naphthylamines is much larger in the S_1 state than in the S_0 state (Mataga, 1963). The degree of charge transfer in the excited state may be affected by the interaction with polar solvent as discussed in section 2.2. Moreover, it was demonstrated that the intramolecular charge transfer in the excited state is much stronger for 1-naphthylamine than for 2-naphthylamine (Mataga, 1963).

The considerable intramolecular charge transfer to the naphthalene ring in the S_1 state suggests that the proton-induced fluorescence quenching occurs due to charge transfer from the naphthalene ring of the excited naphthylamines to the proton.

$$(D^{+\delta}-A^{-\delta})^* + H^+ \rightarrow (D^{+\delta}-A^{-\delta})^* \cdots H^+ \rightarrow (D\dot{-}A)^+ \cdots \dot{H} \rightarrow$$

This mechanism is supported by the experimental result that the quenching rate constant in methanol solution for 1-naphthylamine is much larger than that for 2-naphthylamine (Tsutsumi and Shizuka, 1977) in agreement with the above results of the intramolecular charge transfer in the S_1 state (Mataga, 1963).

Analogous proton-induced fluorescence quenching has been observed also in the case of 9,9'-bianthryl in strongly polar solvents (Shizuka et al., 1977). As discussed in section 2.2, the electronic structure of excited 9,9'-bianthryl in strongly polar solvents is very polar due to the solvent-induced intramolecular charge transfer from one anthracene nucleus to another (Schneider and Lippert, 1968; Lippert, 1975; Nakashima et al., 1976).

Therefore, the proton-induced fluorescence quenching may be due to the charge transfer from excited bianthryl to proton.

$$(A^+-A^-)^* + H^+ \rightarrow (A^+-A^-)^* \cdots H^+ \rightarrow (A\dot{-}A)^+ \cdots \dot{H} \rightarrow$$

In this case, not only the fluorescence quenching, but also the proton-induced formation of the bianthryl triplet state was observed (Shizuka et al., 1977). The mechanism of proton-induced triplet formation might be hyperfine coupling in the geminate radical pair as indicated in equation (12.48).

$$^1(A^+-A^-)^* \cdots H^+ \rightarrow {}^1\{(A\dot{-}A)^+ \cdots \dot{H}\}$$
$$\updownarrow$$
$$^3\{(A\dot{-}A)^+ \cdots \dot{H}\} \rightarrow {}^3(A-A)^* + H^+ \quad (12.48)$$

12.7.2 *Hydrogen bonding interaction in the excited electronic state*

We discuss here briefly hydrogen bonding interactions in the excited electronic state, which have a close relation to the excited-state charge transfer phenomena described in previous sections.

It is well known that the hydrogen bonding interaction of some aromatic hydroxy and amine compounds such as naphthols, naphthylamines, and carbazole with various proton acceptors is stronger in the S_1 state than in the S_0 state (Mataga and Kubota, 1970). Namely, the hydrogen bonding energy is larger in S_1 than S_0 and, moreover, donor–acceptor distance, position of the proton in the hydrogen bond, as well as the configuration of the surrounding solvents may be different in S_1 and S_0.

The relaxation process from the Franck–Condon to the equilibrium state involves rearrangements of the hydrogen bond as well as of the configurations of solvents. Thus the circumstance is similar to the case of excited EDA complexes and exciplexes in the previous sections.

For example, some hydrogen-bonded systems, such as 2-naphthol–triethylamine (TEA) in benzene, form ion-pairs due to proton transfer in the S_1 state, from which the fluorescence is emitted (Mataga and Kaifu, 1962, 1963; Mataga *et al.*, 1964; Beens *et al.*, 1965; Matsuzaki *et al.*, 1974).

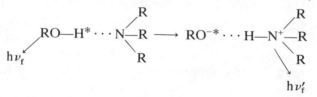

Although the hydrogen bonding shift of the $S_1 \leftarrow S_0$ absorption band of this system is only $370 \, \text{cm}^{-1}$, the corresponding shift of the fluorescence band $(\delta \tilde{\nu}_f = \tilde{\nu}_f - \tilde{\nu}'_f)$ due to ion-pair formation is $4000 \, \text{cm}^{-1}$. The fluorescence band shift becomes larger with increasing solvent polarity due to stabilization of the ion-pair in the equilibrium S_1 state and destabilization in the Franck–Condon S_0 state. For example, it amounts to $4600 \, \text{cm}^{-1}$ and $5000 \, \text{cm}^{-1}$, respectively, in dichloroethane and acetonitrile.

When monobutylamine (MBA) is used as proton acceptor and solvent for 2-napthol, $\delta \tilde{\nu}_f$ amounts to $5700 \, \text{cm}^{-1}$ (Mataga *et al.*, 1964; Beens *et al.*, 1965), although MBA is not so polar. This result was explained as follows. Namely, the proton transfer from excited naphthol to MBA is followed by proton transfer between MBA as indicated in equation (12.49), leading to a more extensive charge separation.

$$
\begin{array}{l}
\text{RO—H}^* \cdots \underset{\substack{| \\ \text{R H}}}{\text{N}}\text{—H} \cdots \underset{\substack{| \\ \text{R H}}}{\text{N}}\text{—H} \cdots \underset{\substack{| \\ \text{R H}}}{\text{N}}\text{—H} \cdots \\[2ex]
\rightarrow \text{RO}^{-*} \cdots \text{H—}\underset{\substack{| \\ \text{R H}}}{\text{N}} \cdots \text{H—}\underset{\substack{| \\ \text{R H}}}{\text{N}^+}\text{—H} \cdots \underset{\substack{| \\ \text{R H}}}{\text{N}}\text{—H} \cdots \quad (12.49)
\end{array}
$$

The stabilization of this large dipole moment in the equilibrium state S_1 by MBA and the large destabilization energy in the Franck–Condon ground state will result in an anomalously large $\delta\tilde{\nu}_f$ value. The proton transfer due to the cooperative process as indicated in equation (12.49) is well known as Grotthus conduction in the case of protolytic reaction in aqueous solution.

In some cases, the fluorescence yield is also affected strongly by hydrogen bonding interaction (Mataga and Kubota, 1970). Fluorescence is quenched almost completely when two conjugate π-electronic systems interact by hydrogen bonding and the hydrogen bond is intimately connected with the conjugate π-electron system as in the cases of naphthol–pyridine, naphthylamine–pyridine (Mataga, 1958), carbazole–pyridine (Mataga et al., 1962), and acridine dye–phenol as well as –naphthol (Mataga and Tsuno, 1957).

It was demonstrated for the carbazole–pyridine system in a rigid state at low temperature that a characteristic of hydrogen bonding quenching of fluorescence is the fact that the rate of the non-radiative deactivation of the S_1 state is extremely enhanced by the hydrogen bonding but this deactivation does not lead to the triplet state (Mataga et al., 1968). That is, a rapid non-radiative process leading to the ground state plays an important role in this quenching process. As a mechanism of this fluorescence quenching, charge transfer interaction between two conjugate π-electronic systems via the hydrogen bond was proposed (Mataga and Tsuno, 1957; Mataga and Kaifu, 1963).

Fluorescence quenching of such systems as naphthol–pyridine and hydroxypyrene–pyridine was examined in acetonitrile solution and it was shown that the quenching rate constants were not very different from the value of the rate constant for diffusion-controlled reaction (Rehm and Weller, 1970). On the other hand, estimation of k_q by means of equation (12.33) for these systems gives values smaller than $10^6 \, M^{-1} \, s^{-1}$, contrary to the experimental results. However, it can be shown that the ΔG^0 value becomes sufficiently negative and the k_q value becomes diffusion controlled if one assumes hydrogen atom transfer of the type (Rehm and Weller, 1970)

$$\text{ROH}^* \cdots \text{N} \hspace{-0.2em} \bigcirc\!\!\!\!\rangle \longrightarrow \text{R}\dot{\text{O}} \cdots \text{H--N} \hspace{-0.2em} \stackrel{\bullet}{\bigcirc}\!\!\!\!\rangle$$

Detailed microsecond flash photolysis studies have been made for some proton donor–acceptor systems such as 2-naphthol (2-N) as well as 1-anthrol (1-A) (proton donors) and several nitrogen heterocycles (proton acceptors) (Kikuchi et al., 1973; Yamamoto et al., 1976a, b, c, 1977). According to these studies, deactivation of triplet 2-N and 1-A by pyridine and/or quinoline occurs efficiently by diffusion-controlled rate and it has been proved that this deactivation is due to the hydrogen atom transfer.

Contrary to this, although radicals, are formed to some extent due to hydrogen atom transfer, during encounter collision in the S_1 state, no radical formation from the excited singlet hydrogen-bonded complex was recognized. The reaction scheme for these systems might be summarized as indicated in equation (12.50).

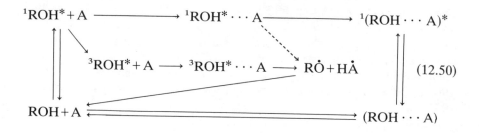

$$(12.50)$$

where $^1(ROH \cdots A)^*$ is the hydrogen-bonded complex in the S_1 state and $^1ROH^* \cdots A$ is a loose encounter complex in the S_1 state.

Therefore, the mechanism of the very efficient fluorescence quenching in hydrogen-bonded complexes is still unclear. Moreover, the hydrogen bonding quenching is not limited to the aromatic hydroxy compound–nitrogen heterocycle pairs but there are many cases of aromatic amine–nitrogen heterocycle pairs. More detailed studies including picosecond laser photolysis may be necessary for the elucidation of the quenching mechanism.

12.8 References

Arimitsu, S., Masuhara, H., Mataga, N., and Tsubomura, H. (1975). *J. Phys. Chem.*, **79**, 1255.

Bakhshiev, N. G. (1972). *Spektroskopiia Mezhmolekulyarnui Bzaimodeistbii (Spectroscopy of Intermolecular Interactions)*. Akad. Nauk. USSR.

Bakhshiev, N. G., Mazurenko, Y. T., and Piterskaya, I. V. (1966). *Opt. Spectrosc.*, **21**, 307.

Beens, H., Grellmann, K. H., Gurr, M., and Weller, A. H. (1965). *Discuss. Faraday Soc.*, **39**, 183.

Beens, H., and Weller, A. (1969). *Chem. Phys. Lett.*, **3**, 666.

Beens, H., and Weller, A. (1975). *Organic Molecular Photophysics* (Ed. Birks J. B.), Vol. 2, Wiley–Interscience, p. 159.

Birks, J. B. (1970). *Photophysics of Aromatic Molecules* Wiley–Interscience.

Birks, J. B. (1975a). *The Exciplex* (Eds Gordon, M. and Ware, W. R.), Academic Press, p. 39.

Birks, J. B. (1975b). *Rep. Progr. Phys.*, **38**, 903.

Bonneau, R., and Joussot-Dubien, J. (1967). *Z. Phys. Chem. (NF)*, **101**, 225.

Brocklehurst, B. (1974). *Chem. Phys. Lett.*, **28**, 357.

Chandross, E. A., and Dempster, C. J. (1970). *J. Amer. Chem. Soc.*, **92**, 3586.

Chandross, E. A., and Thomas, H. T. (1971). *Chem. Phys. Lett.*, **6**, 393.

Chuang, T. J., and Eisenthal, K. B. (1975). *J. Chem. Phys.*, **62**, 2213.

Craig, B. B., and Rodgers, M. A. J. (1976). *J. Chem. Soc. Faraday II*, 72, 1259.

Davidson, R. S. (1975). *Molecular Association* (Ed. Foster, R.), Vol. 1, Academic Press, p. 216.

DeToma, R. P., and Brand, L. (1977). *Chem. Phys. Lett.*, 47, 231.

DeToma, R. P., Easter, J. H., and Brand, L. (1976). *J. Amer. Chem. Soc.*, 98, 5001.

Egawa, K., Nakashima, N., Mataga, N., and Yamanaka, C. (1971). *Chem. Phys. Lett.*, 8, 108; *Bull. Chem. Soc. Japan*, 44, 3287.

Förster, Th. (1965). *Modern Quantum Chemistry* (Ed. Sinanoğlu, O.) Part III, Academic Press, p. 93.

Förster, Th. (1969). *Angew. Chem. Intern. Ed. Engl.*, 8, 333.

Förster, Th. (1972). *Chem. Phys. Lett.*, 17, 309.

Förster, Th. (1975). *The Exciplex* (Eds. Gordon, M. and Ware, W. R.), Academic Press, p. 1.

Fujiwara, H., Nakashima, N., and Mataga, N. (1977). *Chem. Phys. Lett.*, 47, 185.

Goodall, D. M., Orbach, N., and Ottolenghi, M. (1974). *Chem. Phys. Lett.*, 26, 365.

Grätzel, M., and Thomas, J. K. (1976). *Modern Fluorescence Spectroscopy* (Ed. Wehry, E. L.), Vol. 2, Plenum Press, p. 169.

Gronau, B., Lippert, E., and Rapp, W. (1972). *Ber. Bunsenges. Phys. Chem.*, 76, 432.

Hallidy, L. A., and Topp, M. R. (1977). *Chem. Phys. Lett.*, 48, 40.

Hayashi, H., Iwata, S., and Nagakura, S. (1969). *J. Chem. Phys.*, 50, 993.

Hayashi, T., Mataga, N., Sakata, Y., Misumi, S., Morita, M., and Tanaka, J. (1976). *J. Amer. Chem. Soc.*, 98, 5910.

Hayashi, T., Mataga, N., Umemoto, T., Sakata, Y., and Misumi, S. (1977a). *J. Phys. Chem.*, 81, 424.

Hayashi, T., Suzuki, T., Mataga, N., Sakata, Y., and Misumi, S. (1977b). *J. Phys. Chem.*, 81, 420.

Henglein, A., and Scheerer, R. (1978). *Ber. Bunsenges. Phys. Chem.*, 82, 1107.

Hinatsu, J., Masuhara, H., Mataga, N., Sakata, Y., and Misumi, S. (1978). *Bull. Chem. Soc. Japan*, 51, 1032.

Hino, T., Akazawa, H., Masuhara, H., and Mataga, N. (1976a). *J. Phys. Chem.*, 80, 33.

Hino, T., Masuhara, H., and Mataga, N. (1976b). *Bull. Chem. Soc. Japan*, 49, 394.

Hirayama, F. (1965). *J. Chem. Phys.*, 42, 3163.

Holten, D., Gouterman, M., Parson, W. W., Windsor, M. W., and Rockley, M. G. (1976). *Photochem. Photobiol.*, 23, 415.

Hopfield, J. J. (1974). *Proc. Natl. Acad. Sci. U.S.*, 71, 3640.

Hopfield, J. J. (1977). *Biophys. J.*, 18, 311.

Hui, M.-H., and Ware, W. R. (1976). *J. Amer. Chem. Soc.*, 98, 4718.

Huppert, D., Rentzepis, P. M., and Tollin, G. (1976). *Biochim. Biophys. Acta*, 440, 356.

Infelta, P. P. (1979). *Chem. Phys. Lett.*, 61, 88.

Irie, M., Masuhara, H., Hayashi, K., and Mataga, N. (1974). *J. Phys. Chem.*, 78, 341.

Iwata, S., Tanaka, J., and Nagakura, S. (1966). *J. Amer. Chem. Soc.*, 88, 894.

Jortner, J., (1976). *J. Chem. Phys.*, 64, 4860

Katušin-Ražem, B., Wong, M., and Thomas, J. K. (1978). *J. Am. Chem. Soc.*, 100, 1679.

Kaufmann, K. J., Dutton, P. L., Netzel, T. L., Leigh, J. S., and Rentzepis, P. M. (1975). *Science*, 188, 1301.

Kikuchi, K., Tamura, S.-I., Iwanaga, C., Kokubun, H., and Usui, Y. (1977). *Z. Phys. Chem. N. F.*, 106, 17.

Kikuchi, K., Watarai, H., and Koizumi, M. (1973). *Bull. Chem. Soc. Japan*, 46, 749.

Klöpffer, W. (1973). *Organic Molecular Photophysics* (Ed. Birks, J. B.), Vol. 1, Wiley–Interscience, p. 357.

Knibbe, H., Röllig, K., Schäfer, F. P., and Weller, A. (1967). *J. Chem. Phys.*, **47**, 1184.

Kobayashi, T., Yoshihara, K., and Nagakura, S. (1971). *Bull. Chem. Soc. Japan*, **44**, 2603.

Lippert, E. (1955). *Z. Naturforsch.*, **10a**, 541.

Lippert, E. (1957). *Ber. Bunsenges. Phys. Chem.*, **61**, 962.

Lippert, E. (1975). *Organic Molecular Photophysics* (Ed. Birks, J. B.), Vol. 2, Wiley–Interscience, p. 1.

Lippert, E., Lüder, W., and Boos, H. (1962). *Advances in Molecular Spectroscopy*, Pergamon Press, p. 443.

Maestri, M., Infelta, P. P., and Grätzel, M. (1978). *J. Chem. Phys.*, **69**, 1522.

Marcus, R. A. (1956). *J. Chem. Phys.*, **24**, 966.

Masaki, S., Okada, T., Mataga, N., Sakata, Y., and Misumi, S. (1976). *Bull. Chem. Soc. Japan*, **44**, 1277.

Masuhara, H., Hino, T., and Mataga, N. (1975). *J. Phys. Chem.*, **79**, 994.

Masuhara, H., Kaji, K., and Mataga, N. (1977a). *Bull. Chem. Soc. Japan*, **50**, 2084.

Masuhara, H., and Mataga, N. (1970). *Chem. Phys. Lett.*, **6**, 608.

Masuhara, H., and Mataga, N. (1972). *Z. Phys. Chem.* (*NF*), **80**, 113.

Masuhara, H., and Mataga, N. (1973). *Chem. Phys. Lett.*, **22**, 305.

Masuhara, H., Mataga, N., Yoshida, M., Tatemitsu, H., Sakata, Y., and Misumi, S. (1977b). *J. Phys. Chem.*, **80**, 879.

Masuhara, H., Shimada, M., Tsujino, N., and Mataga, N. (1971). *Bull. Chem. Soc. Japan*, **44**, 3310.

Masuhara, H., Tsujino, N., and Mataga, N. (1972). *Chem. Phys. Lett.*, **15**, 357.

Masuhara, H., Tsujino, N., and Mataga, N. (1973). *Bull. Chem. Soc. Japan*, **46**, 1088.

Mataga, N. (1958). *Bull. Chem. Soc. Japan* **31**, 481.

Mataga, N. (1963). *Bull. Chem. Soc. Japan*, **36**, 654.

Mataga, N. (1975). *The Exciplex* (Eds. Gordon, M. and Ware, W. R.), Academic Press, p. 113.

Mataga, N., and Ezumi, K. (1967). *Bull. Chem. Soc. Japan*, **40**, 1355.

Mataga, N., and Kaifu, Y. (1962). *J. Chem. Phys.*, **36**, 2804.

Mataga, N., and Kaifu, Y. (1963). *Mol. Phys.*, **7**, 137.

Mataga, N., Kaifu, Y., and Koizumi, M. (1955). *Bull. Chem. Soc. Japan*, **28**, 690.

Mataga, N., Kaifu, Y., and Koizumi, M. (1956). *Bull. Chem. Soc. Japan*, **29**, 465.

Mataga, N., Kawasaki, Y., and Torihashi, Y. (1964). *Acta Phys. Pol.*, **26**, 477.

Mataga, N., and Kubota, T. (1970). *Molecular Interactions and Electronic Spectra*, Marcel-Dekker.

Mataga, N., Migita, M., and Nishimura, T. (1978). *J. Mol. Struct.*, **47**, 199.

Mataga, N., and Murata, Y. (1969). *J. Amer. Chem. Soc.*, **91**, 3144.

Mataga, N., Okada, T., Masuhara, H., Nakashima, N., Sakata, Y., and Misumi, S. (1976). *J. Luminescence*, **12/13**, 159.

Mataga, N., Okada, T., and Oohari, H. (1966). *Bull. Chem. Soc. Japan*, **39**, 2563.

Mataga, N., Okada, T., and Yamamoto, N. (1967). *Chem. Phys. Lett.*, **1**, 119.

Mataga, N., and Ottolenghi, M. (1979). *Molecular Association* (Ed. Foster, R.), Vol. 2, Academic Press.

Mataga, N., Tanaka, F., and Kato, M. (1968). *Acta Phys. Pol.*, **34**, 733.

Mataga, N., and Tanimoto, O. (1969). *Theor. Chim. Acta*, **15**, 111.

Mataga, N., Torihashi, Y., and Kaifu, Y. (1962). *Z. Phys. Chem.* (*NF*), **34**, 379.

Mataga, N., and Tsuno, S. (1957). *Bull. Chem. Soc. Japan*, **30**, 711.

Matsumoto, S., Nagakura, S., Iwata, S., and Nakamura, J. (1972). *Chem. Phys. Lett.*, **13**, 463; (1973). *Mol. Phys.*, **26**, 1465.

Matsuzaki, A., Yoshihara, K., and Nagakura, S. (1974). *Bull. Chem. Soc. Japan* **47**, 1152.

Michel-Beyerle, M. E., Haberkorn, R., Bube, W., Steffens, E., Schröder, H., Neusser, H. J., Schlag, E. W., and Seidlitz, H. (1976). *Chem. Phys.*, **17**, 139.

Migita, M., Kawai, M., Mataga, N., Sakata, Y., and Misumi, S. (1978). *Chem. Phys. Lett.*, **53**, 67.

Mozumder, A. (1968). *J. Chem. Phys.*, **48**, 1659; (1969). *J. Chem. Phys.*, **50**, 3153, 3162.

Nagakura, S. (1975). *Excited States* (Ed. Lim, E. C.), Vol. 2, Academic Press, p. 321.

Nakashima, N., Inoue, H., Mataga, N., and Yamanaka, C. (1973a). *Bull. Chem. Soc. Japan*, **46**, 2288.

Nakashima, N., Mataga, N., Yamanaka, C., Ide, R., and Misumi, S. (1973b). *Chem. Phys. Lett.*, **18**, 386.

Nakashima, N., Mataga, N., and Yamanaka, C. (1973c). *Int. J. Chem. Kinetics*, **5**, 833.

Nakashima, N., Murakawa, M., and Mataga, N. (1976). *Bull. Chem. Soc. Japan* **49**, 854.

Nakashima, N., Ushio, F., Mataga, N., and Yamanaka, C. (1972). *Z. Phys. Chem.* (*NF*), **79**, 150.

Nishimura, T., Nakashima, N., and Mataga, N. (1977). *Chem. Phys. Lett.*, **46**, 334.

Noyes, R. M. (1961). *Progr. Reaction Kinetics* (Ed. Porter, G.), Vol. 7, Pergamon Press, p. 129.

O'Connor, D. V., and Ware, W. R. (1976). *J. Amer. Chem. Soc.*, **98**, 4706.

Okada, T., and Mataga, N. (1976). *Bull. Chem. Soc. Japan*, **49**, 2190.

Okada, T., Mori, T., and Mataga, N. (1976). *Bull. Chem. Soc. Japan*, **49**, 3398.

Okada, T., Saito, T., Mataga, N., Sakata, Y., and Misumi, S. (1977). *Bull. Chem. Soc. Japan*, **50**, 331.

Orbach, N., and Ottolenghi, M. (1975a). *Chem. Phys. Lett.*, **35**, 175.

Orbach, N., and Ottolenghi, M. (1975b). *The Exciplex* (Eds. Gordon, M. and Ware, W. R.), Academic Press, p. 75.

Ottolenghi, M. (1973). *Acc. Chem. Res.*, **6**, 153.

Potashnik, R., and Ottolenghi, M. (1970). *Chem. Phys. Lett.*, **6**, 525.

Prochorow, J. (1973). *Chem. Phys. Lett.*, **19**, 596.

Prochorow, J., and Bernard, E. (1974). *J. Luminescence*, **8**, 471.

Prochorow, J., and Siegoczyński, R. (1969). *Chem. Phys. Lett.*, **3**, 635.

Rapp, W., Klingenberg, H.-H., and Lessing, H. E. (1971). *Ber. Bunsenges. Phys. Chem.*, **75**, 883.

Rehm, D., and Weller, A. (1969). *Ber. Bunsenges. Phys. Chem.*, **73**, 834.

Rehm, D., and Weller, A. (1970). *Israel J. Chem.*, **8**, 259.

Rockley, M. G., Windsor, M. W., Cogdell, R. J., and Parson, W. W. (1975). *Proc. Natl. Acad. Sci. U.S.*, **72**, 2251.

Rodgers, M. A. J., Wheeler, M. F., and da Silva E. (1978). *Chem. Phys. Lett.*, **53**, 165.

Rotkiewicz, K., Grabowski, Z. R., and Jasny, J. (1975). *Chem. Phys. Lett.*, **34**, 55.

Rotkiewicz, K., Grabowski, Z. R., Krówczyński, A., and Kühnle, W. (1976). *J. Luminescence*, **12/13**, 377.

Scheerer, R., and Grätzel, M. (1977). *J. Am. Chem. Soc.*, **99**, 865.

Schneider, F., and Lippert, E. (1968). *Ber. Bunsenges. Phys. Chem.*, **12**, 1155.

Schulman, S. G., and Liedke, P. (1973). *Z. Phys. Chem. (NF)*, **84,** 317.

Schulten, K., Staerk, H., Weller, A., Werner, H.-J., and Nickel, B. (1976). *Z. Phys. Chem. (NF)*, **101,** 371.

Shimada, M., Masuhara, H., and Mataga, N. (1973). *Bull. Chem. Soc. Japan*, **46,** 1903.

Shizuka, H., Ishii, Y., and Morita, T. (1977). *Chem. Phys. Lett.*, **51,** 40.

Struve, W. S., and Rentzepis, P. M. (1974a). *J. Chem. Phys.*, **60,** 1533.

Struve, W. S., and Rentzepis, P. M. (1974b). *Chem. Phys. Lett.*, **29,** 23.

Taniguchi, Y., and Mataga, N. (1972). *Chem. Phys. Lett.*, **13,** 596.

Taniguchi, Y., Nishina, Y., and Mataga, N. (1972). *Bull. Chem. Soc. Japan*, **45,** 764.

Thomas, J. K. (1977). *Acc. Chem. Res.*, **10,** 133.

Tsujino, N., Masuhara, H., and Mataga, N. (1973). *Chem. Phys. Lett.*, **21,** 301.

Tsutsumi, K., and Shizuka, H. (1977). *Chem. Phys. Lett.*, **52,** 485.

Vogelmann, E., and Kramer, H. E. A. (1976). *Photochem. Photobiol.*, **24,** 595.

Vogelmann, E., Schreiner, S., Rauschen, W., and Kramer, H. E. A. (1976). *Z. Phys. Chem. (NF)*, **101,** 321.

Waka, Y., Hamamoto, K., and Mataga, N. (1978). *Chem. Phys. Lett.*, **53,** 242.

Waka, Y., Hamamoto, K., and Mataga, N. (1979). *Chem. Phys. Lett.*, **62,** 364.

Ware, W. R., Holmes, J. D., and Arnold, D. R. (1974b). *J. Amer. Chem. Soc.*, **96,** 7861.

Ware, W. R., Lee, S. K., Brant, G. J., and Chow, P. P. (1971). *J. Chem. Phys.*, **54,** 4729.

Ware, W. R., and Richter, H. P. (1968). *J. Chem. Phys.*, **48,** 1595.

Ware, W. R., Watt, D., and Holmes, J. D. (1974a). *J. Amer. Chem. Soc.*, **96,** 7853.

Wolff, C., and Grätzel, M. (1977). *Chem. Phys. Lett.*, **52,** 542.

Yamamoto, S., Kikuchi, K., and Kokubun, H. (1976a). *J. Photochem.*, **5,** 469.

Yamamoto, S., Kikuchi, K., and Kokubun, H. (1976b). *Chem. Lett.*, 65.

Yamamoto, S., Kikuchi, K., and Kokubun, H. (1976c). *Bull. Chem. Soc. Japan*, **49,** 2950.

Yamamoto, S., Kikuchi, K., and Kokubun, H. (1977). *J. Photochem.*, **7,** 177.

Yomosa, S. (1973). *J. Phys. Soc. Japan*, **35,** 1501.

Yoshida, M., Tatemitsu, H., Sakata, Y., Misumi, S., Masuhara, H., and Mataga, N. (1976). *J. Chem. Soc. Chem. Commun.*, 587.

Yoshihara, K., Kasuya, T., Inoue, A., and Nagakura, S. (1971). *Chem. Phys. Lett.*, **9,** 469.

Zachariasse, K., and Kühnle, W. (1976). *Z. Phys. Chem. (NF)*, **101,** 267.

Molecular Interactions, Volume 2
Edited by H. Ratajczak and W. J. Orville-Thomas
© 1981 John Wiley & Sons Ltd

Author index

Aano, S., 447, 450
Abe, Y., 431, 436
Accascina, F., 82, 100
Adam, W., 13, 39, 97
Adams, J. M., 213, 223
Adman, E., 217, 223
Aggarwal, S. L., 275, 303
Ainbinder, N. E., 365, 367, 421, 427, 433
Aitken, H. W., 93, 97, 193, 223
Akamatu, H., 35, 103, 452, 464–6, 468, 470–1, 477, 488, 490–2
Akazawa, H., 554, 567
Alagona, G., 205, 223, 227
Alden, R. A., 208–9, 213–4, 223–4, 227
Alexander, C. J., 73, 97
Alexandre, M., 456, 488
Alkatis, S. A., 148, 174
Allen, L., 47, 49, 99, 102
Allen, L. C., 137, 140, 179, 185, 223, 226, 478, 490
Almgren, M., 152, 163, 167–8, 172–4
Almlof, J., 9, 97
Al-Obeidi, F. A., 286, 288, 301
Alymov, I. M., 380, 405, 424, 430, 432, 434
Amano, T., 35, 103, 477, 490
Amick, R. M., 291, 302
Anderson, G. R., 28, 104
Andrade, E. N. da C., 322, 340
Andreeva, A. I., 369–70, 376, 383, 401, 403–5, 407, 409–10, 412, 414, 417, 419, 421–4, 426, 428–9, 433
Andrews, L. J., 35, 102
Anferov, V. P., 365–7, 377–8, 384, 388, 398, 400, 402–4, 408, 414, 416, 418–22, 424, 431, 433, 438
Aniansson, E. A. G., 152, 154, 163, 165, 167–8, 172–4
Anthony, A. A., 332, 340

Ardjomande, S., 356, 360–1, 373, 375–6, 380, 398–400, 402–4, 407–8, 410–1, 414–5, 417–9, 422–6, 429
Arend, H., 247, 269
Argay, G., 186, 226
Arisman, R. K., 205, 223
Armitsu, S., 562, 566
Armstrong, D. R., 507–8
Arnett, E. M., 183, 226
Arnold, D. R., 532, 570
Arnone, A., 213–4, 223
Arshadi, M., 184, 223
Arvidsson, E. D., 211, 223
Ash, R. P., 282, 301
Ashmore, J. P., 216, 226
Askar, A., 476, 492
Asselin, M., 14, 97
Atkinson, G., 112–3, 139
Attig, R., 9, 97
Attwood, D., 174
Aue, D. H., 73, 97
Ault, B. S., 47–8, 97
Azizov, E. O., 416, 423, 433
Azzaro, M., 196, 224

Baba, H., 26, 28, 51, 69, 97, 104, 123, 139
Babushkina, T. A., 344, 351, 357, 388, 403–6, 408, 411, 413–5, 418–23, 427, 429, 433, 436, 438, 498
Bacelon, P., 18, 99
Bader, F., 113, 139
Baer, Y., 493, 508
Baker, A. D., 493, 507–8
Baker, C., 493, 507–8
Bakhshiev, N. G., 522, 566
Baldwin, J. M., 216, 226
Balkanski, M., 256, 269
Balmbra, R., 147, 175

Bambagiotti, M., 14, 20, 104
Bando, M., 452, 488
Bandrauk, A. D., 452, 492
Banerjee, S. K., 212, 223
Bansal, V. M., 148, 175
Baranov, A. I., 248, 269
Baraton, M. I., 31, 91, 97
Barber, M., 501, 508
Barigand, M., 293, 302
Barraque, C., 95, 97
Barrow, G. M., 17, 40, 50, 52, 97, 122, 139
Barrow, M. J., 185, 223
Barthel, J., 80, 97
Basanov, A. G., 127, 141
Basilier, E., 496, 508
Batrakova, L. P., 52, 103
Bauche, F. G. K., 95, 97
Baumuller, W., 168, 175
Baur, W. H., 180, 223
Baxter, R. J., 249, 269
Bayles, J. W., 27, 41, 97, 104
Bechgaard, K., 470, 488
Becker, J., 250, 270
Becker, J. W., 209, 227
Bednarek, J. M., 205, 223
Beens, H., 510, 514, 520, 564, 566
Beier, G., 28, 48, 69, 76, 98, 105
Beilis, Y. I., 37, 103
Belanger, G., 14, 97
Belin, C., 18, 58, 98
Bell, C. L., 27, 40, 50, 52, 86, 98
Bell, J. D., 460, 488
Belostotskaya, I. S., 130, 141
Belozerskaya, L. P., 134, 139
Bender, C. F., 9, 102
Benedetti, E., 206, 223
Benesi, H. A., 447, 450
Bennett, R. A., 388, 402–3, 405–6, 410, 412, 415, 417, 419, 421, 433
Bennion, B. C., 151, 175
Berendsen, H. J. C., 195, 217, 223
Berger, A., 37, 98
Bergman, K., 112–3, 139
Bergmann, E. D., 297, 302
Bergmann, K., 140, 315
Bergmark, T., 493, 508
Bergsten, P. C., 209, 226
Bernander, L., 37, 98
Bernard, E., 545, 569
Bernasconi, C. F., 143, 148–51, 175

Bernstein, H. J., 36, 100, 105
Berry, R. S., 488–9
Bertie, J. E., 341
Bertram, R., 95, 97
Besnaiou, S., 91, 97
Beveridge, D. L., 481, 488, 494, 508
Bhat, S. N., 281, 293–5, 302, 465, 488
Bhat, T. N., 188–9, 223
Bhatnaga, S. S., 275, 302
Bhowmik, S., 93, 103
Biedenkapp, D., 365–6, 369, 377–8, 398, 408–9, 414–5, 419–21, 423, 426–7, 433
Bier, C. J., 213–4, 223
Birchall, T., 37, 98, 100
Birks, J. B., 510, 566
Birktoft, J. J., 208–9, 212–4, 216, 218, 220–1, 223, 227
Birr, E. J., 79, 106
Biryukov, I. P., 344, 347, 433
Bjorvatten, T., 11, 98
Blankenhorn, G., 293, 302
Bledsoe, W., 446, 448, 450
Blinc, R., 231, 233, 237, 240, 247, 252, 254–7, 259–60, 269
Blinck, R., 65, 98
Bloch, A. N., 499, 508
Bloor, J. E., 293–4, 302
Blow, D. M., 208–9, 212–3, 216, 218, 220–1, 223, 225
Blum, H., 474, 490
Bode, W., 207–8, 212, 214–5, 218–21, 223–4
Boeyens, J. C. A., 452, 488
Bogachev, Y. S., 125, 141
Bohlmann, F., 76, 99
Bonneau, R., 554, 566
Bonner, O. D., 205, 223
Bonnet, B., 193, 225
Bonsor, D., 55, 98
Bonsor, D. H., 192, 194, 224
Boos, H., 518, 568
Borah, B., 55–6, 98, 105, 192, 194, 224
Bordewijk, P., 335, 340
Bordua, C. L. Jr., 213, 228
Borgen, B., 11, 98
Borucki, L., 113, 139
Boschov, P., 183, 227
Botchkariova, M. N., 135, 140
Bothnerby, A. A., 212, 225
Bottcher, C. J. F., 308, 312–3, 333, 340

Bowers, M. T., 73, 97
Bowmaker, G. A., 352–4, 375, 384–6, 398, 400, 406–7, 410–2, 415–6, 420–1, 423–4, 430–1, 433
Boyd, D. B., 216, 224
Bradley, J. N., 143, 148, 175
Bragin, J., 115, 120, 140
Brand, L., 523–4, 567
Brant, G. J., 522, 570
Bratos, S., 123, 140
Bratoz, S., 14, 16, 98, 101, 478, 488
Brau, A., 294, 302, 470, 489
Bray, P. J., 344, 400, 404, 430, 436, 438
Bregadze, V. I., 376, 379, 381, 405–10, 417, 420–1, 426, 436–7
Breneman, G. L., 386, 400, 408, 433
Brezinsky, B., 54, 98
Brickenkampf, C. S., 269
Briegleb, G., 14, 98, 274, 302, 453, 488
Bright, H. J., 291, 293, 303
Brill, T. B., 349, 352, 355–8, 378, 400–1, 403, 405, 408, 411–2, 414–6, 419, 422, 424–7, 430, 431, 432, 434, 436
Brocklehurst, B., 553, 566
Brockliss, L. D., 352–3, 400, 423–4, 430, 433
Brodsky, A. I., 110, 136, 139
Bron, J., 28, 41, 42, 98
Brooker, H. R., 347, 434
Brot, C., 339–41
Brout, R. K., 231, 233, 255, 257, 260, 269
Brown, I. D., 470, 488
Brown, J. D., 183, 224
Brown, J. R., 287–8, 303
Brown, T. L., 374, 434
Bruckenstein, S., 16, 93–4, 98, 103
Bruggemann, R., 385, 412, 415–6, 422, 431, 434
Bruice, T. C., 290, 302
Brun, N., 205, 226
Brundle, C. R., 493, 507–8
Brunette, J. P., 382, 406, 418, 426–7, 429, 434
Bryukhova, E. G., 362, 396–7, 405, 409, 411, 420
Bryukhova, E. V., 356–8, 376, 379–81, 383, 386, 401–10, 414, 417, 421, 424–6, 430–2, 434, 436–7
Brzezinski, B., 192, 194, 224

Bube, W., 547, 554, 561, 569
Bubnov, N. N., 126, 130, 141
Buchfelder, J., 44, 106
Budo, A., 321, 340
Bugg, C. E., 273, 303
Bukshpan, S., 446–50
Bukvetsky, B. V., 184, 228
Bunze, K., 30, 75–6, 105
Bunzl, K., 331, 342
Burbelo, V. M., 386, 414, 425, 437
Bureiko, S. F., 125, 130, 133–9, 141
Burgard, M., 373, 379–80, 382, 402, 406, 418, 426–9, 434
Burger, L. L., 38, 102
Burmeister, J. L., 374, 434
Bystrow, D. S., 13, 100

Cabana, A., 14, 98
Cahay, R., 365–6, 384, 400, 405, 408–9, 414, 434, 437
Caldin, E., 122, 139
Caldin, E. F., 41, 98, 125, 139, 143, 148, 175
Calende, M. T., 347, 434
Camerman, A., 206, 224
Camerman, N., 206, 224
Campbell, J. W., 208–11, 218–21, 228
Cannon, C. G., 12, 96
Cardinal, J. R., 153, 174, 176
Carlbom, V., 209, 226
Carnevale, A., 352, 355–6, 363, 376, 378, 400–1, 406–8, 416, 419, 425, 431, 437
Carnochan, P., 465, 467, 488
Carsaro, R. D., 112–3, 139
Carter, C. W. Jr., 208–9, 224
Carter, J. C., 360, 363, 376, 378, 382, 397–9, 405–7, 409–10, 417, 426, 434
Carter, J. M., 314, 341
Cartwright, C. H., 306, 337, 340
Casabella, P. A., 362, 434
Casey, G. J., 44, 106
Castellano, E., 455, 488
Castro, G., 471–2, 491
Chakrabarty, M. R., 38, 98
Chamberlain, J. E., 338, 340
Chance, B., 143, 175
Chandros, E. A., 544, 566
Chang, C. Y., 205, 225

Chang, K. C., 29, 103
Chantooni, M. K., 29, 72, 86, 93–4, 98, 103
Chantry, G. W., 335–8, 340
Chao, G. Y., 11, 98
Chassy, B. M., 291, 302
Chelkowski, A., 330–1, 340–1
Chen, C. S., 186–8, 224
Chen, J., 287–8, 302
Cheng, C. P., 349, 355, 374–5, 400–1, 405, 411–2, 424, 426, 430–2, 434, 436
Chenon, B., 14, 98
Chetwyn, A., 27, 97
Cheung, A. S., 285, 290, 302
Chihara, H., 344, 347, 413, 416, 420–4, 434
Childs, J. D., 286, 288, 302
Chitoku, K., 332, 340
Choi, C. S., 9, 98
Choi, S., 483, 489
Chojnacki, H., 476, 480–3, 486–8
Chothia, C., 212, 225
Chow, P. P., 522, 570
Christian, S. D., 174–5, 286, 288, 302, 304
Cizikov, S., 264, 269
Claesson, S., 143, 148, 175
Clague, A. D. H., 36, 100
Clarke, R. P., 309–10, 324, 340
Claydon, M. F., 15, 98
Clementi, E., 47, 98
Clements, R., 24–5, 55–6, 98, 192, 224
Clerbaux, T., 196, 224
Cleuren, W., 31–3, 69, 73, 98, 101
Clifford, J., 147, 175
Clotman, D., 43, 98
Clunie, J., 147, 175
Coates, J. E., 240, 329
Codding, P. W., 214, 224
Coetzee, J. F., 79, 86, 90, 93–4, 98, 194, 224
Cogdell, R. J., 551, 569
Cohen, J. S., 212, 228
Cole, F. C., 186, 224
Cole, K., 315, 340
Cole, R. H., 208, 315–7, 330, 336, 339–40
Colen, A., 152, 175
Collaer, H., 86, 96, 98
Colman, P. M., 212, 224

Combelas, P., 18, 99
Conner, W. P., 209–10, 324, 340
Connor, J. A., 501, 508
Connors, K. A., 290, 302
Conor, T. M., 36, 99
Constant, E., 338, 340–1
Cook, D., 18–9, 99
Cooke, R., 217, 224
Coop, I. E., 337, 340
Cooper, J. R., 470, 488
Coppens, P., 9, 99, 188, 225, 228, 456, 490
Corbett, J. D., 386, 398, 424, 436
Corkill, J., 147, 175
Cornil, P., 365–6, 384, 400, 405, 408–9, 414, 434
Cornill, P., 400, 414, 437
Cornut, J. C., 19, 103
Cornwell, C. D., 385–6, 424–5, 430–1, 434, 438
Corset, J., 55, 102
Cosani, A., 282, 284, 303
Cotton, F. A., 115, 140, 188, 213–4, 223–4
Coulson, C. A., 478, 488
Courcheene, W. L., 175
Cousseau, J., 344, 387, 417, 430, 434
Cowan, D. O., 499, 508
Cox, B. C., 37, 99
Craig, B. B., 545, 566
Craven, B. M., 204, 225, 460, 488
Creel, J. B., 347, 434
Cronkright, W., 41–2, 106
Crooks, J. E., 27, 41, 69, 99, 125–6, 139
Crossley, J., 332, 334, 340–1
Crowthers, D., 288, 303
Cruege, F., 18, 99
Cruse, K., 95, 97
Cummings, D. L., 24, 99
Cunningham, G., 79, 86, 93–4, 98
Curie, M., 60, 99, 103
Currie, M., 185–6, 223–4
Curtis, A. J., 325, 341
Czerlinski, G., 143, 148–9, 151
Czismadia, I. G., 19, 101

Dadok, J., 212, 225
Dahl, T., 10, 99
Dailey, B. P., 345, 349, 434, 438
Damen, T. C., 256, 269

Damm, E., 10, 99
Darville, J., 347, 434
Das, A. R., 28, 104
Das, T. P., 344, 388, 434
Dasgupta, S., 337, 340
Da Silva, E., 556, 569
Davenport, G., 151–2, 176
Davidson D. W., 316, 340
Davidson, N., 100
Davidson, R. S., 510, 559, 562, 567
Davies, G. J., 339–40
Davies, M., 95–9, 201, 224, 330, 338–40
Davis, M. M., 25–28, 93, 99, 122, 139, 195, 224
Day, V. W., 188, 213–4, 223–4
Dean, R. L., 25, 55–6, 98–9, 192, 194, 224
Debye, P., 306, 316, 320–1, 330, 338, 340
De Carvalho, A. V., 244–6, 255, 269
Deeg, T., 353, 364, 397–8, 417–9, 422–3, 426, 434
Deenadayalan, K. C., 41, 104
Dega–Szafran, Z., 197, 227
De Gennes, P. G., 231, 233, 255, 257, 260, 269
Dehmelt, H. G., 343–5, 362, 369, 397–8, 403, 409, 414, 428, 431, 434, 437
Deisenhofer, J., 208–9, 224
De Jeu, W. H., 39, 99
Delbaere, L. T. J., 187, 214, 224–5
Del Bene, J. E., 14, 19, 99
Delcoigne, V., 86, 93, 99
Delle, H., 14, 98, 453, 488
De Lozé, C., 18, 99
Delplanque, G., 470, 488
De Maeyer, L., 143, 148, 175, 474, 489
Demoss, R. D., 283, 302
Dempster, C. J., 544, 566
Denison, J., 88, 99
Denisov, G. S., 16–7, 43, 53, 69, 93, 99–101, 108, 110, 114, 116–7, 119–21, 123–5, 127, 130, 133–41
Denney, D. J., 335, 340
De Ortiz, C. Z., 13, 39, 97
Depalma, D., 28, 50, 105, 199, 228
Deranleau, D. A., 282, 287–8, 301–2
Derissen, J. L., 185–7, 224
Desplanques, P., 338, 340–1
Dessen, P., 174, 177

Detitta, G. T., 188, 205, 224, 228
De Toma, R. P., 523–4, 567
Detoni, S., 15, 99
Deutch, J. M., 339–40
De Villepin, J., 57, 59, 99
De Waard, H., 449–50
Dewar, M. J. S., 272, 275, 302, 379, 382, 396, 404–10, 414, 416–7, 420, 422, 434, 477, 488
Dhondt, J., 23, 26, 28, 43, 99–100
Diamond, R., 208–11, 218–21, 228
Dicarlo, E. N., 328, 340
Dickerson, R. E., 209, 228
Diekmann, S., 173, 177
Dierckx, A. M., 71, 99
Dill, J., 47, 99
Dillon, K. B., 360, 397–9, 402–7, 409, 414, 417, 422–4, 429, 434
Di Lorenzo, J. V., 372–3, 379, 404, 427, 429, 438
Dimicoli, J. L., 288, 302
Dimroth, K., 75, 99
Ditmarsch, R., 151, 176
Doi, S., 470, 488
Donell, D. O., 41, 98
Donohue, J., 5, 100, 181, 188, 191, 216, 224, 227
Dorokhov, V. A., 135, 140
Dorval, C., 2, 100
Douglass, D. C., 366, 413, 422, 434
Drago, R. S., 91, 102
Drenth, J., 211, 224
Dresvyankin, B. V., 365, 374, 384–5, 388, 400, 402–5, 408, 412, 416, 418–20, 424–5, 431, 434
Drickamer, H. G., 291, 304
Duax, W. L., 205, 228
Duchesne, J., 365–6, 384, 400, 405, 408–9, 414, 434, 437
Duesler, E., 206, 226
Duke, B. J., 480, 489
Dumas, J. M., 479, 489
Dupont, J. P., 26, 28, 43, 100
Dupuis, P., 470, 489
Duterme, P., 196, 224
Dutton, P. L., 551, 567
Dwivedi, P. C., 281, 293, 295, 302
Dzizenko, A. K., 64, 104

Earp, C. D., 353, 424, 430, 433
Easter, J. H., 523, 567

Eastman, D. E., 499, 508
Edelman, G. M., 209, 227
Edmonds, D. T., 355, 435
Edmonds, J. W., 188, 205, 224, 228
Edward, J. T., 327, 340
Edwards, P. A., 386, 398, 424, 436
Egan, W., 125, 140
Egawa, K., 541, 543, 567
Ege, O., 347, 377, 401, 414, 426, 429, 437
Eggers, F., 174, 176
Egorov, V. A., 380, 405, 424, 430, 432, 434
Eia, G., 10, 100
Eigen, M., 112–3, 139, 141, 143, 148, 175, 474, 489
Eisenhardt, R. H., 143, 175
Eisenthal, K. B., 533, 566
Ekwall, P., 147, 175
Eland, J. H. D., 493, 508
El-Bayoumi, M. A., 283, 304
Elegant, L., 196, 224
Eley, D. D., 486, 489
Ellenson, W. D., 247–8, 255, 270
Elliott, M. A., 40, 100
Emin, D., 473, 489
Enderman, H. J., 185–6, 224
Engelhardt, H., 30, 75–6, 105, 331
Epstein, A. J., 499, 508
Erdneev, N. S., 356–8, 383, 401, 409–10, 417, 425, 431, 434
Erlich, B. S., 450, 499
Ermler, W. G., 507–8
Errera, J., 306, 337, 340
Ershov, V. V., 130, 141
Etienne, G., 205, 226
Evans, B., 27, 98
Evans, G. J., 339–40
Evans, J. C., 15, 100, 353, 362, 386–7, 398, 410, 414, 420, 424–6, 430, 435
Evans, J. R. N., 475, 491
Evans, M., 339–40
Eyring, E. M., 151, 175
Eyring, H., 320, 341
Ezell, J. B., 93, 100
Ezumi, K., 526, 559, 568

Fabbrizzi, L., 183, 224
Fabre, J. M., 470, 488
Faingor, B. A., 408, 420–1, 437

Fair, C. K., 9, 105, 224
Falk, M., 183, 224
Faller, L. D., 174, 177
Fang, J. H., 296, 303
Farges, J. P., 294, 302, 470, 489
Faruqui, A. R., 216, 226
Fawcett, J. K., 206, 224
Feder, J., 248, 250, 255, 269–70
Fehlhammer, H., 208, 224
Felix, N., 40, 100
Fendler, J. H., 147, 175
Fenske, J. D., 283, 302
Ferraris, G., 183, 225
Ferraris, J. P., 499, 508
Fersht, A. R., 215, 225
Feshin, V. P., 344, 371, 410, 435, 438
Fichtner, W., 348, 356, 383, 406, 408–9, 415–6, 418, 420, 425, 435
Fidanza, J., 196, 224
Filimonov, V. N., 13, 100
Filipic, C., 269
Finn, F. M., 212, 225
Finney, J. L., 217, 225
Fischer, E., 321, 340
Fischer, S. F., 476, 489
Fisher, M. E., 233, 269
Flanagan, T. B., 475, 492
Fleming, H. C., 384–5, 412, 424, 435
Flora, H. B., 93, 100, 193, 225
Flynn, G. W., 174, 177
Fogg, J., 216, 226
Foglizzo, R., 13, 19–20, 100
Folger, R., 152, 175
Fong, F. K., 332–3, 340
Fordemwalt, J. N., 28, 104
Forest, E., 318, 340
Forneris, R., 21, 28, 104, 106
Forsen, S., 119, 125, 134, 140
Forster, T. H., 510, 529, 562, 567
Foss, R. M., 40, 80, 82, 86, 89, 96, 100–4
Foster, R., 271, 274, 278, 283, 302, 498, 508
Fraenkel, G., 37, 100
Franchini-Angela, M., 183, 225
Franconi, C., 37, 100
Frank, F. C., 309–10, 340
Frankevich, E. L., 468, 489
Franklin, J. L., 47. 101
Franklin, T. C., 148, 175
Franz, M., 32, 69, 100–1

Franzen, J. S., 205, 226
Fratiello, A., 115, 120, 140
Frauenfelder, H., 440, 450
Frausto da Silva, J. J. R., 383, 400, 405, 412, 416, 419, 425, 435
Frazer, B. C., 9, 106, 247–8, 255, 270
Freedman, M. H., 40, 105
Freeman, H. C., 460, 488
Freer, S. T., 208–9, 214, 224, 227
Fridborg, K., 209, 226
Friedman, L., 473, 484, 486, 489
Frohlich, H., 308, 340, 480, 484, 489
Fruton, J. S., 187, 225
Fryar, C. W., 14, 20, 104
Fu, P. P., 283–4, 302
Fuggle, J. C., 372, 375, 381, 388, 404–5, 412, 430, 435
Fujii, S., 151, 177
Fujimori, E., 282, 302
Fujita, H., 481, 489
Fujiwara, F., 97, 104, 125, 140
Fujiwara, H., 543, 546, 553, 561, 567
Fukui, K., 35, 100
Fukushima, K., 463, 489
Fulton, G. P., 297–8, 302
Fultz, W. C., 374, 434
Fung, B. M., 360, 399, 407, 436
Fuoss, R., 199, 225
Furukawa, Y., 347, 364, 367, 374, 377–8, 384–5, 397, 400, 404, 414, 420, 424, 427, 437
Fyfe, C. A., 456, 489

Gasco, M. R., 297, 302
Gaultier, J., 479, 489
Gayles, J. N., 47, 98
Gbryukhova, E. V., 369, 371, 382, 412
Gebbie, H. A., 338, 340
Geffen, T. M., 148, 175
Geiger, W., 76, 104
Geiger, W. E., 296, 303
Gelius, U., 493, 495–6, 508
Gerard, A., 347, 434
Gerasinov, I. V., 114, 140
Gerbier, J., 91, 97
Gettins, J., 164, 167–8, 170–1, 174–5
Gettins, W. J., 143, 148, 174–5
Ghisla, S., 291, 302
Gibb, T. C., 439, 445, 450
Gibson, R. H., 143, 175
Gierer, A., 321, 328, 341

Giger, W., 38, 100
Giglio, E., 479, 489
Gil, V. M. S., 39, 100, 104
Gilkerson, W. R., 86, 90–1, 93, 96–7, 100, 102, 105, 193–5, 223, 225–6
Gillespie, R., 37, 98, 100
Gilson, B. R., 293–4, 302
Gilson, D. F. R., 360, 362, 375, 387, 399, 402, 407–8, 414, 418–9, 421, 435
Ginet, C., 340, 342
Ginn, S. G., 20–1, 24, 100
Ginzburg, V. L., 234–5, 269
Giral, L., 470, 488
Glaeser, R. M., 488–9
Glass, A. M., 269
Glasser, F. D., 507–8
Glasser, L., 334, 340–1
Glasstone, S., 320, 341
Glazunov, V. P., 16, 60, 64, 100, 104, 197, 227
Glowiak, T., 56, 100, 193, 225
Goeppert-Mayer, M., 482, 489
Gold, V., 122, 127, 139–40
Goldanskii, V. I., 439, 445, 450
Goldberg, I., 455–6, 489, 491
Goldshtein, I. P., 35–6, 100
Goldstein, C., 446–8, 450
Golebiewski, A., 482, 488
Golic, L., 185–6, 225
Golovchenko, L. S., 408, 420–1, 437
Golubev, N. S., 108, 116–7, 119–21, 123, 125, 127, 133, 135–6, 138–41
Golubinskaya, L. M., 376, 379, 381, 405–10, 417, 426, 436
Gomaa, C. S., 295–6, 303
Gomel, M., 479, 489
Goodall, D. M., 547, 567
Goodman, J., 147, 175
Goodman, M., 205–6, 223, 226
Goodwin, T. H., 461, 491
Gordeev, A. D., 337–8, 384, 388, 398, 401, 408, 415, 418–25, 427, 435
Gordienko, V. G., 37, 103
Gordon, J. E., 70, 100
Gordon, M. P., 287–8, 302
Gordon, R. G., 339, 341
Gordy, W., 398, 403, 431, 437
Goren, S. D., 349, 435
Goto, K., 40, 104
Gouin, L., 387, 417, 430, 434

Gould, C., 167–8, 170–1, 175
Gouterman, M., 550, 567
Govil, G., 34, 100
Graber, E., 151–2, 175
Grabowski, L. R., 519, 569
Graham, L. L., 205, 225
Gramstead, T., 91–100
Gransch, F., 335, 340
Grant, D. M., 39, 100
Gratzel, M., 148, 174, 555–6, 558, 567–70
Gravatt, C. C., 475, 489
Graybeal, J. D., 354, 373, 435
Grech, E., 55–6, 100, 193, 225
Grechishkin, V. S., 344, 365–7, 374, 377–8, 384–5, 388, 398, 400–6, 408–9, 412, 414–6, 418–25, 427, 431, 433–5, 438
Green, R. D., 191, 225
Greene, L. J., 212, 220, 225
Greenwood, N. N., 439, 445, 450
Greenwood, R. C., 174–5
Gregson, M., 330, 341
Grellmann, K. H., 564, 566
Griffin, J. F., 188, 225
Grimison, A., 13, 39, 97
Grimm, H., 461, 476, 491
Grisaro, V., 284, 303
Grobman, W. D., 499–500, 508
Gronan, B., 524, 567
Gross, P. M., 475, 489
Groth, P., 10, 100
Grunwald, E., 36, 100, 115, 127, 136, 140
Gruverman, I. J., 439, 450
Guest, M. F., 501, 508
Guibe, L., 354, 375, 383, 388, 401–4, 408–10, 412, 414–8, 435–6
Gunnarson, G., 125, 140
Gurr, M., 564, 566
Gurskaya, G. V., 181, 225
Gurudath, K., 281, 293, 295, 302
Guryanova, E. N., 35–6, 100, 344, 356, 362–6, 368–71, 380, 396–7, 399–400, 403–6, 408–16, 418–26, 428, 430, 435–6, 438
Gusakova, G. V., 16, 53, 69, 100–1, 123, 140
Gushkin, S. I., 401, 414, 418, 423, 427, 435
Gutlich, P., 40, 445, 450

Gutmann, V., 76, 83, 101, 104
Gutowski, H., 36, 101
Guttmann, F., 276, 295
Gwynne, E., 95, 99

Ha, T. K., 375, 396, 402–3, 408, 412, 435
Haake, P., 189, 226, 228
Haarer, D., 468, 471–2, 489–91
Haas, R. J., 293–4, 302
Haas, T. E., 387, 410, 424, 430, 435
Haberkorn, R., 547, 554, 561, 569
Hacobian, S., 375, 384–6, 400, 407, 410–2, 415–6, 420, 423–4, 430–1, 433
Hadzi, D., 10, 12, 14–6, 19, 57, 59, 101, 123, 140, 185, 225
Hagen, R., 40, 101
Hagiwara, S., 431, 436
Hague, D. N., 143, 148, 175
Hahn, E. L., 344, 388, 434
Hall, D., 164, 167–8, 170–1, 175
Hall, F. R., 452, 490
Hallidy, L. A., 525, 567
Hamamoto, K., 555–7, 570
Hamilton, W. C., 1, 2, 5–6, 9, 57, 101, 103, 105, 179–80, 185–6, 189–90, 225–6, 228, 463, 478, 489
Hamlin, R., 290, 302
Hammes, G. G., 107, 112, 140, 143, 148, 174–6
Hamori, E., 151–2, 176
Hamrin, K., 493, 508
Hand, E. S., 187, 225
Handloser, C. S., 38, 98
Haney, M. A., 47, 101
Hanna, M. W., 448, 450, 489
Hanson, A. W., 209, 229
Haque, I., 22, 24, 100–1
Harada, H., 386, 405, 410, 417, 420, 422–3, 430–1, 436
Haran, R., 376, 378, 382, 405–7, 409–10, 417, 426, 434
Harbury, H. A., 291, 302
Harned, H. S., 79, 101
Harpool, R. D., 205, 225
Harris, P. M., 9, 105
Harris, W. I., 212, 225
Hart, R. M., 360, 362, 375, 399, 407
Hartley, B. S., 209, 213, 223, 228

Hasegawa, K., 469, 490
Haser, R., 193, 225
Hassel, O., 1, 10–1, 100–1
Hata, T., 188, 228
Haulait, M. C., 79, 86, 89, 93–4, 97, 99, 101–2, 193–4, 200, 225
Hauw, C. H., 479, 489
Hawranek, J. P., 15, 89, 101
Hayakawa, K., 214, 224
Hayashi, H., 543, 567
Hayashi, K., 551, 567
Hayashi, T., 520–1, 544, 567
Hayman, H. J. G., 285, 302
Hazen, E. E. Jr., 188, 213–4, 221, 223–5
Heden, P. F., 493, 508
Hedman, J., 493, 508
Heidner, E. J., 195, 226
Helene, C., 283, 286, 288, 302
Heller, H. A., 506, 508
Henglein, A., 556, 567
Hennelly, E. J., 324–5, 341
Herber, R. H., 439, 445, 450
Herbstein, F. H., 452, 455–6, 488–9
Hercules, D. M., 497–9, 508
Hermann, A. M., 471, 489
Hernandez, G., 75, 102
Herriott, J. R., 207–8, 229, 282, 301
Hertel, E., 453, 489
Heston, W. M. Jr., 336, 341
Hetzer, H. B., 26, 99
Heyn, M. P., 174, 176
Hibbert, F., 42, 76, 101, 201, 225
Higashimura, T., 282, 284, 303
Higasi, K., 189, 226, 309, 332, 340–1
Highsmith, S., 200, 225
Higuchi, T., 191, 228, 290, 302
Hilbrants, W., 449–50
Hildebrand, J. H., 314, 341, 450, 477
Hill, J., 446, 448, 450
Hill, N. E., 307–8, 316, 320, 323, 326, 330, 337–8, 341
Hillier, I. H., 501, 508
Hinatsu, J., 535, 547, 567
Hinmann, L., 287–8, 302
Hino, T., 549, 551–2, 554, 567
Hinton, J. F., 205, 225
Hirayama, F., 544, 567
Hirotsu, K., 191, 228
Hirsch, E., 89, 101
Hofacker, G. L., 476, 489

Hoffman, C. J., 465, 491
Hoffman, R. A., 134, 140
Hoffman, T., 187, 225
Hoffmann, H., 143, 152, 163, 167–76
Hoffmann, R., 13, 39, 97
Hol, W. G. J., 211, 224
Holloway, J. H., 50, 372, 375, 381, 388, 404–5, 412, 430, 435, 449
Holmberg, P., 147, 175
Holmes, J. D., 532, 570
Holmes, L. P., 151, 175
Holstein, T., 473, 489
Holten, D., 550, 567
Honda, T., 148, 175
Honig, B., 213, 229
Hooper, H. O., 365–6, 384, 387–8, 400, 402–6, 410, 412, 414–5, 417, 419, 421, 430, 433, 436
Hope, H., 11, 101
Hopfield, J. J., 530, 567
Hopfinger, A. J., 217, 225
Hopkins, M. P., 73, 97
Hopkinson, A. C., 19, 101
Hopmann, R. F. W., 108, 140
Hornig, D. F., 59, 106
Hsieh, Y. N., 349, 355, 375, 400–1, 405, 411–2, 424, 426, 430–2, 436
Hsu, I. N., 187, 204, 225
Huang, C. W., 287–8, 302
Hubbard, C. D., 174, 176
Huber, R., 207–8, 212, 218–21, 223
Hudson, R. A., 28, 30, 101, 123, 140, 199, 201, 212, 225, 228
Hughus, Z. Z., 42, 357–8, 400–1, 403, 405, 408, 412, 414, 416, 419, 422, 425, 431, 434
Hui, M. H., 532, 567
Huong, P. V., 14, 18–9, 58, 101, 103
Huoslef, J., 1, 101
Huppert, D., 550, 567
Hush, N. S., 285, 290, 302, 479, 489
Hutcheon, W., 214, 224
Huyskens, P., 2, 23, 31–5, 40, 64, 69–70, 74–5, 79, 81, 86, 91, 94, 96, 100–5, 136, 140, 193–6, 200, 202–3, 224–7
Hvidt, A., 205, 226

Ibers, J. A., 5, 6, 57–8, 101–2, 179, 190, 225, 463, 478, 489
Ichika, S., 446–8, 450

Ichikawa, M., 60, 102
Ide, R., 524, 532, 569
Ikeda, R., 374, 417, 421, 437
Ikemoto, I., 35, 103, 458, 477, 490, 492
Imai, K., 216, 226
Imamura, A., 481, 489
Imanishi, H., 282, 284, 303
Immamura, A., 35, 100
Inabe, T., 470, 488
Infelta, P. P., 556, 567-8
Ing, S. D., 354, 435
Inokuchi, H., 464, 486, 488-9
Inoue, H., 518, 531, 569-70
Inoue, N., 17, 104
Inoue, T., 168, 176
Iogansen, A. V., 16, 38-9, 102, 104, 135, 140, 197, 227
Irie, M., 551, 567
Ishihara, H., 362-3, 396, 398-9, 401, 431, 437
Ishii, K., 470, 491
Ishii, Y., 563, 570
Ivin, K. J., 41, 102, 125, 140
Iwamoto, R., 95, 104
Iwanaga, C., 539, 567
Iwasaki, F., 456, 490
Iwata, S., 542-3, 567-8

Jaccarino, V., 446, 450
Jackman, L. M., 38, 102, 115, 140
Jacobs, P. W. M., 474, 489
Jadzyn, J., 32-3, 69, 72, 102, 199, 202, 225
Jakubetz, W., 48, 105
Jallon, J. M., 174, 177
James, C. J., 80, 102
James, M. N. G., 187, 214, 224-5
Janin, J., 212, 220, 225
Janjic, T., 152, 176
Jannakoudakis, D., 28, 102
Janoschek, R., 192, 225
Jansonius, J. N., 211-2, 224
Jarup, L., 209, 226
Jasinski, T., 10, 11, 16-7, 28-9, 41-2, 86, 94, 102, 201, 225
Jasny, J., 519, 569
Jaycock, M. J., 151, 176
Jencks, W. P., 187, 225
Jensen, L. H., 186, 207-8, 217-9, 223, 228-9

Jerome, D., 470, 488
Jesson, J. P., 10, 11, 102
Jnoue, H., 112, 141
Jobling, P. L., 164, 167-8, 170-1, 175
Jodkowski, J. T., 483, 488
Joeston, M. D., 12, 40, 91, 102
Johansson, A., 19, 47, 102, 205, 225
Johansson, G., 493, 508
Johari, G. P., 334, 341
Johnson, S. L., 52, 72, 102, 197, 226
Jone, R. T., 216, 226
Jones, F. M., III, 183, 226
Jones, G. P., 330
Jones, L. V., 387, 417, 430, 434
Jono, I., 481, 488
Jonsson, P. G., 1, 3, 5, 58, 102, 104, 181, 227
Jordan, C. F., 205, 223
Jordan, F., 14, 49, 102
Jortner, I., 530, 567
Jortner, J., 483, 489
Josien, M. L., 55, 102
Jost, J. W., 360, 364, 373, 375, 399, 404, 407, 411, 421, 423-4, 426-7, 429, 436
Joussot-Dubien, J., 554, 566
Jugie, G., 348, 360, 363, 376, 378, 387, 397-9, 405-7, 409-10, 413, 417, 422, 426, 430, 434, 436, 438, 482
Juliano, L., 183, 227
Julien-La Ferriere, S., 19, 104
Junker, M. L., 93, 102, 193-5, 226
Jursek, L., 214, 224
Justile, J. C., 80, 97, 106

Kabachnik, M. I., 123, 126, 130, 141
Kadaba, P., 264, 269
Kaftory, M., 455, 489
Kahlweit, M., 173, 177
Kaino, H., 471, 489
Kallai, O. B., 209, 228
Kallen, R. G., 37, 104
Kallman, O. F., 326-7, 341
Kalman, A., 186, 226
Kamaraev, K. I., 488, 492
Kamb, B., 1, 2, 9, 103
Kamenar, B., 5, 11, 105
Kaminov, I. P., 256, 269
Kanazawa, K. K., 468, 491
Kanduser, A., 247, 269

Kaneko, N., 189, 226
Kannan, K. K., 209, 226
Kanters, J. A., 478, 490
Kapur, P. C., 275, 302
Karabatsos, P. J., 36, 100
Karl, N., 468, 471, 489
Karle, I. L., 205–6, 226
Karle, J., 205, 226
Karlsson, R., 290, 302
Katchalski, E., 284, 303
Katz, J. L., 483, 489
Kautsch, F., 330, 341
Kauzmann, W., 217, 226
Kawada, A., 474, 489
Kazakova, E. M., 130, 139
Kebarle, P., 184, 223
Keder, W. E., 38, 102
Keefer, R. M., 35, 102
Keller, R. A., 488–9
Kepler, R. G., 470, 489
Kerr, K. A., 216, 226
Keyzer, H., 276, 302
Khairutdinov, R. F., 488, 492
Kharlamova, E. N., 36, 100
Khmelnitskii, D. E., 235, 269
Kielmann, I., 152, 163, 167–8, 172–4
Kilmartin, J. V., 216, 226
Kilp, H., 218–9, 341
Kimelfeld, J. M., 21, 102
Kinoshita, M., 465–6, 470, 490–1
Kiriyama, H., 248, 269
Kirkwood, J. G., 307–8, 335, 341
Kirszenbaum, M., 55, 102
Kiselev, S. A., 38, 102
Kitaigorodsky, A. I., 462–3, 477, 489
Kivinen, A., 117, 141
Klaboe, P., 21, 102
Klein, P., 204, 226
Klepanda, T. I., 52, 103
Klotz, I. M., 18, 106, 205, 226
Kludt, J. R., 20, 102
Kniche, W., 143, 148, 176
Knier, B. L., 189, 226
Knop, D., 183, 224
Knox, J. R., 209, 229
Kobayashi, H., 459, 490
Kobayashi, K. K., 231, 255–7, 264, 269
Kobayashi, M., 466, 490
Kobilarov, N., 15, 101
Kobinata, S., 276, 302
Koekoek, R., 211, 224

Koelle, U., 119, 140
Koetzle, T. F., 180–1, 185–6, 189–90, 216, 226
Koetzlee, T. Γ., 190, 228
Kofler, A., 452, 490
Kohler, F., 38, 102, 195–6, 226
Kohot, Z., 16–7, 41, 102
Koizumi, S., 454, 490
Kokado, H., 469, 490
Kokobun, H., 28, 51, 69, 97
Kokubun, J., 123, 139
Koll, A., 32, 105
Kollman, P., 9, 19, 47, 49, 102, 205, 225
Kollman, P. A., 137, 140, 179, 226, 478, 490
Kolomijtsova, T. D., 116, 140
Kolthoff, I. M., 29, 72, 86, 92–4, 103
Koltsov, A. I., 117, 123, 139–40
Kommandeur, J., 452, 490
Komorowski, L., 452, 490
Kosower, E. M., 76, 103
Kotaki, A., 290, 304
Kotov, B. V., 468, 489
Kraft, J., 28, 50, 103
Krajewska, A., 452, 490
Krakower, E., 38, 103
Kramarenko, N. L., 468, 490
Kraus, C. A., 86, 93, 96, 100, 103–4, 106, 195, 226
Kraut, J., 208–9, 213–4, 223–4, 226–7
Kreevoy, M. M., 29, 103, 127, 140
Kresheck, G. C., 151–2, 176, 205, 226
Kretsinger, R. H., 207–9, 226–7
Krishnan, K. S., 308, 342
Kristof, W., 197, 226
Kromhout, R. A., 36, 100
Kronick, P. L., 452, 469, 471, 490
Kroon, J., 478, 490
Kroon, S. G., 338, 341
Kroupa, J., 256, 269
Kubota, T., 25–6, 103–4
Kuhn, S. T., 37, 103
Kumakura, S., 456, 490
Kuntz, E., 283, 304
Kuntz, I. D., 217, 224, 226
Kuopio, R., 117, 141
Kuramshin, I. Ya., 369, 383, 410, 412, 417, 421, 423, 433
Kuroda, H., 35, 103, 454, 457–8, 465–6, 468, 470–1, 477, 488, 490–2
Kustin, K., 143, 148, 176

Kvick, A., 185–6, 226
Kwong, G. Y., 20, 102
Kyogoku, Y., 251, 290–1, 302
Kyuntsel, I. A., 365, 367, 421, 427, 433

Laane, J., 59, 103
Labbe, H. J., 336, 341
Labes, M. M., 452, 469, 471, 474, 489–90
Lacour, T., 213–4, 224
Ladik, J., 481, 488
Ladner, R. C., 195, 218–21, 226, 228
Laidler, K. J., 320, 341
Lam Wee, N., 474, 489
Lamar, G. N., 297–8, 302
Lambeau, Y., 79, 81, 86, 91, 102
Lamberts, L., 34, 104, 196, 200, 227
Lancelot, G., 188, 226
Landolt-Börnstein J., 452, 490
Lang, J., 151–2, 163, 167–8, 172–5
Lange, I. J., 125, 133, 135–6, 138–9
Lange, K. R., 147, 177
Langs, D. A., 205, 228
Lanue, K. F., 291, 302
La Placa, S. J., 1, 2, 9, 103
Larkin, A. I., 235, 269
Larsen, F. K., 456, 490
Larsen, S., 188, 213–4, 224
Lascombe, J., 18, 99
Lassegues, J. C., 19, 103
Lassen, H., 205, 226
Lassier, B., 339–40
Lathan, W. A., 477, 479, 490
Laubengayer, A. W., 506, 508
Lautie, A., 19, 103
Lavrencic, B. B., 256, 264, 269
LeBas, J. M., 19, 104
Le Blanc, O. H., Jr., 480, 490
Lee, B., 209, 229
Lee, S. K., 522, 570
Lee, T. S., 93, 103
Legg. M., 213–4, 224
Lehmann, M. S., 181, 186, 190, 226, 228, 479, 490
Leigh, J. S., 551, 567
Leijonmarck, M., 290, 302
Leiserowitz, L., 191, 226
Leroy, M. J. F., 382, 406, 418, 426–7, 429, 434
Leroy, Y., 338, 340–1

Lessing, H. E., 523, 569
Leuchs, M., 75, 103
Leung, R. C., 256, 269
Le-Van, L., 64, 103
Levstik, A., 264, 269
Levstik, I., 264, 269
Lewis, G. L., 324, 342
Lewis, T. J., 475, 491
Li, N. C., 135, 141
Lieb, E. M., 248–9, 254, 269
Liedke, P., 562, 569
Liler, M., 37, 103
Liljas, A., 209, 226
Lillford, P. J., 112, 140
Limbach, H. H., 136, 141
Lin, D., 115, 120, 140
Lin, L.-N., 286, 288, 302, 304
Lindemann, R., 54, 69, 72, 75, 103, 123, 141, 197–8, 202–3, 226
Lindgren, J., 9, 97
Lindheimer, M., 205, 226
Linnell, R. H., 12, 106, 123, 141
Liotta, C. L., 73, 97
Lipari, N. O., 499, 508
Lippert, E., 478, 491, 510, 512, 518–9, 524, 563, 567–9
Lippincott, E. R., 44, 103
Lipscomb, W. N., 208, 227
Lisichkina, I. N., 386, 414, 425, 437
Little, R. G., 456, 490
Llinas, M., 204, 226
Lloyd, D. R., 506, 508
Lo, G. Y. S., 353, 362, 386–7, 398, 410, 414, 420, 424–6, 430, 435
Loach, P. A., 291, 302
Lobanova, L. A., 362, 396–7, 405, 408, 412, 416, 419, 422–4, 436
Loewenstein, A., 36–7, 98
Lonberg–Holm, K. K., 143, 175
London, F., 312–3
Lord, R. C., 251, 290, 302
Lorentz, H. A., 317
Lorentz, K., 280, 282, 302
Lorenz, K., 470, 487–8, 490
Lorenz, P. M., 28, 106
Lorenzelli, V., 91, 97
Lovas, G., 186, 226
Lovelace, M. I., 27, 99
Lovgren, S., 209, 226
Lowenstein, H., 205, 226
Lower, S. K., 454, 490

Lowndes, R. P., 256, 269
Lucas, J.-P., 375, 388, 402–3, 409–10, 412, 415–8, 436
Lucken, E. A. C., 344, 349, 356, 360–1, 373, 375–6, 379, 380, 382, 398–400, 402–4, 406–8, 410–1, 414–5, 417–9, 422–9, 434, 436
Luder, W., 518, 568
Luganskii, Yu. M., 416, 424, 433
Lumpkin, O., 376, 424, 436
Lundberg, S. K., 174, 177
Lundgren, J. O., 1, 9, 103, 185, 227
Lunegov, V. I., 374, 388, 402–5, 412, 416, 418, 424–5, 434, 485
Lutskii, A. E., 37, 52, 103
Luzzi, L. A., 283–4, 302
Lyerla, J. R., 40, 105
Lynaugh, N., 506, 508
Lynch, R. J., 406, 434
Lyons, L. E., 452, 464, 470, 489

Maas, E. T., 499, 508
Manne, R., 493, 508
Masuhara, H., 567–8
Mataga, N., 510, 512, 514, 517–8, 520–1, 524, 526, 528, 532–7, 540–7, 549, 551–61, 566–70
Matsumoto, S., 543, 569
Matsuzaki, A., 564, 569
Mazurenko, Y. T., 522, 566
McWeeny, R., 493, 508
Metzger, R. M., 499, 508
Michel-Beyerle, M. E., 547, 554, 561, 569
Migita, M., 534–5, 559, 568–9
Misumi, S., 514, 517, 520–1, 524, 532–5, 544, 547, 567–70
Mori, T., 559–60, 569
Morita, M., 544, 567
Morita, T., 563, 570
Mostad, A., 496, 508
Mozumder, A., 552, 569
Mulliken, R. S., 496, 508
Murakawa, M., 520, 563, 569
Murata, Y., 540, 545, 568

Naboikin, J. V., 468, 490
Nagakura, S., 2, 26, 97, 104, 106, 276, 279, 302–3, 510, 531, 540, 542–3, 564, 567–70

Nagato, C., 35, 100
Nagel, R., 168–73, 176
Nagyrevi, A., 69, 106, 197–8, 202, 229
Nahringbauer, I., 185, 227
Nakagawa, T., 152, 176
Nakamura, D., 374, 386, 405, 410, 417, 420–23, 430–1, 436–8
Nakamura, J., 543, 569
Nakamura, N., 344, 347, 413, 416, 420–4, 434
Nakao, A., 398, 414–5, 427, 437
Nakashima, N., 518, 520, 524, 532–3, 537, 541, 543, 546, 553, 561, 563, 567–9
Nakashima, S., 256, 269
Naoi, M., 290, 304
Narasimhamurthy, M. R., 186, 227
Nasielski, J., 52, 104
Nataragan, R., 174–5
Nawrot, C. F., 37, 104
Nead, D. J., 86, 104
Neel, J., 13, 104, 470, 489
Neerinck, D., 34, 104, 196, 227
Negita, H., 347, 362–5, 367, 374, 377–8, 384–5, 396–9, 400–1, 404–5, 414–5, 420, 423–4, 426–7, 429, 431, 437, 446–8, 450
Negran, T. J., 269
Nelson, I. V., 95, 104
Nelson, R. D. Jr., 327, 341
Nesmeyanov, A. N., 386, 408, 414, 420–1, 425, 437
Netzel, T. L., 551, 567
Neusser, H. J., 547, 554, 561, 569
Neven, L., 86, 104
Ng, K. T., 497–9, 508
Nickel, B., 547, 554, 561, 570
Nicola, C. U., 174, 176
Nielsen, P., 499, 508
Nilsson, R., 496, 508
Nishimura, T., 553, 559–61, 568–9
Nishina, Y., 549, 570
Nockolds, C. E., 208–9, 226
Nogaj, B., 197, 227
Nogami, T., 279, 303
Noll, G., 18, 58, 101
Nordling, C., 493, 496, 508
Nouwen, R., 32, 64, 69, 70, 104
Novak, A., 13, 19, 20, 57, 59, 99–101, 103–4, 185, 227
Noyes, R. M., 532, 569

Nrisbin, D. A., 297–8, 302
Nusslein, H., 169, 176

O'Connor, D. V., 532, 569
Odinokov, S. E., 16, 59, 60, 64, 100, 104, 197, 227
O'Donnell, D., 125, 139
O'Dwyer, J. J., 317, 341
Ogawa, S., 385, 398, 401, 410, 431, 436–7
Ohki, M., 188, 227
Ohmasa, M., 470, 491
Ohrt, J. M., 188, 227
Ohta, T., 454, 492
Oja, T., 362, 434
Ojelund, G., 205, 227
Okada, K., 40, 104, 290, 304
Okada, T., 514, 517, 526, 533, 536, 544–6, 559–60, 568–9
Okamoto, B. Y., 291, 304
O'Keeffe, M., 474, 491
Okhlobystin, O. Yu., 408, 420–1, 437
Okonski, S. T., 375, 387, 396, 402–3, 408, 412, 414, 418–21, 435
Okuda, T., 347, 362–5, 367, 377–8, 396–9, 401, 405, 414–5, 420–1, 423, 426–7, 429, 431, 437
O'Leary, B., 480, 489
Olofsson, G., 37, 98
Olovsson, L., 1, 3, 5, 9, 58–9, 103–4, 106, 181, 185, 227
Onaka, R., 474, 491
Oncley, J. L., 321, 324, 341
Onda, S., 374, 417, 421, 437
Ong, N. P., 470, 491
Onsager, L., 74, 80, 104, 233, 270, 307–9, 317, 341
Onuki, H., 474, 491
Oohari, H., 526, 559, 568
Orbach, N., 546–7, 553, 567, 569
O'Reilly, D. E., 384, 400, 414, 436
Orel, B., 57, 59, 101
Orlov, I. G., 349, 364–9, 371–2, 375, 377, 379, 381, 397, 404, 408–9, 412, 415–6, 418–23, 427–30, 437–8
Orszach, J., 293, 302
Orville-Thomas, W. J., 49, 65, 105, 478, 491
Osawa, E., 453, 490
Osawa, R., 453, 490

Osiecki, J., 465, 491
Osokin, D. Ya., 369–70, 376, 383, 401, 403–5, 407, 409–10, 412, 414, 417, 419, 421–4, 426, 428–9, 433
Osterheld, R. K., 269
Oszut, J., 89, 101
Ottenberg, A., 465, 491
Ottersen, T. O., 204, 227
Ottewill, R. H., 151, 176
Ottolenghi, M., 510, 541, 543, 546–7, 552–3, 567–9
Owen, B., 79, 101
Ozawa, T., 290, 304

Paabo, M., 26, 99
Padmanabhan, G. R., 93–4, 98, 194, 224
Paiva, A. C. M., 183, 227
Paivin, V. S., 468, 490
Pajdowska, M., 69, 102
Palacios-Gomes, J., 461, 476, 491
Palumbo, M., 282–4, 303
Paoletti, P., 183, 224
Parbhoo, D. M., 41, 104
Pardoe, G. W. F., 338, 340
Parson, W. W., 550–1, 567, 569
Parthasarathy, R., 180, 186–90, 216, 224, 226–8
Pasternak, M., 445–8, 450
Patel, R. C., 174–5
Paterson, W. G., 135, 141
Pathania, M., 354, 435
Patterson, D. B., 352, 355–6, 363, 376, 378–9, 382, 396, 400–1, 404–10, 414, 416–7, 419–20, 422, 425, 431, 434, 437
Pauwels, H., 2, 32, 104
Pawelka, Z., 64, 69, 106, 197, 199, 227–8
Pawlak, Z., 56, 86, 93–4, 102, 104, 194, 201, 225, 227
Pealstein, J. H., 499, 508
Pearson, R. G., 27, 104
Peat, I. R., 40, 105
Peercy, P. S., 256, 270
Peerdman, A. F., 185–6, 224
Peggion, E., 282–4, 303
Peneau, A., 402, 407, 437
Pepinsky, R., 9, 106
Perepelkova, T. I., 35, 100

Perkins, P. G., 507–8
Perrier-Datin, A., 19, 104
Perrin, F., 321, 341
Perrino, C. T., 474, 491
Person, W. B., 5, 28, 104, 271–2, 300, 303, 447, 450, 476, 479, 491, 496, 508
Perutz, M. F., 194–5, 208, 211–2, 215, 226–7
Peterson, E. M., 384, 400, 414, 436
Peterson, G. E., 355–6, 400–1, 408, 416, 419, 425, 431, 437
Pethig, R., 465, 467, 488
Petif, M., 209, 226
Petrakis, L., 412, 416, 421, 424, 437
Petrosyan, V. S., 369, 371, 382, 401, 403–9, 411–2, 417, 431–2, 437
Petsko, G. A., 212, 228
Petzelt, J., 256, 264, 269
Pfeiffer, H., 192, 225
Piekara, A., 330–1, 341–2
Piekara, B., 330, 342
Pietrzak, J., 197, 227
Pigon, K., 452, 470, 487–8, 490
Pilyugin, V. S., 27, 104
Pimentel, G. C., 12, 25, 47, 97, 104
Pirc, R., 269
Pirson, D., 81, 86, 97, 102, 104
Piterskaya, I. V., 522, 566
Pitt, D. A., 326–8, 342
Plass, K. G., 113, 139
Platz, G., 168–73, 176
Plekhov, V. P., 369–70, 405, 407, 412, 417, 419, 421–2, 428, 437
Podjarny, A., 195, 213, 227, 229
Pogolotti, A. Jr., 212, 223
Poleshchuk, O. Kh., 349, 363–9, 371–2, 375, 377, 379–81, 396–7, 399, 401–6, 408–9, 411–2, 415–23, 427–30, 437–8
Poley, J. P., 337, 342
Pollak, R. A., 499, 508
Pollock, McC. J., 474, 491
Poltev, W. I., 479, 491
Pongs, O., 195, 228
Pople, J. A., 36, 38, 47, 99, 104–5, 494, 508
Porschke, D., 174, 176
Porter, D. J. T., 291, 293, 303
Ports, A. M., 470, 491
Poskin, G., 86, 95, 102, 105

Potashnik, R., 541, 569
Potier, J., 18, 58, 98
Pott, G. T., 498, 508
Powles, J. G., 317, 326, 338, 342
Prakash, A., 1, 2, 9, 103
Prasad, N., 181, 185, 227
Price, A. H., 329, 341
Prince, E., 9, 98
Prochorow, J., 545, 569
Prokofev, A. K., 356–8, 383, 401, 403, 409–10, 417, 425, 431, 434
Prokofiev, A. I., 126, 130, 141
Prout, C. K., 5, 11, 105, 290, 303, 455, 488, 490–1
Pshenichnov, E. A., 49, 105
Puchkova, V. V., 363, 397, 408, 410, 416, 420–3, 435
Pudovik, A. N., 369–70, 376, 383, 401, 403–5, 407, 409–10, 412, 414, 417–9, 421–4, 426, 428–9, 433, 437–8
Pugnuie, R. J., 39, 100
Pullman, A., 205, 223, 227, 476, 491
Purlee, E. L., 36, 100

Quick, A., 56, 105, 193, 227
Quiocho, F. A., 208, 227
Quivoron, C., 18, 99

Rabinovich, D., 195, 227
Raczy, L., 338, 340
Radushnova, I. L., 125, 141
Ragle, J. L., 349, 353, 388, 401–2, 406, 409, 421, 437–8
Rahmann, G. B., 341
Raidt, H., 211, 227
Ralf, E. K., 115, 140
Ralph, E, III., 86, 93, 96, 100, 105
Ramakrishna, J., 344, 437
Ramakrishnan, C., 181, 185, 227
Ramakrishnan, L., 344, 437
Raman, C. V., 309, 342
Ramdas, V., 213, 223
Ramsey, J., 88, 99
Rao, C. N. R., 205, 229, 281, 293, 295, 302, 465, 488
Rao, V. N. M., 356, 358, 401, 408, 410, 416, 420–1, 425, 438
Rapp, W., 523–4, 567, 569
Rassadin, B. V., 16, 38, 102, 104, 135, 140, 197, 227

Rassing, J., 108, 141, 151–2, 163–4, 167–8, 170–2, 174–5
Rast, H. E., Jr., 488–9
Ratajczak, H., 32, 49, 64–5, 69, 103, 105, 196–7, 227, 478, 491
Rauschen, W., 539, 570
Ray, S., 28, 104
Read, M., 365–6, 394, 400, 405, 408–9, 414, 434, 437
Reed, J. W., 9, 105
Reeke, G. N. Jr., 209, 227
Reeve, R. E., 360, 390, 397–8, 402–5, 407, 409, 414, 417, 422–4, 429, 434
Reeves, L. W., 38, 103
Rehm, D., 538–9, 565, 569
Reichardt, C., 76, 105
Reid, C., 44, 105, 301, 303
Reilley, C. N., 40, 105
Reinsborough, V. C., 171, 176
Reiter, F., 385, 412, 415–6, 422, 431, 434
Rentzepis, P. M., 519, 550–1, 567, 570
Requena, Y., 215, 225
Rey-Lafon, M., 18, 99
Reynolds, W. F., 40, 105
Rice, S. A., 483, 489
Rich, A., 100, 251, 290–1, 302, 304
Richards, F. M., 209, 229
Richardson, D. C., 213–4, 223
Richardson, J. S., 213–4, 223
Richter, H. P., 531, 570
Riedel, E. F., 386, 410, 420, 424–5, 437
Rigny, P., 456, 488
Riordan, J. F., 213, 228
Rip, A., 335, 340
Robas, V. I., 349, 437
Robas, V. L., 412, 415, 417–8, 421–2, 438
Roberti, D. M., 315, 340
Roberts, Y. D., 40, 101
Robertson, J. M., 58, 105, 461, 491
Robinson, B. H., 43, 69, 99, 174, 176
Robinson, B. U., 126, 139
Robinson, E. A., 44, 105
Robinson, H. G., 398, 403, 431, 437
Robinson, P. D., 296, 303
Robinson, R. A., 79, 105
Rocard, Y., 338, 342
Rockley, M. G., 550–1, 567, 569

Rode, B. M., 48, 105
Rodgers, M. A. J., 545, 556, 566, 569
Rodriquez, G., 39, 97
Rogers, M. T., 352, 365, 369, 379, 426–9, 437
Rohrer, D. C., 205, 228
Rollefson, B., 37, 105
Rollig, K., 545, 568
Romanaski, H., 28, 43, 51, 59, 105
Romanovska, K., 49, 105
Romm, I. P., 364–6, 397, 419–22, 438
Romming, C., 11, 98, 101
Roos, B., 495, 508
Rose, A., 470, 491
Rosenfield, R. E. Jr., 216, 228
Rosenstein, R. D., 269
Rospenk, M., 32, 105
Ross, I. G., 309, 342
Rothenberg, S., 47, 102, 205, 225
Rotkiewicz, K., 519, 569
Roy, R. S., 59, 105
Rozenberg, Yu. I., 365, 367, 378, 384, 388, 401–3, 406, 408–9, 412, 415, 417–8, 421, 424–5, 427, 435–6
Roziere, J., 193, 225
Rubenacker, G. V., 349, 355, 375, 400–1, 405, 411–2, 424, 426, 430–2, 436
Ruby, S. L., 446, 448, 450
Rumon, K. A., 52, 72, 102, 197, 226
Rundle, R. E., 58, 105
Rupley, J. A., 212, 223, 228
Ruterjans, H., 195, 228
Ryan, J. A., 352, 365, 369, 379, 426–9, 437
Ryltsev, E. V., 130, 133, 137, 139, 141

Saatsazov, V. V., 351, 377, 386, 414, 425, 437
Sabin, J. R., 24, 56, 105, 476, 489
Sabine, T. M., 9, 99, 106
Sack, R. A., 309, 317, 341–2
Sackmann, E., 468, 491
Safarov, N. A., 123, 140
Safin, I. A., 344, 347, 369–70, 376, 383, 401, 403–5, 407, 409–10, 412, 414, 417–9, 421–4, 426, 428–9, 433, 437–8
Sahney, R. C., 275, 303
Saika, A., 36, 101
Saito, A., 14, 98

Saito, G., 453, 490–1
Satio, T., 533, 569
Satio, Y., 456, 490
Sakai, H., 446–8, 450
Sakata, Y., 514, 517, 520–1, 533–5, 544, 547, 567–9
Sakurada, M., 463, 489
Sakurai, N., 188, 228
Sakurai, T., 463, 491
Salinas, S. R., 244–6, 255, 269
Samara, G. A., 256, 260, 270
Samoilenko, A. A., 38, 102
Sams, P. J., 151–2, 163, 171–2, 176
Sandman, D. J., 499, 508
Sandorfy, C., 14, 97–8, 101, 103–5, 107, 141, 181, 226, 331, 342
Sandstrom, J., 91, 100
Sano, T., 113, 141
Santry, D. P., 476, 490
Sarneski, J. E., 40, 58, 105
Sasada, Y., 188, 191, 227–8
Sasane, A., 430–1, 438
Sastry, V. S. S., 344, 437
Satchell, D. N. P., 42, 76, 101, 201, 225
Sato, S., 188, 228
Savelev, V. A., 58, 105
Sawyer, L., 208–11, 218–21, 228
Saya, A., 195, 227
Scaife, D. E., 357, 408, 412, 423, 425–6, 438
Scarborough, F. E., 290–1, 303
Schaad, L. J., 12, 40, 102
Schafer, F. P., 545, 568
Scheerev, R., 556, 558, 567, 569
Scheie, C. E., 384, 400, 414, 436
Schein, L. B., 473, 491
Schempp, E., 344, 438
Scheraga, H. A., 151–2, 176, 216, 228
Schille, B., 18, 99
Schlaak, M., 14, 101
Schlag, E. W., 547, 554, 561, 569
Schleich, T., 37, 105
Schlemperer, E. O., 9, 105, 224
Schmidt, V. H., 232–3, 248, 254–5, 270, 474, 491
Schmueli, U., 455–6, 489, 491
Schneider, F., 519, 563, 569
Schneider, R. F., 364, 372–3, 379, 404, 426–7, 429, 436, 438
Schneider, W. G., 36, 38, 105, 469, 490
Schoenborn, B. P., 209, 218, 228

Schoenes, J., 468, 491
Scholte, P. G., 308, 342
Schreiber, H. D., 44, 105–6
Schreiber, V. M., 17, 20, 43, 69, 99, 121, 123–4, 127, 139–40
Schreiner, S., 539, 570
Schrobilgen, G. J., 449–50
Schroder, H., 547, 554, 561, 569
Schroeder, R., 44, 103
Schuhmann, P. J., 26, 99
Schulman, S. G., 562, 569
Schulten, K., 541, 554, 561, 570
Schuster, P., 48, 101, 103–5, 107, 141, 181, 226, 331, 342
Schuster, R. E., 115, 120, 140
Schvoerer, M., 479, 489
Schwager, P., 208, 212, 214–5, 218–21, 223
Schwartz, D., 349, 353, 388, 401, 409, 438
Schwarz, G., 174, 176
Schwartz, M. E., 495, 508
Schwarz, M., 501, 508
Schwarzenbach, G., 183, 224
Scott, B. A., 499, 508
Scott, R., 28–30, 50, 101, 105, 123, 140, 199, 201, 225, 228
Scrocco, E., 205, 223
Sears, D. S., 506, 508
Seeligloffler, A., 174, 176
Seidlitz, H., 547, 554, 561, 569
Semenova, A. E., 135, 139
Semin, G. K., 344, 349, 351, 356, 358, 362, 365–6, 368–71, 375–6, 379–82, 386, 388, 396–438
Semmingsen, D., 247–8, 250, 255, 270
Sewell, G. L., 480, 484, 489
Shabanov, V. F., 351, 375, 388, 413–5, 436
Shah, D. O., 148, 175
Shamir, J., 450, 499
Shapetko, N. N., 125, 141
Shares, J., 113, 141
Sharon, N., 284, 303
Sharp, D. W. A., 372, 375, 381, 388, 404–5, 412, 430, 435
Sharp, I. H., 471, 491
Shenoy, G. K., 446, 448, 450
Sharples, O., 287–8, 303
Shehepkin, D. N., 116, 140
Sheina, G. G., 52, 103

Sheppard, N., 15, 98
Shibaguchi, T., 474, 491
Shibata, K., 374, 384–5, 400, 404, 424, 437
Shibuya, Y., 168, 176
Shieh, H-S., 290–1, 303
Shifrina, R. R., 362, 396–7, 405, 408, 412, 416, 419, 422–4, 436
Shigorin, D. N., 125, 141
Shimada, A., 191, 228
Shimada, M., 545, 549, 568, 570
Shimanouchi, H., 188, 227
Shimauchi, A., 385, 410, 431, 436
Shimozawa, R., 168, 176
Shindo, H., 212, 228
Shinitsky, M., 284, 303
Shinoda, K., 143, 147, 166, 176
Shipman, L. L., 216, 228
Shirane, G., 247–8, 255, 270
Shiroyama, M., 398, 414–5, 427, 437
Shishkin, V. A., 401, 414, 418, 423, 427, 435
Shizuka, H., 563, 570
Shoshtakovskii, M. F., 349, 365–8, 377, 415–23, 427, 438
Shotton, D. M., 208–11, 214, 218–21, 228
Shteingartz, V. D., 412, 415, 417–8, 412–2, 438
Shurubura, A. K., 133, 137, 141
Siderov, A. N., 297, 303
Siebrand, W., 479, 491
Siegbahn, K., 493, 496, 508
Siegbahn, P., 495, 508
Siegoczyriski, R., 545, 569
Sieker, L. C., 207–8, 229
Sielecki, A., 213, 229
Siepmann, T., 76, 99
Silsbee, H. B., 232–3, 248, 254–5, 270
Silverman, B. D., 500, 508
Sim, G. A., 461, 491
Simmons, E. L., 28, 41–2, 98, 102
Simmons, L. L., 125, 140
Simon, W., 38, 100
Simonov, V. I., 184, 228
Simpson, W. I., 379, 382, 396, 404–10, 414, 416–7, 420, 422, 434
Singh, M., 275, 303
Singh, T. R., 58, 105, 192, 224
Singhabhandhu, A., 296, 303
Sisido, M., 282, 284, 303

Skarzynska-Klentak, T., 28, 41, 102
Sklar, A. L., 482, 489
Skold, R., 205, 227
Slater, J. C., 232–3, 248, 254–5, 270
Slifkin, M. A., 271, 275, 279, 282–6, 289–90, 295–6, 302–3
Small, R., 41, 102, 125, 140
Small, R. W. H., 213, 223
Smerkdy, R., 15, 99
Smilansky, A., 195, 227
Smith, G. D., 205, 228
Smith, I. C., 38, 105
Smith, J. A. S., 344, 347–8, 360–1, 373, 382, 387, 399, 404, 407–11, 413–4, 417–8, 422, 426, 430, 434, 436, 438
Smith, J. W., 17, 105, 309, 342
Smith, L. C., 295, 302
Smolyanski, A. L., 16, 53, 69, 93, 99, 100–1, 123, 133, 138–40
Smyth, C. P., 306–10, 314, 316–9, 324, 326–8, 331–8, 340–2
Snijder, W. R., 44, 105
Snyman, J. A., 452, 456, 489
Sobczyk, L., 15, 28, 30, 32, 43, 51, 55–6, 59, 64, 69, 73, 75–6, 89, 100–2, 104–6, 193, 196–7, 199, 225, 227–8, 331, 341
Sokolik, I. A., 468, 489
Sokolov, N. D., 49, 58, 105, 125, 141
Solodovnikov, S. P., 126, 130, 141
Somorjai, R. L., 59, 106
Sonnino, T., 446–8, 450, 455
Soriano, J., 449–50
Soundararajan, S., 344, 437
Spatz, H. Ch., 474, 489
Speakman, J. C., 185, 223, 228
Spedding, H., 135, 141
Spencer, J. N., 44, 105–6
Spicer, D. P., 471, 492
Springs, B., 189, 228
Stacey, M., 501, 508
Staerk, H., 547, 554, 561, 570
Stallings, W., 188, 224
Stammreich, H., 21, 28, 104, 106
Stanley, H. E., 233, 270
Starosta, Ya., 17, 29, 99, 123, 139
Statz, G., 478, 491
Steele, W. A., 338, 342
Steffens, E., 547, 554, 561, 569
Steigman, J., 28, 41–2, 106

Steigman, W., 208–9, 224
Stenkamp, R. E., 186, 228
Sternhell, S., 38, 102
Stevens, E. D., 188, 228
Stevens, J., 439, 450
Stevens, V., 439, 450
Stiller, H., 461, 476
Stockmayer, W. H., 479, 491
Stokes, R. H., 79, 105
Stoops, W. N., 333, 342
Storms, R. D., 189, 228
Stout, G. H., 207, 228
Strandberg, B., 209, 226
Strauss, H. L., 337, 340
Strehlow, 143, 148, 176
Stromme, K. O., 2, 10–1, 101, 106
Stronski, R. E., 495, 508
Strure, W. S., 519, 570
Sucharda-Sobczyk, A., 73, 106
Suhai, S., 481, 491
Sukhorukov, B. I., 479, 491
Surprenant, H. L., 40, 105
Sussot, R., 269
Sutcliffe, B. T., 493, 508
Sutin, N., 174, 177
Sutton, L. E., 337, 340
Suwelack, D., 250, 270
Suzuki, T., 521, 544, 567
Suzuki, Y., 282, 302
Svensson, S., 496, 508
Svetina, S., 252, 254–5, 269
Swanson, R., 209, 228
Sweet, R. M., 212, 220, 225
Swift, H., 412, 416, 421, 424, 437
Switalski, J. D., 495, 508
Sychev, O. F., 379–80, 402, 429, 438
Szafran, M., 194, 197, 224, 227
Szent-Györgyi, A., 299–300, 303, 451, 491
Szpakowska, M., 95, 106
Szwarc, M., 90, 106

Tabuchi, D., 108, 141
Tachikawa, N., 457, 491
Tagesson, B., 168, 176
Tagusagawa, F., 185–6, 226
Taha, A. M., 295–6, 303
Takagi, H., 282, 284, 303
Takahashi, F., 292–3, 303
Takahashi, H., 189, 226

Takahashi, N., 470, 491
Takano, T., 209, 216, 218, 228
Takeda, K., 152, 176
Takemoto, S., 466, 490
Takenaka, A., 188, 277
Takusagawa, F., 191, 228
Tam, J. W. O., 18, 106
Tamura, C., 188, 191, 228
Tamura, S.-I., 539, 567
Tanaka, F., 291, 304, 565, 568
Tanaka, J., 2, 106, 542, 544, 567
Tanasijehuk, A. S., 123, 140
Taniguchi, Y., 546, 549, 551, 570
Tanimoto, O., 528, 568
Tatemitsu, H., 547, 568, 570
Tatsumoto, N., 112–3, 141
Tavares, Y., 21, 106
Taylor, A. F., 27, 98
Taylor, J. C., 9, 106
Tagenfeldt, J., 9, 97
Tellgren, R., 58–9, 106, 185, 227
Tenzer, L., 9, 106
Terao, H., 347, 377, 401, 414, 426, 429, 437
Terbojevich, M., 282, 284, 303
Teubner, M., 172–3, 177
Tewari, K. C., 135, 141
Thewalt, U., 273, 303
Thiebaut, J. M., 330, 342
Thomas, H., 231, 233, 255, 257, 260, 269
Thomas, H. T., 544, 566
Thomas, J. K., 555, 557–8, 567, 570
Thomas, J. M., 475, 491
Thomas, R., 185–6, 226
Thompson, C. C. Jr., 272, 275, 302, 477, 488
Thompson, H. W., 91, 106
Thusius, D., 174, 177
Tiddy, G. T., 164, 177
Tobin, M. C., 471, 492
Tokhadze, K. G., 110, 114, 133–5, 137–41
Tollin, G., 550, 567
Tolstaya, T. P., 386, 414, 425, 437
Tomasi, J., 205, 223, 227
Tondeau, J. J., 293, 302
Tondre, C., 168, 175
Tong, D. A., 347, 360–1, 372–3, 375, 381–2, 388, 399, 403–5, 407–12, 414, 418, 422, 426, 430, 435, 438

Tong, L. K., 151, 175
Topp, M. R., 525, 567
Topp, W. C., 47, 99
Torihashi, Y., 564–5, 568
Tornberg, N. E., 256, 269
Tournon, J., 283, 304
Townes, C. H., 345, 349, 434, 438
Toyama, M., 347, 424–5, 436
Toyoda, K., 2, 106, 478, 492
Traub, W., 195, 213, 227, 229
Treiner, C., 80, 106
Tremillon, R., 95, 97
Tribel, M. M., 468, 489
Trontelj, Z., 65, 98
Truong, K. D., 452, 492
Trus, B. L., 209, 228
Truscheit, E., 212, 220, 225
Tschesche, H., 212, 220, 225
Tsernoglou, D., 209, 229
Tsernoglu, D., 212, 228
Tsidris, J. C. M., 291, 304
Tsubomura, H., 562, 566
Tsujino, N., 541–2, 545, 549, 554, 568, 570
Tsukada, K., 385, 410, 431, 436
Tsuno, S., 565, 568
Tsutsumi, K., 563, 570
Tufte, T., 11, 101
Tupitsyn, I. F., 133–4, 139
Turner, D. H., 174, 177
Turner, D. W., 493, 507–8

Ubbelohde, A. R., 474, 491
Uchida, T., 452, 592
Uehara, H., 152, 176
Uehling, E. A., 232–3, 248, 254–5, 270
Ulaszkiewicz, Y. V., 127, 141
Ulbricht, W., 152, 163, 167–76
Umemoto, T., 520–1, 544, 567
Umeyama, H., 185, 228
Ushio, F., 537, 569
Usui, Y., 539, 567

Vaccaro, A., 19, 99
Vaks, V. I., 252, 254–5, 270
Van Audenhaege, A., 34, 41, 106, 196, 227
Van Belle, O. C., 335, 340
Van der Donckt, E., 52, 104

Van der Elsken, J., 338, 341
Van der Helm, D., 286, 304
Van der Meulen, P. A., 506, 508
Van Duijneveldt-Van de Rijdt, J. G. C. M., 478, 490
Van Even, V., 89, 101
Van Lerberghe, D., 43, 98
Van Wart, H. E., 216, 228
Vanbrabant-Govaerts, H., 31, 33, 101
Vasilescu, D., 470, 489
Vaughn, W. E., 307, 342
Vedel, J., 95, 97
Veis, A., 37, 104
Velichko, F. K., 401–3, 408–9, 434
Venkatesan, K. 186, 227
Verbist, J. J., 190, 226, 228
Verma, M. R., 275, 302
Versilov, V. S., 365–7, 377–8, 398, 414, 418–22, 438
Versinov, V. S., 377, 398, 414, 419–22, 433
Vidulick, G. A., 115, 120, 140
Vijayan, M., 188–9, 223
Vila Boas. L. F., 383, 400, 405, 412, 416, 419, 425, 435
Vincent, V. M., 452, 469, 492
Vinogradov, S. N., 12, 28–30, 50, 101, 105–6, 123, 140–1, 181–4, 199, 201, 225, 228
Vitoria, M. C., 17, 105
Vliegenthart, J. A., 478, 490
Voet, D., 290–1, 293, 303–4
Vogelmann, E., 539, 554, 570
Vogelsong, D. C., 27, 104
Vogt, B., 197, 226
Voitlander, J., 385, 412, 415–6, 422, 431, 434
Volkov, A., 256, 270
Volkov, A. F., 355, 362–6, 396–7, 405, 408, 410, 412, 416, 419–24, 435, 438
Volkov, V. E., 351, 375, 388, 413–5, 436
Von Hippel, P. H., 37, 105
Voronkov, M. G., 344, 347, 371, 410, 433, 435, 438
Vuylsteke, A., 31, 33, 101

Waara, I., 209, 226
Wachter, R., 80, 97

Waddington, T. C., 360, 390, 397–8, 402–5, 407, 409, 414, 417, 422–3, 429, 434
Wadso, I., 205, 227
Walden, P., 78–9, 106
Walker, J. E., 212, 225
Walker, S., 28, 50, 103, 332, 340
Wall, S. N., 152, 163, 167–8, 172–4
Wallwork, S. C., 275, 304, 457, 492
Walmsley, R. H., 275, 279, 303
Walsh, M. F., 164, 177
Wang, C. H., 189, 228
Warnhoff, E. W., 38, 103
Watenpaugh, K. D., 207–8, 217–9, 223, 229
Watson, H. C., 214, 228
Watson, H. E., 208–11, 218–21, 228
Weast, R. C., 10, 106
Webb, H. M., 73, 97
Weber, G., 291, 304
Weber, K. E., 475, 492
Weeks, C. M., 205, 228
Weger, M., 470, 488
Weidemann, E. G., 192, 225
Weigle, J., 309, 342
Weiner, J. H., 476, 492
Weisbecker, A., 330, 342
Weiss, A., 353, 364–6, 369, 377–8, 397–8, 408–9, 414–5, 417–23, 426–7, 433–4
Weisstuck, A., 147, 177
Wendell, P. L., 208–11, 218–21, 228
Wennerstrom, W., 125, 140
Werner, P. E., 290, 302
Wheland, G. W., 26, 106
Whidby, J. F., 37, 106
Whiffen, D. H., 336, 342
White, D. N. J., 185, 223
Whiting, R., 352–4, 398, 400, 406–7, 420–1, 423–4, 430, 433
Widernikowa, T., 29, 41–2, 102
Whillett, R. D., 386, 400, 408, 433
Williams, D., 56, 105
Williams, D. E., 317, 326, 338, 342
Williams, D. J., 193, 227
Williams, G., 201, 224
Williams, J. M., 7, 9, 97, 103, 185, 227
Williams, R. M., 457, 492
Willie, G., 336–8, 340
Willis, M. R., 486, 489
Winkler, F., 186, 227

Wirtz, K., 321–2, 328, 341
Witschonke, C. R., 86, 93, 106
Witzel, H., 195, 228
Wojtowicz, A., 199, 229
Wolfenden, R., 205, 229
Wong, S. T. K., 188, 224
Wood, J. L., 20–2, 24–5, 55–6, 58, 98, 100–1, 105, 192–4, 224, 229
Woos, J. L., 192, 227
Wright, C. S., 213, 223
Wright, J. D., 290, 303, 452, 454–5, 465, 469, 491–2
Wright, L. D., 291, 304
Wright, N., 15, 100
Wu, F. Y., 248–9, 254, 269
Wyckoff, H. W., 209, 229
Wyn-Jones, E., 143, 148, 150–2, 163–4, 167–8, 170–7

Yada, M., 2, 106
Yagi, K., 290, 304
Yakobson, G. G., 344, 403–6, 408, 411–2, 414–5, 417–23, 427, 429, 438, 498
Yakushi, K., 457–8, 470, 491–2
Yamada, K., 347, 362–3, 374, 384–5, 396, 398–401, 404, 424, 431, 437–8
Yamamoto, N., 536, 545–6, 568
Yamamoto, S., 565, 570
Yamanaka, C., 518, 524, 532, 537, 541, 543, 567, 569
Yamasaki, R. S., 385–6, 424–5, 430–1, 434, 438
Yamdagni, R., 184, 223
Yamoaka, T., 279, 303
Yano, V., 290, 302
Yarwood, J., 273, 304
Yashina, N. S., 369, 371, 382, 401, 403–9, 411–2, 417, 431–2, 437
Yasunaga, K., 151, 177
Yasunaga, T., 112–3, 141, 152, 176
Yomosa, S., 530, 570
Yonath, A., 195, 213–4, 223, 227, 229
Yonezowa, T., 35, 40, 100, 104
Yoshida, M., 547, 568, 570
Yoshihara, K., 279, 303, 465–6, 471, 490, 531, 542, 564, 568–70
Yoshikawa, K., 40, 104
Yu, B. S., 290–1, 302
Yusupov, M. Z., 365–7, 377–8, 384,

388, 398, 400–4, 408, 414–6, 418–
24, 427, 431, 433, 435, 438

Zakirov, D. U., 414, 418, 428, 438
Zana, R., 151–2, 163, 167–8, 172–5
Zboinski, Z., 470–1, 473, 486, 492
Zeegers-Huyskens, T., 2, 12, 14, 17, 23,
    28, 43, 53, 70, 73, 100, 102, 106,
    136, 140, 195–6, 224–5
Zefirov, Y. V., 478, 492
Zeks, B., 231, 233, 237, 240, 254–7,
    259–60, 269
Zelinski, I. V., 127, 141
Zhelonkina, L., 414, 418, 428, 438
Ziegler, J., 471, 489
Zimmerman, H., 40, 106
Zinenko, V. I., 251–5, 270
Zirkle, C. L., 293–4, 302
Znaskar, K. T., 205, 229
Zundel, G., 54, 69, 72, 75, 101, 103–7,
    123, 141, 181, 192, 197–8, 202–3,
    224–7, 229, 331, 342

Molecular Interactions, Volume 2
Edited by H. Ratajczak and W. J. Orville-Thomas
© 1981 John Wiley & Sons Ltd

Subject index

ab initio calculations, 47, 345
Absorption,
  by induced temporary dipoles, 339
  coefficient, 306
  far infrared, 339
  infrared, 332
  intensity, 338
  millimeter and submillimeter, 336
  of non-polar liquids, 337
  rotational type, 339
  transient spectra, 535–36
Accepting orbitals, 374, 385, 388
Acceptor,
  activity, 381, 385
  centre, 351, 377
  hydrogen bond, 181–82
  π, 374
  π inorganic, 375
  π strength, 385
Acceptor and accepting group, 351,
  354–56, 359–60, 363–65, 368–69,
  372–73, 375, 377, 379–81, 383–85
Acetic acid, 112, 114, 122, 132, 136,
  185
Acetonitrile, 292
Acetylcholine, 295
Acid salts,
  pyridine-2,3-dicarboxylate, 185
  sodium hydrogen diacetate, 185
Acids, Lewis, 365, 379
Acridine, 285, 288
Acridine orange, 288
Activation,
  enthalpy of, 320
  entropy of, 320
  free energy of, 320, 332
Activation energy,
  of dissociation, 113, 117
  hydrogen bond dissociation, 113, 117
  hydrogen bond formation, 112
  hydrogen exchange, 134–35

molecular exchange, 120
Activity, acceptor, 381, 385
Acyl-oxygen hydrogen bonds, carboxyl
  groups, 187
Adduct, chloranil-donor, 497
Adenine, 284, 288, 290–91
Aerobic state, 299
Aggregation,
  concentrations, 162
  mean number, 160, 163–64, 166
  number, micelles, 146–47, 152
  numbers, 146, 151, 164, 173
  other phenomena, 173–74
  space, 153–54, 158
  space flow through, 155
Alanine, infrared spectra, 189
L-Alanine, Raman spectra, 189
Alcohols, 129, 131–38, 333
Aldehydes, 299–300
Aliphatic amines, 278
Aliphatic amino acids, 282–83, 285–86
Alkaloids, 296
Alkyl bromides, 324
Alpha effect, 351, 359, 371, 377
Amides,
  group, 183
  hydrogen bonding in, 204
  infrared spectra, 205
  interaction with water, 205, 218
  N—H group, 204–5
  Raman spectra, 205
  self-association, 205
Amines, 123, 125–26, 128, 132–34,
  136–37, 281–82, 293, 299
  complexes with phenols, 196–97
  in monocation complexes, 192
Amino acids, 282–83, 285–86
  aliphatic, 282–83, 285–86
  crystal structure, 181
  hydrogen bonds in crystals, 182
  N—H group, 181, 183

Ammonia, 115
Ammonium carboxylate, interaction in, 182–84, 188–89, 193
Ammonium dihydrogen orthophosphate, 249
Ammonium groups, 181, 183
Ammonium salts, conductance, 193
AMP, 284, 287–88
Ampholytic surfactants, 144
Amplitudes, relaxation, 150–51
  concentration dependence, 172
  volume changes, 173
Anaerobic state, 299
Angle, bond, 343, 350, 361–62, 364, 367, 379
Angular frequency, 314
Aniansson and Wall model, 152–63, 171
  Gaussian form, 159
  Hermite polynomials, 160
  micelles, 153
  mixed micelles, 165
Anionic surfactants, 144
Anions,
  bromide, 182–84
  carboxylate, 182–83
  chloride, 182–84
  nitrate, 182–83
  oxygen, 182–83
  phosphate, 182–83
  sulphate, 182–83
Anisotropy, 309
Anthracene, 452, 454, 485, 488
Anthracene–dimethylpyromellitimide, complex, 469
Anthracene–pyromelliticdianhydride complex, 471–72
Anthracene–S-trinitrobenzene complex, 470–71, 473, 485
Anthracene–TCNQ complex, 456–58
Anthranilic acid, 283
Anti–syn Isomerization, 521
Antiferroelectric, 232, 250, 321
Antishielding factor, 349
Apparent volume, 325
Approximate methods, calculation, 345
Aquacobalamin, 285
Arc, 315–16
  plot, 316
  semicircular, 315
  skewed, 316
Arginine glutamate, 188
Aromatic amino acids, 282–84

Aromatic hydrocarbons, 285, 289
As–Al bond, 378
Aspartic acid, 185, 188
Assignment of quadrupole resonance lines, 352, 366, 385
Association, molecular, 308–9
Asymmetry parameter, 346–47, 350, 358, 363, 369, 374, 378, 380, 388
Atom,
  axial, 370, 372
  bridge, 351, 357–58, 366–67, 383, 388–89
  transfer hydrogen, 558, 656
Atomic charge, 386, 495–96, 502–3, 505
Atomic polarization, 306, 337–38
ATP, 284, 287–88
Atropine, 295
Attractive property, electron, 360, 384
Axes, principal set of, 346
Axial,
  atom, 370, 372
  equatorial NQR frequency splitting, 370
Azulene–s–trinitrobenzene complex, 456

B–Cl bond, 361
Bacillus alvei, 283
Back coordination, 360
Back CT transitions, 546, 560
Bacterial photosynthesis, 551
Band structure calculations, 479
Barrier of proton transfer, 129
Barriers, potential, 384
Bases, Lewis, 365
  organic, 381
Basicity, donor, 357
Bathochromic effect, 26
Behaviour, Debye, 315–16
Bending of bond, 337–367
Benzanthracene, 384
Benzene, 298
Benzenoid, 277
Benzidine, 464
Benzidine–TCNQ complex, 457, 470
Benzidine–trinitrobenzene complex, 457
Benziophene–S–trinitrobenzene complex, 456
Benzoic acid, 112–13
Benzpyrene, 284
Bifluoride, 123

Bifurcated hydrogen bond, 189, 191
Bilayer, 144
Bimaleate ion, 123
Binary interatomic potentials, 477, 479
Binding energy, 380
Biological regulation, 299
Biological systems, 376
Boiling point, 312–13
Bond, 363
   angle, 343, 350, 361–62, 364, 367, 379
   As—Al, 378
   B—Cl, 361
   bending of, 337, 367
   Br—Al, 367
   bridged, 380
   C—Cl, 361
   Cl—Sb, 373
   coordinative, 355–56
   covalent, 372
   double, 361
   Ga—Cl, 378
   He—As, 379
   Hg—Hal, 355
   intramolecular dative, 376
   L—Sb, 379
   length, 379, 386
   metal oxygen, 365
   N—P, 369
   P—Cl, 369
   Sb—Cl, 379
   Sn—Cl, 369
   Sn—Hal, 368
   Sn—Lig, 368
Bonding,
   chemical, 344, 361
   intermolecular, 348, 356, 358
   orbital, 348, 350, 352
   π, 350, 361, 363–64
Boron trifluoride complex, 501
Br—Al bond, 367
Breit–Wigner formula, 441
Bridge,
   atom, 351, 357–58, 366–67, 383, 388–89
   ligand, 357
   structure, 363
Bridged bond, 380
Broken symmetry, solvent-induced, 519
Bromide anions, 182–4
o-Bromoaniline, 452–53
o-Bromoaniline–picric acid complex,

452–53
Bromphenol blue, 126
Brucine, 295

C—H $\cdots$ O,
   interactions, 189
   interactions in glycylglycine monohydrate, 190
   interactions in L-histidine, 190
   interactions in pyridine carboxylic acids, 191
Caesium dihydrogen phosphate, 247
Calculation, 360–61, 371, 373–74, 378, 385
   approximate methods, 345, 494, 502
   band structure, 479
   CNDO/2, 24, 48–49
   theoretical, 361–62, 380, 387
Camphor, 326
Cancerous cell, 300
Caprolactam, 112–13
Carbazole–TCNQ complex, 459
Carbon tetrachloride, 295
Carboncarbonyl, 290, 299
Carbonyl group,
   hydrogen bond acceptor, 182
   hydrogen bond with amide N—H, 204–5
Carboxyl–carboxyl hydrogen bond, 185
Carboxyl groups, 182–83
   acyl oxygen hydrogen bonds, 187
   carboxylate hydrogen bonds, 187
   complexes with nucleic acid bases, 188
   complexes with ureido group, 188
   deformation on crystallization, 187
   O—H groups, 182–83
Carboxylate anions, 182–83
Carboxylate hydrogen bonds, carboxyl groups, 187
Carboxylate interactions, guanidium group, 182, 188
Carboxyl–carboxylate interactions, proteins, 187
Carboxylic acids, 120–22, 129, 133, 138
   complexes with amines, 203
   complexes with amines, proton transfer, 197, 203
   complexes with polypeptides, proton transfer, 197
   hydrogen bonds, 185–86
Carboxypeptidase A, 210, 212

Catalysis, micelles, 147
Catechol, 284, 286
Cationic surfactants, 144
Centre, acceptor, 351, 377
    donor, 351, 361, 381
Centre of,
    complexation, 353, 355, 367–69, 371,
        383
    coordination, 353
Centrosymmetric structure, 387
Chain length, 324
    in surfactants, 167, 171
Change in hybridization, 252, 355, 377,
    382
Character,
    covalent, 353
    donor, 369
    ionic, 354, 361, 373, 376
Characteristic time of proton transfer,
    123
Charge, 360, 371
    amount, 364, 373, 375, 385
    atomic, 386
    degree, 364, 373, 375, 385
    number, 78
    value of, 364, 373, 375, 385
Charge transfer, 351–52, 354–58, 361–
    81, 384–88, 449
    localized, 291
    non-relaxed state, 554
    solvent-induced, 516, 528
Charge-transfer complex, 351, 353, 355,
    447, 496
    conductivity and doping of, 470, 487
    crystal disorder of, 456, 459
    crystal structures of, 455–59
    energy level diagram, 466–68, 483–84
    molecular motion in crystals, 456
    paramagnetism, 465
    photoconductivity, 465, 468–69
    polymorphic transitions in, 452–53,
        456
    stoichiometry, 452–54
    ternary, 454
Charge transfer, extent of, 19, 23
    systems, phase diagrams, 452–54
    transition, 274
Charge-and-proton-transfer   complex,
    452
Chemical bonding, 344, 361
Chemical relaxation, 143

Chloranil, 285, 294–95
Chloranil–donor adduct, 497
Chloride anions, 182–84
Chloroacetic acid, 122
Chloroform, 295, 297–98
o-Chlorophenol, 134
Chlorpromazine, 276, 293–95
Chromatium high potential iron protein,
    208–9
Chrysene–TCNQ complex, 469
Chymotrypsin, 208–9, 213
Circular dichroism, 283
cis–position, 352
cis–trans,
    conformation, 352, 369–70, 389
    frequency shift, 370
    isomerism, 383
Cl—Sb bond, 373
Clusters in proteins, water, 220
CNDO/2 calculations, 24, 48–49
Coefficient, absorption, 306
Coenzymes, 289
Cohesion, 312
    energy of, 312, 314
Cole–Cole plot, 316, 320
    semicircles, 337
Cole–Davidson equation, 316
Collidine, 119
Collisions, molecular, 336
Complexes, 343–45, 351–89
    amine carboxylic acids, 203
    anthracene–dimethylpyromellitimide,
        469
    anthracene–pyromelliticdianhydride,
        471–2
    anthracene–S-trinitrobenzene,   470–
        71, 473, 485
    anthracene–TCNQ, 456–58
    benzidine–S-trinitrobenzene, 458
    benzidine–TCNQ, 457, 470
    benziophene–S-trinitrobenzene, 456
    boron trifluoride, 501
    carbazole–TCNQ, 459
    carboxyl–nucleic acid base, 186, 188
    carboxyl–ureido, 188
    carboxylic acid–amine, 203
    charge transfer, 351, 353, 355, 496
    charge-and-proton-transfer, 452
    chrysene–TCNQ, 469
    dielectric constant of, 315
    donor–acceptor, 351

excited EDA, 513
formation, enthalpy of, 33–36, 60
formation, 329, 350, 365–66, 368, 379, 385
ionic, 344
isomerism, 453
molecular, 344–47, 350–51, 362, 365, 381
n–σ, 344, 364, 367, 378
n–isopropylcarbazole–picryl chloride, 471
NQR, 446
nucleic acid base–carboxyl, 186, 188
o-bromaniline–picric acid, 452–53
o-phenanthroline–TCNQ, 455–56
outer, 5
perylene–chloranil, 465, 467
perylene–TCNQ, 452
phenanthrene–pyromelliticidian-hydride, 471–72
phenol–amine, 196–99
phenothiazine–TCNQ, 459
π–π, 344, 365
π–σ, 344, 364–65, 367
polarity of, 34
poly(N-vinylcarbazole)–iodine, 471–72
poly(N-vinylcarbozole)–trinitrofluore-none, 468
proton exchange in hydrogen bond, 107, 110, 130, 138
pyrene–pyromelliticdianhydride, 456
pyrene–TCNE, 468, 471
pyridine–iodine–halogen, 449
pyridine–iodomonochloride, 496
σ–σ, 344
   stabilization of, 385
strength, 383
tetrabenzonaphthalene–TCNQ, 469
trans, 369
trimethylamine–chloranil, 470, 487
ureido–carboxyl, 188
Van der Waals, 384
weak EDA, 527, 539
Complexation, 345, 348, 351–52, 355–57, 361–63, 367–71, 373–74, 376, 380–81
   centre of, 353, 355, 367–69, 371, 383
   effect of, 361, 368–69, 374, 380–81
   enthalpy of, 363–64
Complex in acid metal cyanides, N—H—
N, 193
Complex N $\cdots$ N distances, N—H—N, 193
Concanavalin, 209
Concentrations, aggregation, 162
Conductance,
   ammonium salts, 193
   uni-univalent electrolytes, 199
Conductivity and doping of charge trans-fer complex, 470, 487
Conductivity,
   ionic and electronic, 473–76
   ligand, 93, 96
   protonic, 474–75
   titration, 294
Configuration, 352, 362, 371
Conformation, 354, 368, 370–71
   effect of, 369
Constant of inner friction, 321
Contact ion-pair, 542, 546
Continuous techniques, 143
Cooperative proton motion, 476
Coordination, 355, 357, 376
   back, 360
   centre of, 353
   hexacoordination, 370–71
   intermolecular, 358, 386
   intramolecular 351, 358, 371
   number, 353, 358
   penta-, 371, 379
   π, 353
   tetra-, 357
   tri-, 353, 357
Coordinative bond, 355–56
Core, electron, 349
Core ionization energy, 495–96, 498, 500–1, 504–5
Correlation, 285, 289
Correlation factor, 329, 333, 335, 339
Correlation of parameter of Kirkwood, 83
Correlation parameter, 307
Coulomb potential, 443
Counter ions, micelles, 155
Coupling processes, micelles, 163
Covalent bond, 372
   character, 353
Critical exponents, 233, 248, 250, 253, 255
Critical micelle concentration, 145, 152, 159, 166–70

hydrocarbon chain length, 165, 167
Critical region, 235
Cryoscopy, 361
Cryptonase, 283
Crystal, 386
  disordered, 349
  effects, 351–52, 362, 368, 378, 381
  electric field, 351
  field effect, 368
  single, 347–48, 377
Crystal disorder of charge transfer, complex, 456, 459
Crystal structures, 356, 358, 366, 389
  amino acids, 181
  charge transfer complex, 455–59
Crystallographic equivalence, unequivalence, 346, 352
Cupric formate tetrahydrate, 248
Cyclic peptides, 206
Cytidine, 286
Cytosine, 284

Debye,
  behaviour, 315–16
  dispersion, 338
  equation, 306–9, 311, 314, 322, 327–28
  model, 441
  temperature, 440–41
Debye–Hückel equation, 82
Definition of,
  dielectric constant, 315
  dipole, 305
  frequency, 314
  moment, 305
  relaxation, 315
Deformation on crystallization, carboxyl groups, 187
Degree,
  amount, 364, 373, 375, 385
  charge, 364, 373, 375, 385
  value of, 364, 373, 375, 385
Denison and Ramsey's equation, 88–90
Density, electron, 374, 380
Deuteration, effect of, 387
Deuterium resonances, 388
d-hybridization, 350, 374, 377, 380
Di-L-leucine hydrochloride, 185
Diamagnetism, 275
Dicarbonyls, 299–300

Dichloroacetic acid, 122
Dielectric constant, 74–75, 83, 88, 95, 236, 246, 248, 255, 305
  complex, 315
  definition of, 315
  infinite frequency, 307
  influence of proton transfer, 74
  optical, 307
  static, 306
  use of, 305
Dielectric loss, 314, 316
  permittivity, 306
  relaxation, 314
  relaxation time, 61, 64
Diethyl ether, 114, 120
Diethylamine, 132–33
Differential overlap, method using neglect of, 494, 505
Diffraction, electron, 363
Dimeric structure, 351, 357–58, 366, 371, 386, 388
Dimerization, 289, 293
Dimers, 287
Dimethyglyoxime, 479
Dimethyl ether, 122
Dimethylformamide, 122
Dimethylsulphoxide, 129, 280–81
Dinitrobenzene, 298
Diphenylamine, 134
Dipole,
  definition of, 305
  –dipole interaction, 329
  electric, 305
  field of, 310
  induced, 306, 312
  induced dipole, 337
  interaction energy of, 312–13
  moment, 30–1, 60, 276, 305–6, 343, 385
  multiple–induced, 337
  orientation, 329
  polarizable point, 312
Dipole increment, 32–36
Dipole moments, phenol–amine complex, 196
Dipropylphosphine, 133
Direction, symmetry, 350
Disordered crystal, 349
Dispersion, Debye, 338
Dispersion forces, 272
Dissociation,

activation energy of, 113, 117
energies, 274
   enthalpy, proton transfer complex, 84–86
hydrogen bonded complex, 112–13
ionic, 529, 537, 545
ion-pairs influence of pK_a, 91
ion-pairs influence of dielectric constants, 83
kinetics of, 115
proton transfer complexes, 72
Dissociation constants, ion-pairs experimental values, 84
Dissociation constants, proton transfer complex, 77, 82, 84–86, 88–89, 92
Dissociation equilibrium constants, ion-pairs, 200
Dissociation yield, ionic, 549–50
Disubstituted quinones, 278
Distortion polarization, 306
Distribution curve, micelles, 152–53, 164, 173
Distribution, electronic charge, 343, 345, 368–69, 381, 388
Distribution minimum, micelles, 172–73
Distribution of relaxation time, 316
DNA, 287–88
Donor, 350–51, 354, 356–57, 360–61, 365–66, 368–70, 372–79, 381–83, 385, 388
  –acceptor formulation, 344
  –acceptor complex, 351
basicity, 357
centre, 351, 361, 381
character, 369
hydrogen bond, 181–82
inorganic, 382
n-donation, 364, 378
n-donor, 367, 372, 377, 384
π-donation, 364
π-donor, 372
power, 378
Doppler, effect, 442
Double bond, 361
strength, 356, 366, 373
Double minimum potential, 44, 46, 49–50, 52, 58–59
Durene, 477

Effect Doppler, 442

Effect of,
  complexation, 361, 368–69, 374, 380–81
conformation, 369
deuteration, 387
  interaction, 387
  solvent polarity, proton transfer, 199
  specific interaction, proton, transfer, 199
Effects, crystal, 351–52, 362, 368, 378, 381
Effects, infrared, 306
Elastase, 208–10, 212, 214
Electric conductivity, 78
ice, 474
Electric dipole, 305, 355
EFG, 345, 353
field gradient EFG, 208–10, 212, 214–350, 355, 379, 389
orientation, 347
Electric field, crystal, 351
Electric field, high, 330
Electric field gradient, 444
Electric quadrupole, interaction, 444
Electrochemistry, 276
Electron,
  $3d_{5/2}$, 380
attractive property, 360, 384
core, 349
density, 374, 380
diffraction, 363
n, 384
pdf, 349–50
π, 347, 365, 367, 377, 384
releasing character, 354
valence, 344–45, 349
Electron acceptor, 272, 277, 293
Electron affinity, 274
Electron donor, 272, 277, 285
Electronegativity, 371
Electronic,
energy, 343
polarization, 306
spectroscopy, 25–27, 50
Electronic charge, 374, 385
charge transfer effect, 358, 362, 377, 384
distribution, 343, 345, 368–69, 381, 388
structure, 343, 362, 376

Electronic structural change,
  solvent induced, 517–19, 548
Electrophoretic effect, 79–80
Electrostatic binding, 277
  forces, 385
  interaction, 311
Ellipsoid, 321
Ellipsoidal molecules, 308
Emission spectra, time-resolved, 522
Empty orbitals, 363–64
Energy,
  activation, 320, 322
  binding, 380
  core ionization, 495–96, 498, 500–
    1, 504–5
  cohesion, 312, 314
  dissociation, 274
  electronic, 343
  levels, 345–47
  level diagram, charge-transfer com-
    plex, 466–68, 483–84
  nuclear, 343
  promotion, 354
Enthalpy of activation, 320
Enthalpy of complex formation, 33–36,
  60, 363–64
Enthalpy of proton transfer, 41–43
  ion-pairs influence of pK_a, 91
  ion-pairs influence of dielectric con-
    stants, 83
  kinetics of, 115
  proton transfer complexes, 72
Entropy of activation, 320
Entropy of fusion, 314
Entropy of proton transfer, 41–43
Entropy changes, 279
Environmental effects, 7, 47, 49
Epinephrine, 286, 288
Equation, Debye, 306–9, 311, 314, 322,
  327–28
Equations for relaxation, 315
Equatorial NQR frequency splitting,
  axial–, 370, 372
Equilibrium constants, 280
Equilibrium constants, micelles, 161
Equilibrium constant of proton transfer,
  125
Equilibrium disturbance, relaxation, 150
Equilibrium excited state, 512, 514, 540
Equilibrium formation constants, 188,
  194–5, 201

enthalpy, 189, 195–6
Equivalent ionic conductance, 79–80
ESCA, 493
ESR, 276, 299–300
Ethane-like structure, 363
Ethylene dichloride, 296
Ethylene glycol, 287–88
Etioporphyrin, 297
Excimer, 509, 544
Excimer state, intramolecular, 521
Exciplex, 509, 547
  intermolecular, 513, 531
  intramolecular, 513, 528, 538, 546
  Experimental data, micelles, 165
  Excited EDA complex, 513
Excited state, 300
  Franck–Condon, 527, 540
Experimental criteria, 277
Experimental data, micelles, 165
Extended Hückel theory, 39
Extent of charge transfer, 19, 23
External heavy atom effect, 551
Extinction coefficient, 28–29

FAD, 290, 292–93
Far infrared absorption, 339
Fast process, micelles, 154–55, 159–61,
  163
Fast time, relaxation, 159–61, 164–66,
  171–73
Fatty acid esters, 325
Fermi resonance, 14, 16, 55
Ferredoxin, 217
Ferredoxin, ion pairs, 211–12
Ferroelastic, 231
Ferroelectric, 231–32, 250
Field,
  effect, crystal, 368
  of dipole, 310
  factor, Lorentz, 316
  gradient EFG, electric dipole, 343–50,
    355, 379, 389
  intense electric, 318
  internal, 307, 317, 319
  local, 307
Flavin co-enzyme, 290–91
Flavins, 283, 289–93
Flavoenzymes, 292
Flavoprotein, 291
Flavoquinones, 293

Florescence, 283
Fluorescence, charge transfer, 517, 540, 545
Fluorescence, time-resolved, 522, 524, 537
Fluoride, 183–84
p-fluorophenol, 120
FMN, 292
Folic acid, 298
Folinic acid, 298
Force constant, friction, 338
Forces, inductive electrostatic, 448
Formation equilibrium constants, ion pairs, 200
Formic acid, 121, 138, 185
Formulation, donor–acceptor, 344
Fourier spectroscopy, 337
Franck–Condon,
    excited state, 527, 540
    ground state, 514, 544
    principle, 510
Free energy of activation, 320, 332
Frequency,
    angular, 314
    definition of, 314
Frequency, relaxation, 150
Friction,
    force constant, 338
    inner, 320
Fuoss equation, 80

Ga—Cl bond, 378
Gamma ray measurements, spectroscopy, 370
Geminate,
    ion-pair, 547, 552–53
    recombination, 553, 561
Glutamic acid, 188
Glutathione, 299
Glyoxyl, 299
GMP, 288–89
Gouy–Chapman layer, 147
Grottus condition, 565
Ground state, Franck–Condon, 514, 544
Group, amides, 183
Group rotation, 334
Guanidine hydrochloride, 188
Guanidium group,
    carboxylate interactions, 182, 188
    phosphate interactions, 213

proton donor, 183
Guanine, 285
Guanosine, 285
Gutmann acceptor number, 76, 83
Gutmann donor number, 76, 82

Haematoporphyrin, 285, 298–99
Haemoglobin, 195, 208, 215
    ion pairs, 215
Half life of reaction, 149
Hamiltonian, 346
Hamiltonian, pseudo-spin, 236, 240, 244, 247, 256–57
He—As bond, 379
Heat of formation, 506
Heat of vaporization, 313
Hermite polynomials, Aniansson and Wall model, 160
Heteroatom, 367
Heteroconjugation, 93
Heteroexcimer, 509, 520, 526, 531–33, 537, 545–16
    intramolecular, 544, 547–18
Hexacoordination, 370–1
Hexacoordination, octahedral complex, 371
Hexadecapole term, 347
Hexafluoro-isopropanol, 120
Hexamethylbenzene, 497
Hexamethylbenzene–TCNE complex, 452
Hexamethylbenzene–TCNQ, 465
Hexamethylphosphorustriamide, 116–17, 119
Highest occupied molecular orbital, 289
Histamine, 284
Histidine, 283
Holotryptophanase, 283
Homoconjugated salts, infrared spectra, 18, 58
Homoconjugation, 93, 95
Homoconjugation constants, 94
Homogeneous recombination, 553
Hybridization, 343, 350–51, 372
    $3d_z^2s$, 353
    change in, 352, 355, 377, 382
    d, 350, 374, 377, 380
    f, 350
    s, 350, 381
    sp^2, 353, 360

sp^3, 360
sp^3d^2, 379
sp, 350
Hydrogen atom transfer, 558, 656
Hydrogen atom positions, 180, 207
Hydrogen bond, 29, 108, 112, 291, 329,
    331, 334, 350, 353, 355, 367, 370,
    375, 387–88
  acceptors, 181–82
  amides, 204
  bifurcated, 189, 191
  carboxyl–carboxyl, 185
  criteria for the formation of, 179
  donors, 181–82
  interaction, 525, 564–65
  intermolecular, 53, 112
  intramolecular, 119, 134
  ion pairs, 187, 195
  length, 183
  monoanion, 185
  monocation, 185, 191
  polarity of, 331
  rate of formation, 112
  quasisymmetrical, 124
  structure, 375
  symmetrical, 185
  types, in proteins, 207
Hydrogen bond complex,
  kinetic study dynamic NMR, 115
  kinetic study ESR, 126, 131
  kinetic study shock-wave technique,
    114
  kinetic study ultrasonic absorption,
    112–14
  kinetics, 111, 115
Hydrogen exchange,
  activation energy, 134–35
  kinetics, 111, 115, 135, 137
  limiting stage, 135, 137–38
Hydrogen halides, 137
Hydrophobic energy, surfactants, 167
Hydrophobic stacking, 283
Hydroquinone, 289
Hydroquinone clathrates, 463
Hydroxyl group, orientation of, 333
Hyperchromic effect, 26
Hyperfine coupling, 553
Hyperfine interactions, 442
Hypothetical volume, 324

Ice, electrical conductivity, 474
Ice rules, 248–49, 254
Imidazole, 284, 459–60, 474–75, 479–
    81, 484
Imidazolium, 284
Imidazolium group, 182–83
IMP, 284, 287
In-plane deformation vibration, 12–13,
    57
Independent processes, relaxation, 315
Indole, 278, 283–84, 289, 296
Indole–S-trinitrobenzene complex, 456
Induced dipole, 306, 312
Induced dipole–dipole, 337
Induced emission probability, 524
Induced moment, 306, 311–12
Inductive effects, 309
Inductive electrostatic forces, 448
Inertial effect, 339
Infinite frequency, dielectic constant,
    307
Infrared,
  absorption, 332
  effects, 306
  refractive index, 337
  spectroscopy, 275
  stopped-flow method, 130
Infrared and Raman, Spectroscopy, 343,
    352–53, 356–57, 361, 364, 375,
    382–83, 387
Infrared spectra, 331
  alanine, 189
  amides, 205
  homoconjugated salts, 18, 58
  monoanion complexes, 192
  monocation complexes, 192
Inner complexes, 5–6
Inner friction, constant of, 321
Inner friction, 320
Inorganic donor, 382
Integrated intensity, vibrational bands,
    12, 15
Intense electric field, 318
Intensity, absorption, 338
Interaction,
  electric quadrupole, 444
  energy of dipole, 312
  C—H $\cdots$ O, 189
  in ammonium-carboxylate, 182–84,
    188–89, 193

in glycylglycine monohydrate, C—H ··· O, 190
in L-histidine, C—H ··· O, 190
in pyridine carboxylic acids, C—H ··· O, 191
Interaction, 352, 360, 367
  dipole–dipole, 329
  effect of, 387
  electrostatic, 311
  hydrogen bond, 525, 564–5
  intermolecular, 343, 351–52, 355, 358, 366, 375
  intramolecular, 371, 375, 377
  molecular, 439
  monopole, 443
  N—H—S, 216
  n–σ and π–σ, 364
  orbital, 359
  π–π, 375
  potential surface proton donor and acceptor, 109–10, 123
  proton-lattice, 255–60, 264, 266
  solid state, 455–56, 476–79
  solute–solvent, 332, 510, 513–14, 522–23
  steric, 294
  strong, 529
  very weak, 529–30
  weak, 529
  in water, 217–18
Interaction energy, 312–13
  dipole, 312–13
  London–Van der Waals, 312–13
Interactions, hyperfine, 442
Interatomic distances, 275
Intermediate species, micelles, 146, 153, 156, 171
Intermolecular,
  exciplex, 513, 531
  hydrogen bond, 54, 112
  interaction, 343, 351–52, 355, 358, 366, 375
  stretching vibration, 12–13, 55
Internal field, 307, 317, 319
Internal refractive index, 307
Internal rotation, 113, 119
Internuclear distance, 1, 3, 46, 60
Interpretation of QCC, 349–51
Intersystem crossing ISC, 552
Intramolecular,

excimer state, 521
exciplex, 513, 528, 538, 546
  heteroexcimer, 544, 547–48
  hydrogen bond, 119, 134
  interaction, 371, 375, 377
Intramolecular N—H ··· Q—C bonds, peptides, 205
Iodanil, 295
Iodine, 285–86, 288, 295, 452, 455, 464–65, 470
Iodine, molecular complexes, 439
Iodine–benzene complex, 447
Ion, 344, 387
  complex, 374
  structure, 387
Ion pairs,
  dissociation equilibrium constants, 200
    experimental values, dissociation constants, 84
  ferredoxin, 211–12
  formation equilibrium constants, 200
  geminate, 547, 552–53
  haemoglobin, 215
  hydrogen bond, 187, 195
  influence of pK_a, dissociation, 91
  influence of dielectric constants, dissociation, 83
  O—H ··· N and N—H ··· O, 195
  roles in proteins, 211
  serine proteases, 213
  staphylococcal nuclease, 213–14
  trypsin–trypsin inhibitor complexes, 212
Ion-pairs, proton transfer, 564
Ionic, 364, 369, 381
  character, 354, 361, 373, 376
  complex, 344
  dissociation, 529, 537, 545
  dissociation yield, 549–50
  micelles, 147
  photodissociation, 548, 551–52, 554
  strength, micelles, 163
  strength and, micelles, 171
  structure, 362, 372, 385
Ionic and electronic conductivity, 473–76
Ionic pair formation, kinetics, 126
Ionicity, 350, 352, 355, 361, 369, 379
Ionization potential, 332, 354, 366, 377, 388

Ionization potentials, 274, 290, 293
Ionogenic processes, 4
Ising,
   model, 233, 244
   transverse tunnelling field, 239, 257
Isoalloxazine, 291
Isomer, 380
Isomer shift, 443
Isomerism, complex, 453
Isomers, rotational, 330
N-Isopropylcarbozole–picryl chloride, complex, 471
Isotope effects, 247, 256, 267
Isotopic ratio, 14, 18–19, 58–60

Kinetic study dynamic NMR, hydrogen bond complex, 115
Kinetic study ESR, hydrogen bond complex, 126, 131
Kinetic study shock-wave technique, hydrogen bond complex, 114
Kinetic study ultrasonic absorption, hydrogen bond complex, 112–14
Kinetics,
   of dissociation, 115
   hydrogen bond complex, 111, 115
   hydrogen exchange, 135, 137
   ionic pair formation, 126
   micelles, 148, 151
   molecular exchange, 121–22
   proton transfer, 125
Kirkwood equation, 307–8
Koopmans theorem, 494, 501, 507
Kosower Z-parameter, 76
Kraft point, 146

L–Sb bond, 379
Landau theory, 234, 250
Laser methods, picosecond, 513, 533, 559
Laser photolysis,
   nanosecond, 535, 546, 552, 562
   picosecond, 552, 566
Lattice dimensionality, 231, 233, 248, 254
Lead diorthoarsenate, 243
Lead diorthophosphate, 243–44, 255–56
Length, bond, 183, 379, 386
   N—H $\cdots$ O bonds, 181–82
Lewis acids, 365, 379

Lewis bases, 365
Levels, energy, 345–47
Libration, 338
Librational absorption, 339
Ligand, 354, 356, 360, 363, 365, 370–71, 379, 382
   bridging, 357
   conductivity, 93, 96
   oxygenated, 364
Ligands, 287
Light scattering, micelles, 153, 161, 166
Limiting stage, hydrogen exchange, 135, 137–38
Linear combination of atomic orbitals, 345
Lippincott–Schroeder potential function, 44
Liquid crystal, lyotropic, 144
Liquid structure, 336
Local field, 307
Localized charge transfer, 291
London–Van der Waals interaction energy, 312–13
Lone pair, 363, 372, 378
   orbitals, 350
Loose heteroexcimer, 535
Lorentz field factor, 316
Loss, dielectric, 314, 316
Loss tangent, 314
Lumiflavin, 290
Luminescence, 283
Lyotropic liquid crystal, 144
Lysozyme, 212

Macroscopic relaxation time, 305–6, 316–18, 324, 331
Macroscopic viscosity, 325
Madelung effect, 49
Madelung potential, 500
Magnetic field, 345, 347
Magnetic resonance shift, proton transfer, 180
Mean number aggregation, 160, 163–64, 166
Mechanism of electrical conductivity, 467, 472–76, 484–88
Melting point, 314
Membranes, 530, 555
Menshutkins complex, 365, 377–78
Mesitylene, 477
Metal, 363, 374, 376, 379

–oxygen bond, 365
–trimethyl molecules, 363
Mctalloporphyrins, 297
Methanes, tetrasubstituted, 325–26
Methanol, 132–34, 136, 138
Methorexate, 298
Methylamine, 115
Methylchloromethanes, 326
Methylene blue, 288
Methylene chloride, 470
Methylglyoxal, 300
Methyl group, hindered rotation, 364
Methylnitroamine, 127–29
N-Methylphenothiazine, 294–95
Micelles, 144, 146, 530, 555
    aggregation number, 146–47, 152
    Aniansson and Wall model, 153
    catalysis, 147
    counter ions, 155
    coupling processes, 163
    critical micelle concentration, 145, 152, 159, 162, 166–67
    critical micelle concentration table, 168–69
    distribution curve, 152–53, 164, 173
    distribution minimum, 172–73
    equilibrium constant, 161
    experimental data, 165
    fast process, 154–55, 159–61, 163
    Gouy–Chapman layer, 147
    intermediate species, 146, 153, 156, 171
    ionic, 147
    ionic strength, 163
    ionic strength and, 171
    kinetic model, 151
    kinetics, 148, 151
    light scattering, 153, 161, 166
    mixed, 164–65
    model for formation, 152
    monomer exchange, 151, 154, 157, 163, 165
    neutral molecule exchange, 164
    pseudo-equilibrium, 157, 159–60
    rate constants, 149, 151, 163–64, 167
    rate constant table of, 168–70
    rate equations, 155–63
    redistribution of, 154
    Sams, Wyn-Jones and Rassing model, 163
    second c.m.c., 147

slow processes, 154–55, 161
solubilization, 147
solutions, 554
spherical shape, 147
step-wise build up, 154
stern layer, 147
thermodynamic properties, 172–73
volume changes, 172–73
water exchange, 164
Microscopic relaxation time, 305, 317, 323–24
Microscopic viscosity, 325
Microsomal oxidations, 299
Microwave spectroscopy, 348
Millimeter and submillimeter absorption, 336
Mixed micelles, Aniansson and Wall model, 164–65
Mixed micelles, relaxation, 164–65
Mixture, 319, 337
Model, Debye, 441
Model for formation, micelles, 152
Model, Ising, 233, 244
Model, potential, 495, 505
Model, Slater-Takagi, 254–55
Model, tunneling and hopping, 486–88
Molar, polarization, 306
Molecular,
    association, 308–9
    complexes, iodine, 439
    interaction, 439
    moment, 305
    motion, 345, 348, 375
    orbitals, 370, 385
    polarizability, 306
    relaxation time, 305, 317–18
    structure, 350, 373, 376, 378
    viscosity, 325
Molecular exchange,
    activation energy, 120
    hydrogen bond complex, 109
    kinetics of, 121
    rate of, 115
Molecular ionic tautomerism, 109, 123–24
Molecular orbital calculation, 494, 502
Moment,
    definition of, 305
    dipole, 276, 305–6, 343
    induced, 306, 311–12
    molecular, 305

Momentum drift, 322
Monoanion, hydrogen bond, 185
Monoanion complexes, infrared spectra, 192
Monocation, hydrogen bond, 185, 191
Monocation complexes, infrared spectra, 192
Mono-substituted quinones, 278, 282
Monomer exchange, micelles, 151, 154, 157, 163, 165
Monomer–dimer relaxation, 112, 114
Monomeric unit, 366
Monomers, 144, 146, 151–55, 158, 160, 164–65, 167, 172
Monomers, thermodynamic properties of, 172
Monopole interaction, 443
Mössbauer spectroscopy, 348, 371, 379–80
Mössbauer effect, 439
Most probable relaxation time, 316
Motion,
    molecular, 345, 348, 375
    orientational, 388
Motional effect, 387
Mulliken atomic population, 25
Multimers, 335–36
Multiplicity, quadrupole resonance spectrum, 347, 353, 357–58, 368–71, 376
Mutual viscosity, 323, 326
Myoglobin, 208, 216, 218

N-Alkylanilines, 134
N—H group,
    amides, 204–5
    amino acids, 181, 183
    peptides, 181, 183
N—H · · ·O bonds,
    lengths, 181–82
    types, 181
N—H · · · S interactions, 216
N—H—N,
    complex in acid metal cyanides, 193
    complex N · · · N distances, 193
N—H—S, interaction, 216
N—H—S, interaction,
    hydrogen bond types, 207
    salt bridges, 187–88
n–σ and π–σ, interaction, 364
Nactryptophanamide, 288

Nanosecond,
    laser photolysis, 535, 546, 552, 562
    time-resolved spectroscopy, 513
Naphthalene–S-trinitrobenzene complex, 456
Naphthalene–tetracyanobenzene complex, 456
Naphthaquinone, 295
Neutral molecule exchange, micelles, 164
Neutral red, 288
Neutron diffraction, 3, 5, 8, 59, 180–81
Nicotinamide, 282
Nicotinamide biscoenzyme, 291
Nitro group, 314
Nitrogen, 278, 299
Nitrophenol, 290
o-Nitrophenol, 117, 119
NMR, 36, 64
No-band wave function, 385
Non-ionic surfactants, 144
Noncentrosymmetric structure, 387
Non-relaxed charge transfer state, 554
Non-polar liquids, 336
    absorption of, 337
NQR, 65, 345, 350, 355–56, 359–64, 366–67, 370, 372, 377–87, 389
    pressure dependence 345, 347–48, 366
    spectroscopy, 343–45, 348, 351, 353, 365–66, 368–69, 371, 375–76, 380, 388
NQR frequencies, temperature dependence, 345, 347, 353–55, 366–67, 369–70, 374–75, 377–78, 384, 386–87
Nuclear gamma resonance, 440
Nuclear magnetic resonance spectra, see NMR,
Nuclear quadrupole resonance, see NQR,
Nuclear relaxtion time, temperature dependence, 345, 367, 377–78, 387
Nuclear site symmetry, 343, 353
Nuclear spin, 345–46, 358
Nuclease, staphylococcal, 213–14, 221
Nucleic acid, 283–88
Nucleic acid bases, complex with carboxyl groups, 186
Number, quadrupole resonance lines, 356

O—H groups,
  carboxyl, 182–83
  water, 182, 184–85, 218
O—H · · · N and N—H · · · O, ion pairs,
  195
Octahedra, 358
Octahedral complex, 369, 371, 377
  hexacoordination, 371
  hybridization, 379
  structure, 370, 372, 381
Octopole moment, 337
Onsager equation, 80, 307–8, 329–30
Optical dielectric constant, 307
Optical spectroscopy, 343
Orbitals, 344
  ab initio MOs, 345
  accepting, 374, 385, 388
  bonding, 348, 350, 352
  empty, 363–64
  interaction, 359
  linear combination of atomic, 345
  lone pair, 350
  molecular, 370, 385
  p, 350, 360, 380, 388
  population, 345, 350, 359–60, 380
  s, 349, 380
  σ, 365, 384
  three centre delocalized σ, 374
  valence, 343, 345, 350
Order parameter, 232, 253
Organic bases, 381
Orientation dipole, 329
Orientation effects, 279
Orientation electric dipole, 347
Orientation of hydroxyl group, 333
Orientational motion, 388
Out-of-plane deformation vibration, 12–
  13, 57
Outer complex, 5
Oxygen, 299–300
Oxygen anions, 182–83
Oxygenated ligand, 364

p-orbitals, 350, 360, 380, 388
Pairing, quadrupole resonance lines, 346
Papain, 195
Paramagnetism, 275–76
  charge transfer complex, 465
Parvalbumin, 208
P—Cl bond, 369
pdf electrons, 349–50

Penta-coordination, 371, 379
Pentafluoropropanol, 114
Peptides,
  cyclic, 206
  hydrogen bonds in crystals, 182
  intramolecular N—H · · · O—C bonds,
  205
  N—H group, 181, 183
  water interactions, 206
Perfluoro-*t*-butanol, 116
Permittivity, dielectric, 306
  effect on proton transfer, 199
Permittivity of the medium, 50
Perylene 452
Perylene–chloranil complex, 465, 467
Perylene–TCNQ complex, 452
Phase diagrams, charge transfer systems,
  452–54
Phase transition, 232, 243, 248, 354,
  363–64, 374, 384
Phenanthrene, 454
Phenanthrene–pyromelliticdianhydride
  complex, 471–72
o-phenanthroline TCNQ complex, 455–
  56
Phenazine, 465
Phenol, 289
Phenol–amine complexes, proton trans-
  fer, 196–99, 202–3,
  dipole moments, 196
Phenols, 117, 119–20, 125, 132–34
Phenosfranine, 288
Phenothiazine, 293–96, 465
Phenothiazine–TCNQ complex, 459
Phenyl, 283
Phenylalanine, 283
p-phenylenediamine, 464
Phosphate, 276
Phosphate anions, 182–83
Phosphate interactions, guanidium group,
  213
Phosphine, 293
Phosphorylating chain, 301
Photoconductivity, charge transfer com-
  plex, 465, 468–69
Photoconductivity transient, 546, 551
Photodissociation, ionic, 548, 551–52,
  554
Photosensitizers, 287
π,
  acceptor, 374
  bonding, 350, 361, 363–64

coordination, 353
donation, donor, 364
donor, donor, 372
electron, 347, 365, 367, 377, 384
inorganic acceptor, 375
π complex, 344, 365
π interaction, 375
σ complex, 344, 364–65, 367
strength, acceptor, 385
Picosecond,
  laser methods, 513, 533, 559
  laser photolysis, 552, 566
Picosecond time-resolved spectroscopy, 513, 540
Picric acid, 282, 452, 454
Picryl chloride, 452–53, 488
Plot, arc, 316
PMR, 332
Polarity of complex, 34
Polarity of hydrogen bond, 331
Polarizable point dipole, 312
Polarizability, 54
  molecular, 306
Polarization,
  atomic 306
  distortion, 306
  electronic, 306
  molar, 306
  solvent-induced, 520–21
  spontaneous, 232–33, 246
  sublattice, 232
Polarization of the solvent, 49
Poly(N-vinylcarbazole)–iodine complex, 471–72
Poly(N-vinylcarbazole--trinitrofluorenone complex, 468
Poly-L-histidine hydrochloride, 283–84
Poly-L-tryptophan, 283–84
Polyacrylamide, 277
Polycrystalline samples, 347, 362
Polycytidylic acid, 286
Polyglycine, 481
Polymer conformation, 282
Polymeric structure, 353, 374
Polymeric substances, 349, 364–65, 368
Polymorphic transitions in charge transfer complex, 452–53, 456
Polymorphism, 363
Polysarcosine, 282

Population, orbitals, 345, 350, 359–60, 380
Porphyrazine, 326–27
Porphyrins, 296–97, 299
Position, *trans* effect, 352, 278
Potassium dihydrogen orthophosphate, 249, 254
Potential barrier, 38, 59, 62
  model, 495, 505
Potentials, binary interatomic, 477, 479
Potential surface proton donor and acceptor, interaction, 109–10, 123
Power, donor, 378
Pressure dependence, NQR, 345, 347–48, 366
Pressure jump, 143, 148, 150–51, 165, 171–72
Primary amines, 278–79
Principle, Franck–Condon, 510
Principle of maximum superposition, 275
Principal set of axes, 346
Proflavine, 288
Proline, 279–81
Promotion energy, 354
Propionic acid, 113
Proteins, 299–300, 324
  carboxyl–carboxylate interactions, 187
Protinfluence affinity on proton transfer, 73
Proton acceptors, 331, 333
  affinity, 44, 46–47, 73
  donor, guanidium group, 183
  donors, 331–32
  exchange in hydrogen bond, complex, 107, 110, 130, 138
  magnetic resonance, 332
  motion, cooperative, 476
Proton transfer, 558
  pK_a influence, 65
  barrier of, 129
  carboxylic acid–amine complexes, 197, 203
  carboxylic acid–polypeptide, complexes, 197
  characteristic time of, 123
  dielectric constant influence, 74
  effect of solvent polarity, 199
  effect of specific interaction, 199
  enthalpy, 41–43

entropy, 41–43
equilibrium constant of, 125
hydrogen bond complex, 107, 109, 122–23, 125, 128
kinetics, 125
magnetic resonance shift, 180
other specific influences on, 74
permittivity effect on, 199
phenol-amine complexes, 196–99, 202–3
proton affinity influence, 73
rate of, 128
relationship to pK_a, 195–96
reversible, 109, 122
synchronous, 136
Proton transfer complex, dissociation enthalpy, 84–86
Proton transfer complexes, dissociation, 72
thermodynamic data, 41–43
Proton tunnelling, 237, 239, 253, 257, 267
Proton–lattice interaction, 255–60, 264, 266
Protonic conductivity, 474–75
Protonic order–disorder transitions, 240, 244, 248, 255
Pseudo-equilibrium, micelles, 157, 159–60
Pseudo one-dimensional systems, 233, 243
Pseudo-spin Hamiltonian, 236, 240, 244, 247, 256–57
Pseudo two-dimensional systems, 233, 248
Psychotropic compounds, 294
Pteridine, 297
Purine, 285–86, 289–90, 296
Pyrene, 452–53, 488
Pyrene–pyromelliticdianhydride complex, 456
Pyrene–TCNE complex, 468, 471
Pyridine, 117, 122, 287
Pyridine-2,3-dicarboxylate, acid salts, 185
Pyridine–iodine–halogen complex, 449
Pyridine–iodomonochloride complex, 496
Pyridoxal-P, 283
Pyrimidine, 286, 289

Pyromelliticdianhydride (PMDA), 452, 464
Pyronine G, 288
Pyrrole, 296

QCC, 344–45, 347–48, 350–53, 355–58, 362–64, 368–69, 374, 376–80, 382, 385–86, 388–89
Quadrupole, 314
Quadrupole moment, 337, 344–46, 352, 381
Quadrupole nucleus, 343, 349, 356, 375, 385
Quadrupole resonance, 348, 351–52, 354–55, 357–58, 360–61, 372–73, 375–80, 384–85, 388
frequency, 344, 347–48, 353, 355–64, 366–69, 371, 373–76, 378–79, 381, 383, 386–87
frequency shift, 351–53, 356, 360, 362, 364, 371, 377–80, 383–85, 387–88
lines, 346–48, 352, 361, 388
assignment of, 352, 366, 385
number, 356
pairing, 346
shape and width, 345, 347, 349, 362, 375
spectrum, 345–46, 348, 352, 356–57, 366, 368–71, 373–74, 380, 386
multiplicity, 347, 353, 357–58, 368–71, 376, 378–79
Quadrupole splitting, 445
Quasisymmetrical hydrogen bond, 124
Quinhydrone, 279
Quinone, 190, 275, 278–82, 284, 289, 293

Radii, Van der Waals, 4, 46, 216, 325, 358
Radii N, O, C, H, Van der Waals, 179
Raman spectra, 22–23, 57, 180
L-alanine, 189
amides, 205
Rate constants, 279
Rate constant, micelles, 149, 151, 163–64, 167–70
Rate equations, micelles, 155–63

Rate of,
   hydrogen bond formation, 112
   hydrogen exchange, 133–34
   molecular exchange, 115
   proton transfer, 128
Rate theory, 320
Reaction field, 312
Recoil energy, 440
Recoilless fraction, 440
Recombination, geminate, 553, 561
Redistribution of micelles, 154
Reduced relaxation time, 332
Reflection coefficient, 306
Refractive index, 306
   infrared, 337
   internal, 307
Reichardt E_T parameter, 76
Relationship to pK_a, proton transfer, 195–96
Releasing character, electron, 354
Relative weights, 315
Relaxation,
   amplitudes, 150–51
   amplitudes concentration dependence, 172
   amplitudes volume changes, 173
   chemical, 143
   definition of, 315
   dielectric, 314
   equations for, 315
   equilibrium disturbance, 150
   fast time, 159–61, 164–66, 171–73
   frequency, 150
   independent processes, 315
   methods, 148, 158
   mixed micelles, 164–65
   slow time, 152, 156, 159, 161, 172–73
   solvation, 528, 530, 537
   vibrational, 511–12, 528–29
Relaxation effect, 79
Relaxation time, 150–53, 159, 165, 173, 345, 348–49, 367, 377–78, 384
   distribution of, 316
   macroscopic, 305–6, 316–18, 324, 331
   microscopic, 305, 317, 323–24
   molecular, 305, 317–18
   most probable, 316
   reduced, 332
Reorientation relaxation, 511–12, 522
Repulsion forces, 46
Resonant nucleus, 343–45, 349, 370, 385

Respiratory transport chain, 299
Reststrahlen, 306
Reversibility, 279
Reversible proton transfer, 109, 122
Ribonuclease, 195
Roles in proteins, ion pairs, 211
Roles of proteins, water, 220
Rotamer, 31
Rotation, group, 334
Rotation of methyl group, 362
Rotation spectra, 348
   freedom, 314
   frequencies, 378
   isomers, 330
Rotational type, absorption, 339
Rubredoxin, 207–8, 217–18

s-hybridization, 350, 381
s-orbitals, 349, 380
$S_1 \rightarrow S_n$ spectra, 542
Salicylic aldehyde, 112, 118
Salt bridges, N—H—S interaction, 187–88
Sams, Wyn-Jones and Rassing model, 163–65
Sandwich heteroexcimer, 535
Sb—Cl bond, 379
Second c.m.c., micelles, 147
Self-association, amides, 205
Semicircular arc, 315
Semiconductors, 299
Semi-empirical calculation, 362
Semiquinone, 289
Serine proteases, ion pairs, 213
Serotonin, 295
Shape and width, quadrupole resonance lines, 345, 347, 349, 362, 375
Shells, valence, 347, 365, 372
Shock-wave technique, 143, 148
σ-orbitals, 365, 384
σ–σ complex, 344
Single crystal, 347–48, 377
Single minimum potential, 25, 28
Sites, 346, 370, 380
Skeletal vibrations, 13, 21, 23–25
Skewed arc, 316
Slater–Takagi model, 254–55
Slow processes, micelles, 154–55, 161
Slow time, relaxation, 152, 156, 159, 161, 172–73
Sn—Cl bond, 369

Sn—Hal bond, 368
Sn—Lig bond, 368
Soap films, 144
Sodium cacodylate, 288
Sodium hydrogen diacetate, acid salts, 185
Sodium trihydrogen selenite, 248, 255
Soft mode, 240–41, 243, 256, 264–66
Solid state, 276
Solid state effects, 351, 366
Solubility, 287, 312, 314
Solubilization, 147
Solute–solvent interaction, 332, 510, 513–14, 522–23
Solutions, micelles, 554
Solvation, 279
  relaxation, 528, 530, 537
Solvation effects, 297
Solvent, specific interaction of, 91, 93
Solvent effect, 309–11
Solvent polarity, 21, 28, 50–52
Solvent-induced,
  broken symmetry, 519
  charge transfer, 516, 528
  electronic structural change, 517–19, 548
  polarization, 520–21
sp^2 hybridization, 353, 360
sp^3 hybridization, 360
sp^3d^2 hybridization, 379
sp hybridization, 350
Space, aggregation, 153–54, 158
Space flow through, aggregation, 155
Specific bonds, 49, 74–75
Specific interaction of solvent, 91, 93
Spectra,
  $S_1 \rightarrow S_n$, 542
  $T_1 \rightarrow T_n$, 543
  valence photoelectron, 506
Spectrometers, 348–49, 354
Spectroscopy,
  electronic, 25–27, 50
  ESR, 296, 299–300
  Fourier, 337
  gamma ray measurements, 370
  IR, 275
  IR and Raman, 343, 352–53, 356–57, 361, 364, 375, 382–83, 287
  microwave, 348
  Mössbauer, 348, 371, 379–80
  NMR, 36, 64
  NQR, 343–45, 348, 351, 353, 365–66,

368–69, 371, 375–76, 380, 388
  optical, 343
  picosecond time-resolved, 513, 540
  UV photoclectron, 493
  vibrational, 7, 21, 52, 57
Spherical shape, micelles, 147
Spontaneous polarization, 232–33, 246
Spontaneous transfer, 300
Squaric acid, 248, 250–51, 255–56
Stabilization of complex, 385
Stacking, 291
Standard free energy of transfer, 65
Stannous chloride dihydrate, 248
Staphylococcal nuclease, ion pairs, 213–14
State, 343
Static dielectric constant, 306
Step-wise build up of micelles, 154
Steric,
  hindrance, 275
  interaction, 294
Steric effect, 362, 365, 377, 382, 386
  hindrance, 368, 370
Stern layer, 147
Steroids, 285
Stiochiometry, charge transfer complex, 452–54
Stopped-flow method in IR spectroscopy, 130
Stopped-flow techniques, 143, 148, 150–51, 172
Strength,
  complex, 383
  donor, 356, 366, 373
Stretching vibration, 12, 52, 55, 57
  intermolecular, 12–13, 55
Strong interaction, 529
Structural arrangement, 369
Structural properties, 381
Structure, 356–57, 365, 369, 371–73, 375, 379–81, 383, 386
  bridge, 363
  centrosymmetric, 387
  crystal, 356, 358, 366, 389
  dimeric, 351, 357–58, 366, 371, 386, 388
  electronic charge, 343, 362, 376
  ethane like, 363
  hydrogen bond, 375
  ion, 387
  ionic, 362, 372, 385
  liquid, 336

molecular, 350, 373, 376, 378
non-centrosymmetric, 387
octahedral complex, 370, 372, 381
polymeric, 353
Sublattice, polarization, 232
Substitution effect, 374
Subtilisin, 211
Sulphate anions, 182–83
Sulphite, 292–93
Sulphur, 299
Sulphydryl, 300
Surfactants,
  ampholytic, 144
  anionic, 144
  cationic, 144
  chain length and, 167, 171
  hydrophobic energy, 167
  non-ionic, 144
  properties of, 143
  table of, 168–70
Symmetrical hydrogen bond, 185
Symmetry,
  direction, 350
  nuclear site, 343, 353
  tetrahedral, 360
Synchronous proton transfer, 136
System, 365–66, 384

$T_1 \rightarrow T_n$ spectra, 543
Table of surfactants, 168–70
Tafts constant, 359
Tautomerism, 44–45, 53, 55, 60, 62
TCNB, 295
TCNE, 295
TCNQ, 295
TCNQ tetracyanoquinodimethane, 499
Temperature coefficient NQR frequencies, 348, 356, 364
Temperature, Debye, 440–51
Temperature dependence,
  NQR frequencies, 345, 347, 353–55, 362–3, 366–7, 369–70, 374–75, 377–78, 384, 386–87
  nuclear relaxation time, 345, 367, 377–78, 387
Temperature-jump, 143, 148, 150–51, 172
Terminal atoms, 357–58, 366–67, 374, 383, 386, 388–89
Ternary charge–transfer complex, 454

Tetrabenzonaphthalene–TCNQ complex, 469
Tetra-coordination, 357
Tetracyanoethylene, 477
Tetracyanoquinodimethane (TCNQ), 452, 464–65, 469
Tetrahedral symmetry, 360
Tetrahydrofurane, 121
Tetramethylpheryldiamine, 298
Tetraphenylporphyrin, 296
Tetrasubstituted methanes, 325–26
Tetrathiofulvalene (TTF), 464, 499
Tetrathiotetracene (TTT), 464–65
Theorem, Koopmans, 494, 501, 507
Theoretical calculation, 361, 380, 387
Theory, Landau calculation, 234, 250
Theory, extended Hückel, 39
Thermal motion, 6, 351, 355
Thermochemical data, 343, 356, 364, 387
Thermodynamic data, proton transfer complexes, 41–43
Thermodynamic parameters, 279
Thermodynamic properties of micelles, 172–73
Thermodynamic properties of monomers, 172
Thermolysin, 212
Thiols, 131–33, 138
Thionine, 288
Thiophenol, 128–29
Three centre delocalized σ orbitals, 374
Three-dimensional systems, 233, 254
Thymine, 284, 288
Time-resolved,
  emission spectra, 522
  fluorescence, 522, 524, 537
  spectroscopy, nanosecond, 513
TMP, 284, 287
TMPD N,N,N,N-tetramethyl-p-phenylenediamine, 498
Toluidine blue, 288
Townes and Dailey's theory, 345, 349, 354, 360, 362, 372–74
Tranquilizers, 293
Trans effect, 378
  complex, 369
  position, 352, 378
Transfer constant, 53, 65–67, 70–74
Transient,
  absorption spectra, 535–36
  photoconductivity, 546, 551

Transition, 343, 345–47
   phase, 354, 363–64, 374, 384
Transmission windows, 15–16, 58
Transverse tunnelling field, Ising, 239, 257
Tri-coordination, 353, 357
Trichloroacetic acid, 122
Triethylamine, 116, 120, 127–28
Trifluoroacetic acid, 114, 121–22, 127
Trifluoroethanol, 134
Trifluoromethane, 116
Trimethylamine, 114–15
Trimethylamine–chloranil complex, 470–487
Trinitrobenzene, 295–96, 298, 454, 464
Trinitrofluorenone, 296, 298
Triphenylphosphine, 292
Triple ions, 82, 96
Trouton's law, 313
Trypsin, 208, 214–15, 218
Trypsin–trypsin inhibitor complexes, ion pairs, 212
Trypsin inhibitor pancreatic, 208, 218, 220
Trypsinogen, 207–8, 215
Tryptophan, 282, 284, 286–88, 292–93
Tryptophan picrate, 282
TTF(tetrathiofulvalenium), 499
Tunnelling and Hopping model, 486–88
Tunnelling transition, 55

Ultrasonic technique, 143, 148, 151, 165, 171
Ultraviolet photoelectron spectroscopy, 493
Unequivalence, crystallographic equivalence, 346, 352
Uni-univalent electrolytes, conductance, 199
Uracil, 284, 288
Urea, 204
UV spectroscopy, 343, 388

Valence bond, 272
Valence bond model, 369
Valence electron, 344–45, 349
Valence orbitals, 343, 345, 350
Valence photoelectron spectra, 506
Valence shells, 347, 365, 372
Value of,

charge, 364, 373, 375, 385
   degree, 364, 373, 375, 385
Van der Waals,
   complex, 384
   radii, 4, 46, 216, 325, 358
   radii N, O, C, H, 179
Van't Hoff isochore, 148
Very weak interaction, 529–30
Vibrational,
   bands, integrated intensity, 12, 15
   relaxation, 511–12, 528–29
   spectroscopy, 7, 21, 52, 57
Viscosity, 314, 317, 319–20, 324, 327
   macroscopic, 325
   microscopic, 325
   molecular, 325
   mutual, 323, 326
Viscous flow, 325
Volume,
   apparent, 325
   changes, micelles, 172–73
   hypothetical, 324

Walden product, 79, 81
Water, 115, 132–33, 135, 138, 333
   clusters in proteins, 220
   exchange, micelles, 164
   influence on transfer constant, 75
   interaction with amide group, 218
   interaction with carbonyl group, 218
   interaction with carboxyl groups 218
   interactions, peptides, 206
   interaction with proteins, 217
   molecules internal in proteins, 219–20
   O—H groups, 182, 184–85, 218
   roles of proteins, 220
   types of molecules, 217
Wave function, no bond, 385
Weak interaction, 529
Weak EDA complex, 527, 539

X-ray crystallography 275, 285, 291, 296
X-ray diffraction, 5–6, 10–11, 21, 56, 343, 352, 356–57, 367, 373, 379, 383, 388–89
X-ray structural analysis, 282
o-xylene, 477

Zeeman effect, 347, 362, 367, 377
Zinc tetraphenylporphyrin, 296

Molecular Interactions, Volume 2
Edited by H. Ratajczak and W. J. Orville-Thomas
© 1981 John Wiley & Sons Ltd

Subject index – Volume 1

For the reader's information and for reference purposes, the subject index for Volume 1 is included here in an expanded form.

A, B, C, bands, 318–19, 339–40
ab initio,
  band structure of DNA, 174
  band structure of proteins, 180
  calculations, 6, 281–82, 291
  hydrogen bond studies, 74
  methods, 5, 68
  MO calculations, 11, 128
  MO–SCF studies, 9, 118
  perturbation scheme, 3, 9
  SCF calculations, 15, 137
  SCF LCAO crystal orbital method, 152
  SCF studies, 9, 136, 205, 209, 211–12, 215, 217, 227
Absorption, infrared, 215, 220
Acceptor, proton, 11
Acetic acid,
  crystals, 330–31
  in matrices, 324–5
  in solution, 293, 317–18, 333
Acetonitrile, 314, 316, 319
Acid salts, 337–38
Acid salts of carboxylic acids, 341–43
Acids, H-bonds in carboxylic, 239, 241
Acids, stretching bands of carboxylic, 240–44, 251, 253
Acoustic branch, 356
Adiabatic approximation, 352
Aggregation, 274, 280, 283–85, 289–95
Alcohol–amine complexes, 281
Alcohols, 274, 309, 311, 314–15, 322, 327
Alkali bifluoride, 218, 222–27
Alkene–halogen complexes, 294–96
Alkenes, 287
Alternative perturbation theory, 3

Amines, 274
Ammonia, 210, 279, 281–82, 287, 293
  complex of, 324
  dimer, 282, 293
Ammonia–chlorine complex, 296–97
Analysis,
  back, 31, 34, 46, 50
  configuration, 28
  donor–acceptor complexes, 84
  energy and charge decomposition, 21, 24
  hydrogen, 44, 49–50, 52, 83
  hydrogen bonded systems, 83
  vibrational, 210
Anharmonic,
  interactions, Fermi resonances in H-bonds, 261
  interactions in H-bonds, 235, 248, 253, 255
  potential, 219
Anharmonicity, 218, 220
  electrical, 215, 220, 227, 302, 306–7, 311–12, 314, 326
  in crystals, 381
  mechanical, 220, 302, 306–7, 311–12, 314, 326
Antarafacial, 107
Anticooperativity, 118
Antisymmetrized wave functions, 2
Approach, supermolecule, 5
Approximation, ZDQ, 118
Argon matrices, 273–96
Associations,
  intermolecular, 120
  molecular, 118
Asymmetric hydrogen bond, 341
Atom–atom potential, 370

Atomic clustering, 128, 135
Autocorrelation function, 213, 220
Autocorrelation function, dipole moment, 215
Axilrod–Teller,
 forces, 143
 terms, 145

Back charge transfer, 31, 34, 46, 50
Band charge transfer, 7, 13
 infrared, 215
 transition, 13
Band shape mechanisms, 302, 304–12
Band structure of DNA,
 ab initio, 174
 semi-empirical, 173
Band structure of proteins,
 ab initio, 180
 semi-empirical, 179
Band structure of $(SN)_x$, 181
Base temperature, 276–77
Basis of the representation, 372
Basis set dependency, 45
Bending vibrations, 329, 331, 336, 339
Benzene, 296
Benzene–bromine complex, 296
Benzene–chlorine complex, 296–97
Benzene–iodine complex, 296
Benzene–iodine monochloride complex, 296
Bifluoride, alkali, 218, 222–27
Bifurcated hydrogen bond, 321
Bond, hydrogen, 2, 6–7, 11–13, 15, 139, 208, 210, 212
Bond, intramolecular hydrogen, 13
Born–Oppenheimer approximation, 119–20, 135, 302, 304
$BrH \cdots O$, hydrogen bond, 323
Buckingham potential function, 370

Calculations, *ab initio*, 281–82, 291
 first-order perturbational, 128
 quantum mechanical, 12, 145
Carbon monoxide matrix, 289
Carboxylic acids, 280, 295, 317–20, 324–25, 330–31, 333–34
Carboxylic acids, H-bonds in, 239, 241
Carboxylic acids, stretching bands of, 240–44, 251, 253

Centred hydrogen bond, 337, 341, 343
Centrifugal distortion constants, 291
Character of the matrix, 373
Character tables, 373
Charge–bond order matrix, 154
Charge densities, 139
Charge transfer, 2, 4, 6–13, 31, 34, 46, 50, 273, 275, 294–97
 band, 8, 13
 complex, 6–7, 12, 44, 93, 120
 electrostatic, 82, 84
 in charge transfer complexes, 10
 in donor acceptor complexes, 84
 in H-bond complexes, 11–13
 in hydrogen bonds, 84
 interaction, 9, 12, 123
 interaction energy, 26
 theory, 7
 transition band, 12
$CH \cdots O$, hydrogen bond, 315
Chemical shift, NMR, 6
$Cl—D \cdots O$, hydrogen bond, 324
$Cl—H \cdots N$, hydrogen bond, 324
$Cl—H \cdots O$, hydrogen bond, 313–14, 316–17, 321, 323–24
$Cl-H \cdots$ diether complexes, 234
Closed-cycle cryostat, 277
Cluster,
 expansion, 117
 interactions, 117
 molecular, 118
CNDO/2 calculations, 296
CNDO crystal orbital method, 164
CNDO method, 69
Coefficients, virial, 143, 209
Coherent potential approximation, 170, 181
Cohesive energy, 364
Combination bands, 314, 316–17, 319, 327, 329, 331, 336, 339
Combination frequencies, 381
Combined symmetry operation, 155–56
Complex,
 alcohol–amine, 281
 alkene–halogen, 294–96
 ammonia, 324
 ammonia–chlorine, 296–77
 benzene–bromine, 296
 benzene–chlorine, 296–97
 benzene–iodine, 296
 benzene–iodine monochloride, 296

charge–transfer, 6, 12, 44, 93, 120
Cl—H · · · diether, 234
component, 34–36, 39
donor–acceptor, 9, 12
donor–acceptor complexes, 84
electron, 27, 34
electron donor–acceptor, 30, 42, 52
electrostatic, 82, 84
energy, 24, 40, 42, 45, 49, 52, 57, 62
ethene–chlorine, 295–97
ethene–iodine, 295
formaldehyde, 291
heptyne, 315
hydrogen bonded, 119
hydrogen bromide, 287–88, 321, 323
hydrogen chloride, 285–88, 303, 313–
    14, 316–17, 321, 323–24
  –alcohol, 286
  –alkene, 286
  –alkyne, 286
  –benzene, 286
  –carbon dioxide, 286
  –carbon monoxide, 286
  –chloroalkane, 286
  –cyclopropane 286
  –diethyl ether, 286
  –dimethyl sulphide, 286
  –hydrogen bromide, 286
  –hydrogen fluoride, 282
  –hydrogen iodide 286
  –methylamine, 286
  –nitrogen, 286
  –pyridine, 286
  –sulphur dioxide, 286
  –trimethylamine, 286
  –water, 281, 286–88
hydrogen cyanide–ammonia, 281
hydrogen fluoride, 321, 323
  –ammonia, 282
  –ethene, 282
  –hydrogen cyanide, 281–82
  –water, 281–82
hydrogen halide, 285–88
hydrogen halide–amine, 285
hydrogen halide–nitrogen, 285
hydrogen sulphide, 282
hydrogen sulphide–water, 282
intermolecular, 118, 123–24
iodine, 296
ion-molecule, 142
molecular, 2, 7, 12, 273–89, 291–97

n–σ type, 44
phenol, 314–16
π, 44
pyrrolc, 315
space group, 351
water, 74, 291–92
water–ammonia, 282, 291
water–ammonia–carbon dioxide, 291
water–ammonia–chlorine, 296–97
water–ammonia–dimethyl ether, 29
X—H · · · N–, 233
X—H · · · O–, 233
Component, density map, 34–36, 39
Components, 3
Computer simulation, 208, 303
Configuration,
  analysis, 28, 91–93
  interaction, 5, 146
Constant, diffusion, 208
Constants, nuclear coupling, 6
Contribution, dispersion, 8, 9
Contributions, three-body, 142
Cooperativity, 117–18, 143
Coordination number, 207, 212
Correction, quantum, 206, 227
Correlation,
  dynamic, 117
  electron, 5
  energy, 22, 31, 45
  intermolecular, 22
  intramolecular, 22
  molecular crystals, 349
  time, 215, 313
Correlation field splitting, 312, 326,
    329–30
Correlation function, 206, 305–7
  dipole, 209
  pair, 206, 208
  quantum, 208
  radial, 206
  time, 207–8
  vibrational, 216
Correlation tables, 374
Coulomb interaction, 94, 96, 108
Coulomb operators, 154, 161
Coulomb term, 3
Coupling, 8
Coupling, Davidov resonant, 311–12,
    326–27
Coupling n, N, 302–3, 305, 308, 311,
    314, 316, 318, 320, 324, 326, 336

Cryostats, 276–77
Crystal basis, 350
Crystal-field polarization, 128
Crystal orbital methods, 151
Crystal, pyrrole, 327–28
Crystals, 326
Crystals, acetic acid, 330–31
Cyclic,
    dimer, 283, 293–94, 329
    dimers H-bonded, 239
    tetramer, 284
    trimer, 284–85
Cyclic conditions, 355

De Broglie wavelength, 206
Debye forces, 366
Deformation energy 29, 32, 54, 56, 62
    modes, 329, 331, 336, 339
    tensor, 365
Degeneracy, lifting of in matrix, 278
Delocalization,
    electron, 11
    interaction, 94–96, 100–2, 108–10
Dense packing, 349
Density, electron, 8
Density of states, 309–10
Density of states in $(SN)_x$, 182
Deposition methods, 278
Deuterated compounds, 316–17, 319,
    324–25, 329–43
Diaquahydrogen ion, 308
Dibutylether, 315
Dicarboxylic acids, H-bonds in, 247
Dielectric constant of DNA, 169
Diffusion, 207
Diffusion constant, 208
Diffusion in matrices, 274
Dimers,
    ammonia, 282, 293
    cyclic, 283, 293–94, 329
    formic acid, 319
    hydrazoic acid, 294–95
    hydrogen bonding, 80, 384
    hydrogen bromide, 285
    hydrogen cyanide, 281–82, 289–90
    hydrogen fluoride, 282–83
    hydrogen iodide, 285
    hydrogen peroxide, 294–95
    H-bonded, cyclic, 239
    inductive, 278
    interaction, 23, 29

ionic, 57
MINDO, 69
nitric acid, 294–95
open chain, 282–83, 285, 289, 292,
    294
water, 3, 14, 282–83, 289–90
Dimethylamine, 287
Dimethylarsenic acid, 339–40
Dimethyl ether, 303, 313–14, 316–17,
    321
Dimethylformamide, 315
Dimethylsulphoxide, 210–14, 318–19
Dioxane, 318–19
Dipole,
    autocorrelation function, 215
    correlation function, 209
    moment, 7, 13, 215, 220, 282, 291
Dislocation site, 280
Dispersion, 3
    contribution, 8
    energy, 4, 22, 125
    forces, 13
Dispersion energy contribution, 3
Dispersive interaction, 273, 278
Dissociation energy, 211
Distance, hydrogen bond, 326, 329–30,
    333–34, 336–37, 339, 341
Distribution function,
    four body, 206
    radial, 206, 209
    triple, 206
DNA models, 173
DODS crystal orbital method, 157
Donor–acceptor,
    complexes, 9, 12
    electron, 2, 6
    proton, 11
Donor-acceptor complexes,
    charge-transfer in, 84
    electrostatic picture of, 84
    theoretical studies of, 83
Doping experiment, 281
Double Dewar cryostat, 276
Double perturbation theory, 122
Dynamic correlation, 117
Dynamic matrix, 355
Dynamical correlation field effect, 357
Dynamical coupling, 362

EDA complex interactions, 5
EDA interactions, 2

Effective electron model, 128
Effects, many-body, 128, 135
Eigenvector, 355
Electric deflection measurements, 291
Electric resonance spectroscopy, 281–82
Electrical anharmonicity, 215, 220, 227, 302, 306–7, 311–12, 314, 326
Electron,
  correlation, 5, 124
  delocalization, 11
  density, 8, 27, 34
  donor-acceptor, 2, 7
  interaction, 6, 15
  properties, 11
Electron density function, 128
Electron donor–acceptor complex, 30, 42, 52
Electron polaron model, 168, 177
Electronic excited states, 16
Electronic spectra, 7
Electrostatic,
  charge transfer, 82, 84
  complexes, 82, 84
  component, 9
  coulomb energy, 366
  energy contribution, 3, 9
  first-order, 3
  forces, 125
  interaction, 9, 13, 24, 273, 278
  interaction energy, 24
  model, 49–50
  molecular potential model, 15
Energy,
  charge transfer interaction, 26
  component, 24, 40, 42, 45, 49, 52, 57, 62
  correlation, 22, 31, 45
  deformation, 29, 32, 54, 56, 62
  dispersion, 4, 22, 125
  electrostatic interaction, 24
  exchange interaction, 3, 26
  hydrogen-bonded systems, 4, 80–81
  interaction, 3, 5, 8
  intermolecular, 121, 142
  ionization, 125
  polarization interaction, 25
  potential functions, 80
  SCF interaction, 24
  semi-empirical, 74, 83
  space group, 351
  superposition, 23
  three-body, 129, 132, 139

  three-body dispersion, 127
Energy and charge decomposition, analysis, 21, 24
Energy partitioning, 3, 9, 81, 121, 141
Energy partitioning, intermolecular, 121
Energy relaxation, 227
Ethanethiol, 292–93
Ethanol, 291
Ethene–chlorine complex, 295–97
Ethene–iodine complex, 295
Ethene matrix, 279
Evans hole effect, 304, 307–8, 333, 343
Exchange,
  energy, 3
  forces, 2
  interaction, 94–96, 98, 100–1, 108–10
  interaction energy, 26
  polarization, 3
  repulsion, 8
  repulsion, terms, 13
    energy contribution, 3, 9
    operator, 154, 161
    overlap forces, 367
Exchange term, three-body, 139
Excited state, hydrogen bonding, 80
Excitons, 352
Expansion, cluster, 117
Extended basis set, 3
Extended Hückel theory, 69, 71
External vibrations, 357, 360, 374

F—H $\cdots$ N—C—H, infrared spectra of, 237
FH $\cdots$ O, hydrogen bond, 319
Face-centred cubic lattice, 274
Factor group splitting, 357
Far infrared spectra, 335–36
Far infrared spectroscopy, 274, 280, 290
Fermi resonance, 304, 307, 311–12, 318–19, 323, 326–27, 329
First Brillouin zone, 356
First-order perturbational calculation, 128
First-order term, electrostatic, 3
Fluctuation dissipation theorem, 207, 219
Fluorescence, 277
Fluoroalcohols, 291
Fluoromethane, 279
Fock matrix, 154, 162
Force constants, 289–90, 293

Forces,
   Axilrod–Teller, 143
   dispersion, 13
   electrostatic, 125
   exchange, 2
   intermolecular, 117–21, 125
   many-body, 127–28
   polarization, 125
   repulsion, 279
   three-body, 127–28, 130–31, 133, 135–37, 139, 143, 145
Formic acid dimers, 319
Four body distribution function, 206
Frequencies, vibrational, 211
   density distribution, vibrational, 220
Frequency bandwidth correlation, 326
   dispersion, 355
   distance correlation, 326
   excited state, 80
Frontier electron density, 103–5
Frontier orbital, interaction, 106, 108
Functions,
   electron density, 128
   potential, 117
   stochastic, 306–7

Gaussian infrared band shape profile, 306
Gaussian orbitals, 68
Geometric structure, 5
Grain boundary, 280
Greens function, 309–10

Hamiltonian, 2, 119, 306–7, 311–12
Hamiltonian, intermolecular, 125
Hamiltonian, stochastic, 306–7, 311–12
Hard core potential, 204
Harmonic approximation, 354
Harmonic potential, 210
Hartree–Fock model, 68, 117, 129
Hartree–Fock–Roothaan method, 6, 9, 151
Heptyne complex, 315
Hetero-associated complexes, 285
HF · · · F, hydrogen bond, 320, 343
Highest-occupied molecular orbital, 103
Hückel model, 68–69
Hydrated ions, 140
Hydrates, 328–29
Hydrazoic acid, dimer, 294–95

Hydrogen, 2, 6, 9–13, 15, 44, 49, 52, 136, 139, 208, 278–79
Hydrogen bonds, 7, 44, 49, 52, 136, 139, 273–75, 281–94
   ab initio methods, 74
   anharmonic interactions in, 235, 248, 253, 255
   asymmetric, 341
   bending modes, 274, 284, 289, 291
   bifurcated, 321
   BrH · · · O, 323
   CH · · · O, 315
   charge transfer in, 83–84
   Cl—D · · · O, 324
   Cl—H · · · N, 324
   Cl—H · · · O, 313–14, 316–17, 321, 323–24
   compared with D-bonds, 254, 256
   complexes, 119
   cyclic dimers, 239
   in dicarboxylic acids, 247
   directionality, 50
   distance, 326, 329–30, 333–34, 336–37, 339, 341
   electrical anharmonicity in, 258–59, 264
   electrostatic picture, 82
   energy, 4, 74
   energy partitioning, 81
   enthalpy of, 236, 245, 252
   excited state, 80
   far infrared spectra of, 238, 246, 267
   Fermi resonances in, 261
   FH · · · O, 319
   HF · · · F, 320, 343
   in crystals, 384
   in gases, 232
   interactions, 4, 6, 10, 15
   isolated, 236
   linear complexes, 237
   medium strong, 314, 316, 323–24, 329
   near infrared spectra of, 246
   NH · · · N, 334, 336
   NH · · · O, 315
   NH · · · π, 327–28
   OD · · · O, 318–19, 329, 331–34, 339, 341, 343
   OH · · · N, 316, 319–20
   OH · · · O, 314–16, 319–20, 322, 324–26, 329–30, 333–34, 337, 339, 341

polarizable, 307
potential functions, 80
predissociation in, 265
rotational envelope of infrared bands, 256, 265
single isolated, 305, 307–8
stretching bands of, 233–34, 240, 247, 255
stretching mode, 274, 289, 291
strong, 316, 337
symmetric, 337, 341, 343
system, 203, 208–9, 211, 215, 217
trimer, 137
vibrations, 302, 331, 336
weak, 314, 323, 326
Hydrogen bromide,
  complex, 321, 323
  complexes, 287–88
  dimer, 285
  trimer, 285
Hydrogen chloride, 279, 281, 283, 285
  complex, 285–88, 303, 313–14, 316–17, 321, 323–24
  tetramer, 284–85
  trimer, 284–85
Hydrogen chloride–alcohol complexes, 286
  –alkene complexes, 286
  –alkyne complexes, 286
  –ammonia complex, 281, 286–87
  –benzene complex, 286
  –carbon dioxide complex, 286
  –carbon monoxide complex, 286
  –chloroalkane complexes, 286
  –cyclopropane complex, 286
  –diethyl ether complex, 286
  –dimethyl sulphide complex, 286
  –hydrogen bromide complex, 286
  –hydrogen fluoride complex, 282
  –hydrogen iodide complex, 286
  –methylamine complex, 286
  –nitrogen complex, 286
  –pyridine complex, 286
  –sulphur dioxide complex, 286
  –trimethylamine complex, 286
  –water complex, 281, 286–88
Hydrogen cyanide, 281, 289–90
  dimer, 281–82, 289–90
  trimer, 289–90
Hydrogen cyanide–ammonia complex, 281

Hydrogen fluoride, 209–15, 280–81, 283
  complex, 321, 323
  dimer, 282–83
Hydrogen fluoride–ammonia complex, 282
  –ethene complex, 282
  –hydrogen cyanide complex, 281–82
  –water complex, 281–82
Hydrogen halide complexes, 285–88
  –amine complexes, 285
  –nitrogen complexes, 285
Hydrogen halides, 274, 279, 282–89
Hydrogen iodide,
  dimer, 285
  trimer, 285
Hydrogen peroxide dimer, 294–95
Hydrogen sulphide–water complex, 282

Ices, 329
Imidazole, 334–36
Induction, 3
Induction energy, 366
Induction energy contribution, 3
Inductive interaction, 278
Infrared,
  absorption, 215, 220
  band, 215
Infrared absolute intensities, 6
Infrared absorption intensities, 376–77
Infrared band intensity, 12
Infrared band shape profile, 301–8, 316–43
  Gaussian, 306
  of a single hydrogen bond, 308
  of gases, 302
  of liquids, 304, 312
  of matrix isolated species, 322
Infrared spectroscopy, 273–87, 297
Infrared spectrum, 375
Intensities, 376
Intensities, infrared absolute, 6
Intensities of the Raman bands, 6, 377
Intensity, infrared band, 12
Interaction,
  charge transfer, 9, 12, 26, 123
  configuration, 5
  cluster, 117
  cooperative molecular, 118
  coulomb, 94, 96, 108
  degeneracy, 278

delocalization, 94–96, 100–2, 108–10
dispersive, 273, 278
EDA, 2
EDA complex, 6
electron donor–acceptor, 6, 15
electrostatic, 9, 13, 24, 273, 278
energy, 3, 5, 9–10, 23, 29
Fermi resonances in H-bonds,
    anharmonic, 261
frontier orbital, 106, 108
hydrogen bond, 4, 6, 10, 15, 76, 79
in H-bonds, anharmonic, 235, 248,
    253, 255
intermolecular, 3, 118, 140
long-range, 2
many-body, 119
molecular, 1–2, 5, 7–9, 15–16
nearest neighbour, 274
next nearest neighbour, 274
orbital, 89, 106–8
polarization, 95–96, 100–2, 108, 111
reactive, 64
short-range, 2, 6, 128
solute–matrix, 279, 281
solvent, 273
specific, 278, 281
three-body, 139
Interaction energy, 139
between molecules, 185
between polymers, 196
polarization, 25
Interatomic potentials, 127
Intermolecular,
associations, 120
clusters, 136
complexes, 118, 123–24
cooperativity, 143
correlation, 22
electron correlation, 124
energies, 121, 142
energy partitioning, 121
forces, 117–21, 125
Hamiltonians, 125
hydrogen bond complex, 13
interactions, 3, 118, 140
perturbation theory, 121–22, 124
potential, 122, 128
potential in crystals, 368
vibrational modes, 283, 290
Internal vibrations, 358, 374

Interstitial site, 280
Intramolecular correlation, 22
Intramolecular forces, 348
Intramolecular hydrogen bond, 13
Intramolecular vibrational modes, 283,
    290
Intrinsic reaction coordinate, 110
Inversion of ammonia, 293
Inversion temperature, 276
Iodine, 278, 294
Ion–molecule complexes, 142
Ionic hydrogen bonding, 57
Ionization energies, 125
Ionization potential, 46, 60
Irreducible representation, 373
Isopropanol, 314–15
Isotope effect, 306–7, 310, 314–16, 331,
    336, 341–42
Isotopic splitting pattern, 293
Isotropic potential, 204

Jahn–Teller effect, 304
Joule–Thomson principle, 276

Keesom forces, 366
Krypton matrix, 279

Langevin equation, 306
Lattice modes, 309, 326, 336
Lennard–Jones potential function, 205,
    210, 370
Libration, 278, 357, 361
Librational modes, 284, 291
Lippincott Schroeder potential, 210–1,
    219, 221
Liquid helium cryostat, 276
Liquid hydrogen cryostat, 276
Liquid nitrogen trap, 277
    acetic acid, 333
Lithium fluoride, 279
Lithium hydrogen oxalate, 341–42
London dispersion forces, 366
Long-range interactions, 2
Long-range forces, 365
Lowest unoccupied orbital, 103

Magnesium sulphate hexahydrate, 329
Maleimide, 281

Many-body,
  effects, 128, 135
  forces, 127–28
  interactions, 119
Matrix effects, 278–81
  isolation, 273–80, 296–97
  shift, 279
  to solute ratio, 273–74
Mechanics,
  quantum, 1, 8–9, 117, 121
  statistical, 1
Mechanical anharmonicity, 220, 302, 306–7, 311–12, 314, 326
Medium strong hydrogen bond, 314, 316, 323–24, 329
Methane, 279
Methanethiol, 292–93
Methanol, 274–75, 291, 309, 316, 322
Method,
  ab initio SCF LCAO crystal orbital, 152
  CNDO crystal orbital, 164
  crystal orbital, 151
  DODS crystal orbital, 157
  Hartree–Fock–Roothaan, 151
  MINDO crystal orbital, 165, 179
  molecular dynamics, 203–5, 209, 217, 221
  Monte Carlo, 203, 209
  mutually consistent field, 186, 188
  NDDO, 69
  OAO, 167
  open shell crystal orbital, 160
  Pauli, 26, 32
  PPP crystal orbital, 146
  PRODDO, 70
  SCF, 124
  SCF LCAO, 151
  SCF, *ab initio*, 1, 5
  space group, 351
  variational, 22, 28
  water, 78
Methylamine, 280, 287, 293
Microwave spectroscopy, 281–82, 289
MINDO method, 69
MINDO crystal orbital method, 165, 179
Minimum basis set, 5
MO calculations, *ab initio*, 118, 128
Model,
  effective electron, 128

electrostatic, 15, 49–50
Hartree–Fock, 68, 117
Hückel, 68–69
donor-acceptor complexes, 83
charge, 31, 34, 46, 50
Watson–Crick, 4
Modes,
  n, 302, 304, 307–8, 314, 316, 318–19, 323–24, 326, 329, 331, 336
  N, 307–8, 316, 318, 326, 329
Molecular,
  associations, 118
  beam electric resonance spectroscopy, 281–82
  complex, 2, 6–7, 11, 96, 273–89, 291–95, 297
  clusters, 118
  crystals, definition of, 349
  dynamics method, 203–5, 209, 217, 221
  interactions, 1, 2, 5, 7–9, 15–16, 117
  orbital theory, 5
  polarizability, 6, 125
  wavefunctions, 2
Molecule–molecule potential, 369
Moment, dipole, 6, 12
Moment, quadrupole, 6
Monte Carlo method, 203, 209
Mulliken charge transfer theory, 7, 11–12
  populations, 139
Multiple trapping sites, 278, 280
Multiplet band structure, 278–80
Multipole potential, 205
Mutually consistent field method, 186, 188

N coupling N, 302–3, 305, 308, 311, 314, 316, 318, 320, 324, 326, 336
n modes, 302, 304, 307–8, 314, 316, 318–19, 323–24, 326, 329, 331, 336
N modes, 307–8, 316, 318, 326, 329
NH $\cdots$ N, hydrogen bond, 334, 336
NH $\cdots$ O, hydrogen bond, 315
NH $\cdots$ π, hydrogen bond, 327–28
n–σ type complex, 44
NDDO method, 69
Nearest neighbour interactions, 274
Needle valve, 278
Neutron scattering, 208

Newton equation, 204
Next nearest neighbour interactions, 274
Nitric acid dimer, 294–95
Nitrogen matrices, 273–97
NMR chemical shift, 6
Nodes of the lattice, 350
Normal coordinate analysis, 289, 293
Normal coordinates, 358
Nuclear coupling constants, 6
Nuclear spin conversion, 289, 293

OAO method, 167
OD $\cdots$ O, hydrogen bond, 318–19, 329, 331–34, 339, 341, 343
OH $\cdots$ O, hydrogen bond, 314–16, 319–20, 322, 324–26, 329–30, 333–34, 337, 339, 341
OH $\cdots$ N, hydrogen bond, 316, 319–20
Oil diffusion pump, 277
One-electron systems, 3
Open chain,
  dimer, 282–3, 285, 289, 292, 294
  tetramer, 284, 291
  trimer, 284, 291
Open cycle cryostat, 276
Open shell crystal orbital method, 160
Optical branch, 356
Orbital interaction, 25, 89, 106–8
Orbital phase continuity, 105
Orbital symmetry, 107
Order–disorder in crystals, 385
Oriented gas model, 379
Overtones, 381
Oxalic acid, 333–34

Pair, correlation function, 206, 208
Pair potential, 365
Partitioning energy, 141
Pauli exclusion principle, 2, 7, 9, 26, 32
Periodic boundary conditions, 153, 205
Perturbation theory, 2, 71, 125, 127–28, 136, 302
  intermolecular, 121–22, 124
Perturbational interaction calculations, 186
Phase transitions, 385
Phenol, 1,4-chloro, 327
Phenol complex, 314–16
Phenomena, quantum, 206
Photolysis, 289

π-complex, 44
Plastic ODIC crystals, 353, 386
Polarization, 8, 13
  crystal-field, 128
  exchange, 3
  forces, 125
  interaction, 95–96, 100–2, 108, 111
  interaction energy, 25
Polarization energy, 366
  three-body, 143
    energy contribution, 3, 9
    forces, three-body, 146
    interaction energy, 25
Polarizability, molecular, 6
Polarizable hydrogen bond, 307
Polychroism of crystals, 379
Polyethylene window, 277
Polymorphism, 385
Potassium hydrogen carbonate, 302
Potassium hydrogen oxalate, 341–42
Potential,
  anharmonic, 219
  functions, 117
  hard core, 204
  harmonic, 210
  intermolecular, 122, 128
  isotropic, 204
  Lennard–Jones, 205, 210
  Lippincott Schroeder, 210–11, 219, 221
  multipole, 205
  semi-empirical, 205
  stochastic, 307
  three-body, 130, 133, 135, 137, 143
  two-body, 140
Potential functions for hydrogen bonds, 80
PPP crystal orbital method, 146
Primitive cell, 350
Principal limiting frequencies of the crystal, 356, 373
Principle of molecular packing, 368
PRODDO method, 70
Profile, Raman band shape, 301–43
Profile liquids, Raman band shape, 315, 322
Profile solids, Raman band shape, 327–28, 330–34, 336–40
Profile solutions, Raman band shape, 316–17, 321
Proton,
  acceptor, 11
  donor, 11

Proton affinity, 58, 286–88
Pulsed matrix deposition, 278
Pyrrole,
  complex, 315
  crystal, 327–28
  vitreous, 327–28

Quadrupole moment, 6
Quality of basis set, 23
Quantum correction, 206, 227
Quantum correlation function, 208
Quantum mechanical calculations, 12, 145
Quantum mechanics, 1, 8–9, 117, 121
Quantum phenomena, 206

Radial correlation function, 206
Radial distribution function, 206, 209
Radii, Van der Waals, 11
Raman absolute intensities, 6
Raman spectroscopy, 210, 273, 277–78, 280, 292–95
Raman band shape profile, 301–43
Raman band shape, profile liquids, 315, 322
Raman band shape, profile solids, 327–28, 330–34, 336–40
Raman profile solutions, 316–17, 321
Raman spectrum, 375
Rayleigh–Schrödinger, 2
Reaction field, 209
Reactive interaction, 64
Relaxation time, 227
Representation of the symmetry operation, 371
Repulsion exchange, 8
Repulsion forces, 279
Resonance Raman spectroscopy, 294
Rotary pump, 277
Rotation, 206, 213, 215, 278–80, 289–90, 292–93
Rotation–vibration spectrum, 281
Rotational barrier, 34
Rotational structure, 274

S-tetrazine, 289
Sample deposition, 278
SCF, interaction energy, 24
  ab initio methods, 3, 15, 136–37

method, 124
interaction energy, 24
LCAO method, 151
Scheme, Hartree–Fock–Roothaan, 6, 9
Schrödinger equation, 119
Second-order term, 3
  polarization term, 3
  spectra, 383
Secular equation, 355
Selection rules for vibrational spectra, 375
Self-association, 274, 283–95
Semi-empirical methods, 68
  band structure of DNA, 173
  band structure of proteins, 179
  in donor–acceptor complexes, 83
  in hydrogen bonded systems, 74
  LCAO–MO methods, 118
  quantum mechanical methods, 5
Semi-empirical potential, 205
Set, extended basis, 3
  minimum basis, 5
Shift, NMR chemical, 7
Short-range, interactions, 2, 6, 128
  forces, 367
Simultaneous deposition technique, 274, 278
Single crystals, 335–36, 339
Single determinant theory, 8
Single isolated hydrogen bond, 305, 307–8
  dispersion, 4, 125
  exchange, 3
Spectroscopy,
  infrared, 210, 217, 273–87, 297,
  microwave, 281–82, 289
  molecular beam electric resonance, 281–82
  hydrogen iodide, 285
  Raman, 210, 273, 277–78, 280, 292–95
  resonance Raman, 294
  ultraviolet-visible, 294–96
  vibrational, 5, 221
Spectrum density, vibrational, 221
States, electronic excited, 16
Static-field site effect, 357
Statistical mechanics, 1
Statistics, 274
Stereoselection, 102, 105
Steric hindrance, 26, 32, 61
Stirling refrigeration cycle, 277

Stochastic,
  functions, 306–7
  Hamiltonian, 306–7, 311–12
  potential, 307
Stretching vibration, 215, 219
Strong coupling theory, 301
Strong hydrogen bond, 316, 337
Structure, geometric, 5
Structural disorder, 311, 326–27, 333, 339
  solids, 309, 326
  solutions, 312
Sub-bands submaxima, 316, 327, 330, 333, 336–37, 339–41
Substituent effect, 59
Substitutional site, 280
Sulphur dioxide, 279
Supermolecule approach, 5
Supermolecule calculation, 22
Superposition error, 23
Symmetric hydrogen bond, 337, 341, 343
Symmetry of molecular vibrations, 371
Symmetry of vibrations in molecular, crystals, 373
Symmetry types species, 373
Symmetry-adapted perturbation theory, 3
Symmetry-forbidden, 105, 107–8
Systems,
  hydrogen bond, 203, 208–9, 211, 215, 217
  one-electron, 3
  two-electron, 3

Temperature effect dependence, 307, 314–16, 330–33, 336–37, 340–43
Temperature effects on infrared bands, 236, 241
Terbutanol, 314
Term,
  Axilrod–Teller, 145
  coulomb, 3
  exchange repulsion, 13
  first-order, 3
  second-order, 3
  second-order polarization, 3
  three-body, 139–40, 142–3
Tetramer, cyclic, 284
Tetramer, hydrogen chloride, 284–85

Tetramer, open chain, 284, 291
Theory,
  ab initio perturbation, 9
  ab initio SCF molecular orbital, 8
  alternative perturbation, 3
  charge transfer, 7
  double perturbation, 122
  intermolecular perturbation, 122, 124
  molecular orbital, 6
  Mulliken charge transfer, 7, 12–13
  perturbation, 2, 125, 127–28, 136
  semi-empirical exchange perturbation, 4
  single determinant, 8
  symmetry-adapted perturbation, 3
  valence-bond, 11
  zero-overlap, 3
Three-body,
  contributions, 142
  dispersion energy, 127
  energy, 129, 132, 139
  exchange term, 139
  forces, 127–28, 130–31, 133, 135–37, 139, 143, 145
  interaction, 139
  polarization energy, 143
  polarization forces, 146
  potential, 130, 133, 135, 137, 143
  term, 139–40, 142–43
Time,
  correlation, 213, 215, 313
  correlation function, 207–8
  relaxation, 227
Total energy per unit cell, 159
Transfer charge, 2, 4
Transition probabilities of infrared bands, 235–36, 242, 245, 257, 264
Translation, 206, 213
Trapping sites, 278–80
Triazole, 1, 2, 4
Trichloroacetic acid, 318–20
Trifluoroacetic aicd, 324
Trimer,
  cyclic, 284–85
  hydrogen bond, 137
  hydrogen bromide, 285
  hydrogen chloride, 284–85
  hydrogen cyanide, 289–90
  open chain, 284, 291
Trimethylamine, 287
Triple distribution function, 206

Two-body potentials, 140
Two-electron systems, 3

UHF equations, 164
Ultraviolet, vacuum, 12
Ultraviolet–visible spectroscopy, 294–96

Vacuum ultraviolet, 12
Vacuum shroud, 277
Valence-active, 97, 99
Valence-bond theory, 11
Valence-inactive, 97
Van der Waals radius, 2, 9, 349
Variational model, 22, 28
Vibration, 206, 213, 215, 217
Vibration, stretching, 215, 219
Vibrations, hydrogen bond, 302, 331, 336
Vibrational,
  analysis, 210
  approach, 6
  correlation diagram, 285
  correlation function, 216
  frequencies, 211
  frequency density distribution, 220
  spectrum, 5, 221, 374
  spectrum density, 221

Virial coefficient, 143, 209
Vitreous pyrrole, 327–28

Water, 208–9, 274, 279, 281–82, 289–90, 316, 323
  complexes, 291–2
  dimer, 3, 14, 74, 282–83, 289–90
  oligomers, 78
Water–ammonia complex, 282, 291
  carbon dioxide complex, 291
  chlorine complex, 296–97
  dimethyl ether complex, 29
  formaldehyde complex, 291
  hydrogen sulphide complex, 282
  iodine complex, 296
Watson–Crick model, 4
Wavefunctions, molecular, 2
Weak hydrogen bond, 314, 323, 326
Woodward–Hoffmann rule, 108

X—H · · · N-complexes, 233
X—H · · · O-complexes, 233
X-ray scattering, 208

ZDO approximation, 118
Zero-overlap theory, 3